Material Property Relations

Poisson's ratio

$$\nu = -\frac{\epsilon_{lat}}{\epsilon_{long}}$$

Generalized Hooke's Law

$$\epsilon_x = \frac{1}{E}[\sigma_x - \nu(\sigma_y + \sigma_z)]$$

$$\epsilon_y = \frac{1}{E}[\sigma_y - \nu(\sigma_x + \sigma_z)]$$

$$\epsilon_z = \frac{1}{E}[\sigma_z - \nu(\sigma_x + \sigma_y)]$$

$$\gamma_{xy} = \frac{1}{G}\tau_{xy}, \quad \gamma_{yz} = \frac{1}{G}\tau_{yz}, \quad \gamma_{zx} = \frac{1}{G}\tau_{zx}$$

where

$$G = \frac{E}{2(1+\nu)}$$

Relations Between *w, V, M*

$$\frac{dV}{dx} = -w(x), \quad \frac{dM}{dx} = V$$

Elastic Curve

$$\frac{1}{\rho} = \frac{M}{EI}$$

$$EI\frac{d^4v}{dx^4} = -w(x)$$

$$EI\frac{d^3v}{dx^3} = V(x)$$

$$EI\frac{d^2v}{dx^2} = M(x)$$

Buckling
Critical axial load

$$P_{cr} = \frac{\pi^2 EI}{(KL)^2}$$

Critical stress

$$\sigma_{cr} = \frac{\pi^2 E}{(KL/r)^2}, \quad r = \sqrt{I/A}$$

Secant formula

$$\sigma_{max} = \frac{P}{A}\left[1 + \frac{ec}{r^2}\sec\left(\frac{L}{2r}\sqrt{\frac{P}{EA}}\right)\right]$$

Energy Methods
Conservation of energy

$$U_e = U_i$$

Strain energy

$$U_i = \frac{N^2 L}{2AE} \quad \text{constant axial load}$$

$$U_i = \int_0^L \frac{M^2 dx}{EI} \quad \text{bending moment}$$

$$U_i = \int_0^L \frac{f_s V^2 dx}{2GA} \quad \text{transverse shear}$$

$$U_i = \int_0^L \frac{T^2 dx}{2GJ} \quad \text{torsional moment}$$

Geometric Properties of Area Elements

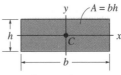

$A = bh$

$I_x = \frac{1}{12}bh^3$

$I_y = \frac{1}{12}hb^3$

Rectangular area

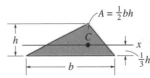

$A = \frac{1}{2}bh$

$I_x = \frac{1}{36}bh^3$

Triangular area

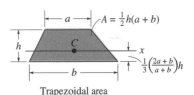

$A = \frac{1}{2}h(a+b)$

$\frac{1}{3}\left(\frac{2a+b}{a+b}\right)h$

Trapezoidal area

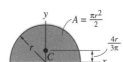

$A = \frac{\pi r^2}{2}$

$\frac{4r}{3\pi}$

$I_x = \frac{1}{8}\pi r^4$

$I_y = \frac{1}{8}\pi r^4$

Semicircular area

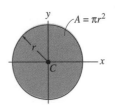

$A = \pi r^2$

$I_x = \frac{1}{4}\pi r^4$

$I_y = \frac{1}{4}\pi r^4$

Circular area

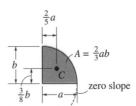

$\frac{2}{5}a$

$A = \frac{2}{3}ab$

zero slope

$\frac{3}{8}b$

Semiparabolic area

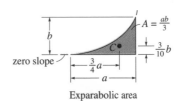

$A = \frac{ab}{3}$

$\frac{3}{10}b$

zero slope

$\frac{3}{4}a$

Exparabolic area

MECHANICS OF MATERIALS

FIFTH EDITION

MECHANICS OF MATERIALS

R. C. Hibbeler

Prentice
Hall

Pearson Education, Inc.
Upper Saddle River, New Jersey 07458

Library of Congress Cataloging-in-Publication Data

Hibbeler, R. C.
    Mechanics of materials / R. C. Hibbeler.—5th ed.
        p.   cm.
    Includes bibliographical references and index.
    ISBN 0-13-008181-7
    1. Strength of materials.   2. Structural analysis (Engineering) I. Title.

TA405.H47   2002
620.1'12—dc21   2002032958
                                                                                    CIP

Vice President and Editorial Director, ECS: Marcia J. Horton
Vice President and Director of Production and Manufacturing, ESM: David W. Riccardi
Editorial Assistant: Brian Hoel
Executive Managing Editor: Vince O'Brien
Managing Editor: David A. George
Director of Creative Services: Paul Belfanti
Manager of Formatting: Jim Sullivan
Formatting: Judith R. Wilkens and Allyson Graesser
Creative Director: Carole Anson
Art Director and Cover Designer: Wanda España
Art Editor: Xiaohong Zhu
Manufacturing Manager: Trudy Pisciotti
Manufacturing Buyer: Lisa McDowell
Marketing Manager: Holly Stark

About the Cover: Cover image courtesy of Vishay Measurements Group, Inc.,
                Raleigh, NC 27611

© 2003 by R. C. Hibbeler
Published by Pearson Education, Inc.
Pearson Education, Inc.
Upper Saddle River, New Jersey 07458

The author and publisher of this book have used their best efforts in preparing this book. These efforts include the development, research, and testing of the theories and programs to determine their effectiveness. The author and publisher make no warranty of any kind, expressed or implied, with regard to these programs or the documentation contained in this book. The author and publisher shall not be liable in any event for incidental or consequential damages with, or arising out of, the furnishing, performance, or use of these programs.

Printed in the United States of America.

10 9 8 7 6 5 4 3 2 1

ISBN 0-13-008181-7

Pearson Education Ltd., *London*
Pearson Education Australia Pty. Ltd., *Sydney*
Pearson Education Singapore, Pte. Ltd.
Pearson Education North Asia Ltd., *Hong Kong*
Pearson Education Canada, Inc., *Toronto*
Pearson Educación de Mexico, S.A. de C.V.
Pearson Education—Japan, *Tokyo*
Pearson Education Malaysia, Pte. Ltd.
Pearson Education, Inc., *Upper Saddle River, New Jersey*

CONTENTS

v

4

AXIAL LOAD   117

5

TORSION   177

6

BENDING   255

7

TRANSVERSE SHEAR   363

8

COMBINED LOADINGS   409

9

STRESS TRANSFORMATION   439

10

STRAIN TRANSFORMATION   489

11

DESIGN OF BEAMS AND SHAFTS   539

12

DEFLECTIONS OF BEAMS AND SHAFTS   569

13

BUCKLING OF COLUMNS   649

14

ENERGY METHODS 705

A

GEOMETRIC PROPERTIES OF AN AREA 775

B

GEOMETRICAL PROPERTIES OF STRUCTURAL SHAPES 792

C

SLOPES AND DEFLECTIONS OF BEAMS 800

D

REVIEW FOR THE FUNDAMENTALS OF ENGINEERING EXAM 802

ANSWERS 822

INDEX 844

PREFACE

This book is intended to provide the student with a clear and thorough presentation of both the theory and application of the fundamental principles of mechanics of materials. Understanding is based on the explanation of the physical behavior of materials under load and then modeling this behavior to develop the theory. Emphasis is placed on the importance of satisfying equilibrium, compatibility of deformation, and material behavior requirements.

Unique Features

The following is a list of some of the more important features of the text.

• *Summaries.* The "procedure for analysis" and "important points," sections provide a guide for problem solving and a summary of the concepts.

• *Photographs.* Many photographs are used throughout the book to explain how the principles of mechanics of materials apply to real-world situations. In some sections they show how materials deform or fail under load in order to provide a conceptual understanding of the terms and concepts.

• *Problems.* The problems provide a balance between easy, medium, and difficult applications. In addition, some problems require solution by computer. Extra care has been taken in the presentation and solution of the problems, and all the problem sets have been reviewed and the solutions checked and rechecked to ensure both their clarity and numerical accuracy.

Contents

The subject matter is organized into 14 chapters. Chapter 1 begins with a review of the important concepts of statics, followed by a formal definition of both normal and shear stress, and a discussion of normal stress in axially loaded members and average shear stress caused by direct shear. In Chapter 2 normal and shear strain are defined, and in Chapter 3 a discussion of some of the important mechanical properties of materials is given.

Separate treatments of axial load, torsion, and bending are presented in Chapters 4, 5, and 6, respectively. In each of these chapters, both linear-elastic and plastic behavior of the material are considered. Also, topics related to stress concentrations and residual stress are included. Transverse shear is discussed in Chapter 7, along with a discussion of thin-walled tubes, shear flow, and the shear center. Chapter 8 provides a partial review of the material covered in the previous chapters, in which the state of stress resulting from combined loadings is discussed. In Chapter 9 the concepts for transforming multiaxial states of stress are presented. In a similar manner, Chapter 10 discusses the methods for strain transformation, including the application of various theories of failure. Chapter 11 provides a means for a further summary and review of previous material by covering design applications of beams and shafts. In Chapter 12 various methods for computing deflections of beams and shafts are covered. Also included is a discussion for finding the reactions on these members if they are statically indeterminate. Chapter 13 provides a discussion of column buckling, and lastly, in Chapter 14 the problem of impact and the application of various energy methods for computing deflections are considered.

Sections of the book that contain more advanced material are indicated by a star (∗). Time permitting, some of these topics may be included in the course. Furthermore, this material provides a suitable reference for basic principles when it is covered in other courses, and it can be used as a basis for assigning special projects.

Alternative Method of Coverage. Some instructors prefer to cover stress and strain transformations *first,* before discussing specific applications of axial load, torsion, bending, and shear. One possible method for doing this would be first to cover stress and its transformation, Chapter 1 and Chapter 9, followed by strain and its transformation, Chapter 2 and the first part of Chapter 10. The discussion and example problems in these later chapters have been styled so that this is possible. Also, the problem sets have been subdivided so that this material can be covered without prior knowledge of the intervening chapters. Chapters 3 through 8 can then be covered with no loss in continuity.

Special Features

Organization and Approach.   In order to aid both the instructor and the student, the contents of each chapter are organized into well-defined sections. Selected groups of sections contain an explanation of specific topics, followed by illustrative example problems and a set of homework problems. The topics within each section are often placed in subgroups denoted by boldface titles. The purpose of this is to present a structured method for introducing each new definition or concept and to make the book convenient for later reference and review. Furthermore, important terms in the chapter have been highlighted in boldface to provide a convenient means for review.

Chapter Contents.   Each chapter begins with a photo to illustrate a broad range application of the material within the chapter. The "chapter objectives" are then provided to give a general overview of the material that will be covered.

Procedures for Analysis.   Found in many sections of the book, this unique feature provides the student with a logical and orderly method to follow when applying the theory. The example problems are then solved using this outlined method in order to clarify its numerical application. It is to be understood, however, that once the relevant principles have been mastered and enough confidence and judgment have been acquired, the student can then develop his or her own procedures for solving problems.

Important Points.   This feature provides a review or summary of the most important concepts in a section and highlights the most significant points that should be realized when applying the theory to solve problems.

Conceptual Understanding.   Through the use of photographs placed throughout the book, examples of the theory are provided in order to illustrate some of its more important conceptual features and instill the physical meaning of many of the terms used in the equations.

Example Problems.   All the example problems are presented in a concise manner and in a style that is easy to understand. New examples have been added throughout the text, and some from the previous edition have been shortened.

Homework Problems.   Numerous problems in the book depict realistic situations encountered in engineering practice. It is hoped that this realism will both stimulate the student's interest in the subject and provide a means for developing the skill to reduce any such problem from its physical description to a model or symbolic representation to which the principles may be applied.

Throughout the book there is an approximate balance of problems using either SI or FPS units. Furthermore, in any set, an attempt has been made to arrange the problems in order of increasing difficulty. The answers to all but every fourth problem are listed in the back of the book. To alert the user to a problem without a reported answer, an asterisk (*) is placed before the problem number. Answers are reported to three significant figures, even though the data for material properties may be known with less accuracy. Although this might appear to be poor practice, it is done simply to be consistent and to allow the student a better chance to validate his or her solution. All the problems and their solutions have been independently checked for accuracy. A solid square (■) is used to identify problems that require a numerical analysis or computer application.

Appendices.   The appendices of the book provide a source for review and a listing of tabular data. Appendix A provides information on the centroid and the moment of inertia of an area. Appendices B and C list tabular data for structural shapes, and the deflection and slopes of various types of beams and shafts. Appendix D, which is titled "Review for the Fundamentals of Engineering Exam," contains typical problems, along with their partial solutions, that are commonly used on *FE exams*. These problems may also be used for review and practice in preparing for class examinations.

Accuracy Checking.   The edition has undergone rigorous accuracy checking and proofing of pages. Besides the author's review of all art pieces and pages, Karim Nohra of the University of South Florida and Scott Hendricks of Virginia Polytechnic Institute rechecked the page proofs twice, and together reviewed the entire Solutions Manual.

Supplements.   Several instructor's supplements are available with this text. An instructor's solutions manual was prepared and typeset by the author. It has been completely checked for accuracy. In addition, an Instructor's Resource CD is available containing PowerPoint slides and pdf files of text art, as well as PowerPoint slides of text examples.

Acknowledgments

Over the years, this text has been shaped by the suggestions and comments of many of my colleagues in the teaching profession. Their encouragement and willingness to provide constructive criticism are very much appreciated and it is hoped that they will accept this anonymous recognition.

A note of thanks is given to our reviewers: Patrick Kwon of Michigan State University, Cliff Lissenden of Penn State University, Dahsin Liu of Michigan State University, Ting-Wen Wu of the University of Kentucky, Javad Hashemi of Texas Tech University, and Assimina Pelegri of Rutgers— The State University of New Jersey. A particular note of thanks is given to Karim Nohra of the University of South Florida and Scott Hendricks of Virginia Polytechnic Institute, who rigorously checked both the text and the Solutions Manual. I would also like to thank all my students who have used the previous edition and have made comments to improve its contents. Lastly, I should like to acknowledge the assistance of my wife, Cornelie (Conny), during the time it has taken to prepare the manuscript for publication.

I would greatly appreciate hearing from you if at any time you have any comments or suggestions regarding the contents of this edition.

RUSSELL CHARLES HIBBELER
hibbeler@bellsouth.net

The bolts used for the connections of this steel framework are subjected to stress. In this chapter we will discuss how engineers design these connections and their fasteners.

1 STRESS

CHAPTER OBJECTIVES

In this chapter we will review some of the important principles of statics and show how they are used to determine the internal resultant loadings in a body. Afterwards the concepts of normal and shear stress will be introduced, and specific applications of the analysis and design of members subjected to an axial load or direct shear will be discussed.

1.1 INTRODUCTION

Mechanics of materials is a branch of mechanics that studies the relationships between the *external* loads applied to a deformable body and the intensity of *internal* forces acting within the body. This subject also involves computing the *deformations* of the body, and it provides a study of the body's *stability* when the body is subjected to external forces.

In the design of any structure or machine, it is *first* necessary to use the principles of statics to determine the forces acting both on and within its various members. The size of the members, their deflection, and their stability depend not only on the internal loadings, but also on the type of material from which the members are made. Consequently, an accurate determination and fundamental understanding of *material behavior* will be of vital importance for developing the necessary equations used in mechanics of materials. Realize that many formulas and rules for design, as defined in engineering codes and used in practice, are based on the fundamentals of mechanics of materials, and for this reason an understanding of the principles of this subject is very important.

Historical Development. The origin of mechanics of materials dates back to the beginning of the seventeenth century, at which time, Galileo performed experiments to study the effects of loads on rods and beams made of various materials. For a proper understanding, however, it was necessary to establish accurate experimental descriptions of a material's mechanical properties. Methods for doing this were remarkably improved at the beginning of the eighteenth century. At that time both experimental and theoretical studies in this subject were undertaken primarily in France by such notables as Saint-Venant, Poisson, Lamé, and Navier. Because their efforts were based on material-body applications of mechanics, they called this study "strength of materials." Currently, however, it is usually referred to as "mechanics of deformable bodies" or simply "mechanics of materials."

Over the years, after many of the fundamental problems of mechanics of materials had been solved, it became necessary to use advanced mathematical and computer techniques to solve more complex problems. As a result, this subject expanded into other subjects of advanced mechanics such as the *theory of elasticity* and the *theory of plasticity*. Research in these fields is ongoing, not only to meet the demands for solving advanced engineering problems, but to justify further use and the limitations upon which the fundamental theory of mechanics of materials is based.

1.2 EQUILIBRIUM OF A DEFORMABLE BODY

Since statics plays an important role in both the development and application of mechanics of materials, it is very important to have a good grasp of its fundamentals. For this reason we will review some of the main principles of statics that will be used throughout the text.

External Loads. A body can be subjected to several different types of external loads; however, any one of these can be classified as either a surface force or a body force, Fig. 1-1.

Surface Forces. As the name implies, **surface forces** are caused by the direct contact of one body with the surface of another. In all cases these forces are distributed over the *area* of contact between the bodies. If this area is small in comparison with the total surface area of the body, then the surface force can be *idealized* as a single **concentrated force**, which is applied to a *point* on the body. For example, the force of the ground on the wheels of a bicycle can be considered as a concentrated force when studying the loading on the bicycle. If the surface loading is applied along a narrow area, the loading can be *idealized* as a **linear distributed load**, $w(s)$. Here the loading is measured as having an intensity of force/length along the area and is represented graphically by a series of arrows along the line s. *The* *resultant force F_R of $w(s)$ is equivalent to the area under the distributed* *loading curve, and this resultant acts through the centroid C or geometric* *center of this area.* The loading along the length of a beam is a typical example of where this idealization is often applied.

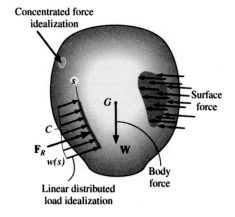

Concentrated force idealization

Linear distributed load idealization

Fig. 1–1

Body Force. A ***body force*** is developed when one body exerts a force on another body without direct physical contact between the bodies. Examples include the effects caused by the earth's gravitation or its electromagnetic field. Although body forces affect each of the particles composing the body, these forces are normally represented by a single concentrated force acting on the body. In the case of gravitation, this force is called the ***weight*** of the body and acts through the body's center of gravity.

Support Reactions. The surface forces that develop at the supports or points of contact between bodies are called ***reactions***. For two-dimensional problems, i.e., bodies subjected to coplanar force systems, the supports most commonly encountered are shown in Table 1–1. Note carefully the symbol used to represent each support and the type of reactions it exerts on its contacting member. In general, one can always determine the type of support reaction by imagining the attached member as being translated or rotated in a particular direction. *If the support prevents translation in a given direction, then a force must be developed on the member in that direction. Likewise, if rotation is prevented, a couple moment must be exerted on the member.* For example, a roller support can only prevent translation in the contact direction, perpendicular or normal to the surface. Hence, the roller exerts a normal force **F** on the member at the point of contact. Since the member can freely rotate about the roller, a couple moment cannot be developed on the member.

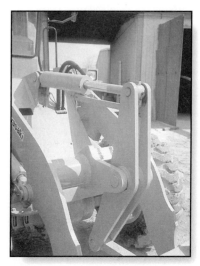

Many machine elements are pin connected in order to enable free rotation at their connections. These supports exert a force on a member, but no moment.

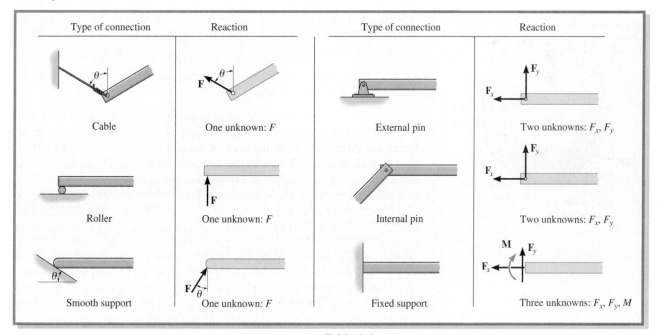

Table 1–1

Equations of Equilibrium. Equilibrium of a body requires both a *balance of forces*, to prevent the body from translating or having accelerated motion along a straight or curved path, and a *balance of moments*, to prevent the body from rotating. These conditions can be expressed mathematically by the two vector equations

$$\Sigma \mathbf{F} = \mathbf{0}$$
$$\Sigma \mathbf{M}_O = \mathbf{0}$$

(1–1)

Here, $\Sigma \mathbf{F}$ represents the sum of all the forces acting on the body, and $\Sigma \mathbf{M}_O$ is the sum of the moments of all the forces about any point O either on or off the body. If an x, y, z coordinate system is established with the origin at point O, the force and moment vectors can be resolved into components along the coordinate axes and the above two equations can be written in scalar form as six equations, namely,

$$\Sigma F_x = 0 \qquad \Sigma F_y = 0 \qquad \Sigma F_z = 0$$
$$\Sigma M_x = 0 \qquad \Sigma M_y = 0 \qquad \Sigma M_z = 0$$

(1–2)

Often in engineering practice the loading on a body can be represented as a system of *coplanar forces*. If this is the case, and the forces lie in the x–y plane, then the conditions for equilibrium of the body can be specified by only three scalar equilibrium equations; that is,

$$\Sigma F_x = 0$$
$$\Sigma F_y = 0$$
$$\Sigma M_O = 0$$

(1–3)

In this case, if point O is the origin of coordinates, then moments will always be directed along the z axis, which is perpendicular to the plane that contains the forces.

Successful application of the equations of equilibrium requires complete specification of all the known and unknown forces that act *on* the body. *The best way to account for these forces is to draw the body's free-body diagram.* Obviously, if the free-body diagram is drawn correctly, the effects of all the applied forces and couple moments can be accounted for when the equations of equilibrium are written.

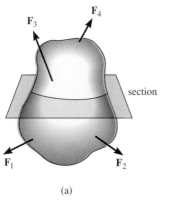

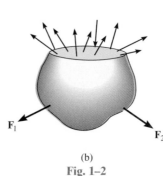

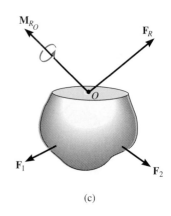

(a)

(b)

(c)

Fig. 1–2

Internal Resultant Loadings.   One of the most important applications of statics in the analysis of mechanics of materials problems is to be able to determine the resultant force and moment acting *within* a body, which are necessary to hold the body together when the body is subjected to external loads. For example, consider the body shown in Fig. 1–2*a*, which is held in equilibrium by the four external forces.* In order to obtain the *internal loadings* acting on a specific region within the body, it is necessary to use the **method of sections**. This requires that an imaginary section or "cut" be made through the region where the internal loadings are to be determined. The two parts of the body are then separated, and a free-body diagram of one of the parts is drawn, Fig. 1–2*b*. Here it can be seen that there is actually a *distribution* of internal force acting on the "exposed" area of the section. These forces represent the effects of the material of the top part of the body acting on the adjacent material of the bottom part.

Although the exact distribution of internal loading may be *unknown*, we can use the equations of equilibrium to relate the external forces on the body to the distribution's *resultant force and moment*, $\mathbf{F}_R$ and $\mathbf{M}_{R_O}$, *at any specific point O* on the sectioned area, Fig. 1–2*c*. When doing so, note that $\mathbf{F}_R$ acts through point O, although its computed value will *not* depend on the location of this point. On the other hand, $\mathbf{M}_{R_O}$, does depend on this location, since the moment arms must extend from O to the line of action of each external force on the free-body diagram. It will be shown in later portions of the text that point O is most often chosen at the *centroid* of the sectioned area, and so we will always choose this location for O, unless otherwise stated. Also, if a member is long and slender, as in the case of a rod or beam, the section to be considered is generally taken *perpendicular* to the longitudinal axis of the member. This section is referred to as the **cross section**.

---

*The body's weight is not shown, since it is assumed to be quite small, and therefore negligible compared with the other loads.

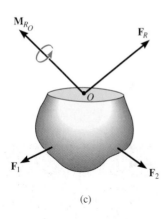

(c)

Fig. 1–2

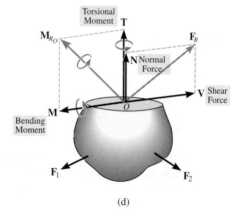

(d)

Three Dimensions.   Later in this text we will show how to relate the resultant loadings, $\mathbf{F}_R$ and $\mathbf{M}_{R_O}$, to the *distribution of force* on the sectioned area, and thereby develop equations that can be used for analysis and design. To do this, however, the components of $\mathbf{F}_R$ and $\mathbf{M}_{R_O}$, acting both normal or perpendicular to the sectioned area and within the plane of the area, must be considered, Fig. 1–2*d*. Four different types of resultant loadings can then be defined as follows:

Normal force, **N.**   This force acts perpendicular to the area. It is developed whenever the external loads tend to push or pull on the two segments of the body.

Shear force, **V.** The shear force lies in the plane of the area and is developed when the external loads tend to cause the two segments of the body to slide over one another.

Torsional moment or torque, **T.** This effect is developed when the external loads tend to twist one segment of the body with respect to the other.

Bending moment, **M.** The bending moment is caused by the external loads that tend to bend the body about an axis lying within the plane of the area.

In this text, note that graphical representation of a moment or torque is shown in three dimensions as a vector with an associated curl. By the *right-hand rule*, the thumb gives the arrowhead sense of the vector and the fingers or curl indicate the tendency for rotation (twist or bending). Using an *x, y, z* coordinate system, each of the above loadings can be determined directly from the six equations of equilibrium applied to either segment of the body.

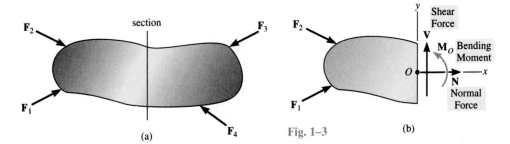

(a)

Fig. 1–3

(b)

Coplanar Loadings. If the body is subjected to a *coplanar system of forces*, Fig. 1–3a, then only normal-force, shear, and bending-moment components will exist at the section, Fig. 1–3b. If we use the *x, y, z* coordinate axes, with origin established at point *O*, as shown on the left segment, then a direct solution for **N** can be obtained by applying $\Sigma F_x = 0$, and **V** can be obtained directly from $\Sigma F_y = 0$. Finally, the bending moment $\mathbf{M}_O$ can be determined directly by summing moments about point *O* (the *z* axis), $\Sigma M_O = 0$, in order to eliminate the moments caused by the unknowns **N** and **V**.

In order to design the members of this building frame, it is first necessary to find the internal loadings at various points along their length.

IMPORTANT POINTS

- *Mechanics of materials* is a study of the relationship between the external loads on a body and the intensity of the internal loads within the body.

- External forces can be applied to a body as *distributed* or *concentrated surface loadings*, or as *body forces* which act throughout the volume of the body.

- Linear distributed loadings produce a *resultant force* having a *magnitude* equal to the *area* under the load diagram, and having a *location* that passes through the *centroid* of this area.

- A support produces a *force* in a particular direction on its attached member if it *prevents translation* of the member in that direction, and it produces a *couple moment* on the member if it *prevents rotation*.

- The equations of equilibrium $\Sigma \mathbf{F} = 0$ and $\Sigma \mathbf{M} = 0$ must be satisfied in order to prevent a body from translating with accelerated motion and from rotating.

- When applying the equations of equilibrium, it is important to first draw the body's free-body diagram in order to account for all the terms in the equations.

- The method of sections is used to determine the internal resultant loadings acting on the surface of the sectioned body. In general, these resultants consist of a normal force, shear force, torsional moment, and bending moment.

PROCEDURE FOR ANALYSIS

The method of sections is used to determine the resultant *internal* loadings at a point located on the section of a body. To obtain these resultants, application of the method of sections requires the following steps.

Support Reactions.

• First decide which segment of the body is to be considered. If the segment has a support or connection to another body, then *before* the body is sectioned, it will be necessary to determine the reactions acting on the chosen segment of the body. Draw the free-body diagram for the *entire body* and then apply the necessary equations of equilibrium to obtain these reactions.

Free-Body Diagram.

• Keep all external distributed loadings, couple moments, torques, and forces acting on the body in their *exact locations*, then pass an imaginary section through the body at the point where the resultant internal loadings are to be determined.

• If the body represents a member of a structure or mechanical device, the section is often taken *perpendicular* to the longitudinal axis of the member.

• Draw a free-body diagram of one of the "cut" segments and indicate the unknown resultants N, V, M, and T at the section. These resultants are normally placed at the point representing the geometric center or *centroid* of the sectioned area.

• If the member is subjected to a *coplanar* system of forces, only N, V, and M act at the centroid.

• Establish the x, y, z coordinate axes with origin at the centroid and show the resultant components acting along the axes.

Equations of Equilibrium.

• Moments should be summed at the section, about each of the coordinate axes where the resultants act. Doing this eliminates the unknown forces N and V and allows a direct solution for M (and T).

• If the solution of the equilibrium equations yields a negative value for a resultant, the assumed *directional sense* of the resultant is *opposite* to that shown on the free-body diagram.

The following examples illustrate this procedure numerically and also provide a review of some of the important principles of statics.

EXAMPLE 1–1

Determine the resultant internal loadings acting on the cross section at C of the beam shown in Fig. 1–4a.

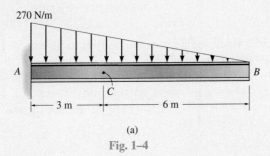

(a)

Fig. 1–4

SOLUTION

Support Reactions.   This problem can be solved in the most direct manner by considering segment CB of the beam, since then the support reactions at A do not have to be computed.

Free-Body Diagram.   Passing an imaginary section perpendicular to the longitudinal axis of the beam yields the free-body diagram of segment CB shown in Fig. 1–4b. It is important to keep the distributed loading exactly where it is on the segment until *after* the section is made. Only then should this loading be replaced by a single resultant force. Notice that the intensity of the distributed loading at C is found by proportion, i.e., from Fig.1–4a, $w/6\text{ m} = (270\text{ N/m})/9\text{m}$, $w = 180\text{ N/m}$. The magnitude of the resultant of the distributed load is equal to the area under the loading curve (triangle) and acts through the centroid of this area. Thus, $F = \frac{1}{2}(180\text{ N/m})(6\text{ m}) = 540\text{ N}$, which acts $1/3(6\text{ m}) = 2\text{ m}$ from C as shown in Fig. 1–4b.

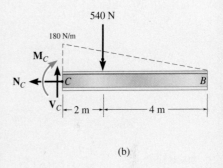

(b)

Equations of Equilibrium.   Applying the equations of equilibrium we have

$\xrightarrow{+} \Sigma F_x = 0;$ $-N_C = 0$

$N_C = 0$ *Ans.*

$+\uparrow \Sigma F_y = 0;$ $V_C - 540\text{ N} = 0$

$V_C = 540\text{ N}$ *Ans.*

$\zeta + \Sigma M_C = 0;$ $-M_C - 540\text{ N}(2\text{ m}) = 0$

$M_C = -1080\text{ N} \cdot \text{m}$ *Ans.*

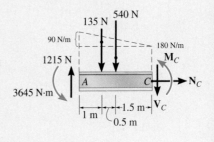

(c)

The negative sign indicates that $\mathbf{M}_C$ acts in the opposite direction to that shown on the free-body diagram. Try solving this problem using segment AC, by first obtaining the support reactions at A, which are given in Fig. 1–4c.

EXAMPLE 1–2

Determine the resultant internal loadings acting on the cross section at *C* of the machine shaft shown in Fig. 1–5*a*. The shaft is supported by bearings at *A* and *B*, which exert only vertical forces on the shaft.

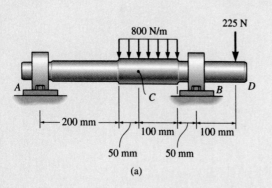

(a)

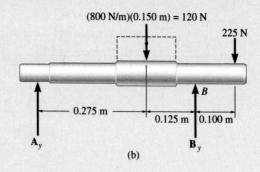

(b)

Fig. 1–5

SOLUTION

We will solve this problem using segment *AC* of the shaft.

Support Reactions.   A free-body diagram of the entire shaft is shown in Fig. 1–5*b*. Since segment *AC* is to be considered, only the reaction at *A* has to be determined. Why?

$$\zeta^+ \Sigma M_B = 0; \quad -A_y(0.400 \text{ m}) + 120 \text{ N}(0.125 \text{ m}) - 225 \text{ N}(0.100 \text{ m}) = 0$$
$$A_y = -18.75 \text{ N}$$

The negative sign for A_y indicates that $\mathbf{A}_y$ acts in the *opposite sense* to that shown on the free-body diagram.

Free-Body Diagram.   Passing an imaginary section perpendicular to the axis of the shaft through *C* yields the free-body diagram of segment *AC* shown in Fig. 1–5*c*.

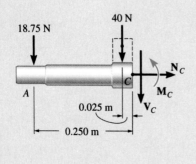

(c)

Equations of Equilibrium.

$$\xrightarrow{+} \Sigma F_x = 0; \qquad\qquad\qquad N_C = 0 \qquad\qquad\qquad Ans.$$
$$+\uparrow \Sigma F_y = 0; \qquad -18.75 \text{ N} - 40 \text{ N} - V_C = 0$$
$$V_C = -58.8 \text{ N} \qquad\qquad Ans.$$
$$\zeta^+ \Sigma M_C = 0; \quad M_C + 40 \text{ N}(0.025 \text{ m}) + 18.75 \text{ N}(0.250 \text{ m}) = 0$$
$$M_C = -5.69 \text{ N} \cdot \text{m} \qquad\qquad Ans.$$

What do the negative signs for V_C and M_C indicate? As an exercise, calculate the reaction at *B* and try to obtain the same results using segment *CBD* of the shaft.

EXAMPLE 1-3

The hoist in Fig. 1–6*a* consists of the beam *AB* and attached pulleys, the cable, and the motor. Determine the resultant internal loadings acting on the cross section at *C* if the motor is lifting the 500-lb load *W* with constant velocity. Neglect the weight of the pulleys and beam.

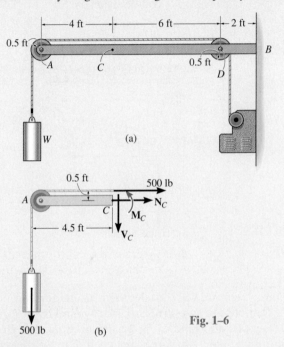

Fig. 1–6

SOLUTION

The most direct way to solve this problem is to section both the cable and the beam at *C* and then consider the entire left segment.

Free-Body Diagram.   See Fig. 1–6*b*.

Equations of Equilibrium.

$$\xrightarrow{+} \Sigma F_x = 0; \qquad 500 \text{ lb} + N_C = 0 \qquad N_C = -500 \text{ lb} \qquad \textit{Ans.}$$

$$+\uparrow \Sigma F_y = 0; \qquad -500 \text{ lb} - V_C = 0 \qquad V_C = -500 \text{ lb} \qquad \textit{Ans.}$$

$$\zeta^+ \Sigma M_C = 0; \qquad 500 \text{ lb}(4.5 \text{ ft}) - 500 \text{ lb}(0.5 \text{ ft}) + M_C = 0$$

$$M_C = -2000 \text{ lb} \cdot \text{ft} \qquad \textit{Ans.}$$

As an exercise, try obtaining these same results by considering just the beam segment *AC*, i.e., remove the pulley at *A* from the beam and show the 500-lb force components of the pulley acting on the beam segment *AC*. Also, this problem can be worked by first finding the reactions at *B*, ($B_x = 0$, $B_y = 1000$ lb, $M_B = 7000$ lb $\cdot$ ft) and then considering segment *CB*.

EXAMPLE 1-4

Determine the resultant internal loadings acting on the cross section at G of the wooden beam shown in Fig. 1–7a. Assume the joints at A, B, C, D, and E are pin connected.

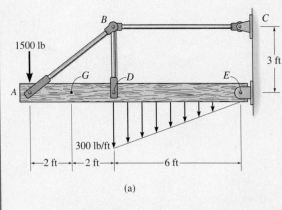

1500 lb

300 lb/ft

2 ft | 2 ft | 6 ft

(a)

Fig. 1–7

$F_{BC} = 6200$ lb

3 ft

1500 lb

$E_x = 6200$ lb

$E_y = 2400$ lb

6 ft

$\frac{2}{3}$ (6 ft) = 4 ft

$\frac{1}{2}$ (6 ft)(300 lb/ft) = 900 lb

(b)

SOLUTION

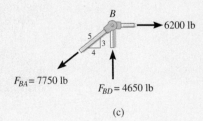

B

6200 lb

$F_{BA} = 7750$ lb

$F_{BD} = 4650$ lb

(c)

Support Reactions. Here we will consider segment AG for the analysis. A free-body diagram of the *entire* structure is shown in Fig. 1–7b. Verify the computed reactions at E and C. In particular, note that BC is a *two-force member* since only two forces act on it. For this reason the reaction at C must be horizontal as shown.

Since BA and BD are also two-force members, the free-body diagram of joint B is shown in Fig. 1–7c. Again, verify the magnitudes of the computed forces $\mathbf{F}_{BA}$ and $\mathbf{F}_{BD}$.

1500 lb   7750 lb

A   G

$\mathbf{N}_G$

$\mathbf{M}_G$

2 ft

$\mathbf{V}_G$

(d)

Free-Body Diagram. Using the result for $\mathbf{F}_{BA}$, the left section AG of the beam is shown in Fig. 1–7d.

Equations of Equilibrium. Applying the equations of equilibrium to segment AG, we have

$$\xrightarrow{+} \Sigma F_x = 0; \quad 7750 \text{ lb}(\tfrac{4}{5}) + N_G = 0 \quad N_G = -6200 \text{ lb} \qquad Ans.$$

$$+\uparrow \Sigma F_y = 0; \quad -1500 \text{ lb} + 7750 \text{ lb}(\tfrac{3}{5}) - V_G = 0$$

$$V_G = 3150 \text{ lb} \qquad Ans.$$

$$\zeta + \Sigma M_G = 0; \quad M_G - (7750 \text{ lb})(\tfrac{3}{5})(2 \text{ ft}) + 1500 \text{ lb}(2 \text{ ft}) = 0$$

$$M_G = 6300 \text{ lb} \cdot \text{ft} \qquad Ans.$$

As an exercise, compute these same results using segment GE.

EXAMPLE 1-5

Determine the resultant internal loadings acting on the cross section at B of the pipe shown in Fig. 1–8a. The pipe has a mass of 2 kg/m and is subjected to both a vertical force of 50 N and a couple moment of 70 N · m at its end A. It is fixed to the wall at C.

SOLUTION

The problem can be solved by considering segment AB, which does *not* involve the support reactions at C.

Free-Body Diagram. The x, y, z axes are established at B and the free-body diagram of segment AB is shown in Fig. 1–8b. The resultant force and moment components at the section are assumed to act in the positive coordinate directions and to pass through the *centroid* of the cross-sectional area at B. The weight of each segment of pipe is calculated as follows:

$$W_{BD} = (2 \text{ kg/m})(0.5 \text{ m})(9.81 \text{ N/kg}) = 9.81 \text{ N}$$
$$W_{AD} = (2 \text{ kg/m})(1.25 \text{ m})(9.81 \text{ N/kg}) = 24.525 \text{ N}$$

These forces act through the center of gravity of each segment.

Equations of Equilibrium. Applying the six scalar equations of equilibrium, we have*

$$\Sigma F_x = 0; \qquad\qquad (F_B)_x = 0 \qquad\qquad \textit{Ans.}$$
$$\Sigma F_y = 0; \qquad\qquad (F_B)_y = 0 \qquad\qquad \textit{Ans.}$$
$$\Sigma F_z = 0; \quad (F_B)_z - 9.81 \text{ N} - 24.525 \text{ N} - 50 \text{ N} = 0$$
$$\qquad\qquad\qquad (F_B)_z = 84.3 \text{ N} \qquad\qquad \textit{Ans.}$$
$$\Sigma (M_B)_x = 0; \quad (M_B)_x + 70 \text{ N} \cdot \text{m} - 50 \text{ N}(0.5 \text{ m}) - 24.525 \text{ N}(0.5 \text{ m})$$
$$\qquad\qquad\qquad\qquad - 9.81 \text{ N}(0.25 \text{ m}) = 0$$
$$\qquad\qquad\qquad (M_B)_x = -30.3 \text{ N} \cdot \text{m} \qquad\qquad \textit{Ans.}$$
$$\Sigma (M_B)_y = 0; \quad (M_B)_y + 24.525 \text{ N}(0.625 \text{ m}) + 50 \text{ N}(1.25 \text{ m}) = 0$$
$$\qquad\qquad\qquad (M_B)_y = -77.8 \text{ N} \cdot \text{m} \qquad\qquad \textit{Ans.}$$
$$\Sigma (M_B)_z = 0; \qquad\qquad (M_B)_z = 0 \qquad\qquad \textit{Ans.}$$

What do the negative signs for $(M_B)_x$ and $(M_B)_y$ indicate? Note that the normal force $N_B = (F_B)_y = 0$, whereas the shear force is $V_B = \sqrt{(0)^2 + (84.3)^2} = 84.3 \text{ N}$. Also, the torsional moment is $T_B = (M_B)_y = 77.8 \text{ N} \cdot \text{m}$ and the bending moment is $M_B = \sqrt{(30.3)^2 + (0)^2} = 30.3 \text{ N} \cdot \text{m}$.

---

*The *magnitude* of each moment about an axis is equal to the magnitude of each force times the perpendicular distance from the axis to the line of action of the force. The *direction* of each moment is determined using the right-hand rule, with positive moments (thumb) directed along the positive coordinate axes.

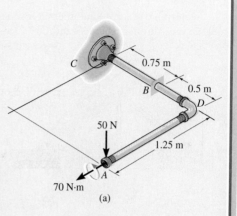

(a)

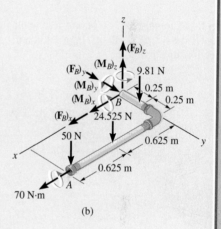

(b)

Fig. 1–8

PROBLEMS

1–1. Determine the resultant internal torque acting on the cross sections through points B and C.

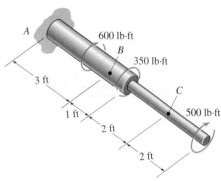

Prob. 1–1

1–2. A force of 80 N is supported by the bracket as shown. Determine the resultant internal loadings acting on the section through point A.

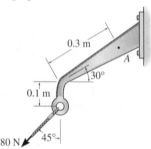

Prob. 1–2

1–3. The beam AB is fixed to the wall and has a uniform weight of 80 lb/ft. If the trolley supports a load of 1500 lb, determine the resultant internal loadings acting on the cross sections through points C and D.

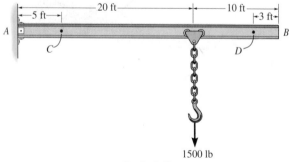

Prob. 1–3

*1–4. Determine the resultant internal torque acting on the cross sections through points C and D of the shaft. The shaft is fixed at B.

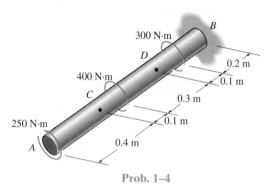

Prob. 1–4

1–5. The beam supports the distributed load shown. Determine the resultant internal loadings on the cross section through point C. Assume the reactions at the supports A and B are vertical.

1–6. The beam supports the distributed load shown. Determine the resultant internal loadings on the cross sections through points D and E. Assume the reactions at the supports A and B are vertical.

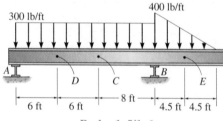

Probs. 1–5/1–6

1–7. Determine the resultant internal loadings on the cross section through point D on member AB.

*1–8. Determine the resultant internal loadings at cross sections through points E and F on the assembly.

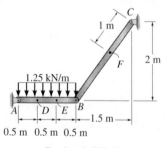

Probs. 1–7/1–8

1–9. The boom *DF* of the jib crane and the column *DE* have a uniform weight of 50 lb/ft. If the hoist and load weigh 300 lb, determine the resultant internal loadings in the crane on cross sections through points *A*, *B*, and *C*.

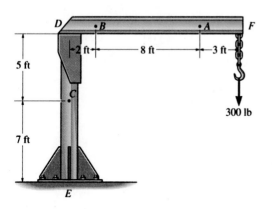

Prob. 1–9

1–10. Determine the resultant internal loadings acting on the cross section at point *C*. The cooling unit has a total weight of 52 kip and a center of gravity at *G*.

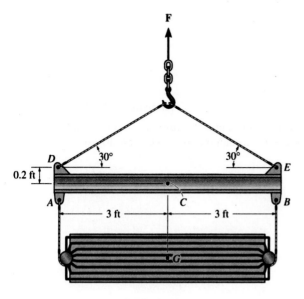

Prob. 1–10

1–11. Determine the resultant internal loadings acting on the cross section at point *B*.

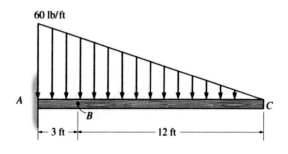

Prob. 1–11

***1–12.** Determine the resultant internal loading on the cross section through point *C* of the pliers. There is a pin at *A*, and the jaws at *B* are smooth.

1–13. Determine the resultant internal loading on the cross section through point *D* of the pliers.

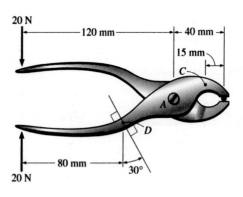

Probs. 1–12/1–13

1–14. The 800-lb load is being hoisted at a constant speed using the motor *M*, which has a weight of 90 lb. Determine the resultant internal loadings acting on the cross section through point *B* in the beam. The beam has a weight of 40 lb/ft and is fixed to the wall at *A*.

1–15. Determine the resultant internal loadings acting on the cross section through points *C* and *D* of the beam in Prob. 1–14.

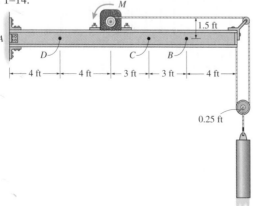

Prob. 1–14/1–15

***1–16.** Determine the normal force, shear force, and moment at a section through point *C*. Take *P* = 8 kN.

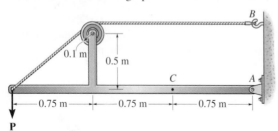

Prob. 1–16

1–17. The cable will fail when subjected to a tension of 2 kN. Determine the largest vertical load *P* the frame will support and calculate the internal normal force, shear force, and moment at the cross section through point *C* for this loading.

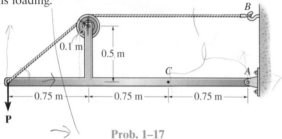

Prob. 1–17

1–18. The sky hook is used to support the cable of a scaffold over the side of a building. If it consists of a smooth rod that contacts the parapet of a wall at points *A*, *B*, and *C*, determine the normal force, shear force, and moment on the cross section at points *D* and *E*.

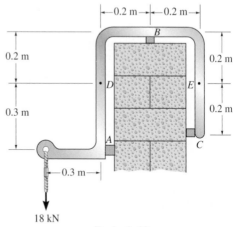

Prob. 1–18

1–19. The serving tray *T* used on an airplane is supported on *each side* by an arm. The tray is pin connected to the arm at *A*, and at *B* there is a smooth pin. (The pin can move within the slot in the arms to permit folding the tray against the front passenger seat when not in use.) Determine the resultant internal loadings acting on the cross section of the arm through point *C* when the tray arm supports the loads shown.

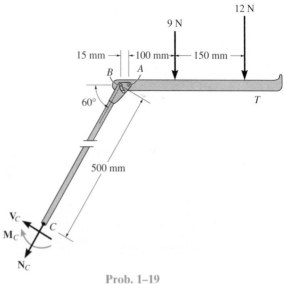

Prob. 1–19

*1–20. The metal stud punch is subjected to a force of 120 N on the handle. Determine the magnitude of the reactive force at the pin A and in the short link BC. Also, determine the internal resultant loadings acting on the cross section passing through the handle arm at D.

1–21. Solve Prob. 1–20 for the resultant internal loadings acting on the cross section passing through the handle arm at E and at a cross section of the short link BC.

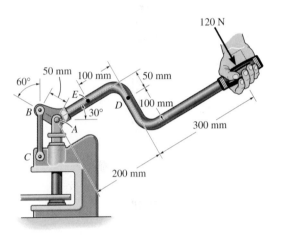

Probs. 1–20/1–21

1–22. Determine the resultant internal loadings in the beam at cross sections through points D and E. Point E is just to the right of the 3-kip load.

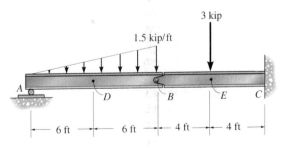

Prob. 1–22

1–23. Determine the resultant internal loadings acting on section a–a through the centroid, point C on the beam.

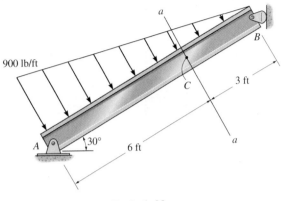

Prob. 1–23

*1–24. The wishbone construction of the power pole supports the three lines, each exerting a force of 800 lb on the bracing struts. If the struts are pin connected at A, B, and C, determine the resultant internal loadings at cross sections through points D, E, and F.

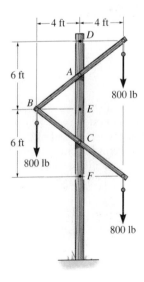

Prob. 1–24

1–25. Determine the resultant internal loadings acting on section b–b through the centroid, point C on the beam.

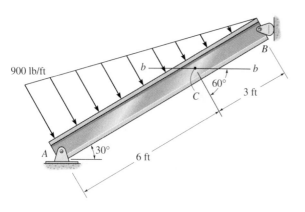

900 lb/ft

60°

3 ft

30°

6 ft

Prob. 1–25

1–26. The shaft is supported at its ends by two bearings A and B and is subjected to the forces applied to the pulleys fixed to the shaft. Determine the resultant internal loadings acting on the cross section located at point C. The 300-N forces act in the –z direction and the 500-N forces act in the +x direction. The journal bearings at A and B exert only x and z components of force on the shaft.

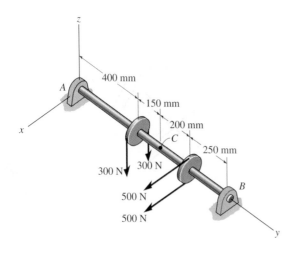

400 mm

150 mm

200 mm

C

250 mm

300 N

300 N

500 N

500 N

B

Prob. 1–26

1–27. The shaft is supported at its ends by two bearings A and B and is subjected to the forces applied to the pulleys fixed to the shaft. Determine the resultant internal loadings acting on the cross section through point D. The 400-N forces act in the −z direction and the 200-N and 80-N forces act in the +y direction. The journal bearings at A and B exert only y and z components of force on the shaft.

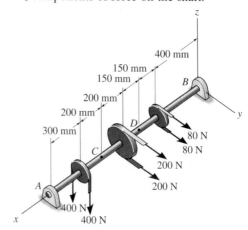

400 mm

150 mm

150 mm

200 mm

200 mm

300 mm

D

B

80 N

80 N

200 N

200 N

A

400 N

400 N

Prob. 1–27

*1–28. The shaft is supported at its ends by two bearings A and B and is subjected to the forces applied to the pulleys fixed to the shaft. Determine the resultant internal loadings acting on the cross section through point C. The 400-N forces act in the −z direction and the 200-N and 80-N forces act in the +y direction. The journal bearings at A and B exert only y and z components of force on the shaft.

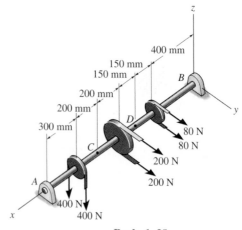

400 mm

150 mm

150 mm

200 mm

200 mm

300 mm

D

B

80 N

80 N

200 N

200 N

C

A

400 N

400 N

Prob. 1–28

1–29. The sign has a weight of 1500 lb and center of gravity at *G*. If it is subjected to the uniform wind load of 60 lb/ft², determine the resultant internal loadings acting on the cross section of the post at *A*. The post has a weight of 100 lb/ft.

1–31. The curved rod has a radius *r* and is fixed to the wall at *B*. Determine the resultant internal loadings acting on the cross section through *A* which is located at an angle *θ* from the horizontal.

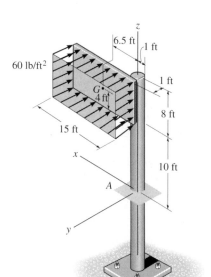

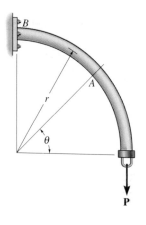

Prob. 1–29

Prob. 1–31

1–30. The curved rod *AD* of radius *r* has a weight per length of *w*. If it lies in the horizontal plane, determine the resultant internal loadings acting on the cross section through point *B*. *Hint:* The distance from the centroid *C* of segment *AB* to point *O* is *CO* = 0.9745*r*.

*1–32. A differential element taken from a curved bar is shown in the figure. Show that $dN/d\theta = V$, $dV/d\theta = -N$, $dM/d\theta = -T$, and $dT/d\theta = M$.

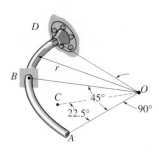

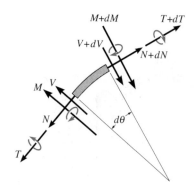

Prob. 1–30

Prob. 1–32

1.3   STRESS

It was stated in Sec. 1.2 that the force and moment acting at a specified point on the sectioned area of a body, Fig. 1–9, represents the resultant effects of the actual *distribution of force* acting over the sectioned area, Fig. 1–9*b*. Obtaining this *distribution* of internal loading is of primary importance in mechanics of materials. To solve this problem it is necessary to establish the concept of stress.

Consider the sectioned area to be subdivided into small areas, such as ΔA shown dark shaded in Fig. 1–10*a*. As we reduce ΔA to a smaller and smaller size, we must make two assumptions regarding the properties of the material. We will consider the material to be **continuous**, that is, to consist of a *continuum* or uniform distribution of matter having no voids, rather than being composed of a finite number of distinct atoms or molecules. Furthermore, the material must be **cohesive**, meaning that all portions of it are connected together, rather than having breaks, cracks, or separations. A typical finite yet very small force $\Delta \mathbf{F}$, acting on its associated area ΔA, is shown in Fig. 1–10*a*. This force, like all the others, will have a unique direction, but for further discussion we will replace it by its *three components*, namely, $\Delta \mathbf{F}_x$, $\Delta \mathbf{F}_y$, and $\Delta \mathbf{F}_z$, which are taken tangent and normal to the area, respectively. As the area ΔA approaches zero, so do the force $\Delta \mathbf{F}$ and its components; however, the quotient of the force and area will, in general, approach a finite limit. This quotient is called *stress*, and as noted, it describes the *intensity of the internal force* on a *specific plane* (area) passing through a point.

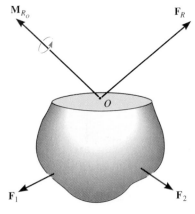

Fig. 1–9

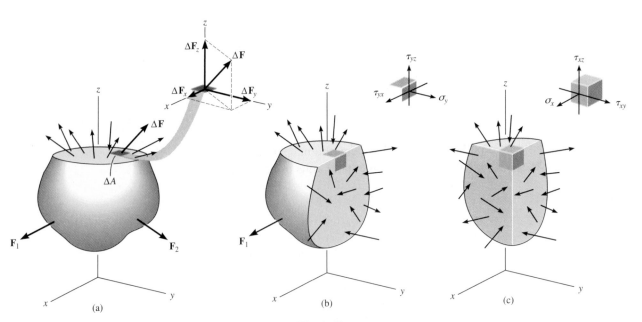

Fig. 1–10

Normal Stress. The *intensity* of force, or force per unit area, acting normal to ΔA is defined as the **normal stress**, σ (sigma). Since ΔF_z is normal to the area then

$$\sigma_z = \lim_{\Delta A \to 0} \frac{\Delta F_z}{\Delta A} \qquad (1\text{--}4)$$

If the normal force or stress "pulls" on the area element ΔA as shown in Fig. 1–10a, it is referred to as *tensile stress*, whereas if it "pushes" on ΔA it is called *compressive stress*.

Shear Stress. The intensity of force, or force per unit area, acting tangent to ΔA is called the **shear stress**, τ (tau). Here we have shear stress components,

$$\tau_{zx} = \lim_{\Delta A \to 0} \frac{\Delta F_x}{\Delta A}$$

$$\tau_{zy} = \lim_{\Delta A \to 0} \frac{\Delta F_y}{\Delta A} \qquad (1\text{--}5)$$

Note that the subscript notation z in σ_z is used to reference the *direction* of the outward normal line, which specifies the orientation of the area ΔA, Fig. 1–11. Two subscripts are used for the shear-stress components, τ_{zx} and τ_{zy}. The z axis specifies the orientation of the area, and x and y refer to the direction lines for the shear stresses.

General State of Stress. If the body is further sectioned by planes parallel to the x–z plane, Fig. 1–10b, and the y–z plane, Fig. 1–10c, we can then "cut out" a cubic volume element of material that represents the **state of stress** acting around the chosen point in the body, Fig. 1–12. This state of stress is then characterized by three components acting on each face of the element. These stress components describe the state of stress at the point only for the element orientated along the x, y, z axes. Had the body been sectioned into a cube having some other orientation, then the state of stress would be defined using a different set of stress components.

Units. In the International Standard or SI system, the magnitudes of both normal and shear stress are specified in the basic units of newtons per square meter (N/m^2). This unit, called a *pascal* $(1\ Pa = 1\ N/m^2)$ is rather small, and in engineering work prefixes such as kilo- (10^3), symbolized by k, mega- (10^6), symbolized by M, or giga- (10^9), symbolized by G, are used to represent larger, more realistic values of stress.* Likewise, in the U.S. Customary or Foot-Pound-Second system of units, engineers usually express stress in pounds per square inch (psi) or kilopounds per square inch (ksi), where 1 kilopound (kip) = 1000 lb.

---

*Sometimes stress is expressed in units of N/mm^2, where $1\ mm = 10^{-3} m$. However, in the SI system, prefixes are not allowed in the denominator of a fraction and therefore it is better to use the equivalent $1\ N/mm^2 = 1\ MN/m^2 = 1\ MPa$.

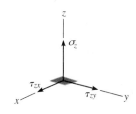

Fig. 1–11

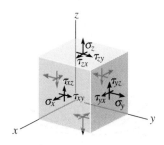

General state of stress

Fig. 1–12

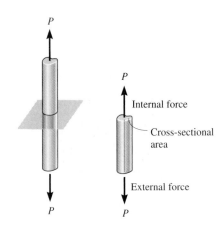

P

P

Internal force

Cross-sectional area

External force

P

P

(a)   (b)

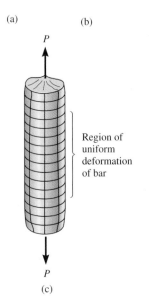

P

Region of uniform deformation of bar

P

(c)

Fig. 1–13

1.4   AVERAGE NORMAL STRESS IN AN AXIALLY LOADED BAR

Frequently structural or mechanical members are made long and slender. Also, they are subjected to axial loads that are usually applied to the ends of the member. Truss members, hangers, and bolts are typical examples. In this section we will determine the average stress distribution acting on the cross section of an axially loaded bar, such as the one having the general form shown in Fig. 1–13*a*. This section defines the **cross-sectional area** of the bar, and since all such cross sections are the same, the bar is referred to as being **prismatic**. If we neglect the weight of the bar and section it as indicated, then, for equilibrium of the bottom segment, Fig. 1–13*b*, the internal resultant force acting on the cross-sectional area must be equal in magnitude, opposite in direction, and collinear to the external force acting at the bottom of the bar.

Assumptions.   Before we determine the average stress distribution acting over the bar's cross-sectional area, it is necessary to make two simplifying assumptions concerning the material description and the specific application of the load.

1. It is necessary that the bar remains straight both before and after the load is applied, and also, the cross section should remain flat or plane during the deformation, that is, during the time the bar changes its volume and shape. If this occurs, then horizontal and vertical grid lines inscribed on the bar will *deform uniformly* when the bar is subjected to the load, Fig. 1–13*c*. Here we will not consider regions of the bar near its ends, where application of the external loads can cause *localized distortions*. Instead we will focus only on the stress distribution within the bar's midsection.

2. In order for *the bar to undergo uniform deformation*, it is necessary that **P** be applied along the *centroidal axis* of the cross section, and the material must be homogeneous and isotropic. **Homogeneous material** has the same physical and mechanical properties throughout its volume, and **isotropic material** has these same properties in all directions. Many engineering materials may be approximated as being both homogeneous and isotropic as assumed here. Steel, for example, contains thousands of randomly oriented crystals in each cubic millimeter of its volume, and since most problems involving this material have a physical size that is much larger than a single crystal, the above assumption regarding its material composition is quite realistic. It should be mentioned, however, that steel can be made anisotropic by cold-rolling, i.e., rolling or forging it at subcritical temperatures. **Anisotropic materials** have different properties in different directions, and although this is the case, if the anisotropy is oriented along the bar's axis, then the bar will also deform uniformly when subjected to an axial load. For example, timber, due to its grains or fibers of wood, is an engineering material that is homogeneous and anisotropic and is therefore suited for the following analysis.

Average Normal Stress Distribution.   Provided the bar is subjected to a constant uniform deformation as noted, then this deformation is the result of a *constant* normal stress σ, Fig. 1–13d. As a result, each area ΔA on the cross section is subjected to a force $\Delta F = \sigma \Delta A$, and the *sum* of these forces acting over the entire cross-sectional area must be equivalent to the internal resultant force P at the section. If we let $\Delta A \to dA$ and therefore $\Delta F \to dF$, then, recognizing σ is *constant*, we have

$$+\uparrow F_{Rz} = \Sigma F_z; \qquad \int dF = \int_A \sigma \, dA$$
$$P = \sigma A$$

$$\boxed{\sigma = \frac{P}{A}}$$

(1–6)

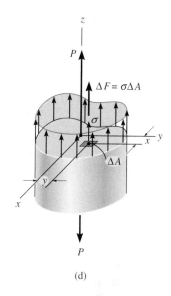

Fig. 1–13

Here

σ =   average normal stress at any point on the cross-sectional area

P =   internal resultant normal force, which is applied through the *centroid* of the cross-sectional area. P is determined using the method of sections and the equations of equilibrium.

A =   cross-sectional area of the bar

The internal load P must pass through the centroid of the cross-section since the uniform stress distribution will produce zero moments about any x and y axes passing through this point, Fig. 1–13d. When this occurs,

$$(M_R)_x = \Sigma M_x; \qquad 0 = \int_A y \, dF = \int_A y\sigma \, dA = \sigma \int_A y \, dA$$

$$(M_R)_y = \Sigma M_y; \qquad 0 = -\int_A x \, dF = -\int_A x\sigma \, dA = -\sigma \int_A x \, dA$$

These equations are indeed satisfied, since by definition of the centroid, $\int y \, dA = 0$ and $\int x \, dA = 0$. (See Appendix A.)

Equilibrium.   It should be apparent that only a normal stress exists on any volume element of material located at each point on the cross section of an axially loaded bar. If we consider vertical equilibrium of the element, Fig. 1-14, then applying the equation of force equilibrium,

$$\Sigma F_z = 0; \qquad \sigma (\Delta A) - \sigma'(\Delta A) = 0$$
$$\sigma = \sigma'$$

In other words, the two normal stress components on the element must be equal in magnitude but opposite in direction. This is referred to as *uniaxial stress.*

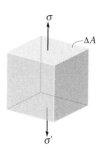

Fig. 1–14

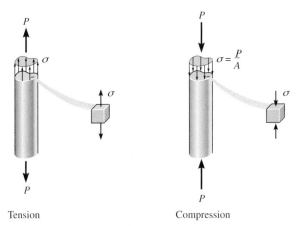

Tension   Compression

Fig. 1–15

The previous analysis applies to members subjected to either tension or compression, as shown in Fig. 1–15. As a graphical interpretation, the **magnitude** of the internal resultant force **P** is **equivalent** to the **volume** under the stress diagram; that is, $P = \sigma A$ (volume = height $\times$ base). Furthermore, as a consequence of the balance of moments, **this resultant passes through the centroid of this volume**.

Although we have developed this analysis for *prismatic* bars, this assumption can be relaxed somewhat to include bars that have a *slight taper*. For example, it can be shown, using the more exact analysis of the theory of elasticity, that for a tapered bar of rectangular cross section, for which the angle between two adjacent sides is 15°, the average normal stress, as calculated by $\sigma = P/A$, is only 2.2% *less* than its value found from the theory of elasticity.

Maximum Average Normal Stress. In our analysis both the internal force P and the cross-sectional area A were *constant* along the longitudinal axis of the bar, and as a result the normal stress $\sigma = P/A$ is also *constant* throughout the bar's length. Occasionally, however, the bar may be subjected to *several* external loads along its axis, or a change in its cross-sectional area may occur. As a result, the normal stress within the bar could be different from one section to the next, and, if the *maximum* average normal stress is to be determined, then it becomes important to find the location where the ratio P/A is a *maximum*. To do this it is necessary to determine the internal force P at various sections along the bar. Here it may be helpful to show this variation by drawing an **axial or normal force diagram**. Specifically, this diagram is a plot of the normal force P versus its position x along the bar's length. As a sign convention, P will be positive if it causes tension in the member, and negative if it causes compression. Once the internal loading throughout the bar is known, the maximum ratio of P/A can then be identified.

This steel tie rod is used to suspend a portion of a staircase, and as a result it is subjected to tensile stress.

IMPORTANT POINTS

• When a body that is subjected to an external load is sectioned, there is a distribution of force acting over the sectioned area which holds each segment of the body in equilibrium. The intensity of this internal force at a point in the body is referred to as *stress*.

• Stress is the limiting value of force per unit area, as the area approaches zero. For this definition, the material at the point is considered to be continuous and cohesive.

• In general, there are six independent components of stress at each point in the body, consisting of *normal stress,* σ_x, σ_y, σ_z, and *shear stress,* τ_{xy}, τ_{yz}, τ_{xz}.

• The magnitude of these components depends upon the type of loading acting on the body, and the orientation of the element at the point.

• When a prismatic bar is made from homogeneous and isotropic material, and is subjected to axial force acting through the centroid of the cross-sectioned area, then the material within the bar is subjected *only to normal stress.* This stress is assumed to be uniform or *averaged* over the cross-sectional area.

PROCEDURE FOR ANALYSIS

The equation $\sigma = P/A$ gives the *average* normal stress on the cross-sectional area of a member when the section is subjected to an internal resultant normal force **P**. For axially loaded members, application of this equation requires the following steps.

Internal Loading.

• Section the member *perpendicular* to its longitudinal axis at the point where the normal stress is to be determined and use the necessary free-body diagram and equation of force equilibrium to obtain the internal axial force **P** at the section.

Average Normal Stress.

• Determine the member's cross-sectional area at the section and compute the average normal stress $\sigma = P/A$.

• It is suggested that σ be shown acting on a small volume element of the material located at a point on the section where stress is calculated. To do this, first draw σ on the face of the element coincident with the sectioned area A. Here σ acts in the *same direction* as the internal force **P** since all the normal stresses on the cross section act in this direction to develop this resultant. The normal stress σ acting on the opposite face of the element can be drawn in its appropriate direction.

EXAMPLE 1–6

The bar in Fig. 1–16a has a constant width of 35 mm and a thickness of 10 mm. Determine the maximum average normal stress in the bar when it is subjected to the loading shown.

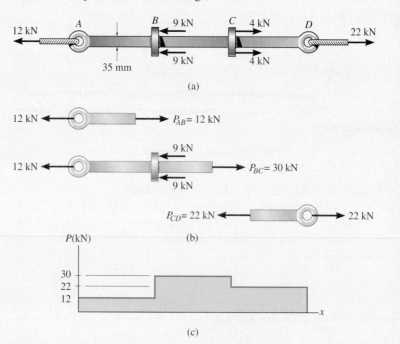

(a)

(b)

(c)

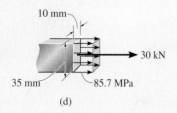

(d)

Fig. 1–16

SOLUTION

Internal Loading. By inspection, the internal axial forces in regions *AB*, *BC*, and *CD* are all constant yet have different magnitudes. Using the method of sections, these loadings are determined in Fig. 1–16b; and the normal force diagram which represents these results graphically is shown in Fig. 1–16c. By inspection, the largest loading is in region *BC*, where $P_{BC} = 30$ kN. Since the cross-sectional area of the bar is *constant*, the largest average normal stress also occurs within this region of the bar.

Average Normal Stress. Applying Eq. 1–6, we have

$$\sigma_{BC} = \frac{P_{BC}}{A} = \frac{30(10^3)\ \text{N}}{(0.035\ \text{m})(0.010\ \text{m})} = 85.7\ \text{MPa} \qquad \textit{Ans.}$$

The stress distribution acting on an arbitrary cross section of the bar within region *BC* is shown in Fig. 1–16d. Graphically the *volume* (or "block") represented by this distribution of stress is equivalent to the load of 30 kN; that is, 30 kN = (85.7 MPa)(35 mm)(10 mm).

EXAMPLE 1-7

The 80-kg lamp is supported by two rods AB and BC as shown in Fig. 1–17a. If AB has a diameter of 10 mm and BC has a diameter of 8 mm, determine the average normal stress in each rod.

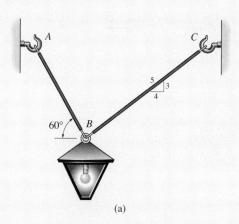

(a)

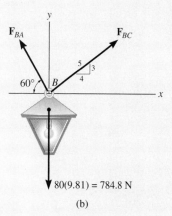

(b)

Fig. 1–17

SOLUTION

Internal Loading. We must first determine the axial force in each rod. A free-body diagram of the lamp is shown in Fig. 1–17b. Applying the equations of force equilibrium yields

$$\xrightarrow{+} \Sigma F_x = 0; \qquad F_{BC}(\tfrac{4}{5}) - F_{BA} \cos 60° = 0$$
$$+\uparrow \Sigma F_y = 0; \quad F_{BC}(\tfrac{3}{5}) + F_{BA} \sin 60° - 784.8 \text{ N} = 0$$
$$F_{BC} = 395.2 \text{ N}, \qquad F_{BA} = 632.4 \text{ N}$$

By Newton's third law of action, equal but opposite reaction, these forces subject the rods to tension throughout their length.

Average Normal Stress. Applying Eq. 1–6, we have

$$\sigma_{BC} = \frac{F_{BC}}{A_{BC}} = \frac{395.2 \text{ N}}{\pi(0.004 \text{ m})^2} = 7.86 \text{ MPa} \qquad Ans.$$

$$\sigma_{BA} = \frac{F_{BA}}{A_{BA}} = \frac{632.4 \text{ N}}{\pi(0.005 \text{ m})^2} = 8.05 \text{ MPa} \qquad Ans.$$

The average normal stress distribution acting over a cross section of rod AB is shown in Fig. 1–17c, and at a point on this cross section, an element of material is stressed as shown in Fig. 1–17d.

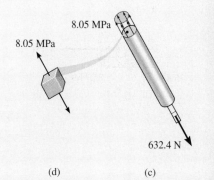

(d)   (c)

EXAMPLE 1-8

The casting shown in Fig. 1–18a is made of steel having a specific weight of $\gamma_{st} = 490$ lb/ft³. Determine the average compressive stress acting at points A and B.

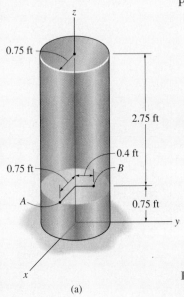

(a)

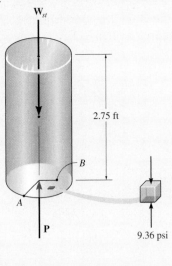

Fig. 1–18   (b)   (c)

SOLUTION

Internal Loading.   A free-body diagram of the top segment of the casting where the section passes through points A and B is shown in Fig. 1–18b. The weight of this segment is determined from $W_{st} = \gamma_{st} V_{st}$. Thus the internal axial force P at the section is

$$+\uparrow \Sigma F_z = 0; \qquad\qquad P - W_{st} = 0$$
$$P - (490 \text{ lb/ft}^3)(2.75 \text{ ft})\pi(0.75 \text{ ft})^2 = 0$$
$$P = 2381 \text{ lb}$$

Average Compressive Stress.   The cross-sectional area at the section is $A = \pi(0.75 \text{ ft})^2$, and so the average compressive stress becomes

$$\sigma = \frac{P}{A} = \frac{2381 \text{ lb}}{\pi(0.75 \text{ ft})^2}$$
$$= 1347.5 \text{ lb/ft}^2 = 1347.5 \text{ lb/ft}^2(1 \text{ ft}^2/144 \text{ in}^2)$$
$$= 9.36 \text{ psi} \qquad\qquad\qquad\qquad\qquad\qquad ***Ans.***$$

The stress shown on the volume element of material in Fig. 1–18c is representative of the conditions at either point A or B. Notice that this stress acts *upward* on the bottom or shaded face of the element since this face forms part of the bottom surface area of the cut section, and on this surface, the resultant internal force **P** is pushing upward.

EXAMPLE 1-9

Member *AC* shown in Fig. 1–19*a* is subjected to a vertical force of 3 kN. Determine the position *x* of this force so that the average compressive stress at the smooth support *C* is equal to the average tensile stress in the tie rod *AB*. The rod has a cross-sectional area of 400 mm^2 and the contact area at *C* is 650 mm^2.

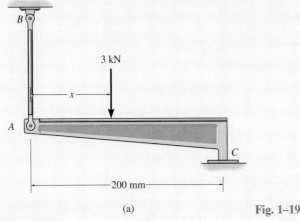

(a) Fig. 1–19 (b)

F_C

SOLUTION

Internal Loading. The forces at *A* and *C* can be related by considering the free-body diagram for member *AC*, Fig. 1–19*b*. There are three unknowns, namely, F_{AB}, F_C, and *x*. To solve this problem we will work in units of newtons and millimeters.

$$+\uparrow \Sigma F_y = 0; \qquad\qquad F_{AB} + F_C - 3000 \text{ N} = 0 \qquad (1)$$
$$\downarrow^+ \Sigma M_A = 0; \qquad -3000 \text{ N}(x) + F_C (200 \text{ mm}) = 0 \qquad (2)$$

Average Normal Stress. A necessary third equation can be written that requires the tensile stress in the bar *AB* and the compressive stress at *C* to be equivalent, i.e.,

$$\sigma = \frac{F_{AB}}{400 \text{ mm}^2} = \frac{F_C}{650 \text{ mm}^2}$$
$$F_C = 1.625 F_{AB}$$

Substituting this into Eq. 1, solving for F_{AB}, then solving for F_C, we obtain

$$F_{AB} = 1143 \text{ N}$$
$$F_C = 1857 \text{ N}$$

The position of the applied load is determined from Eq. 2,

$$x = 124 \text{ mm} \qquad\qquad \textit{Ans.}$$

Note that $0 < x < 200$ mm, as required.

1.5   AVERAGE SHEAR STRESS

Shear stress has been defined in Sec. 1–3 as the stress component that acts *in the plane* of the sectioned area. In order to show how this stress can develop, we will consider the effect of applying a force **F** to the bar in Fig. 1–20*a*. If the supports are considered rigid, and **F** is large enough, it will cause the material of the bar to deform and fail along the planes identified by *AB* and *CD*. A free-body diagram of the unsupported center segment of the bar, Fig. 1–20*b*, indicates that the shear force $V = F/2$ must be applied at each section to hold the segment in equilibrium. The *average shear stress* distributed over each sectioned area that develops this shear force is defined by

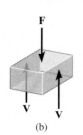

(a)

$$\tau_{avg} = \frac{V}{A}$$

(1–7)

Here

τ_{avg} = average shear stress at the section, which is assumed to be the *same* at each point located on the section

V = internal resultant shear force at the section determined from the equations of equilibrium

A = area at the section

(b)

The distribution of average shear stress is shown acting over the sections in Fig. 1–20*c*. Notice that τ_{avg} is in the *same direction* as **V**, since the shear stress must create associated forces all of which contribute to the internal resultant force **V** at the section.

The loading case discussed in Fig. 1–20 is an example of *simple or direct shear*, since the shear is caused by the *direct action* of the applied load **F**. This type of shear often occurs in various types of simple connections that use bolts, pins, welding material, etc. In all these cases, however, application of Eq. 1–7 is *only approximate*. A more precise investigation of the shear-stress distribution over the critical section often reveals that much larger shear stresses occur in the material than those predicted by this equation. Although this may be the case, application of Eq. 1–7 is generally acceptable for many problems in engineering design and analysis. For example, engineering codes allow its use when considering design sizes for fasteners such as bolts and for obtaining the bonding strength of joints subjected to shear loadings. In this regard, two types of shear frequently occur in practice, which deserve separate treatment.

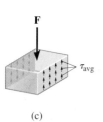

(c)

Fig. 1–20

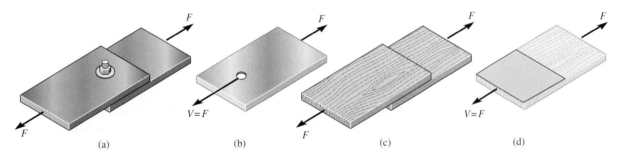

Fig. 1–21

Single Shear.   The steel and wood joints shown in Fig. 1–21*a* and 1–21*c*, respectively, are examples of **single-shear connections** and are often referred to as *lap joints*. Here we will assume that the members are thin and that the nut in Fig. 1–21*a* is not tightened to any great extent so friction between the members can be neglected. Passing a section between the members yields the free-body diagrams shown in Fig. 1–21*b* and 1–21*d*. Since the members are thin, we can neglect the moment created by the force *F*. Hence for equilibrium the cross-sectional area of the bolt in Fig. 1–21*b* and the bonding surface between the members in Fig. 1–21*d* are subjected only to a *single shear force V = F*. This force is used in Eq. 1–7 to determine the average shear stress acting on the colored section of Fig. 1–21*d*.

The pin on this tractor is subjected to double shear.

Double Shear.   When the joint is constructed as shown in Fig. 1–22*a* or 1–22*c*, two shear surfaces must be considered. These types of connections are often called *double lap joints*. If we pass a section between each of the members, the free-body diagrams of the center member are shown in Fig. 1–22*b* and 1–22*d*. Here we have a condition of **double shear**. Consequently, $V = F/2$ acts on *each* sectioned area and this shear must be considered when applying $\tau_{avg} = V/A$.

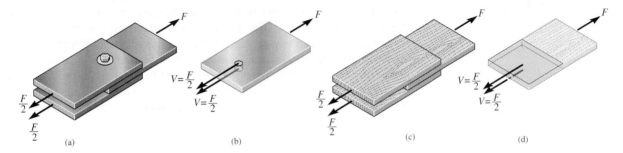

Fig. 1–22

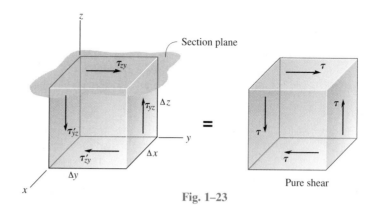

Fig. 1–23

Equilibrium.   Consider a volume element of material taken at a point located on the surface of any sectioned area on which the average shear stress acts, Fig. 1–23a. If we consider force equilibrium in the *y* direction, then

$$\Sigma F_y = 0; \qquad \overset{\displaystyle\text{force}}{\overbrace{\underset{\text{stress}}{\underbrace{\tau_{zy}}}(\underset{\text{area}}{\underbrace{\Delta x \Delta y}})}} - \tau'_{zy}\Delta x \Delta y = 0$$

$$\tau_{zy} = \tau'_{zy}$$

And in a similar manner, force equilibrium in the *z* direction yields $\tau_{yz} = \tau'_{yz}$. Finally, taking moments about the *x* axis,

$$\Sigma M_x = 0; \qquad -\underset{\text{stress}}{\underbrace{\tau_{zy}}}(\underset{\text{area}}{\underbrace{\Delta x \Delta y}})\underset{\text{arm}}{\Delta z} + \tau_{yz}(\Delta x \Delta z)\Delta y = 0$$

$$\tau_{zy} = \tau_{yz}$$

so that

$$\tau_{zy} = \tau'_{zy} = \tau_{yz} = \tau'_{yz} = \tau$$

In other words, force and moment equilibrium requires the shear stress acting on the top face of the element, to be accompanied by shear stress acting on three other faces, Fig. 1-23b. Here ***all four shear stresses must have equal magnitude and be directed either toward or away from each other at opposite edges of the element.*** This is referred to as the *complementary property of shear,* and under the conditions shown in Fig. 1-23, the material is subjected to *pure shear*.

Although we have considered here a case of simple shear as caused by the *direct* action of a load, in later chapters we will show that shear stress can also arise *indirectly* due to the action of other types of loading.

IMPORTANT POINTS

• If two parts which are *thin or small* are joined together, the applied loads can cause shearing of the material with negligible bending. If this is the case, it is generally suitable for engineering analysis to assume that an *average shear stress* acts over the cross-sectional area.

• Oftentimes fasteners, such as nails and bolts, are subjected to shear loads. The magnitude of a shear force on the fastener is greatest along a plane which passes through the surfaces being joined. A carefully drawn free-body diagram of a segment of the fastener will enable one to obtain the magnitude and direction of this force.

PROCEDURE FOR ANALYSIS

The equation $\tau_{avg} = V/A$ is used to compute only the *average shear stress* in the material. Application requires the following steps.

Internal Shear.

• Section the member at the point where the average shear stress is to be determined.

• Draw the necessary free-body diagram, and calculate the internal shear force **V** acting at the section that is necessary to hold the part in equilibrium.

Average Shear Stress.

• Determine the sectioned area A, and compute the average shear stress $\tau_{avg} = V/A$.

• It is suggested that τ_{avg} be shown on a small volume element of material located at a point on the section where it is determined. To do this, first draw τ_{avg} on the face of the element, coincident with the sectioned area A. This shear stress acts in the same direction as **V**. The shear stresses acting on the three adjacent planes can then be drawn in their appropriate directions following the scheme shown in Fig. 1–23.

EXAMPLE 1–10

The bar shown in Fig. 1–24a has a square cross section for which the depth and thickness are 40 mm. If an axial force of 800 N is applied along the centroidal axis of the bar's cross-sectional area, determine the average normal stress and average shear stress acting on the material along (a) section plane a–a and (b) section plane b–b.

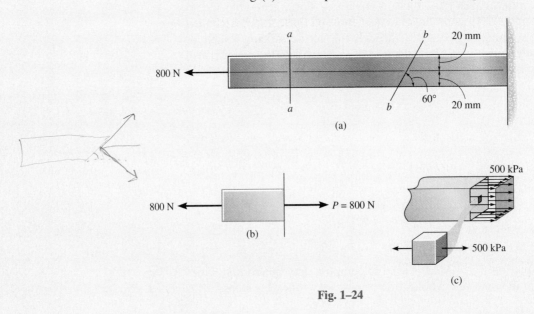

Fig. 1–24

SOLUTION

Part (a)
Internal Loading. The bar is sectioned, Fig. 1–24b, and the internal resultant loading consists only of an axial force for which $P = 800$ N.

Average Stress. The average normal stress is determined from Eq.1–6.

$$\sigma = \frac{P}{A} = \frac{800 \text{ N}}{(0.04 \text{ m})(0.04 \text{ m})} = 500 \text{ kPa} \qquad \textit{Ans.}$$

No shear stress exists on the section, since the shear force at the section is zero.

$$\tau_{avg} = 0 \qquad \textit{Ans.}$$

The distribution of average normal stress over the cross section is shown in Fig. 1–24c.

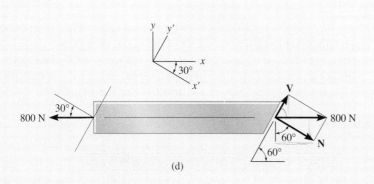

(d)

Part (b)

Internal Loading.   If the bar is sectioned along *b–b*, the free-body diagram of the left segment is shown in Fig. 1–24*d*. Here both a normal force (**N**) and shear force (**V**) act on the sectioned area. Using *x*, *y* axes, we require

$\xrightarrow{+} \Sigma F_x = 0;$ $-800 \text{ N} + N \sin 60° + V \cos 60° = 0$

$+\uparrow \Sigma F_y = 0;$ $V \sin 60° - N \cos 60° = 0$

or, more directly, using *x′*, *y′* axes,

$+\searrow \Sigma F_{x'} = 0;$ $N - 800 \text{ N} \cos 30° = 0$

$+\nearrow \Sigma F_{y'} = 0;$ $V - 800 \text{ N} \sin 30° = 0$

Solving either set of equations,

$$N = 692.8 \text{ N}$$
$$V = 400 \text{ N}$$

Average Stresses.   In this case the sectioned area has a thickness and depth of 40 mm and 40 mm/sin 60° = 46.19 mm, respectively, Fig. 1–24*a*. Thus the average normal stress is

$$\sigma = \frac{N}{A} = \frac{692.8 \text{ N}}{(0.04 \text{ m})(0.04619 \text{ m})} = 375 \text{ kPa} \qquad \textbf{\textit{Ans.}}$$

and the average shear stress is

$$\tau_{avg} = \frac{V}{A} = \frac{400 \text{ N}}{(0.04 \text{ m})(0.04619 \text{ m})} = 217 \text{ kPa} \qquad \textbf{\textit{Ans.}}$$

The stress distribution is shown in Fig. 1–24*e*.

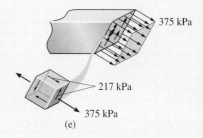

375 kPa

217 kPa

375 kPa

(e)

EXAMPLE 1-11

The wooden strut shown in Fig. 1–25a is suspended from a 10-mm-diameter steel rod, which is fastened to the wall. If the strut supports a vertical load of 5 kN, compute the average shear stress in the rod at the wall and along the two shaded planes of the strut, one of which is indicated as *abcd*.

SOLUTION

Internal Shear.   As shown on the free-body diagram in Fig. 1–25b, the rod resists a shear force of 5 kN where it is fastened to the wall. A free-body diagram of the sectioned segment of the strut that is in contact with the rod is shown in Fig. 1–25c. Here the shear force acting along each shaded plane is 2.5 kN.

Average Shear Stress.   For the rod,

$$\tau_{avg} = \frac{V}{A} = \frac{5000 \text{ N}}{\pi(0.005 \text{ m})^2} = 63.7 \text{ MPa} \qquad \textit{Ans.}$$

For the strut,

$$\tau_{avg} = \frac{V}{A} = \frac{2500 \text{ N}}{(0.04 \text{ m})(0.02 \text{ m})} = 3.12 \text{ MPa} \qquad \textit{Ans.}$$

The average-shear-stress distribution on the sectioned rod and strut segment is shown in Fig. 1–25d and 1–25e, respectively. Also shown with these figures is a typical volume element of the material taken at a point located on the surface of each section. Note carefully how the shear stress must act on each shaded face of these elements and then on the adjacent faces of the elements.

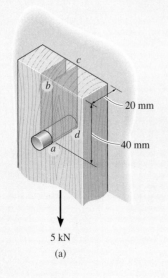

20 mm

40 mm

5 kN

(a)

force of strut on rod

5 kN

$V = 5$ kN

(b)

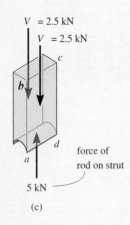

$V = 2.5$ kN

$V = 2.5$ kN

force of rod on strut

5 kN

(c)

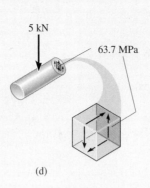

5 kN

63.7 MPa

(d)

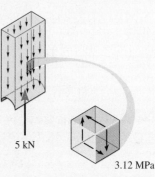

5 kN

3.12 MPa

(e)

Fig. 1–25

EXAMPLE 1-12

The inclined member in Fig. 1–26a is subjected to a compressive force of 600 lb. Determine the average compressive stress along the smooth areas of contact defined by AB and BC, and the average shear stress along the horizontal plane defined by EDB.

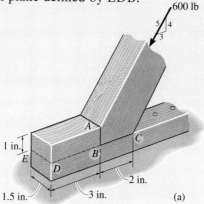

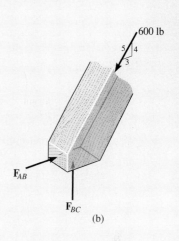

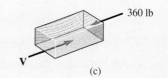

(b)

Fig. 1–26

SOLUTION

Internal Loadings. The free-body diagram of the inclined member is shown in Fig. 1–26b. The compressive forces acting on the areas of contact are

$\xrightarrow{+} \Sigma F_x = 0;$ $F_{AB} - 600 \text{ lb}(\frac{3}{5}) = 0$ $F_{AB} = 360 \text{ lb}$

$+\uparrow \Sigma F_y = 0$ $F_{BC} - 600 \text{ lb}(\frac{4}{5}) = 0$ $F_{BC} = 480 \text{ lb}$

Also, from the free-body diagram of the top segment of the bottom member, Fig. 1–26c, the shear force acting on the sectioned horizontal plane EDB is

$\xrightarrow{+} \Sigma F_x = 0;$ $V = 360 \text{ lb}$

Average Stress. The average compressive stresses along the horizontal and vertical planes of the inclined member are

$$\sigma_{AB} = \frac{360 \text{ lb}}{(1 \text{ in.})(1.5 \text{ in.})} = 240 \text{ psi} \qquad \textit{Ans.}$$

$$\sigma_{BC} = \frac{480 \text{ lb}}{(2 \text{ in.})(1.5 \text{ in.})} = 160 \text{ psi} \qquad \textit{Ans.}$$

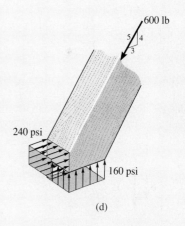

(d)

These stress distributions are shown in Fig. 1–26d.

The average shear stress acting on the horizontal plane defined by EDB is

$$\tau_{\text{avg}} = \frac{360 \text{ lb}}{(3 \text{ in.})(1.5 \text{ in.})} = 80.0 \text{ psi} \qquad \textit{Ans.}$$

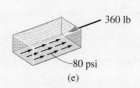

(e)

This stress is shown distributed over the sectioned area in Fig. 1–26e.

PROBLEMS

1–33. The column is subjected to an axial force of 8 kN at its top. If the cross-sectional area has the dimensions shown in the figure, determine the average normal stress acting at section *a–a*. Show this distribution of stress acting over the area's cross section.

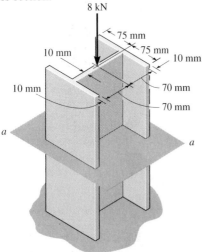

Prob. 1–33

1–34. The cinder block has the dimensions shown. If the material fails when the average normal stress reaches 120 psi, determine the largest centrally applied vertical load **P** it can support.

1–35. The cinder block has the dimensions shown. If it is subjected to a centrally applied force of **P** = 800 lb, determine the average normal stress in the material. Show the result acting on a differential volume element of the material.

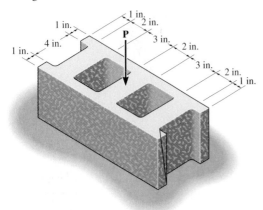

Probs. 1–34/1–35

*1–36. The 50-lb lamp is supported by two steel rods connected by a ring at *A*. Determine which rod is subjected to the greater average normal stress and compute its value. Take $\theta = 60°$. The diameter of each rod is given in the figure.

1–37. Solve Prob. 1–36 for $\theta = 45°$.

1–38. The 50-lb lamp is supported by two steel rods connected by a ring at *A*. Determine the angle of orientation θ of *AC* such that the average normal stress in rod *AC* is twice the average normal stress in rod *AB*. What is the magnitude of this stress in each rod? The diameter of each rod is given in the figure.

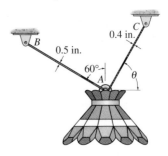

Probs. 1–36/1–37/1–38

1–39. The thrust bearing is subjected to the loads shown. Determine the average normal stress developed on cross sections through points *B*, *C*, and *D*. Sketch the results on a differential volume element located at each section.

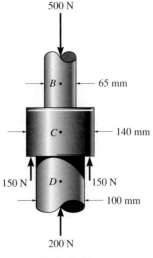

Prob. 1–39

*1–40. The shaft is subjected to the axial force of 30 kN. If the shaft passes through the 53-mm diameter hole in the fixed support A, determine the bearing stress acting on the collar C. Also, what is the average shear stress acting along the inside surface of the collar where it is fixed connected to the 52-mm diameter shaft?

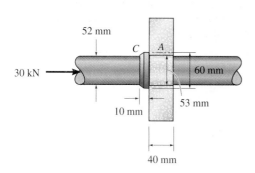

Prob. 1–40

1–41. The small block has a thickness of 5 mm. If the stress distribution at the support developed by the load varies as shown, determine the force **F** applied to the block, and the distance d to where it is applied.

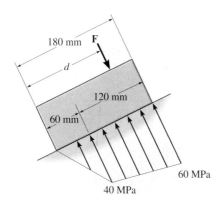

Prob. 1–41

1–42. The supporting wheel on a scaffold is held in place on the leg using a 4-mm-diameter pin as shown. If the wheel is subjected to a normal force of 3 kN, determine the average shear stress developed in the pin. Neglect friction between the inner scaffold puller leg and the tube used on the wheel.

Prob. 1–42

1–43. The pins on the frame at B and C each have a diameter of 0.25 in. If these pins are subjected to *double shear*, determine the average shear stress in each pin.

*1–44. Solve Prob. 1–43 assuming that pins B and C are subjected to *single shear*.

1–45. The pins on the frame at D and E each have a diameter of 0.25 in. If these pins are subjected to *double shear*, determine the average shear stress in each pin.

1–46. Solve Prob. 1–45 assuming that pins D and E are subjected to *single shear*.

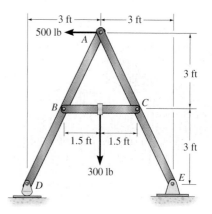

Probs. 1–43/1–44/1–45/1–46

1–47. The pedestal has a triangular cross section as shown. If it is subjected to a compressive force of 500 lb, specify the x and y coordinates for the location of point $P(x, y)$, where the load must be applied on the cross section, so that the average normal stress is uniform. Compute the stress and sketch its distribution acting on the cross section at a location removed from the point of load application.

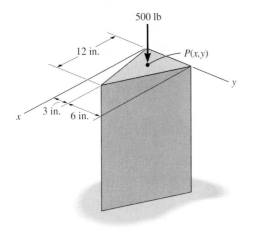

Prob. 1–47

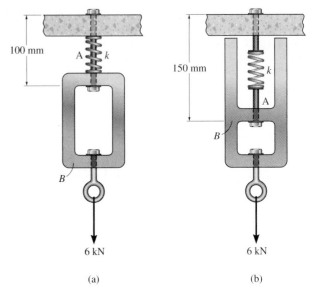

Probs. 1–48/1–49

*1–48.** Two designs for a shock support are shown in the figure. The spring has a stiffness of $k = 15$ kN/m and in (a) it is uncompressed, whereas in (b) it is originally stretched 0.2 m. Determine the average normal stress in the 5-mm-diameter bolt shank at A when the 6-kN load is applied. In (b) the bracket B is not connected to the support.

1–49. Two designs for a shock support are shown in the figure. The spring has a stiffness of $k = 15$ kN/m and in (a) it is uncompressed. Determine the maximum amount which the spring in (b) must be originally stretched so that the average normal stress in the 5-mm bolt shank at A is equivalent for both designs when the load of 6 kN is applied. In (b) the bracket B is not connected to the support. What is the bolt stress in both cases?

1–50. The yoke is subjected to the force and couple moment. Determine the average shear stress in the bolt acting on the cross sections through A and B. The bolt has a diameter of 0.25 in. *Hint:* The couple moment is resisted by a set of couple forces developed in the shank of the bolt.

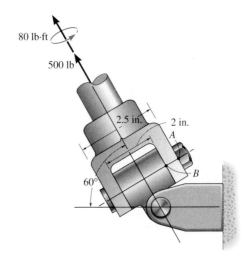

Prob. 1–50

1–51. The built-up shaft consists of a pipe AB and solid rod BC. The pipe has an inner diameter of 20 mm and outer diameter of 28 mm. The rod has a diameter of 12 mm. Determine the average normal stress at points D and E and represent the stress on a volume element located at each of these points.

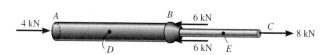

Prob. 1–51

*1–52. The anchor bolt was pulled out of the concrete wall and the failure surface formed part of a frustum and cylinder. This indicates a shear failure occurred along the cylinder BC and tension failure along the frustum AB. If the shear and normal stresses along these surfaces have the magnitudes shown, determine the force **P** that must have been applied to the bolt.

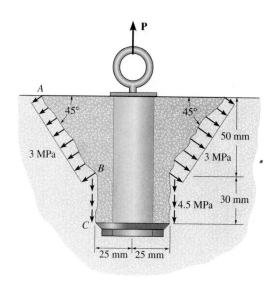

Prob. 1–52

1–53. The plastic block is subjected to an axial compressive force of 600 N. Assuming that the caps at the top and bottom distribute the load uniformly throughout the block, determine the average normal and average shear stress acting along section a–a.

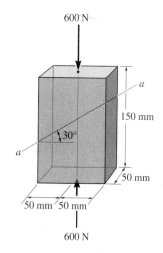

Prob. 1–53

1–54. The crimping tool is used to crimp the end of the wire E. If a force of 20 lb is applied to the handles, determine the average shear stress in-the pin at A. The pin is subjected to double shear and has a diameter of 0.2 in. Only a vertical force is exerted on the wire.

1–55. Solve Prob. 1–54 for pin B. The pin is subjected to double shear and has a diameter of 0.2 in.

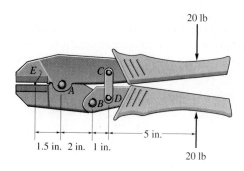

Probs. 1–54/1–55

*1–56.   The joint is subjected to the axial member force of 6 kip. Determine the average normal stress acting on sections AB and BC. Assume the member is smooth and is 1.5 in. thick.

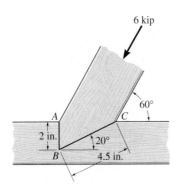

Prob. 1–56

1–57.   The driver of the sports car applies the rear brakes and causes the tires to slip. If the normal force on each rear tire is 400 lb and the coefficient of kinetic friction between the tires and the pavement is $\mu_k = 0.5$, determine the average shear stress developed by the friction force on the tires. Assume the rubber of the tires is flexible and each tire is filled with an air pressure of 32 psi.

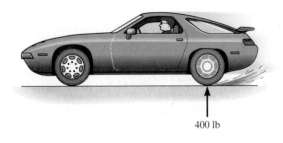

Prob. 1–57

1–58.   The bar has a cross-sectional area A and is subjected to the axial load P. Determine the average normal and average shear stresses acting over the shaded section, which is oriented at θ from the horizontal. Plot the variation of these stresses as a function of θ $(0 \le \theta \le 90°)$.

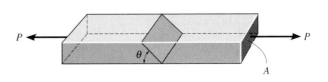

Prob. 1–58

1–59.   The jaw clutch is used to transmit a torque of 450 lb · ft in only one direction. If each shaft has two teeth placed around its circumference as shown, determine the average shear stress along the root AB of each of the teeth.

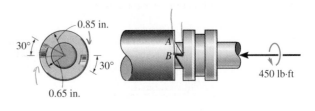

Prob. 1–59

*1–60. The bars of the truss each have a cross-sectional area of 1.25 in². Determine the average normal stress in each member due to the loading $P = 8$ kip. State whether the stress is tensile or compressive.

1–61. The bars of the truss each have a cross-sectional area of 1.25 in². If the maximum average normal stress in any bar is not to exceed 20 ksi, determine the maximum magnitude P of the loads that can be applied to the truss.

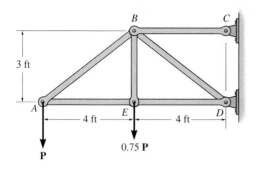

Probs. 1–60/1–61

1–62. The prismatic bar has a cross-sectional area A. If it is subjected to a distributed axial loading that increases linearly from $w = 0$ at $x = 0$ to $w = w_0$ at $x = a$, and then decreases linearly to $w = 0$ at $x = 2a$, determine the average normal stress in the bar as a function of x for $0 \le x < a$.

1–63. The prismatic bar has a cross-sectional area A. If it is subjected to a distributed axial loading that increases linearly from $w = 0$ at $x = 0$ to $w = w_0$ at $x = a$, and then decreases linearly to $w = 0$ at $x = 2a$, determine the average normal stress in the bar as a function of x for $a < x \le 2a$.

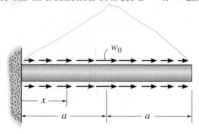

Probs. 1–62/1–63

*1–64. The bar is subjected to a uniform distributed axial loading of 10 kN/m and a concentrated force of 1.5 kN at its midpoint as shown. Determine the maximum average normal stress in the bar and its location x.

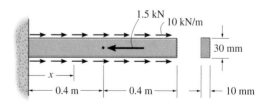

Prob. 1–64

1–65. The bar has a cross-sectional area of $400 \, (10^{-6})$ m². If it is subjected to a uniform axial distributed loading along its length and to two concentrated loads as shown, determine the average normal stress in the bar as a function of x for $0 < x \le 0.5$ m.

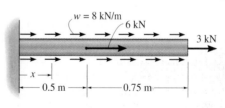

Prob. 1–65

1–66. The tapered rod has a radius of $r = (2 - x/6)$ in. and is subjected to the distributed loading of $w = (60 + 40x)$ lb /in. Determine the average normal stress at the center of the rod.

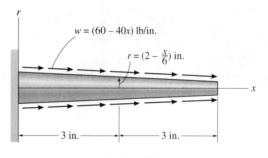

Prob. 1–66

1–67. The truss is made from three pin-connected members having the cross-sectional areas shown in the figure. Determine the average normal stress developed in each member when the truss is subjected to the load shown. State whether the stress is tensile or compressive.

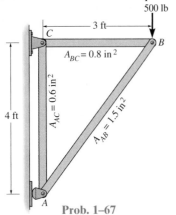

Prob. 1–67

*1–68. Rods AB and BC have diameters of 25 mm and 18 mm, respectively. If a load of 6 kN is applied to the ring at B, determine the average normal stress in each rod if $\theta = 60°$.

1–69. Rods AB and BC each have a diameter of 4 mm. If a load of 6 kN is applied to the ring at B, determine the smallest angle θ of rod BC so that the average normal stress in each rod is equivalent.

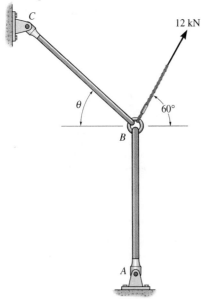

Probs. 1–68/ 1–69

1–70. The beam is supported by a pin at A and a short link BC. If $P = 15$ kN, determine the average shear stress developed in the pins at A, B, and C. All pins are in double shear as shown, and each has a diameter of 18 mm.

1–71. The beam is supported by a pin at A and a short link BC. Determine the maximum magnitude P of the loads the beam will support if the average shear stress in each pin is not to exceed 80 MPa. All pins are in double shear as shown, and each has a diameter of 18 mm.

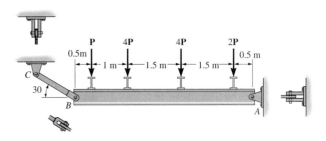

Probs. 1–70/1–71

*1–72. The pedestal in the shape of a frustum of a cone is made of concrete having a specific weight of γ. Determine the average normal stress acting at its base. *Hint:* The volume of a cone of radius r and height h is $V = \frac{1}{3}\pi r^2 h$.

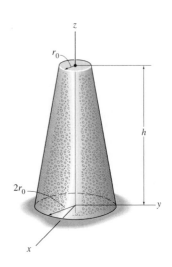

Prob. 1–72

1–73. The shape has a radius that is defined by $r = 0.4\cos(\pi y/3)$ m. Determine the average normal stress at the support if the material has a density of $\rho = 3$ Mg/m^3.

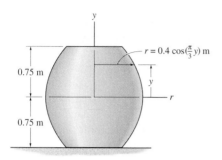

$r = 0.4\cos(\frac{\pi}{3}y)$ m

0.75 m

0.75 m

Prob. 1–73

1–74. The radius of the pedestal is defined by $r = (0.5e^{-0.08y^2})$ m, where y is given in meters. If the material has a density of 2.5 Mg/m^3, determine the average normal stress at the support.

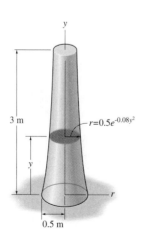

3 m

$r = 0.5e^{-0.08y^2}$

0.5 m

Prob. 1–74

1–75. The column is made of concrete having a density of 2.30 Mg/m^3. At its top B it is subjected to an axial compressive force of 15 kN. Determine the average normal stress in the column as a function of the distance z measured from its base. *Note:* The result will be useful only for finding the average normal stress at a section removed from the ends of the column, because of localized deformation at the ends.

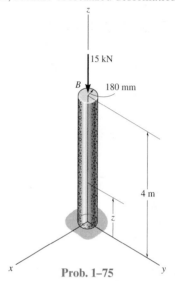

15 kN

B 180 mm

4 m

Prob. 1–75

*1–76. The pedestal supports a load **P** at its center. If the material has a mass density ρ, determine the radial dimension r as a function of z so that the average normal stress in the pedestal remains constant. The cross section is circular.

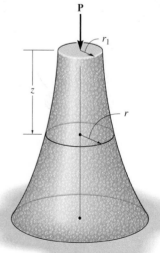

P

r_1

r

Prob. 1–76

1.6   ALLOWABLE STRESS

An engineer in charge of the *design* of a structural member or mechanical element must restrict the stress in the material to a level that will be safe. Furthermore, a structure or machine that is currently in use may, on occasion, have to be *analyzed* to see what additional loadings its members or parts can support. So again it becomes necessary to perform the calculations using a safe or allowable stress.

Appropriate factors of safety must be considered when designing cranes and cables used to transfer heavy loads.

To ensure safety, it is necessary to choose an allowable stress that restricts the applied load to one that is *less* than the load the member can fully support. There are several reasons for this. For example, the load for which the member is designed may be different from actual loadings placed on it. The intended measurements of a structure or machine may not be exact due to errors in fabrication or in the assembly of its component parts. Unknown vibrations, impact, or accidental loadings can occur that may not be accounted for in the design. Atmospheric corrosion, decay, or weathering tend to cause materials to deteriorate during service. And lastly, some materials, such as wood, concrete, or fiber-reinforced composites, can show high variability in mechanical properties.

One method of specifying the allowable load for the design or analysis of a member is to use a number called the factor of safety. The *factor of safety* (F.S.) is a ratio of the failure load F_{fail} divided by the allowable load, F_{allow}. Here F_{fail} is found from experimental testing of the material, and the factor of safety is selected based on experience so that the above mentioned uncertainties are accounted for when the member is used under similar conditions of loading and geometry. Stated mathematically,

$$\text{F.S.} = \frac{F_{\text{fail}}}{F_{\text{allow}}} \tag{1–8}$$

If the load applied to the member is *linearly related* to the stress developed within the member, as in the case of using $\sigma = P/A$ and $\tau_{\text{avg}} = V/A$, then we can express the factor of safety as a ratio of the failure stress σ_{fail} (or τ_{fail}) to the allowable stress σ_{allow} (or τ_{allow});* that is,

$$\text{F.S.} = \frac{\sigma_{\text{fail}}}{\sigma_{\text{allow}}} \tag{1–9}$$

or

$$\text{F.S.} = \frac{\tau_{\text{fail}}}{\tau_{\text{allow}}} \tag{1–10}$$

---

*In some cases, such as columns, the applied load is *not* linearly related to stress and therefore only Eq. 1–8 can be used to determine the factor of safety. See Chapter 13.

In any of these equations, the factor of safety is chosen to be *greater* than 1 in order to avoid the potential for failure. Specific values depend on the types of materials to be used and the intended purpose of the structure or machine. For example, the F.S. used in the design of aircraft or space-vehicle components may be close to 1 in order to reduce the weight of the vehicle. On the other hand, in the case of a nuclear power plant, the factor of safety for some of its components may be as high as 3 since there may be uncertainties in loading or material behavior. In general, however, factors of safety and therefore the allowable loads or stresses for both structural and mechanical elements have become well standardized, since their design uncertainties have been reasonably evaluated. Their values, which can be found in design codes and engineering handbooks, are intended to form a balance of insuring public and environmental safety and providing a reasonable economic solution to design.

1.7   DESIGN OF SIMPLE CONNECTIONS

By making simplifying assumptions regarding the behavior of the material, the equations $\sigma = P/A$ and $\tau_{avg} = V/A$ can often be used to analyze or design a simple connection or a mechanical element. In particular, if a member is subjected to a *normal force* at a section, its required area at the section is determined from

$$A = \frac{P}{\sigma_{allow}} \qquad (1\text{--}11)$$

On the other hand, if the section is subjected to a *shear force*, then the required area at the section is

$$A = \frac{V}{\tau_{allow}} \qquad (1\text{--}12)$$

As discussed in Sec. 1.6, the allowable stress used in each of these equations is determined either by applying a factor of safety to a specified normal or shear stress or by finding these stresses directly from an appropriate design code.

We will now discuss four common types of problems for which the above equations can be used for design.

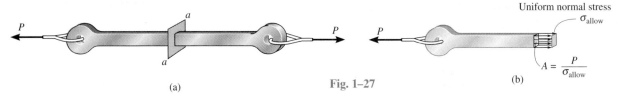

(a)

Fig. 1–27

$A = \dfrac{P}{\sigma_{\text{allow}}}$

Uniform normal stress
σ_{allow}

(b)

Cross-Sectional Area of a Tension Member.

The cross-sectional area of a prismatic member subjected to a tension force can be determined *provided* the force has a line of action that passes through the centroid of the cross section. For example, consider the "eye bar" shown in Fig. 1–27a. At the intermediate section a–a, the stress distribution is uniform over the cross section and the required area A is determined, as shown in Fig. 1–27b.

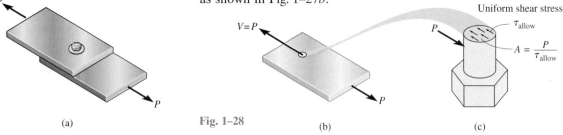

$V = P$

Uniform shear stress
τ_{allow}

$A = \dfrac{P}{\tau_{\text{allow}}}$

(a)

Fig. 1–28

(b)

(c)

Cross-Sectional Area of a Connector Subjected to Shear.

Often bolts or pins are used to connect plates, boards, or several members together. For example, consider the lap joint shown in Fig. 1–28a. If the bolt is loose or the clamping force of the bolt is unknown, it is safe to assume that any frictional force *between* the plates is negligible. As a result, the free-body diagram for a section passing *between* the plates and through the bolt is shown in Fig. 1–28b. The bolt is subjected to a resultant internal shear force of $V = P$ at this cross section. Assuming that the shear stress causing this force is *uniformly distributed* over the cross section, the required A is determined as shown in Fig. 1–28c.

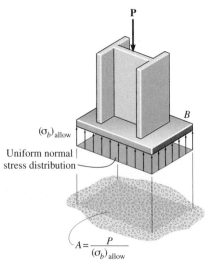

B

$(\sigma_b)_{\text{allow}}$

Uniform normal stress distribution

$A = \dfrac{P}{(\sigma_b)_{\text{allow}}}$

Fig. 1–29

Required Area to Resist Bearing.

A normal stress that is produced by the compression of one surface against another is called a ***bearing stress***. If this stress becomes large enough, it may crush or locally deform one or both of the surfaces. Hence, in order to prevent failure it is necessary to determine the proper bearing area for the material using an allowable bearing stress. For example, the area A of the column base plate B shown in Fig. 1–29 is determined from the allowable bearing stress of the concrete using $A = P/(\sigma_b)_{\text{allow}}$. This assumes, of course, that the allowable bearing stress for the concrete is smaller than that of the base plate material, and furthermore the bearing stress is uniformly distributed between the plate and the concrete as shown in the figure.

Required Area to Resist Shear Caused by Axial Load. Occasionally rods or other members will be supported in such a way that shear stress can be developed in the member even though the member may be subjected to an axial load. An example of this situation would be a steel rod whose end is encased in concrete and loaded as shown in Fig. 1–30a. A free-body diagram of the rod, Fig. 1–30b, shows that *shear stress* acts over the area of contact of the rod with the concrete. This area is $(\pi d)l$, where d is the rod's diameter and l is the length of embedment. Although the actual shear-stress distribution along the rod would be difficult to determine, if we assume it is *uniform*, we can use $A = V/\tau_{\text{allow}}$ to calculate l, provided we know d and τ_{allow}, Fig. 1–30b.

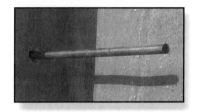

IMPORTANT POINTS

• Design of a member for strength is based on selecting an allowable stress that will enable it to safely support its intended load. There are many unknown factors that can influence the actual stress in a member and so, depending upon the intended uses of the member, a *factor of safety* is applied to obtain the allowable load the member can support.

• The four cases illustrated in this section represent just a few of the many applications of the average normal and shear stress formulas used for engineering design and analysis. Whenever these equations are applied, however, it is important to be aware that the stress distribution is assumed to be *uniformly distributed* or "averaged" over the section.

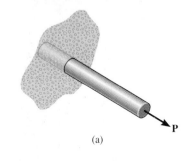

(a)

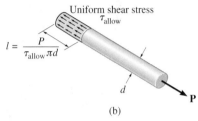

$$l = \frac{P}{\tau_{\text{allow}}\,\pi d}$$

Uniform shear stress
τ_{allow}

d

P

(b)

Fig. 1–30

PROCEDURE FOR ANALYSIS

When solving problems using the average normal and shear stress equations, a careful consideration should first be made as to the section over which the critical stress is acting. Once this section is made, the member must then be designed to have a sufficient area at the section to resist the stress that acts on it. To determine this area, application requires the following steps.

Internal Loading.

• Section the member through the area and draw a free-body diagram of a segment of the member. The internal resultant force at the section is then determined using the equations of equilibrium.

Required Area.

• Provided the allowable stress is known or can be determined, the required area needed to sustain the load at the section is then computed from $A = P/\sigma_{\text{allow}}$ or $A = V/\tau_{\text{allow}}$.

EXAMPLE 1–13

The two members are pinned together at B as shown in Fig. 1–31a. Top views of the pin connections at A and B are also given in the figure. If the pins have an allowable shear stress of $\tau_{allow} = 12.5$ ksi and the allowable tensile stress of rod CB is $(\sigma_t)_{allow} = 16.2$ ksi, determine to the nearest $\frac{1}{16}$ in. the smallest diameter of pins A and B and the diameter of rod CB necessary to support the load.

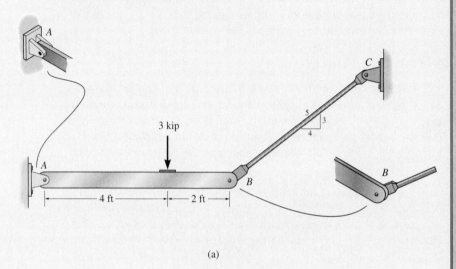

(a)

Fig. 1–31

SOLUTION

Recognizing CB to be a two-force member, the free-body diagram of member AB along with the computed reactions at A and B is shown in Fig. 1–31b. As an exercise, verify the computations and notice that the *resultant force* at A must be used for the design of pin A, since this is the shear force the pin resists.

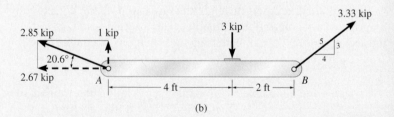

(b)

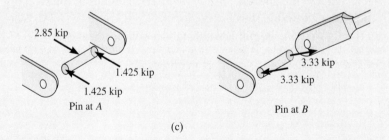

Pin at A   Pin at B

(c)

Diameter of the Pins.   From Fig. 1–31*a* and the free-body diagrams of the sectioned portion of each pin in contact with member *AB*, Fig. 1–31*c*, it is seen that pin A is subjected to double shear, whereas pin *B* is subjected to single shear. Thus,

$$A_A = \frac{V_A}{\tau_{\text{allow}}} = \frac{1.425 \text{ kip}}{12.5 \text{ kip/in}^2} = 0.1139 \text{ in}^2 = \pi\left(\frac{d_A^2}{4}\right) \quad d_A = 0.381 \text{ in.}$$

$$A_B = \frac{V_B}{\tau_{\text{allow}}} = \frac{3.333 \text{ kip}}{12.5 \text{ kip/in}^2} = 0.2667 \text{ in}^2 = \pi\left(\frac{d_B^2}{4}\right) \quad d_B = 0.583 \text{ in.}$$

Although these values represent the *smallest* allowable pin diameters, a *fabricated* or available pin size will have to be chosen. We will choose a size *larger* to the nearest $\frac{1}{16}$ in. as required.

$$d_A = \tfrac{7}{16} \text{ in.} = 0.4375 \text{ in.} \qquad\qquad \textit{Ans.}$$
$$d_B = \tfrac{5}{8} \text{ in.} = 0.625 \text{ in.} \qquad\qquad \textit{Ans.}$$

Diameter of Rod.   The required diameter of the rod throughout its midsection is thus,

$$A_{BC} = \frac{P}{(\sigma_t)_{\text{allow}}} = \frac{3.333 \text{ kip}}{16.2 \text{ kip/in}^2} = 0.2058 \text{ in}^2 = \pi\left(\frac{d_{BC}^2}{4}\right)$$
$$d_{BC} = 0.512 \text{ in.}$$

We will choose

$$d_{BC} = \tfrac{9}{16} \text{ in.} = 0.5625 \text{ in.} \qquad\qquad \textit{Ans.}$$

EXAMPLE 1-14

The control arm is subjected to the loading shown in Fig. 1–32a. Determine to the nearest $\frac{1}{4}$ in. the required diameter of the steel pin at C if the allowable shear stress for the steel is $\tau_{\text{allow}} = 8$ ksi. Note in the figure that the pin is subjected to double shear.

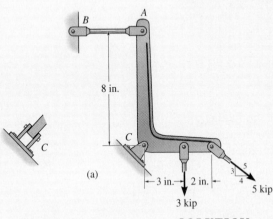

(a)

Fig. 1–32

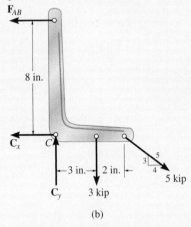

(b)

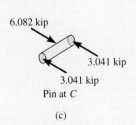

Pin at C

(c)

SOLUTION

Internal Shear Force.   A free-body diagram of the arm is shown in Fig. 1–32b. For equilibrium we have

$$\zeta^+ \ \Sigma M_C = 0; \qquad F_{AB}(8 \text{ in.}) - 3 \text{ kip}(3 \text{ in.}) - 5 \text{ kip}(\tfrac{3}{5})(5 \text{ in.}) = 0$$

$$F_{AB} = 3 \text{ kip}$$

$$\xrightarrow{+} \Sigma F_x = 0; \qquad -3 \text{ kip} - C_x + 5 \text{ kip}(\tfrac{4}{5}) = 0 \qquad C_x = 1 \text{ kip}$$

$$+\uparrow \Sigma F_y = 0; \qquad C_y - 3 \text{ kip} - 5 \text{ kip}(\tfrac{3}{5}) = 0 \qquad C_y = 6 \text{ kip}$$

The pin at C resists the resultant force at C. Therefore,

$$F_C = \sqrt{(1 \text{ kip})^2 + (6 \text{ kip})^2} = 6.082 \text{ kip}$$

Since the pin is subjected to double shear, a shear force of 3.041 kip acts over its cross-sectional area *between* the arm and each supporting leaf for the pin, Fig. 1–32c.

Required Area.   We have

$$A = \frac{V}{\tau_{\text{allow}}} = \frac{3.041 \text{ kip}}{8 \text{ kip/in}^2} = 0.3802 \text{ in}^2$$

$$\pi \left(\frac{d}{2}\right)^2 = 0.3802 \text{ in}^2$$

$$d = 0.696 \text{ in.}$$

Use a pin having a diameter of

$$d = \tfrac{3}{4} \text{ in.} = 0.750 \text{ in.} \qquad\qquad \textit{Ans.}$$

EXAMPLE 1–15

The suspender rod is supported at its end by a fixed-connected circular disk as shown in Fig. 1–33a. If the rod passes through a 40-mm-diameter hole, determine the minimum required diameter of the rod and the minimum thickness of the disk needed to support the 20-kN load. The allowable normal stress for the rod is $\sigma_{\text{allow}} = 60$ MPa, and the allowable shear stress for the disk is $\tau_{\text{allow}} = 35$ MPa.

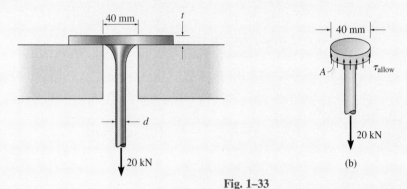

Fig. 1–33

SOLUTION

Diameter of Rod. By inspection, the axial force in the rod is 20 kN. Thus the required cross-sectional area of the rod is

$$A = \frac{P}{\sigma_{\text{allow}}} = \frac{20(10^3) \text{ N}}{60(10^6) \text{ N/m}^2} = 0.3333(10^{-3}) \text{ m}^2$$

So that

$$A = \pi(\frac{d^2}{4}) = 0.3333(10^{-3}) \text{ m}^2$$

$$d = 0.0206 \text{ m} = 20.6 \text{ mm} \qquad \textbf{Ans.}$$

Thickness of Disk. As shown on the free-body diagram of the core section of the disk, Fig. 1–33b, the material at the sectioned area must resist shear stress to prevent movement of the disk through the hole. If this shear stress is assumed to be distributed uniformly over the sectioned area, then, since $V = 20$ kN, we have

$$A = \frac{V}{\tau_{\text{allow}}} = \frac{20(10^3) \text{ N}}{35(10^6) \text{ N/m}^2} = 0.5714(10^{-3}) \text{ m}^2$$

Since the sectioned area $A = 2\pi(0.02 \text{ m})(t)$, the required thickness of the disk is

$$t = \frac{0.5714(10^{-3}) \text{ m}^2}{2\pi(0.02 \text{ m})} = 4.55(10^{-3}) \text{ m} = 4.55 \text{ mm} \qquad \textbf{Ans.}$$

EXAMPLE 1–16

An axial load on the shaft shown in Fig. 1–34a is resisted by the collar at C, which is attached to the shaft and located on the right side of the bearing at B. Determine the largest value of P for the two axial forces at E and F so that the stress in the collar does not exceed an allowable bearing stress at C of $(\sigma_b)_{\text{allow}} = 75$ MPa and the average normal stress in the shaft does not exceed an allowable tensile stress of $(\sigma_t)_{\text{allow}} = 55$ MPa.

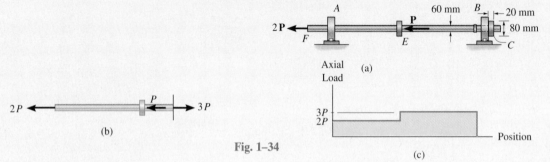

Fig. 1–34

SOLUTION

To solve the problem we will determine P for each possible failure condition. Then we will choose the *smallest* value. Why?

Normal Stress.   Using the method of sections, the axial load within region *FE* of the shaft is 2P, whereas the *largest* axial load, 3P, occurs within region *EC*, Fig. 1–34b. The variation of the internal loading is clearly shown on the normal-force diagram, Fig. 1–34c. Since the cross-sectional area of the entire shaft is constant, region *EC* will be subjected to the maximum average normal stress. Applying Eq. 1–11, we have

$$\sigma_{\text{allow}} = \frac{P}{A} \qquad 55(10^6) \text{ N/m}^2 = \frac{3P}{\pi(0.03 \text{ m})^2}$$

$$P = 51.8 \text{ kN}$$

Bearing Stress.   As shown on the free-body diagram in Fig. 1–34d, the collar at C must resist the load of 3P, which acts over a bearing area of $A_b = [\pi(0.04 \text{ m})^2 - \pi(0.03 \text{ m})^2] = 2.199(10^{-3}) \text{ m}^2$. Thus,

$$A = \frac{P}{\sigma_{\text{allow}}}; \qquad 75(10^6) \text{ N/m}^2 = \frac{3P}{2.199(10^{-3}) \text{ m}^2}$$

$$P = 55.0 \text{ kN}$$

By comparison, the largest load that can be applied to the shaft is P = 51.8 kN, since any load larger than this will cause the allowable normal stress in the shaft to be exceeded.

EXAMPLE 1-17

The rigid bar AB shown in Fig. 1–35a is supported by a steel rod AC having a diameter of 20 mm and an aluminum block having a cross-sectional area of 1800 mm². The 18-mm-diameter pins at A and C are subjected to *single shear*. If the failure stress for the steel and aluminum is $(\sigma_{st})_{\text{fail}} = 680$ MPa and $(\sigma_{al})_{\text{fail}} = 70$ MPa, respectively, and the failure shear stress for each pin is $\tau_{\text{fail}} = 900$ MPa, determine the largest load P that can be applied to the bar. Apply a factor of safety of F.S. = 2.

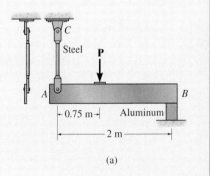

(a)

SOLUTION

Using Eqs. 1–9 and 1–10, the allowable stresses are

$$(\sigma_{st})_{\text{allow}} = \frac{(\sigma_{st})_{\text{fail}}}{\text{F.S.}} = \frac{680 \text{ MPa}}{2} = 340 \text{ MPa}$$

$$(\sigma_{al})_{\text{allow}} = \frac{(\sigma_{al})_{\text{fail}}}{\text{F.S.}} = \frac{70 \text{ MPa}}{2} = 35 \text{ MPa}$$

$$\tau_{\text{allow}} = \frac{\tau_{\text{fail}}}{\text{F.S.}} = \frac{900 \text{ MPa}}{2} = 450 \text{ MPa}$$

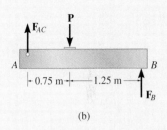

(b)

Fig. 1–35

The free-body diagram for the bar is shown in Fig. 1–35b. There are three unknowns. Here we will apply the equations of equilibrium so as to express F_{AC} and F_B in terms of the applied load P. We have

$$\zeta^+ \Sigma M_B = 0; \qquad P(1.25 \text{ m}) - F_{AC}(2 \text{ m}) = 0 \qquad (1)$$
$$\zeta^+ \Sigma M_A = 0; \qquad F_B(2 \text{ m}) - P(0.75 \text{ m}) = 0 \qquad (2)$$

We will now determine each value of P that creates the allowable stress in the rod, block, and pins, respectively.

Rod AC. This requires

$$F_{AC} = (\sigma_{st})_{\text{allow}}(A_{AC}) = 340(10^6) \text{ N/m}^2[\pi(0.01 \text{ m})^2] = 106.8 \text{ kN}$$

Using Eq. 1,

$$P = \frac{(106.8 \text{ kN})(2 \text{ m})}{1.25 \text{ m}} = 171 \text{ kN}$$

Block B. In this case,

$$F_B = (\sigma_{al})_{\text{allow}} A_B = 35(10^6) \text{ N/m}^2[1800 \text{ mm}^2(10^{-6}) \text{ m}^2/\text{mm}^2] = 63.0 \text{ kN}$$

Using Eq. 2,

$$P = \frac{(63.0 \text{ kN})(2 \text{ m})}{0.75 \text{ m}} = 168 \text{ kN}$$

Pin A or C. Here

$$V = F_{AC} = \tau_{\text{allow}} A = 450(10^6) \text{ N/m}^2[\pi(0.009 \text{ m})^2] = 114.5 \text{ kN}$$

From Eq. 1,

$$P = \frac{114.5 \text{ kN}(2 \text{ m})}{1.25 \text{ m}} = 183 \text{ kN}$$

By comparison, when P reaches its *smallest value* (168 kN), it develops the allowable normal stress in the aluminum block. Hence,

$$P = 168 \text{ kN} \qquad \qquad \textit{Ans.}$$

PROBLEMS

1–77. Member B is subjected to a compressive force of 800 lb. If A and B are both made of wood and are $\frac{3}{8}$ in. thick, determine to the nearest $\frac{1}{4}$ in. the smallest dimension h of the support so that the average shear stress does not exceed $\tau_{\text{allow}} = 300$ psi.

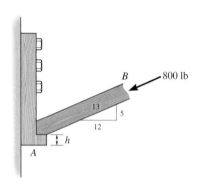

Prob. 1–77

1–78. The lever is attached to the shaft A using a key that has a width d and length of 25 mm. If the shaft is fixed and a vertical force of 200 N is applied perpendicular to the handle, determine the dimension d if the allowable shear stress for the key is $\tau_{\text{allow}} = 35$ MPa.

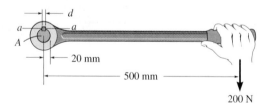

Prob. 1–78

1–79. The eye bolt is used to support the load of 5 kip. Determine its diameter d to the nearest $\frac{1}{8}$ in. and the required thickness h to the nearest $\frac{1}{8}$ in. of the support so that the washer will not penetrate or shear through it. The allowable normal stress for the bolt is $\sigma_{\text{allow}} = 21$ ksi and the allowable shear stress for the supporting material is $\tau_{\text{allow}} = 5$ ksi.

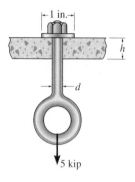

Prob. 1–79

***1–80.** Member A of the timber step joint for a truss is subjected to a compressive force of 5 kN. Determine the required diameter d of the steel rod at C and the height h of member B if the allowable normal stress for the steel is $(\sigma_{\text{allow}})_{\text{st}} = 157$ MPa and the allowable normal stress for the wood is $(\sigma_{\text{allow}})_{\text{w}} = 2$ MPa. Member B is 50 mm thick.

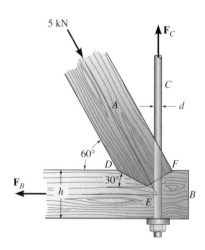

Prob. 1–80

1–81. The fillet weld size a is determined by computing the average shear stress along the shaded plane, which has the smallest cross section. Determine the smallest size a of the two welds if the force applied to the plate is $P = 20$ kip. The allowable shear stress for the weld material is $\tau_{allow} = 14$ ksi.

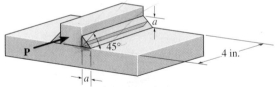

Prob. 1–81

1–82. The fillet weld size $a = 0.25$ in. If the joint is assumed to fail by shear on both sides of the block along the shaded plane, which is the smallest cross section, determine the largest force P that can be applied to the plate. The allowable shear stress for the weld material is $\tau_{allow} = 14$ ksi.

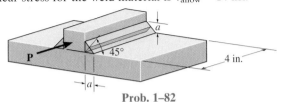

Prob. 1–82

1–83. The hanger is supported using the rectangular pin. Determine the magnitude of the allowable suspended load **P** if the allowable bearing stress is $(\sigma_b)_{allow} = 220$ MPa, the allowable tensile stress is $(\sigma_t)_{allow} = 150$ MPa, and the allowable shear stress is $\tau_{allow} = 130$ MPa. Take $t = 6$ mm, $a = 5$ mm, and $b = 25$ mm.

*1–84. The hanger is supported using the rectangular pin. Determine the required thickness t of the hanger, and dimensions a and b if the suspended load is $P = 60$ kN. The allowable tensile stress is $(\sigma_t)_{allow} = 150$ MPa, the allowable bearing stress is $(\sigma_b)_{allow} = 290$ MPa, and the allowable shear stress is $\tau_{allow} = 125$ MPa.

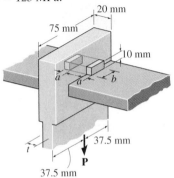

Probs. 1–83/1–84

1–85. The frame is subjected to the load of 1.5 kip. Determine the required diameter of the pins at A and B if the allowable shear stress for the material is $\tau_{allow} = 6$ ksi. Pin A is subjected to double shear, whereas pin B is subjected to single shear.

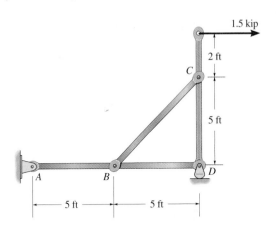

Prob. 1–85

1–86. Determine the required cross-sectional area of member BC and the diameter of the pins at A and B if the allowable normal stress is $\sigma_{allow} = 3$ ksi and the allowable shear stress is $\tau_{allow} = 4$ ksi.

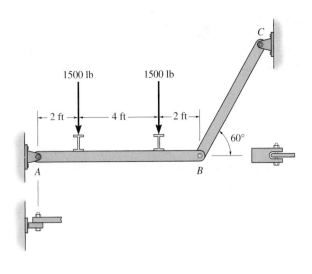

Prob. 1–86

1–87. The connection is made using a bolt and nut and two washers. If the allowable bearing stress for the boards is $(\sigma_b)_{allow} = 2$ ksi, and the allowable tensile stress for the bolt shank S is $(\sigma_t)_{allow} = 18$ ksi, determine the maximum allowable tension in the bolt shank. The bolt shank has a diameter of 0.31 in., and the washers have an outer diameter of 0.75 in. and inner diameter (hole) of 0.50 in.

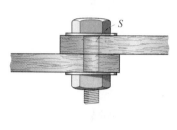

Prob. 1–87

*1–88. The hangers support the joist uniformly, so that it is assumed the four nails on each hanger carry an equal portion of the load. If the joist is subjected to the loading shown, determine the average shear stress in each nail of the hanger at ends A and B. Each nail has a diameter of 0.25 in. The hangers only support vertical loads.

1–89. The hangers support the joist uniformly, so that it is assumed the four nails on each hanger carry an equal portion of the load. Determine the smallest diameter of the nails at A and at B if the allowable shear stress for the nails is $\tau_{allow} = 4$ ksi. The hangers only support vertical loads.

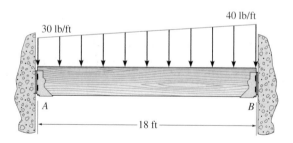

Probs. 1–88/1–89

1–90. The aluminum bracket A is used to support the centrally applied load of 8 kip. If it has a constant thickness of 0.5 in., determine the smallest height h in order to prevent a shear failure. The failure shear stress is $\tau_{fail} = 23$ ksi. Use a factor of safety for shear of F.S. = 2.5.

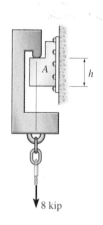

8 kip

Prob. 1–90

1–91. Determine the smallest dimensions of the circular shaft and circular end cap if the load it is required to support is $P = 150$ kN. The allowable tensile stress, bearing stress, and shear stress is $(\sigma_t)_{allow} = 175$ MPa, $(\sigma_b)_{allow} = 275$ MPa, and $\tau_{allow} = 115$ MPa.

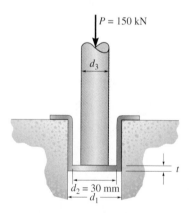

Prob. 1–91

*1–92. The hanger assembly is used to support a distributed loading of $w = 0.8$ kip/ft. Determine the average shear stress in the 0.40-in.-diameter bolt at A and the average tensile stress in rod AB, which has a diameter of 0.5 in. If the yield shear stress for the bolt is $\tau_y = 25$ ksi, and the yield tensile stress for the rod is $\sigma_y = 38$ ksi, determine the factor of safety with respect to yielding in each case.

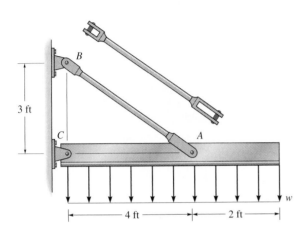

3 ft

4 ft

2 ft

Prob. 1–92

1–93. Determine the intensity w of the maximum distributed load that can be supported by the hanger assembly so that an allowable shear stress of $\tau_{allow} = 13.5$ ksi is not exceeded in the 0.40-in.-diameter bolts at A and B, and an allowable tensile stress of $\sigma_{allow} = 22$ ksi is not exceeded in the 0.5-in.-diameter rod AB.

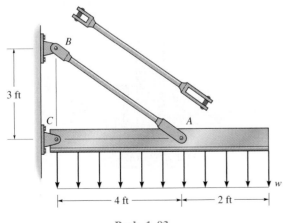

3 ft

4 ft

2 ft

Prob. 1–93

1–94. The wood specimen is subjected to the pull of 10 kN in a tension testing machine. If the allowable normal stress for the wood is $(\sigma_t)_{allow} = 12$ MPa and the allowable shear stress is $\tau_{allow} = 1.2$ MPa, determine the required dimensions b and t so that the specimen reaches these stresses simultaneously. The specimen has a width of 25 mm.

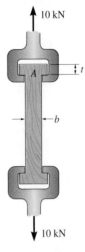

10 kN

10 kN

Prob. 1–94

1–95. The compound wooden beam is connected together by a bolt at B. Assuming that the connections at A, B, C, and D exert only vertical forces on the beam, determine the required diameter of the bolt at B and the required outer diameter of its washers if the allowable tensile stress for the bolt is $(\sigma_t)_{allow} = 150$ MPa and the allowable bearing stress for the wood is $(\sigma_b)_{allow} = 28$ MPa. Assume that the hole in the washers has the same diameter as the bolt.

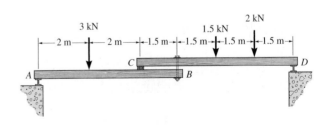

3 kN   1.5 kN   2 kN

2 m   2 m   1.5 m   1.5 m   1.5 m   1.5 m

Prob. 1–95

*1–96. If the allowable bearing stress for the material under the support at A and B is $(\sigma_b)_{\text{allow}} = 400$ psi, determine the maximum load **P** that can be applied to the beam. The bearing plates A' and B' have square cross sections of 2 in. $\times$ 2 in. and 4 in. $\times$ 4 in., respectively.

1–97. If the allowable bearing stress for the material under the supports at A and B is $(\sigma_b)_{\text{allow}} = 400$ psi, deter-mine the size of *square* bearing plates A' and B' required to support the load. Dimension the plates to the nearest $\frac{1}{2}$ in. The reactions at the supports are vertical. Take $P = 1500$ lb.

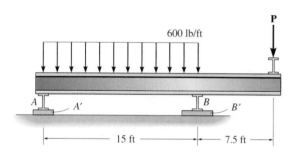

Probs. 1–96/1–97

1–98. The lapbelt assembly is to be subjected to a force of 800 N. Determine (a) the required thickness t of the belt if the allowable tensile stress for the material is $(\sigma_t)_{\text{allow}} = 10$ MPa, (b) the required lap length d_l if the glue can sustain an allowable shear stress of $(\tau_{\text{allow}})_g = 0.75$ MPa, and (c) the required diameter d_r of the pin if the allowable shear stress for the pin is $(\tau_{\text{allow}})_p = 30$ MPa.

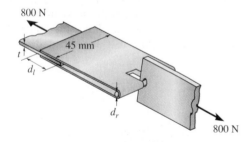

Prob. 1–98

1–99. The pin is subjected to double shear since it is used to connect the three links together. Due to wear, the load is distributed over the top and bottom of the pin as shown on the free-body diagram. Determine the diameter d of the pin if the allowable shear stress is $\tau_{\text{allow}} = 10$ ksi and the load $P = 8$ kip. Also, determine the load intensities w_1 and w_2.

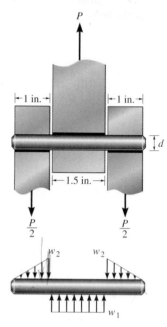

Prob. 1–99

*1–100. The thrust bearing consists of a circular collar A fixed to the shaft B. Determine the maximum axial force P that can be applied to the shaft so that it does not cause the shear stress along a cylindrical surface a or b to exceed an allowable shear stress of $\tau_{\text{allow}} = 170$ MPa.

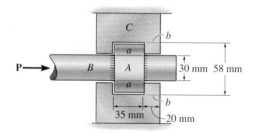

Prob. 1–100

1–101. The spring mechanism is used as a shock absorber for a load applied to the drawbar AB. Determine the force in each spring when the 50-kN force is applied. Each spring is originally unstretched and the drawbar slides along the smooth guide posts CG and EF. The ends of all springs are attached to their respective members. Also, what is the required diameter of the shank of bolts CG and EF if the allowable stress for the bolts is $\sigma_{\text{allow}} = 150$ MPa.

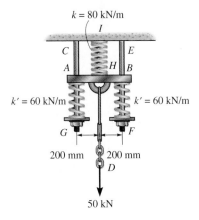

Prob. 1–101

1–102. The beam AB is pin supported at A and supported by a cable BC. A *separate* cable CG is used to hold up the frame. If AB weighs 120 lb/ft and the column FC has a weight of 180 lb/ft, determine the resultant internal loadings acting on cross sections located at points D and E. Neglect the thickness of both the beam and column in the calculation.

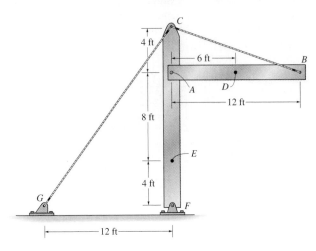

Prob. 1–102

1–103. The assembly is used to support the distributed loading of $w = 500$ lb/ft. Determine the factor of safety with respect to yielding for the steel rod BC and the pins at B and C if the yield stress for the steel in tension is $\sigma_Y = 36$ ksi and in shear $\tau_Y = 18$ ksi. The rod has a diameter of 0.4 in., and the pins each have a diameter of 0.30 in.

*1–104. If the allowable shear stress for each of the 0.3-in.-diameter steel pins at A, B, and C is $\tau_{\text{allow}} = 12.5$ ksi, and the allowable normal stress for the 0.40-in.-diameter rod BC is $\sigma_{\text{allow}} = 22$ ksi, determine the largest intensity w of the uniform distributed load that can be supported by the beam.

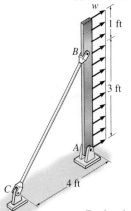

Probs. 1–103/1–104

1–105. The bar is supported by the pin. If the allowable tensile stress for the bar is $(\sigma_t)_{\text{allow}} = 21$ ksi, and the allowable shear stress for the pin is $\tau_{\text{allow}} = 12$ ksi, determine the diameter of the pin for which the load P will be a maximum. What is this maximum load? Assume the hole in the bar has the same diameter d as the pin. Take $t = \frac{1}{4}$ in. and $w = 2$ in.

1–106. The bar is connected to the support using a pin having a diameter of $d = 1$ in. If the allowable tensile stress for the bar is $(\sigma_t)_{\text{allow}} = 20$ ksi, and the allowable bearing stress between the pin and the bar is $(\sigma_b)_{\text{allow}} = 30$ ksi, determine the dimensions w and t such that the gross area of the cross section is $wt = 2$ in^2 and the load P is a maximum. What is this maximum load? Assume the hole in the bar has the same diameter as the pin.

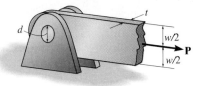

Probs. 1–105/1–106

REVIEW PROBLEMS

1–107. A force of 8 kN is applied at the *center* of the wooden post. If the post is placed at the corner of its base plate *B*, can the bearing stress that the base plate exerts on the slab *S* be assumed uniformly distributed? Why or why not? What is the average compressive stress in the wooden post?

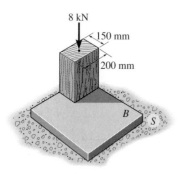

Prob. 1–107

*1–108. The column is made of concrete having a density of $2.30 \ \text{Mg/m}^3$. At its top *B* it is subjected to an axial compressive force of 15 kN. Determine the compressive stress in the column as a function of the distance *z* measured from its base.

1–109. The column is made of concrete having a density of $2.30 \ \text{Mg/m}^3$. At its top *B* it is subjected to an axial compressive force of 15 kN. Determine the compressive stress in the column at $z = 1$ m, 2 m, and 3 m.

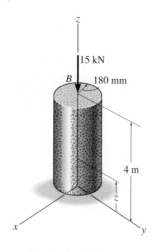

Probs. 1–108/1–109

1–110. The pulley is held fixed to the 20-mm-diameter shaft using a key that fits within a groove cut into the pulley and shaft. If the suspended load has a mass of 50 kg, determine the average shear stress in the key along section *a–a*. The key is 5 mm by 5 mm square and 12 mm long.

Prob. 1–110

1–111. The circular punch *B* exerts a force of 2 kN on the top of the plate *A*. Determine the average shear stress in the plate due to this loading.

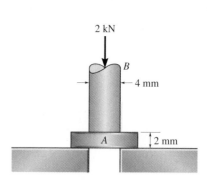

Prob. 1–111

*1–112. The two aluminum rods support the vertical force of $P = 20$ kN. Determine their required diameters if the allowable tensile stress for the aluminum is $\sigma_{allow} = 150$ MPa.

1–114. Member B is subjected to a compressive force of 650 lb. If A and B are both made of wood and are $\frac{3}{8}$ in. thick, determine to the nearest $\frac{1}{4}$ in. the smallest dimension d of the support so that the average shear stress along section C does not exceed $\tau_{allow} = 300$ psi.

1–115. Member B is subjected to a compressive force of 650 lb. If A and B are both made of wood and are $\frac{3}{8}$ in. thick, and $d = 6$ in., determine the average shear stress along section C.

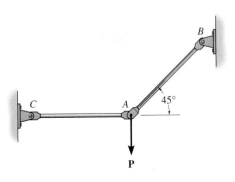

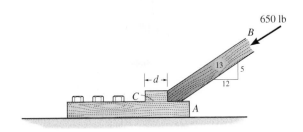

Prob. 1–112

Probs. 1–114/1–115

1–113. The cable has a specific weight γ (weight/volume) and cross-sectional area A. If the sag s is small, so that its length is approximately L and its weight can be distributed uniformly along the horizontal axis, determine the average normal stress in the cable at its lowest point C.

*1–116. The 3-Mg concrete pipe is suspended by the three wires. If BD and CD have a diameter of 10 mm and AD has a diameter of 7 mm, determine the average normal stress in each wire.

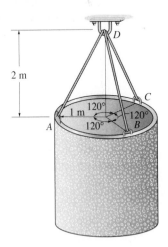

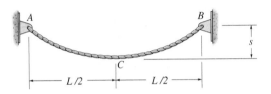

Prob. 1–113

Prob. 1–116

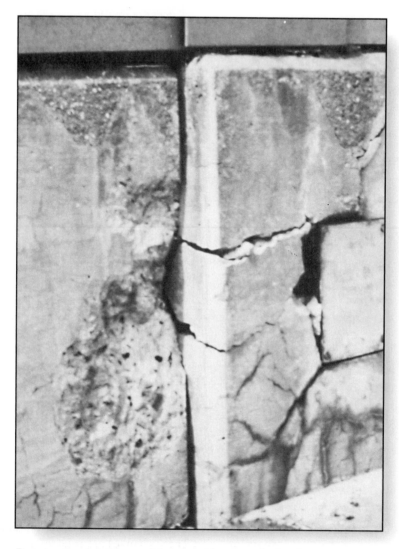

Excessive stress in brittle materials such as this concrete bridge abutment has caused it to strain until it fractured. By making measurements of strain, engineers can then predict the stress in the material.

2 STRAIN

Chapter Objectives

In engineering the deformation of a body is specified using the concept of normal and shear strain. In this chapter we will define these quantities and show how they can be determined for various types of problems.

2.1 DEFORMATION

Whenever a force is applied to a body, it will tend to change the body's shape and size. These changes are referred to as *deformation*, and they may be either highly visible or practically unnoticeable, without the use of equipment to make precise measurements. For example, a rubber band will undergo a very large deformation when stretched. On the other hand, only slight deformations of structural members occur when a building is occupied with people walking about. Deformation of a body can also occur when the temperature of the body is changed. A typical example is the thermal expansion or contraction of a roof caused by the weather.

Note the before and after positions of three different line segments on this rubber membrane which is subjected to tension. The vertical line is lengthened, the horizontal line is shortened, and the inclined line changes its length and rotates.

In the general sense, the deformation of a body will not be uniform throughout its volume, and so the change in geometry of any line segment within the body may vary along its length. For example, one portion of the line may elongate, whereas another portion may contract. As shorter and shorter line segments are considered, however, they remain straighter after the deformation, and so to study deformational changes in a more uniform manner, we will consider the lines to be very short and located in the neighborhood of a point. In doing so, realize that any line segment located at one point in the body will change by a different amount from one located at some other point. Furthermore, these changes will also depend on the orientation of the line segment at the point. For example, a line segment may elongate if it is oriented in one direction, whereas it may contract if it is oriented in another direction.

2.2   STRAIN

In order to describe the deformation by changes in length of line segments and the changes in the angles between them, we will develop the concept of strain. Measurements of strain are actually made by experiments, and once the strains are obtained, it will be shown in the next section how they can be related to the applied loads, or stresses, acting within the body.

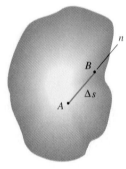

Undeformed body

(a)

Normal Strain.   The elongation or contraction of a line segment per unit of length is referred to as **normal strain**. To develop a formalized definition of normal strain, consider the line AB, which is contained within the undeformed body shown in Fig. 2–1a. This line lies along the n axis and has an original length of Δs. After deformation, points A and B are displaced to A' and B', and the line becomes a curve having a length of $\Delta s'$, Fig. 2–1b. The change in length of the line is therefore $\Delta s' - \Delta s$. If we define the *average normal strain* using the symbol ϵ_{avg} (epsilon), then

$$\epsilon_{\text{avg}} = \frac{\Delta s' - \Delta s}{\Delta s} \tag{2–1}$$

As point B is chosen closer and closer to point A, the length of the line becomes shorter and shorter, such that $\Delta s \to 0$. Also, this causes B' to approach A', such that $\Delta s' \to 0$. Consequently, in the limit the normal strain at *point A* and in the direction of n is

$$\epsilon = \lim_{B \to A \text{ along } n} \frac{\Delta s' - \Delta s}{\Delta s} \tag{2–2}$$

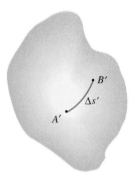

Deformed body

(b)

Fig. 2–1

If the normal strain is known, we can use this equation to obtain the approximate final length of a *short* line segment in the direction of n after it is deformed. We have

$$\Delta s' \approx (1 + \epsilon)\, \Delta s \tag{2–3}$$

Hence, when ϵ is positive the initial line will elongate, whereas if ϵ is negative the line contracts.

Units.   Note that normal strain is a *dimensionless quantity*, since it is a ratio of two lengths. Although this is the case, it is common practice to state it in terms of a ratio of length units. If the SI system is used, then the basic units will be meters/meter (m/m). Ordinarily, for most engineering applications ϵ will be very small, so measurements of strain are in micrometers per meter (μm/m), where $1\ \mu$m $= 10^{-6}$ m. In the Foot-Pound-Second system, strain can be stated in units of inches per inch (in./in.). Sometimes for experimental work, strain is expressed as a percent, e.g., 0.001 m/m $= 0.1\%$. As an example, a normal strain of $480(10^{-6})$ can be reported as $480(10^{-6})$ in./in., 480 μm/m, or 0.0480%. Also, one can state this answer as simply 480 μ (480 "micros").

Shear Strain.   The change in angle that occurs between two line segments that were originally *perpendicular* to one another is referred to as **shear strain**. This angle is denoted by γ (gamma) and is measured in radians (rad). To show how it is developed, consider the line segments AB and AC originating from the same point A in a body, and directed along the perpendicular n and t axes, Fig. 2–2a. After deformation, the ends of the lines are displaced, and the lines themselves become curves, such that the angle between them at A is θ', Fig. 2–2b. Hence we define the shear strain at point A that is associated with the n and t axes as

$$\gamma_{nt} = \frac{\pi}{2} - \lim_{\substack{B \to A \text{ along } n \\ C \to A \text{ along } t}} \theta' \qquad (2\text{–}4)$$

Notice that if θ' is smaller than $\pi/2$ the shear strain is positive, whereas if θ' is larger than $\pi/2$ the shear strain is negative.

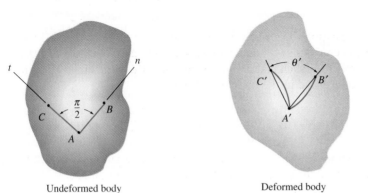

Undeformed body

(a)

Deformed body

(b)

Fig. 2–2

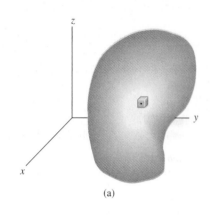

(a)

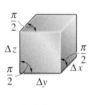

Undeformed
element

(b)

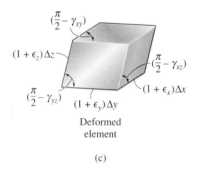

Deformed
element

(c)

Fig. 2–3

Cartesian Strain Components.   Using the above definitions of normal and shear strain, we will now show how they can be used to describe the deformation of the body, Fig. 2–3a. To do so, imagine the body to be subdivided into small elements such as the one shown in Fig. 2–3b. This element is rectangular, has undeformed dimensions Δx, Δy, and Δz, and is located in the neighborhood of a point in the body, Fig. 2–3a. Assuming that the element's dimensions are very small, the deformed shape of the element will be a parallelepiped, Fig. 2–3c, since very small line segments will remain approximately straight after the body is deformed. In order to achieve this deformed shape, we must first consider how the normal strain changes the lengths of the sides of the rectangular element, and then how the shear strain changes the angles of each side. Hence, using Eq. 2–3, $\Delta s' \approx (1 + \epsilon)\Delta s$, in reference to the lines Δx, Δy, and Δz, the approximate lengths of the sides of the parallelepiped are

$$(1 + \epsilon_x)\,\Delta x \qquad (1 + \epsilon_y)\,\Delta y \qquad (1 + \epsilon_z)\,\Delta z$$

And the approximate angles between the sides, again originally defined by the sides Δx, Δy, and Δz, are

$$\frac{\pi}{2} - \gamma_{xy} \qquad \frac{\pi}{2} - \gamma_{yz} \qquad \frac{\pi}{2} - \gamma_{xz}$$

In particular, notice that the *normal strains cause a change in volume* of the rectangular element, whereas the *shear strains cause a change in its shape*. Of course, both of these effects occur simultaneously during the deformation.

In summary, then, the *state of strain* at a point in a body requires specifying three normal strains, ϵ_x, ϵ_y, ϵ_z, and three shear strains, γ_{xy}, γ_{yz}, γ_{xz}. These strains completely describe the deformation of a rectangular volume element of material located at the point and oriented so that its sides are originally parallel to the x, y, z axes. Once these strains are defined at all points in the body, the deformed shape of the body can then be described. It should also be added that by knowing the state of strain at a point, defined by its six components, it is possible to determine the strain components on an element oriented at the point in any other direction. This is discussed in Chapter 10.

Small Strain Analysis. Most engineering design involves applications for which only *small deformations* are allowed. For example, almost all structures and machines appear to be rigid, and the deformations that occur during use are hardly noticeable. Furthermore, even if the deflection of a member such as a thin plate or slender rod may appear to be large, the material from which it is made may only be subjected to very small deformations. In this text, therefore, we will assume that the deformations that take place within a body are almost infinitesimal, so that the *normal strains* occurring within the material are *very small* compared to 1, that is, $\epsilon \ll 1$. This assumption, which is based on the magnitude of the strain, has wide practical application in engineering, and it is often referred to as a *small strain analysis*. For example, it allows us to approximate $\sin \theta \approx \theta$, $\cos \theta \approx 1$ and $\tan \theta \approx \theta$ provided θ is very small.

The rubber bearing support under this concrete bridge girder is subjected to both normal and shear strain. The normal strain is caused by the weight and bridge loads on the girder, and the shear strain is caused by the horizontal movement of the girder due to temperature changes.

IMPORTANT POINTS

• Loads will cause all material bodies to deform and, as a result, points in the body will undergo *displacements or changes in position*.

• *Normal strain* is a measure of the elongation or contraction of a small line segment in the body, whereas *shear strain* is a measure of the change in angle that occurs between two small line segments that are originally perpendicular to one another.

• The state of strain at a point is characterized by six strain components: three normal strains ϵ_x, ϵ_y, ϵ_z and three shear strains γ_{xy}, γ_{yz}, γ_{xz}. These components depend upon the orientation of the line segments and their location in the body.

• Strain is the geometrical quantity that is measured using experimental techniques. Once obtained, the stress in the body can then be determined from material property relations.

• Most engineering materials undergo small deformations, and so the normal strain $\epsilon \ll 1$. This assumption of "small strain analysis" allows the calculations for normal strain to be simplified since first-order approximations can be made about their size.

EXAMPLE 2-1

The slender rod shown in Fig. 2–4 is subjected to an increase of temperature along its axis, which creates a normal strain in the rod of $\epsilon_z = 40(10^{-3})z^{1/2}$, where z is given in meters. Determine (a) the displacement of the end B of the rod due to the temperature increase, and (b) the average normal strain in the rod.

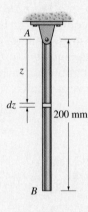

Fig. 2–4

SOLUTION

Part (a).   Since the normal strain is reported at each point along the rod, a differential segment dz, located at position z, Fig. 2–4, has a deformed length that can be determined from Eq. 2–3; that is,

$$dz' = [1 + 40(10^{-3})z^{1/2}]\,dz$$

The sum total of these segments along the axis yields the *deformed length* of the rod, i.e.,

$$z' = \int_0^{0.2\ \text{m}} [1 + 40(10^{-3})z^{1/2}]\,dz$$

$$= z + 40(10^{-3})(\tfrac{2}{3}z^{3/2}) \Big|_0^{0.2\ \text{m}}$$

$$= 0.20239\ \text{m}$$

The displacement of the end of the rod is therefore

$$\Delta_B = 0.20239\ \text{m} - 0.2\ \text{m} = 0.00239\ \text{m} = 2.39\ \text{mm} \downarrow \qquad Ans.$$

Part (b).   The average normal strain in the rod is determined from Eq. 2–1, which assumes that the rod or "line segment" has an original length of 200 mm and a change in length of 2.39 mm. Hence,

$$\epsilon_{\text{avg}} = \frac{\Delta s' - \Delta s}{\Delta s} = \frac{2.39\ \text{mm}}{200\ \text{mm}} = 0.0119\ \text{mm/mm} \qquad Ans.$$

EXAMPLE 2-2

A force acting on the grip of the lever arm shown in Fig. 2–5a causes the arm to rotate clockwise through an angle of $\theta = 0.002$ rad. Determine the average normal strain developed in the wire BC.

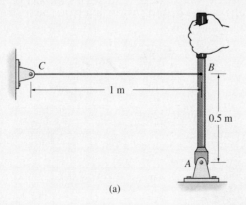

(a)

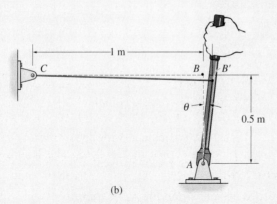

(b)

Fig. 2–5

SOLUTION

Since $\theta = 0.002$ rad is small, the stretch in the wire CB, Fig. 2–5b, is $BB' = \theta\,(0.5\text{ m}) = (0.002\text{ rad})(0.5\text{ m}) = 0.001$ m. The average normal strain in the wire is therefore,

$$\epsilon_{avg} = \frac{BB'}{CB} = \frac{0.001}{1\text{ m}} = 0.001\text{ m/m} \qquad Ans.$$

EXAMPLE 2-3

The plate is deformed into the dashed shape shown in Fig. 2–6a. If in this deformed shape horizontal lines on the plate remain horizontal and do not change their length, determine (a) the average normal strain along the side AB, and (b) the average shear strain in the plate relative to the x and y axes.

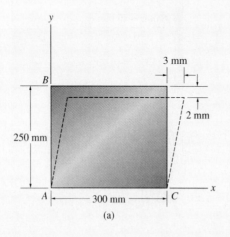

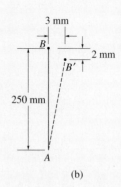

Fig. 2–6

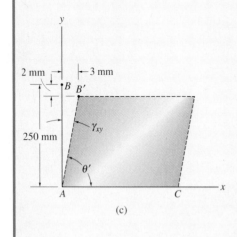

SOLUTION

Part (a). Line AB, coincident with the y axis, becomes line AB′ after deformation, as shown in Fig. 2–6b. The length of this line is

$$AB' = \sqrt{(250 - 2)^2 + (3)^2} = 248.018 \text{ mm}$$

The average normal strain for AB is therefore

$$(\epsilon_{AB})_{\text{avg}} = \frac{AB' - AB}{AB} = \frac{248.018 \text{ mm} - 250 \text{ mm}}{250 \text{ mm}}$$

$$= -7.93(10^{-3}) \text{ mm/mm} \qquad \textbf{Ans.}$$

The negative sign indicates the strain causes a contraction of AB.

Part (b). As noted in Fig. 2–6c, the once 90° angle BAC between the sides of the plate, referenced from the x, y axes, changes to θ′ due to the displacement of B to B′. Since $\gamma_{xy} = \pi/2 - \theta'$, then γ_{xy} is the angle shown in the figure. Thus,

$$\gamma_{xy} = \tan^{-1}\left(\frac{3 \text{ mm}}{250 \text{ mm} - 2 \text{ mm}}\right) = 0.0121 \text{ rad} \qquad \textbf{Ans.}$$

EXAMPLE 2–4

The plate shown in Fig. 2–7a is held in the rigid horizontal guides at its top and bottom, *AD* and *BC*. If its right side *CD* is given a uniform horizontal displacement of 2 mm, determine (*a*) the average normal strain along the diagonal *AC*, and (*b*) the shear strain at *E* relative to the *x*, *y* axes.

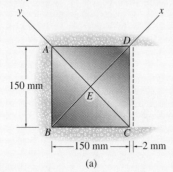

(a)

Fig. 2–7

(b)

SOLUTION

Part (a).   When the plate is deformed, the diagonal *AC* becomes *AC′*, Fig. 2–7b. The length of diagonals *AC* and *AC′* can be found from the Pythagorean theorem. We have

$$AC = \sqrt{(0.150)^2 + (0.150)^2} = 0.21213 \text{ m}$$
$$AC' = \sqrt{(0.150)^2 + (0.152)^2} = 0.21355 \text{ m}$$

Therefore the average normal strain along the diagonal is

$$(\epsilon_{AC})_{\text{avg}} = \frac{AC' - AC}{AC} = \frac{0.21355 \text{ m} - 0.21213 \text{ m}}{0.21213 \text{ m}}$$
$$= 0.00669 \text{ mm/mm} \qquad \textit{Ans.}$$

Part (b).   To find the shear strain at *E* relative to the *x* and *y* axes, it is first necessary to find the angle *θ′*, which specifies the angle between these axes after deformation, Fig. 2–7b. We have

$$\tan\left(\frac{\theta'}{2}\right) = \frac{76 \text{ mm}}{75 \text{ mm}}$$
$$\theta' = 90.759° = \frac{\pi}{180°}(90.759°) = 1.58404 \text{ rad}$$

Applying Eq. 2–4, the shear strain at *E* is therefore

$$\gamma_{xy} = \frac{\pi}{2} - 1.58404 \text{ rad} = -0.0132 \text{ rad} \qquad \textit{Ans.}$$

According to the sign convention, the *negative sign* indicates that the angle *θ′* is *greater than* 90°. Note that if the *x* and *y* axes were horizontal and vertical, then due to the deformation $\gamma_{xy} = 0$ at point *E*.

PROBLEMS

2–1. The center portion of the rubber balloon has a diameter of $d = 4$ in. If the air pressure within it causes the balloon's diameter to become $d = 5$ in., determine the average normal strain in the rubber.

Prob. 2–1

2–2. A rubber band has an unstretched length of 10 in. If it is stretched around a pole having an outer diameter of 6 in., determine the average normal strain in the band.

2–3. The rigid bar ABC is originally in a horizontal position. If loads cause the end A to be displaced downwards $\Delta_A = 0.002$ in. and the bar rotates $\theta = 0.2°$, determine the average normal strain in the rods AD, BE, and CF.

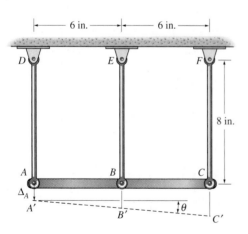

Prob. 2–3

*2–4. The rigid beam is supported by a pin at A and wires BD and CE. If the load **P** on the beam is displaced 10 mm downward, determine the normal strain developed in wires CE and BD.

2–5. The rigid beam is supported by a pin at A and wires BD and CE. If the maximum allowable normal strain in each wire is $\epsilon_{max} = 0.002$ mm/mm, determine the maximum vertical displacement of the load **P**.

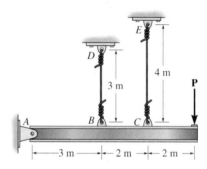

Probs. 2–4/2–5

2–6. Determine the approximate average normal strain in the wire AB as a function of the rotation θ of the rigid bar CA by assuming θ is small. What is this value if $\theta = 2°$?

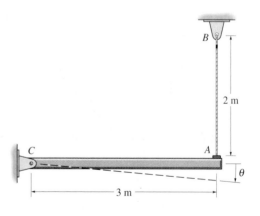

Prob. 2–6

2–7. The two wires are connected together at A. If the force **P** causes point A to be displaced vertically 3 mm, determine the normal strain developed in each wire.

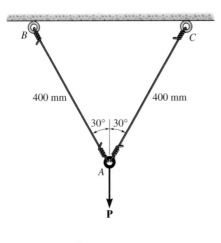

400 mm   400 mm

30° | 30°

A

P

Prob. 2–7

2–9. Two bars are used to support a load **P**. When unloaded, AB is 5 in. long, AC is 8 in. long, and the ring at A has coordinates $(0, 0)$. If a load is applied to the ring at A, so that it moves it to the coordinate position (0.25 in., -0.73 in.), determine the normal strain in each bar.

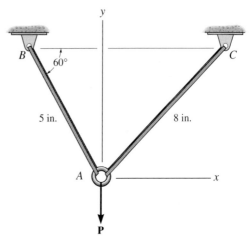

y

B 60° C

5 in.   8 in.

A x

P

Prob. 2–9

*2–8. Two bars are used to support a load. When unloaded, AB is 5 in. long, AC is 8 in. long, and the ring at A has coordinates $(0,0)$. If a load **P** acts on the ring at A, the normal strain in AB becomes $\epsilon_{AB} = 0.02$ in./in., and the normal strain in AC becomes $\epsilon_{AC} = 0.035$ in./in. Determine the coordinate position of the ring due to the load.

2–10. The guy wire AB of a building frame is originally unstretched. Due to an earthquake, the two columns of the frame tilt $\theta = 2°$. Determine the approximate normal strain in the wire when the frame is in this position. Assume the columns are rigid and rotate about their lower supports.

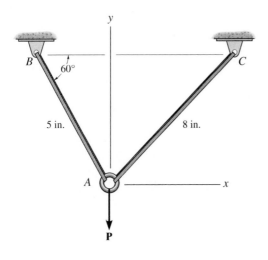

y

B 60° C

5 in.   8 in.

A x

P

Prob. 2–8

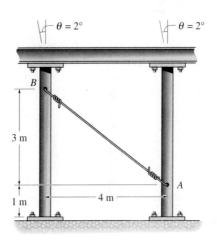

$\theta = 2°$ $\theta = 2°$

B

3 m

A

1 m   4 m

Prob. 2–10

2–11. Due to its weight, the rod is subjected to a normal strain that varies along its length such that $\epsilon = kz$, where k is a constant. Determine the displacement ΔL of its end B when it is suspended as shown.

Prob. 2–11

2–13. The rectangular plate is subjected to the deformation shown by the dashed line. Determine the average shear strain γ_{xy} of the plate.

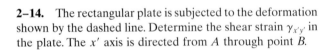

Prob. 2–13

*2–12. The rectangular membrane has an unstretched length L_1 and width L_2. If the sides are increased by small amounts ΔL_1 and ΔL_2, determine the normal strain along the diagonal AB.

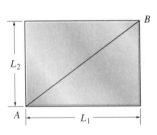

Prob. 2–12

2–14. The rectangular plate is subjected to the deformation shown by the dashed line. Determine the shear strain $\gamma_{x'y'}$ in the plate. The x' axis is directed from A through point B.

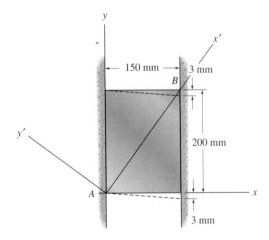

Prob. 2–14

2–15. The rectangular plate is subjected to the deformation shown by the dashed line. Determine the normal strains ϵ_x, ϵ_y, $\epsilon_{x'}$ $\epsilon_{y'}$.

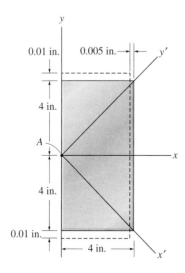

Prob. 2–15

*2–16. The rectangular plate is subjected to the deformation shown by the dashed line. Determine the shear strains γ_{xy} and $\gamma_{x'y'}$ developed at point A.

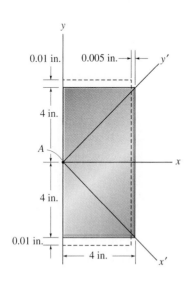

Prob. 2–16

2–17. The piece of plastic is originally rectangular. Determine the shear strain γ_{xy} at corners A and B if the plastic distorts as shown by the dashed lines.

2–18. The piece of plastic is originally rectangular. Determine the shear strain γ_{xy} at corners D and C if the plastic distorts as shown by the dashed lines.

2–19. The piece of plastic is originally rectangular. Determine the average normal strain that occurs along the diagonals AC and DB.

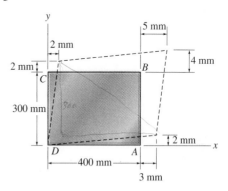

Probs. 2–17/2–18/2–19

*2–20. A square piece of material is deformed into the dashed position. Determine the shear strain γ_{xy} at A.

2–21. A square piece of material is deformed into the dashed parallelogram. Determine the average normal strain that occurs along the diagonals AC and BD.

2–22. A square piece of material is deformed into the dashed position. Determine the shear strain γ_{xy} at C.

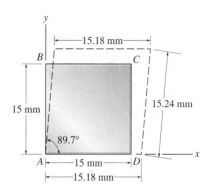

Probs. 2–20/2–21/2–22

2–23. The square deforms into the position shown by the dashed lines. Determine the average normal strain along each diagonal, *AB* and *CD*. Side *DB* remains horizontal.

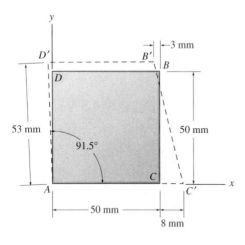

Prob. 2–23

2–25. The block is deformed into the position shown by the dashed lines. Determine the average normal strain along line *AB*.

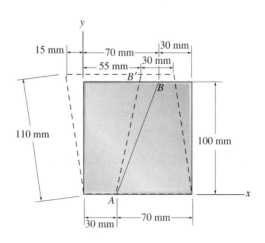

Prob. 2–25

*2–24. The square deforms into the position shown by the dashed lines. Determine the shear strain at each of its corners, *A* and *C*. Side *DB* remains horizontal.

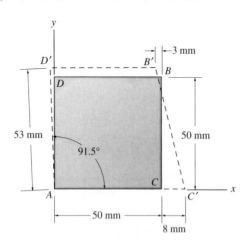

Prob. 2–24

2–26. The block is deformed into the position shown by the dashed lines. Determine the shear strain at corner *C*.

Prob. 2–26

2–27.   The bar is originally 300 mm long when it is flat. If it is subjected to a shear strain defined by $\gamma_{xy} = 0.02x$, where x is in millimeters, determine the displacement Δy at the end of its bottom edge. It is distorted into the shape shown, where no elongation of the bar occurs in the x direction.

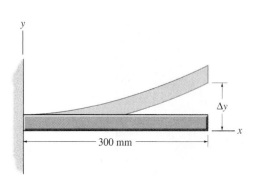

Prob. 2–27

*2–28.   The rubber band AB has an unstretched length of 1 ft. If it is fixed at B and attached to the surface at point A', determine the average normal strain in the band. The surface is defined by the function $y = (x^2)$ ft, where x is in feet.

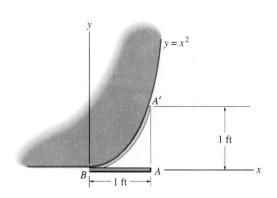

Prob. 2–28

2–29.   A thin wire is wrapped along a surface having the form $y = 0.5x^2$, where x and y are in inches. Originally the end B is at $x = 10$ in. If the wire undergoes a normal strain along its length of $\epsilon = 0.005x$, determine the change in length of the wire. *Hint*: For the curve, $y = f(x)$, $ds = \sqrt{1 + (dy/dx)^2}\, dx$.

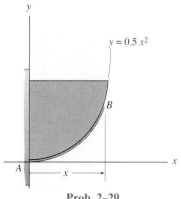

Prob. 2–29

2–30.   The fiber AB has a length L and orientation θ. If its ends A and B undergo very small displacements u_A and v_B, respectively, determine the normal strain in the fiber when it is in position $A'B'$.

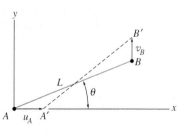

Prob. 2–30

2–31.   If the normal strain is defined in reference to the final length, that is,

$$\epsilon_n' = \lim_{P \to P'} \left(\frac{\Delta s' - \Delta s}{\Delta s'} \right)$$

instead of in reference to the original length, Eq. 2–2, show that the difference in these strains is represented as a second-order term, namely, $\epsilon_n - \epsilon_n' = \epsilon_n \epsilon_n'$.

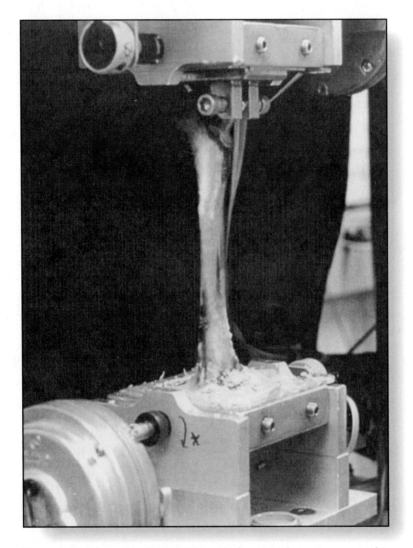

The mechanical properties of a material must be known so that engineers can relate the measured strain in a material to its associated stress. Here the mechanical properties of bone are determined from a compression test.

3 MECHANICAL PROPERTIES OF MATERIALS

CHAPTER OBJECTIVES

Having discussed the basic concepts of stress and strain, we will in this chapter show how stress can be related to strain by using experimental methods to determine the stress–strain diagram for a specific material. The behavior described by this diagram will then be discussed for materials that are commonly used in engineering. Also, mechanical properties and other tests that are related to the development of mechanics of materials will be discussed.

3.1 THE TENSION AND COMPRESSION TEST

The strength of a material depends on its ability to sustain a load without undue deformation or failure. This property is inherent in the material itself and must be determined by *experiment*. One of the most important tests to perform in this regard is the ***tension or compression test***. Although many important mechanical properties of a material can be determined from this test, it is used primarily to determine the relationship between the average normal stress and average normal strain in many engineering materials such as metals, ceramics, polymers, and composites.

To perform the tension or compression test a specimen of the material is made into a "standard" shape and size. Before testing, two small punch marks are identified along the specimen's length. These marks are located away from both ends of the specimen because the stress distribution at the ends is somewhat complex due to gripping at the connections where the load is applied. Measurements are taken of both the specimen's initial cross-sectional area, A_0, and the **gauge-length** distance L_0 between the punch marks. For example, when a metal specimen is used in a tension test it generally has an initial diameter of $d_0 = 0.5$ in. (13 mm) and a gauge length of $L_0 = 2$ in. (50 mm), Fig. 3–1. In order to apply an axial load with no bending of the specimen, the ends are usually seated into ball-and-socket joints. A testing machine like the one shown in Fig. 3–2 is then used to stretch the specimen at a very slow, constant rate until it reaches the breaking point. The machine is designed to read the load required to maintain this uniform stretching.

At frequent intervals during the test, data is recorded of the applied load P, as read on the dial of the machine or taken from a digital readout. Also, the elongation $\delta = L - L_0$ between the punch marks on the specimen may be measured using either a caliper or a mechanical or optical device called an **extensometer**. This value of δ (delta) is then used to calculate the average normal strain in the specimen. Sometimes, however, this measurement is not taken, since it is also possible to read the strain *directly* by using an **electrical-resistance strain gauge**, which looks like the one shown in Fig. 3–3. The operation of this gauge is based on the change in electrical resistance of a very thin wire or piece of metal foil under strain. Essentially the gauge is cemented to the specimen in a specified direction. If the cement is very strong in comparison to the gauge, then the gauge is in effect an integral part of the specimen, so that when the specimen is strained in the direction of the gauge, the wire and specimen will experience the same strain. By measuring the electrical resistance of the wire, the gauge may be calibrated to read values of normal strain directly.

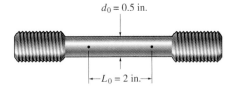

$d_0 = 0.5$ in.

$L_0 = 2$ in.

Fig. 3–1

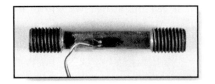

Typical steel specimen with attached strain gauge.

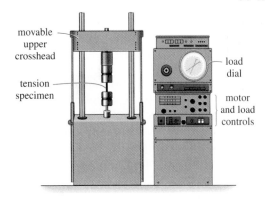

movable upper crosshead

tension specimen

load dial

motor and load controls

Fig. 3–2

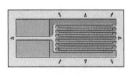

Electrical–resistance strain gauge

Fig. 3–3

3.2   THE STRESS–STRAIN DIAGRAM

From the data of a tension or compression test, it is possible to compute various values of the stress and corresponding strain in the specimen and then plot the results. The resulting curve is called the **stress–strain diagram**, and there are two ways in which it is normally described.

Conventional Stress–Strain Diagram.   Using the recorded data, we can determine the **nominal or engineering stress** by dividing the applied load P by the specimen's *original* cross-sectional area A_0. This calculation assumes that the stress is *constant* over the cross section and throughout the region between the gauge points. We have

$$\sigma = \frac{P}{A_0} \qquad\qquad (3\text{–}1)$$

Likewise, the **nominal or engineering strain** is found directly from the strain gauge reading, or by dividing the change in the specimen's gauge length, δ, by the specimen's original gauge length L_0. Here the strain is assumed to be constant throughout the region between the gauge points. Thus,

$$\epsilon = \frac{\delta}{L_0} \qquad\qquad (3\text{–}2)$$

If the corresponding values of σ and ϵ are plotted as a graph, for which the ordinate is the stress and the abscissa is the strain, the resulting curve is called a **conventional stress–strain diagram**. This diagram is very important in engineering since it provides the means for obtaining data about a material's tensile (or compressive) strength *without* regard for the material's physical size or shape, i.e., its geometry. Realize, however, that no two stress–strain diagrams for a particular material will be *exactly* the same, since the results depend on such variables as the material's composition, microscopic imperfections, the way it is manufactured, the rate of loading, and the temperature during the time of the test.

We will now discuss the characteristics of the conventional stress–strain curve as it pertains to *steel*, a commonly used material for fabricating both structural members and mechanical elements. Using the method described above, the characteristic stress–strain diagram for a steel specimen is shown in Fig. 3–4. From this curve we can identify four different ways in which the material behaves, depending on the amount of strain induced in the material.

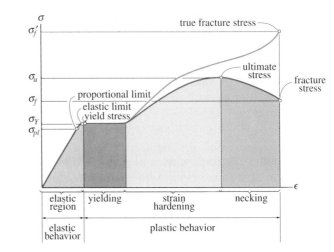

Conventional and true stress-strain diagrams
for ductile material (steel) (not to scale)

Fig. 3–4

Elastic Behavior.   Elastic behavior of the material occurs when the strains in the specimen are within the lightly shaded region shown in Fig. 3–4. It can be seen that the curve is actually a *straight line* throughout most of this region, so that stress is *proportional* to the strain. In other words, the material is said to be *linearly elastic*. The upper stress limit to this linear relationship is called the ***proportional limit***, σ_{pl}. If the stress slightly exceeds the proportional limit, the material may still respond elastically; however, the curve tends to bend and flatten out as shown. This continues until the stress reaches the ***elastic limit***. Upon reaching this point, if the load is removed the specimen will still return back to its original shape. Normally for steel, however, the elastic limit is seldom determined, since it is very close to the proportional limit and therefore rather difficult to detect.

Yielding.   A slight increase in stress above the elastic limit will result in a breakdown of the material and cause it to *deform permanently*. This behavior is called ***yielding***, and it is indicated by the dark-shaded region of the curve. The stress that causes yielding is called the ***yield stress or yield point***, σ_Y, and the deformation that occurs is called ***plastic deformation***. Although not shown in Fig. 3–4, for low-carbon steels or those that are hot rolled, the yield point is often distinguished by two values. The ***upper yield point*** occurs first, followed by a sudden decrease in load-carrying capacity to a ***lower yield point***. Once the yield point is reached, however, then as shown in Fig. 3–4, the specimen will continue to elongate (strain) *without any increase in load*. Realize that this figure is not drawn to scale. If it was, the induced strains due to yielding would be about 10 to 40 times greater than those produced up to the elastic limit. When the material is in this state, it is often referred to as being ***perfectly plastic***.

Strain hardening. When yielding has ended, a further load can be applied to the specimen, resulting in a curve that rises continuously but becomes flatter until it reaches a maximum stress referred to as the **ultimate stress**, σ_u. The rise in the curve in this manner is called **strain hardening**, and it is identified in Fig. 3–4 as the region in light-shaded color. Throughout the test, while the specimen is elongating, its cross-sectional area will decrease. This decrease in area is fairly *uniform* over the specimen's entire gauge length, even up to the strain corresponding to the ultimate stress.

Necking. At the ultimate stress, the cross-sectional area begins to decrease in a *localized* region of the specimen, instead of over its entire length. This phenomenon is caused by slip planes formed within the material, and the actual strains produced are caused by shear stress (see Sec. 10.7). As a result, a constriction or "neck" gradually tends to form in this region as the specimen elongates further, Fig. 3–5a. Since the cross-sectional area within this region is continually decreasing, the smaller area can only carry an ever-decreasing load. Hence the stress–strain diagram tends to curve downward until the specimen breaks at the **fracture stress**, σ_f, Fig. 3–5b. This region of the curve due to necking is indicated in dark color in Fig. 3–4.

True Stress–Strain Diagram. Instead of always using the *original* cross-sectional area and specimen length to calculate the (engineering) stress and strain, we could have used the *actual* cross-sectional area and specimen length at the *instant* the load is measured. The values of stress and strain computed from these measurements are called *true stress* and *true strain*, and a plot of their values is called the **true stress–strain diagram**. When this diagram is plotted it has a form shown by the light-colored curve in Fig. 3–4. Note that both the conventional and true σ–ϵ diagrams are practically coincident when the strain is small. The differences between the diagrams begin to appear in the strain-hardening range, where the magnitude of strain becomes more significant. In particular, there is a large divergence within the necking region. Here it can be seen from the conventional σ–ϵ diagram that the specimen *actually* supports a *decreasing load*, since A_0 is constant when calculating engineering stress, $\sigma = P/A_0$. However, from the true σ–ϵ diagram, the actual area A within the necking region is always decreasing until fracture, $\sigma_{f'}$, and so the material actually sustains *increasing stress*, since $\sigma = P/A$.

Typical necking pattern which has occurred on this steel specimen just before fracture.

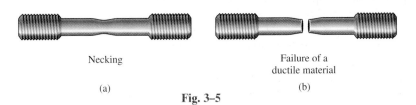

Necking	Failure of a ductile material
(a)	(b)

Fig. 3–5

Although the true and conventional stress–strain diagrams are different, most engineering design is done within the elastic range, since the distortion of the material is generally not severe within this range. Provided the material is "stiff," like most metals, the strain up to the elastic limit will remain small and the error in using the engineering values of σ and ϵ is very small (about 0.1%) compared with their true values. This is one of the primary reasons for using conventional stress–strain diagrams.

The above concepts can be summarized with reference to Fig. 3–6, which shows an actual conventional stress–strain diagram for a mild steel specimen. In order to enhance the details, the elastic region of the curve has been shown in light color to an exaggerated strain scale, also shown in light color. Tracing the behavior, the proportional limit is reached at $\sigma_{pl} = 35$ ksi (241 MPa), where $\epsilon_{pl} = 0.0012$ in./in. This is followed by an upper yield point of $(\sigma_Y)_u = 38$ ksi (262 MPa), then suddenly a lower yield point of $(\sigma_Y)_l = 36$ ksi (248 MPa). The end of yielding occurs at a strain of $\epsilon_Y = 0.030$ in./in., which is 25 times greater than the strain at the proportional limit! Continuing, the specimen is strain hardened until it reaches the ultimate stress of $\sigma_u = 63$ ksi (435 MPa), then it begins to neck down until failure occurs, $\sigma_f = 47$ ksi (324 MPa). By comparison, the strain at failure, $\epsilon_f = 0.380$ in./in., is 317 times greater than ϵ_{pl}!

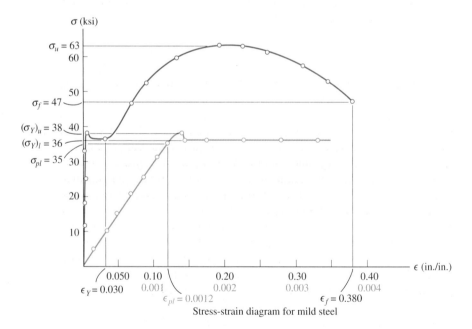

Stress-strain diagram for mild steel

Fig. 3–6

3.3   STRESS–STRAIN BEHAVIOR OF DUCTILE AND BRITTLE MATERIALS

Materials can be classified as either being ductile or brittle, depending on their stress–strain characteristics. Each will now be given separate treatment.

Ductile Materials.   Any material that can be subjected to large strains before it ruptures is called a ***ductile material***. Mild steel, as discussed previously, is a typical example. Engineers often choose ductile materials for design because these materials are capable of absorbing shock or energy, and if they become overloaded, they will usually exhibit large deformation before failing.

One way to specify the ductility of a material is to report its percent elongation or percent reduction in area at the time of fracture. The ***percent elongation*** is the specimen's fracture strain expressed as a percent. Thus, if the specimen's original gauge-mark length is L_0 and its length at fracture is L_f, then

$$\text{Percent elongation} = \frac{L_f - L_0}{L_0}(100\%) \qquad (3\text{--}3)$$

As seen in Fig. 3–6, since $\epsilon_f = 0.380$, this value would be 38% for a mild steel specimen.

The ***percent reduction in area*** is another way to specify ductility. It is defined within the region of necking as follows:

$$\text{Percent reduction of area} = \frac{A_0 - A_f}{A_0}(100\%) \qquad (3\text{--}4)$$

Here A_0 is the specimen's original cross-sectional area and A_f is the area at fracture. Mild steel has a typical value of 60%.

Besides steel, other metals such as brass, molybdenum, and zinc may also exhibit ductile stress–strain characteristics similar to steel, whereby they undergo elastic stress–strain behavior, yielding at constant stress, strain hardening, and finally necking until rupture. In most metals, however, constant yielding will *not occur* beyond the elastic range. One metal for which this is the case is aluminum. Actually, this metal often does not have a well-defined *yield point*, and consequently it is standard practice to define a **yield strength** for aluminum using a graphical procedure called the ***offset method***. Normally a 0.2% strain (0.002 in./in.) is chosen, and from this point on the ϵ axis, a line parallel to the initial straight-line portion of the stress–strain diagram is drawn. The point where this line intersects the curve defines the yield strength. An example of the construction for determining the yield strength for an aluminum alloy is shown in Fig. 3–7. From the graph, the yield strength is $\sigma_{YS} = 51$ ksi (352 MPa).

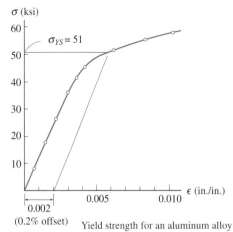

Yield strength for an aluminum alloy

Fig. 3–7

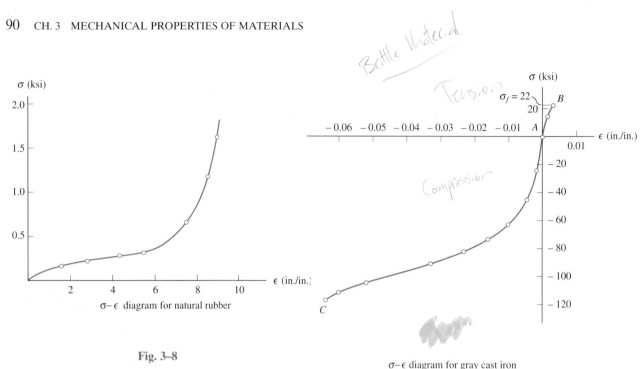

$\sigma - \epsilon$ diagram for natural rubber

Fig. 3–8

$\sigma - \epsilon$ diagram for gray cast iron

Fig. 3–9

Realize that the yield strength is not a physical property of the material, since it is a stress that caused a *specified* permanent strain in the material. In this text, however, we will assume that the yield strength, yield point, elastic limit, and proportional limit all *coincide* unless otherwise stated. An exception would be natural rubber, which in fact does not even have a proportional limit, since stress and strain are *not* linearly related, Fig. 3–8. Instead, this material, which is known as a polymer, exhibits *nonlinear elastic behavior.*

Wood is a material that is often moderately ductile, and as a result it is usually designed to respond only to elastic loadings. The strength characteristics of wood vary greatly from one species to another, and for each species they depend on the moisture content, age, and the size and arrangement of knots in the wood. Since wood is a fibrous material, its tensile or compressive characteristics will differ greatly when it is loaded either parallel or perpendicular to its grain. Specifically, wood splits easily when it is loaded in tension perpendicular to its grain, and consequently tensile loads are almost always intended to be applied parallel to the grain of wood members.

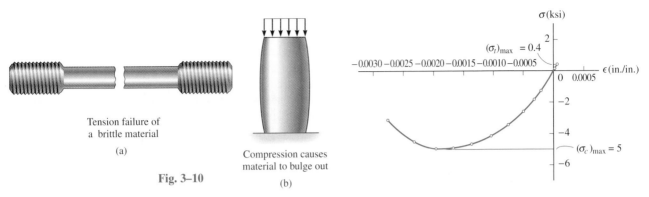

Tension failure of
a brittle material

(a)

Fig. 3–10

Compression causes
material to bulge out

(b)

σ-ϵ diagram for typical concrete mix

Fig. 3–11

Brittle Materials. Materials that exhibit little or no yielding before failure are referred to as ***brittle materials***. Gray cast iron is an example, having a stress–strain diagram in tension as shown by portion AB of the curve in Fig. 3–9. Here fracture at $\sigma_f = 22$ ksi (152 MPa) took place initially at an imperfection or microscopic crack and then spread rapidly across the specimen, causing complete fracture. As a result of this type of failure, brittle materials do not have a well-defined tensile fracture stress, since the appearance of initial cracks in a specimen is quite random. Instead the *average* fracture stress from a set of observed tests is generally reported. A typical failed specimen is shown in Fig. 3–10a.

Compared with their behavior in tension, brittle materials, such as gray cast iron, exhibit a much higher resistance to axial compression, as evidenced by portion AC of the curve in Fig. 3–9. For this case any cracks or imperfections in the specimen tend to close up, and as the load increases the material will generally bulge out or become barrel shaped as the strains become larger, Fig. 3–10b.

Like gray cast iron, concrete is classified as a brittle material, and it also has a low strength capacity in tension. The characteristics of its stress–strain diagram depend primarily on the mix of concrete (water, sand, gravel, and cement) and the time and temperature of curing. A typical example of a "complete" stress–strain diagram for concrete is given in Fig. 3–11. By inspection, its maximum compressive strength is almost 12.5 times greater than its tensile strength, $(\sigma_c)_{max} = 5$ ksi (34.5 MPa) versus $(\sigma_t)_{max} = 0.40$ ksi (2.76 MPa). For this reason, concrete is almost always reinforced with steel bars or rods whenever it is designed to support tensile loads.

It can generally be stated that most materials exhibit both ductile and brittle behavior. For example, steel has brittle behavior when it contains a high carbon content, and it is ductile when the carbon content is reduced. Also, at low temperatures materials become harder and more brittle, whereas when the temperature rises they become softer and more ductile. This effect is shown in Fig. 3–12 for a methacrylate plastic.

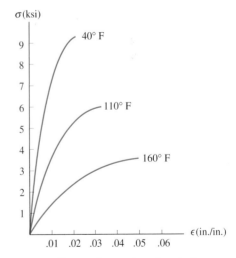

Steel rapidly loses its strength when heated. For this reason engineers often require main structural members to be insulated in case of fire.

σ-ϵ diagrams for a methacrylate plastic

Fig. 3–12

3.4   HOOKE'S LAW

As noted in the previous section, the stress–strain diagrams for most engineering materials exhibit a *linear relationship* between stress and strain within the elastic region. Consequently, an increase in stress causes a proportionate increase in strain. This fact was discovered by Robert Hooke in 1676 using springs and is known as *Hooke's law*. It may be expressed mathematically as

$$\sigma = E\epsilon$$

(3–5)

Here E represents the constant of proportionality, which is called the **modulus of elasticity** or **Young's modulus,** named after Thomas Young, who published an account of it in 1807.

Equation 3–5 actually represents the equation of the *initial straight-lined portion* of the stress–strain diagram up to the proportional limit. Furthermore, the modulus of elasticity represents the *slope* of this line. Since strain is dimensionless, from Eq. 3–5, E will have units of stress, such as psi, ksi, or pascals. As an example of its calculation, consider the stress–strain diagram for steel shown in Fig. 3–6. Here $\sigma_{pl} = 35$ ksi and $\epsilon_{pl} = 0.0012$ in./in., so that

$$E = \frac{\sigma_{pl}}{\epsilon_{pl}} = \frac{35 \text{ ksi}}{0.0012 \text{ in./in.}} = 29(10^3) \text{ ksi}$$

As shown in Fig. 3–13, the proportional limit for a particular type of steel depends on its alloy content; however, most grades of steel, from the softest rolled steel to the hardest tool steel, have about the same modulus of elasticity, generally accepted to be $E_{st} = 29(10^3)$ ksi or 200 GPa. Common values of E for other engineering materials are often tabulated in engineering code and reference books. Representative values are also listed on the inside back cover of this book. It should be noted that the modulus of elasticity is a mechanical property that indicates the *stiffness* of a material. Materials that are very stiff, such as steel, have large values of E [$E_{st} = 29(10^3)$ ksi or 200 GPa], whereas spongy materials such as vulcanized rubber may have low values [$E_r = 0.10(10^3)$ ksi or 0.70 MPa].

The modulus of elasticity is one of the most important mechanical properties used in the development of equations presented in this text. It must always be remembered, though, that E can be used only if a material has *linear-elastic* behavior. Also, if the stress in the material is *greater* than the proportional limit, the stress–strain diagram ceases to be a straight line and Eq. 3–5 is no longer valid.

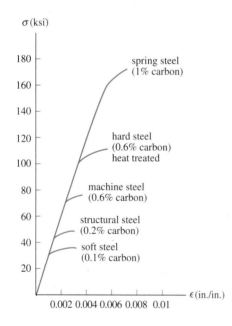

Fig. 3–13

Strain Hardening.   If a specimen of ductile material, such as steel, is loaded into the *plastic region* and then unloaded, *elastic strain is recovered* as the material returns to its equilibrium state. The *plastic strain remains*, however, and as a result the material is subjected to a **permanent set**. For example, a wire when bent (plastically) will spring back a little (elastically) when the load is removed; however, it will not fully return to its original position. This behavior can be illustrated on the stress–strain diagram shown in Fig. 3–14a. Here the specimen is first loaded beyond its yield point A to the point A′. Since interatomic forces have to be overcome to elongate the specimen *elastically*, then these same forces pull the atoms back together when the load is removed, Fig. 3–14a. Consequently, the modulus of elasticity, E, is the same, and therefore the slope of line O′A′ is the same as line OA.

If the load is reapplied, the atoms in the material will again be displaced until yielding occurs at or near the stress A′, and the stress–strain diagram continues along the same path as before, Fig. 3–14b. It should be noted, however, that this new stress–strain diagram, defined by O′A′B, now has a *higher* yield point (A′), a consequence of strain-hardening. In other words, the material now has a *greater elastic region*; however, it has *less ductility*, a smaller plastic region, than when it was in its original state.

In the true sense some heat or *energy* may be *lost* as the specimen is unloaded from A′ and then again loaded to this same stress. As a result, slight curves in the paths A′ to O′ and O′ to A′ will occur during a carefully measured cycle of loading. This is shown by the dashed curves in Fig. 3–14b. The colored area between these curves represents lost energy and is called **mechanical hysteresis**. It becomes an important consideration when selecting materials to serve as dampers for vibrating structural or mechanical equipment, although its effects will not be considered in this text.

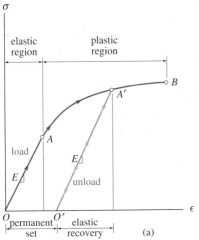

(a)

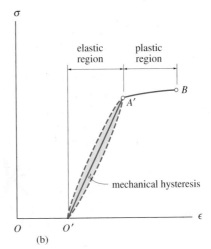

(b)

Fig. 3–14

3.5   STRAIN ENERGY

As a material is deformed by an external loading, it tends to store energy *internally* throughout its volume. Since this energy is related to the strains in the material, it is referred to as **strain energy**. For example, when a tension-test specimen is subjected to an axial load, a volume element of the material is subjected to uniaxial stress as shown in Fig. 3–15. This stress develops a force $\Delta F = \sigma \Delta A = \sigma(\Delta x \, \Delta y)$ on the top and bottom faces of the element *after* the element undergoes a vertical displacement $\epsilon \, \Delta z$. By definition, *work* is determined by the product of the force and displacement in the direction of the force. Since the force is increased uniformly from zero to its final magnitude ΔF when the displacement $\epsilon \, \Delta z$ is attained, the work done on the element by the force is equal to the *average* force magnitude $(\Delta F/2)$ times the displacement $\epsilon \, \Delta z$. This "external work" is equivalent to the "internal work" or strain energy stored in the element—assuming that no energy is lost in the form of heat. Consequently, the strain energy ΔU is $\Delta U = (1/2 \, \Delta F) \, \epsilon \, \Delta z = (1/2 \, \sigma \, \Delta x \, \Delta y) \, \epsilon \, \Delta z$. Since the volume of the element is $\Delta V = \Delta x \, \Delta y \, \Delta z$, then $\Delta U = 1/2 \, \sigma \epsilon \, \Delta V$.

It is sometimes convenient to formulate the strain energy per unit volume of material. This is called the **strain-energy density**, and it can be expressed as

$$u = \frac{\Delta U}{\Delta V} = \frac{1}{2}\sigma\epsilon \tag{3–6}$$

If the material behavior is *linear elastic*, then Hooke's law applies, $\sigma = E\epsilon$, and therefore we can express the strain-energy density in terms of the uniaxial stress as

$$u = \frac{1}{2}\frac{\sigma^2}{E} \tag{3–7}$$

Modulus of Resilience. In particular, when the stress σ reaches the proportional limit, the strain-energy density, as calculated by Eq. 3–6 or 3–7, is referred to as the **modulus of resilience**, i.e.,

$$\boxed{u_r = \frac{1}{2}\sigma_{pl}\epsilon_{pl} = \frac{1}{2}\frac{\sigma_{pl}^2}{E}} \tag{3–8}$$

From the elastic region of the stress–strain diagram, Fig. 3–16a, notice that u_r is equivalent to the shaded *triangular area* under the diagram. Physically a material's resilience represents the ability of the material to absorb energy without any permanent damage to the material.

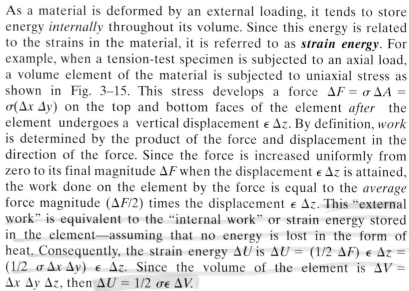

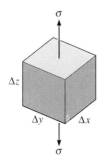

Fig. 3–15

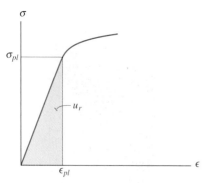

Modulus of resilience u_r

(a)

Fig. 3–16

Modulus of Toughness.   Another important property of a material is the *modulus of toughness*, u_t. This quantity represents the *entire area* under the stress–strain diagram, Fig. 3–16b, and therefore it indicates the strain-energy density of the material just before it fractures. This property becomes important when designing members that may be accidentally overloaded. Materials with a high modulus of toughness will distort greatly due to an overloading; however, they may be preferable to those with a low value, since materials having a low u_t may suddenly fracture without warning of an approaching failure. Alloying metals can also change their resilience and toughness. For example, by changing the percentage of carbon in steel, the resulting stress–strain diagrams in Fig. 3–17 show how the degrees of resilience and toughness can be changed.

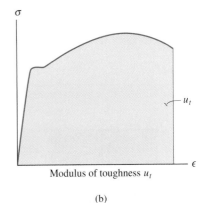

Modulus of toughness u_t

(b)

Fig. 3–16

IMPORTANT POINTS

- A *conventional stress–strain diagram* is important in engineering since it provides a means for obtaining data about a material's tensile or compressive strength without regard for the material's physical size or shape.

- *Engineering stress and strain* are calculated using the *original* cross-sectional area and gauge length of the specimen.

- A *ductile material*, such as mild steel, has four distinct behaviors as it is loaded. They are *elastic behavior, yielding, strain hardening,* and *necking*.

- A material is *linear elastic* if the stress is proportional to the strain within the elastic region. This is referred to as *Hooke's law*, and the slope of the curve is called the *modulus of elasticity, E*.

- Important points on the stress–strain diagram are the *proportional limit, elastic limit, yield stress, ultimate stress,* and the *fracture stress*.

- The *ductility* of a material can be specified by the specimen's *percent elongation* or the *percent reduction in area*.

- If a material does not have a distinct yield point, a *yield strength* can be specified using a graphical procedure such as the *offset method*.

- *Brittle materials*, such as gray cast iron, have very little or no yielding and fracture suddenly.

- *Strain hardening* is used to establish a higher yield point for a material. This is done by straining the material beyond the elastic limit, then releasing the load. The modulus of elasticity remains the same; however, the material's ductility *decreases*.

- *Strain energy* is energy that is stored in a material due to its deformation. This energy per unit volume is called *strain-energy density*. If it is measured up to the proportional limit, it is referred to as the *modulus of resilience*, and if it is measured up to the point of fracture, it is called the *modulus of toughness*.

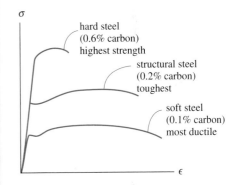

Fig. 3–17

This nylon specimen exhibits a high degree of toughness as noted by the large amount of necking that has occurred just before fracture.

EXAMPLE 3-1

A tension test for a steel alloy results in the stress–strain diagram shown in Fig. 3–18. Calculate the modulus of elasticity and the yield strength based on a 0.2% offset. Identify on the graph the ultimate stress and the fracture stress.

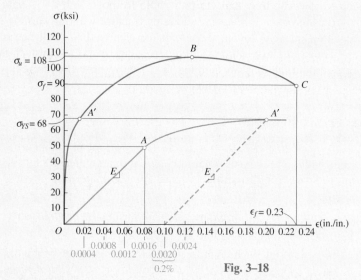

Fig. 3–18

SOLUTION

Modulus of Elasticity. We must calculate the *slope* of the initial straight-line portion of the graph. Using the magnified curve and scale shown in color, this line extends from point O to an estimated point A, which has coordinates of approximately (0.0016 in./in., 50 ksi). Therefore,

$$E = \frac{50 \text{ ksi}}{0.0016 \text{ in./in.}} = 31.2(10^3) \text{ ksi} \qquad \textit{Ans.}$$

Note that the equation of the line OA is thus $\sigma = 31.2(10^3)\epsilon$.

Yield Strength. For a 0.2% offset, we begin at a strain of 0.2% or 0.0020 in./in. and graphically extend a (dashed) line parallel to OA until it intersects the σ–ϵ curve at A'. The yield strength is approximately

$$\sigma_{YS} = 68 \text{ ksi} \qquad \textit{Ans.}$$

Ultimate Stress. This is defined by the peak of the σ–ϵ graph, point B in Fig. 3–18.

$$\sigma_u = 108 \text{ ksi} \qquad \textit{Ans.}$$

Fracture Stress. When the specimen is strained to its maximum of $\epsilon_f = 0.23$ in./in., it fractures at point C. Thus,

$$\sigma_f = 90 \text{ ksi} \qquad \textit{Ans.}$$

EXAMPLE 3-2

The stress–strain diagram for an aluminum alloy that is used for making aircraft parts is shown in Fig. 3–19. If a specimen of this material is stressed to 600 MPa, determine the permanent strain that remains in the specimen when the load is released. Also, compute the modulus of resilience both before and after the load application.

SOLUTION

Permanent Strain.   When the specimen is subjected to the load, it strain-hardens until point B is reached on the σ–ϵ diagram, Fig. 3–19. The strain at this point is approximately 0.023 mm/mm. When the load is released, the material behaves by following the straight line BC, which is parallel to line OA. Since both lines have the same slope, the strain at point C can be determined analytically. The slope of line OA is the modulus of elasticity, i.e.,

$$E = \frac{450 \text{ MPa}}{0.006 \text{ mm/mm}} = 75.0 \text{ GPa}$$

From triangle CBD, we require

$$E = \frac{BD}{CD} = \frac{600(10^6) \text{ Pa}}{CD} = 75.0(10^9) \text{ Pa}$$

$$CD = 0.008 \text{ mm/mm}$$

This strain represents the amount of *recovered elastic strain*. The permanent strain, ϵ_{OC}, is thus

$$\epsilon_{OC} = 0.023 \text{ mm/mm} - 0.008 \text{ mm/mm}$$
$$= 0.0150 \text{ mm/mm} \qquad\qquad Ans.$$

Note: If gauge marks on the specimen were originally 50 mm apart, then after the load is *released* these marks will be 50 mm + (0.0150) (50 mm) = 50.75 mm apart.

Modulus of Resilience.   Applying Eq. 3–8, we have*

$$(u_r)_{\text{initial}} = \frac{1}{2}\sigma_{pl}\epsilon_{pl} = \frac{1}{2}(450 \text{ MPa})(0.006 \text{ mm/mm})$$
$$= 1.35 \text{ MJ/m}^3 \qquad\qquad Ans.$$

$$(u_r)_{\text{final}} = \frac{1}{2}\sigma_{pl}\epsilon_{pl} = \frac{1}{2}(600 \text{ MPa})(0.008 \text{ mm/mm})$$
$$= 2.40 \text{ MJ/m}^3 \qquad\qquad Ans.$$

By comparison, the effect of strain-hardening the material has caused an increase in the modulus of resilience; however, note that the modulus of toughness for the material has decreased since the area under the original curve, $OABF$, is larger than the area under curve CBF.

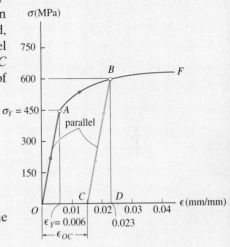

Fig. 3–19

---

*Work in the SI system of units is measured in joules, where 1 J = 1 N · m.

EXAMPLE 3–3

An aluminum rod shown in Fig. 3–20a has a circular cross section and is subjected to an axial load of 10 kN. If a portion of the stress–strain diagram for the material is shown in Fig. 3–20b, determine the approximate elongation of the rod when the load is applied. If the load is removed, what is the permanent elongation of the rod? Take E_{al} = 70 GPa.

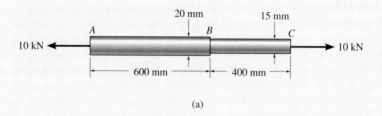

(a)

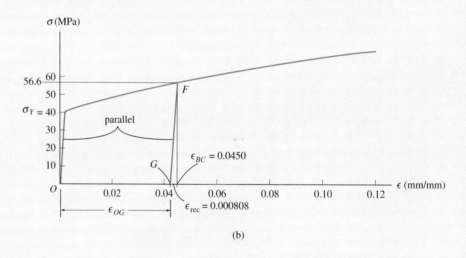

(b)

Fig. 3–20

SOLUTION

For the analysis we will neglect the *localized deformations* at the point of load application and where the rod's cross-sectional area suddenly changes. (These effects will be discussed in Secs. 4.1 and 4.7.) Throughout the midsection of each segment the normal stress and deformation are uniform.

EXAMPLE 3–3 (*cont.*)

In order to study the deformation of the rod, we must obtain the strain. This is done by first calculating the stress, then using the stress–strain diagram to obtain the strain. The normal stress within each segment is

$$\sigma_{AB} = \frac{P}{A} = \frac{10(10^3)\ N}{\pi\,(0.01\ m)^2} = 31.83\ MPa$$

$$\sigma_{BC} = \frac{P}{A} = \frac{10(10^3)\ N}{\pi\,(0.0075\ m)^2} = 56.59\ MPa$$

From the stress–strain diagram, the material in region *AB* is strained *elastically* since $\sigma_Y = 40\ MPa > 31.83\ MPa$. Using Hooke's law,

$$\epsilon_{AB} = \frac{\sigma_{AB}}{E_{al}} = \frac{31.83(10^6)\ Pa}{70(10^9)\ Pa} = 0.0004547\ mm/mm$$

The material within region *BC* is strained plastically, since $\sigma_Y = 40\ MPa < 56.59\ MPa$. From the graph, for $\sigma_{BC} = 56.59\ MPa$,

$$\epsilon_{BC} \approx 0.045\ mm/mm$$

The approximate elongation of the rod is therefore

$$\delta = \Sigma \epsilon L = 0.0004547(600\ mm) + 0.045(400\ mm)$$

$$= 18.3\ mm \qquad \textbf{Ans.}$$

When the 10-kN load is removed, segment *AB* of the rod will be restored to its original length. Why? On the other hand, the material in segment *BC* will recover elastically along line *FG*, Fig. 3–20*b*. Since the slope of *FG* is E_{al}, the elastic strain recovery is

$$\epsilon_{rec} = \frac{\sigma_{BC}}{E_{al}} = \frac{56.59(10^6)\ Pa}{70(10^9)\ Pa} = 0.000808\ mm/mm$$

The remaining plastic strain in segment *BC* is then

$$\epsilon_{OG} = 0.0450 - 0.000808 = 0.0442\ mm/mm$$

Therefore, when the load is removed the rod remains elongated by an amount

$$\delta' = \epsilon_{OG}L_{BC} = 0.0442(400\ mm) = 17.7\ mm \qquad \textbf{Ans.}$$

PROBLEMS

3–1. A concrete cylinder having a diameter of 6.00 in. and gauge length of 12 in. is tested in compression. The results of the test are reported in the table as load versus contraction. Draw the stress–strain diagram using scales of 1 in. = 0.5 ksi and 1 in. = $0.2(10^{-3})$ in./in. From the diagram, determine approximately the modulus of elasticity.

Load (kip)	Contraction (in.)
0	0
5.0	0.0006
9.5	0.0012
16.5	0.0020
20.5	0.0026
25.5	0.0034
30.0	0.0040
34.5	0.0045
38.5	0.0050
46.5	0.0062
50.0	0.0070
53.0	0.0075

Prob. 3–1

3–2. Data taken from a stress–strain test for a ceramic are given in the table. The curve is linear between the origin and the first point. Plot the diagram, and determine the modulus of elasticity and the modulus of resilience.

3–3. Data taken from a stress–strain test for a ceramic are given in the table. The curve is linear between the origin and the first point. Plot the diagram, and determine approximately the modulus of toughness. The rupture stress is $\sigma_r = 53.4$ ksi.

σ (ksi)	ϵ (in./in.)
0	0
33.2	0.0006
45.5	0.0010
49.4	0.0014
51.5	0.0018
53.4	0.0022

Probs. 3–2/3–3

*3–4. Data taken from a stress–strain test are given in the table. The curve is linear between the origin and the first point. Plot the diagram, and determine the modulus of elasticity and the modulus of resilience.

σ (ksi)	ϵ (in./in.)
0	0
32.0	0.0016
33.5	0.0018
40.0	0.0030
41.2	0.0050

Prob. 3–4

3–5. A tension test was performed on a specimen having an original diameter of 12.5 mm and a gauge length of 50 mm. The data are listed in the table. Plot the stress–strain diagram, and determine approximately the modulus of elasticity, the ultimate stress, and the fracture stress. Use a scale of 20 mm = 50 MPa and 20 mm = 0.05 mm/mm. Redraw the linear-elastic region, using the same stress scale but a strain scale of 20 mm = 0.001 mm/mm.

3–6. A tension test was performed on a steel specimen having an original diameter of 12.5 mm and gauge length of 50 mm. Using the data listed in the table, plot the stress–strain diagram, and determine approximately the modulus of toughness. Use a scale of 20 mm = 50 MPa and 20 mm = 0.05 mm/mm.

Load (kN)	Elongation (mm)
0	0
11.1	0.0175
31.9	0.0600
37.8	0.1020
40.9	0.1650
43.6	0.2490
53.4	1.0160
62.3	3.0480
64.5	6.3500
62.3	8.8900
58.8	11.9380

Probs. 3–5/3–6

3–7. The stress–strain diagram for a steel alloy having an original diameter of 0.5 in. and a gauge length of 2 in. is given in the figure. If the specimen is loaded until it is stressed to 70 ksi, determine the approximate amount of elastic recovery and the permanent increase in the gauge length after it is unloaded.

*3–8. The stress–strain diagram for a steel alloy having an original diameter of 0.5 in. and a gauge length of 2 in. is given in the figure. Determine approximately the modulus of elasticity for the material, the load on the specimen that causes yielding, and the ultimate load the specimen will support.

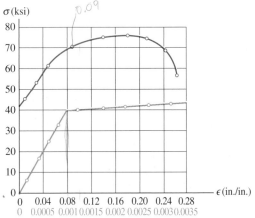

Probs. 3–7/3–8

3–9. The stress–strain diagram for a steel alloy having an original diameter of 0.5 in. and a gauge length of 2 in. is given in the figure. Determine approximately the modulus of resilience and the modulus of toughness for the material.

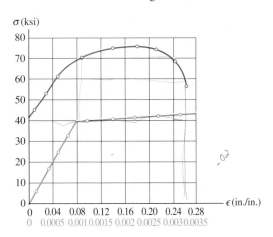

Prob. 3–9

3–10. The stress–strain diagram for a bar of steel alloy is shown in the figure. Determine approximately the modulus of elasticity, the proportional limit, the ultimate stress, and the modulus of resilience. If the bar is loaded until it is stressed to 360 MPa, determine the elastic strain recovery and the permanent set or strain in the bar when it is unloaded.

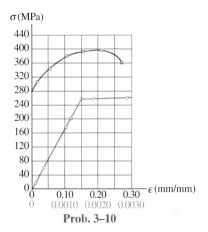

Prob. 3–10

3–11. An A-36 steel bar has a length of 50 in. and cross-sectional area of 0.7 in². Determine the length of the bar if it is subjected to an axial tension of 5000 lb. The material has linear-elastic behavior.

*3–12. The stress–strain diagram for polyethylene, which is used to sheath coaxial cables, is determined from testing a specimen that has a gauge length of 10 in. If a load P on the specimen develops a strain of $\epsilon = 0.024$ in./in., determine the approximate length of the specimen, measured between the gauge points, when the load is removed. Assume the specimen recovers elastically.

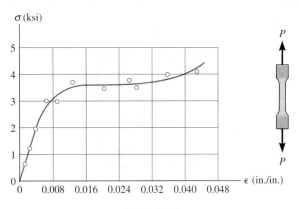

Prob. 3–12

3–13. The change in weight of an airplane is determined from reading the strain gauge A mounted in the plane's aluminum wheel strut. *Before* the plane is loaded, the strain-gauge reading in a strut is $\epsilon_1 = 0.00100$ in./in., whereas after loading $\epsilon_2 = 0.00243$ in./in. Determine the change in the force on the strut if the cross-sectional area of the strut is 3.5 in^2. $E_{al} = 10(10^3)$ ksi.

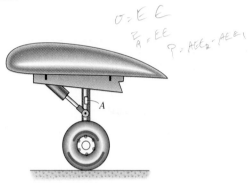

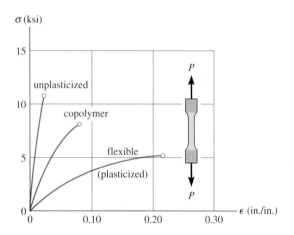

Prob. 3–13

3–14. By adding plasticizers to polyvinyl chloride, it is possible to reduce its stiffness. The stress–strain diagrams for three types of this material showing this effect are given below. Specify the type that should be used in the manufacture of a rod having a length of 5 in. and a diameter of 2 in., that is required to support at least an axial load of 20 kip and also be able to stretch at most $\frac{1}{4}$ in.

Prob. 3–14

3–15. A bar having a length of 5 in. and cross-sectional area of 0.7 in^2 is subjected to an axial force of 8000 lb. If the bar stretches 0.002 in., determine the modulus of elasticity of the material. The material has linear-elastic behavior.

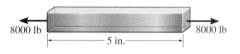

8000 lb   5 in.   8000 lb

Prob. 3–15

*3–16.** A specimen is originally 1 ft long, has a diameter of 0.5 in., and is subjected to a force of 500 lb. When the force is increased from 500 lb to 1800 lb, the specimen elongates 0.009 in. Determine the modulus of elasticity for the material if it remains linear elastic.

3–17. A structural member in a nuclear reactor is made of a zirconium alloy. If an axial load of 4 kip is to be supported by the member, determine its required cross-sectional area. Use a factor of safety of 3 relative to yielding. What is the load on the member if it is 3 ft long and its elongation is 0.02 in.? $E_{zr} = 14(10^3)$ ksi, $\sigma_Y = 57.5$ ksi. The material has elastic behavior.

3–18. The steel wires AB and AC support the 200-kg mass. If the allowable axial stress for the wires is $\sigma_{allow} = 130$ MPa, determine the required diameter of each wire. Also, what is the new length of wire AB after the load is applied? Take the unstretched length of AB to be 750 mm. $E_{st} = 200$ GPa.

Prob. 3–18

3–19. The bar *DBA* is rigid and is originally held in the horizontal position. When the weight *W* is supported from *D*, it causes the end *D* to displace downward 0.025 in. Determine the strain in wires *CD* and *BE*. Also, if the wires are made of A-36 steel and have a cross-sectional area of 0.002 in², determine the weight *W*.

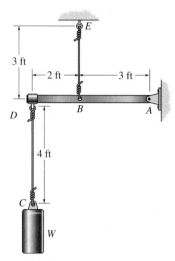

Prob. 3–19

*3–20. The stress–strain diagram for many metal alloys can be described analytically using the Ramberg-Osgood three parameter equation $\epsilon = \sigma/E + k\sigma^n$, where *E*, *k*, and *n* are determined from measurements taken from the diagram. Using the stress–strain diagram shown in the figure, take $E = 30(10^3)$ ksi and determine the other two parameters *k* and *n*, and thereby obtain an analytical expression for the curve.

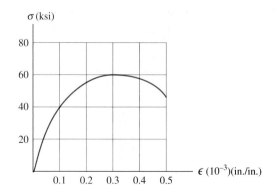

Prob. 3–20

3–21. Direct tension indicators are sometimes used instead of torque wrenches to insure that a bolt has a prescribed tension when used for connections. If a nut on the bolt is tightened so that the six heads of the indicator that were originally 3 mm high, are crushed 0.3 mm, leaving a contact area on each head of 1.5 mm², determine the tension in the bolt shank. The material has the stress–strain diagram shown.

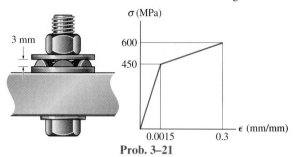

Prob. 3–21

3–22. The two bars are made of polystyrene, which has the stress–strain diagram shown. If the cross-sectional area of bar *AB* is 1.5 in² and *BC* is 4 in², determine the largest force *P* that can be supported before any member ruptures. Assume that buckling does not occur.

3–23. The two bars are made of polystyrene, which has the stress–strain diagram shown. Determine the cross-sectional area of each bar so that the bars rupture simultaneously when the load $P = 3$ kip. Assume that buckling does not occur.

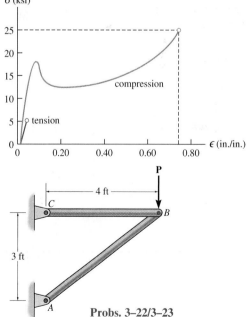

Probs. 3–22/3–23

3.6 POISSON'S RATIO

When a deformable body is subjected to an axial tensile force, not only does it elongate but it also contracts laterally. For example, if a rubber band is stretched, it can be noted that both the thickness and width of the band are decreased. Likewise, a compressive force acting on a body causes it to contract in the direction of the force and yet its sides expand laterally. These two cases are illustrated in Fig. 3–21 for a bar having an original radius r and length L.

When the load **P** is applied to the bar, it changes the bar's length by an amount δ and its radius by an amount δ'. Strains in the longitudinal or axial direction and in the lateral or radial direction are, respectively,

$$\epsilon_{\text{long}} = \frac{\delta}{L} \quad \text{and} \quad \epsilon_{\text{lat}} = \frac{\delta'}{r}$$

In the early 1800s, the French scientist S. D. Poisson realized that within the *elastic range* the *ratio* of these strains is a *constant*, since the deformations δ and δ' are proportional. This constant is referred to as **Poisson's ratio,** ν (nu), and it has a numerical value that is unique for a particular material that is both *homogeneous and isotropic*. Stated mathematically it is

$$\nu = -\frac{\epsilon_{\text{lat}}}{\epsilon_{\text{long}}} \tag{3–9}$$

When the rubber block is compressed (negative strain) its sides will expand (positive strain). The ratio of these strains is constant.

The negative sign is used here since *longitudinal elongation* (positive strain) causes *lateral contraction* (negative strain), and vice versa. Notice that this lateral strain is the *same* in all lateral (or radial) directions. Furthermore, this strain is caused only by the axial or longitudinal force; i.e., no force or stress acts in a lateral direction in order to strain the material in this direction.

Poisson's ratio is seen to be *dimensionless*, and for most nonporous solids it has a value that is generally between $\frac{1}{4}$ and $\frac{1}{3}$. Typical values of ν for common materials are listed on the inside back cover. In particular, an ideal material having no lateral movement when it is stretched or compressed will have $\nu = 0$. Furthermore, it will be shown in Sec. 10.6 that the *maximum* possible value for Poisson's ratio is 0.5. Therefore $0 \le \nu \le 0.5$.

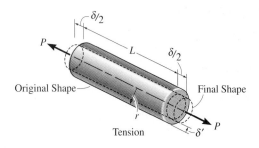

Tension

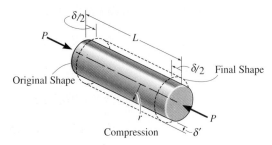

Compression

Fig. 3–21

EXAMPLE 3-4

A bar made of A-36 steel has the dimensions shown in Fig. 3–22. If an axial force of $P = 80$ kN is applied to the bar, determine the change in its length and the change in the dimensions of its cross section after applying the load. The material behaves elastically.

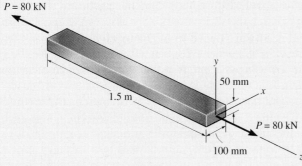

$P = 80$ kN

y

50 mm

x

1.5 m

$P = 80$ kN

100 mm

z

Fig. 3–22

SOLUTION

The normal stress in the bar is

$$\sigma_z = \frac{P}{A} = \frac{80(10^3)\ \text{N}}{(0.1\ \text{m})(0.05\ \text{m})} = 16.0(10^6)\ \text{Pa}$$

From the table on the inside back cover, for A-36 steel, $E_{st} = 200$ GPa, and so the strain in the z direction is

$$\epsilon_z = \frac{\sigma_z}{E_{st}} = \frac{16.0(10^6)\ \text{Pa}}{200(10^9)\ \text{Pa}} = 80(10^{-6})\ \text{mm/mm}$$

The axial elongation of the bar is therefore

$$\delta_z = \epsilon_z L_z = [80(10^{-6})](1.5\ \text{m}) = 120\ \mu\text{m} \qquad \textit{Ans.}$$

Using Eq. 3–9, where $\nu_{st} = 0.32$ as found from the inside back cover, the contraction strains in *both* the x and y directions are

$$\epsilon_x = \epsilon_y = -\nu_{st}\epsilon_z = -0.32[80(10^{-6})] = -25.6\ \mu\text{m/m}$$

Thus the changes in the dimensions of the cross section are

$$\delta_x = \epsilon_x L_x = -[25.6(10^{-6})](0.1\ \text{m}) = -2.56\ \mu\text{m} \qquad \textit{Ans.}$$

$$\delta_y = \epsilon_y L_y = -[25.6(10^{-6})](0.05\ \text{m}) = -1.28\ \mu\text{m} \qquad \textit{Ans.}$$

3.7   THE SHEAR STRESS–STRAIN DIAGRAM

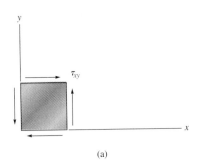

In Sec. 1.5 it was shown that when an element of material is subjected to *pure shear*, equilibrium requires that equal shear stresses must be developed on four faces of the element. These stresses must be directed either toward or away from diagonally opposite corners of the element, Fig. 3–23a. Furthermore, if the material is *homogeneous* and *isotropic*, then this shear stress will distort the element uniformly, Fig. 3–23b. As mentioned in Sec. 2.2, the shear strain γ_{xy} measures the angular distortion of the element relative to the sides originally along the x and y axes.

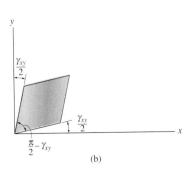

(a)

(b)

Fig. 3–23

The behavior of a material subjected to pure shear can be studied in a laboratory by using specimens in the shape of thin circular tubes and subjecting them to a torsional loading. If measurements are made of the applied torque and the resulting angle of twist, then by the methods to be explained in Chapter 5, the data can be used to determine the shear stress and shear strain, and a shear stress–strain diagram plotted. An example of such a diagram for a ductile material is shown in Fig. 3–24. Like the tension test, this material when subjected to shear will exhibit linear-elastic behavior and it will have a defined *proportional limit* τ_{pl}. Also, strain hardening will occur until an *ultimate shear stress* τ_u is reached. And finally, the material will begin to lose its shear strength until it reaches a point where it fractures, τ_f.

For most engineering materials, like the one just described, the elastic behavior is *linear*, and so Hooke's law for shear can be written as

$$\tau = G\gamma \tag{3–10}$$

Here G is called the **shear modulus of elasticity** or the **modulus of rigidity**. Its value can be measured as the slope of the line on the τ–γ diagram, that is, $G = \tau_{pl}/\gamma_{pl}$. Typical values for common engineering materials are listed on the inside back cover. Notice that the units of measurement for G will be the *same* as those for E (Pa or psi), since γ is measured in radians, a dimensionless quantity.

It will be shown in Sec. 10.6 that the three material constants, E, ν, and G are actually *related* by the equation

$$G = \frac{E}{2(1 + \nu)} \tag{3–11}$$

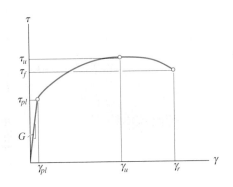

Fig. 3–24

Provided E and G are known, the value of ν can be determined from this equation rather than through experimental measurement. For example, in the case of A-36 steel, $E_{st} = 29(10^3)$ ksi and $G_{st} = 11.0(10^3)$ ksi, so that, from Eq. 3–11, $\nu_{st} = 0.32$.

EXAMPLE 3-5

A specimen of titanium alloy is tested in torsion and the shear stress–strain diagram is shown in Fig. 3–25a. Determine the shear modulus G, the proportional limit, and the ultimate shear stress. Also, determine the maximum distance d that the top of a block of this material, shown in Fig. 3–25b, could be displaced horizontally if the material behaves elastically when acted upon by a shear force **V**. What is the magnitude of **V** necessary to cause this displacement?

SOLUTION

Shear Modulus. This value represents the slope of the straight-line portion OA of the τ–γ diagram. The coordinates of point A are (0.008 rad, 52 ksi). Thus,

$$G = \frac{52 \text{ ksi}}{0.008 \text{ rad}} = 6500 \text{ ksi} \qquad Ans.$$

The equation of line OA is therefore $\tau = 6500\gamma$, which is Hooke's law for shear.

Proportional Limit. By inspection, the graph ceases to be linear at point A. Thus,

$$\tau_{pl} = 52 \text{ ksi} \qquad Ans.$$

Ultimate Stress. This value represents the maximum shear stress, point B. From the graph,

$$\tau_u = 73 \text{ ksi} \qquad Ans.$$

Maximum Elastic Displacement and Shear Force. Since the maximum elastic shear strain is 0.008 rad, a very small angle, the top of the block in Fig. 3–25b will be displaced horizontally:

$$\tan(0.008 \text{ rad}) \approx 0.008 \text{ rad} = \frac{d}{2 \text{ in.}}$$

$$d = 0.016 \text{ in.} \qquad Ans.$$

The corresponding *average* shear stress in the block is $\tau_{pl} = 52$ ksi. Thus, the shear force V needed to cause the displacement is

$$\tau_{avg} = \frac{V}{A}; \qquad 52 \text{ ksi} = \frac{V}{(3 \text{ in.})(4 \text{ in.})}$$

$$V = 624 \text{ kip} \qquad Ans.$$

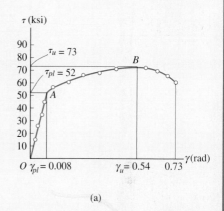

(a)

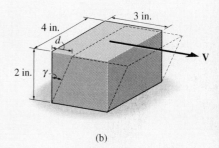

(b)

Fig. 3–25

EXAMPLE 3-6

An aluminum specimen shown in Fig. 3–26 has a diameter of $d_0 = 25$ mm and a gauge length of $L_0 = 250$ mm. If a force of 165 kN elongates the gauge length 1.20 mm, determine the modulus of elasticity. Also, determine by how much the force causes the diameter of the specimen to contract. Take $G_{al} = 26$ GPa and $\sigma_Y = 440$ MPa.

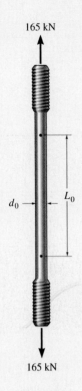

165 kN

d_0 —|← →| L_0

165 kN

Fig. 3–26

SOLUTION

Modulus of Elasticity.   The average normal stress in the specimen is

$$\sigma = \frac{P}{A} = \frac{165(10^3) \text{ N}}{(\pi/4)(0.025 \text{ m})^2} = 336.1 \text{ MPa}$$

and the average normal strain is

$$\epsilon = \frac{\delta}{L} = \frac{1.20 \text{ mm}}{250 \text{ mm}} = 0.00480 \text{ mm/mm}$$

Since $\sigma < \sigma_Y = 440$ MPa, the material behaves elastically. The modulus of elasticity is

$$E_{al} = \frac{\sigma}{\epsilon} = \frac{336.1(10^6) \text{ Pa}}{0.00480} = 70.0 \text{ GPa} \qquad \textit{Ans.}$$

Contraction of Diameter.   First we will determine Poisson's ratio for the material using Eq. 3–11.

$$G = \frac{E}{2(1 + \nu)}$$

$$26 \text{ GPa} = \frac{70.0 \text{ GPa}}{2(1 + \nu)}$$

$$\nu = 0.346$$

Since $\epsilon_{long} = 0.00480$ mm/mm, then by Eq. 3–9,

$$\nu = -\frac{\epsilon_{lat}}{\epsilon_{long}}$$

$$0.346 = -\frac{\epsilon_{lat}}{0.00480 \text{ mm/mm}}$$

$$\epsilon_{lat} = -0.00166 \text{ mm/mm}$$

The contraction of the diameter is therefore

$$\delta' = (0.00166)(25 \text{ mm})$$

$$= 0.0415 \text{ mm} \qquad \textit{Ans.}$$

*3.8 FAILURE OF MATERIALS DUE TO CREEP AND FATIGUE

The mechanical properties of a material have up to this point been discussed only for a static or slowly applied load at constant temperature. In some cases, however, a member may have to be used in an environment for which loadings must be sustained over long periods of time at elevated temperatures, or in other cases, the loading may be repeated or cycled. We will not consider these effects in this book, although we will briefly mention how one determines a material's strength for these conditions since they are given special treatment in design.

Creep. When a material has to support a load for a very long period of time, it may continue to deform until a sudden fracture occurs or its usefulness is impaired. This time-dependent permanent deformation is known as *creep*. Normally creep is considered when metals and ceramics are used for structural members or mechanical parts that are subjected to high temperatures. For some materials, however, such as polymers and composite materials—including wood or concrete—temperature is *not* an important factor, and yet creep can occur strictly from long-term load application. As a typical example, consider the fact that a rubber band will not return to its original shape after being released from a stretched position in which it was held for a very long period of time. In the general sense, therefore, both *stress and/or temperature* play a significant role in the *rate* of creep.

For practical purposes, when creep becomes important, a material is usually designed to resist a specified creep strain for a given period of time. In this regard, an important mechanical property that is used for the design of members subjected to creep is *creep strength*. This value represents the highest initial stress the material can withstand during a specified time without causing a given amount of creep strain. The creep strength will vary with temperature, and for design, a given temperature, duration of loading, and allowable creep strain must all be specified. For example, a creep strain of 0.1% per year has been suggested for steel in bolts and piping, and 0.25% per year for lead sheathing on cables.

Several methods exist for determining an allowable creep strength for a particular material. One of the simplest involves testing several specimens simultaneously at a constant temperature, but with each subjected to a *different axial stress*. By measuring the length of time needed to produce either an allowable strain or the rupture strain for each specimen, a curve of stress versus time can be established. Normally these tests are run to a maximum of 1000 hours. An example of the results for stainless steel at a temperature of 1200°F and prescribed creep strain of 1% is shown in Fig. 3–27. As noted this material has a yield strength of 40 ksi (276 MPa) at room temperature (0.2% offset) and the creep strength at 1000 h is found to be approximately $\sigma_c = 20$ ksi (138 MPa).

The long-term application of the cable loading on this pole has caused it to deform due to creep.

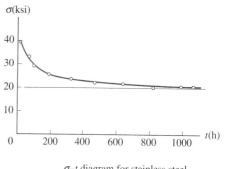

σ–t diagram for stainless steel at 1200°F and creep strain at 1%

Fig. 3–27

In general, the creep strength will *decrease* for *higher temperatures* or for *higher applied stresses*. For longer periods of time, extrapolations from the curves must be made. To do this usually requires a certain amount of experience with creep behavior, and some supplementary knowledge as to the creep properties of the material to be used. Once the material's creep strength has been determined, however, a factor of safety is applied to obtain an appropriate allowable stress for design.

Fatigue.   When a metal is subjected to repeated *cycles* of stress or strain, it causes its structure to break down, ultimately leading to fracture. This behavior is called **fatigue**, and it is usually responsible for a large percentage of failures in connecting rods and crankshafts of engines; steam or gas turbine blades; connections or supports for bridges, railcar wheels, and axles; and other parts subjected to cyclic loading. In all these cases, fracture will occur at a stress that is *less* than the material's yield stress.

The design of amusement park rides requires careful consideration of loadings that can cause fatigue.

The nature of this failure apparently results from the fact that there are microscopic regions, usually on the surface of the member, where the localized stress becomes *much greater* than the average stress acting over the cross section. As this higher stress is cycled, it leads to the formation of minute cracks. Occurrence of these cracks causes a further increase of stress at their tips or boundaries, which in turn causes a further extension of the cracks into the material as the stress continues to be cycled. Eventually the cross-sectional area of the member is reduced to the point where the load can no longer be sustained, and as a result sudden fracture occurs. The material, even though known to be ductile, behaves as if it were *brittle*.

In order to specify a safe strength for a metallic material under repeated loading, it is necessary to determine a limit below which no evidence of failure can be detected after applying a load for a specified number of cycles. This limiting stress is called the *endurance* or *fatigue limit*. Using a testing machine for this purpose, a series of specimens are each subjected to a specified maximum stress and cycled to failure. The results are plotted as a graph representing the stress S (or σ) as the ordinate and the number of cycles-to-failure N as the abscissa. This graph is called an *S–N diagram* or *stress–cycle diagram*, and most often the values of N are plotted on a logarithmic scale since they are generally quite large.

Examples of *S–N* diagrams for two common engineering metals are shown in Fig. 3–28. The endurance limit is that stress for which the *S–N* graph becomes horizontal or asymptotic. As noted, it has a well-defined value of $(S_{el})_{st} = 27$ ksi (186 MPa) for steel. For aluminum, however, the endurance limit is not well defined, and so it is normally specified as the stress having a limit of 500 million cycles, $(S_{el})_{al} = 19$ ksi (131 MPa). Typical values of endurance limits for various engineering

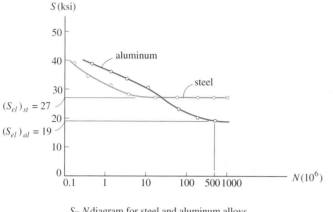

S–N diagram for steel and aluminum alloys
(N axis has a logarithmic scale)

Fig. 3–28

materials are usually reported in handbooks. Once a particular value is obtained, it is often assumed that for any stress below this value the fatigue life is infinite, and therefore the number of cycles to failure is no longer given consideration.

IMPORTANT POINTS

• *Poisson's ratio*, v, is a measure of the lateral strain of a homogeneous and isotropic material versus its longitudinal strain. These strains are generally of opposite signs, that is, if one is an elongation, the other will be a contraction.

• The *shear stress–strain diagram* is a plot of the shear stress versus the shear strain. If the material is homogeneous and isotropic, and is also linear elastic, the slope of the curve within the elastic region is called the modulus of rigidity or the shear modulus, G.

• There is a mathematical relationship between G, E, and v.

• *Creep* is the time related deformation of a material for which stress and/or temperature play an important role. Members are designed to resist the effects of creep based on their creep strength, which is the highest initial stress a material can withstand during a specified time without causing a specified creep strain.

• *Fatigue* occurs in metals when the stress or strain is cycled. It causes brittle fracture to occur. Members are designed to resist fatigue by insuring that the stress in the member does not exceed its *endurance* or *fatigue limit*. This value is determined from an *S–N* diagram as the maximum stress the member can resist when subjected to a specified number of cycles of loading.

PROBLEMS

*3–24. The plastic rod is made of Kevlar 49 and has a diameter of 10 mm. If an axial load of 80 kN is applied to it, determine the change in its length and the change in diameter.

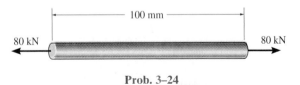

Prob. 3–24

3–25. The support consists of three rigid plates, which are connected together using two symmetrically placed rubber pads. If a vertical force of 50 N is applied to plate A, determine the approximate vertical displacement of this plate due to shear strains in the rubber. Each pad has cross-sectional dimensions of 30 mm and 20 mm. $G_r = 0.20$ MPa.

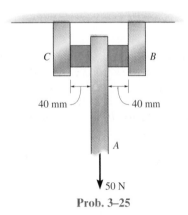

Prob. 3–25

3–26. A short cylindrical block of bronze C86100, having an original diameter of 1.5 in. and a length of 3 in., is placed in a compression machine and squeezed until its length becomes 2.98 in. Determine the new diameter of the block.

3–27. A short cylindrical block of 6061-T6 aluminum, having an original diameter of 20 mm and a length of 75 mm, is placed in a compression machine and squeezed until the axial load applied is 5 kN. Determine (a) the decrease in its length and (b) its new diameter.

*3–28. The elastic portion of the stress–strain diagram for a steel alloy is shown in the figure. The specimen from which it was obtained had an original diameter of 13 mm and a gauge length of 50 mm. When the applied load on the specimen is 50 kN, the diameter is 12.99265 mm. Determine Poisson's ratio for the material.

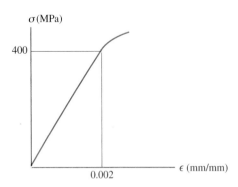

Prob. 3–28

3–29. The block is made of titanium Ti-6A1-4V and is subjected to a compression of 0.06 in. along the y axis, and its shape is given a tilt of $\theta = 89.7°$. Determine ϵ_x, ϵ_y, and γ_{xy}.

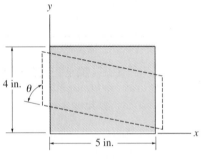

Prob. 3–29

3–30. The elastic portion of the stress–strain diagram for a steel alloy is shown in the figure. The specimen from which it was obtained had an original diameter of 13 mm and a gauge length of 50 mm. If a load of $P = 20$ kN is applied to the specimen, determine its diameter and gauge length. Take $\nu = 0.4$.

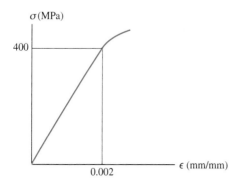

Prob. 3–30

3–31. The shear stress–strain diagram for a steel alloy is shown in the figure. If a bolt having a diameter of 0.25 in. is made of this material and used in the lap joint, determine the modulus of elasticity E and the force P required to cause the material to yield. Take $\nu = 0.3$.

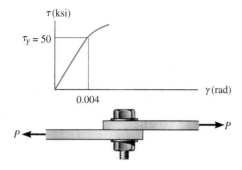

Prob. 3–31

*3–32. A shear spring is made from two blocks of rubber, each having a height h, width b, and thickness a. The blocks are bonded to three plates as shown. If the plates are rigid and the shear modulus of the rubber is G, determine the displacement of plate A if a vertical load $\mathbf{P}$ is applied to this plate. Assume that the displacement is small so that $\delta = a \tan \gamma \approx a\gamma$.

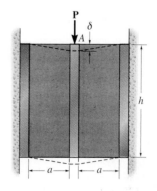

Prob. 3–32

3–33. A shear spring is made by bonding the rubber annulus to a rigid fixed ring and a plug. When an axial load $\mathbf{P}$ is placed on the plug, show that the slope at point A in the rubber is $dy/dr = -\tan \gamma = -\tan(P/(2\pi h Gr))$. For small angles we can write $dy/dr = -P/(2\pi h Gr)$. Integrate this expression and evaluate the constant of integration using the condition that $y = 0$ at $r = r_o$. From the result compute the deflection $y = \delta$ of the plug.

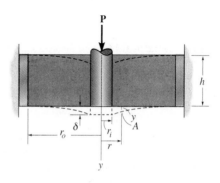

Prob. 3–33

REVIEW PROBLEMS

3–34. The head *H* is connected to the cylinder of a compressor using six steel bolts. If the clamping force in each bolt is 800 lb, determine the normal strain in the bolts. Each bolt has a diameter of $\frac{3}{16}$ in. If $\sigma_Y = 40$ ksi and $E_{st} = 29(10^3)$ ksi, what is the strain in each bolt when the nut is unscrewed so that the clamping force is released?

Prob. 3–34

3–35. The stone has a mass of 800 kg and center of gravity at *G*. It rests on a pad at *A* and a roller at *B*. The pad is fixed to the ground and has a compressed height of 30 mm, a width of 140 mm, and a length of 150 mm. If the coefficient of static friction between the pad and the stone is $\mu_s = 0.8$, determine the approximate horizontal displacement of the stone, caused by the shear strains in the pad, before the stone begins to slip. Assume the normal force at *A* acts 1.5 m from *G* as shown. The pad is made from a material having $E = 4$ MPa and $\nu = 0.35$.

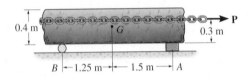

Prob. 3–35

*3–36. The rigid pipe is supported by a pin at *C* and an A-36 steel guy wire *AB*. If the wire has a diameter of 0.2 in., determine how much it stretches when a load of $P = 300$ lb acts on the pipe. The material remains elastic.

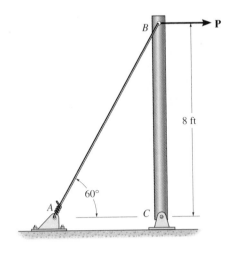

Prob. 3–36

3–37. The rigid pipe is supported by a pin at *C* and an A-36 guy wire *AB*. If the wire has a diameter of 0.2 in., determine the load *P* if the end *B* is displaced 0.10 in. to the right. $E_s = 29(10^3)$ ksi.

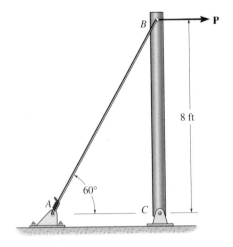

Prob. 3–37

3–38. The 8-mm-diameter bolt is made of an aluminum alloy. It fits through a magnesium sleeve that has an inner diameter of 12 mm and an outer diameter of 20 mm. If the original lengths of the bolt and sleeve are 80 mm and 50 mm, respectively, determine the strains in the sleeve and the bolt if the nut on the bolt is tightened so that the tension in the bolt is 8 kN. Assume the material at A is rigid. $E_{al} = 70$ GPa, $E_{mg} = 45$ GPa.

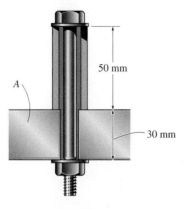

50 mm

A

30 mm

Prob. 3–38

3–39. A tension test was performed on a steel specimen having an original diameter of 0.503 in. and gauge length of 2.00 in. The data are listed in the table below. Plot the stress–strain diagram and determine approximately the modulus of elasticity, the yield stress, the ultimate stress, and the rupture stress. Use a scale of 1 in. = 20 ksi and 1 in. = 0.05 in./in. Redraw the elastic region, using the same stress scale but a strain scale of 1 in. = 0.001 in./in.

Load (kip)	Elongation (in.)
0	0
1.50	0.0005
4.60	0.0015
8.00	0.0025
11.00	0.0035
11.80	0.0050
11.80	0.0080
12.00	0.0200
16.60	0.0400
20.00	0.1000
21.50	0.2800
19.50	0.4000
18.50	0.4600

Prob. 3–39

*3–40. The stress–strain diagram for a polyester resin is given in the figure. If the rigid beam is supported by a link AB and post CD, both made of this material, and subjected to a load of $P = 80$ kN, determine the angle of tilt of the beam when the load is applied. The diameter of the link is 40 mm and the diameter of the post is 80 mm.

3–41. The stress–strain diagram for a polyester resin is given in the figure. If the rigid beam is supported by a link AB and post CD made of this material, determine the largest load P that can be applied to the beam before either AB or CD fractures. The diameter of the link is 12 mm and the diameter of the post is 40 mm.

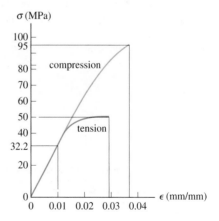

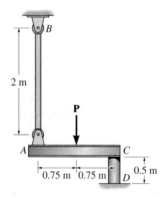

Probs. 3–40/3–41

The string of drill pipe suspended from this traveling block on an oil rig is subjected to extremely large loadings and axial deformations.

4 AXIAL LOAD

CHAPTER OBJECTIVES

In Chapter 1 we developed the method for finding the normal stress in axially loaded members. In this chapter we will discuss how to determine the deformation of these members, and we will also develop a method for finding the support reactions when these reactions cannot be determined strictly from the equations of equilibrium. An analysis of the effects of thermal stress, stress concentrations, inelastic deformations, and residual stress will also be discussed.

4.1 SAINT-VENANT'S PRINCIPLE

In the previous chapters we have developed the concept of *stress* as a means of measuring the force distribution within a body and *strain* as a means of measuring a body's deformation. We have also shown that the mathematical relationship between stress and strain depends on the type of material from which the body is made. In particular, if the material behaves in a linear-elastic manner, then Hooke's law applies, and there is a proportional relationship between stress and strain.

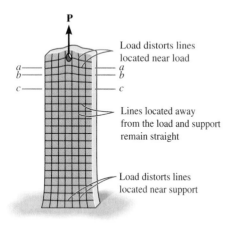

P

Load distorts lines located near load

a
b

a
b

c

c

Lines located away from the load and support remain straight

Load distorts lines located near support

(a)

Fig. 4–1

Using this idea, consider the manner in which a rectangular bar will deform elastically when the bar is subjected to a force **P** applied along its centroidal axis, Fig. 4–1*a*. Here the bar is fixed-connected at one end, with the force applied through a hole at its other end. Due to the loading, the bar deforms as indicated by the distortions of the once horizontal and vertical grid lines drawn on the bar. Notice the *localized deformation* that occurs at each end. This effect tends to *decrease* as measurements are taken farther and farther away from the ends. Furthermore, the deformations even out and become uniform throughout the midsection of the bar.

Since the deformation or strain is related to stress within the bar, we can state that stress will be distributed more uniformly throughout the cross-sectional area if the section is taken farther and farther from the point where the external load is applied. For example, consider a profile of the variation of the stress distribution acting at sections *a–a*, *b–b*, and *c–c*, each of which is shown in Fig. 4–1*b*. By comparison, the stress *almost* reaches a uniform value at section *c–c*, which is sufficiently removed from the end. In other words, section *c–c* is far enough away from the application of **P** so that the localized deformation caused by **P** *vanishes*. The minimum distance from the bar's end where this occurs can be determined using a mathematical analysis based on the theory of elasticity.

However, as a *general rule*, which applies as well to many other cases of loading and member geometry, we can consider this distance to be at least equal to the *largest dimension* of the loaded cross section. Hence, for the bar in Fig. 4–1*b*, section *c–c* should be located at a distance at least equal to the width (not the thickness) of the bar.* This rule is based on *experimental observation of material behavior*, and only in special cases, like the one discussed here, has it been validated mathematically. It should be noted, however, that this rule does not apply to every type of member and loading case. For example, members made from thin-walled elements, and subjected to loadings that cause large deflections, may create localized stresses and deformations that have an influence a considerable distance away from the point of application of loading.

---

*When section *c–c* is so located, the theory of elasticity predicts the maximum stress to be $\sigma_{max} = 1.02\sigma_{avg}$.

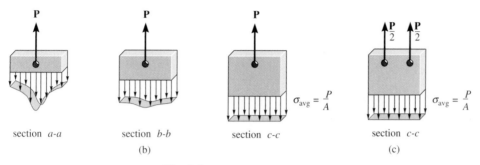

section *a-a*   section *b-b*   section *c-c* $\sigma_{\text{avg}} = \dfrac{P}{A}$ section *c-c* $\sigma_{\text{avg}} = \dfrac{P}{A}$

(b)   (c)

Fig. 4–1

At the support, in Fig. 4–1*a* notice how the bar is prevented from decreasing its width, which should occur due to the bar's lateral elongation—a consequence of the "Poisson effect," discussed in Sec. 3–6. By the same arguments given above, however, we could demonstrate that the stress distribution at the support will also even out and become uniform over the cross section at a short distance from the support; and furthermore, the magnitude of the resultant force created by this stress distribution must also equal *P*.

The fact that stress and deformation behave in this manner is referred to as *Saint-Venant's principle*, since it was first noticed by the French scientist Barré de Saint-Venant in 1855. Essentially it states that the stress and strain produced at points in a body sufficiently removed from the region of load application will be the *same* as the stress and strain produced by *any applied loadings* that have the same statically equivalent resultant and are applied to the body within the same region. For example, if two symmetrically applied forces *P*/2 act on the bar, Fig. 4–1*c*, the stress distribution at section *c–c*, which is sufficiently removed from the localized effects of these loads, will be uniform and therefore equivalent to $\sigma_{\text{avg}} = P/A$ as before.

To summarize, then, we do not have to consider the somewhat complex stress distributions that may actually develop at points of load application, or at supports, when studying the stress distribution in a body at sections *sufficiently removed* from the points of load application. Saint-Venant's principle claims that the localized effects caused by any load acting on the body will dissipate or smooth out within regions that are sufficiently removed from the location of the load. Furthermore, the resulting stress distribution at these regions will be the *same* as that caused by any other statically equivalent load applied to the body within the same localized area.

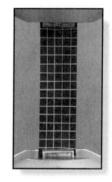

Notice how the lines on this rubber membrane distort after it is stretched. The localized distortions at the grips smooth out as expected. This is due to Saint-Venant's principle.

4.2   Elastic Deformation of an Axially Loaded Member

Using Hooke's law and the definitions of stress and strain, we will now develop an equation that can be used to determine the elastic deformation of a member subjected to axial loads. To generalize the development, consider the bar shown in Fig. 4–2a, which has a cross-sectional area that *gradually* varies along its length *L*. The bar is subjected to concentrated loads at its ends and a variable external load distributed along its length. This distributed load could, for example, represent the weight of a vertical bar, or friction forces acting on the bar's surface. Here we wish to find the ***relative displacement*** δ (delta) of one end of the bar with respect to the other end as caused by this loading. In the following analysis we will neglect the localized deformations that occur at points of concentrated loading and where the cross section suddenly changes. As noted in Sec. 4.1, these effects occur within small regions of the bar's length and will therefore have only a slight effect on the final result. For the most part, the bar will deform uniformly, so the normal stress will be uniformly distributed over the cross section.

Using the method of sections, a differential element (or wafer) of length *dx* and cross-sectional area $A(x)$ is isolated from the bar at the arbitrary position *x*. The free-body diagram of this element is shown in Fig. 4–2b. The resultant internal axial force is represented as $P(x)$, since the external loading will cause it to vary along the length of the bar. This load, $P(x)$, will deform the element into the shape indicated by the dashed outline, and therefore the displacement of one end of the element with respect to the other end is *dδ*. The stress and strain in the element are

$$\sigma = \frac{P(x)}{A(x)} \qquad \text{and} \qquad \epsilon = \frac{d\delta}{dx}$$

Provided these quantities do not exceed the proportional limit, we can relate them using Hooke's law; i.e.,

$$\sigma = E\epsilon$$
$$\frac{P(x)}{A(x)} = E\left(\frac{d\delta}{dx}\right)$$
$$d\delta = \frac{P(x)\ dx}{A(x)\ E}$$

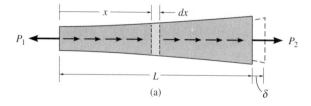

(a)

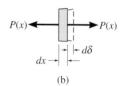

(b)

Fig. 4–2

For the entire length L of the bar, we must integrate this expression to find the required end displacement. This yields

$$\delta = \int_0^L \frac{P(x)\,dx}{A(x)\,E} \qquad (4\text{--}1)$$

where

δ = displacement of one point on the bar relative to another point
L = distance between the points
$P(x)$ = internal axial force at the section, located a distance x from one end
$A(x)$ = cross-sectional area of the bar, expressed as a function of x
E = modulus of elasticity for the material

Constant Load and Cross-Sectional Area.   In many cases the bar will have a constant cross-sectional area A; and the material will be homogeneous, so E is constant. Furthermore, if a constant external force is applied at each end, Fig. 4–3, then the internal force P throughout the length of the bar is also constant. As a result, Eq. 4–1 can be integrated to yield

$$\delta = \frac{PL}{AE} \qquad (4\text{--}2)$$

If the bar is subjected to several different axial forces, or the cross-sectional area or modulus of elasticity changes abruptly from one region of the bar to the next, the above equation can be applied to each *segment* of the bar where these quantities are all *constant*. The displacement of one end of the bar with respect to the other is then found from the *algebraic addition* of the end displacements of each segment. For this general case,

$$\delta = \sum \frac{PL}{AE} \qquad (4\text{--}3)$$

The vertical displacement at the top of these building columns depends upon the loading applied on the roof and from the floor attached to their midpoint.

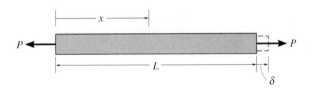

Fig. 4–3

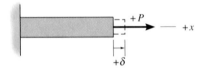

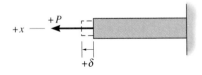

Positive sign convention for P and δ

Fig. 4–4

Sign Convention. In order to apply Eq. 4–3, we must develop a sign convention for the internal axial force and the displacement of one end of the bar with respect to the other end. To do so, we will consider both the force and displacement to be positive if they cause tension and elongation, respectively, Fig. 4–4; whereas a negative force and displacement will cause compression and contraction, respectively.

For example, consider the bar shown in Fig. 4–5a. The *internal axial forces "P,"* are determined by the method of sections for each segment, Fig. 4-5b. They are $P_{AB} = +5$ kN, $P_{BC} = -3$ kN, $P_{CD} = -7$ kN. This variation in axial load is shown on the axial (or normal) force diagram for the bar, Fig. 4–5c. Applying Eq. 4–3 to obtain the displacement of end A relative to end D, we have

$$\delta_{A/D} = \sum \frac{PL}{AE} = \frac{(5 \text{ kN}) L_{AB}}{AE} + \frac{(-3 \text{ kN}) L_{BC}}{AE} + \frac{(-7 \text{ kN}) L_{CD}}{AE}$$

If the other data are substituted and a positive answer is computed, it means that end A will move away from end D (the bar elongates), whereas a negative result would indicate that end A moves toward end D (the bar shortens). The double subscript notation is used to indicate this relative displacement $(\delta_{A/D})$; however, if the displacement is to be determined relative to a *fixed point*, then only a single subscript will be used. For example, if D is located at a *fixed* support, then the computed displacement will be denoted as simply δ_A.

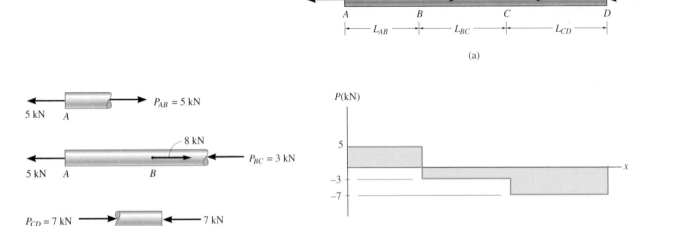

Fig. 4–5

IMPORTANT POINTS

• *Saint-Venant's principle* states that both the localized deformation and stress which occur within the regions of load application or at the supports tend to "even out" at a distance sufficiently removed from these regions.

• The displacement of an axially loaded member is determined by relating the applied load to the stress using $\sigma = P/A$ and relating the displacement to the strain using $\epsilon = d\delta/dx$. Finally these two equations are combined using Hooke's law, $\sigma = E\epsilon$, which yields Eq. 4–1.

• Since Hooke's law has been used in the development of the displacement equation, it is important that the loads do not cause yielding of the material and that the material is homogeneous and behaves in a linear–elastic manner.

PROCEDURE FOR ANALYSIS

The relative displacement between two points A and B on an axially loaded member can be determined by applying Eq. 4–1 (or Eq. 4–2). Application requires the following steps.

Internal Force.

• Use the method of sections to determine the internal axial force P in the member.

• If this force varies along the member's length, a section should be made at the arbitrary location x from one end of the member and the force represented as a function of x, i.e., $P(x)$.

• If several *constant external forces* act on the member, the internal force in each *segment* of the member, between any two external forces, must then be determined.

• For any segment, an internal *tensile force* is *positive* and an internal *compressive force* is *negative*. For convenience, the results of the internal loading can be shown graphically by constructing the normal-force diagram.

Displacement.

• When the member's cross-sectional area *varies* along its axis, the area must be expressed as a function of its position x, i.e., $A(x)$.

• If the cross-sectional area, the modulus of elasticity, or the internal loading *suddenly changes,* then Eq. 4–2 should be applied to each segment for which these quantities are constant.

• When substituting the data into Eqs. 4–1 through 4–3, be sure to account for the proper sign for P, tensile loadings are positive and compressive loadings are negative. Also, use a consistent set of units. For any segment, if the computed result is a *positive* numerical quantity, it indicates *elongation*; if it is *negative*, it indicates a *contraction*.

EXAMPLE 4-1

The composite A-36 steel bar shown in Fig. 4–6a is made from two segments, AB and BD, having cross-sectional areas of $A_{AB} = 1$ in^2 and $A_{BD} = 2$ in^2. Determine the vertical displacement of end A and the displacement of B relative to C.

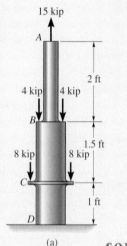

(a)

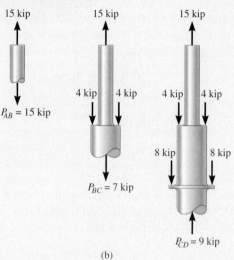

(b)

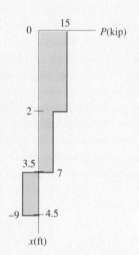

(c)

Fig. 4–6

SOLUTION

Internal Force.   Due to the application of the external loadings, the *internal axial forces* in regions AB, BC, and CD will all be *different*. These forces are obtained by applying the method of sections and the equation of vertical force equilibrium as shown in Fig. 4–6b. This variation is plotted in Fig. 4–6c.

Displacement.   From the inside back cover, $E_{st} = 29(10^3)$ ksi. Using the sign convention, i.e., the internal tensile forces are positive and the compressive forces are negative, the vertical displacement of A relative to the *fixed* support D is

$$\delta_A = \sum \frac{PL}{AE} = \frac{[+15 \text{ kip}](2 \text{ ft})(12 \text{ in./ft})}{(1 \text{ in}^2)[29(10^3) \text{ kip/in}^2]} + \frac{[+7 \text{ kip}](1.5 \text{ ft})(12 \text{ in./ft})}{(2 \text{ in}^2)[29(10^3) \text{ kip/in}^2]}$$

$$+ \frac{[-9 \text{ kip}](1 \text{ ft})(12 \text{ in./ft})}{(2 \text{ in}^2)[29(10^3) \text{ kip/in}^2]}$$

$$= +0.0127 \text{ in.} \qquad\qquad\qquad\qquad\qquad Ans.$$

Since the result is *positive*, the bar *elongates* and so the displacement at A is upward.

Applying Eq. 4–2 between points B and C, we obtain,

$$\delta_{B/C} = \frac{P_{BC}L_{BC}}{A_{BC}E} = \frac{[+7 \text{ kip}](1.5 \text{ ft})(12 \text{ in./ft})}{(2 \text{ in}^2)[29(10^3) \text{ kip/in}^2]} = +0.00217 \text{ in.} \qquad Ans.$$

Here B moves away from C, since the segment elongates.

EXAMPLE 4-2

The assembly shown in Fig. 4–7a consists of an aluminum tube AB having a cross-sectional area of 400 mm². A steel rod having a diameter of 10 mm is attached to a rigid collar and passes through the tube. If a tensile load of 80 kN is applied to the rod, determine the displacement of the end C of the rod. Take $E_{st} = 200$ GPa, $E_{al} = 70$ GPa.

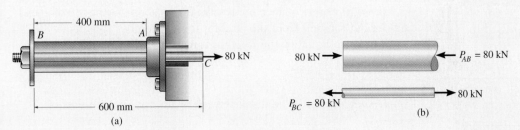

(a)

(b)

Fig. 4–7

SOLUTION

Internal Force   The free-body diagram of the tube and rod, Fig. 4–7b, shows that the rod is subjected to a tension of 80 kN and the tube is subjected to a compression of 80 kN.

Displacement.   We will first determine the displacement of end C with respect to end B. Working in units of newtons and meters, we have

$$\delta_{C/B} = \frac{PL}{AE} = \frac{[+80(10^3)\ \text{N}](0.6\ \text{m})}{\pi(0.005\ \text{m})^2[200(10^9)\ \text{N/m}^2]} = +0.003056\ \text{m} \rightarrow$$

The positive sign indicates that end C moves *to the right* relative to end B, since the bar elongates.

The displacement of end B with respect to the *fixed* end A is

$$\delta_B = \frac{PL}{AE} = \frac{[-80(10^3)\ \text{N}](0.4\ \text{m})}{[400\ \text{mm}^2(10^{-6})\ \text{m}^2/\text{mm}^2][70(10^9)\ \text{N/m}^2]}$$

$$= -0.001143\ \text{m} = 0.001143\ \text{m} \rightarrow$$

Here the negative sign indicates that the tube shortens, and so B moves to the *right* relative to A.

Since both displacements are to the right, the resultant displacement of C relative to the fixed end A is therefore

$$(\overset{+}{\rightarrow}) \qquad \delta_C = \delta_B + \delta_{C/B} = 0.001143\ \text{m} + 0.003056\ \text{m}$$

$$= 0.00420\ \text{m} = 4.20\ \text{mm} \rightarrow \qquad \qquad \textit{Ans.}$$

EXAMPLE 4-3

A *rigid beam AB* rests on the two short posts shown in Fig. 4–8a. *AC* is made of steel and has a diameter of 20 mm, and *BD* is made of aluminum and has a diameter of 40 mm. Determine the displacement of point *F* on *AB* if a vertical load of 90 kN is applied over this point. Take E_{st} = 200 GPa, E_{al} = 70 GPa.

SOLUTION

Internal Force. The compressive forces acting at the top of each post are determined from the equilibrium of member *AB*, Fig. 4–8b. These forces are equal to the internal forces in each post, Fig. 4–8c.

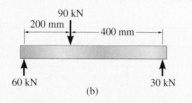

(a)

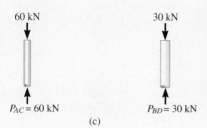

(b)

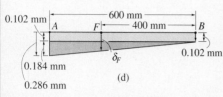

(c)

0.102 mm

0.184 mm

0.286 mm

(d)

Fig. 4–8

Displacement. The displacement of the top of each post is

Post *AC*:

$$\delta_A = \frac{P_{AC}L_{AC}}{A_{AC}E_{st}} = \frac{[-60(10^3)\ \text{N}](0.300\ \text{m})}{\pi(0.010\ \text{m})^2[200(10^9)\ \text{N/m}^2]} = -286(10^{-6})\ \text{m}$$

$$= 0.286\ \text{mm} \downarrow$$

Post *BD*:

$$\delta_B = \frac{P_{BD}L_{BD}}{A_{BD}E_{al}} = \frac{[-30(10^3)\ \text{N}](0.300\ \text{m})}{\pi(0.020\ \text{m})^2[70(10^9)\ \text{N/m}^2]} = -102(10^{-6})\ \text{m}$$

$$= 0.102\ \text{mm} \downarrow$$

A diagram showing the centerline displacements at points *A, B,* and *F* on the beam is shown in Fig. 4–8d. By proportion of the shaded triangle, the displacement of point *F* is therefore

$$\delta_F = 0.102\ \text{mm} + (0.184\ \text{mm})\left(\frac{400\ \text{mm}}{600\ \text{mm}}\right) = 0.225\ \text{mm} \downarrow \qquad \textit{Ans.}$$

EXAMPLE 4-4

A member is made from a material that has a specific weight γ and modulus of elasticity E. If it is formed into a *cone* having the dimensions shown in Fig. 4–9a, determine how far its end is displaced due to gravity when it is suspended in the vertical position.

SOLUTION

Internal Force.   The internal axial force varies along the member since it is dependent on the weight $W(y)$ of a segment of the member below any section, Fig. 4–9b. Hence, to calculate the displacement, we must use Eq. 4–1. At the section located at a distance y from its bottom end, the radius x of the cone as a function of y is determined by proportion; i.e.,

$$\frac{x}{y} = \frac{r_0}{L}; \qquad x = \frac{r_0}{L}y$$

The volume of a cone having a base of radius x and height y is

$$V = \frac{\pi}{3}yx^2 = \frac{\pi r_0^2}{3L^2}y^3$$

Since $W = \gamma V$, the internal force at the section becomes $W = \rho g V$

$$+\uparrow \Sigma F_y = 0; \qquad P(y) = \frac{\gamma \pi r_0^2}{3L^2}y^3$$

Displacement.   The area of the cross section is also a function of position y, Fig. 4–9b. We have

$$A(y) = \pi x^2 = \frac{\pi r_0^2}{L^2}y^2$$

Applying Eq. 4–1 between the limits of $y = 0$ and $y = L$ yields

$$\delta = \int_0^L \frac{P(y)\, dy}{A(y)\, E} = \int_0^L \frac{[(\gamma \pi r_0^2/3L^2)\, y^3]\, dy}{[(\pi r_0^2/L^2)\, y^2]\, E}$$

$$= \frac{\gamma}{3E}\int_0^L y\, dy$$

$$= \frac{\gamma L^2}{6E} \qquad\qquad\qquad\qquad\qquad\qquad\text{Ans.}$$

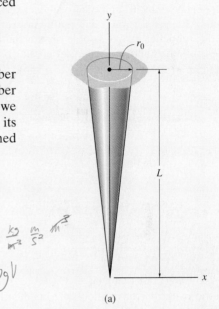

(a)

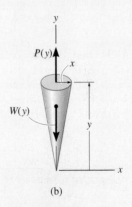

(b)

Fig. 4–9

As a partial check of this result, notice how the units of the terms, when canceled, give the displacement in units of length as expected.

PROBLEMS

4–1. The ship is pushed through the water using an A-36 steel propeller shaft that is 8 m long, measured from the propeller to the thrust bearing D at the engine. If it has an outer diameter of 400 mm and a wall thickness of 50 mm, determine the amount of axial contraction of the shaft when the propeller exerts a force on the shaft of 5 kN. The bearings at B and C are journal bearings.

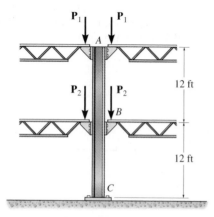

Prob. 4–1

4–2. The A-36 steel column is used to support the symmetric loads from the two floors of a building. Determine the vertical displacement of its top A if $P_1 = 40$ kip, $P_2 = 62$ kip, and the column has a cross-sectional area of 23.4 in².

4–3. The A-36 steel column is used to support the symmetric loads from the two floors of a building. Determine the loads P_1 and P_2 if A moves downward 0.12 in. and B moves downward 0.09 in. when the loads are applied. The column has a cross-sectional area of 23.4 in².

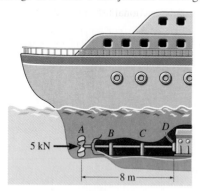

Probs. 4–2/4–3

*4–4. The bronze C86100 shaft is subjected to the axial loads shown. Determine the displacement of end A with respect to end D if the diameters of each segment are $d_{AB} = 0.75$ in., $d_{BC} = 2$ in., and $d_{CD} = 0.5$ in.

4–5. Determine the displacement of end A with respect to end C of the shaft in Prob. 4–4.

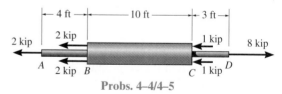

Probs. 4–4/4–5

4–6 The assembly consists of an A-36 steel rod CB and a 6061-T6 aluminum rod BA, each having a diameter of 1 in. If the rod is subjected to the axial loading $P_1 = 12$ kip at A and $P_2 = 18$ kip at the coupling B, determine the displacement of the coupling B and the end A. The unstretched length of each segment is shown in the figure. Neglect the size of the connections at B and C, and assume that they are rigid.

4–7. The assembly consists of an A-36 steel rod CB and a 6061-T6 aluminum rod BA, each having a diameter of 1 in. Determine the applied loads P_1 and P_2 if A is displaced 0.08 in. to the right and B is displaced 0.02 in. to the left when the loads are applied. The unstretched length of each segment is shown in the figure. Neglect the size of the connections at B and C, and assume that they are rigid.

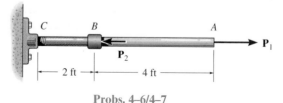

Probs. 4–6/4–7

*4–8. The joint is made from three A-36 steel plates that are bonded together at their seams. Determine the displacement of end A with respect to end D when the joint is subjected to the axial loads shown. Each plate has a thickness of 6 mm.

Prob. 4–8

4–9. The assembly consists of two rigid bars that are originally horizontal. They are supported by pins and 0.25-in.-diameter A-36 steel rods *FC* and *EB*. If the vertical load of 5 kip is applied to the bottom bar *AB*, determine the displacement at *C*, *B*, and *E*.

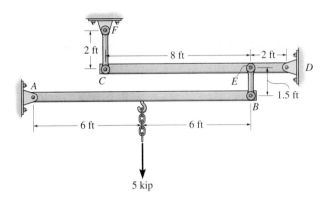

Prob. 4–9

4–10. The truss is made from three A-36 steel members, each having a cross-sectional area of 400 mm². Determine the vertical displacement of the roller at *C* when the truss supports the load of $P = 10$ kN.

4–11. The truss is made from three A-36 steel members, each having a cross-sectional area of 400 mm². Determine the load *P* required to displace the roller downward 0.2 mm.

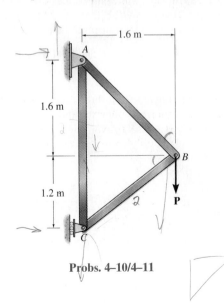

Probs. 4–10/4–11

4–12. The load is supported by the four 304 stainless steel wires that are connected to the rigid members *AB* and *DC*. Determine the vertical displacement of the 500-lb load if the members were horizontal when the load was originally applied. Each wire has a cross-sectional area of 0.025 in².

4–13. The load is supported by the four 304 stainless steel wires that are connected to the rigid members *AB* and *DC*. Determine the angle of tilt of each member after the 500-lb load is applied. The members were originally horizontal, and each wire has a cross-sectional area of 0.025 in².

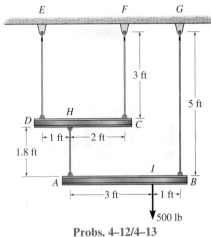

Probs. 4–12/4–13

4–14. The linkage is made from three pin-connected 304 stainless steel members, each having a cross-sectional area of 0.75 in². If a horizontal force of $P = 6$ kip is applied to the end *B* of member *AB*, determine the horizontal displacement of point *B*.

4–15. The linkage is made from three pin-connected 304 stainless steel members, each having a cross-sectional area of 0.75 in². Determine the magnitude of the force **P** needed to displace point *B* 0.08 in. to the right.

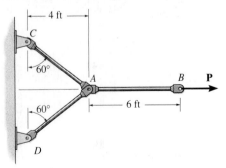

Probs. 4–14/4–15

*4–16. A spring-supported pipe hanger consists of two springs which are originally unstretched and have a stiffness of $k = 60$ kN/m, three 304 stainless steel rods, AB and CD, which have a diameter of 5 mm and EF, which has a diameter of 12 mm, and a rigid beam GH. If the pipe and the fluid it carries have a total weight of 4 kN, determine the displacement of the pipe when it is attached to the support.

4–17. A spring-supported pipe hanger consists of two springs, which are originally unstretched and have a stiffness of $k = 60$ kN/m, three 304 stainless steel rods, AB and CD which have a diameter of 5 mm and EF which has a diameter of 12 mm, and a rigid beam GH. If the pipe is displaced 82 mm when it is filled with fluid, determine the weight of the fluid.

*4–20. The rigid beam is supported at its ends by two A-36 steel tie rods. If the allowable stress for the steel is $\sigma_{allow} = 16.2$ ksi, the load $w = 3$ kip/ft, and $x = 4$ ft, determine the diameter of each rod so that the beam remains in the horizontal position when it is loaded.

4–21. The rigid beam is supported at its ends by two A-36 steel tie rods. The rods have diameters $d_{AB} = 0.5$ in. and $d_{CD} = 0.3$ in. If the allowable stress for the steel is $\sigma_{allow} = 16.2$ ksi, determine the intensity of the distributed load w and its length x on the beam so that the beam remains in the horizontal position when it is loaded.

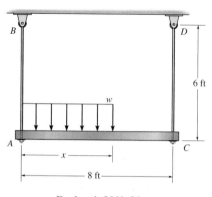

Probs. 4–20/4–21

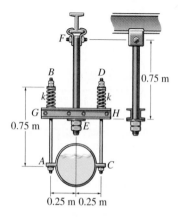

Probs. 4–16/4–17

4–18. The assembly consists of three titanium (Ti-6A1-4V) rods and a rigid bar AC. The cross-sectional area of each rod is given in the figure. If a force of 6 kip is applied to the ring, determine the horizontal displacement of point F.

4–19. The assembly consists of three titanium (Ti-6A1-4V) rods and a rigid bar AC. The cross-sectional area of each rod is given in the figure. If a force of 6 kip is applied to the ring F, determine the angle of tilt in radians of bar AC.

4–22. The truss consists of three members, each made from A-36 steel and having a cross-sectional area of 0.75 in². Determine the greatest load P that can be applied so that the roller support at B is not displaced more than 0.03 in.

4–23. Solve Prob. 4-22 when the load P acts vertically downward at C.

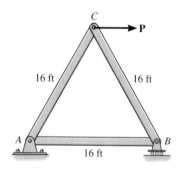

Probs. 4–22/4–23

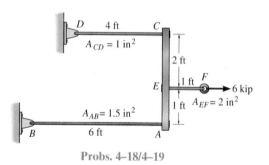

Probs. 4–18/4–19

*4–24.   The segments of pipe and couplings used for drilling an oil well 15 000 ft deep are made of A-36 steel weighing 20 lb/ft. They have an outer diameter of 5.50 in. and an inner diameter of 4.75 in. In order to prevent buckling or sidesway of the pipe due to its own weight, it is partially supported at its top by the drawworks of the rig. If this force is $P = 299$ kip, determine the force **F** of the ground on the drill pipe and the elongation of the pipe for this condition.

4–25.   The segments of pipe and couplings used for drilling an oil well 15 000 ft deep are made of A-36 steel weighing 20 lb/ft. They have an outer diameter of 5.50 in. and an inner diameter of 4.75 in. Determine the force **P** needed to remove the pipe, excluding friction along its sides and requiring $F = 0$. Also, what is the elongation of the pipe as it just begins to be lifted out?

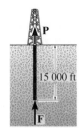

Probs. 4–24/4–25

4–26.   The post is made of Douglas fir and has a diameter of 60 mm. If it is subjected to the load of 20 kN and the soil provides a frictional resistance that is uniformly distributed along its sides of $w = 4$ kN/m, determine the force **F** at its bottom needed for equilibrium. Also, what is the displacement of the top of the post A with respect to its bottom B? Neglect the weight of the post.

4–27.   The post is made of Douglas fir and has a diameter of 60 mm. If it is subjected to the load of 20 kN and the soil provides a frictional resistance that is distributed along its length and varies linearly from $w = 0$ at $y = 0$ to $w = 3$ kN/m at $y = 2$ m, determine the force **F** at its bottom needed for equilibrium. Also, what is the displacement of the top of the post A with respect to its bottom B? Neglect the weight of the post.

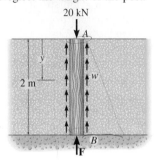

Probs. 4–26/4–27

*4–28.   The rod has a slight taper and length L. It is suspended from the ceiling and supports a load **P** at its end. Show that the displacement of its end due to this load is $\delta = PL/(\pi E r_2 r_1)$. Neglect the weight of the material. The modulus of elasticity is E.

4–29.   Solve Prob. 4–28 by including the weight of the material, considering its specific weight to be γ (weight/volume).

Probs. 4–28/4–29

4–30.   Determine the elongation of the aluminum strap when it is subjected to an axial force of 30 kN. $E_{al} = 70$ GPa.

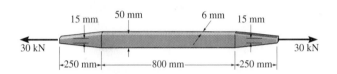

Prob. 4–30

4–31. The bar has a cross-sectional area of $A = 3$ in², and $E = 35(10^3)$ ksi. Determine the displacement of its end A when it is subjected to the distributed loading.

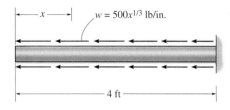

$w = 500x^{1/3}$ lb/in.

x

4 ft

Prob. 4–31

*4–32.** Bone material has a stress–strain diagram that can be defined by the relation $\sigma = kE[\epsilon/(1 + E\epsilon)]$, where k and E are constants. Determine the compression within the length L of the bone, where it is assumed the cross-sectional area A of the bone is constant.

P

L

P

Prob. 4–32

4–33. The support is made by cutting off the two opposite sides of a sphere that has a radius r_0. If the original height of the support is $r_0/2$, determine how far it shortens when it supports a load **P**. The modulus of elasticity is E.

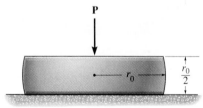

P

r_0

$\dfrac{r_0}{2}$

Prob. 4–33

4–34. The casting is made of a material that has a specific weight γ and modulus of elasticity E. If it is formed into a pyramid having the dimensions shown, determine how far its end is displaced due to gravity when it is suspended in the vertical position.

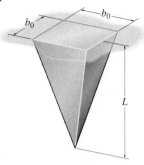

b_0

b_0

L

Prob. 4–34

4–35. Two posts support the rigid beam, each having a width d, a thickness d, and a length L. If the modulus of elasticity for material A is E_A, and for material B it is E_B, determine the distance x for placement of the force **P** so that the beam remains horizontal.

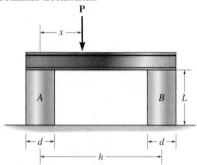

P

x

A　　　B　L

d　　　　d

h

Prob. 4–35

■*4–36.** Consider the general problem of a bar made from m segments, each having a constant cross-sectional area A_m and length L_m. If there are n loads on the bar as shown, write a computer program that can be used to determine the displacement of the bar at any specified location x. Show an application of the program using the values $L_1 = 4$ ft, $d_1 = 2$ ft, $P_1 = 400$ lb, $A_1 = 3$ in², $L_2 = 2$ ft, $d_2 = 6$ ft, $P_2 = -300$ lb, $A_2 = 1$ in².

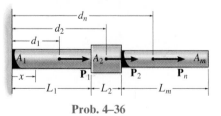

d_n

d_2

d_1

A_1　　　A_2　　　A_m

x

P_1　P_2　　P_n

L_1　L_2　L_m

Prob. 4–36

4.3   PRINCIPLE OF SUPERPOSITION

The principle of superposition is often used to determine the stress or displacement at a point in a member when the member is subjected to a complicated loading. By subdividing the loading into components, the **principle of superposition** states that the resultant stress or displacement at the point can be determined by first finding the stress or displacement caused by each component load acting *separately* on the member. The resultant stress or displacement is then determined by algebraically adding the contributions caused by each of the components.

The following two conditions must be valid if the principle of superposition is to be applied.

1. ***The loading must be linearly related to the stress or displacement that is to be determined.*** For example, the equations $\sigma = P/A$ and $\delta = PL/AE$ involve a linear relationship between P and σ or δ.

2. ***The loading must not significantly change the original geometry or configuration of the member.*** If significant changes do occur, the direction and location of the applied forces and their moment arms will change, and consequently, application of the equilibrium equations will yield different results. For example, consider the slender rod shown in Fig. 4–10a, which is subjected to the load **P**. In Fig. 4–10b, **P** is replaced by two of its components, $\mathbf{P} = \mathbf{P}_1 + \mathbf{P}_2$. If **P** causes the rod to deflect a large amount, as shown, the moment of the load about its support, Pd, will *not* equal the sum of the moments of its component loads, $Pd \neq P_1 d_1 + P_2 d_2$, because $d_1 \neq d_2 \neq d$.

Most of the equations involving load, stress, and displacement that are developed in this text consist of linear relationships between these quantities. Also, members or bodies that are to be considered will be such that the loading will produce deformations that are so small that the change in position and direction of the loading will be insignificant and can be neglected. One exception to this rule, however, will be discussed in Chapter 13. It consists of a column that carries an axial load that is equivalent to the critical or buckling load. It will be shown that when this load increases only slightly, it will cause the column to have a large lateral deflection, even if the material remains linear-elastic. These deflections, associated with the components of any axial load, *cannot* be superimposed.

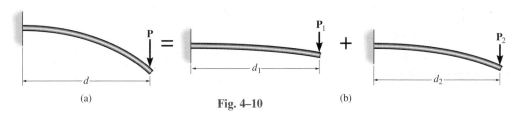

(a)

Fig. 4–10

(b)

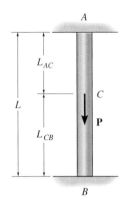

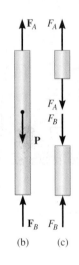

(a)

(b)   (c)

Fig. 4–11

4.4   STATICALLY INDETERMINATE AXIALLY LOADED MEMBER

When a bar is fixed-supported at only one end and is subjected to an axial force, the force equilibrium equation applied along the axis of the bar is *sufficient* to find the reaction at the fixed support. A problem such as this, where the reactions can be determined strictly from the equations of equilibrium, is called *statically determinate*. If the bar is fixed at *both ends*, however, as in Fig. 4–11*a*, then two unknown axial reactions occur, Fig. 4–11*b*, and the force equilibrium equation becomes

$$+\uparrow \ \Sigma F = 0; \qquad\qquad F_B + F_A - P = 0$$

In this case the bar is called **statically indeterminate**, since the equilibrium equation(s) are not sufficient to determine the reactions.

In order to establish an additional equation needed for solution, it is necessary to consider the geometry of the deformation. Specifically, an equation that specifies the conditions for displacement is referred to as a **compatibility** or **kinematic condition**. A suitable compatibility condition would require the relative displacement of one end of the bar with respect to the other end to be equal to zero, since the end supports are fixed. Hence, we can write

$$\delta_{A/B} = 0$$

This equation can be expressed in terms of the applied loads by using a *load–displacement relationship*, which depends on the material behavior. For example, if linear-elastic behavior occurs, $\delta = PL/AE$ can be used. Realizing that the internal force in segment AC is $+F_A$, and in segment CB the internal force is $-F_B$, Fig. 4–11*c* the compatibility equation can be written as

$$\frac{F_A L_{AC}}{AE} - \frac{F_B L_{CB}}{AE} = 0$$

Assuming that AE is constant, we can solve the above two equations for the reactions, which gives

$$F_A = P\left(\frac{L_{CB}}{L}\right) \quad \text{and} \quad F_B = P\left(\frac{L_{AC}}{L}\right)$$

Both of these results are positive, so the reactions are shown correctly on the free-body diagram.

IMPORTANT POINTS

• The *principle of superposition* is sometimes used to simplify stress and displacement problems having complicated loadings. This is done by subdividing the loading into components, then algebraically adding the results.

• Superposition requires that the loading be linearly related to the stress or displacement, and the loading does not significantly change the original geometry of the member.

• A member is *statically indeterminate* if the equations of equilibrium are not sufficient to determine the reactions on a member.

• *Compatibility conditions* specify the displacement constraints that occur at the supports or other points on a member.

PROCEDURE FOR ANALYSIS

The unknown forces in statically indeterminate problems are determined by satisfying equilibrium, compatibility, and force-displacement requirements for the member.

Equilibrium.

• Draw a free–body diagram of the member in order to identify all the forces that act on it.

• The problem can be classified as statically indeterminate if the number of unknown reactions on the free-body diagram is greater than the number of available equations of equilibrium.

• Write the equations of equilibrium for the member.

Compatibility.

• To write the compatibility equations, consider drawing a displacement diagram in order to investigate the way the member will elongate or contract when subjected to the external loads.

• Express the compatibility conditions in terms of the displacements caused by the forces.

• Use a load–displacement relation, such as $\delta = PL/AE$, to relate the unknown displacements to the unknown reactions.

• Solve the equilibrium and compatibility equations for the unknown reactive forces. If any of the magnitudes has a negative numerical value, it indicates that this force acts in the opposite sense of direction to that indicated on the free-body diagram.

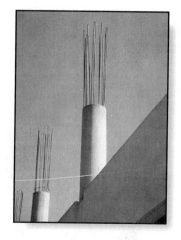

Most concrete columns are reinforced with steel rods; and since these two materials work together in supporting the applied load, the column becomes statically indeterminate.

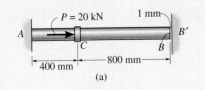

(a)

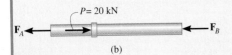

(b)

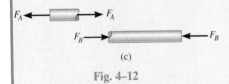

(c)

Fig. 4–12

EXAMPLE 4–5

The steel rod shown in Fig. 4–12a has a diameter of 5 mm. It is attached to the fixed wall at A, and before it is loaded, there is a gap between the wall at B' and the rod of 1 mm. Determine the reactions at A and B' if the rod is subjected to an axial force of P = 20 kN as shown. Neglect the size of the collar at C. Take E_{st} = 200 GPa.

SOLUTION

Equilibrium. As shown on the free-body diagram, Fig. 4–12b, we will assume that the force P is large enough to cause the rod's end B to contact the wall at B'. The problem is statically indeterminate since there are two unknowns and only one equation of equilibrium. Equilibrium of the rod requires

$$\xrightarrow{+} \Sigma F_x = 0; \qquad -F_A - F_B + 20(10^3) \text{ N} = 0 \qquad (1)$$

Compatibility. The loading causes point B to move to B', with no further displacement. Therefore the compatibility condition for the rod is

$$\delta_{B/A} = 0.001 \text{ m}$$

This displacement can be expressed in terms of the unknown reactions by using the load–displacement relationship, Eq. 4–2, applied to segments AC and CB, Fig. 4–12c. Working in units of newtons and meters, we have

$$\delta_{B/A} = 0.001 \text{ m} = \frac{F_A L_{AC}}{AE} - \frac{F_B L_{CB}}{AE}$$

$$0.001 \text{ m} = \frac{F_A(0.4 \text{ m})}{\pi(0.0025 \text{ m})^2[200(10^9) \text{ N/m}^2]}$$

$$- \frac{F_B(0.8 \text{ m})}{\pi(0.0025 \text{ m})^2[200(10^9) \text{ N/m}^2]}$$

or

$$F_A(0.4 \text{ m}) - F_B(0.8 \text{ m}) = 3927.0 \text{ N} \cdot \text{m} \qquad (2)$$

Solving Eqs. 1 and 2 yields

$$F_A = 16.6 \text{ kN} \qquad F_B = 3.39 \text{ kN}$$

Ans.

Since the answer for F_B is *positive*, indeed the end B contacts the wall at B' as originally assumed. On the other hand, if F_B were a negative quantity, the problem would be statically determinate, so that $F_B = 0$ and $F_A = 20$ kN.

EXAMPLE 4-6

The aluminum post shown in Fig. 4–13a is reinforced with a brass core. If this assembly supports a resultant axial compressive load of $P =$ 9 kip, applied to the rigid cap, determine the average normal stress in the aluminum and the brass. Take $E_{al} = 10(10^3)$ ksi and $E_{br} = 15(10^3)$ ksi.

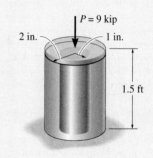

(a)

SOLUTION

Equilibrium. The free-body diagram of the post is shown in Fig. 4–13b. Here the resultant axial force at the base is represented by the unknown components carried by the aluminum, $\mathbf{F}_{al}$, and brass, $\mathbf{F}_{br}$. The problem is statically indeterminate. Why?

Vertical force equilibrium requires

$$+\uparrow \Sigma F_y = 0; \qquad -9 \text{ kip} + F_{al} + F_{br} = 0 \qquad (1)$$

Compatibility. The rigid cap at the top of the post causes both the aluminum and brass to displace the same amount. Therefore,

$$\delta_{al} = \delta_{br}$$

Using the load–displacement relationships,

$$\frac{F_{al}L}{A_{al}E_{al}} = \frac{F_{br}L}{A_{br}E_{br}}$$

$$F_{al} = F_{br}\left(\frac{A_{al}}{A_{br}}\right)\left(\frac{E_{al}}{E_{br}}\right)$$

$$F_{al} = F_{br}\left[\frac{\pi[(2 \text{ in.})^2 - (1 \text{ in.})^2]}{\pi(1 \text{ in.})^2}\right]\left[\frac{10(10^3) \text{ ksi}}{15(10^3) \text{ ksi}}\right]$$

$$F_{al} = 2F_{br} \qquad (2)$$

(b)

Solving Eqs. 1 and 2 simultaneously yields

$$F_{al} = 6 \text{ kip} \qquad F_{br} = 3 \text{ kip}$$

Since the results are positive, indeed the stress will be compressive.
The average normal stress in the aluminum and brass is therefore

$$\sigma_{al} = \frac{6 \text{ kip}}{\pi[(2 \text{ in.})^2 - (1 \text{ in.})^2]} = 0.637 \text{ ksi} \qquad \textit{Ans.}$$

$$\sigma_{br} = \frac{3 \text{ kip}}{\pi(1 \text{ in.})^2} = 0.955 \text{ ksi} \qquad \textit{Ans.}$$

The stress distributions are shown in Fig. 4–13c.

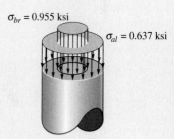

(c)

Fig. 4–13

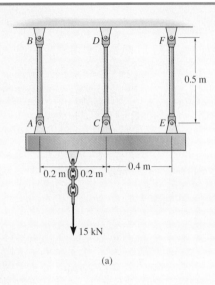

(a)

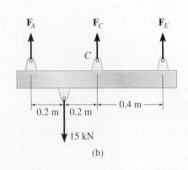

(b)

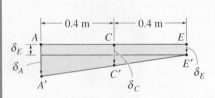

(c)

Fig. 4–14

EXAMPLE 4-7

The three A-36 steel bars shown in Fig. 4–14*a* are pin connected to a *rigid* member. If the applied load on the member is 15 kN, determine the force developed in each bar. Bars *AB* and *EF* each have a cross-sectional area of 25 mm², and bar *CD* has a cross-sectional area of 15 mm².

SOLUTION

Equilibrium. The free-body diagram of the rigid member is shown in Fig. 4–14*b*. This problem is statically indeterminate since there are three unknowns and only two available equilibrium equations. These equations are

$$+\uparrow \Sigma F_y = 0; \qquad F_A + F_C + F_E - 15 \text{ kN} = 0 \qquad (1)$$
$$\zeta + \Sigma M_C = 0; \quad -F_A(0.4 \text{ m}) + 15 \text{ kN}(0.2 \text{ m}) + F_E(0.4 \text{ m}) = 0 \quad (2)$$

Compatibility. The applied load will cause the horizontal line *ACE* shown in Fig. 4–14*c* to move to the inclined line *A′C′E′*. The displacements of points *A*, *C*, and *E* can be related by proportional triangles. Thus the compatibility equation for these displacements is

$$\frac{\delta_A - \delta_E}{0.8 \text{ m}} = \frac{\delta_C - \delta_E}{0.4 \text{ m}}$$

$$\delta_C = \frac{1}{2}\delta_A + \frac{1}{2}\delta_E$$

Using the load–displacement relationship, Eq. 4–2, we have

$$\frac{F_C L}{(15 \text{ mm}^2)E_{st}} = \frac{1}{2}\left[\frac{F_A L}{(25 \text{ mm}^2)E_{st}}\right] + \frac{1}{2}\left[\frac{F_E L}{(25 \text{ mm}^2)E_{st}}\right]$$

$$F_C = 0.3F_A + 0.3F_E \qquad (3)$$

Solving Eqs. 1–3 simultaneously yields

$$F_A = 9.52 \text{ kN} \qquad \qquad \textit{Ans.}$$
$$F_C = 3.46 \text{ kN} \qquad \qquad \textit{Ans.}$$
$$F_E = 2.02 \text{ kN} \qquad \qquad \textit{Ans.}$$

EXAMPLE 4–8

The bolt shown in Fig. 4–15a is made of 2014-T6 aluminum alloy and is tightened so it compresses a cylindrical tube made of Am 1004-T61 magnesium alloy. The tube has an outer radius of $\frac{1}{2}$ in., and it is assumed that both the inner radius of the tube and the radius of the bolt are $\frac{1}{4}$ in. The washers at the top and bottom of the tube are considered to be rigid and have a negligible thickness. Initially the nut is hand-tightened slightly; then, using a wrench, the nut is further tightened one-half turn. If the bolt has 20 threads per inch, determine the stress in the bolt.

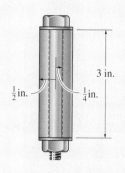

3 in.

$\frac{1}{2}$ in. $\frac{1}{4}$ in.

(a)

SOLUTION

Equilibrium. The free-body diagram of a section of the bolt and the tube, Fig. 4–15b, is considered in order to relate the force in the bolt F_b to that in the tube, F_t. Equilibrium requires

$$+\uparrow \Sigma F_y = 0; \qquad\qquad F_b - F_t = 0 \qquad\qquad (1)$$

The problem is statically indeterminate since there are two unknowns in this equation.

Compatibility. When the nut is tightened on the bolt, the tube will shorten δ_t, and the bolt will *elongate* δ_b, Fig. 4-15c. Since the nut undergoes one-half turn, it advances a distance of $(\frac{1}{2})(\frac{1}{20}$ in.$) = 0.025$ in. along the bolt. Thus, the compatibility of these displacements requires

$$(+\uparrow) \qquad\qquad \delta_t = 0.025 \text{ in.} - \delta_b$$

Taking the modulus of elasticity from the table on the inside back cover, and applying Eq. 4–2, yields

$$\frac{F_t(3 \text{ in.})}{\pi[(0.5 \text{ in.})^2 - (0.25 \text{ in.})^2][6.48(10^3) \text{ ksi}]} = 0.025 \text{ in.} - \frac{F_b(3 \text{ in.})}{\pi(0.25 \text{ in.})^2[10.6(10^3) \text{ ksi}]}$$

$$0.78595F_t = 25 - 1.4414F_b \qquad\qquad (2)$$

Solving Eqs. 1 and 2 simultaneously, we get

$$F_b = F_t = 11.22 \text{ kip}$$

The stresses in the bolt and tube are therefore

$$\sigma_b = \frac{F_b}{A_b} = \frac{11.22 \text{ kip}}{\pi(0.25 \text{ in.})^2} = 57.2 \text{ ksi} \qquad\qquad Ans.$$

$$\sigma_s = \frac{F_t}{A_t} = \frac{11.22 \text{ kip}}{\pi[(0.5 \text{ in.})^2 - (0.25 \text{ in.})^2]} = 19.1 \text{ ksi}$$

These stresses are less than the reported yield stress for each material, $(\sigma_Y)_{al} = 60$ ksi and $(\sigma_Y)_{mg} = 22$ ksi (see the inside back cover), and therefore this "elastic" analysis is valid.

F_t

F_b

(b)

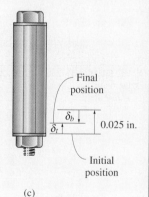

Final position

δ_b

δ_t

0.025 in.

Initial position

(c)

Fig. 4–15

4.5 THE FORCE METHOD OF ANALYSIS FOR AXIALLY LOADED MEMBERS

It is also possible to solve statically indeterminate problems by writing the compatibility equation using the superposition of the forces acting on the free-body diagram. This method of solution is often referred to as the *flexibility* or *force method of analysis*. To show how it is applied, consider again the bar in Fig. 4–11a. In order to write the necessary equation of compatibility, we will first choose any one of the two supports as "redundant" and temporarily remove its effect on the bar. The word *redundant,* as used here, indicates that the support is not needed to hold the bar in stable equilibrium, so that when it is removed, the bar becomes statically determinate. Here we will choose the support at B as redundant. By using the principle of superposition, the bar in Fig. 4–16a, is then equivalent to the bar subjected only to the external load P, Fig. 4–16b, plus the bar subjected only to the unknown redundant load F_B, Fig. 4–16c.

If the load P causes B to be displaced *downward* by an amount δ_P, the reaction F_B must displace the end B of the bar *upward* by an amount δ_B, such that no displacement occurs at B when the two loadings are superimposed. Thus,

$$(+\downarrow) \qquad\qquad 0 = \delta_P - \delta_B$$

This equation represents the compatibility equation for displacements at point B, for which we have assumed that displacements are positive downward.

Applying the load–displacement relationship to each case, we have $\delta_P = PL_{AC}/AE$ and $\delta_B = F_B L/AE$. Consequently,

$$0 = \frac{PL_{AC}}{AE} - \frac{F_B L}{AE}$$

$$F_B = P\left(\frac{L_{AC}}{L}\right)$$

From the free-body diagram of the bar, Fig. 4–11b, the reaction at A can now be determined from the equation of equilibrium,

$$+\uparrow \ \Sigma F_y = 0; \qquad P\left(\frac{L_{AC}}{L}\right) + F_A - P = 0$$

Since $L_{CB} = L - L_{AC}$, then

$$F_A = P\left(\frac{L_{CB}}{L}\right)$$

These results are the same as those obtained in Sec. 4.4, except that here we have applied the condition of compatibility *and then* the equilibrium condition to obtain the solution. Also note that the principle of superposition can be used here since the displacement and the load are linearly related ($\delta = PL/AE$), which assumes, of course, that the material behaves in a linear-elastic manner.

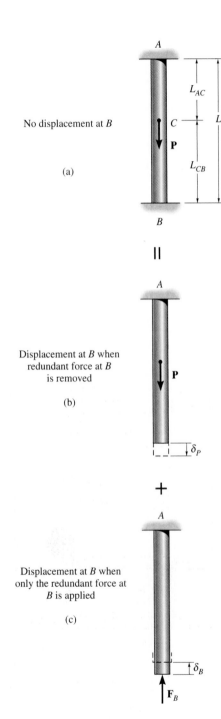

No displacement at B

(a)

$||$

Displacement at B when redundant force at B is removed

(b)

$+$

Displacement at B when only the redundant force at B is applied

(c)

Fig. 4–16

PROCEDURE FOR ANALYSIS

The force method of analysis requires the following steps.

Compatibility.

• Choose one of the supports as redundant and write the equation of compatibility. To do this, the known displacement at the redundant support, which is usually zero, is equated to the displacement at the support caused *only* by the external loads acting on the member *plus* (vectorially) the displacement at the support caused *only* by the redundant reaction acting on the member.

• Express the external load and redundant displacements in terms of the loadings by using a load–displacement relationship, such as $\delta = PL/AE$.

• Once established, the compatibility equation can then be solved for the magnitude of the redundant force.

Equilibrium.

• Draw a free-body diagram and write the appropriate equations of equilibrium for the member using the calculated result for the redundant. Solve these equations for the other reactions.

EXAMPLE 4–9

The A-36 steel rod shown in Fig. 4–17a has a diameter of 5 mm. It is attached to the fixed wall at A, and before it is loaded there is a gap between the wall at B' and the rod of 1 mm. Determine the reactions at A and B'.

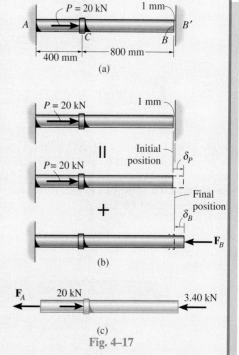

SOLUTION

Compatibility.   Here we will consider the support at B' as redundant. Using the principle of superposition, Fig. 4–17b, we have

$$(\overset{+}{\rightarrow}) \qquad\qquad 0.001 \text{ m} = \delta_P - \delta_B \qquad\qquad (1)$$

The deflections δ_P and δ_B are determined from Eq. 4–2.

$$\delta_P = \frac{PL_{AC}}{AE} = \frac{[20(10^3) \text{ N}](0.4 \text{ m})}{\pi(0.0025 \text{ m})^2[200(10^9) \text{ N/m}^2]} = 0.002037 \text{ m}$$

$$\delta_B = \frac{F_B L_{AB}}{AE} = \frac{F_B(1.20 \text{ m})}{\pi(0.0025 \text{ m})^2[200(10^9) \text{ N/m}^2]} = 0.3056(10^{-6})F_B$$

Substituting into Eq. 1, we get

$$0.001 \text{ m} = 0.002037 \text{ m} - 0.3056(10^{-6})F_B$$

$$F_B = 3.40(10^3) \text{ N} = 3.40 \text{ kN} \qquad\qquad Ans.$$

Equilibrium.   From the free-body diagram, Fig. 4–17c,

$$\overset{+}{\rightarrow} \Sigma F_x = 0; \quad -F_A + 20 \text{ kN} - 3.40 \text{ kN} = 0 \quad F_A = 16.6 \text{ kN} \qquad Ans.$$

Fig. 4–17

PROBLEMS

4–37. The A-36 steel column, having a cross-sectional area of 18 in², is encased in high-strength concrete as shown. If an axial force of 60 kip is applied to the column, determine the average compressive stress in the concrete and in the steel. How far does the column shorten? It has an original length of 8 ft.

4–38. The A-36 steel column is encased in high-strength concrete as shown. If an axial force of 60 kip is applied to the column, determine the required area of the steel so that the force is shared equally between the steel and concrete. How far does the column shorten? It has an original length of 8 ft.

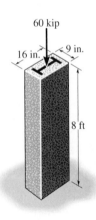

Probs. 4–37/4–38

4–39. The A-36 steel pipe has a 6061-T6 aluminum core. It is subjected to a tensile force of 200 kN. Determine the average normal stress in the aluminum and the steel due to this loading. The pipe has an outer diameter of 80 mm and an inner diameter of 70 mm.

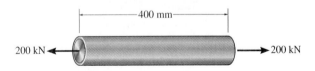

Prob. 4–39

***4–40.** The 304 stainless steel post A has a diameter of $d = 2$ in. and is surrounded by a red brass C83400 tube B. Both rest on the rigid surface. If a force of 5 kip is applied to the rigid cap, determine the average normal stress developed in the post and the tube.

4–41. The 304 stainless steel post A is surrounded by a red brass C83400 tube B. Both rest on the rigid surface. If a force of 5 kip is applied to the rigid cap, determine the required diameter d of the steel post so that the load is shared equally between the post and tube.

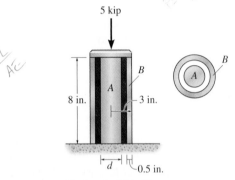

Probs. 4–40/4–41

4–42. The concrete column is reinforced using four steel reinforcing rods, each having a diameter of 18 mm. Determine the average normal stress in the concrete and the steel if the column is subjected to an axial load of 800 kN. $E_{st} = 200$ GPa, $E_c = 25$ GPa.

4–43. The column is constructed from high-strength concrete and four A-36 steel reinforcing rods. If it is subjected to an axial force of 800 kN, determine the required diameter of each rod so that one-fourth of the load is carried by the steel and three-fourths by the concrete.

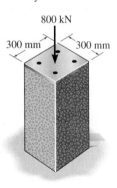

Probs. 4–42/4–43

*4–44. The uniform bar is subjected to the load **P** at collar B. Determine the reactions at the pins A and C. Neglect the size of the collar.

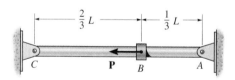

Prob. 4–44

4–45. The two pipes are made of the same material and are connected as shown. If the cross-sectional area of BC is A and that of CD is $2A$, determine the reactions at B and D when a force **P** is applied at the junction C.

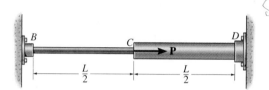

Prob. 4–45

4–46. The bolt AB has a diameter of 20 mm and passes through a sleeve that has an inner diameter of 40 mm and an outer diameter of 50 mm. The bolt and sleeve are made of A-36 steel and are secured to the rigid brackets as shown. If the bolt length is 220 mm and the sleeve length is 200 mm, determine the tension in the bolt when a force of 50 kN is applied to the brackets.

Prob. 4–46

4–47. The load of 1500 lb is to be supported by the two vertical A-36 steel wires. If, originally, wire AB is 50 in. long and wire AC is 50.1 in. long, determine the force developed in each wire after the load is suspended. Each wire has a cross-sectional area of 0.02 in².

*4–48. The load of 15 kip is to be supported by the two vertical A-36 steel wires. If, originally, wire AB is 50 in. long and wire AC is 50.1 in. long, determine the cross-sectional area of AB if the load is to be shared equally between both wires. Wire AC has a cross-sectional area of 0.02 in².

Probs. 4–47/4–48

4–49. The rigid link is supported by a pin at A, a steel wire BC having an unstretched length of 200 mm and cross-sectional area of 22.5 mm², and a short aluminum block having an unloaded length of 50 mm and cross-sectional area of 40 mm². If the link is subjected to the vertical load shown, determine the average normal stress in the wire and the block. $E_{st} = 200$ GPa, $E_{al} = 70$ GPa.

4–50. The rigid link is supported by a pin at A, a steel wire BC having an unstretched length of 200 mm and cross-sectional area of 22.5 mm², and a short aluminum block having an unloaded length of 50 mm and cross-sectional area of 40 mm². If the link is subjected to the vertical load shown, determine the rotation of the link about the pin A. Report the answer in radians. $E_{st} = 200$ GPa, $E_{al} = 70$ GPa.

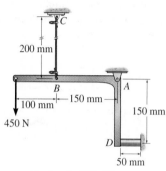

Probs. 4–49/4–50

4–51. The three A-36 steel wires each have a diameter of 2 mm and unloaded lengths of $L_{AC} = 1.60$ m and $L_{AB} = L_{AD} = 2.00$ m. Determine the force in each wire after the 150-kg mass is suspended from the ring at A.

*4–52. The A-36 steel wires AB and AD each have a diameter of 2 mm and the unloaded lengths of each wire are $L_{AC} = 1.60$ m and $L_{AB} = L_{AD} = 2.00$ m. Determine the required diameter of wire AC so that each wire is subjected to the same force caused by the 150-kg mass suspended from the ring at A.

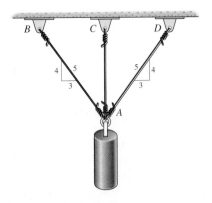

Probs. 4–51/4–52

4–53. The center post B of the assembly has an original length of 124.7 mm, whereas posts A and C have a length of 125 mm. If the caps on the top and bottom can be considered rigid, determine the average normal stress in each post. The posts are made of aluminum and have a cross-sectional area of 400 mm². $E_{al} = 70$ GPa.

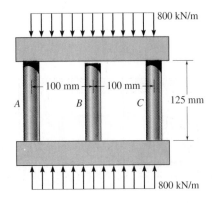

Prob. 4–53

4–54. The assembly consists of an A-36 steel bolt and a C83400 red brass tube. If the nut is drawn up snug against the tube so that $L = 75$ mm, then turned an additional amount so that it advances 0.02 mm on the bolt, determine the force in the bolt and the tube. The bolt has a diameter of 7 mm and the tube has a cross-sectional area of 100 mm².

4–55. The assembly consists of an A-36 steel bolt and a C83400 red brass tube. The nut is drawn up snug against the tube so that $L = 75$ mm. Determine the maximum additional amount of advance of the nut on the bolt so that none of the material will yield. The bolt has a diameter of 7 mm and the tube has a cross-sectional area of 100 mm².

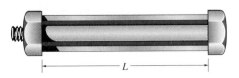

Probs. 4–54/4–55

*4–56. The specimen represents a filament-reinforced matrix system made from plastic (matrix) and glass (fiber). If there are n fibers, each having a cross-sectional area of A_f, embedded in a matrix having a cross-sectional area of A_m, determine the stress in the matrix and each fiber when the force P is imposed on the specimen.

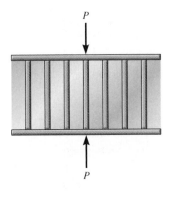

Prob. 4–56

4–57. The three suspender bars are made of the same material and have equal cross-sectional areas A. Determine the average normal stress in each bar if the rigid beam ACE is subjected to the force $\mathbf{P}$.

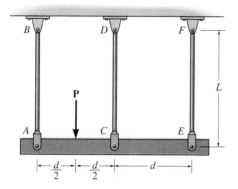

Prob. 4–57

4–58. The bar is pinned at A and supported by two aluminum rods, each having a diameter of 1 in. and a modulus of elasticity $E_{al} = 10(10^3)$ ksi. If the bar is assumed to be rigid and initially vertical, determine the displacement of the end B when the force of 2 kip is applied.

4–59. The bar is pinned at A and supported by two aluminum rods, each having a diameter of 1 in. and a modulus of elasticity $E_{al} = 10(10^3)$ ksi. If the bar is assumed to be rigid and initially vertical, determine the force in each rod when the 2-kip load is applied.

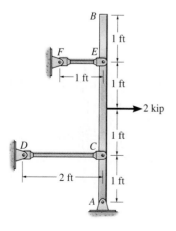

Probs. 4–58/4–59

*4–60. The rigid bar is supported by the two short wooden posts and a spring. If each of the posts has an unloaded length of 500 mm and a cross-sectional area of 800 mm², and the spring has a stiffness of $k = 1.8$ MN/m and an unstretched length of 520 mm, determine the force in each post after the load is applied to the bar. $E_w = 11$ GPa.

4–61. The rigid bar is supported by the two short wooden posts and a spring. If each of the posts has an unloaded length of 500 mm and a cross-sectional area of 800 mm², and the spring has a stiffness of $k = 1.8$ MN/m and an unstretched length of 520 mm, determine the vertical displacement of A and B after the load is applied to the bar. $E_w = 11$ GPa.

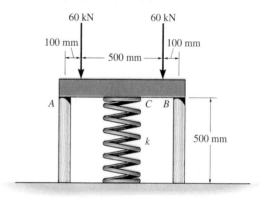

Probs. 4–60/4–61

4–62. The assembly consists of two posts made from material 1 having a modulus of elasticity of E_1 and a cross-sectional area A_1 and a material 2 having a modulus of elasticity E_2 and cross-sectional area A_2. If a central load $\mathbf{P}$ is applied to the rigid cap, determine the force in each post. The support is also rigid.

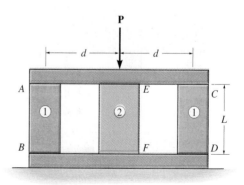

Prob. 4–62

4–63. The assembly consists of two posts AB and CD made from material 1 having a modulus of elasticity of E_1 and a cross-sectional area A_1 and a central post EF made from material 2 having a modulus of elasticity E_2 and a cross-sectional area A_2. If posts AB and CD are to be replaced by those having a material 2, determine the required cross-sectional area of these new posts so that both assemblies deform the same amount when loaded. The support is also rigid.

*4–64. The assembly consists of two posts AB and CD made from material 1 having a modulus of elasticity of E_1 and a cross-sectional area A_1 and a central post EF made from material 2 having a modulus of elasticity E_2 and a cross-sectional area A_2. If post EF is to be replaced by one having a material 1, determine the required cross-sectional area of this new post so that both assemblies deform the same amount when loaded. The support is also rigid.

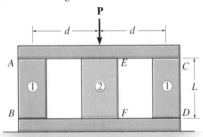

Probs. 4–63/4–64

4–65. The assembly consists of a 6061-T6-aluminum member and a C83400-red-brass member that rest on the rigid plates. Determine the distance d where the vertical load **P** should be placed on the plates so that the plates remain horizontal when the materials deform. Each member has a width of 8 in. and they are not bonded together.

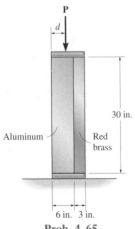

Prob. 4–65

4–66. The bracket is held to the wall using three A-36 steel bolts at B, C, and D. Each bolt has a diameter of 0.5 in. and an unstretched length of 2 in. If a force of 800 lb is placed on the bracket as shown, determine how far, s, the top bracket at bolt D moves away from the wall. For the calculation, assume that the bolts carry no shear; rather, the vertical force of 800 lb is supported by the toe at A. Also, assume that the wall and bracket are rigid. A greatly exaggerated deformation of the bolts is shown.

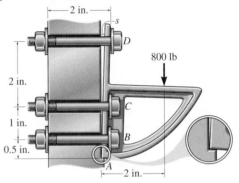

Prob. 4–66

4–67. The post is made of 6061-T6 aluminum and has a diameter of 50 mm. It is fixed supported at A and B, and at its center C, there is a coiled spring attached to the rigid collar which is fixed to the post. If the spring is originally uncompressed, determine the reactions at A and B when the force $P = 40$ kN is applied to the collar.

*4–68. The post is made of 6061-T6 aluminum and has a diameter of 50 mm. It is fixed supported at A and B, and at its center C, there is a coiled spring attached to the rigid collar which is fixed to the post. If the spring is originally uncompressed, determine the compression in the spring when the load of $P = 50$ kN is applied to the collar.

Probs. 4–67/4–68

4–69. The rigid bar is supported by the two short white spruce wooden posts and a spring. If each of the posts has an unloaded length of 1 m and a cross-sectional area of 600 mm², and the spring has a stiffness of $k = 2$ MN/m and an unstretched length of 1.02 m, determine the force in each post after the load is applied to the bar.

4–70. The rigid bar is supported by the two short white spruce wooden posts and a spring. If each of the posts has an unloaded length of 1 m and a cross-sectional area of 600 mm², and the spring has a stiffness of $k = 2$ MN/m and an unstretched length of 1.02 m, determine the vertical displacement of A and B after the load is applied to the bar.

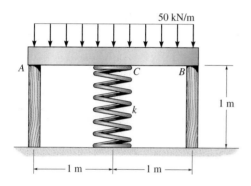

50 kN/m

1 m

1 m   1 m

Probs. 4–69/4–70

4–71. The tapered member is fixed connected at its ends A and B and is subjected to a load $P = 7$ kip at $x = 30$ in. Determine the reactions at the supports. The material is 2 in. thick and is made from 2014-T6 aluminum.

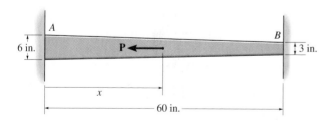

6 in.

P

3 in.

x

60 in.

Prob. 4–71

*4–72. The bolt is made of A-36 steel and the single threaded screw on the bolt has a lead of 0.100 mm. Determine the maximum average normal stress in the bolt if the nut is given one full turn after it first fits snug against the rigid plate. Also, assume the washer, the bolt head, and nut are rigid. *Note:* The lead represents the distance the screw advances along its axis for one complete turn of the screw.

20 mm

30 mm

90 mm

45 mm

Prob. 4–72

4–73. The spring has an unstretched length of 250 mm and stiffness $k = 400$ kN/m. If it is compressed and placed over the 200-mm-long portion AC of the aluminum bar AB and released, determine the force that the bar exerts on the wall at A. Before loading there is a gap of 0.1 mm between the bar and the wall at B. The bar is fixed to the wall at A. Neglect the thickness of the rigid plate at C. $E_{al} = 70$ GPa.

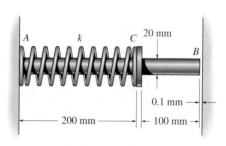

A   k   C   20 mm

B

0.1 mm

200 mm   100 mm

Prob. 4–73

4.6   THERMAL STRESS

A change in temperature can cause a material to change its dimensions. If the temperature increases, generally a material expands, whereas if the temperature decreases, the material will contract. Ordinarily this expansion or contraction is *linearly* related to the temperature increase or decrease that occurs. If this is the case, and the material is homogeneous and isotropic, it has been found from experiment that the deformation of a member having a length L can be calculated using the formula

$$\boxed{\delta_T = \alpha \Delta T L} \tag{4-4}$$

where

Most traffic bridges are designed with expansion joints to accommodate the thermal movement of the deck and thus avoid thermal stress.

α = a property of the material, referred to as the ***linear coefficient of thermal expansion***. The units measure strain per degree of temperature. They are 1/°F (Fahrenheit) in the Foot-Pound-Second or FPS system, and 1/°C (Celsius) or 1/°K (Kelvin) in the SI system. Typical values are given on the inside back cover

ΔT = the algebraic change in temperature of the member

L = the original length of the member

δ_T = the algebraic change in length of the member

If the change in temperature varies throughout the length of the member, i.e., $\Delta T = \Delta T(x)$, or if α varies along the length, then Eq. 4–4 applies for each segment having a length dx. In this case the change in the member's length is

$$\delta_T = \int_0^L \alpha \, \Delta T \, dx \tag{4-5}$$

The change in length of a *statically determinate* member can readily be computed using Eq. 4–4 or 4–5, since the member is free to expand or contract when it undergoes a temperature change. However, in a *statically indeterminate* member, these thermal displacements can be constrained by the supports, producing ***thermal stresses*** that must be considered in design.

Computations of these thermal stresses can be made using the methods outlined in the previous sections. The following examples illustrate some applications.

EXAMPLE 4-10

The A-36 steel bar shown in Fig. 4–18 is constrained to just fit between two fixed supports when $T_1 = 60°F$. If the temperature is raised to $T_2 = 120°F$, determine the average normal thermal stress developed in the bar.

SOLUTION

Equilibrium. The free-body diagram of the bar is shown in Fig. 4–18b. Since there is no external load, the force at A is equal but opposite to the force acting at B; that is,

$$+\uparrow \Sigma F_y = 0; \qquad F_A = F_B = F$$

The problem is statically indeterminate since this force cannot be determined from equilibrium.

Compatibility. Since $\delta_{A/B} = 0$, the thermal displacement δ_T at A that would occur, Fig. 4–18c, is counteracted by the force **F** that would be required to push the bar δ_F back to its original position. The compatibility condition at A becomes

$$(+\uparrow) \qquad \delta_{A/B} = 0 = \delta_T - \delta_F$$

Applying the thermal and load–displacement relationships, we have

$$0 = \alpha \Delta TL - \frac{FL}{AE}$$

Thus, from the data on the inside back cover,

$$\begin{aligned}
F &= \alpha \Delta TAE \\
&= [6.60(10^{-6})/°F](120°F - 60°F)(0.5\ \text{in.})^2[29(10^3)\ \text{kip/in}^2] \\
&= 2.87\ \text{kip}
\end{aligned}$$

From the magnitude of **F**, it should be apparent that changes in temperature can cause large reaction forces in statically indeterminate members.

Since **F** also represents the internal axial force within the bar, the average normal compressive stress is thus

$$\sigma = \frac{F}{A} = \frac{2.87\ \text{kip}}{(0.5\ \text{in.})^2} = 11.5\ \text{ksi} \qquad \qquad Ans.$$

0.5 in.

0.5 in.

A

2 ft

B

(a)

F

F

(b)

δ_T

δ_F

(c)

Fig. 4–18

EXAMPLE 4-11

A 2014-T6 aluminum tube having a cross-sectional area of 600 mm^2 is used as a sleeve for an A-36 steel bolt having a cross-sectional area of 400 mm^2, Fig. 4–19a. When the temperature is $T_1 = 15°C$, the nut holds the assembly in a snug position such that the axial force in the bolt is negligible. If the temperature increases to $T_2 = 80°C$, determine the average normal stress in the bolt and sleeve.

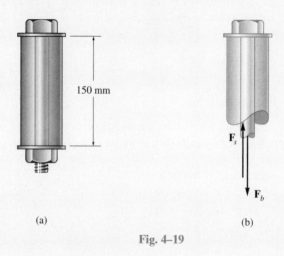

(a)   (b)

Fig. 4–19

SOLUTION

Equilibrium.   A free-body diagram of a sectioned segment of the assembly is shown in Fig. 4–19b. The forces F_b and F_s are produced since the sleeve has a higher coefficient of thermal expansion than the bolt, and therefore the sleeve will expand more when the temperature is increased. The problem is statically indeterminate since these forces cannot be determined from equilibrium. However, it is required that

$$+\uparrow \Sigma F_y = 0; \qquad\qquad F_s = F_b \qquad\qquad (1)$$

Compatibility.   The temperature increase causes the sleeve and bolt to expand $(\delta_s)_T$ and $(\delta_b)_T$, Fig. 4–19c. However, the redundant forces F_b and F_s elongate the bolt and shorten the sleeve. Consequently, the end of the assembly reaches a final position, which is not the same as the initial position. Hence, the compatibility condition becomes

$$(+\downarrow) \qquad\qquad \delta = (\delta_b)_T + (\delta_b)_F = (\delta_s)_T - (\delta_s)_F$$

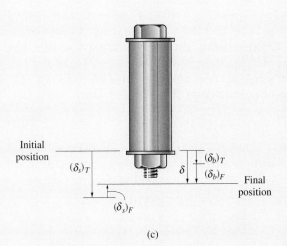

Initial position $(\delta_s)_T$ $(\delta_b)_T$
δ $(\delta_b)_F$ Final position
$(\delta_s)_F$

(c)

Applying Eqs. 4–2 and 4–4, and using the mechanical properties from the table on the inside back cover, we have

$$[12(10^{-6})/°C](80°C - 15°C)(0.150 \text{ m})$$
$$+ \frac{F_b(0.150 \text{ m})}{(400 \text{ mm}^2)(10^{-6} \text{ m}^2/\text{mm}^2)[200(10^9) \text{ N/m}^2]}$$
$$=$$
$$[23(10^{-6})/°C](80°C - 15°C)(0.150 \text{ m})$$
$$- \frac{F_s(0.150 \text{ m})}{600 \text{ mm}^2(10^{-6} \text{ m}^2/\text{mm}^2)[73.1(10^9) \text{ N/m}^2]}$$

Using Eq. 1 and solving gives

$$F_s = F_b = 20.26 \text{ kN}$$

The average normal stress in the bolt and sleeve is therefore

$$\sigma_b = \frac{20.26 \text{ kN}}{400 \text{ mm}^2(10^{-6} \text{ m}^2/\text{mm}^2)} = 50.6 \text{ MPa} \qquad \textit{Ans.}$$

$$\sigma_s = \frac{20.26 \text{ kN}}{600 \text{ mm}^2(10^{-6} \text{ m}^2/\text{mm}^2)} = 33.8 \text{ MPa} \qquad \textit{Ans.}$$

Since linear–elastic material behavior was assumed in this analysis, the calculated stresses should be checked to make sure that they do not exceed the proportional limits for the material.

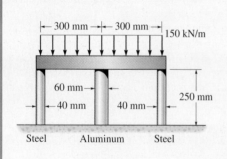

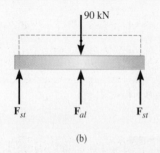

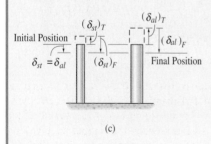

Fig. 4–20

EXAMPLE 4–12

The rigid bar shown in Fig. 4–20a is fixed to the top of the three posts made of A-36 steel and 2014-T6 aluminum. The posts each have a length of 250 mm when no load is applied to the bar, and the temperature is $T_1 = 20°C$. Determine the force supported by each post if the bar is subjected to a uniform distributed load of 150 kN/m and the temperature is raised to $T_2 = 80°C$.

SOLUTION

Equilibrium.   The free-body diagram of the bar is shown in Fig. 4–20b. Moment equilibrium about the bar's center requires the forces in the steel posts to be equal. Summing forces on the free-body diagram, we have

$$+\uparrow \Sigma F_y = 0; \qquad 2F_{st} + F_{al} - 90(10^3)\ \text{N} = 0 \qquad (1)$$

Compatibility.   Due to load, geometry, and material symmetry, the top of each post is displaced by an equal amount. Hence,

$$(+\downarrow) \qquad\qquad \delta_{st} = \delta_{al} \qquad\qquad (2)$$

The final position of the top of each post is equal to its displacement caused by the temperature increase, plus its displacement caused by the internal axial compressive force, Fig. 4–20c. Thus, for a steel and aluminum post, we have

$$(+\downarrow) \qquad\qquad \delta_{st} = -(\delta_{st})_T + (\delta_{st})_F$$
$$(+\downarrow) \qquad\qquad \delta_{al} = -(\delta_{al})_T + (\delta_{al})_F$$

Applying Eq. 2 gives

$$-(\delta_{st})_T + (\delta_{st})_F = -(\delta_{al})_T + (\delta_{al})_F$$

Using Eqs. 4–2 and 4–4 and the material properties on the inside back cover, we get

$$-[12(10^{-6})/°C](80°C - 20°C)(0.250\ \text{m}) + \frac{F_{st}(0.250\ \text{m})}{\pi(0.020\ \text{m})^2[200(10^9)\ \text{N/m}^2]}$$

$$= -[23(10^{-6})/°C](80°C - 20°C)(0.250\ \text{m}) + \frac{F_{al}(0.250\ \text{m})}{\pi(0.03\ \text{m})^2[73.1(10^9)\ \text{N/m}^2]}$$

$$F_{st} = 1.216F_{al} - 165.9(10^3) \qquad (3)$$

To be *consistent*, all numerical data has been expressed in terms of newtons, meters, and degrees Celsius. Solving Eqs. 1 and 3 simultaneously yields

$$F_{st} = -16.4\ \text{kN} \qquad F_{al} = 123\ \text{kN} \qquad\qquad Ans.$$

The negative value for F_{st} indicates that this force acts opposite to that shown in Fig. 4–20b. In other words, the steel posts are in tension and the aluminum post is in compression.

PROBLEMS

4–74. Three bars each made of different materials are connected together and placed between two walls when the temperature is $T_1 = 12°C$. Determine the force exerted on the (rigid) supports when the temperature becomes $T_2 = 18°C$. The material properties and cross-sectional area of each bar are given in the figure.

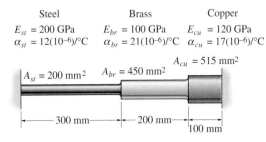

Steel	Brass	Copper
$E_{st} = 200$ GPa	$E_{br} = 100$ GPa	$E_{cu} = 120$ GPa
$\alpha_{st} = 12(10^{-6})/°C$	$\alpha_{br} = 21(10^{-6})/°C$	$\alpha_{cu} = 17(10^{-6})/°C$

$A_{cu} = 515$ mm^2

$A_{st} = 200$ mm^2 $A_{br} = 450$ mm^2

— 300 mm — — 200 mm — 100 mm

Prob. 4–74

4–75. A 6-ft-long steam pipe is made of A-36 steel and is connected directly to two turbines A and B as shown. The pipe has an outer diameter of 4 in. and a wall thickness of 0.25 in. The connection was made at $T_1 = 70°F$. If the turbines' points of attachment are assumed rigid, determine the force the pipe exerts on the turbines when the steam and thus the pipe reach a temperature of $T_2 = 275°F$.

*4–76. A 6-ft-long steam pipe is made of A-36 steel and is connected directly to two turbines A and B as shown. The pipe has an outer diameter of 4 in. and a wall thickness of 0.25 in. The connection was made at $T_1 = 70°F$. If the turbines' points of attachment are assumed to have a stiffness of $k = 80(10^3)$ kip/in., determine the force the pipe exerts on the turbines when the steam and thus the pipe reach a temperature of $T_2 = 275°F$.

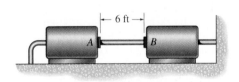

— 6 ft —

A B

Probs. 4–75/4–76

4–77. The 40-ft-long A-36 steel rails on a train track are laid with a small gap between them to allow for thermal expansion. Determine the required gap δ so that the rails just touch one another when the temperature is increased from $T_1 = -20°F$ to $T_2 = 90°F$. Using this gap, what would be the axial force in the rails if the temperature were to rise to $T_3 = 110°F$? The cross-sectional area of each rail is 5.10 in^2.

δ δ

— 40 ft —

Prob. 4–77

4–78. The 0.4-in.-diameter A-36 steel bolt is used to hold the (rigid) assembly together. Determine the clamping force that must be provided by the bolt when $T_1 = 90°F$ so that the clamping force it exerts when $T_2 = 175°F$ is 500 lb.

4–79. The 0.4-in.-diameter A-36 steel bolt is used to hold the (rigid) assembly together. If the nut is snug (no axial force in the bolt) when $T = 90°F$, determine the clamping force it exerts on the assembly when $T = 20°F$.

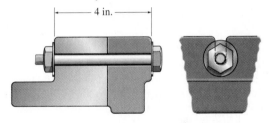

— 4 in. —

Probs. 4–78/4–79

*4–80. A thermo gate consists of a 6061-T6-aluminum plate AB and an Am-1004-T61-magnesium plate CD, each having a width of 15 mm and fixed supported at their ends. If the gap between them is 1.5 mm when the temperature is $T_1 = 25°C$, determine the temperature required to just close the gap. Also, what is the axial force in each plate if the temperature becomes $T_2 = 100°C$? Assume bending or buckling will not occur.

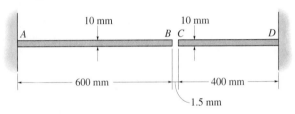

	10 mm			10 mm	
A			B C		D

— 600 mm — — 400 mm —

1.5 mm

Prob. 4–80

4–81. The A-36 steel pipe having a cross-sectional area of 0.5 in^2 is connected to fixed supports and carries a liquid that causes the pipe to be subjected to a temperature drop of $\Delta T = (-0.2x^{3/2})$ °F, where x is in inches. Determine the axial force developed in the pipe.

4–82. The A-36 steel pipe having a cross-sectional area of 0.5 in^2 is connected to fixed supports and carries a liquid that causes the pipe to be subjected to a temperature drop of $\Delta T = (-0.2x^{1/2})$ °F, where x is in inches. Determine the maximum and minimum normal strain.

4–83. The A-36 steel pipe having a cross-sectional area of 0.5 in^2 is connected to fixed supports and carries a liquid that causes the pipe to be subjected to a temperature variation of $\Delta T = (20 - x)$ °F, where x is in inches. Determine the average normal stress developed in the pipe.

Probs. 4–81/4–82/4–83

*4–84. The C83400-red-brass rod AB and 2014-T6-aluminum rod BC are joined at the collar B and fixed connected at their ends. If there is no load in the members when $T_1 = 50$°F, determine the average normal stress in each member when $T_2 = 120$°F. Also, how far will the collar be displaced? The cross-sectional area of each member is 1.75 in^2.

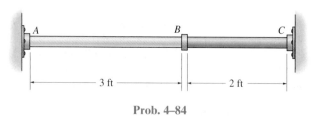

Prob. 4–84

4–85. The two circular rod segments, one of aluminum and the other of copper, are fixed to the rigid walls such that there is a gap of 0.008 in. between them when $T_1 = 60$°F. What larger temperature T_2 is required in order to just close the gap? Each rod has a diameter of 1.25 in., $\alpha_{al} = 13(10^{-6})$/°F, $E_{al} = 10(10^3)$ ksi, $\alpha_{cu} = 9.4(10^{-6})$/°F, $E_{cu} = 18(10^3)$ ksi. Determine the average normal stress in each rod if $T_2 = 200$°F.

4–86. The two circular rod segments, one of aluminum and the other of copper, are fixed to the rigid walls such that there is a gap of 0.008 in. between them when $T_1 = 60$°F. Each rod has a diameter of 1.25 in., $\alpha_{al} = 13(10^{-6})$/°F, $E_{al} = 10(10^3)$ ksi, $\alpha_{cu} = 9.4(10^{-6})$/°F, $E_{cu} = 18(10^3)$ ksi. Determine the average normal stress in each rod if $T_2 = 300$°F, and also calculate the new length of the aluminum segment.

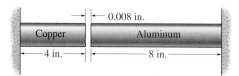

Probs. 4–85/4–86

4–87. The pipe is made of A-36 steel and is connected to the collars at A and B. When the temperature is 60°F, there is no axial load in the pipe. If hot gas traveling through the pipe causes its temperature to rise by $\Delta T = (40 + 15x)$°F, where x is in feet, determine the average normal stress in the pipe. The inner diameter is 2 in., the wall thickness is 0.15 in.

*4–88. The bronze 86100 pipe has an inner radius of 0.5 in. and a wall thickness of 0.2 in. If the gas flowing through it changes the temperature of the pipe uniformly from $T_A = 200$°F at A to $T_B = 60$°F at B, determine the axial force it exerts on the walls. The pipe was fitted between the walls when $T = 60$°F.

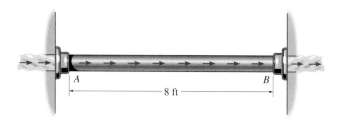

Probs. 4–87/4–88

4–89. The rigid block has a weight of 80 kip and is to be supported by posts A and B, which are made of A-36 steel, and the post C, which is made of C83400 red brass. If all the posts have the same original length before they are loaded, determine the average normal stress developed in each post when post C is heated so that its temperature is increased by 20°F. Each post has a cross-sectional area of 8 in^2.

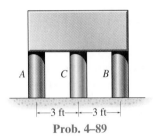

Prob. 4–89

4–90. The wires AB and AC are made of steel, and wire AD is made of copper. Before the 150-lb force is applied, AB and AC are each 60 in. long and AD is 40 in. long. If the temperature is increased by 80°F, determine the force in each wire needed to support the load. Take $E_{st} = 29(10^3)$ ksi, $E_{cu} = 17(10^3)$ ksi, $\alpha_{st} = 8(10^{-6})/°F$, $\alpha_{cu} = 9.60(10^{-6})/°F$. Each wire has a cross-sectional area of 0.0123 in².

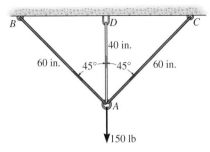

Prob. 4–90

4–91. The steel bolt has a diameter of 7 mm and fits through an aluminum sleeve as shown. The sleeve has an inner diameter of 8 mm and an outer diameter of 10 mm. The nut at A is adjusted so that it just presses up against the sleeve. If the assembly is originally at a temperature of $T_1 = 20°C$ and then is heated to a temperature of $T_2 = 100°C$, determine the average normal stress in the bolt and the sleeve. $E_{st} = 200$ GPa, $E_{al} = 70$ GPa, $\alpha_{st} = 14(10^{-6})/°C$, $\alpha_{al} = 23(10^{-6})/°C$.

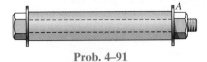

Prob. 4–91

*4–92. The aluminum 2014-T6 pipe CD is placed within the clamp and the screws on the clamp are tightened snug. If the assembly experiences a temperature increase of $\Delta T = 50°C$, determine the average normal stress developed within the pipe and screw. Assume the heads on the clamp are rigid and the screws are made of A-36 steel. The screws have a diameter of 14 mm, and the pipe has an outer diameter of 35 mm and a wall thickness of 2 mm.

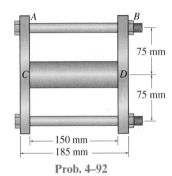

Prob. 4–92

4–93. The 50-mm-diameter cylinder is made from Am 1004-T61 magnesium and is placed in the clamp when the temperature is $T_1 = 15°C$. If the two 304-stainless-steel carriage bolts of the clamp each have a diameter of 10 mm, and they hold the cylinder snug with negligible force against the rigid jaws, determine the temperature at which the average normal stress in either the magnesium or steel becomes 12 MPa.

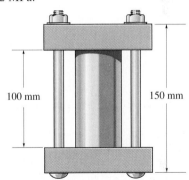

Prob. 4–93

4–94. The bar has a cross-sectional area A, length L, modulus of elasticity E, and coefficient of thermal expansion α. The temperature of the bar changes uniformly along its length from an original temperature of T_A at A to T_B at B so that at any point x along the bar $T = T_A + x(T_B - T_A)/L$. Determine the force the bar exerts on the rigid walls. Initially no axial force is in the bar.

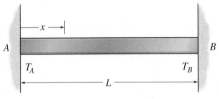

Prob. 4–94

4.7 STRESS CONCENTRATIONS

In Sec. 4.1 it was pointed out that when an axial force is applied to a member, it creates a complex stress distribution within a localized region of the point of load application. Such typical stress distributions are shown in Fig. 4–1. Not only do complex stress distributions arise just under a concentrated loading, they also arise at sections where the member's cross-sectional area changes. For example, consider the bar in Fig. 4–21a, which is subjected to an axial force P. Here it can be seen that the once horizontal and vertical grid lines deflect into an irregular pattern around the hole centered in the bar. The maximum normal stress in the bar occurs on section a–a, which is taken through the bar's *smallest* cross-sectional area. Provided the material behaves in a linear-elastic manner, the stress distribution acting on this section can be determined either from a mathematical analysis, using the theory of elasticity, or experimentally by measuring the strain normal to section a–a and then calculating the stress using Hooke's law, $\sigma = E\epsilon$. Regardless of the method used, the general shape of the stress distribution will be like that shown in Fig. 4–21b. In a similar manner, if the bar has a reduction in its cross section, achieved using shoulder fillets as in Fig. 4–22a, then again the maximum normal stress in the bar will occur at the *smallest* cross-sectional area, section a–a, and the stress distribution will be like that shown in Fig. 4–22b.

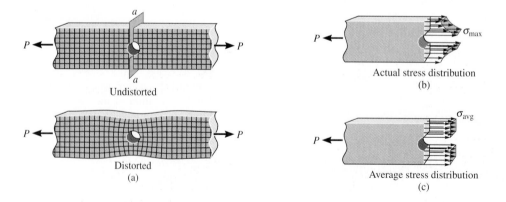

Undistorted

Actual stress distribution
(b)

Distorted
(a)

Average stress distribution
(c)

Fig. 4–21

In both of these cases, *force equilibrium* requires the magnitude of the *resultant force* developed by the stress distribution to be equal to P. In other words,

$$P = \int_A \sigma \, dA \qquad (4\text{--}6)$$

As stated in Sec. 1.4, this integral *graphically* represents the *volume* under each of the stress-distribution diagrams shown in Fig. 4–21*b* or 4–22*b*. Furthermore, moment equilibrium requires each stress distribution to be symmetrical over the cross section, so that **P** must pass through the *centroid* of each *volume*.

In engineering practice, though, the actual stress distribution does *not* have to be determined. Instead, only the *maximum stress* at these sections must be known, and the member is then designed to resist this stress when the axial load **P** is applied. In cases where a member's cross-sectional area changes, such as those discussed above, specific values of the maximum normal stress at the critical section can be determined by experimental methods or by advanced mathematical techniques using the theory of elasticity. The results of these investigations are usually reported in graphical form using a **stress-concentration factor** K. We define K as a ratio of the maximum stress to the average stress acting at the smallest cross section; i.e.,

$$K = \frac{\sigma_{max}}{\sigma_{avg}} \qquad (4\text{--}7)$$

Provided K is known, and the average normal stress has been calculated from $\sigma_{avg} = P/A$, where A is the *smallest* cross-sectional area, Figs. 4–21*c* and 4–22*c*, then from the above equation the maximum stress at the cross section is $\sigma_{max} = K(P/A)$.

Stress concentrations often arise at sharp corners on heavy machinery. Engineers can mitigate this effect by using stiffeners welded to the corners.

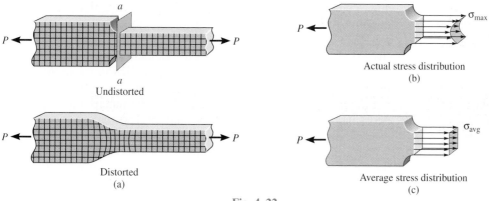

Undistorted

Distorted
(a)

Actual stress distribution
(b)

Average stress distribution
(c)

Fig. 4–22

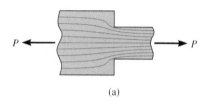

(a)

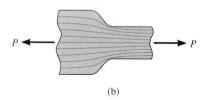

(b)

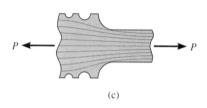

(c)

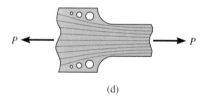

(d)

Fig. 4–23

Cracking of the concrete has occurred at all the corners of this slab due to shrinkage of the concrete while it cured. These stress concentrations can be avoided by making the hole circular.

Specific values of K are generally reported in graphical form in handbooks related to stress analysis. Examples of these graphs are given in Figs. 4–24 and 4–25, respectively.* In particular, note that K is independent of the bar's material properties; rather, it *depends* only on the bar's *geometry* and the type of discontinuity. As the size r of the discontinuity is *decreased*, the stress concentration is increased. For example, if a bar requires a change in cross section, theoretically it has been determined that a sharp corner, Fig. 4–23a, produces a stress-concentration factor greater than 3. In other words, the maximum normal stress will be three times greater than the average normal stress on the smallest cross section. However, this can be reduced to, say, 1.5 by introducing a fillet, Fig. 4–23b. A further reduction can be made by means of small grooves or holes placed at the transition, Fig. 4–23c and 4–23d. In all of these cases the designs help to reduce the rigidity of the material surrounding the corners, so that both the strain and the stress are more evenly spread throughout the bar.

The stress-concentration factors given in Figs. 4–24 and 4–25 were determined on the basis of a static loading, with the assumption that the stress in the material does not exceed the proportional limit. If the material is *very brittle*, the proportional limit may be at the rupture stress, and so for this material, failure will begin *at* the point of stress concentration when the proportional limit is reached. Essentially a crack begins to form at this point, and a higher stress concentration will develop at the *tip* of this crack. This, in turn, causes the crack to progress over the cross section, resulting in sudden fracture. For this reason, it is very important to use stress-concentration factors in design when using brittle materials. On the other hand, if the material is ductile and subjected to a static load, designers usually neglect using stress-concentration factors since any stress that exceeds the proportional limit will not result in a crack. Instead, the material will have reserve strength due to yielding and strain-hardening. In the next section we will discuss the effects caused by this phenomenon.

Stress concentrations are also responsible for many failures of structural members or mechanical elements subjected to *fatigue loadings*. For these cases, a stress concentration will cause the material to crack if the stress exceeds the material's endurance limit, whether or not the material is ductile or brittle. Here, the material *localized* at the tip of the crack remains in a *brittle state*, and so the crack continues to grow, leading to a progressive fracture. Consequently, engineers involved in the design of such members must seek ways to limit the amount of damage that can be caused by fatigue.

---

*See Lipson, C. and Juvinall, R. C., *Handbook of Stress and Strength,* Macmillan, 1963.

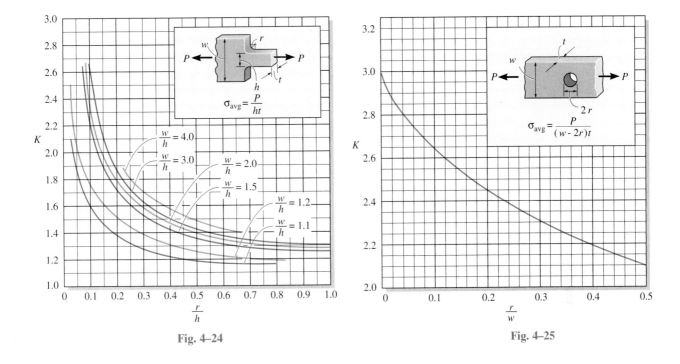

Fig. 4–24

Fig. 4–25

IMPORTANT POINTS

• *Stress concentrations* occur at sections where the cross-sectional area suddenly changes. The more severe the change, the larger the stress concentration.

• For design or analysis, it is only necessary to determine the maximum stress acting on the smallest cross-sectional area. This is done using a *stress concentration factor, K,* that has been determined through experiment and is only a function of the geometry of the specimen.

• Normally the stress concentration in a ductile specimen that is subjected to a static loading will *not* have to be considered in design; however, if the material is *brittle,* or subjected to *fatigue* loadings, then stress concentrations become important.

EXAMPLE 4–13

A steel bar has the dimensions shown in Fig. 4–26. If the allowable stress is $\sigma_{allow} = 16.2$ ksi, determine the largest axial force P that the bar can carry.

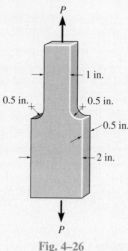

0.5 in.

1 in.

0.5 in.

0.5 in.

2 in.

Fig. 4–26

SOLUTION

Because there is a shoulder fillet, the stress-concentration factor can be determined using the graph in Fig. 4–24. Calculating the necessary geometric parameters yields

$$\frac{r}{h} = \frac{0.5 \text{ in.}}{1 \text{ in.}} = 0.50$$

$$\frac{w}{h} = \frac{2 \text{ in.}}{1 \text{ in.}} = 2$$

Thus, from the graph,

$$K = 1.4$$

Computing the average normal stress at the *smallest* cross section, we have

$$\sigma_{avg} = \frac{P}{(1 \text{ in.})(0.5 \text{ in.})} = 2P$$

Applying Eq. 4–7 with $\sigma_{allow} = \sigma_{max}$ yields

$$\sigma_{allow} = K\sigma_{avg}$$
$$16.2 \text{ ksi} = 1.4(2P)$$
$$P = 5.79 \text{ kip}$$

Ans.

EXAMPLE 4-14

The steel strap shown in Fig. 4–27 is subjected to an axial load of 80 kN. Determine the maximum normal stress developed in the strap and the displacement of one end of the strap with respect to the other end. The steel has a yield stress of $\sigma_Y = 700$ MPa, and $E_{st} = 200$ GPa.

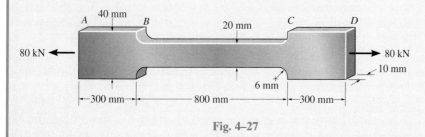

Fig. 4–27

SOLUTION

Maximum Normal Stress.   By inspection, the maximum normal stress occurs at the smaller cross section, where the shoulder fillet begins at B or C. The stress-concentration factor is determined from Fig. 4–23. We require

$$\frac{r}{h} = \frac{6 \text{ mm}}{20 \text{ mm}} = 0.3; \qquad \frac{w}{h} = \frac{40 \text{ mm}}{20 \text{ mm}} = 2$$

Thus, $K = 1.6$.

The maximum stress is therefore

$$\sigma_{\max} = K\frac{P}{A} = 1.6\left[\frac{80(10^3) \text{ N}}{(0.02 \text{ m})(0.01 \text{ m})}\right] = 640 \text{ MPa} \qquad \textit{Ans.}$$

Notice that the material remains elastic, since 640 MPa $< \sigma_Y = 700$ MPa.

Displacement.   Here we will neglect the localized deformations surrounding the applied load and at the sudden change in cross section of the shoulder fillet (Saint-Venant's principle). We have

$$\delta_{A/D} = \sum\frac{PL}{AE} = 2\left\{\frac{80(10^3) \text{ N}(0.3 \text{ m})}{(0.04 \text{ m})(0.01 \text{ m})[200(10^9) \text{ N/m}^2]}\right\}$$

$$+ \left\{\frac{80(10^3) \text{ N}(0.8 \text{ m})}{(0.02 \text{ m})(0.01 \text{ m})[200(10^9) \text{ N/m}^2]}\right\}$$

$$\delta_{A/D} = 2.20 \text{ mm} \qquad \textit{Ans.}$$

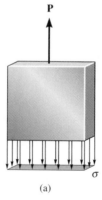

Failure of this steel pipe in tension occurred at its smallest cross-sectional area, which is through the hole. Notice how the material yielded around the fractured surface before it fractured.

*4.8   INELASTIC AXIAL DEFORMATION

Up to this point we have only considered loadings that cause the material of a member to behave elastically. Sometimes, however, a member may be designed so that the loading causes the material to yield and thereby permanently deform. Such members are often made from a highly ductile metal such as annealed low-carbon steel having a stress–strain diagram that is similar to that of Fig. 3–6 and can be *modeled* as shown in Fig. 4–28b. A material that exhibits this behavior is referred to as being *elastic perfectly plastic or elastoplastic*.

To illustrate physically how such a material behaves, consider the bar in Fig. 4–28a, which is subjected to the axial load **P**. If the load causes an *elastic stress* $\sigma = \sigma_1$ to be developed in the bar, then applying Eq. 4–6, *equilibrium* requires $P = \int \sigma_1 \, dA = \sigma_1 A$. Furthermore, the stress σ_1 causes the bar to strain ϵ_1 as indicated on the stress–strain diagram, Fig. 4–28b. If P is now increased to P_p such that it causes yielding of the material, that is, $\sigma = \sigma_Y$, then again $P_p = \int \sigma_Y \, dA = \sigma_Y A$. The load P_p is called the *plastic load* since it represents the maximum load that can be supported by an elastoplastic material. For this case, the strains are *not* uniquely defined. Instead, at the instant σ_Y is attained, the bar is *first* subjected to the yield strain ϵ_Y, Fig. 4–28b, after which the bar will *continue to yield* (or elongate) such that the strains ϵ_2, then ϵ_3, etc., are generated. Since our "model" of the material exhibits perfectly plastic material behavior, this elongation will continue indefinitely with no increase in load. In reality, however, the material will, after some yielding, actually begin to strain-harden so that the extra strength it attains will *stop* any further straining. As a result, any design based on this behavior will be safe, since strain-hardening provides the potential for the material to support an *additional* load if necessary.

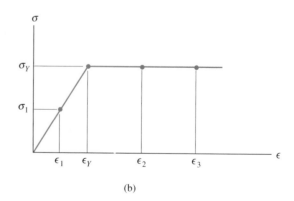

(a)   (b)

Fig. 4–28

Consider now the case of a bar having a hole through it as shown in Fig. 4–29a. As the magnitude of **P** is increased, a stress concentration occurs in the material near the hole, along section a–a. The stress here will reach a maximum value of, say, $\sigma_{max} = \sigma_1$ and have a corresponding *elastic strain* of ϵ_1, Fig. 4–29b. The stresses and corresponding strains at other points along the cross section will be smaller, as indicated by the stress distribution shown in Fig. 4–29c. As expected, equilibrium requires $P = \int \sigma \, dA$. In other words, P is geometrically equivalent to the "volume" contained within the stress distribution. If the load is now increased to P', so that $\sigma_{max} = \sigma_Y$, then the material will begin to yield outward from the hole, until the equilibrium condition $P' = \int \sigma \, dA$ is satisfied, Fig. 4–29d. As shown, this produces a stress distribution that has a geometrically *greater* "volume" than that shown in Fig. 4–29c. An even further increase in load will cause the material over the *entire cross section* to yield eventually, until *no greater load* can be sustained by the bar. This *plastic load* P_p is shown in Fig. 4–29e. It can be calculated from the equilibrium condition

$$P_p = \int_A \sigma_Y \, dA = \sigma_Y A \qquad (4\text{–}7)$$

Here σ_Y is the yield stress and A is the bar's cross-sectional area at section a–a.

The following examples illustrate numerically how these concepts apply to other types of problems for which the material has elastoplastic behavior.

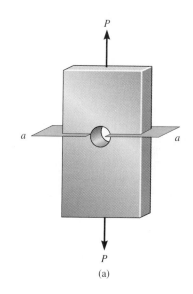

(a)

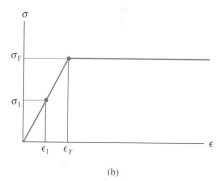

(b)

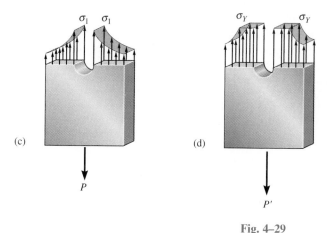

(c)

(d)

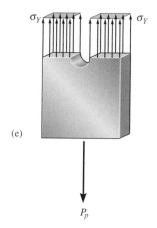

(e)

Fig. 4–29

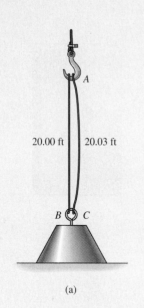

20.00 ft   20.03 ft

(a)

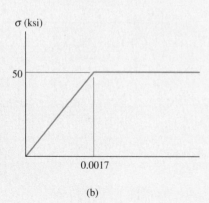

(b)

Fig. 4–30

EXAMPLE 4–15

Two steel wires are used to lift the weight of 3 kip, Fig. 4–30a. Wire AB has an unstretched length of 20.00 ft and wire AC has an unstretched length of 20.03 ft. If each wire has a cross-sectional area of 0.05 in^2, and the steel can be considered elastic perfectly plastic as shown by the σ–ϵ graph in Fig. 4–30b, determine the force in each wire and its elongation.

SOLUTION

By inspection, wire AB begins to carry the weight when the hook is lifted. However, if this wire stretches more than 0.03 ft, the load is then carried by both wires. For this to occur, the strain in wire AB must be

$$\epsilon_{AB} = \frac{0.03 \text{ ft}}{20 \text{ ft}} = 0.0015$$

which is less than the maximum elastic strain, $\epsilon_Y = 0.0017$, Fig. 4–30b. Furthermore, the stress in wire AB when this happens can be determined from Fig. 4–30b by proportion; i.e.,

$$\frac{0.0017}{50 \text{ ksi}} = \frac{0.0015}{\sigma_{AB}}$$

$$\sigma_{AB} = 44.12 \text{ ksi}$$

As a result, the force in the wire is thus

$$F_{AB} = \sigma_{AB} A = (44.12 \text{ ksi})(0.05 \text{ in}^2) = 2.21 \text{ kip}$$

Since the weight to be supported is 3 kip, we can conclude that both wires must be used for support.

Once the weight is supported, the stress in the wires depends on the corresponding strain. There are three possibilities, namely, the strains in both wires are elastic, wire AB is plastically strained while wire AC is elastically strained, or both wires are plastically strained. We will begin by assuming that both wires remain *elastic*. Investigation of the free-body diagram of the suspended weight, Fig. 4–30c, indicates

$\mathbf{T}_{AB}$ $\mathbf{T}_{AC}$

3 kip   (c)

that the problem is statically indeterminate. The equation of equilibrium is

$$+\uparrow \Sigma F_y = 0; \qquad T_{AB} + T_{AC} - 3 \text{ kip} = 0 \qquad (1)$$

Since AC is 0.03 ft longer than AB, then from Fig. 4–30d, compatibility of displacement of the ends B and C requires that

$$\delta_{AB} = 0.03 \text{ ft} + \delta_{AC} \qquad (2)$$

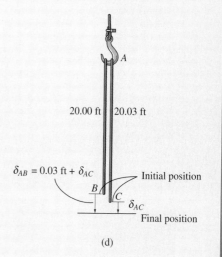

The modulus of elasticity, Fig. 4–30b, is $E_{st} = 50\text{ksi}/0.0017 = 29.4(10^3)$ ksi. Since this is a linear–elastic analysis, the load–displacement relationship is $\delta = PL/AE$, and therefore

$$\frac{T_{AB}(20.00 \text{ ft})(12 \text{ in./ft})}{(0.05 \text{ in}^2)[29.4(10^3) \text{ ksi}]} = 0.03 \text{ ft}(12 \text{ in./ft}) + \frac{T_{AC}(20.03 \text{ ft})(12 \text{ in./ft})}{(0.05 \text{ in}^2)[29.4(10^3) \text{ ksi}]}$$

$$20.00 T_{AB} = 44.11 + 20.03 T_{AC} \qquad (3)$$

Solving Eqs. 1 and 3, we have

$$T_{AB} = 2.60 \text{ kip}$$
$$T_{AC} = 0.400 \text{ kip}$$

The stress in wire AB is thus

$$\sigma_{AB} = \frac{2.60 \text{ kip}}{0.05 \text{ in}^2} = 52.0 \text{ ksi}$$

This stress is greater than the maximum elastic stress allowed ($\sigma_Y = 50$ ksi), and therefore wire AB becomes plastically strained and supports its maximum load of

$$T_{AB} = 50 \text{ ksi } (0.05 \text{ in}^2) = 2.50 \text{ kip} \qquad \textit{Ans.}$$

From Eq. 1,

$$T_{AC} = 0.500 \text{ kip} \qquad \textit{Ans.}$$

Note that wire AC remains elastic since the stress in the wire is $\sigma_{AC} = 0.500 \text{ kip}/0.05 \text{ in}^2 = 10 \text{ ksi} < 50$ ksi. The corresponding elastic strain is determined by proportion, Fig. 4–30b; i.e.,

$$\frac{\epsilon_{AC}}{10 \text{ ksi}} = \frac{0.0017}{50 \text{ ksi}}$$

$$\epsilon_{AC} = 0.000340$$

The elongation of AC is thus

$$\delta_{AC} = (0.000340)(20.03 \text{ ft}) = 0.00681 \text{ ft} \qquad \textit{Ans.}$$

Applying Eq. 2, the elongation of AB is then

$$\delta_{AB} = 0.03 \text{ ft} + 0.00681 \text{ ft} = 0.0368 \text{ ft} \qquad \textit{Ans.}$$

EXAMPLE 4–16

The bar in Fig. 4–31a is made of steel that is assumed to be elastic perfectly plastic, with $\sigma_Y = 250$ MPa. Determine (a) the maximum value of the applied load P that can be applied without causing the steel to yield and (b) the maximum value of P that the bar can support. Sketch the stress distribution at the critical section for each case.

SOLUTION

Part (a). *When the material behaves* elastically, we must use a stress-concentration factor determined from Fig. 4–23 that is unique for the bar's geometry. Here

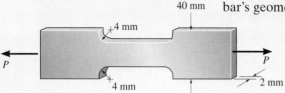

$$\frac{r}{h} = \frac{4 \text{ mm}}{(40 \text{ mm} - 8 \text{ mm})} = 0.125$$

$$\frac{w}{h} = \frac{40 \text{ mm}}{(40 \text{ mm} - 8 \text{ mm})} = 1.25$$

The maximum load, without causing yielding, occurs when $\sigma_{\max} = \sigma_Y$. The average normal stress is $\sigma_{\text{avg}} = P/A$. Using Eq. 4–7, we have

$$\sigma_{\max} = K\sigma_{\text{avg}}; \qquad \sigma_Y = K\left(\frac{P_Y}{A}\right)$$

$$250(10^6) \text{ Pa} = 1.75\left[\frac{P_Y}{(0.002 \text{ m})(0.032 \text{ m})}\right]$$

$$P_Y = 9.14 \text{ kN} \qquad \qquad Ans.$$

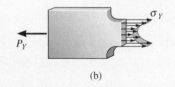

(b)

This load has been calculated using the *smallest* cross section. The resulting stress distribution is shown in Fig. 4–31b. For equilibrium, the "volume" contained within this distribution must equal 9.14 kN.

Part (b). The maximum load sustained by the bar causes *all the material* at the smallest cross section to yield. Therefore, as P is increased to the *plastic load* P_p, it gradually changes the stress distribution from the elastic state shown in Fig. 4–31b to the plastic state shown in Fig. 4–31c. We require

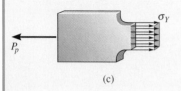

(c)

Fig. 4–31

$$\sigma_Y = \frac{P_p}{A}$$

$$250(10^6) \text{ Pa} = \frac{P_p}{(0.002 \text{ m})(0.032 \text{ m})}$$

$$P_p = 16.0 \text{ kN} \qquad \qquad Ans.$$

Here P_p equals the "volume" contained within the stress distribution, which in this case is $P_p = \sigma_Y A$.

*4.9 RESIDUAL STRESS

If an axially loaded member or group of such members forms a statically indeterminate system that can support both tensile and compressive loads, then excessive external loadings, which cause yielding of the material, will create *residual stresses* in the members when the loads are removed. The reason for this has to do with the elastic recovery of the material that occurs during unloading. For example, consider a prismatic member made from an elastoplastic material having the stress–strain diagram OAB, as shown in Fig. 4–32. If an axial load produces a stress σ_Y in the material and a corresponding plastic strain ϵ_C, then when the load is *removed*, the material will respond elastically and follow the line CD in order to recover some of the plastic strain. A full recovery to zero stress at point O' will be possible only if the member is statically determinate, since the support reactions for the member must be zero when the load is removed. Under these circumstances the member will be permanently deformed so that the permanent set or strain in the member is $\epsilon_{O'}$. If the member is *statically indeterminate*, however, removal of the external load will cause the support forces to respond to the elastic recovery CD. Since these forces will constrain the member from full recovery, they will induce **residual stresses** in the member.

To solve a problem of this kind, the complete cycle of loading and then unloading of the member can be considered as the *superposition* of a positive load (loading) on a negative load (unloading). The loading, O to C, results in a plastic stress distribution, whereas the unloading, along CD, results only in an elastic stress distribution. Superposition requires the loads to cancel; however, the stress distributions will not cancel, and so residual stresses will remain.

The following example illustrates these concepts numerically.

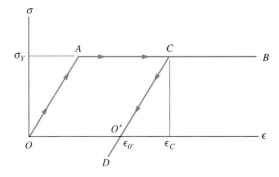

Fig. 4–32

EXAMPLE 4-17

The rod shown in Fig. 4–33a has a radius of 5 mm and is made from an elastic-perfectly plastic material for which $\sigma_Y = 420$ MPa, $E = 70$ GPa, Fig. 4–33b. If a force of $P = 60$ kN is applied to the rod and then removed, determine the residual stress in the rod and the permanent displacement of the collar at C.

SOLUTION

The free-body diagram of the rod is shown in Fig. 4–33b. By inspection, the rod is statically indeterminate. Application of the load **P** will cause one of three possibilities, namely, both segments AC and CB remain elastic, AC is plastic while CB is elastic, or both AC and CB are plastic.*

An *elastic analysis*, similar to that discussed in Sec. 4.4, will produce $F_A = 45$ kN and $F_B = 15$ kN at the supports. However, this results in a stress of

$$\sigma_{AC} = \frac{45 \text{ kN}}{\pi(0.005 \text{ m})^2} = 573 \text{ MPa (compression)} > \sigma_Y = 420 \text{ MPa}$$

in segment AC, and

$$\sigma_{CB} = \frac{15 \text{ kN}}{\pi(0.005 \text{ m})^2} = 191 \text{ MPa (tension)}$$

in segment CB. Since the material in segment AC will yield, we will assume that AC becomes plastic, while CB remains elastic.

For this case, the maximum possible force developed in AC is

$$(F_A)_Y = \sigma_Y A = 420(10^3) \text{ kN/m}^2[\pi(0.005 \text{ m})^2]$$
$$= 33.0 \text{ kN}$$

and from the equilibrium of the rod, Fig. 4–33b,

$$F_B = 60 \text{ kN} - 33.0 \text{ kN} = 27.0 \text{ kN}$$

The stress in each segment of the rod is therefore

$$\sigma_{AC} = \sigma_Y = 420 \text{ MPa (compression)}$$

$$\sigma_{CB} = \frac{27.0 \text{ kN}}{\pi(0.005 \text{ m})^2} = 344 \text{ MPa (tension)} < 420 \text{ MPa (OK)}$$

Residual Stress. In order to obtain the residual stress, it is also necessary to know the strain in each segment due to the loading. Since CB responds elastically,

$$\delta_C = \frac{F_B L_{CB}}{AE} = \frac{(27.0 \text{ kN})(0.300 \text{ m})}{\pi(0.005 \text{ m})^2[70(10^6) \text{ kN/m}^2]} = 0.001474 \text{ m}$$

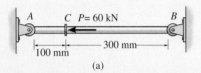

$P = 60$ kN

100 mm　　300 mm

(a)

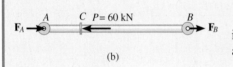

F_A　A　C　$P = 60$ kN　B　F_B

(b)

Fig. 4–33

Thus,

$$\epsilon_{CB} = \frac{\delta_C}{L_{CB}} = \frac{0.001474 \text{ m}}{0.300 \text{ m}} = +0.004913$$

Also, since δ_C is known, the strain in AC is

$$\epsilon_{AC} = \frac{\delta_C}{L_{AC}} = -\frac{0.001474 \text{ m}}{0.100 \text{ m}} = -0.01474$$

Therefore, when **P** is *applied*, the stress–strain behavior for the material in segment CB moves from O to A', Fig. 4–33c, and the stress–strain behavior for the material in segment AC moves from O to B'. If the load **P** is applied in the reverse direction, in other words, the load is removed, then an elastic response occurs and a reverse force of $F_A = 45$ kN and $F_B = 15$ kN must be applied to each segment. As calculated previously, these forces produce stresses $\sigma_{AC} = 573$ MPa (tension) and $\sigma_{CB} = 191$ MPa (compression), and as a result the residual stress in each member is

$$(\sigma_{AC})_r = -420 \text{ MPa} + 573 \text{ MPa} = 153 \text{ MPa} \qquad \textit{Ans.}$$
$$(\sigma_{CB})_r = 344 \text{ MPa} - 191 \text{ MPa} = 153 \text{ MPa} \qquad \textit{Ans.}$$

This tensile stress is the *same* for both segments, which is to be expected. Also note that the stress–strain behavior for segment AC moves from B' to D' in Fig. 4–33c, while the stress–strain behavior for the material in segment CB moves from A' to C'.

Permanent Displacement. From Fig. 4–33c, the residual strain in CB is

$$\epsilon'_{CB} = \frac{\sigma}{E} = \frac{153(10^6) \text{ Pa}}{70(10^9) \text{ Pa}} = 0.002185$$

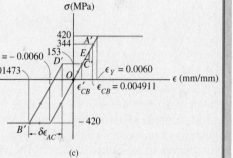

(c)

so that the permanent displacement of C is

$$\delta_C = \epsilon'_{CB} L_{CB} = 0.002185(300 \text{ mm}) = 0.656 \text{ mm} \leftarrow \qquad \textit{Ans.}$$

We can also obtain this result by determining the residual strain ϵ'_{AC} in AC, Fig. 4–33c. Since line $B'D'$ has a slope of E, then

$$\delta\epsilon_{AC} = \frac{\delta\sigma}{E} = \frac{(420 + 153)10^6 \text{ Pa}}{70(10^9) \text{ Pa}} = 0.008185$$

Therefore

$$\epsilon'_{AC} = \epsilon_{AC} + \delta\epsilon_{AC} = -0.01474 + 0.008185 = -0.006555$$

Finally,

$$\delta_C = \epsilon'_{AC} L_{AC} = -0.006555(100 \text{ mm}) = 0.656 \text{ mm} \leftarrow \qquad \textit{Ans.}$$

---

*The possibility of CB becoming plastic before AC will not occur because when point C deforms, the *strain* in AC (since it is shorter) will always be larger than the strain in CB.

PROBLEMS

4–95. If the allowable normal stress for the bar is $\sigma_{allow} = 120$ MPa, determine the maximum axial force P that can be applied to the bar.

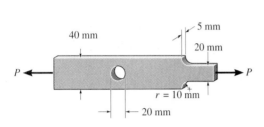

Prob. 4–95

*4–96. The member is to be made from a steel plate that is 0.25 in. thick. If a 1-in. hole is drilled through its center, determine the approximate width w of the plate so that it can support an axial force of 3350 lb. The allowable stress is $\sigma_{allow} = 22$ ksi.

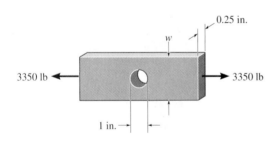

Prob. 4–96

4–97. Determine the maximum normal stress developed in the bar when it is subjected to a tension of $P = 2$ kip.

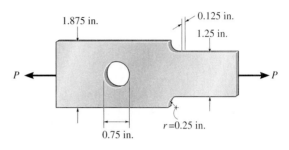

Prob. 4–97

4–98. Determine the maximum normal stress developed in the bar when it is subjected to a tension of $P = 8$ kN.

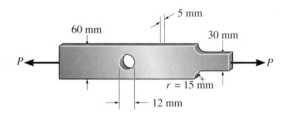

Prob. 4–98

4–99. If the allowable normal stress for the bar is $\sigma_{allow} = 120$ MPa, determine the maximum axial force P that can be applied to the bar.

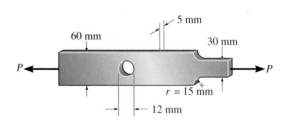

Prob. 4–99

*4–100.   The resulting stress distribution along the section *AB* of the bar is shown in the figure. From this distribution, determine the approximate resultant axial force *P* applied to the bar. Also, what is the stress-concentration factor for this geometry?

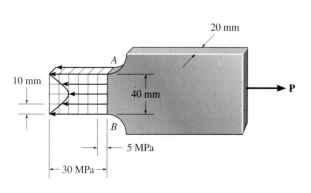

Prob. 4–100

4–101.   The A-36 steel plate has a thickness of 12 mm. If there are shoulder fillets at *B* and *C*, and $\sigma_{allow} = 150$ MPa, determine the maximum axial load *P* that it can support. Compute its elongation neglecting the effect of the fillets.

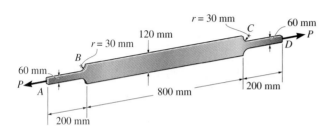

Prob. 4–101

4–102.   The resulting stress distribution along section *AB* for the bar is shown. From this distribution, determine the approximate resultant axial force *P* applied to the bar. Also, what is the stress-concentration factor for this geometry?

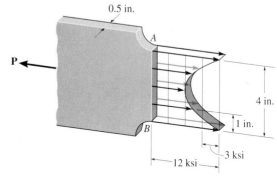

Prob. 4–102

4–103.   The 300-kip weight is slowly set on the top of a post made of 2014-T6 aluminum with an A-36 steel core. If both materials can be considered elastic perfectly plastic, determine the stress in each material.

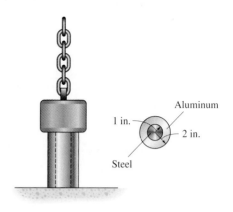

Prob. 4–103

*4–104. The weight is suspended from steel and aluminum wires, each having the same initial length of 3 m and cross-sectional area of 4 mm². If the materials can be assumed to be elastic perfectly plastic, with $(\sigma_Y)_{st}$ = 120 MPa and $(\sigma_Y)_{al}$ = 70 MPa, determine the force in each wire if the weight is (a) 600 N and (b) 720 N. E_{al} = 70 GPa, E_{st} = 200 GPa.

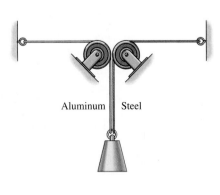

Aluminum Steel

Prob. 4–104

4–105. The bar has a cross-sectional area of 0.5 in² and is made of a material that has a stress–strain diagram that can be approximated by the two line segments shown. Determine the elongation of the bar due to the applied loading.

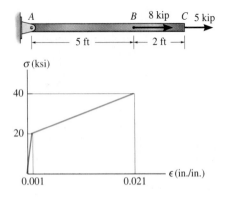

A   B 8 kip C 5 kip

|← 5 ft →|← 2 ft →|

σ (ksi)

40

20

0.001   0.021 ϵ (in./in.)

Prob. 4–105

4–106. The wire BC has a diameter of 0.125 in. and the material has the stress–strain characteristics shown in the figure. Determine the vertical displacement of the handle at D if the pull at the grip is slowly increased and reaches a magnitude of (a) P = 450 lb, (b) P = 600 lb.

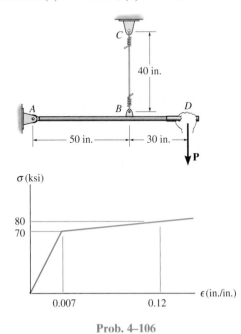

C

40 in.

A   B   D

|← 50 in. →|← 30 in. →|

$\downarrow$ **P**

σ (ksi)

80
70

0.007   0.12 ϵ (in./in.)

Prob. 4–106

4–107. The rigid beam is supported by the three rods A, B, and C of equal length. Rods A and C have a diameter of 25 mm and are made of aluminum, for which E_{al} = 70 GPa and $(\sigma_Y)_{al}$ = 20 MPa. Rod B has a diameter of 10 mm and is made of brass, for which E_{br} = 100 GPa and $(\sigma_Y)_{br}$ = 590 MPa. Determine the smallest magnitude of **P** so that (a) only rods A and C yield and (b) all the rods yield.

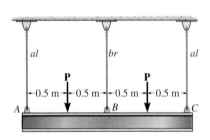

al   br   al

|←0.5 m→|←0.5 m→|←0.5 m→|←0.5 m→|

A **P** B **P** C

Prob. 4–107

*4–108.   The rigid beam is supported by the three posts A, B, and C of equal length. Posts A and C have a diameter of 75 mm and are made of aluminum, for which $E_{al} = 70$ GPa and $(\sigma_Y)_{al} = 20$ MPa. Post B has a diameter of 20 mm and is made of brass, for which $E_{br} = 100$ GPa and $(\sigma_Y)_{br} = 590$ MPa. Determine the smallest magnitude of **P** so that (a) only posts A and C yield and (b) all the posts yield.

4–110.   The rigid bar is supported by a pin at A and two steel wires, each having a diameter of 4 mm. If the yield stress for the wires is $\sigma_Y = 530$ MPa, and $E_{st} = 200$ GPa, determine the intensity of the distributed load w that can be placed on the beam and will just cause wire EB to yield. What is the displacement of point G for this case? For the calculation, assume that the steel is elastic perfectly plastic.

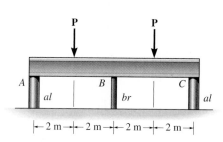

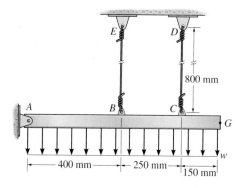

Prob. 4–108

Prob. 4–110

4–109.   A material has a stress–strain diagram that can be described by the curve $\sigma = c\epsilon^{1/2}$. Determine the deflection δ of the end of a rod made from this material if it has a length L, cross-sectional area A, and a specific weight γ.

4–111.   The rigid bar is supported by a pin at A and two steel wires, each having a diameter of 4 mm. If the yield stress for the wires is $\sigma_Y = 530$ MPa, and $E_{st} = 200$ GPa, determine a) the intensity of the distributed load w that can be placed on the beam that will cause only one of the wires to start to yield and b) the smallest intensity of the distributed load that will cause both wires to yield. For the calculation, assume that the steel is elastic perfectly plastic.

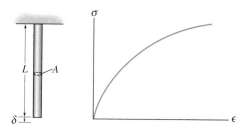

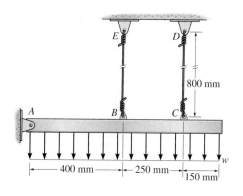

Prob. 4–109

Prob. 4–111

REVIEW PROBLEMS

*4–112. A 0.25-in.-diameter steel rivet having a temperature of 1500°F is secured between two plates such that at this temperature it is 2 in. long and exerts a clamping force of 250 lb between the plates. Determine the approximate clamping force between the plates when the rivet cools to 70°F. For the calculation, assume that the heads of the rivet and the plates are rigid. Take $\alpha_{st} = 8(10^{-6})/°F$, $E_{st} = 29(10^3)$ ksi. Is the result a conservative estimate of the actual answer? Why or why not?

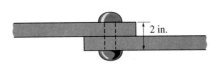

2 in.

Prob. 4–112

4–113. The 50-mm-diameter cylinder is made of Am 1004-T61 magnesium and is placed in the clamp when the temperature is $T_1 = 15°C$. The two 304-stainless-steel carriage bolts of the clamp each have a diameter of 10 mm, and they hold the cylinder snug with negligible force against the rigid jaws. Determine the temperature at which the average normal stress in either the magnesium or steel does not exceed $\sigma_{allow} = 12$ MPa.

100 mm   150 mm

Prob. 4–113

4–114. The assembly consists of a 2014-T6-aluminum cylinder having an outer diameter of 200 mm and inner diameter of 150 mm, together with a concentric solid inner cylinder of Am 1004-T61 magnesium, having a diameter of 125 mm. If the clamping force in the bolts AB and CD is 4 kN when the temperature is $T_1 = 16°C$, determine the clamping force in the bolts when the temperature becomes $T_2 = 48°C$. Assume the bolts and the restraining bars are rigid.

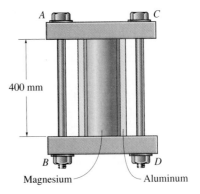

A   C

400 mm

B   D
Magnesium   Aluminum

Prob. 4–114

4–115. The assembly consists of two bars AB and CD of the same material having a modulus of elasticity E_1 and coefficient of thermal expansion α_1, and a bar EF having a modulus of elasticity E_2 and coefficient of thermal expansion α_2. All the bars have the same length L and cross-sectional area A. If the rigid beam is originally horizontal at temperature T_1, determine the angle it makes with the horizontal when the temperature is increased to T_2.

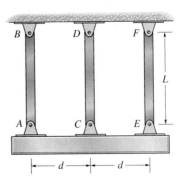

B   D   F

L

A   C   E

d   d

Prob. 4–115

*4–116. The composite shaft, consisting of aluminum, copper, and steel sections, is subjected to the loading shown. Determine the displacement of end A with respect to end D and the average normal stress in each section. The cross-sectional area and modulus of elasticity for each section are shown. Neglect the size of the collars at B and C.

4–117. Determine the displacement of B with respect to C of the composite shaft in Prob. 4–116.

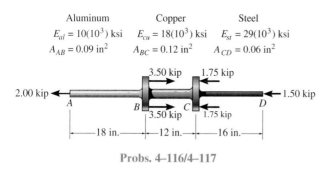

Probs. 4–116/4–117

4–118. The assembly consists of a 30-mm-diameter aluminum bar ABC with fixed collar at B and a 10-mm-diameter steel rod CD. Determine the displacement of point D when the assembly is loaded as shown. Neglect the size of the collar at B and the connection at C. $E_{st} = 200$ GPa, $E_{al} = 70$ GPa.

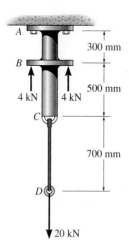

Prob. 4–118

4–119. The brass plug is force-fitted into the rigid casting. The uniform normal bearing pressure on the plug is estimated to be 15 MPa. If the coefficient of static friction between the plug and casting is $\mu_s = 0.3$, determine the axial force P needed to pull the plug out. Also, calculate the displacement of end B relative to end A just before the plug starts to slip out. $E_{br} = 98$ GPa.

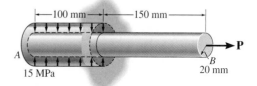

Prob. 4–119

*4–120. The assembly consists of two A-36 steel suspender rods AC and BD attached to the 100-lb uniform rigid beam AB. Determine the position x for the 300-lb loading so that the beam remains in a horizontal position both before and after the load is applied. Each rod has a diameter of 0.5 in.

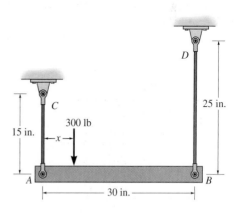

Prob. 4–120

The torsional stress developed within the drive shaft of this condensation fan depends upon the output of the motor.

5 TORSION

CHAPTER OBJECTIVES

In this chapter we will discuss the effects of applying a torsional loading to a long straight member such as a shaft or tube. Initially we will consider the member to have a circular cross section. We will show how to determine both the stress distribution within the member and the angle of twist when the material behaves in a linear-elastic manner and also when it is inelastic. Statically indeterminate analysis of shafts and tubes will also be discussed, along with special topics that include those members having noncircular cross sections. Lastly, stress concentrations and residual stress caused by torsional loadings will be given special consideration.

5.1 TORSIONAL DEFORMATION OF A CIRCULAR SHAFT

Torque is a moment that tends to twist a member about its longitudinal axis. Its effect is of primary concern in the design of axles or drive shafts used in vehicles and machinery. We can illustrate physically what happens when a torque is applied to a circular shaft by considering the shaft to be made of a highly deformable material such as rubber, Fig. 5–1a. When the torque is applied, the circles and longitudinal grid lines originally marked on the shaft tend to distort into the pattern shown in Fig. 5–1b. By inspection, twisting causes the circles to *remain circles*, and each longitudinal grid line deforms into a helix that intersects the circles at equal angles. Also, the cross sections at the *ends* of the shaft remain *flat*—that is, they do not warp or bulge in or out—and radial lines on these ends *remain straight* during the deformation, Fig. 5–1b. From these observations we can assume that if the angle of rotation is *small*, the *length of the shaft* and its *radius* will *remain unchanged*.

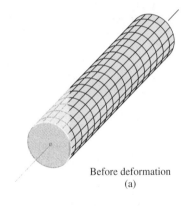

Before deformation
(a)

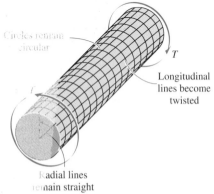

Circles remain
circular

T

Longitudinal
lines become
twisted

Radial lines
remain straight

After deformation
(b)

Fig. 5–1

*Notice the deformation of the rectangular element
when this rubber bar is subjected to a torque.*

If the shaft is fixed at one end and a torque is applied to its other end, the shaded plane in Fig. 5–2 will distort into a skewed form as shown. Here a radial line located on the cross section at a distance x from the fixed end of the shaft will rotate through an angle $\phi(x)$. The angle $\phi(x)$, so defined, is called the *angle of twist*. It depends on the position x and will vary along the shaft as shown.

In order to understand how this distortion strains the material, we will now isolate a small element located at a radial distance ρ (rho) from the axis of the shaft, Fig. 5–3. Due to the deformation as noted in Fig. 5–2, the front and rear faces of the element will undergo a rotation. The back face by $\phi(x)$, and the front face by $\phi(x) + \Delta\phi$. As a result, the *difference* in these rotations, $\Delta\phi$, causes the element to be subjected to a *shear strain*. To calculate this strain, note that before deformation the angle between the edges AB and AC is 90°; after deformation, however, the edges of the element are AD and AC and the angle between them is θ'. From the definition of shear strain, Eq. 2–4, we have

$$\gamma = \frac{\pi}{2} - \lim_{\substack{C \to A \text{ along } CA \\ B \to A \text{ along } BA}} \theta'$$

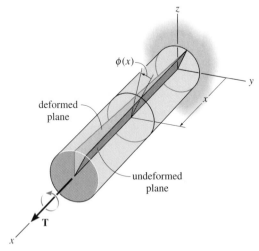

The angle of twist $\phi(x)$ increases as x increases.

Fig. 5–2

This angle, γ, is indicated on the element. It can be related to the length Δx of the element and the difference in the angle of rotation, $\Delta\phi$, between the shaded faces. If we let $\Delta x \rightarrow dx$ and $\Delta\phi \rightarrow d\phi$, we have

$$BD = \rho\, d\phi = dx\, \gamma$$

Therefore,

$$\gamma = \rho \frac{d\phi}{dx} \qquad (5\text{–}1)$$

Since dx and $d\phi$ are the *same* for *all elements* located at points on the cross section at x, then $d\phi/dx$ is constant, and Eq. 5–1 states that the magnitude of the shear strain for any of these elements varies only with its radial distance ρ from the axis of the shaft. In other words, the shear strain within the shaft varies linearly along any radial line, from zero at the axis of the shaft to a maximum γ_{max} at its outer boundary, Fig. 5–4. Since $d\phi/dx = \gamma/\rho = \gamma_{max}/c$, then

$$\gamma = \left(\frac{\rho}{c}\right)\gamma_{max} \qquad (5\text{–}2)$$

The results obtained here are also valid for circular tubes. They depend only on the assumptions regarding the deformations mentioned above.

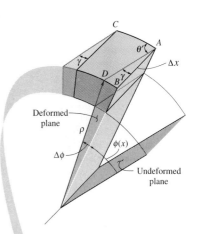

Deformed plane

Undeformed plane

Shear strain of element

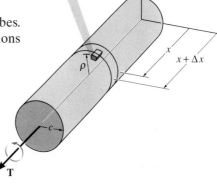

Fig. 5–3

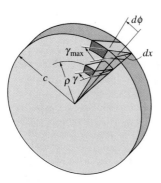

The shear strain for the material increases linearly with ρ, i.e., $\gamma = (\rho/c)\gamma_{max}$

Fig. 5–4

5.2   THE TORSION FORMULA

When an external torque is applied to a shaft it creates a corresponding internal torque within the shaft. In this section, we will develop an equation that relates this internal torque to the shear stress distribution on the cross section of a circular shaft or tube.

If the material is linear-elastic, then Hooke's law applies, $\tau = G\,\gamma$, and consequently a ***linear variation in shear strain***, as noted in the previous section, leads to a corresponding ***linear variation in shear stress*** along any radial line on the cross section. Hence, like the shear-strain variation, for a solid shaft, τ will vary from zero at the shaft's longitudinal axis to a maximum value, τ_{max}, at its outer surface. This variation is shown in Fig. 5–5 on the front faces of a selected number of elements, located at an intermediate radial position ρ and at the outer radius c. Due to the proportionality of triangles, or by using Hooke's law ($\tau = G\,\gamma$) and Eq. 5–2 [$\gamma = (\rho/c)\gamma_{max}$], we can write

$$\tau = \left(\frac{\rho}{c}\right)\tau_{max} \qquad (5\text{–}3)$$

This equation expresses the shear-stress distribution as a *function* of the radial position ρ of the element; in other words, it defines the stress distribution over the cross section in terms of the geometry of the shaft. Using it, we will now apply the condition that requires the torque produced by the stress distribution over the entire cross section to be equivalent to the resultant internal torque T at the section, which holds

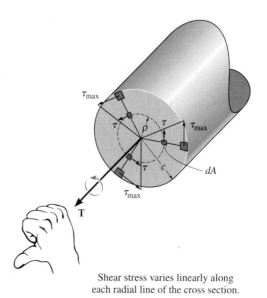

Shear stress varies linearly along each radial line of the cross section.

Fig. 5–5

the shaft in equilibrium, Fig. 5–5. Specifically, each element of area dA, located at ρ, is subjected to a force of $dF = \tau \, dA$. The torque produced by this force is $dT = \rho(\tau \, dA)$. We therefore have for the entire cross section

$$T = \int_A \rho(\tau \, dA) = \int_A \rho \left(\frac{\rho}{c} \right) \tau_{max} \, dA \qquad (5\text{–}4)$$

Since τ_{max}/c is constant,

$$T = \frac{\tau_{max}}{c} \int_A \rho^2 \, dA \qquad (5\text{–}5)$$

The integral in this equation depends only on the geometry of the shaft. It represents the **polar moment of inertia** of the shaft's cross-sectional area computed about the shaft's longitudinal axis. We will symbolize its value as J, and therefore the above equation can be written in a more compact form, namely,

$$\boxed{\tau_{max} = \frac{Tc}{J}} \qquad (5\text{–}6)$$

where

τ_{max} = the maximum shear stress in the shaft, which occurs at the outer surface

T = the resultant internal torque acting at the cross section. Its value is determined from the method of sections and the equation of moment equilibrium applied about the shaft's longitudinal axis

J = the polar moment of inertia of the cross-sectional area

c = the outer radius of the shaft

Using Eqs. 5–3 and 5–6, the shear stress at the intermediate distance ρ can be determined from a similar equation:

$$\boxed{\tau = \frac{T\rho}{J}} \qquad (5\text{–}7)$$

Either of the above two equations is often referred to as the **torsion formula**. Recall that it is used only if the shaft is circular and the material is homogeneous and behaves in a linear-elastic manner, since the derivation is based on the fact that the shear stress is proportional to the shear strain.

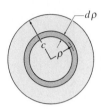

Fig. 5–6

Solid Shaft.   If the shaft has a solid circular cross section, the polar moment of inertia J can be determined using an area element in the form of a *differential ring* or annulus having a thickness $d\rho$ and circumference $2\pi\rho$, Fig. 5–6. For this ring, $dA = 2\pi\rho\,d\rho$, so

$$J = \int_A \rho^2\,dA = \int_0^c \rho^2\,(2\pi\rho\,d\rho) = 2\pi \int_0^c \rho^3\,d\rho = 2\pi\left(\frac{1}{4}\right)\rho^4\,\bigg|_0^c$$

$$\boxed{J = \frac{\pi}{2}c^4} \tag{5–8}$$

Note that J is a *geometric property* of the circular area and is always positive. Common units used for its measurement are mm^4 or in^4.

The shear stress has been shown to vary linearly along each radial line of the cross section of the shaft. However, if a volume element of material on the cross section is isolated, then due to the complementary property of shear, equal shear stresses must also act on four of its adjacent faces as shown in Fig. 5–7a. Hence, *not only does the internal torque T develop a linear distribution of shear stress along each radial line in the plane of the cross-sectional area, but also an associated shear-stress distribution is developed along an axial plane*, Fig. 5–7b. It is interesting to note that because of this axial distribution of shear stress, shafts made from wood tend to *split* along the axial plane when subjected to excessive torque, Fig. 5–8. This is because wood is an anisotropic material. Its shear resistance parallel to its grains or fibers, directed along the axis of the shaft, is much less than its resistance perpendicular to the fibers, directed in the plane of the cross section.

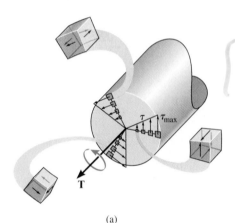

(a)

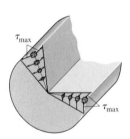

Shear stress varies linearly along each radial line of the cross section.

(b)

Fig. 5–7

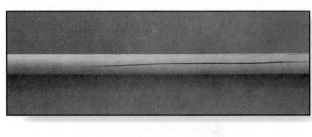

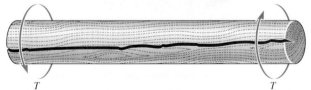

Failure of a wooden shaft due to torsion.

Fig. 5–8

Tubular Shaft.   If a shaft has a tubular cross section, with inner radius c_i and outer radius c_o, then from Eq. 5–8 we can determine its polar moment of inertia by subtracting J for a shaft of radius c_i from that determined for a shaft of radius c_o. The result is

$$J = \frac{\pi}{2}(c_o^4 - c_i^4) \qquad (5\text{–}9)$$

This tubular drive shaft for a truck was subjected to an overload resulting in failure caused by yielding of the material.

Like the solid shaft, the shear stress distributed over the tube's cross-sectional area varies linearly along any radial line, Fig. 5–9a. Furthermore, the shear stress varies along an axial plane in this same manner, Fig. 5–9b. Examples of the shear stress acting on typical volume elements are shown in Fig. 5–9a.

Absolute Maximum Torsional Stress.   At any given cross section of the shaft the maximum shear stress occurs at the outer surface. However, if the shaft is subjected to a series of external torques, or the radius (polar moment of inertia) changes, then the maximum torsional stress within the shaft could be different from one section to the next. If the absolute maximum torsional stress is to be determined, then it becomes important to find the location where the ratio Tc/J is a maximum. In this regard, it may be helpful to show the variation of the internal torque T at each section along the axis of the shaft by drawing a *torque diagram*. Specifically, this diagram is a plot of the internal torque T versus its position x along the shaft's length. As a sign convention, T will be positive if by the right-hand rule the thumb is directed outward from the shaft when the fingers curl in the direction of twist as caused by the torque, Fig. 5–5. Once the internal torque throughout the shaft is determined, the maximum ratio of Tc/J can then be identified.

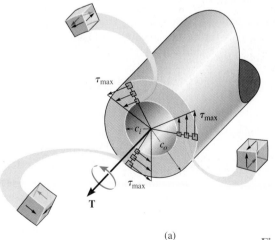

(a)

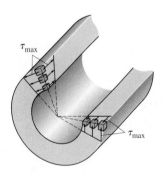

Shear stress varies linearly along
each radial line of the cross section.

(b)

Fig. 5–9

IMPORTANT POINTS

• When a shaft having a *circular cross section* is subjected to a torque, the cross section *remains plane* while radial lines rotate. This causes a *shear strain* within the material that *varies linearly* along any radial line, from zero at the axis of the shaft to a maximum at its outer boundary.

• For linearly elastic homogeneous material, due to Hooke's law, the *shear stress* along any radial line of the shaft also *varies linearly*, from zero at its axis to a maximum at its outer boundary. This maximum shear stress *must not* exceed the proportional limit.

• Due to the complementary property of shear, the linear shear stress distribution within the plane of the cross section is also distributed along an adjacent axial plane of the shaft.

• The torsion formula is based on the requirement that the resultant torque on the cross section is equal to the torque produced by the linear shear stress distribution about the longitudinal axis of the shaft. It is necessary that the shaft or tube have a *circular* cross section and that it is made of *homogeneous* material which has *linear-elastic* behavior.

PROCEDURE FOR ANALYSIS

The torsion formula can be applied using the following procedure.

Internal Loading.

• Section the shaft perpendicular to its axis at the point where the shear stress is to be determined, and use the necessary free-body diagram and equations of equilibrium to obtain the internal torque at the section.

Section Property.

• Compute the polar moment of inertia of the cross-sectional area. For a solid section of radius c, $J = \pi c^4/2$, and for a tube of outer radius c_o and inner radius c_i, $J = \pi(c_o^4 - c_i^4)/2$.

Shear Stress.

• Specify the radial distance ρ, measured from the center of the cross section to the point where the shear stress is to be found. Then apply the torsion formula $\tau = T\rho/J$, or if the maximum shear stress is to be determined use $\tau_{max} = Tc/J$. When substituting the data, make sure to use a consistent set of units.

• The shear stress acts on the cross section in a direction that is always perpendicular to ρ. The force it creates must contribute a torque about the axis of the shaft that is in the *same direction* as the internal resultant torque **T** acting on the section. Once this direction is established, a volume element located at the point where τ is determined can be isolated, and the direction of τ acting on the remaining three adjacent faces of the element can be shown.

EXAMPLE 5-1

The stress distribution in a solid shaft has been plotted along three arbitrary radial lines as shown in Fig. 5–10a. Determine the resultant internal torque at the section.

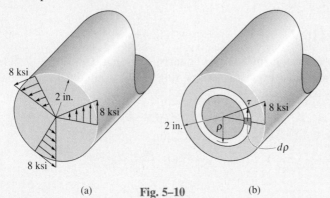

(a)   **Fig. 5–10**   (b)

SOLUTION I

The polar moment of inertia for the cross-sectional area is

$$J = \frac{\pi}{2}(2 \text{ in.})^4 = 25.13 \text{ in}^4$$

Applying the torsion formula, with $\tau_{max} = 8$ ksi, Fig. 5–10a, we have

$$\tau_{max} = \frac{Tc}{J}; \qquad 8 \text{ kip/in}^2 = \frac{T(2 \text{ in.})}{(25.13 \text{ in}^4)}$$

$$T = 101 \text{ kip} \cdot \text{in.} \qquad \textit{Ans.}$$

SOLUTION II

The same result can be obtained by finding the torque produced by the stress distribution about the centroidal axis of the shaft. First we must express $\tau = f(\rho)$. Using proportional triangles, Fig. 5–10b, we have

$$\frac{\tau}{\rho} = \frac{8 \text{ ksi}}{2 \text{ in.}}$$

$$\tau = 4\rho$$

This stress acts on all portions of the differential ring element that has an area $dA = 2\pi\rho \, d\rho$. Since the force created by τ is $dF = \tau \, dA$, the torque is

$$dT = \rho \, dF = \rho(\tau dA) = \rho(4\rho)2\pi\rho \, d\rho = 8\pi\rho^3 \, d\rho$$

For the entire area over which τ acts, we require

$$T = \int_0^2 8\pi\rho^3 \, d\rho = 8\pi\left(\frac{1}{4}\rho^4\right)\Big|_0^2 = 101 \text{ kip} \cdot \text{in.} \qquad \textit{Ans.}$$

EXAMPLE 5-2

The *solid* shaft of radius c is subjected to a torque **T**, Fig. 5–11a. Determine the fraction of T that is resisted by the material contained within the outer region of the shaft, which has an inner radius of $c/2$ and outer radius c.

SOLUTION

The stress in the shaft varies linearly, such that $\tau = (\rho/c)\tau_{max}$, Eq. 5–3. Therefore, the torque dT' on the ring (area) located within the lighter-shaded region, Fig. 5–11b, is

$$dT' = \rho(\tau \, dA) = \rho(\rho/c)\tau_{max}(2\pi\rho \, d\rho)$$

For the entire lighter-shaded area the torque is

$$T' = \frac{2\pi\tau_{max}}{c} \int_{c/2}^{c} \rho^3 \, d\rho$$

$$= \frac{2\pi\tau_{max}}{c} \frac{1}{4}\rho^4 \Big|_{c/2}^{c}$$

So that

$$T' = \frac{15\pi}{32}\tau_{max}c^3 \tag{1}$$

This torque T' can be expressed in terms of the applied torque T by first using the torsion formula to determine the maximum stress in the shaft. We have

$$\tau_{max} = \frac{Tc}{J} = \frac{Tc}{(\pi/2)c^4}$$

or

$$\tau_{max} = \frac{2T}{\pi c^3}$$

Substituting this into Eq. 1 yields

$$T' = \frac{15}{16}T \qquad \text{Ans.}$$

Here, approximately 94% of the torque is resisted by the lighter-shaded region, and the remaining 6% of T (or $\frac{1}{16}$) is resisted by the inner "core" of the shaft, $\rho = 0$ to $\rho = c/2$. As a result, the material located at the *outer region* of the shaft is highly effective in resisting torque, which justifies the use of tubular shafts as an efficient means for transmitting torque, and thereby saves material.

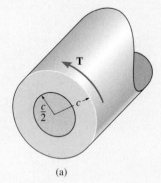

(a)

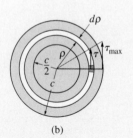

(b)

Fig. 5–11

EXAMPLE 5-3

The shaft shown in Fig. 5–12a is supported by two bearings and is subjected to three torques. Determine the shear stress developed at points A and B, located at section $a-a$ of the shaft, Fig. 5–12b.

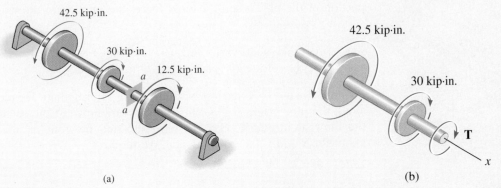

(a)

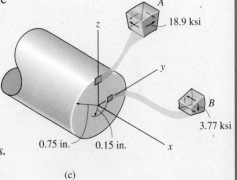

(b)

Fig. 5–12

SOLUTION

Internal Torque. The bearing reactions on the shaft are zero, provided the shaft's weight is neglected. Furthermore, the applied torques satisfy moment equilibrium about the shaft's axis.

The internal torque at section $a-a$ will be determined from the free-body diagram of the left segment, Fig. 5–12b. We have

$$\Sigma M_x = 0; \quad 42.5 \text{ kip} \cdot \text{in.} - 30 \text{ kip} \cdot \text{in.} - T = 0 \quad T = 12.5 \text{ kip} \cdot \text{in.}$$

Section Property. The polar moment of inertia for the shaft is

$$J = \frac{\pi}{2}(0.75 \text{ in.})^4 = 0.497 \text{ in}^4$$

Shear Stress. Since point A is at $\rho = c = 0.75$ in.,

$$\tau_A = \frac{Tc}{J} = \frac{(12.5 \text{ kip} \cdot \text{in.})(0.75 \text{ in.})}{(0.497 \text{ in}^4)} = 18.9 \text{ ksi} \quad \textit{Ans.}$$

Likewise for point B, at $\rho = 0.15$ in., we have

$$\tau_B = \frac{T\rho}{J} = \frac{(12.5 \text{ kip} \cdot \text{in.})(0.15 \text{ in.})}{(0.497 \text{ in}^4)} = 3.77 \text{ ksi} \quad \textit{Ans.}$$

The directions of these stresses on each element at A and B, Fig. 5–12c, are established from the direction of the resultant internal torque **T**, shown in Fig. 5–12b. Note carefully how the shear stress acts on the planes of each of these elements.

EXAMPLE 5-4

The pipe shown in Fig. 5–13a has an inner diameter of 80 mm and an outer diameter of 100 mm. If its end is tightened against the support at A using a torque wrench at B, determine the shear stress developed in the material at the inner and outer walls along the central portion of the pipe when the 80-N forces are applied to the wrench.

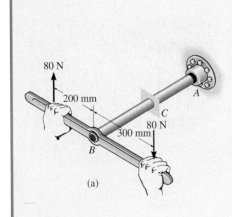

80 N

200 mm

C

80 N

300 mm

A

B

(a)

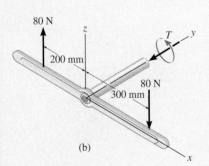

80 N

z

T

y

200 mm

80 N

300 mm

(b)

x

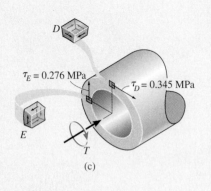

D

$\tau_E = 0.276$ MPa

$\tau_D = 0.345$ MPa

E

T

(c)

Fig. 5–13

SOLUTION

Internal Torque. A section is taken at an intermediate location C along the pipe's axis, Fig. 5–13b. The only unknown at the section is the internal torque **T**. Force equilibrium and moment equilibrium about the x and z axes are satisfied. We require

$$\Sigma M_y = 0; \qquad 80 \text{ N}(0.3 \text{ m}) + 80 \text{ N}(0.2 \text{ m}) - T = 0$$
$$T = 40 \text{ N} \cdot \text{m}$$

Section Property. The polar moment of inertia for the pipe's cross-sectional area is

$$J = \frac{\pi}{2}[(0.05 \text{ m})^4 - (0.04 \text{ m})^4] = 5.80(10^{-6}) \text{ m}^4$$

Shear Stress. For any point lying on the outside surface of the pipe, $\rho = c_o = 0.05$ m, we have

$$\tau_o = \frac{Tc_o}{J} = \frac{40 \text{ N} \cdot \text{m}(0.05 \text{ m})}{5.80(10^{-6}) \text{ m}^4} = 0.345 \text{ MPa} \qquad \textit{Ans.}$$

And for any point located on the inside surface, $\rho = c_i = 0.04$ m, so that

$$\tau_i = \frac{Tc_i}{J} = \frac{40 \text{ N} \cdot \text{m}(0.04 \text{ m})}{5.80(10^{-6}) \text{ m}^4} = 0.276 \text{ MPa} \qquad \textit{Ans.}$$

To show how these stresses act at representative points D and E on the cross-sectional area, we will first view the cross section from the front of segment CA of the pipe, Fig. 5–13a. On this section, Fig. 5–13c, the resultant internal torque is equal but opposite to that shown in Fig. 5–13b. The shear stresses at D and E contribute to this torque and therefore act on the shaded faces of the elements in the directions shown. As a consequence, notice how the shear-stress components act on the other three faces. Furthermore, since the top face of D and the inner face of E are in stress-free regions taken from the pipe's outer and inner walls, no shear stress can exist on these faces or on the other corresponding faces of the elements.

5.3   Power Transmission

Shafts and tubes having circular cross sections are often used to transmit power developed by a machine. When used for this purpose, they are subjected to torques that depend on the power generated by the machine and the angular speed of the shaft. *Power* is defined as the work performed per unit of time. The work transmitted by a rotating shaft equals the torque applied times the angle of rotation. Therefore, if during an instant of time dt an applied torque $\mathbf{T}$ causes the shaft to rotate $d\theta$, then the instantaneous power is

$$P = \frac{T\,d\theta}{dt}$$

Since the shaft's angular velocity $\omega = d\theta/dt$, we can also express the power as

$$\boxed{P = T\omega} \tag{5–10}$$

The drive shaft of this cutting machine must be designed to meet the power requirements of its motor.

In the SI system, power is expressed in *watts* when torque is measured in newton-meters (N · m) and ω is in radians per second (rad/s) (1 W = 1 N · m/s). In the Foot-Pound-Second or FPS system, the basic units of power are foot-pounds per second (ft · lb/s); however, horsepower (hp) is often used in engineering practice, where

$$1 \text{ hp} = 550 \text{ ft} \cdot \text{lb/s}$$

For machinery, the *frequency* of a shaft's rotation, f, is often reported. This is a measure of the number of revolutions or cycles the shaft makes per second and is expressed in hertz (1 Hz = 1 cycle/s). Since 1 cycle $= 2\pi$ rad, then $\omega = 2\pi f$, and the above equation for power becomes

$$\boxed{P = 2\pi f T} \tag{5–11}$$

Shaft Design.   When the power transmitted by a shaft and its frequency of rotation are known, the torque developed in the shaft can be determined from Eq. 5–11, that is, $T = P/2\pi f$. Knowing T and the allowable shear stress for the material, τ_{allow}, we can determine the size of the shaft's cross section using the torsion formula, provided the material behavior is linear-elastic. Specifically, the design or geometric parameter J/c becomes

$$\frac{J}{c} = \frac{T}{\tau_{\text{allow}}} \tag{5–12}$$

For a *solid shaft*, $J = (\pi/2)c^4$, and thus, upon substitution, a *unique value* for the shaft's radius c can be determined. If the shaft is *tubular*, so that $J = (\pi/2)(c_o^4 - c_i^4)$, design permits a wide range of possibilities for the solution. This is because an *arbitrary choice* can be made for either c_o or c_i and the other radius can then be determined from Eq. 5–12.

EXAMPLE 5-5

A solid steel shaft AB shown in Fig. 5–14 is to be used to transmit 5 hp from the motor M to which it is attached. If the shaft rotates at $\omega = 175$ rpm and the steel has an allowable shear stress of $\tau_{allow} = 14.5$ ksi, determine the required diameter of the shaft to the nearest $\frac{1}{8}$ in.

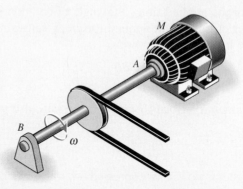

Fig. 5–14

SOLUTION

The torque on the shaft is determined from Eq. 5–10, that is, $P = T\omega$. Expressing P in foot-pounds per second and ω in radians/second, we have

$$P = 5 \text{ hp}\left(\frac{550 \text{ ft} \cdot \text{lb/s}}{1 \text{ hp}}\right) = 2750 \text{ ft} \cdot \text{lb/s}$$

$$\omega = \frac{175 \text{ rev}}{\text{min}}\left(\frac{2\pi \text{ rad}}{1 \text{ rev}}\right)\left(\frac{1 \text{ min}}{60 \text{ s}}\right) = 18.33 \text{ rad/s}$$

Thus,

$$P = T\omega; \qquad 2750 \text{ ft} \cdot \text{lb/s} = T(18.33 \text{ rad/s})$$
$$T = 150.1 \text{ ft} \cdot \text{lb}$$

Applying Eq. 5–12 yields

$$\frac{J}{c} = \frac{\pi}{2}\frac{c^4}{c} = \frac{T}{\tau_{allow}}$$

$$c = \left(\frac{2T}{\pi\tau_{allow}}\right)^{1/3} = \left(\frac{2(150.1 \text{ ft} \cdot \text{lb})(12 \text{ in./ft})}{\pi(14\,500 \text{ lb/in}^2)}\right)^{1/3}$$

$$c = 0.429 \text{ in.}$$

Since $2c = 0.858$ in., select a shaft having a diameter of

$$d = \frac{7}{8} \text{ in.} = 0.875 \text{ in.} \qquad \textit{Ans.}$$

EXAMPLE 5-6

A tubular shaft, having an inner diameter of 30 mm and an outer diameter of 42 mm, is to be used to transmit 90 kW of power. Determine the frequency of rotation of the shaft so that the shear stress will not exceed 50 MPa.

SOLUTION

The maximum torque that can be applied to the shaft is determined from the torsion formula.

$$\tau_{max} = \frac{Tc}{J}$$

$$50(10^6) \text{ N/m}^2 = \frac{T(0.021 \text{ m})}{(\pi/2)[(0.021 \text{ m})^4 - (0.015 \text{ m})^4]}$$

$$T = 538 \text{ N} \cdot \text{m}$$

Applying Eq. 5–11, the frequency of rotation is

$$P = 2\pi f T$$

$$90(10^3) \text{ N} \cdot \text{m/s} = 2\pi f(538 \text{ N} \cdot \text{m})$$

$$f = 26.6 \text{ Hz} \qquad \qquad \textit{Ans.}$$

PROBLEMS

5–1. A shaft is made of a steel alloy having an allowable shear stress of $\tau_{allow} = 12$ ksi. If the diameter of the shaft is 1.5 in., determine the maximum torque **T** that can be transmitted. What would be the maximum torque **T′** if a 1-in.-diameter hole is bored through the shaft? Sketch the shear-stress distribution along a radial line in each case.

5–2. The solid shaft of radius r is subjected to a torque **T**. Determine the radius r' of the inner core of the shaft that resists one-quarter of the applied torque ($T/4$). Solve the problem two ways: (a) by using the torsion formula, (b) by finding the resultant of the shear-stress distribution.

Prob. 5–1

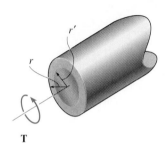

Prob. 5–2

5–3. The shaft has an outer diameter of 1.25 in. and an inner diameter of 1 in. If it is subjected to the applied torques as shown, determine the absolute maximum shear stress developed in the shaft. The smooth bearings at A and B do not resist torque.

*5–4. The shaft has an outer diameter of 1.25 in. and an inner diameter of 1 in. If it is subjected to the applied torques as shown, plot the shear-stress distribution acting along a radial line lying within region EA of the shaft. The smooth bearings at A and B do not resist torque.

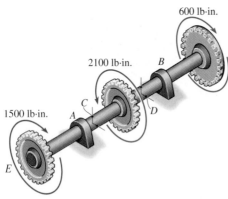

Probs. 5–3/5–4

5–5. The solid 30-mm-diameter shaft is used to transmit the torques applied to the gears. Determine the shear stress developed in the shaft at points C and D. Indicate the shear stress on volume elements located at these points.

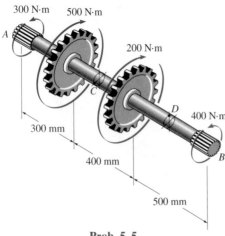

Prob. 5–5

5–6. The assembly consists of two sections of galvanized steel pipe connected together using a reducing coupling at B. The smaller pipe has an outer diameter of 0.75 in. and an inner diameter of 0.68 in., whereas the larger pipe has an outer diameter of 1 in. and an inner diameter of 0.86 in. If the pipe is tightly secured to the wall at C, determine the maximum shear stress developed in each section of the pipe when the couple shown is applied to the handles of the wrench.

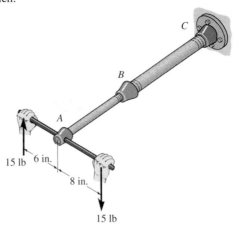

Prob. 5–6

5–7. The solid aluminum shaft has a diameter of 50 mm and an allowable shear stress of $\tau_{\text{allow}} = 6$ MPa. Determine the largest torque $\mathbf{T}_1$ that can be applied to the shaft if it is also subjected to the other torsional loadings. It is required that $\mathbf{T}_1$ act in the direction shown. Also, determine the maximum shear stress within regions CD and DE.

*5–8. The solid aluminum shaft has a diameter of 50 mm. Determine the absolute maximum shear stress in the shaft and sketch the shear-stress distribution along a radial line of the shaft where the shear stress is maximum. Set $T_1 = 20$ N $\cdot$ m.

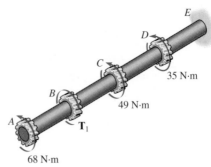

Probs. 5–7/5–8

5–9. The solid shaft has a diameter of 0.75 in. If it is subjected to the torques shown, determine the maximum shear stress developed in regions *BC* and *DE* of the shaft. The bearings at *A* and *F* allow free rotation of the shaft.

5–10. The solid shaft has a diameter of 0.75 in. If it is subjected to the torques shown, determine the maximum shear stress developed in regions *CD* and *EF* of the shaft. The bearings at *A* and *F* allow free rotation of the shaft.

■*5–12.** Consider the general problem of a circular shaft made from m segments each having a radius of c_m. If there are n torques on the shaft as shown, write a computer program that can be used to determine the maximum shear stress at any specified location x along the shaft. Show an application of the program using the values $L_1 = 2$ ft, $c_1 = 2$ in., $L_2 = 4$ ft, $c_2 = 1$ in., $T_1 = 800$ lb · ft, $d_1 = 0$, $T_2 = -600$ lb · ft, $d_2 = 5$ ft.

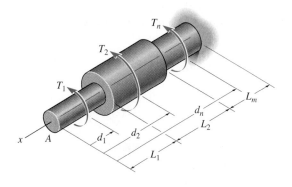

Prob. 5–12

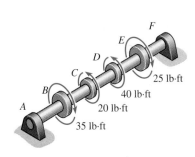

Probs. 5–9/5–10

5–11. A steel tube having an outer diameter of 2.5 in. is used to transmit 35 hp when turning at 2700 rev/min. Determine the inner diameter d of the tube to the nearest $\frac{1}{8}$ in. if the allowable shear stress is $\tau_{allow} = 10$ ksi.

5–13. The copper pipe has an outer diameter of 2.50 in. and an inner diameter of 2.30 in. If it is tightly secured to the wall at *C* and a uniformly distributed torque is applied to it as shown, determine the shear stress developed at points *A* and *B*. These points lie on the pipe's outer surface. Sketch the shear stress on volume elements located at *A* and *B*.

5–14. The copper pipe has an outer diameter of 2.50 in. and an inner diameter of 2.30 in. If it is tightly secured to the wall at *C* and it is subjected to the uniformly distributed torque along its entire length, determine the absolute maximum shear stress in the pipe. Discuss the validity of this result.

Prob. 5–11

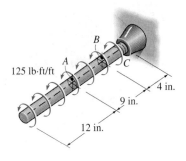

Probs. 5–13/5–14

5–15. The 60-mm-diameter solid shaft is subjected to the distributed and concentrated torsional loadings shown. Determine the shear stress at points A and B, and sketch the shear stress on volume elements located at these points.

*5–16. The 60-mm diameter solid shaft is subjected to the distributed and concentrated torsional loadings shown. Determine the absolute maximum and minimum shear stresses in the shaft and specify their locations, measured from the fixed end.

5–17. The solid shaft is subjected to the distributed and concentrated torsional loadings shown. Determine the required diameter d of the shaft if the allowable shear stress for the material is $\tau_{\text{allow}} = 175$ MPa.

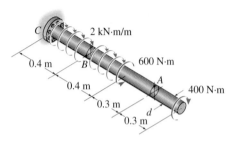

Probs. 5–15/5–16/5–17

5–18. The steel shaft is subjected to the torsional loading shown. Determine the shear stress developed at points A and B and sketch the shear stress on volume elements located at these points. The shaft where A and B are located has an outer radius of 60 mm.

5–19. The steel shaft is subjected to the torsional loading shown. Determine the absolute maximum shear stress in the shaft and sketch the shear-stress distribution along a radial line where it is maximum.

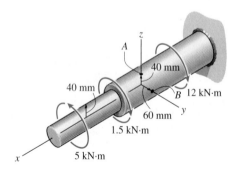

Probs. 5–18/5–19

*5–20. The solid shaft is subjected to the distributed and concentrated torsional loadings shown. Determine the shear stress at points A and B and sketch the shear stress on volume elements located at these points. The distributed torque from D to C varies from zero to 900 N · m/m. Take $d = 40$ mm.

5–21. The solid shaft is subjected to the distributed and concentrated torsional loadings shown. Determine the absolute maximum shear stress in the shaft and specify its location, measured from the fixed end C. Take $d = 40$ mm.

5–22. The solid shaft is subjected to the distributed and concentrated torsional loadings shown. Determine the required diameter d of the shaft if the allowable shear stress for the material is $\tau_{\text{allow}} = 150$ MPa.

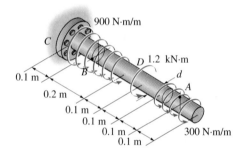

Probs. 5–20/5–21/5–22

5–23. Determine to the nearest $\frac{1}{8}$ in. the diameter of a solid shaft that is required to transmit 150 hp at 4000 rev/min. The material has an allowable shear stress of $\tau_{\text{allow}} = 8$ ksi.

*5–24. The drilling pipe on an oil rig is made from steel pipe having an outer diameter of 4.5 in. and a thickness of 0.25 in. If the pipe is turning at 650 rev/min while being powered by a 15-hp motor, determine the maximum shear stress in the pipe.

5–25. The gear motor can develop 1/10 hp when it turns at 300 rev/min. If the shaft has a diameter of $\frac{1}{2}$ in., determine the maximum shear stress that will be developed in the shaft.

5–26. The gear motor can develop 1/10 hp when it turns at 80 rev/min. If the allowable shear stress for the shaft is $\tau_{allow} = 4$ ksi, determine the smallest diameter of the shaft to the nearest $\frac{1}{8}$ in. that can be used.

Probs. 5–25/5–26

5–27. The coupling is used to connect the two shafts together. Assuming that the shear stress in the bolts is *uniform*, determine the number of bolts necessary to make the maximum shear stress in the shaft equal to the shear stress in the bolts. Each bolt has a diameter d.

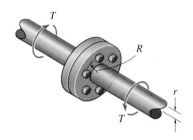

Prob. 5–27

*5–28. The steel shafts are connected together using a fillet weld as shown. Determine the average shear stress in the weld along section *a–a* if the torque applied to the shafts is $T = 60$ N · m. *Note:* The critical section where the weld fails is along section *a–a*.

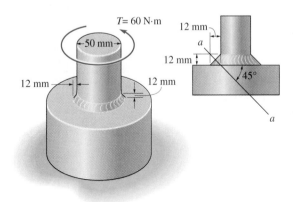

Prob. 5–28

5–29. A motor delivers 500 hp to the steel shaft, which is tubular and has an outer diameter of 2 in. If the shaft is rotating at 200 rad/s, determine its largest inner diameter to the nearest $\frac{1}{8}$ in. if the allowable shear stress for the material is $\tau_{allow} = 25$ ksi.

5–30. The pump operates using the motor that has a power of 85 W. If the impeller at B is turning at 150 rev/min, determine the maximum shear stress developed in the 20-mm-diameter transmission shaft at A.

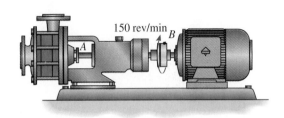

Prob. 5–30

5–31. A steel tube having an outer diameter of $d_1 = 2.5$ in. is used to transmit 35 hp when turning at 2700 rev/min. Determine the inner diameter d_2 of the tube to the nearest $\frac{1}{8}$ in. if the allowable shear stress is $\tau_{allow} = 10$ ksi.

Prob. 5–31

5–34. The shaft is subjected to a distributed torque along its length of $t = (10x^2)$ N · m/m, where x is in meters. If the maximum stress in the shaft is to remain constant at 80 MPa, determine the required variation of the radius c of the shaft for $0 \le x \le 3$ m.

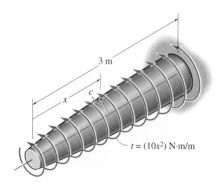

$t = (10x^2)$ N·m/m

Prob. 5–34

*5–32. A steel tube having an outer diameter of $d_1 = 2.5$ in. and an inner diameter of $d_2 = 2$ in. is used to transmit 45 hp. Determine its maximum rate of rotation if the allowable shear stress is $\tau_{allow} = 12$ ksi.

Prob. 5–32

5–33. A ship has a propeller drive shaft that is turning at 1500 rev/min while developing 1800 hp. If it is 8 ft long and has a diameter of 4 in., determine the maximum shear stress in the shaft caused by torsion.

■5–35. The shaft has a diameter of 80 mm and due to friction at its surface within the hole, it is subjected to a variable torque described by the function $t = (25xe^{x^2})$ N · m/m, where x is in meters. Determine the minimum torque T_0 needed to overcome friction and cause it to twist. Also, determine the absolute maximum stress in the shaft.

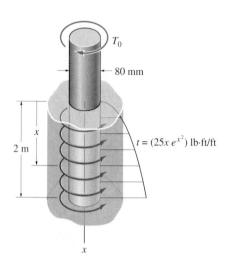

T_0

80 mm

2 m

$t = (25x\,e^{x^2})$ lb·ft/ft

x

Prob. 5–35

*5–36. Determine the diameter d of a solid shaft required to transmit a power P if the allowable shear stress is τ_{allow} and the angular velocity of the shaft is ω.

5–39. The drive shaft of an automobile is to be designed as a thin-walled tube. The engine delivers 150 hp when the shaft is turning at 1500 rev/min. Determine the minimum thickness of the tube's wall if the outer diameter is 2.5 in. The material has an allowable shear stress of $\tau_{\text{allow}} = 7$ ksi.

5–37. The solid shaft has a linear taper from r_A at one end to r_B at the other. Derive an equation that gives the maximum shear stress in the shaft at a location x along the shaft's axis.

*5–40. The motor delivers 50 hp while turning at a constant rate of 1350 rpm at A. Using the belt and pulley system this loading is delivered to the steel blower shaft BC. Determine to the nearest $\frac{1}{8}$ in. the smallest diameter of this shaft if the allowable shear stress for the steel is $\tau_{\text{allow}} = 12$ ksi.

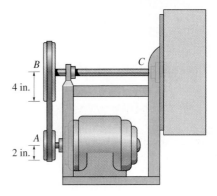

Prob. 5–40

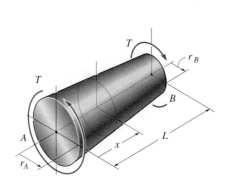

Prob. 5–37

5–41. A cylindrical spring consists of a rubber annulus bonded to a rigid ring and shaft. If the ring is held fixed and a torque **T** is applied to the shaft, determine the maximum shear stress in the rubber.

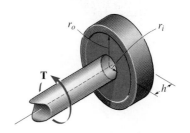

5–38. The drive shaft of an automobile is made of a steel tube having an allowable shear stress of $\tau_{\text{allow}} = 8$ ksi. If the outer diameter is 2.5 in. and the engine delivers 200 hp to the shaft when it is turning at 1140 rev/min, determine the minimum required thickness of the shaft wall.

Prob. 5–41

5.4   ANGLE OF TWIST

Oil wells are commonly drilled to depths exceeding a thousand meters. As a result, the total angle of twist of a string of drill pipe can be substantial and must be computed.

Occasionally the design of a shaft depends on restricting the amount of rotation or twist that may occur when the shaft is subjected to a torque. Furthermore, being able to compute the angle of twist for a shaft is important when analyzing the reactions on statically indeterminate shafts.

In this section we will develop a formula for determining the *angle of twist* ϕ (phi) of one end of a shaft with respect to its other end. The shaft is assumed to have a circular cross section that can gradually vary along its length, Fig. 5–15a, and the material is assumed to be homogeneous and to behave in a linear-elastic manner when the torque is applied. As in the case of an axially loaded bar, we will neglect the localized deformations that occur at points of application of the torques and where the cross section changes abruptly. By Saint-Venant's principle, these effects occur within small regions of the shaft's length and generally have only a slight effect on the final result.

Using the method of sections, a differential disk of thickness dx, located at position x, is isolated from the shaft, Fig. 5–15b. The internal resultant torque is represented as $T(x)$, since the external loading may cause it to vary along the axis of the shaft. Due to $T(x)$, the disk will twist, such that the *relative rotation* of one of its faces with respect to the other face is $d\phi$, Fig. 5–15b. As a result an element of material located at an arbitrary radius ρ within the disk will undergo a shear strain γ. The values of γ and $d\phi$ are related by Eq. 5–1, namely,

$$d\phi = \gamma \frac{dx}{\rho} \qquad (5\text{–}13)$$

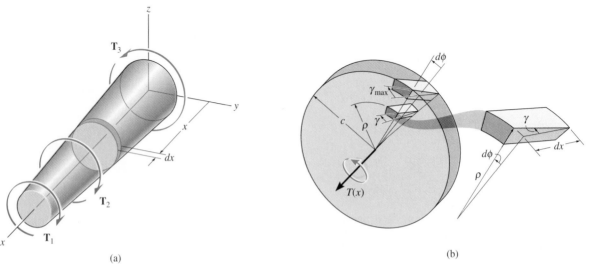

(a)

(b)

Fig. 5–15

Since Hooke's law, $\gamma = \tau/G$, applies and the shear stress can be expressed in terms of the applied torque using the torsion formula $\tau = T(x)\rho/J(x)$, then $\gamma = T(x)\rho/J(x)G$. Substituting this into Eq. 5–13, the angle of twist for the disk is

$$d\phi = \frac{T(x)}{J(x)G}\, dx$$

Integrating over the entire length L of the shaft, we obtain the angle of twist for the entire shaft, namely,

$$\phi = \int_0^L \frac{T(x)dx}{J(x)G} \qquad (5\text{--}14)$$

Here

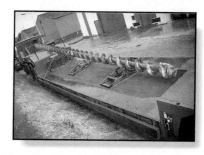

When computing both the stress and the angle of twist of this soil auger, it is necessary to consider the variable loading which acts along its length.

ϕ = the angle of twist of one end of the shaft with respect to the other end, measured in radians

$T(x)$ = the internal torque at the arbitrary position x, found from the method of sections and the equation of moment equilibrium applied about the shaft's axis

$J(x)$ = the shaft's polar moment of inertia expressed as a function of position x

G = the shear modulus of elasticity for the material

Constant Torque and Cross-Sectional Area.   Usually in engineering practice the material is homogeneous so that G is constant. Also, the shaft's cross-sectional area and the applied torque are constant along the length of the shaft, Fig. 5–16. If this is the case, the internal torque $T(x) = T$, the polar moment of inertia $J(x) = J$, and Eq. 5–14 can be integrated, which gives

$$\phi = \frac{TL}{JG} \qquad (5\text{--}15)$$

The similarities between the above two equations and those for an axially loaded bar ($\delta = \int P(x)\, dx/A(x)E$ and $\delta = PL/AE$) should be noted.

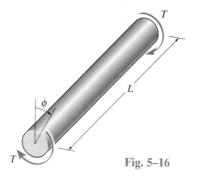

Fig. 5–16

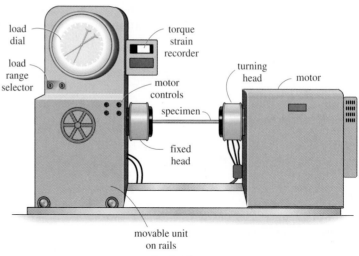

load
dial

load
range
selector

torque
strain
recorder

motor
controls

turning
head

motor

specimen

fixed
head

movable unit
on rails

Fig. 5–17

We can use Eq. 5–15 to determine the shear modulus of elasticity G of the material. To do so, a specimen of known length and diameter is placed in a torsion testing machine like the one shown in Fig. 5–17. The applied torque T and angle of twist ϕ are then measured between a gauge length L. Using Eq. 5–15, $G = TL/J\phi$. Usually, to obtain a more reliable value of G, several of these tests are performed and the average value is used.

If the shaft is subjected to several different torques, or the cross-sectional area or shear modulus changes abruptly from one region of the shaft to the next, Eq. 5–15 can be applied to each segment of the shaft where these quantities are all constant. The angle of twist of one end of the shaft with respect to the other is then found from the vector addition of the angles of twist of each segment. For this case,

$$\phi = \sum \frac{TL}{JG}$$

(5–16)

Sign Convention. In order to apply the above equation, we must develop a sign convention for the internal torque and the angle of twist of one end of the shaft with respect to the other end. To do this, we will use the right-hand rule, whereby both the torque and angle will be *positive*, provided the *thumb* is directed *outward* from the shaft when the fingers curl to give the tendency for rotation, Fig. 5–18.

To illustrate the use of this sign convention, consider the shaft shown in Fig. 5–19*a*, which is subjected to four torques. The angle of twist of end A with respect to end D is to be determined. For this problem, three segments of the shaft must be considered, since the internal torque

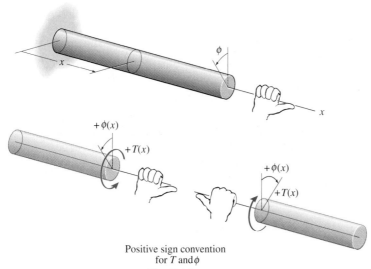

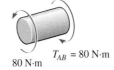

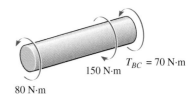

Positive sign convention
for T and ϕ

Fig. 5–18

changes at B and C. Using the method of sections, the internal torques are found for each segment, Fig. 5–19b. By the right-hand rule, with positive torques directed away from the *sectioned end* of the shaft, we have $T_{AB} = +80 \text{ N} \cdot \text{m}$, $T_{BC} = -70 \text{ N} \cdot \text{m}$, and $T_{CD} = -10 \text{ N} \cdot \text{m}$. These results are also shown on the *torque diagram* for the shaft, Fig. 5–19c. Applying Eq. 5–16, we have

$$\phi_{A/D} = \frac{(+80 \text{ N} \cdot \text{m}) \, L_{AB}}{JG} + \frac{(-70 \text{ N} \cdot \text{m}) \, L_{BC}}{JG} + \frac{(-10 \text{ N} \cdot \text{m}) \, L_{CD}}{JG}$$

If the other data is substituted and the answer is found as a *positive* quantity, it means that end A will rotate as indicated by the curl of the right-hand fingers when the thumb is directed *away* from the shaft, Fig. 5–19a. The double subscript notation is used to indicate this relative angle of twist ($\phi_{A/D}$); however, if the angle of twist is to be determined relative to a *fixed point*, then only a single subscript will be used. For example, if D is located at a fixed support, then the computed angle of twist will be denoted as ϕ_A.

(b)

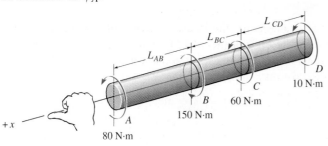

(a)

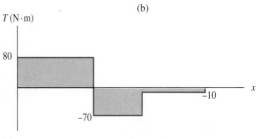

(c)

Fig. 5–19

IMPORTANT POINTS

• The angle of twist is determined by relating the applied torque to the shear stress using the torsion formula, $\tau = T\rho/J$, and relating the relative rotation to the shear strain using $d\phi = \gamma \, dx/\rho$. Finally these equations are combined using Hooke's law, $\tau = G\gamma$ which yields Eq. 5–14.

• Since Hooke's law is used in the development of the formula for the angle of twist, it is important that the applied torques do not cause yielding of the material and that the material is homogeneous and behaves in a linear-elastic manner.

PROCEDURE FOR ANALYSIS

The angle of twist of one end of a shaft or tube with respect to the other end can be determined by applying Eqs. 5–14 through 5–16.

Internal Torque.

• The internal torque is found at a point on the axis of the shaft by using the method of sections and the equation of moment equilibrium, applied along the shaft's axis.

• If the torque varies along the shaft's length, a section should be made at the arbitrary position x along the shaft and the torque represented as a function of x, i.e., $T(x)$.

• If several constant external torques act on the shaft between its ends, the internal torque in each *segment* of the shaft, between any two external torques, must be determined. The results can be represented as a torque diagram.

Angle of Twist.

• When the circular cross-sectional area varies along the shaft's axis, the polar moment of inertia must be expressed as a function of its position x along the axis, $J(x)$.

• If the polar moment of inertia or the internal torque *suddenly changes* between the ends of the shaft, then $\phi = \int (T(x)/J(x)G) \, dx$ or $\phi = TL/JG$ must be applied to *each segment* for which J, G, and T are continuous or constant.

• When the internal torque in each segment is determined, be sure to use a consistent sign convention for the shaft, such as the one discussed above. Also make sure that a consistent set of units is used when substituting numerical data into the equations.

EXAMPLE 5-7

The gears attached to the fixed-end steel shaft are subjected to the torques shown in Fig. 5–20a. If the shear modulus of elasticity is $G = 80$ GPa and the shaft has a diameter of 14 mm, determine the displacement of the tooth P on gear A. The shaft turns freely within the bearing at B.

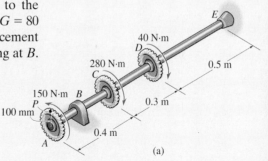

$T_{CD} = 130$ N·m

$T_{AC} = 150$ N·m

150 N·m

150 N·m

150 N·m B

100 mm

280 N·m

(a)

SOLUTION

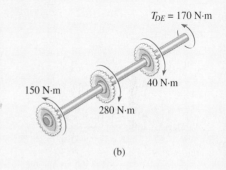

$T_{DE} = 170$ N·m

150 N·m

40 N·m

280 N·m

(b)

Internal Torque.   By inspection, the torques in segments AC, CD, and DE are different yet *constant* throughout each segment. Free-body diagrams of appropriate segments of the shaft along with the calculated internal torques are shown in Fig. 5–20b. Using the right-hand rule and the established sign convention that positive torque is directed away from the sectioned end of the shaft, we have

$$T_{AC} = +150 \text{ N} \cdot \text{m} \qquad T_{CD} = -130 \text{ N} \cdot \text{m} \qquad T_{DE} = -170 \text{ N} \cdot \text{m}$$

These results are also shown on the torque diagram, Fig. 5–20c.

Angle of Twist.   The polar moment of inertia for the shaft is

$$J = \frac{\pi}{2}(0.007 \text{ m})^4 = 3.77(10^{-9}) \text{ m}^4$$

Applying Eq. 5–16 to each segment and adding the results algebraically, we have

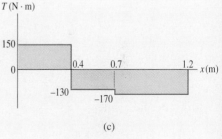

T (N · m)

(c)

$$\phi_A = \sum \frac{TL}{JG} = \frac{(+ 150 \text{ N} \cdot \text{m})(0.4 \text{ m})}{3.77(10^{-9}) \text{ m}^4[80(10^9) \text{ N/m}^2]}$$
$$+ \frac{(-130 \text{ N} \cdot \text{m})(0.3 \text{ m})}{3.77(10^{-9}) \text{ m}^4[80(10^9) \text{ N/m}^2)]}$$
$$+ \frac{(-170 \text{ N} \cdot \text{m})(0.5 \text{ m})}{3.77(10^{-9}) \text{ m}^4[80(10^9) \text{ N/m}^2)]} = -0.212 \text{ rad}$$

Since the answer is negative, by the right-hand rule the thumb is directed *toward* the end E of the shaft, and therefore gear A will rotate as shown in Fig. 5–20d.

The displacement of tooth P on gear A is

$$s_P = \phi_A r = (0.212 \text{ rad})(100 \text{ mm}) = 21.2 \text{ mm} \qquad \textit{Ans.}$$

Remember that this analysis is valid only if the shear stress does not exceed the proportional limit of the material.

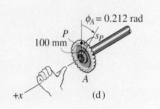

$\phi_A = 0.212$ rad

(d)

Fig. 5–20

EXAMPLE 5-8

The two solid steel shafts shown in Fig. 5–21a are coupled together using the meshed gears. Determine the angle of twist of end A of shaft AB when the torque $T = 45$ N · m is applied. Take $G = 80$ GPa. Shaft AB is free to rotate within bearings E and F, whereas shaft DC is fixed at D. Each shaft has a diameter of 20 mm.

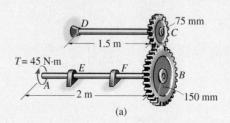

(a)

SOLUTION

Internal Torque.   Free-body diagrams for each shaft are shown in Fig. 5–21b and 5–21c. Summing moments along the x axis of shaft AB yields the tangential reaction between the gears of $F = 45$ N · m/0.15 m = 300 N. Summing moments about the x axis of shaft DC, this force then creates a torque of $(T_D)_x = 300$ N(0.075 m) = 22.5 N · m on shaft DC.

Angle of Twist.   To solve the problem, we will first calculate the rotation of gear C due to the torque of 22.5 N · m in shaft DC, Fig. 5–21b. This angle of twist is

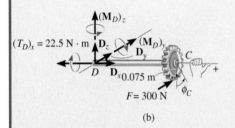

(b)

Fig. 5–21

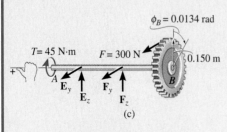

(c)

$$\phi_C = \frac{TL_{DC}}{JG} = \frac{(+22.5 \text{ N} \cdot \text{m})(1.5 \text{ m})}{(\pi/2)(0.010 \text{ m})^4[80(10^9) \text{ N/m}^2]} = +0.0269 \text{ rad}$$

Since the gears at the end of the shaft are in mesh, the rotation ϕ_C of gear C causes gear B to rotate ϕ_B, Fig. 5–21c, where

$$\phi_B(0.15 \text{ m}) = (0.0269 \text{ rad})(0.075 \text{ m})$$

$$\phi_B = 0.0134 \text{ rad}$$

We will now determine the angle of twist of end A with respect to end B of shaft AB caused by the 45 N · m torque, Fig. 5–21c. We have

$$\phi_{A/B} = \frac{T_{AB}L_{AB}}{JG} = \frac{(+45 \text{ N} \cdot \text{m})(2 \text{ m})}{(\pi/2)(0.010 \text{ m})^4[80(10^9) \text{ N/m}^2]} = +0.0716 \text{ rad}$$

The rotation of end A is therefore determined by adding ϕ_B and $\phi_{A/B}$, since both angles are in the *same direction*, Fig. 5–21c. We have

$$\phi_A = \phi_B + \phi_{A/B} = 0.0134 \text{ rad} + 0.0716 \text{ rad} = +0.0850 \text{ rad}   \textit{Ans.}$$

EXAMPLE 5-9

The 2-in.-diameter solid cast-iron post shown in Fig. 5–22a is buried 24 in. in soil. If a torque is applied to its top using a rigid wrench, determine the maximum shear stress in the post and the angle of twist at its top. Assume that the torque is about to turn the post, and the soil exerts a uniform torsional resistance of t lb · in./in. along its 24-in. buried length. $G = 5.5(10^3)$ ksi.

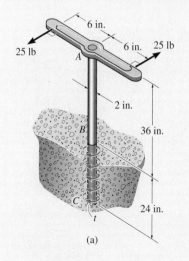

SOLUTION

Internal Torque.   The internal torque in segment AB of the post is constant. From the free-body diagram, Fig. 5–22b, we have

$$\Sigma M_z = 0; \qquad T_{AB} = 25 \text{ lb}(12 \text{ in.}) = 300 \text{ lb} \cdot \text{in.}$$

The magnitude of the uniform distribution of torque along the buried segment BC can be determined from equilibrium of the entire post, Fig. 5–22c. Here

$$\Sigma M_z = 0; \qquad 25 \text{ lb}(12 \text{ in.}) - t(24 \text{ in.}) = 0$$
$$t = 12.5 \text{ lb} \cdot \text{in./in.}$$

Hence, from a free-body diagram of a section of the post located at the position x within region BC, Fig. 5–22d, we have

$$\Sigma M_z = 0; \qquad T_{BC} - 12.5x = 0$$
$$T_{BC} = 12.5x$$

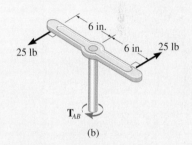

Maximum Shear Stress.   The largest shear stress occurs in region AB, since the torque is largest there and J is constant for the post. Applying the torsion formula, we have

$$\tau_{\max} = \frac{T_{AB}c}{J} = \frac{(300 \text{ lb} \cdot \text{in.})(1 \text{ in.})}{(\pi/2)(1 \text{ in.})^4} = 191 \text{ psi} \qquad \textit{Ans.}$$

Angle of Twist.   The angle of twist at the top can be determined relative to the bottom of the post, since it is fixed and yet is about to turn. Both segments AB and BC twist, and so in this case we have

$$\phi_A = \frac{T_{AB}L_{AB}}{JG} + \int_0^{L_{BC}} \frac{T_{BC} \, dx}{JG}$$

$$= \frac{(300 \text{ lb} \cdot \text{in.})(36 \text{ in.})}{JG} + \int_0^{24 \text{ in.}} \frac{12.5x \, dx}{JG}$$

$$= \frac{10\,800 \text{ lb} \cdot \text{in}^2}{JG} + \frac{12.5[(24)^2/2] \text{ lb} \cdot \text{in}^2}{JG}$$

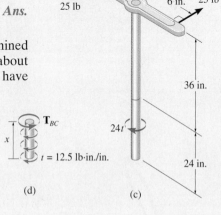

$$= \frac{14\,400 \text{ lb} \cdot \text{in}^2}{(\pi/2)(1 \text{ in.})^4 5500(10^3) \text{ lb/in}^2} = 0.00167 \text{ rad} \qquad \textit{Ans.}$$

Fig. 5–22

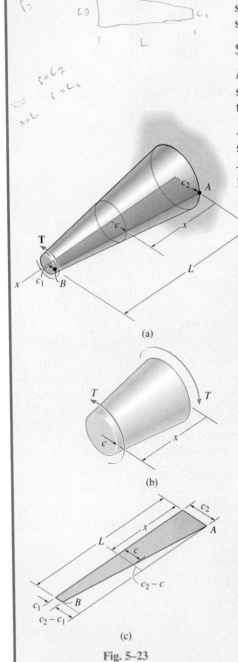

Fig. 5–23

EXAMPLE 5-10

The tapered shaft shown in Fig. 5–23a is made of a material having a shear modulus G. Determine the angle of twist of its end B when subjected to the torque.

SOLUTION

Internal Torque.   By inspection or from the free-body diagram of a section located at the arbitrary position x, Fig. 5–23b, the internal torque is T.

Angle of Twist.   Here the polar moment of inertia varies along the shaft's axis and therefore we must express it in terms of the coordinate x. The radius c of the shaft at x can be determined in terms of x by proportion of the slope of line AB in Fig. 5–23c. We have

$$\frac{c_2 - c_1}{L} = \frac{c_2 - c}{x}$$

$$c = c_2 - x\left(\frac{c_2 - c_1}{L}\right)$$

Thus, at x,

$$J(x) = \frac{\pi}{2}\left[c_2 - x\left(\frac{c_2 - c_1}{L}\right)\right]^4$$

Applying Eq. 5–14, we have

$$\phi = \int_0^L \frac{T\,dx}{\left(\frac{\pi}{2}\right)\left[c_2 - x\left(\frac{c_2 - c_1}{L}\right)\right]^4 G} = \frac{2T}{\pi G}\int_0^L \frac{dx}{\left[c_2 - x\left(\frac{c_2 - c_1}{L}\right)\right]^4}$$

Performing the integration using an integral table, the result becomes

$$\phi = \left(\frac{2T}{\pi G}\right)\frac{1}{3\left(\frac{c_2 - c_1}{L}\right)\left[c_2 - x\left(\frac{c_2 - c_1}{L}\right)\right]^3}\Bigg|_0^L$$

$$= \frac{2T}{\pi G}\left(\frac{L}{3(c_2 - c_1)}\right)\left(\frac{1}{c_1^3} - \frac{1}{c_2^3}\right)$$

Rearranging terms yields

$$\phi = \frac{2TL}{3\pi G}\left(\frac{c_2^2 + c_1 c_2 + c_1^2}{c_1^3 c_2^3}\right) \qquad Ans.$$

To partially check this result, note that when $c_1 = c_2 = c$, then

$$\phi = \frac{TL}{[(\pi/2)c^4]G} = \frac{TL}{JG}$$

which is Eq. 5–15.

PROBLEMS

5–42. The propellers of a ship are connected to a solid A-36 steel shaft that is 60 m long and has an outer diameter of 340 mm and inner diameter of 260 mm. If the power output is 4.5 MW when the shaft rotates at 20 rad/s, determine the maximum torsional stress in the shaft and its angle of twist.

5–43. A shaft is subjected to a torque **T**. Compare the effectiveness of using the tube shown in the figure with that of a solid section of radius *c*. To do this, compute the percent increase in torsional stress and angle of twist per unit length for the tube versus the solid section.

Prob. 5–43

*5–44. The hydrofoil boat has an A-36 steel propeller shaft that is 100 ft long. It is connected to an in-line diesel engine that delivers a maximum power of 2500 hp and causes the shaft to rotate at 1700 rpm. If the outer diameter of the shaft is 8 in. and the wall thickness is $\frac{3}{8}$ in., determine the maximum shear stress developed in the shaft. Also, what is the "wind up," or angle of twist in the shaft at full power?

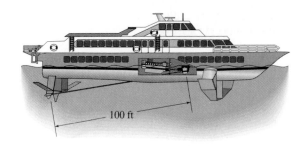

Prob. 5–44

5–45. The splined ends and gears attached to the A-36 steel shaft are subjected to the torques shown. Determine the angle of twist of end *B* with respect to end *A*. The shaft has a diameter of 40 mm.

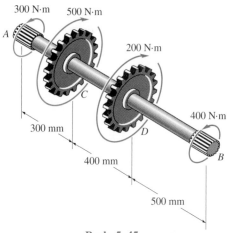

Prob. 5–45

5–46. The A-36 steel axle is made from tubes *AB* and *CD* and a solid section *BC*. It is supported on smooth bearings that allow it to rotate freely. If the ends are subjected to 85 N · m torques, determine the angle of twist of end *A* relative to end *D*. The tubes have an outer diameter of 30 mm and an inner diameter of 20 mm. The solid section has a diameter of 40 mm.

5–47. The A-36 steel axle is made from tubes *AB* and *CD* and a solid section *BC*. It is supported on smooth bearings that allow it to rotate freely. If ends *A* and *D* are subjected to 85 N · m torques, determine the angle of twist of end *B* of the solid section relative to end *C*. The tubes have an outer diameter of 30 mm and an inner diameter of 20 mm. The solid section has a diameter of 40 mm.

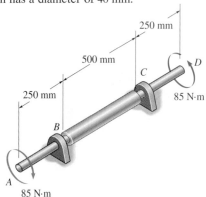

Probs. 5–46/5–47

*5–48. The gears attached to the 304 stainless steel shaft are subjected to the torques shown. Determine the angle of twist of end A with respect to end B. The shaft has a diameter of 1.5 in.

5–49. The gears attached to the 304 stainless steel shaft are subjected to the torques shown. Determine the angle of twist of gear C with respect to gear B. The shaft has a diameter of 1.5 in.

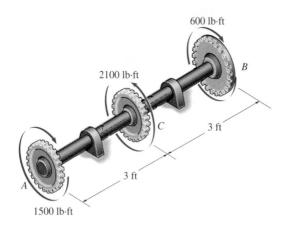

Probs. 5–48/5–49

5–50. The rotating flywheel-and-shaft, when brought to a sudden stop at D, begins to oscillate clockwise- counterclockwise such that a point A on the outer edge of the flywheel is displaced through a 6-mm arc. Determine the maximum shear stress developed in the tubular A-36 steel shaft due to this oscillation. The shaft has an inner diameter of 24 mm and an outer diameter of 32 mm. The bearings at B and C allow the shaft to rotate freely, whereas the support at D holds the shaft fixed.

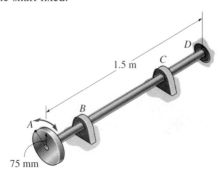

Prob. 5–50

5–51. The motor delivers 40 hp to the 304 stainless steel shaft while it rotates at 20 Hz. The shaft is supported on smooth bearings at A and B, which allow free rotation of the shaft. The gears C and D fixed to the shaft remove 25 hp and 15 hp, respectively. Determine the diameter of the shaft to the nearest $\frac{1}{8}$ in. if the allowable shear stress is $\tau_{\text{allow}} = 8$ ksi and the allowable angle of twist of C with respect to D is 0.20°.

*5–52. The motor delivers 40 hp to the 304 stainless steel solid shaft while it rotates at 20 Hz. The shaft has a diameter of 1.5 in. and is supported on smooth bearings at A and B, which allow free rotation of the shaft. The gears C and D fixed to the shaft remove 25 hp and 15 hp, respectively. Determine the absolute maximum stress in the shaft and the angle of twist of gear C with respect to gear D.

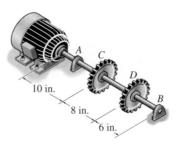

Probs. 5–51/5–52

5–53. The A-36 steel shaft is 2 m long and has an outer diameter of 40 mm. When it is rotating at 80 rad/s, it transmits 32 kW of power from the engine E to the generator G. Determine the smallest thickness of the shaft if the allowable shear stress is $\tau_{\text{allow}} = 140$ MPa and the shaft is restricted not to twist more than 0.05 rad.

5–54. The A-36 solid steel shaft is 3 m long and has a diameter of 50 mm. It is required to transmit 35 kW of power from the engine E to the generator G. Determine the smallest angular velocity the shaft can have if it is restricted not to twist more than 1°.

Probs. 5–53/5–54

5–55. The A-36 steel shaft rotates at $\omega = 125$ rad/s and transmits the power shown. Determine the absolute maximum shear stress in the shaft and the angle of twist of C with respect to F. The inner and outer diameters of the shaft are $d_i = 30$ mm and $d_o = 40$ mm. The journal bearings at B and G are smooth.

*5–56. The A-36 steel shaft rotates at $\omega = 125$ rad/s and transmits the power shown. Determine the inner and outer diameters of the shaft if $d_i/d_o = 0.6$, the allowable shear stress is $\tau_{allow} = 150$ MPa, and the allowable relative angle of twist is $\phi_{allow} = 2°/$m. The journal bearings at B and G are smooth.

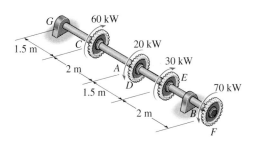

Probs. 5–55/5–56

5–57. The A-36 steel assembly consists of a tube having an outer radius of 1 in. and a wall thickness of 0.125 in. Using a rigid plate at B, it is connected to the solid 1-in.-diameter shaft AB. Determine the rotation of the tube's end C if a torque of 200 lb · in. is applied to the tube at this end. The end A of the shaft is fixed-supported.

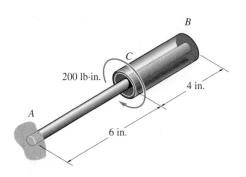

Prob. 5–57

5–58. The two shafts are made of A-36 steel. Each has a diameter of 1 in., and they are supported by bearings at A, B, and C, which allow free rotation. If the support at D is fixed, determine the angle of twist of end B when the torques are applied to the assembly as shown.

5–59. The two shafts are made of A-36 steel. Each has a diameter of 1 in., and they are supported by bearings at A, B, and C, which allow free rotation. If the support at D is fixed, determine the angle of twist of end A when the torques are applied to the assembly as shown.

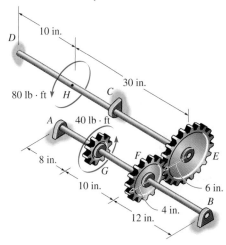

Probs. 5–58/5–59

*5–60. The 6-in.-diameter L-2 steel shaft on the turbine is supported on journal bearings at A and B. If C is held fixed and the turbine blades create a torque on the shaft that increases linearly from zero at C to 2000 lb · ft at D, determine the angle of twist of the shaft of end D relative to end C. Also, compute the absolute maximum shear stress in the shaft. Neglect the size of the blades.

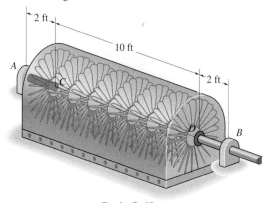

Prob. 5–60

5–61. The assembly is made of A-36 steel and consists of a solid rod 15 mm in diameter connected to the inside of a tube using a rigid disk at *B*. Determine the angle of twist at *A*. The tube has an outer diameter of 30 mm and wall thickness of 3 mm.

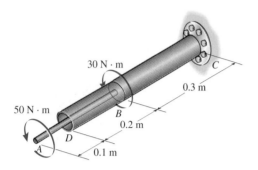

Prob. 5–61

5–62. The device shown is used to mix soils in order to provide in-situ stabilization. If the mixer is connected to an A-36 steel tubular shaft that has an inner diameter of 3 in. and an outer diameter of 4.5 in., determine the angle of twist of the shaft of *A* relative to *C* if each mixing blade is subjected to the torques shown.

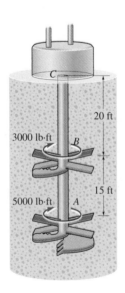

Prob. 5–62

5–63. The device serves as a compact torsional spring. It is made of A-36 steel and consists of a solid inner shaft *CB* which is surrounded by and attached to a tube *AB* using a rigid ring at *B*. The ring at *A* can also be assumed rigid and is fixed from rotating. If a torque of $T = 2$kip · in. is applied to the shaft, determine the angle of twist at the end *C* and the maximum shear stress in the tube and shaft.

*5–64. The device serves as a compact torsion spring. It is made of A-36 steel and consists of a solid inner shaft *CB* which is surrounded by and attached to a tube *AB* using a rigid ring at *B*. The ring at *A* can also be assumed rigid and is fixed from rotating. If the allowable shear stress for the material is $\tau_{allow} = 12$ ksi and the angle of twist at *C* is limited to $\phi_{allow} = 3°$, determine the maximum torque *T* that can be applied at the end *C*.

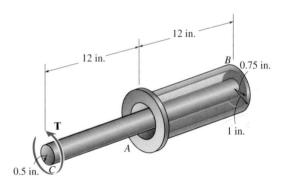

Probs. 5–63/5–64

5–65. The contour of the surface of the shaft is defined by the equation $y = e^{ax}$, where *a* is a constant. If the shaft is subjected to a torque *T* at its ends, determine the angle of twist of end *A* with respect to end *B*. The shear modulus is *G*.

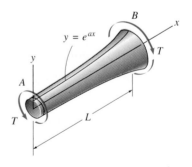

Prob. 5–65

5–66. The tapered shaft is made of 2014-T6 aluminum alloy and has a radius which can be described by the function $r = 0.02(1 + x^{3/2})$ m, where x is in meters. Determine the angle of twist of its end A if it is subjected to a torque of 450 N · m.

*5–68. The A-36 steel shaft has a diameter of 50 mm and is subjected to the distributed and concentrated loadings shown. Determine the absolute maximum shear stress in the shaft and plot a graph of the angle of twist of the shaft in radians versus x.

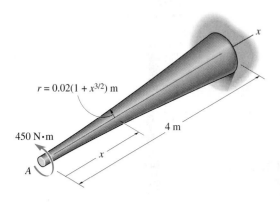

Prob. 5–66

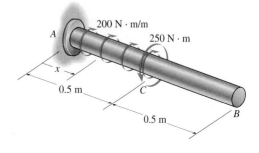

Prob. 5–68

5–67. The shaft of radius c is subjected to a distributed torque t, measured as torque/length of shaft. Determine the angle of twist at end A. The shear modulus is G.

5–69. The glass tube is confined within a rubber stopper, so that when the tube is twisted at constant angular velocity the stopper creates a *constant distribution* of frictional torque along the contacting length AB of the tube. If the tube has an inner diameter of 2 mm and an outer diameter of 4 mm, determine the shear stress developed at a point located at its inner and outer walls at a section through level C. Show the shear-stress distribution acting along a radial line segment at this section. Also, determine the angle of twist at A with respect to B. $G_g = 10$ GPa.

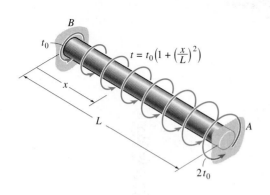

Prob. 5–67

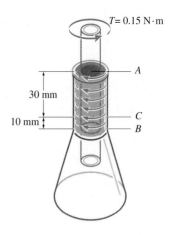

Prob. 5–69

5–70. The 60-mm-diameter solid shaft is made of A-36 steel and is subjected to the distributed and concentrated torsional loadings shown. Determine the angle of twist at the free end *A* of the shaft due to these loadings.

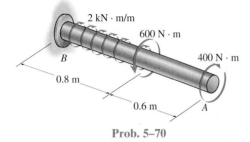

Prob. 5–70

*5–71. The tapered shaft has a length *L* and a radius *r* at end *A* and 2*r* at end *B*. If it is fixed at end *B* and is subjected to a torque *T*, determine the angle of twist of end *A*. The shear modulus is *G*.

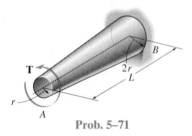

Prob. 5–71

5–72. The solid steel shaft consists of two tapered ends, *AB* and *CD*, and a central portion *BC* having a constant diameter. It is supported by two smooth bearings, which allow it to rotate freely. If the gears fixed to its ends are subjected to counterbalancing torques of 1700 lb · in., determine the angle of twist of one end of the shaft relative to the other end. *Hint:* Use the results of Prob. 5–71. $G_{st} = 12(10^3)$ ksi.

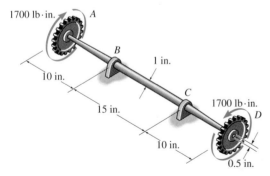

Prob. 5–72

5–73. Show that when the bar is subjected to the torque *T*, the length *L* shortens by an amount $L(1 - \sqrt{1 - (2T/\pi c^3 G)^2})$. The shaft is made of material having a shear modulus *G*.

Prob. 5–73

5–74. A cylindrical spring consists of a rubber annulus bonded to a rigid ring and shaft. If the ring is held fixed and a torque *T* is applied to the rigid shaft, determine the angle of twist of the shaft. The shear modulus of the rubber is *G*. *Hint:* As shown in the figure, the deformation of the element at radius *r* can be determined from $r \, d\theta = dr \, \gamma$. Use this expression along with $\tau = T/(2\pi r^2 h)$, from Prob. 5–41, to obtain the result.

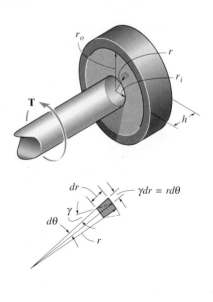

Prob. 5–74

5.5 STATICALLY INDETERMINATE TORQUE-LOADED MEMBERS

A torsionally loaded shaft may be classified as statically indeterminate if the moment equation of equilibrium, applied about the axis of the shaft, is not adequate to determine the unknown torques acting on the shaft. An example of this situation is shown in Fig. 5–24a. As shown on the free-body diagram, Fig. 5–24b, the reactive torques at the supports A and B are unknown. We require that

$$\Sigma M_x = 0; \qquad\qquad T - T_A - T_B = 0$$

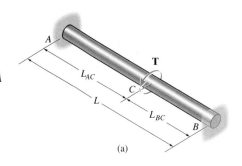

(a)

Since only one equilibrium equation is relevant and there are two unknowns, this problem is statically indeterminate. In order to obtain a solution, we will use the method of analysis discussed in Sec. 4.4.

The necessary condition of compatibility, or the kinematic condition, requires the angle of twist of one end of the shaft with respect to the other end to be equal to zero, since the end supports are fixed. Therefore,

$$\phi_{A/B} = 0$$

In order to write this equation in terms of the unknown torques, we will assume that the material behaves in a linear-elastic manner, so that the load–displacement relationship is expressed by $\phi = TL/JG$. Realizing that the internal torque in segment AC is $+T_A$ and that in segment CB the internal torque is $-T_B$, Fig. 5–24c, the above compatibility equation can be written as

$$\frac{T_A L_{AC}}{JG} - \frac{T_B L_{BC}}{JG} = 0$$

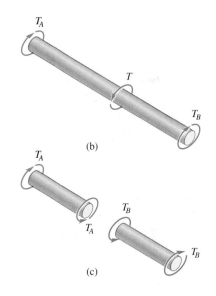

(b)

(c)

Here JG is assumed to be constant.

Solving the above two equations for the reactions, realizing that $L = L_{AC} + L_{BC}$, we get

$$T_A = T\left(\frac{L_{BC}}{L}\right)$$

Fig. 5–24

and

$$T_B = T\left(\frac{L_{AC}}{L}\right)$$

Note that each of these reactive torques increases or decreases linearly with the placement L_{AC} or L_{BC} of the applied torque.

Procedure for Analysis

The unknown torques in statically indeterminate shafts are determined by satisfying equilibrium, compatibility, and torque-displacement requirements for the shaft.

Equilibrium.

• Draw a free-body diagram of the shaft in order to identify all the torques that act on it. Then write the equations of moment equilibrium about the axis of the shaft.

Compatibility.

• To write the compatibility equation, investigate the way the shaft will twist when subjected to the external loads, and give consideration as to how the supports constrain the shaft when it is twisted.

• Express the compatibility condition in terms of the rotational displacements caused by the reactive torques, and then use a torque-displacement relation, such as $\phi = TL/JG$, to relate the unknown torques to the unknown displacements.

• Solve the equilibrium and compatibility equations for the unknown reactive torques. If any of the magnitudes have a negative numerical value, it indicates that this torque acts in the opposite sense of direction to that indicated on the free-body diagram.

EXAMPLE 5-11

The solid steel shaft shown in Fig. 5–25a has a diameter of 20 mm. If it is subjected to the two torques, determine the reactions at the fixed supports A and B.

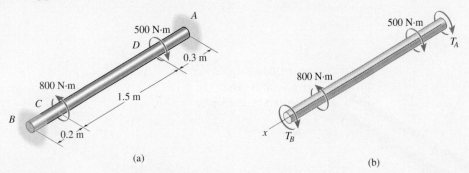

(a)

(b)

SOLUTION

Equilibrium.   By inspection of the free-body diagram, Fig. 5–25b, it is seen that the problem is statically indeterminate since there is only *one* available equation of equilibrium, whereas $\mathbf{T}_A$ and $\mathbf{T}_B$ are unknown. We require

$$\Sigma M_x = 0; \qquad -T_B + 800 \text{ N} \cdot \text{m} - 500 \text{ N} \cdot \text{m} - T_A = 0 \qquad (1)$$

Compatibility.   Since the ends of the shaft are fixed, the angle of twist of one end of the shaft with respect to the other must be zero. Hence, the compatibility equation can be written as

$$\phi_{A/B} = 0$$

This condition can be expressed in terms of the unknown torques by using the load–displacement relationship, $\phi = TL/JG$. Here there are three regions of the shaft where the internal torque is constant, BC, CD, and DA. On the free-body diagrams in Fig. 5–25c we have shown the internal torques acting on segments of the shaft which are sectioned in each of these regions. Using the sign convention established in Sec. 5.4, we have

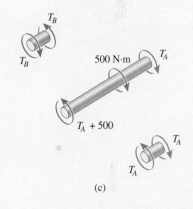

(c)

Fig. 5–25

$$\frac{-T_B(0.2 \text{ m})}{JG} + \frac{(T_A + 500 \text{ N} \cdot \text{m})(1.5 \text{ m})}{JG} + \frac{T_A(0.3 \text{ m})}{JG} = 0$$

or

$$1.8T_A - 0.2T_B = -750 \qquad (2)$$

Solving Eqs. 1 and 2 yields

$$T_A = -345 \text{ N} \cdot \text{m} \qquad T_B = 645 \text{ N} \cdot \text{m} \qquad \textbf{\textit{Ans.}}$$

The negative sign indicates that $\mathbf{T}_A$ acts in the opposite direction of that shown in Fig. 5–25b.

EXAMPLE 5–12

The shaft shown in Fig. 5–26a is made from a steel tube, which is bonded to a brass core. If a torque of $T = 250$ lb · ft is applied at its end, plot the shear-stress distribution along a radial line of its cross-sectional area. Take $G_{st} = 11.4(10^3)$ ksi, $G_{br} = 5.20(10^3)$ ksi.

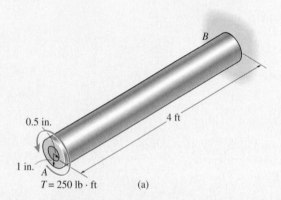

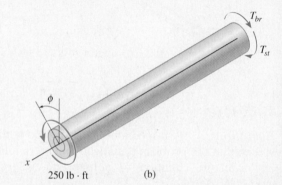

Fig. 5–26

SOLUTION

Equilibrium. A free-body diagram of the shaft is shown in Fig. 5–26b. The reaction at the wall has been represented by the unknown amount of torque resisted by the steel, T_{st}, and by the brass, T_{br}. Working in units of pounds and inches, equilibrium requires

$$-T_{st} - T_{br} + 250 \text{ lb} \cdot \text{ft(12 in./ft)} = 0 \qquad (1)$$

Compatibility. We require the angle of twist of end A to be the same for both the steel and brass since they are bonded together. Thus,

$$\phi = \phi_{st} = \phi_{br}$$

Applying the load–displacement relationship, $\phi = TL/JG$, we have

$$\frac{T_{st}L}{(\pi/2)[(1 \text{ in.})^4 - (0.5 \text{ in.})^4]11.4(10^3) \text{ kip/in}^2} =$$

$$\frac{T_{br}L}{(\pi/2)(0.5 \text{ in.})^4 5.20(10^3) \text{ kip/in}^2}$$

$$T_{st} = 32.88T_{br} \qquad (2)$$

Solving Eqs. 1 and 2, we get

$$T_{st} = 2911.0 \text{ lb} \cdot \text{in.} = 242.6 \text{ lb} \cdot \text{ft}$$
$$T_{br} = 88.5 \text{ lb} \cdot \text{in.} = 7.38 \text{ lb} \cdot \text{ft}$$

These torques act throughout the entire length of the shaft, since no external torques act at intermediate points along the shaft's axis. The shear stress in the brass core varies from zero at its center to a maximum at the interface where it contacts the steel tube. Using the torsion formula,

$$(\tau_{br})_{max} = \frac{(88.5 \text{ lb} \cdot \text{in.})(0.5 \text{ in.})}{(\pi/2)(0.5 \text{ in.})^4} = 451 \text{ psi}$$

For the steel, the minimum shear stress is also at this interface,

$$(\tau_{st})_{min} = \frac{(2911.0 \text{ lb} \cdot \text{in.})(0.5 \text{ in.})}{(\pi/2)[(1 \text{ in.})^4 - (0.5 \text{ in.})^4]} = 988 \text{ psi}$$

and the maximum shear stress is at the outer surface,

$$(\tau_{st})_{max} = \frac{(2911.0 \text{ lb} \cdot \text{in.})(1 \text{ in.})}{(\pi/2)[(1 \text{ in.})^4 - (0.5 \text{ in.})^4]} = 1977 \text{ psi}$$

The results are plotted in Fig. 5–26c. Note the discontinuity of *shear stress* at the brass and steel interface. This is to be expected, since the materials have different moduli of rigidity; i.e., steel is stiffer than brass ($G_{st} > G_{br}$) and thus it carries more shear stress at the interface. Although the shear stress is discontinuous here, the *shear strain* is not. Rather, the shear strain is the *same* for both the brass and the steel. This can be shown by using Hooke's law, $\gamma = \tau/G$. At the interface, Fig. 5–26d, the shear strain is

$$\gamma = \frac{\tau}{G} = \frac{451 \text{ psi}}{5.2(10^6)\text{psi}} = \frac{988 \text{ psi}}{11.4(10^6)\text{psi}} = 0.0867(10^{-3})\text{rad}$$

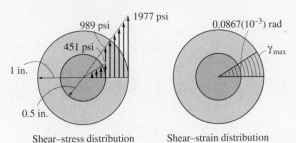

Shear–stress distribution

(c)

Shear–strain distribution

(d)

PROBLEMS

5–75. The steel shaft has a diameter of 40 mm and is fixed at its ends A and B. If it is subjected to the couple, determine the maximum shear stress in regions AC and CB of the shaft. $G_{st} = 10.8(10^3)$ ksi.

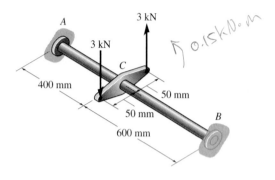

Prob. 5–75

*5–76. The steel shaft is made from two segments: AC has a diameter of 0.5 in., and CB has a diameter of 1 in. If it is fixed at its ends A and B and subjected to a torque of 500 lb · ft, determine the maximum shear stress in the shaft. $G_{st} = 10.8(10^3)$ ksi.

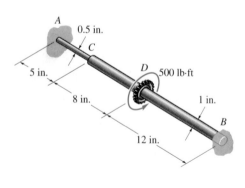

Prob. 5–76

5–77. The bronze C86100 pipe has an outer diameter of 1.5 in. and a thickness of 0.125 in. The coupling on it at C is being tightened using a wrench. If the torque developed at A is 125 lb · in., determine the magnitude F of the couple forces. The pipe is fixed supported at end B.

5–78. The bronze C86100 pipe has an outer diameter of 1.5 in. and a thickness of 0.125 in. The coupling on it at C is being tightened using a wrench. If the applied force is $F = 20$ lb, determine the maximum shear stress in the pipe.

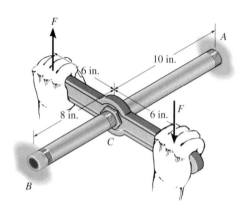

Probs. 5–77/5–78

5–79. A rod is made from two segments: AB is A-36 steel and has a diameter of 30 mm and BD is C83400 red brass and has a diameter of 50 mm. It is fixed at its ends and subjected to a torque of $T = 500$ N · m. Determine the torsional reactions at the walls A and D.

*5–80. Determine the absolute maximum shear stress in the shaft of Prob. 5–79.

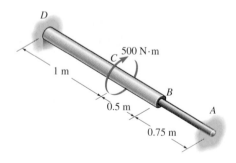

Probs. 5–79/5–80

5–81. The A-36 steel shaft is made from two segments: AC has a diameter of 1 in. and CB has a diameter of 2 in. If it is fixed at its ends A and B and subjected to a torque of $T = 500$ lb · ft, determine the absolute maximum shear stress in the shaft.

5–82. Determine the torsional reactions at the ends A and B of the shaft in Prob. 5–81.

5–83. The A-36 steel shaft is made from two segments: AC has a diameter of 1 in. and CB has a diameter of 2 in. If it is fixed at its ends A and B, determine the magnitude of the applied torque T if the reaction at A is 50 lb · ft.

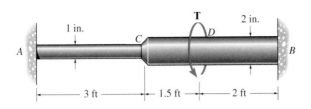

Probs. 5–81/5–82/5–83

5–85. The two shafts are made of A-36 steel. Each has a diameter of 25 mm and they are connected using the gears fixed to their ends. Their other ends are attached to fixed supports at A and B. They are also supported by journal bearings at C and D, which allow free rotation of the shafts along their axes. If a torque of 500 N · m is applied to the gear at E as shown, determine the reactions at A and B.

5–86. Determine the rotation of the gear at E in Prob. 5–85.

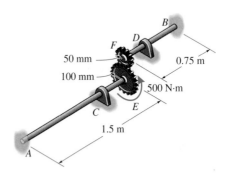

Probs. 5–85/5–86

*5–84.** The composite shaft consists of a mid-section that includes the 1-in.-diameter solid shaft and a tube that is welded to the rigid flanges at A and B. Neglect the thickness of the flanges and determine the angle of twist of end C of the shaft relative to end D. The shaft is subjected to a torque of 800 lb · ft. The material is A-36 steel.

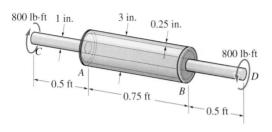

Prob. 5–84

5–87. The shaft is made from a solid steel section AB and a tubular portion made of steel and having a brass core. If it is fixed to a rigid support at A, and a torque of $T = 50$ lb · ft is applied to it at C, determine the angle of twist that occurs at C and compute the maximum shear and maximum shear strain in the brass and steel. Take $G_{st} = 11.5(10^3)$ ksi, $G_{br} = 5.6(10^3)$ ksi.

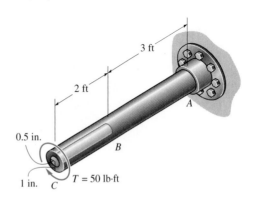

Prob. 5–87

*5–88. The shaft of radius c is subjected to a distributed torque t, measured as torque/length of shaft. Determine the reactions at the fixed supports A and B.

5–89. The tapered shaft is confined by the fixed supports at A and B. If a torque **T** is applied at its midpoint, determine the reactions at the supports.

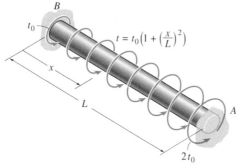

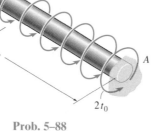

$$t = t_0\left(1 + \left(\frac{x}{L}\right)^2\right)$$

Prob. 5–88

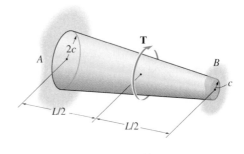

Prob. 5–89

*5.6 SOLID NONCIRCULAR SHAFTS

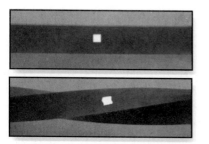

Notice the deformation that occurs to the square element when this rubber bar is subjected to a torque.

It was demonstrated in Sec. 5.1 that when a torque is applied to a shaft having a circular cross section—that is, one that is axisymmetric—the shear strains vary linearly from zero at its center to a maximum at its outer surface. Furthermore, due to the uniformity of the shear strain at all points on the same radius, the cross section does not deform, but rather remains plane after the shaft has twisted. Shafts that have a noncircular cross section, however, are *not* axisymmetric, and because the shear stress over their cross section is distributed in a very complex manner, their cross sections will **bulge** or **warp** when the shaft is twisted. Evidence of this can be seen from the way grid lines deform on a shaft having a square cross section when the shaft is twisted, Fig. 5–27. As a consequence of this deformation the torsional analysis of *noncircular* shafts becomes considerably complicated and will not be considered in this text.

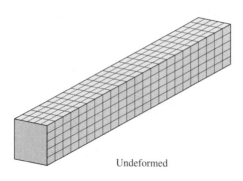

Undeformed

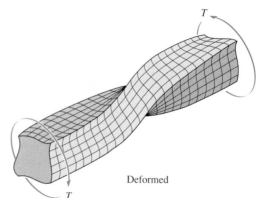

Deformed

Fig. 5–27

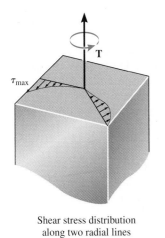

Shear stress distribution
along two radial lines

(a)

Warping of
cross-sectional area

(b)

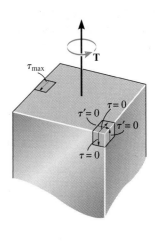

(c)

Fig. 5–28

Using a mathematical analysis based on the theory of elasticity, however, it is possible to determine the shear-stress distribution within a shaft of square cross section. Examples of how this shear stress varies along two radial lines of the shaft are shown in Fig. 5–28a. Because these shear-stress distributions vary in a complex manner, the shear strains they create will *warp* the cross section as shown in Fig. 5–28b. Notice that the corner points of the shaft will be subjected to zero shear stress and therefore zero shear strain. The reason for this can be shown by considering an element of material located at one of these points, Fig. 5–28c. One would expect the top face of this element to be subjected to a shear stress in order to aid in resisting the applied torque **T**. This, however, is *not* the case, since the shear stresses τ and τ', acting on the *outer surface* of the shaft, must be *zero*, which in turn implies that the corresponding shear-stress components τ and τ' on the top face must also be equal to zero.

The results of the analysis for square cross sections, along with other results from the theory of elasticity, for shafts having triangular and elliptical cross sections, are reported in Table 5–1. In all cases the *maximum shear stress* occurs at a point on the edge of the cross section that is *closest to* the center axis of the shaft. In Table 5–1 these points are indicated as "dots" on the cross sections. Also given are formulas for the angle of twist of each shaft. By extending these results to a shaft having an *arbitrary* cross section, it can also be shown that a shaft having a *circular* cross section is most efficient, since it is subjected to both a *smaller* maximum shear stress and a *smaller* angle of twist than a corresponding shaft having the same cross sectional area, but having a noncircular cross section and subjected to the same torque.

Shape of cross section	τ_{max}	ϕ
Square	$\dfrac{4.81\,T}{a^3}$	$\dfrac{7.10\,TL}{a^4 G}$
Equilateral triangle	$\dfrac{20\,T}{a^3}$	$\dfrac{46\,TL}{a^4 G}$
Ellipse	$\dfrac{2\,T}{\pi a b^2}$	$\dfrac{(a^2 + b^2)TL}{\pi a^3 b^3 G}$

Table 5–1

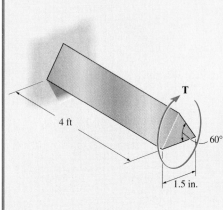

4 ft

1.5 in.

60°

T

Fig. 5–29

EXAMPLE 5–13

The 6061-T6 aluminum shaft shown in Fig. 5–29 has a cross-sectional area in the shape of an equilateral triangle. Determine the largest torque **T** that can be applied to the end of the shaft if the allowable shear stress is $\tau_{allow} = 8$ ksi and the angle of twist at its end is restricted to $\phi_{allow} = 0.02$ rad. How much torque can be applied to a shaft of circular cross section made from the same amount of material?

SOLUTION

By inspection, the resultant internal torque at any cross section along the shaft's axis is also **T**. Using the formulas for τ_{max} and ϕ in Table 5–1, we require

$$\tau_{allow} = \frac{20T}{a^3}; \qquad 8(10^3) \text{ lb/in}^2 = \frac{20T}{(1.5 \text{ in.})^3}$$

$$T = 1350 \text{ lb} \cdot \text{in.}$$

Also,

$$\phi_{allow} = \frac{46TL}{a^4 G_{al}}; \quad 0.02 \text{ rad} = \frac{46T(4 \text{ ft})(12 \text{ in./ft})}{(1.5 \text{ in.})^4 [3.7(10^6) \text{ lb/in}^2]}$$

$$T = 170 \text{ lb} \cdot \text{in.} \qquad \textit{Ans.}$$

By comparison, the torque is limited due to the angle of twist.

Circular Cross Section. If the same amount of aluminum is to be used in making the same length of shaft having a circular cross section, then the radius of the cross section can be calculated. We have

$$A_{circle} = A_{triangle}; \qquad \pi c^2 = \frac{1}{2}(1.5 \text{ in.})(1.5 \sin 60°)$$

$$c = 0.557 \text{ in.}$$

The limitations of stress and angle of twist then require

$$\tau_{allow} = \frac{Tc}{J}; \qquad 8(10^3) \text{ lb/in}^2 = \frac{T(0.557 \text{ in.})}{(\pi/2)(0.557 \text{ in.})^4}$$

$$T = 2170 \text{ lb} \cdot \text{in.}$$

$$\phi_{allow} = \frac{TL}{JG_{al}}; \qquad 0.02 \text{ rad} = \frac{T(4 \text{ ft})(12 \text{ in./ft})}{(\pi/2)(0.557 \text{ in.})^4 [3.7(10^6) \text{ lb/in}^2]}$$

$$T = 233 \text{ lb} \cdot \text{in.} \qquad \textit{Ans.}$$

Again, the angle of twist limits the applied torque.

Comparing this result (233 lb · in.) with that given above (170 lb · in.), it is seen that a shaft of circular cross section can support 37% more torque than the one having a triangular cross section.

*5.7   THIN-WALLED TUBES HAVING CLOSED CROSS SECTIONS

Thin-walled tubes of noncircular shape are often used to construct light-weight frameworks such as those used in aircraft. In some applications, they may be subjected to a torsional loading. In this section we will analyze the effects of applying a torque to a thin-walled tube having a *closed* cross section, that is, a tube that does not have any breaks or slits along its length. Such a tube, having a constant yet arbitrary cross-sectional shape, is shown in Fig. 5–30a. For the analysis we will assume that the walls have a variable thickness t. Since the walls are thin, we will be able to obtain an approximate solution for the shear stress by assuming that this stress is *uniformly distributed* across the thickness of the tube. In other words, we will be able to determine the *average shear stress* in the tube at any given point. Before we do this, however, we will first discuss some preliminary concepts regarding the action of shear stress over the cross section.

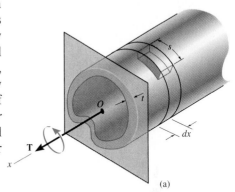

(a)

Shear Flow.   Shown in Figs. 5–30a and 5–30b is a small element of the tube having a finite length s and differential width dx. At one end the element has a thickness t_A, and at the other end the thickness is t_B. Due to the applied torque $\mathbf{T}$, shear stress is developed on the front face of the element. Specifically, at end A the shear stress is τ_A, and at end B it is τ_B. These stresses can be related by noting that equivalent shear stresses τ_A and τ_B must also act on the longitudinal sides of the element, shown shaded in Fig. 5–30b. Since these sides have *constant* thicknesses t_A and t_B, the forces acting on them are $dF_A = \tau_A(t_A\ dx)$ and $dF_B = \tau_B(t_B\ dx)$. Force equilibrium requires these forces to be of equal magnitude but opposite direction, so that

$$\tau_A t_A = \tau_B t_B$$

This important result states that ***the product of the average longitudinal shear stress times the thickness of the tube is the same at each point on the tube's cross-sectional area.*** This product is called ***shear flow,**** q, and in general terms we can express it as

$$\boxed{q = \tau_{\text{avg}} t} \qquad (5\text{–}17)$$

Since q is constant over the cross section, the *largest* average shear stress will occur where the tube's thickness is the *smallest*.

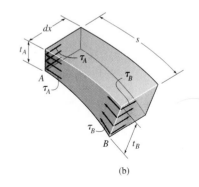

(b)

Fig. 5–30

---

*The terminology "flow" is used since q is analogous to water flowing through an open channel of rectangular cross section having a constant depth and variable width w. Although the water's velocity v at each point along the channel will be different (like τ_{avg}), the flow $q = vy$ will be constant.

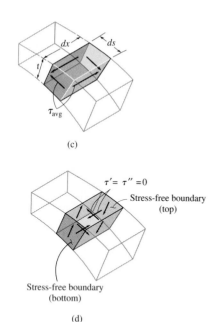

(c)

(d)

Stress-free boundary
(bottom)

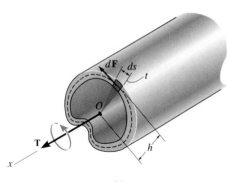

(e)

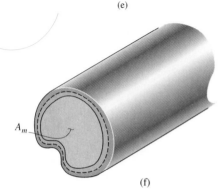

(f)

Fig. 5–30

If a differential element having a thickness t, length ds, and width dx is isolated from the tube, Fig. 5–30c, it is seen that the colored area over which the average shear stress acts is $dA = t\ ds$. Hence, $dF = \tau_{avg}t\ ds = q\ ds$, or $q = dF/ds$. In other words, *the shear flow, which is constant over the cross-sectional area, measures the force per unit length along the tube's cross-sectional area.*

It is important to note that the shear-stress components shown in Fig. 5–30c are the only ones acting on the tube. Components acting in the other direction, as shown in Fig. 5–30d, cannot exist. This is because the top and bottom faces of the element are at the inner and outer walls of the tube, and these boundaries must be free of stress. Instead, as noted above, the applied torque causes *the shear flow and the average stress always to be directed tangent to the wall of the tube, such that it contributes to the resultant torque* **T**.

Average Shear Stress. The average shear stress, τ_{avg}, acting on the shaded area $dA = t\ ds$ of the differential element shown in Fig. 5–30c, can be related to the torque T by considering the torque produced by the shear stress about a selected point O within the tube's boundary, Fig. 5–30e. As shown, the shear stress develops a force $dF = \tau_{avg}\ dA = \tau_{avg}(t\ ds)$ on the element. This force acts tangent to the centerline of the tube's wall, and since the moment arm is h, the torque is

$$dT = h(dF) = h(\tau_{avg}t\ ds)$$

For the entire cross section, we require

$$T = \oint h\ \tau_{avg}t\ ds$$

Here the "line integral" indicates that integration is performed *around* the entire boundary of the area. Since the shear flow $q = \tau_{avg}t$ is *constant*, these terms together can be factored out of the integral, so that

$$T = \tau_{avg}t \oint h\ ds$$

A graphical simplification can be made for evaluating the integral by noting that the *mean area*, shown by the colored triangle in Fig. 5–30e, is $dA_m = (1/2)h\ ds$. Thus,

$$T = 2\tau_{avg}t \int dA_m = 2\tau_{avg}tA_m$$

Solving for τ_{avg}, we have

$$\tau_{avg} = \frac{T}{2tA_m} \qquad (5\text{--}18)$$

Here

τ_{avg} = the average shear stress acting over the thickness of the tube
T = the resultant internal torque at the cross section, which is found using the method of sections and the equations of equilibrium
t = the thickness of the tube where τ_{avg} is to be determined
A_m = the mean area enclosed within the boundary of the *centerline* of the tube's thickness. A_m is shown shaded in Fig. 5–30f.

Since $q = \tau_{avg}t$, we can determine the shear flow throughout the cross section using the equation

$$q = \frac{T}{2A_m} \qquad (5\text{--}19)$$

Angle of Twist. The angle of twist of a thin-walled tube of length L can be determined using energy methods, and the development of the necessary equation is given as a problem later in the text.* If the material behaves in a linear-elastic manner and G is the shear modulus, then this angle ϕ, given in radians, can be expressed as

$$\phi = \frac{TL}{4A_m^2 G} \oint \frac{ds}{t} \qquad (5\text{--}20)$$

Here the integration must be performed around the entire boundary of the tube's cross-sectional area.

IMPORTANT POINTS

• Shear flow q is the product of the tube's thickness and the average shear stress. This value is *constant* at all points along the tube's cross section. As a result, the *largest* average shear stress on the cross section occurs where the tube's thickness is *smallest*.

• Both shear flow and the average shear stress act *tangent* to the wall of the tube at all points and in a direction so as to contribute to the resultant torque.

---

*See Prob. 14–19.

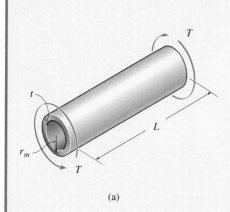

(a)

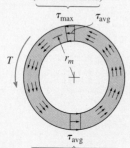

Actual shear–stress
distribution
(torsion formula)

$\tau_{\max}$　τ_{avg}

T　r_m

τ_{avg}

Average shear–stress
distribution
(thin-wall approximation)

(b)

Fig. 5–31

EXAMPLE 5–14

Calculate the average shear stress in a thin-walled tube having a circular cross section of mean radius r_m and thickness t, which is subjected to a torque T, Fig. 5–31a. Also, what is the relative angle of twist if the tube has a length L?

SOLUTION

Average Shear Stress. The mean area for the tube is $A_m = \pi r_m^2$. Applying Eq. 5–18 gives

$$\tau_{\text{avg}} = \frac{T}{2tA_m} = \frac{T}{2\pi tr_m^2} \qquad \textit{Ans.}$$

We can check the validity of this result by applying the torsion formula. In this case, using Eq. 5–9, we have

$$J = \frac{\pi}{2}(r_o^4 - r_i^4)$$

$$= \frac{\pi}{2}(r_o^2 + r_i^2)(r_o^2 - r_i^2)$$

$$= \frac{\pi}{2}(r_o^2 + r_i^2)(r_o + r_i)(r_o - r_i)$$

Since $r_m \approx r_o \approx r_i$ and $t = r_o - r_i$, $J = \frac{\pi}{2}(2r_m^2)(2r_m)t = 2\pi r_m^3 t$

so that
$$\tau_{\text{avg}} = \frac{Tr_m}{J} = \frac{Tr_m}{2\pi r_m^3 t} = \frac{T}{2\pi tr_m^2} \qquad \textit{Ans.}$$

which agrees with the previous result.

The average shear-stress distribution acting throughout the tube's cross section is shown in Fig. 5–31b. Also shown is the shear-stress distribution acting on a radial line as calculated using the torsion formula. Notice how each τ_{avg} acts in a direction such that it contributes to the resultant torque $\mathbf{T}$ at the section. As the tube's thickness decreases, the shear stress throughout the tube becomes more uniform.

Angle of Twist. Applying Eq. 5–20, we have

$$\phi = \frac{TL}{4A_m^2 G}\oint \frac{ds}{t} = \frac{TL}{4(\pi r_m^2)^2 Gt}\oint ds$$

The integral represents the length around the centerline boundary, which is $2\pi r_m$. Substituting, the final result is

$$\phi = \frac{TL}{2\pi r_m^3 Gt} \qquad \textit{Ans.}$$

Show that one obtains this same result using Eq. 5–15

EXAMPLE 5-15

The tube is made of C86100 bronze and has a rectangular cross section as shown in Fig. 5–32a. If it is subjected to the two torques, determine the average shear stress in the tube at points A and B. Also, what is the angle of twist of end C? The tube is fixed at E.

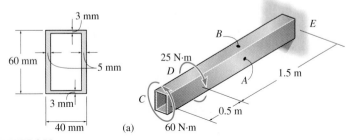

(a)

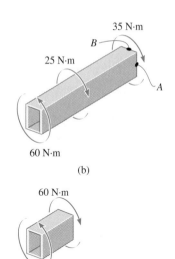

(b)

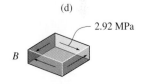

(c)

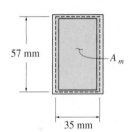

(d)

SOLUTION

Average Shear Stress.   If the tube is sectioned through points A and B, the resulting free-body diagram is shown in Fig. 5–32b. The internal torque is 35 N · m. As shown in Fig. 5–32d, the area A_m is

$$A_m = (0.035 \text{ m})(0.057 \text{ m}) = 0.00200 \text{ m}^2$$

Applying Eq. 5–18 for point A, $t_A = 5$ mm, so that

$$\tau_A = \frac{T}{2tA_m} = \frac{35 \text{ N} \cdot \text{m}}{2(0.005 \text{ m})(0.00200 \text{ m}^2)} = 1.75 \text{ MPa} \qquad \textit{Ans.}$$

And for point B, $t_B = 3$ mm, and therefore

$$\tau_B = \frac{T}{2tA_m} = \frac{35 \text{ N} \cdot \text{m}}{2(0.003 \text{ m})(0.00200 \text{ m}^2)} = 2.92 \text{ MPa} \qquad \textit{Ans.}$$

These results are shown on elements of material located at points A and B, Fig. 5–32e. Note carefully how the 35-N · m torque in Fig. 5–32b creates these stresses on the color-shaded faces of each element.

Angle of Twist.   From the free-body diagrams in Fig. 5–32b and 5–32c, the internal torques in regions DE and CD are 35 N · m and 60 N · m, respectively. Following the sign convention outlined in Sec. 5.4, these torques are both positive. Thus, Eq. 5–20 becomes

$$\phi = \sum \frac{TL}{4A_m^2 G} \oint \frac{ds}{t}$$

$$= \frac{60 \text{ N} \cdot \text{m}(0.5 \text{ m})}{4(0.00200 \text{ m}^2)^2(38(10^9) \text{ N/m}^2)} \left[2\left(\frac{57 \text{ mm}}{5 \text{ mm}} \right) + 2\left(\frac{35 \text{ mm}}{3 \text{ mm}} \right) \right]$$

$$+ \frac{35 \text{ N} \cdot \text{m}(1.5 \text{ m})}{4(0.00200 \text{ m}^2)^2(38(10^9) \text{ N/m}^2)} \left[2\left(\frac{57 \text{ mm}}{5 \text{ mm}} \right) + 2\left(\frac{35 \text{ mm}}{3 \text{ mm}} \right) \right]$$

$$= 6.29(10^{-3}) \text{ rad} \qquad \textit{Ans.}$$

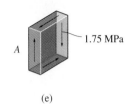

(e)

Fig. 5–32

EXAMPLE 5-16

A square aluminum tube has the dimensions shown in Fig. 5–33a. Determine the average shear stress in the tube at point A if it is subjected to a torque of 85 lb · ft. Also compute the angle of twist due to this loading. Take $G_{al} = 3.80(10^3)$ ksi.

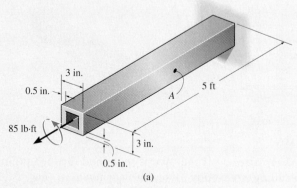

(a)

SOLUTION

Average Shear Stress. By inspection, the internal resultant torque at the cross section where point A is located is $T = 85$ lb · ft. From Fig. 5–33b, the area A_m, shown shaded, is

$$A_m = (2.5 \text{ in.})(2.5 \text{ in.}) = 6.25 \text{ in}^2$$

Applying Eq. 5–18,

$$\tau_{avg} = \frac{T}{2tA_m} = \frac{85 \text{ lb} \cdot \text{ft}(12 \text{ in./ft})}{2(0.5 \text{ in.})(6.25 \text{ in}^2)} = 163 \text{ psi} \qquad \textbf{\textit{Ans.}}$$

Since t is constant except at the corners, the average shear stress is the same at all points on the cross section. It is shown acting on an element located at point A in Fig. 5–33c. Note that τ_{avg} acts upward on the color-shaded face, since it contributes to the internal resultant torque **T** at the section.

Angle of Twist. The angle of twist caused by T is determined from Eq. 5–20; i.e.,

$$\phi = \frac{T L}{4A_m^2 G} \oint \frac{ds}{t} = \frac{85 \text{ lb} \cdot \text{ft}(12 \text{ in./ft})(5 \text{ ft})(12 \text{ in./ft})}{4(6.25 \text{ in}^2)^2[3.80(10^6) \text{ lb/in}^2]} \oint \frac{ds}{(0.5 \text{ in.})}$$

$$= 0.206(10^{-3}) \text{ in}^{-1} \oint ds$$

Here the integral represents the *length* around the centerline boundary of the tube, Fig. 5–33b. Thus,

$$\phi = 0.206(10^{-3}) \text{ in}^{-1}[4(2.5 \text{ in.})] = 2.06(10^{-3}) \text{ rad} \qquad \textbf{\textit{Ans.}}$$

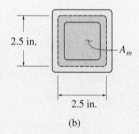

2.5 in.

2.5 in.

A_m

(b)

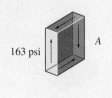

163 psi

A

(c)

Fig. 5–33

EXAMPLE 5–17

A thin tube is made from three 5-mm-thick A-36 steel plates such that it has a cross section that is triangular as shown in Fig. 5–34a. Determine the maximum torque T to which it can be subjected, if the allowable shear stress is $\tau_{\text{allow}} = 90$ MPa and the tube is restricted to twist no more than $\phi = 2(10^{-3})$ rad.

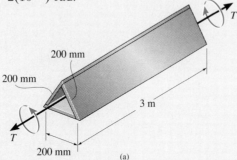

200 mm

200 mm

3 m

200 mm

(a)

SOLUTION

The area A_m is shown shaded in Fig. 5–34b. It is

$$A_m = \frac{1}{2}(200 \text{ mm})(200 \text{ mm sin } 60°)$$

$$= 17.32(10^3) \text{ mm}^2(10^{-6} \text{ m}^2/\text{mm}^2) = 17.32(10^{-3}) \text{ m}^2$$

The greatest average shear stress occurs at points where the tube's thickness is smallest, which is along the sides and not at the corners. Applying Eq. 5–18, with $t = 0.005$ m, yields

$$\tau_{\text{avg}} = \frac{T}{2tA_m}; \quad 90(10^6) \text{ N/m}^2 = \frac{T}{2(0.005 \text{ m})(17.32(10^{-3}) \text{ m}^2)}$$

$$T = 15.6 \text{ kN} \cdot \text{m}$$

Also, from Eq. 5–20, we have

$$\phi = \frac{TL}{4A_m^2 G} \oint \frac{ds}{t}$$

$$0.002 \text{ rad} = \frac{T(3 \text{ m})}{4(17.32(10^{-3}) \text{ m})^2[75(10^9) \text{ N/m}^2]} \oint \frac{ds}{(0.005 \text{ m})}$$

$$300.0 = T \oint ds$$

The integral represents the sum of the dimensions along the three sides of the center-line boundary. Thus,

$$300.0 = T[3(0.20 \text{ m})]$$

$$T = 500 \text{ N} \cdot \text{m} \qquad \qquad \textit{Ans.}$$

By comparison, the application of torque is restricted due to the angle of twist.

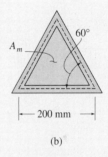

A_m

60°

200 mm

(b)

Fig. 5–34

PROBLEMS

5–90. The aluminum rod has a square cross section of 10 mm by 10 mm. If it is 8 m long, determine the torque T that is required to rotate one end relative to the other end by 90°. $G_{al} = 28$ GPa, $(\tau_Y)_{al} = 240$ MPa.

*5–92. The shaft is made of red brass C83400 and has an elliptical cross section. If it is subjected to the torsional loading shown, determine the maximum shear stress within regions AC and BC, and the angle of twist ϕ of end B relative to end A.

5–93. Solve Prob. 5–92 for the maximum shear stress within regions AC and BC, and the angle of twist ϕ of end B relative to C.

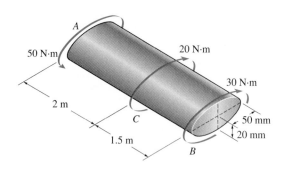

Probs. 5–92/5–93

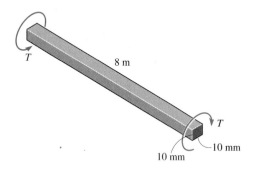

Prob. 5–90

5–91. The 6061-T6 aluminum bar has a square cross section of 25 mm by 25 mm. If it is 2 m long, determine the maximum shear stress in the bar and the rotation of one end relative to the other end.

5–94. The square shaft is used at the end of a drive cable in order to register the rotation of the cable on a gauge. If it has the dimensions shown and is subjected to a torque of 8 N · m, determine the shear stress in the shaft at point A. Sketch the shear stress on a volume element located at this point.

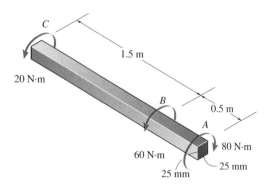

Prob. 5–91

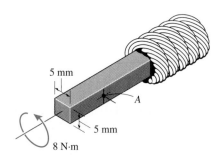

Prob. 5–94

5–95. The uniform C86100 bronze rod has an elliptical cross section. If it is subjected to the five torques shown, determine the dimension a of its cross section so that the shear stress in the rod does not exceed $\tau_{\text{allow}} = 85$ MPa. Also, what is the rotation of end A with respect to end B?

*5–96. The uniform C86100 bronze rod has an elliptical cross section. If it is subjected to the five torques shown, determine the maximum shear stress in the rod and the angle of twist of end A with respect to end B. Take $a = 20$ mm.

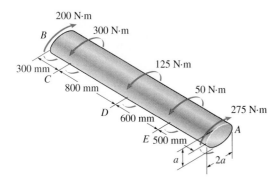

Probs. 5–95/5–96

5–97. The aluminum strut is fixed between the two walls at A and B. If it has a 2 in. by 2 in. square cross section, and it is subjected to the torque of 80 lb · ft at C, determine the reactions at the fixed supports. Also, what is the angle of twist at C? $G_{al} = 3.8(10^3)$ ksi.

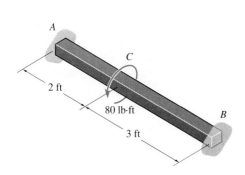

Prob. 5–97

5–98. The brass wire has a triangular cross section, 2 mm on a side. If the yield stress for brass is $\tau_Y = 205$ MPa, determine the maximum torque T to which it can be subjected so that the wire will not yield. If this torque is applied to a segment 4 m long, determine the greatest angle of twist of one end of the wire relative to the other end that will not cause permanent damage to the wire. $G_{br} = 37$ GPa.

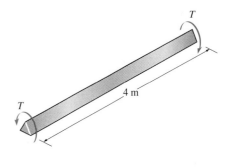

Prob. 5–98

5–99. The plastic tube is subjected to a torque of 150 N · m. Determine the mean dimension a of its sides if the allowable shear stress is $\tau_{\text{allow}} = 60$ MPa. Each side has a thickness of $t = 3$ mm. Neglect stress concentrations at the corners.

*5–100. The plastic tube is subjected to a torque of 150 N · m. Determine the average shear stress in the tube if the mean dimension $a = 200$ mm. Each side has a thickness of $t = 3$ mm. Neglect stress concentrations at the corners.

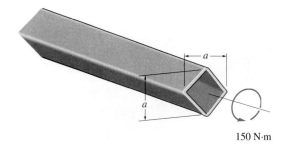

Probs. 5–99/5–100

5–101. A torque of 2 kip · in. is applied to the tube. If the wall thickness is 0.1 in., determine the average shear stress in the tube.

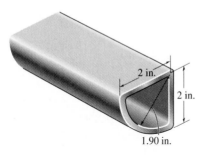

Prob. 5–101

5–102. The 304 stainless steel tube has a thickness of 10 mm. If the allowable shear stress is $\tau_{allow} = 80$ MPa, determine the maximum torque T that it can transmit. Also, what is the angle of twist of one end of the tube with respect to the other if the tube is 4 m long? Neglect the stress concentrations at the corners. The mean dimensions are shown.

5–103. The 304 stainless steel tube has a thickness of 10 mm. If the applied torque is $T = 50$ N · m, determine the average shear stress in the tube. Neglect the stress concentrations at the corners. The mean dimensions are shown.

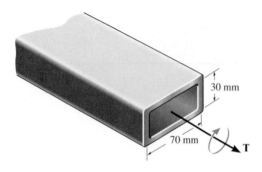

Probs. 5–102/5–103

***5–104.** The tube is made of plastic, is 5 mm thick, and has the mean dimensions shown. Determine the average shear stress at points A and B if it is subjected to the torque of $T = 5$ N · m. Show the shear stress on volume elements located at these points.

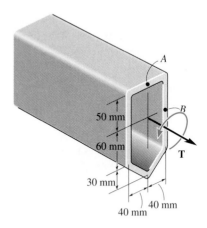

Prob. 5–104

5–105. The steel tube has an elliptical cross section of the mean dimensions shown and a constant thickness of $t = 0.2$ in. If the allowable shear stress is $\tau_{allow} = 8$ ksi, determine the necessary dimension b needed to resist the torque shown. The mean area A_m for the ellipse is $\pi b(0.5b)$.

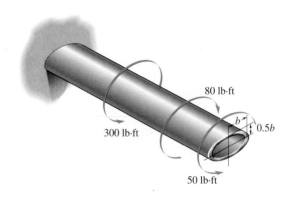

Prob. 5–105

5–106. The tube is made of plastic, is 5 mm thick, and has the mean dimensions shown. Determine the average shear stress at points A and B if the tube is subjected to the torque of $T = 500\ \text{N} \cdot \text{m}$. Show the shear stress on volume elements located at these points. Neglect stress concentrations at the corners.

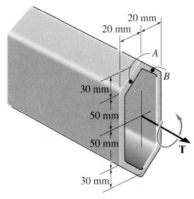

20 mm
20 mm
A
B
30 mm
50 mm
50 mm
T
30 mm

Prob. 5–106

5–107. For a given average shear stress, determine the factor by which the torque-carrying capacity is increased if the half-circular sections are reversed from the dashed-line positions to the section shown. The tube is 0.1 in. thick.

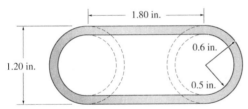

1.80 in.
0.6 in.
1.20 in.
0.5 in.

Prob. 5–107

*5–108. Due to a fabrication error the inner circle of the tube is eccentric with respect to the outer circle. By what percentage is the torsional strength reduced when the eccentricity e is one-fourth of the difference in the radii?

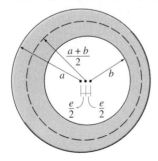

$\dfrac{a+b}{2}$
a b
$\dfrac{e}{2}$ $\dfrac{e}{2}$

Prob. 5–108

5–109. The symmetric tube is made from a high-strength steel, having the mean dimensions shown and a thickness of 5 mm. If it is subjected to a torque of $T = 40\ \text{N} \cdot \text{m}$, determine the average shear stress developed at points A and B. Indicate the shear stress on volume elements located at these points.

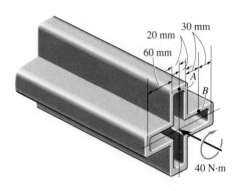

30 mm
20 mm
60 mm
A
B
40 N·m

Prob. 5–109

5–110. The plastic hexagonal tube is subjected to a torque of 150 N · m. Determine the mean dimension a of its sides if the allowable shear stress is $\tau_{\text{allow}} = 60$ MPa. Each side has a thickness of $t = 3$ mm.

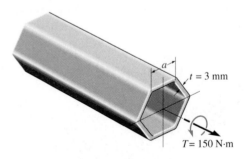

a
$t = 3$ mm
$T = 150$ N·m

Prob. 5–110

5.8 Stress Concentration

The torsion formula, $\tau_{max} = Tc/J$, can be applied to regions of a shaft having a circular cross section that is constant or tapers slightly. When sudden changes arise in the cross section, both the shear-stress and shear-strain distributions in the shaft become complex and can be obtained only by using experimental methods or possibly by a mathematical analysis based on the theory of elasticity. Three common discontinuities of the cross section that occur in practice are shown in Fig. 5–35. They are at *couplings*, which are used to connect two collinear shafts together, Fig. 5–35*a*, *keyways*, used to connect gears or pulleys to a shaft, Fig. 5–35*b*, and *shoulder fillets*, used to fabricate a single collinear shaft from two shafts having different diameters, Fig. 5–35*c*. In each case the maximum shear stress will occur at the point (dot) indicated on the cross section.

In order to eliminate the necessity for the engineer to perform a complex stress analysis at a shaft discontinuity, the maximum shear stress can be determined for a specified geometry using a **torsional stress-concentration factor**, *K*. As in the case of axially loaded members, Sec. 4.7, *K* is usually taken from a graph. An example, for the shoulder-fillet shaft, is shown in Fig. 5–36. To use this graph, one first

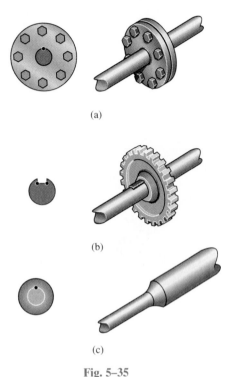

(a)

(b)

(c)

Fig. 5–35

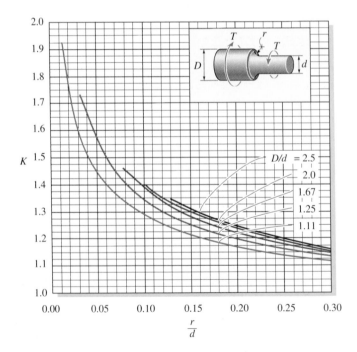

Fig. 5–36

finds the geometric ratio D/d to define the appropriate curve, and then once the abscissa r/d is calculated, the value of K is found along the ordinate. The maximum shear stress is then determined from the equation

$$\tau_{max} = K \frac{Tc}{J}$$

(5–21)

Here the torsion formula is applied to the *smaller* of the two connected shafts, since τ_{max} occurs at the base of the fillet, Fig. 5–35c.

It can be noted from the graph that an *increase* in fillet radius r causes a *decrease* in K. Hence the maximum shear stress in the shaft can be reduced by *increasing* the fillet radius. Also, if the diameter of the larger shaft is reduced, the D/d ratio will be lower and so the value of K and therefore τ_{max} will be lower.

Like the case of axially loaded members, torsional stress concentration factors should *always* be used when designing shafts made from *brittle materials*, or when designing shafts that will be subjected to *fatigue or cyclic torsional loadings*. These conditions give rise to the formation of cracks at the stress concentration, and this can often lead to a sudden failure of the shaft. Also realize that if a large *static* torsional loading is applied to a shaft made from *ductile material*, then *inelastic strains* may develop within the shaft. As a result of yielding, the stress distribution will become more *evenly distributed* throughout the shaft, so that the maximum stress that results will not be limited at the stress concentration. This phenomenon will be discussed further in the next section.

Stress concentrations can arise at the coupling of these shafts, and this must be taken into account when the coupling is designed.

IMPORTANT POINTS

• *Stress concentrations* in shafts occur at points of sudden cross-sectional change, such as couplings, keyways, and at shoulder fillets. The more severe the change the larger the stress concentration.

• For design or analysis, it is not necessary to know the exact shear-stress distribution on the cross section. Instead, it is possible to obtain the maximum shear stress using a stress concentration factor, K, that has been determined through experiment, and is only a function of the geometry of the shaft.

• Normally the stress concentration in a ductile shaft subjected to a static torque will *not* have to be considered in design, however, if the material is *brittle*, or subjected to *fatigue* loadings, then stress concentrations become important.

EXAMPLE 5–18

The stepped shaft shown in Fig. 5–37a is supported by bearings at A and B. Determine the maximum stress in the shaft due to the applied torques. The fillet at the junction of each shaft has a radius of $r = 6$ mm.

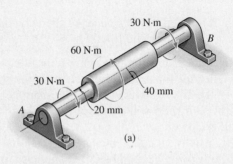

(a)

Fig. 5–37

SOLUTION

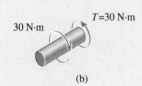

(b)

Internal Torque.   By inspection, moment equilibrium about the axis of the shaft is satisfied. Since the maximum shear stress occurs at the rooted ends of the *smaller* diameter shafts, the internal torque (30 N · m) can be found there by applying the method of sections, Fig. 5–37b.

Maximum Shear Stress.   The stress-concentration factor can be determined by using Fig. 5–36. From the shaft geometry we have

$$\frac{D}{d} = \frac{2(40 \text{ mm})}{2(20 \text{ mm})} = 2$$

$$\frac{r}{d} = \frac{6 \text{ mm}}{2(20 \text{ mm})} = 0.15$$

Thus, the value of $K = 1.3$ is obtained.
Applying Eq. 5–21, we have

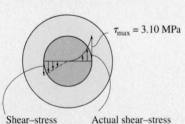

Shear–stress
distribution
predicted by
torsion formula

Actual shear–stress
distribution caused
by stress concentration

(c)

$$\tau_{max} = K \frac{Tc}{J}; \qquad \tau_{max} = 1.3 \left[\frac{30 \text{ N} \cdot \text{m}(0.020 \text{ m})}{(\pi/2)(0.020 \text{ m})^4} \right] = 3.10 \text{ MPa} \qquad \textit{Ans.}$$

From experimental evidence, the actual stress distribution along a radial line of the cross section at the critical section looks similar to that shown in Fig. 5–37c. Notice how this compares with the linear stress distribution found from the torsion formula.

*5.9   INELASTIC TORSION

The equations of stress and deformation developed thus far are valid only if the applied torque causes the material to behave in a linear-elastic manner. If the torsional loadings are excessive, however, the material may yield, and, consequently, a "plastic analysis" must then be used to determine the shear-stress distribution and the angle of twist. To perform this analysis, it is necessary to meet the conditions of both deformation and equilibrium for the shaft.

It was shown in Sec. 5.1 that the shear strains that develop in the material must vary *linearly* from zero at the center of the shaft to a maximum at its outer boundary, Fig. 5–38a. This conclusion was based entirely on geometric considerations and not the material's behavior. Also, the resultant torque at the section must be equivalent to the torque caused by the entire shear-stress distribution over the cross section. This condition can be expressed mathematically by considering the shear stress τ acting on an element of area dA located a distance ρ from the center of the shaft, Fig. 5–38b. The force produced by this stress is $dF = \tau\, dA$, and the torque produced is $dT = \rho\, dF = \rho\tau\, dA$. For the entire shaft we require

$$T = \int_A \rho\tau\, dA \qquad (5\text{--}22)$$

If the area dA over which τ acts can be defined as a *differential ring* having an area of $dA = 2\pi\rho\, d\rho$, Fig. 5–38c, then the above equation can be written as

$$T = 2\pi \int_A \tau\rho^2\, d\rho \qquad (5\text{--}23)$$

These conditions of geometry and loading will now be used to determine the shear-stress distribution in a shaft when the shaft is subjected to three types of torque.

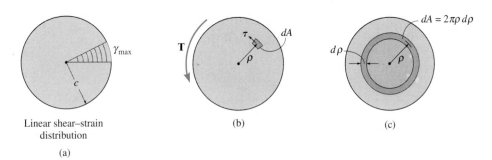

Linear shear–strain
distribution

(a)

(b)

(c)

Fig. 5–38

Maximum Elastic Torque.   If the torque produces the maximum *elastic* shear strain γ_Y, at the outer boundary of the shaft, then the shear-strain distribution along a radial line of the shaft will look like that shown in Fig. 5–39b. To establish the shear-stress distribution, we must either use Hooke's law or find the corresponding values of shear stress from the material's $\tau - \gamma$ diagram, Fig. 5–39a. For example, a shear strain γ_Y produces the shear stress τ_Y at $\rho = c$. Likewise, at $\rho = \rho_1$ the shear strain is $\gamma_1 = (\rho_1/c)\gamma_Y$. From the $\tau - \gamma$ diagram γ_1 produces τ_1. When these stresses and others like them are plotted at $\rho = c$, $\rho = \rho_1$, etc., the expected *linear* shear-stress distribution in Fig. 5–39c results. Since this shear-stress distribution can be described mathematically as $\tau = \tau_Y(\rho/c)$, the maximum elastic torque can be determined from Eq. 5–23; i.e.,

$$T_Y = 2\pi \int_0^c \tau_Y \left(\frac{\rho}{c}\right) \rho^2 \, d\rho$$

or

$$T_Y = \frac{\pi}{2} \tau_Y c^3 \tag{5–24}$$

This same result can of course be obtained in a more direct manner using the torsion formula; $\tau_Y = T_Y c/[(\pi/2)c^4]$. Furthermore, the angle of twist can be determined from Eq. 5–13, namely,

$$d\phi = \gamma \frac{dx}{\rho} \tag{5–25}$$

As noted in Sec. 5.4, this equation results in $\phi = TL/JG$, when the shaft is subjected to a constant torque and has a constant cross-sectional area.

Elastic-Plastic Torque.   Let us now consider the material in the shaft to exhibit an elastic perfectly plastic behavior. As shown in Fig. 5–40a,

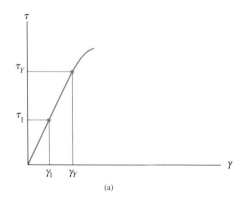

(a)

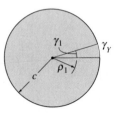

Shear–strain distribution

(b)

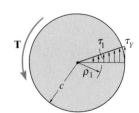

Shear–stress distribution

(c)

Fig. 5–39

this is characterized by a shear stress–strain diagram for which the material undergoes an increasing amount of shear strain when the shear stress in the material reaches the yield point τ_Y. Thus, as the applied torque increases in magnitude above T_Y, it will begin to cause yielding. First at the outer boundary of the shaft, $\rho = c$ and then, as the maximum shear strain increases to, say, γ', the yielding boundary will progress inward toward the shaft's center, Fig. 5–40b. As shown, this produces an *elastic core*, where, by proportion, the outer radius of the core is $\rho_Y = (\gamma_Y/\gamma')c$. Also, the outer portion of the shaft forms a *plastic annulus* or ring, since the shear strains γ are greater than γ_Y within this region. The corresponding shear-stress distribution along a radial line of the shaft is shown in Fig. 5–40c. It was established by taking successive points on the shear-strain distribution and finding the corresponding value of shear stress from the $\tau - \gamma$ diagram. For example, at $\rho = c$, γ' gives τ_Y, and at $\rho = \rho_Y$, γ_Y also gives τ_Y; etc.

Since τ can now be established as a function of ρ, we can apply Eq. 5–23 to determine the torque. As a general formula for elastic-plastic material behavior, we have

$$T = 2\pi \int_0^c \tau \rho^2 \, d\rho$$

$$= 2\pi \int_0^{\rho_Y} \left(\tau_Y \frac{\rho}{\rho_Y}\right) \rho^2 \, d\rho + 2\pi \int_{\rho_Y}^c \tau_Y \rho^2 \, d\rho$$

$$= \frac{2\pi}{\rho_Y} \tau_Y \int_0^{\rho_Y} \rho^3 \, d\rho + 2\pi\tau_Y \int_{\rho_Y}^c \rho^2 \, d\rho$$

$$= \frac{\pi}{2\rho_Y} \tau_Y \rho_Y^4 + \frac{2\pi}{3} \tau_Y(c^3 - \rho_Y^3)$$

$$= \frac{\pi\tau_Y}{6}(4c^3 - \rho_Y^3) \tag{5–26}$$

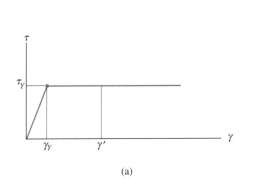

(a)

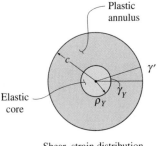

Elastic core

Plastic annulus

Shear–strain distribution

(b)

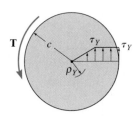

Shear–stress distribution

(c)

Fig. 5–40

Plastic Torque. Further increases in T tend to shrink the radius of the elastic core until all the material will yield, i.e., $\rho_Y \to 0$, Fig. 5–40b. The material of the shaft is then subjected to *perfectly plastic behavior* and the shear-stress distribution is constant, Fig. 5–40d. Since then $\tau = \tau_Y$, we can apply Eq. 5–23 to determine the *plastic torque* T_p, which represents the largest possible torque the shaft will support.

$$T_p = 2\pi \int_0^c \tau_Y \rho^2 \, d\rho$$

$$= \frac{2\pi}{3} \tau_Y c^3 \qquad (5\text{–}27)$$

By comparison with the maximum elastic torque T_Y, Eq. 5–24, it can be seen that

$$T_p = \frac{4}{3} T_Y$$

In other words, the plastic torque is 33% greater than the maximum elastic torque.

The angle of twist ϕ for the shear-stress distribution in Fig. 5–40d *cannot* be uniquely defined. This is because $\tau = \tau_Y$ does not correspond to any unique value of shear strain $\gamma \geq \gamma_Y$. As a result, once $\mathbf{T}_p$ is applied, the shaft will continue to deform or twist with no corresponding increase in shear stress.

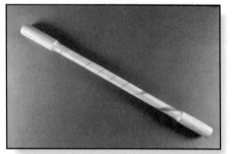

Severe twist of an aluminum specimen caused by the application of a plastic torque.

Ultimate Torque. In the general case, most engineering materials will have a shear stress–strain diagram as shown in Fig. 5–41a. Consequently, if T is increased so that the maximum shear strain in the shaft becomes $\gamma = \gamma_u$, Fig. 5–41b, then, by proportion γ_Y occurs at $\rho_Y = (\gamma_Y/\gamma_u)c$. Likewise, the shear strains at, say, $\rho = \rho_1$ and $\rho = \rho_2$, can be found by proportion, i.e., $\gamma_1 = (\rho_1/c)\gamma_u$ and $\gamma_2 = (\rho_2/c)\gamma_u$. If corresponding values of τ_1, τ_Y, τ_2, and τ_u are taken from the $\tau - \gamma$ diagram and plotted,

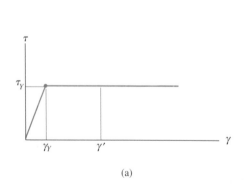

(a)

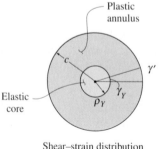

Shear–strain distribution

(b)

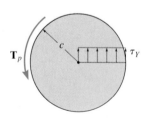

Fully plastic torque

(d)

Fig. 5–40

we obtain the shear-stress distribution, which acts over a radial line on the cross section, Fig. 5–41c. The torque produced by this stress distribution is called the **ultimate torque**, T_u, since any further increase in shear strain would cause the maximum shear stress at the outer boundary of the shaft to be less than τ_u, and therefore the torque produced by the resulting shear-stress distribution would be *less* than T_u.

The magnitude of $\mathbf{T}_u$ can be determined by "graphically" integrating Eq. 5–23. To do this, the cross-sectional area of the shaft is segmented into a finite number of rings, such as the one shown shaded in Fig. 5–41d. The area of this ring, $\Delta A = 2\pi\rho\,\Delta\rho$, is multiplied by the shear stress τ that acts on it, so that the force $\Delta F = \tau\,\Delta A$ can be determined. The torque created by this force is then $\Delta T = \rho\,\Delta F = \rho(\tau\,\Delta A)$. The addition of all the torques for the entire cross section, as determined in this manner, gives the ultimate torque T_u; that is, Eq. 5–23 becomes $T_u \approx 2\pi\Sigma\tau\rho^2\,\Delta\rho$. On the other hand, if the stress distribution can be expressed as an analytical function, $\tau = f(\rho)$, as in the elastic and plastic torque cases, then the integration of Eq. 5–23 can be carried out directly.

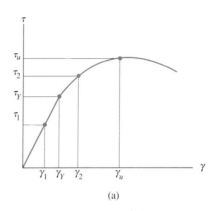

(a)

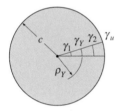

Ultimate shear–strain distribution

(b)

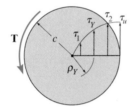

Ultimate shear–stress distribution

(c)

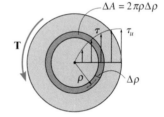

(d)

Fig. 5–41

IMPORTANT POINTS

- The *shear-strain distribution* over a radial line on a shaft is based on geometric considerations, and it is found to *always remain linear*. The shear–stress distribution, however, depends upon the applied torque and must therefore be determined from the material behavior, or shear stress-strain diagram.

- Once the shear-stress distribution for the shaft is established, it produces a torque about the axis of the shaft that is equivalent to the resultant torque acting on the cross section.

- *Perfectly plastic behavior* assumes, the shear-stress distribution is *constant*, and the shaft will continue to twist with no increase in torque. This torque is called the *plastic torque*.

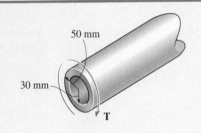

50 mm

30 mm

T

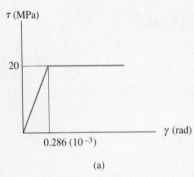

τ (MPa)

20

γ (rad)

0.286 (10^{-3})

(a)

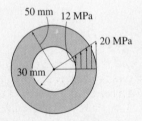

50 mm 12 MPa

20 MPa

30 mm

Elastic shear–stress distribution

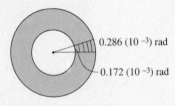

0.286 (10^{-3}) rad

0.172 (10^{-3}) rad

Elastic shear–strain distribution

(b)

Fig. 5–42

EXAMPLE 5–19

The tubular shaft in Fig. 5–42a is made of an aluminum alloy that is assumed to have an elastic-plastic $\tau - \gamma$ diagram as shown. Determine (a) the maximum torque that can be applied to the shaft without causing the material to yield, (b) the maximum torque or plastic torque that can be applied to the shaft. What should the minimum shear strain at the outer radius be in order to develop a plastic torque?

SOLUTION

Maximum Elastic Torque.   We require the shear stress at the outer fiber to be 20 MPa. Using the torsion formula, we have

$$\tau_Y = \frac{T_Y c}{J}; \quad 20(10^6) \text{ N/m}^2 = \frac{T_Y(0.05 \text{ m})}{(\pi/2)[(0.05 \text{ m})^4 - (0.03 \text{ m})^4]}$$

$$T_Y = 3.42 \quad \text{kN} \cdot \text{m} \qquad \textit{Ans.}$$

The shear-stress and shear-strain distributions for this case are shown in Fig. 5–42b. The values at the tube's inner wall are obtained by proportion.

Plastic Torque.   The shear-stress distribution in this case is shown in Fig. 5–42c. Application of Eq. 5–23 requires $\tau = \tau_Y$. We have

$$T_p = 2\pi \int_{0.03 \text{ m}}^{0.05 \text{ m}} [20(10^6) \text{ N/m}^2]\rho^2 \, d\rho = 125.66(10^6) \frac{1}{3}\rho^3 \Big|_{0.03 \text{ m}}^{0.05 \text{ m}}$$

$$= 4.10 \quad \text{kN} \cdot \text{m} \qquad \textit{Ans.}$$

For this tube T_p represents a 20% increase in torque capacity compared with the elastic torque T_Y.

Outer Radius Shear Strain.   The tube becomes fully plastic when the shear strain at the *inner wall* becomes $0.286(10^{-3})$ rad, as shown in Fig. 5–42c. Since the shear strain *remains linear* over the cross section, the plastic strain at the outer fibers of the tube in Fig. 5–42c is determined by proportion;

$$\frac{\gamma_o}{50 \text{ mm}} = \frac{0.286(10^{-3}) \text{ rad}}{30 \text{ mm}}$$

$$\gamma_o = 0.477(10^{-3}) \text{ rad} \qquad \textit{Ans.}$$

20 MPa

0.477 (10^{-3}) rad

0.286 (10^{-3}) rad

Plastic shear–stress distribution   Initial plastic shear–strain distribution

(c)

EXAMPLE 5-20

A solid circular shaft has a radius of 20 mm and length of 1.5 m. The material has an elastic-plastic $\tau - \gamma$ diagram as shown in Fig. 5–43a. Determine the torque needed to twist the shaft $\phi = 0.6$ rad.

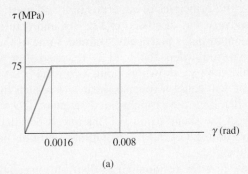

(a)

SOLUTION

To solve the problem, we will first obtain the shear-strain distribution, then establish the shear-stress distribution. Once this is known, the applied torque can be determined.

The maximum shear strain occurs at the surface of the shaft, $\rho = c$. Since the angle of twist is $\phi = 0.6$ rad for the entire 1.5-m length of shaft, then using Eq. 5–25 for the entire length, we have

$$\phi = \gamma \frac{L}{\rho}; \qquad 0.6 = \frac{\gamma_{max}(1.5 \text{ m})}{(0.02 \text{ m})}$$

$$\gamma_{max} = 0.008 \text{ rad}$$

The shear-strain distribution, which always varies linearly, is shown in Fig. 5–43b. Note that yielding of the material occurs since $\gamma_{max} > \gamma_Y = 0.0016$ rad in Fig. 5–43a. The radius of the elastic core, ρ_Y, can be obtained by proportion. From Fig. 5–43b,

$$\frac{\rho_Y}{0.0016} = \frac{0.02 \text{ m}}{0.008}$$

$$\rho_Y = 0.004 \text{ m} = 4 \text{ mm}$$

Based on the shear-strain distribution, the shear-stress distribution, plotted over a radial line segment, is shown in Fig. 5–43c. The torque can now be obtained using Eq. 5–26. Substituting in the numerical data yields

$$T = \frac{\pi \tau_Y}{6}(4c^3 - \rho_Y^3)$$

$$= \frac{\pi [75(10^6) \text{ N/m}^2]}{6} [4(0.02 \text{ m})^3 - (0.004 \text{ m})^3]$$

$$= 1.25 \text{ kN} \cdot \text{m} \qquad\qquad Ans.$$

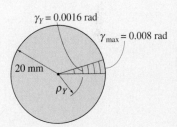

Shear–strain distribution

(b)

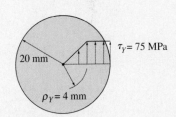

Shear–stress distribution

(c)

Fig. 5–43

*5.10   RESIDUAL STRESS

When a shaft is subjected to plastic shear strains caused by torsion, removal of the torque will cause some shear stress to remain in the shaft. This stress is referred to as **residual stress**, and its distribution can be calculated using the principles of superposition and elastic recovery.

Elastic recovery was discussed in Sec. 3.4, and it refers to the fact that whenever a material is plastically strained, some of the strain in the material is recovered when the load is released. For example, if a material is strained to γ_1, shown as point C on the $\tau - \gamma$ curve in Fig. 5–44, the release will cause a reverse shear stress, such that the material behavior will follow the straight-lined segment CD, creating some *elastic recovery* of the shear strain γ_1. This line is parallel to the initial straight-lined portion AB of the $\tau - \gamma$ diagram, and thus both lines have a slope G as indicated.

To illustrate how the residual-stress distribution can be determined in a shaft, we will first consider the shaft to be subjected to a plastic torque $\mathbf{T}_p$. As explained in Sec. 5.9, $\mathbf{T}_p$ creates the shear-stress distribution

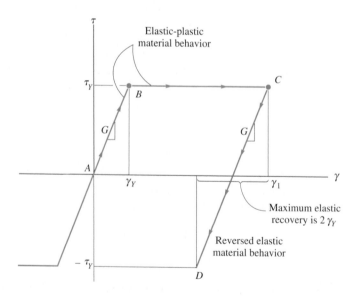

Fig. 5–44

shown in Fig. 5–45a. We will assume that this distribution is a consequence of straining the material at the shaft's outer boundary to γ_1 in Fig. 5–44. Also, γ_1 is large enough so that the radius of the elastic core is assumed to approach zero, that is, $\gamma_1 \gg \gamma_Y$. If $\mathbf{T}_p$ is removed, the material tends to recover *elastically*, following along line *CD*. Since elastic behavior occurs, we can superimpose on the stress distribution in Fig. 5–45a a *linear stress distribution* caused by applying the plastic torque $\mathbf{T}_p$ in the *opposite* direction, Fig. 5–45b. Here the maximum shear stress τ_r, computed for this stress distribution, is called the *modulus of rupture* for torsion. It is determined from the torsion formula,* which gives

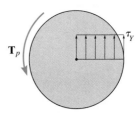

$$\tau_r = \frac{T_p c}{J} = \frac{T_p c}{(\pi/2)c^4}$$

Using Eq. 5–27,

$$\tau_r = \frac{[(2/3)\pi\tau_Y c^3]c}{(\pi/2)c^4} = \frac{4}{3}\tau_Y$$

Plastic torque applied causing plastic shear strains throughout the shaft

(a)

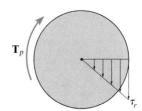

Plastic torque reversed causing elastic shear strains throughout the shaft

(b)

Note that reversed application of $\mathbf{T}_p$ using the linear shear-stress distribution in Fig. 5–45b is possible here, since the maximum recovery for the elastic shear strain is $2\gamma_Y$ as noted in Fig. 5–44. This corresponds to a maximum applied shear stress of $2\tau_Y$, which is *greater* than the maximum shear stress of $\frac{4}{3}\tau_Y$ computed above. Hence, by superimposing the stress distributions involving applications and then removal of the plastic torque, we obtain the residual-shear-stress distribution in the shaft as shown in Fig. 5–45c. It should be noted from this diagram that the shear stress at the center of the shaft, shown as τ_Y, must actually be *zero*, since the material along the axis of the shaft is not strained. The reason this is not zero is because we assumed that *all* the material of the shaft was strained beyond the yield point in order to determine the plastic torque, Fig. 5–45a To be more realistic, an elastic-plastic torque should be considered when modeling the material behavior. Doing so leads to the superposition of the stress distribution shown in Fig. 5–45d.

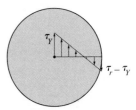

Residual–shear–stress distribution in shaft

(c)

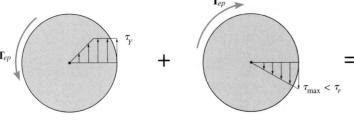

+

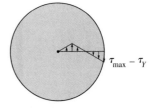

=

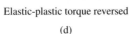

Elastic-plastic torque applied   Elastic-plastic torque reversed   Residual shear stress distribution in shaft

(d)

Fig. 5–45

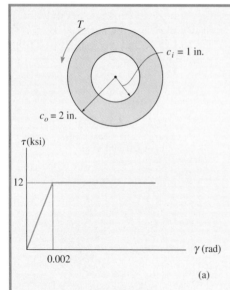

(a)

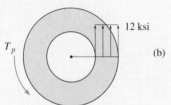

Plastic torque applied

(b)

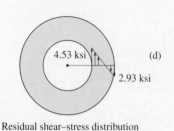

Plastic torque reversed

(c)

Residual shear–stress distribution

(d)

Fig. 5–46

EXAMPLE 5-21

A tube is made from a brass alloy having a length of 5 ft and cross-sectional area shown in Fig. 5–46a. The material has an elastic-plastic $\tau - \gamma$ diagram, also shown in Fig. 5–46a. Determine the plastic torque T_p. What are the residual-shear-stress distribution and permanent twist of the tube that remain if $\mathbf{T}_p$ is removed *just after* the tube becomes fully plastic?

SOLUTION

Plastic Torque. The plastic torque $\mathbf{T}_p$ will strain the tube such that all the material yields. Hence the stress distribution will appear as shown in Fig. 5–46b. Applying Eq. 5–23, we have

$$T_p = 2\pi \int_{c_i}^{c_o} \tau_Y \rho^2 \, d\rho = \frac{2\pi}{3}\tau_Y(c_o^3 - c_i^3)$$

$$= \frac{2\pi}{3}(12(10^3) \text{ lb/in}^2)[(2 \text{ in.})^3 - (1 \text{ in.})^3] = 175.9 \text{ kip} \cdot \text{in.} \quad \textit{Ans.}$$

When the tube just becomes fully plastic, yielding has started at the inner radius, i.e., at $c_i = 1$ in., $\gamma_Y = 0.002$ rad, Fig. 5–46a. The angle of twist that occurs can be determined from Eq. 5–25, which for the entire tube becomes

$$\phi_p = \gamma_Y \frac{L}{c_i} = \frac{(0.002)(5 \text{ ft})(12 \text{ in./ft})}{(1 \text{ in.})} = 0.120 \text{ rad} \; \text{↑}$$

Then $\mathbf{T}_p$ is *removed*, or in effect reapplied in the opposite direction, then the "fictitious" linear shear-stress distribution shown in Fig. 5–46c must be superimposed on the one shown in Fig. 5–46b. In Fig. 5–46c the maximum shear stress or the modulus of rupture is computed from the torsion formula

$$\tau_r = \frac{T_p c_o}{J} = \frac{(175.9 \text{ kip} \cdot \text{in.})(2 \text{ in.})}{(\pi/2)[(2 \text{ in.})^4 - (1 \text{ in.})^4]} = 14.93 \text{ ksi}$$

Also, at the inner wall of the tube the shear stress is

$$\tau_i = (14.93 \text{ ksi})\left(\frac{1 \text{ in.}}{2 \text{ in.}}\right) = 7.47 \text{ ksi}$$

From Fig. 5–46a, $G = \tau_Y/\gamma_Y = 12$ ksi/(0.002 rad) $= 6000$ ksi, so that the corresponding angle of twist ϕ_p' upon removal of $\mathbf{T}_p$ is therefore

$$\phi_p' = \frac{T_p L}{JG} = \frac{(175.9 \text{ kip} \cdot \text{in.})(5 \text{ ft})(12 \text{ in./ft})}{(\pi/2)[(2 \text{ in.})^4 - (1 \text{ in.})^4]6000 \text{ kip/in}^2} = 0.0747 \text{ rad} \; \text{↓}$$

The resulting *residual-shear-stress distribution* is therefore shown in Fig. 5–46d. The permanent rotation of the tube after T_p is removed is

$$\phi = 0.120 - 0.0747 = 0.0453 \text{ rad} \; \text{↑} \quad \textit{Ans.}$$

PROBLEMS

5–111. The steel shaft is made from two segments AB and BC, which are connected using a fillet weld having a radius of 2.8 mm. Determine the maximum shear stress developed in the shaft.

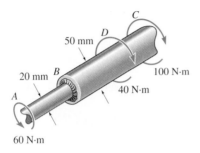

Prob. 5–111

*5–112. The shaft is used to transmit 9 hp while turning at 600 rpm. Determine the maximum shear stress in the shaft. The segments are connected using a fillet weld having a radius of 0.15 in.

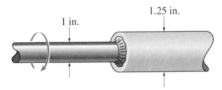

Prob. 5–112

5–113. The members are connected with a fillet weld of radius $r = 4$ mm. Determine the maximum shear stress in the shaft if $T = 10$ N · m.

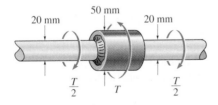

Prob. 5–113

5–114. The assembly is subjected to a torque of 710 lb · in. If the allowable shear stress for the material is $\tau_{allow} = 12$ ksi, determine the radius of the smallest size fillet that can be used to transmit the torque.

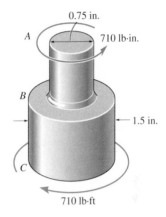

Prob. 5–114

5–115. The steel used for the shaft has an allowable shear stress of $\tau_{allow} = 8$ MPa. If the members are connected together with a fillet weld of radius $r = 2.25$ mm, determine the maximum torque T that can be applied.

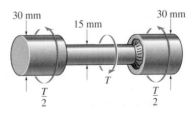

Prob. 5–115

*5–116. A solid shaft is subjected to the torque **T**, which causes the material to yield. If the material is elastic perfectly plastic, show that the torque can be expressed in terms of the angle of twist ϕ of the shaft as $T = \frac{4}{3}T_Y(1 - \phi_Y^3/4\phi^3)$, where T_Y and ϕ_Y are the torque and angle of twist when the material begins to yield.

5–117. The assembly is subjected to a torque of 1800 lb · in. If the allowable shear stress for the material is $\tau_{allow} = 12$ ksi, determine the radius of the smallest size fillet that can be used to transmit the torque.

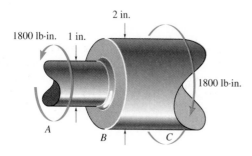

2 in.

1800 lb·in.   1 in.

1800 lb·in.

A

B

C

Prob. 5–117

5–118. A bar having a circular cross section of 3 in. diameter is subjected to a torque of 100 kip · in. If the material is elastic perfectly plastic, with $\tau_Y = 16$ ksi, determine the radius of the elastic core.

5–119. A solid shaft having a diameter of 2 in. is made of elastic-perfectly plastic material having a yield stress of $\tau_Y = 16$ ksi and shear modulus of $G = 12(10^3)$ ksi. Determine the torque required to develop an elastic core in the shaft having a diameter of 1 in. Also, what is the plastic torque?

*5–120. The 2-m-long tube is made from an elastic-plastic material as shown. Determine the applied torque T, which subjects the material of the tube's outer edge to a shearing strain of $\gamma_{max} = 0.008$ rad. What would be the permanent angle of twist of the tube when the torque is removed? Sketch the residual stress distribution of the tube.

45 mm

T

40 mm

τ (MPa)

240

0.003

γ (rad)

Prob. 5–120

5–121. Determine the torque needed to twist a short 3-mm-diameter steel wire through several revolutions if it is made of steel assumed to be elastic perfectly plastic and having a yield stress of $\tau_Y = 80$ MPa. Assume that the material becomes fully plastic.

5–122. A solid shaft has a diameter of 40 mm and length of 1 m. It is made of an elastic perfectly plastic material having a yield stress of $\tau_Y = 100$ MPa. Determine the maximum elastic torque T_Y and the corresponding angle of twist. What is the angle of twist if the torque is increased to $T = 1.2T_Y$? $G = 80$ GPa.

5–123. The stepped shaft is subjected to a torque T that produces yielding on the surface of the larger diameter segment. Determine the radius of the elastic core produced in the smaller diameter segment. Neglect the stress concentration at the fillet.

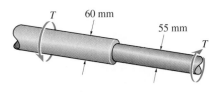

60 mm

55 mm

T T

Prob. 5–123

5–125. The shear stress–strain diagram for a solid 50-mm diameter shaft can be approximated as shown in the figure. Determine the torque required to cause a maximum shear stress in the shaft of 125 MPa. If the shaft is 3 m long, what is the corresponding angle of twist?

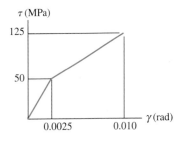

τ (MPa)

125

50

0.0025   0.010 γ (rad)

Prob. 5–125

*5–124. The shaft is made from a strain-hardening material having a τ–γ diagram as shown. Determine the torque T that must be applied to the shaft in order to create an elastic core in the shaft having a radius of $\rho_c = 0.5$ in.

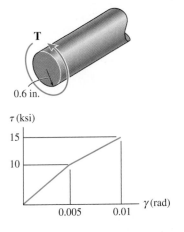

T

0.6 in.

τ (ksi)

15

10

0.005   0.01 γ (rad)

Prob. 5–124

5–126. The 2-m-long tube is made of an elastic perfectly plastic material as shown. Determine the applied torque T that subjects the material at the tube's outer edge to a shear strain of $\gamma_{max} = 0.006$ rad. What would be the permanent angle of twist of the tube when this torque is removed? Sketch the residual stress distribution in the tube.

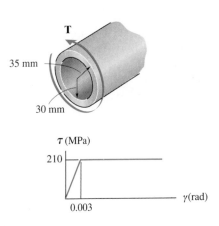

T

35 mm

30 mm

τ (MPa)

210

0.003 γ (rad)

Prob. 5–126

5–127. The solid shaft is made of an elastic-perfectly plastic material as shown. Determine the torque T needed to form an elastic core in the shaft having a radius of $\rho_Y = 20$ mm. If the shaft is 3 m long, through what angle does one end of the shaft twist with respect to the other end? When the torque is removed, determine the residual stress distribution in the shaft and the permanent angle of twist.

τ(MPa)

160

0.004

γ(rad)

Prob. 5–127

*5–128. The shaft consists of two sections that are rigidly connected. If the material is elastic perfectly plastic as shown, determine the largest torque T that can be applied to the shaft. Also, draw the shear-stress distribution over a radial line for each section. Neglect the effect of stress concentration.

τ(ksi)

10

0.002

γ(rad)

Prob. 5–128

5–129. The shaft is made of an elastic-perfectly plastic material as shown. Plot the shear-stress distribution acting along a radial line if it is subjected to a torque of $T = 2$ kN · m. What is the residual stress distribution in the shaft when the torque is removed?

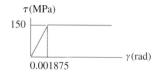

20 mm

τ(MPa)

150

0.001875

γ(rad)

Prob. 5–129

5–130. The shaft is made of an elastic-perfectly plastic material as shown. Determine the torque that the shaft can transmit if the allowable angle of twist is 0.375 rad. Also, determine the permanent angle of twist once the torque is removed. The shaft is 2 m long.

20 mm

τ(MPa)

150

0.001875

γ(rad)

Prob. 5–130

5–131. The shear stress–strain diagram for a solid 50-mm-diameter shaft can be approximated as shown in the figure. Determine the torque T required to cause a maximum shear stress in the shaft of 125 MPa. If the shaft is 1.5 m long, what is the corresponding angle of twist?

*5–132. A torque is applied to the shaft of radius r. If the material has a shear stress–strain relation of $\tau = k\gamma^{1/6}$, where k is a constant, determine the maximum shear stress in the shaft.

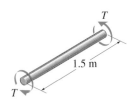

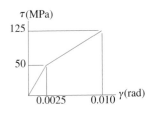

Prob. 5–131

Prob. 5–132

REVIEW PROBLEMS

5–133. The 304 stainless steel shaft is 3 m long and has an outer diameter of 60 mm. When it is rotating at 60 rad/s, it transmits 30 kW of power from the engine E to the generator G. Determine the smallest thickness of the shaft if the allowable shear stress is $\tau_{allow} = 150$ MPa and the shaft is restricted not to twist more than 0.08 rad.

5–134. The 304 stainless solid steel shaft is 3 m long and has a diameter of 50 mm. It is required to transmit 40 kW of power from the engine E to the generator G. Determine the smallest angular velocity the shaft can have if it is restricted not to twist more than 1.5°.

Prob. 5–133

Prob. 5–134

5–135. If the solid shaft AB to which the valve handle is attached is made of C83400 red brass and has a diameter of 10 mm, determine the maximum couple forces F that can be applied to the handle just before the material starts to fail. Take $\tau_{allow} = 40$ MPa. What is the angle of twist of the handle? The shaft is fixed at A.

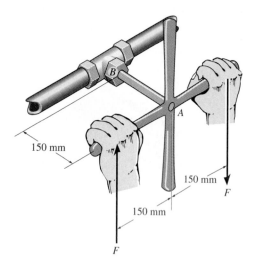

150 mm

150 mm

F

150 mm

A

B

F

Prob. 5–135

*5–136. If the solid shaft AB to which the valve handle is attached is made of C83400 red brass, determine the smallest diameter of the handle so that the angle of twist does not exceed 0.5° and the shear stress does not exceed 40 MPa when $F = 25$ N.

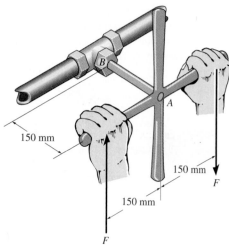

150 mm

150 mm

F

150 mm

A

B

F

Prob. 5–136

5–137. The device shown is used to mix soils in order to provide in-situ stabilization. If the mixer is connected to an A-36 steel tubular shaft that has an inner diameter of 3 in. and an outer diameter of 4.5 in., determine the angle of twist of the shaft of A relative to C if each mixing blade is subjected to the torques shown.

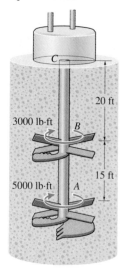

C

20 ft

3000 lb·ft

B

15 ft

5000 lb·ft

A

Prob. 5–137

5–138. The rotating flywheel and shaft is brought to a sudden stop at D when the bearing freezes. This causes the flywheel to oscillate clockwise–counterclockwise, so that a point A on the outer edge of the flywheel is displaced through a 10-mm arc. Determine the maximum shear stress developed in the tubular 304 stainless steel shaft due to this oscillation. The shaft has an inner diameter of 25 mm and an outer diameter of 35 mm. The journal bearings at B and C allow the shaft to rotate freely.

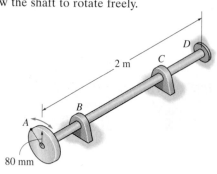

2 m

C

D

B

A

80 mm

Prob. 5–138

5–139. The 60-mm-diameter shaft rotates at 300 rev/min. This motion is caused by the unequal belt tensions on the pulley of 800 N and 450 N. Determine the power transmitted and the maximum shear stress developed in the shaft.

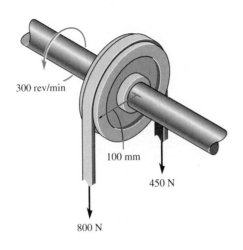

300 rev/min

100 mm

450 N

800 N

Prob. 5–139

5–141. The coupling consists of two disks fixed to separate shafts, each 25 mm in diameter. The shafts are supported on journal bearings that allow free rotation. In order to limit the torque **T** that can be transmitted, a "shear pin" *P* is used to connect the disks together. If this pin can sustain an *average* shear force of 550 N before it fails, determine the maximum constant torque **T** that can be transmitted from one shaft to the other. Also, what is the maximum shear stress in each shaft when the "shear pin" is about to fail?

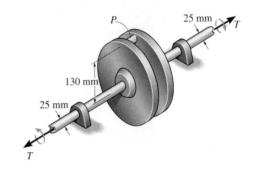

P   25 mm   T

130 mm

25 mm

T

Prob. 5–141

*5–140. The motor *A* develops a torque at gear *B* of 500 lb · ft, which is applied along the axis of the 2-in.-diameter A-36 steel shaft *CD*. This torque is to be transmitted to the pinion gears at *E* and *F*. If these gears are temporarily fixed, determine the maximum shear stress in segments *CB* and *BD* of the shaft. Also, what is the angle of twist of each of these segments? The bearings at *C* and *D* only exert force reactions on the shaft.

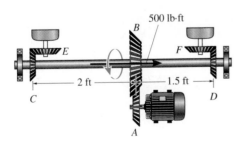

500 lb·ft

B

E

F

2 ft

1.5 ft

C

D

A

Prob. 5–140

5–142. The aluminum tube has a thickness of 5 mm and the outer cross-sectional dimensions shown. Determine the maximum average shear stress in the tube. If the tube has a length of 5 m, determine the angle of twist. $G_{al} = 28$ GPa.

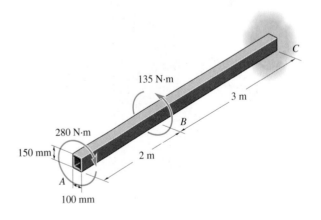

135 N·m

C

3 m

B

280 N·m

150 mm

2 m

A

100 mm

Prob. 5–142

Beams are important structural members used in building construction. Their design is often based upon their ability to resist bending stress, which forms the subject matter of this chapter.

6 BENDING

CHAPTER OBJECTIVES

Beams and shafts are important structural and mechanical elements in engineering. In this chapter we will determine the stress in these members caused by bending. The chapter begins with a discussion of how to establish the shear and moment diagrams for a beam or shaft. Like the normal-force and torque diagrams, the shear and moment diagrams provide a useful means for determining the largest shear and moment in a member, and they specify where these maximums occur. Once the internal moment at a section is determined, the bending stress can then be calculated. First we will consider members that are straight, have a symmetric cross section, and are made of homogeneous linear-elastic material. Afterward we will discuss special cases involving unsymmetric bending and members made of composite materials. Consideration will also be given to curved members, stress concentrations, inelastic bending, and residual stresses.

6.1 SHEAR AND MOMENT DIAGRAMS

Members that are slender and support loadings that are applied perpendicular to their longitudinal axis are called *beams*. In general, beams are long, straight bars having a constant cross-sectional area. Often they are classified as to how they are supported. For example, a *simply supported beam* is pinned at one end and roller-supported at the other, Fig. 6–1, a *cantilevered beam* is fixed at one end and free at the other, and an *overhanging beam* has one or both of its ends freely extended over the supports. Certainly beams may be considered among the most important of all structural elements. Examples include members used to support the floor of a building, the deck of a bridge, or the wing of an aircraft. Also, the axle of an automobile, the boom of a crane, even many of the bones of the body act as beams.

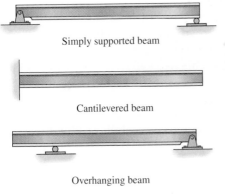

Simply supported beam

Cantilevered beam

Overhanging beam

Fig. 6–1

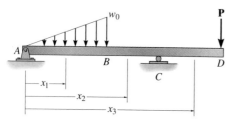

Fig. 6–2

Because of the applied loadings, beams develop an internal shear force and bending moment that, in general, vary from point to point along the axis of the beam. In order to properly design a beam it is first necessary to determine the maximum shear and moment in the beam. One way to do this is to express V and M as functions of the arbitrary position x along the beam's axis. These *shear and moment functions* can then be plotted and represented by graphs called **shear and moment diagrams**. The maximum values of V and M can then be obtained from these graphs. Also, since the shear and moment diagrams provide detailed information about the *variation* of the shear and moment along the beam's axis, they are often used by engineers to decide where to place reinforcement materials within the beam or how to proportion the size of the beam at various points along its length.

In Sec. 1.2 we used the method of sections to find the internal loading in a member at a *specific point*. However, if we must determine the internal V and M as functions of x along a beam, then it is necessary to locate the imaginary section or cut at an *arbitrary distance* x from the end of the beam and formulate V and M in terms of x. In this regard, the choice for the origin and the positive direction for any selected x is *arbitrary*. Most often, however, the origin is located at the left end of the beam and the positive direction is to the right.

In general, the internal shear and bending-moment functions obtained as a function of x will be *discontinuous*, or their slope will be discontinuous, at points where a distributed load changes or where concentrated forces or couples are applied. Because of this, shear and bending-moment functions must be determined for *each region* of the beam located *between* any two discontinuities of loading. For example, coordinates x_1, x_2, and x_3 will have to be used to describe the variation of V and M throughout the length of the beam in Fig. 6–2. These coordinates will be valid *only* within the regions from A to B for x_1, from B to C for x_2, and from C to D for x_3.

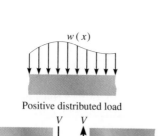

Positive distributed load

Positive internal shear

Positive internal moment

Beam sign convention

Fig. 6–3

Beam Sign Convention.   Before presenting a method for determining the shear and moment as functions of x and later plotting these functions (shear and moment diagrams), it is first necessary to establish a *sign convention* so as to define "positive" and "negative" internal shear force and bending moment. Although the choice of a sign convention is arbitrary, here we will use the one often used in engineering practice and shown in Fig. 6–3. The *positive directions* are as follows: the *distributed load* acts *downward* on the beam; the internal *shear force* causes a *clockwise* rotation of the beam segment on which it acts; and the internal *moment* causes *compression* in the *top fibers* of the segment such that it bends the segment so that it holds water. Loadings that are opposite to these are considered negative.

IMPORTANT POINTS

• *Beams* are long straight members that carry loads perpendicular to their longitudinal axis. They are classified according to the way they are supported, e.g., simply supported, cantilevered, or overhanging.

• In order to properly design a beam, it is important to know the *variation* of the shear and moment along its axis in order to find the points where these values are a maximum.

• By establishing a sign convention for positive shear and moment, the shear and moment in the beam can be determined as a function of its position x and these values plotted to form the shear and moment diagrams.

PROCEDURE FOR ANALYSIS

The shear and moment diagrams for a beam can be constructed using the following procedure.

Support Reactions.

• Determine all the reactive forces and couple moments acting on the beam, and resolve all the forces into components acting perpendicular and parallel to the beam's axis.

Shear and Moment Functions.

• Specify separate coordinates x having an origin at the beam's *left end* and extending to regions of the beam between concentrated forces and/or couple moments, or where there is no discontinuity of distributed loading.

• Section the beam perpendicular to its axis at each distance x, and draw the free-body diagram of one of the segments. Be sure $\mathbf{V}$ and $\mathbf{M}$ are shown acting in their positive sense, in accordance with the sign convention given in Fig. 6–3.

• The shear is obtained by summing forces perpendicular to the beam's axis.

• The moment is obtained by summing moments about the sectioned end of the segment.

Shear and Moment Diagrams.

• Plot the shear diagram (V versus x) and the moment diagram (M versus x). If numerical values of the functions describing V and M are *positive*, the values are plotted above the x axis, whereas negative values are plotted below the axis.

• Generally it is convenient to show the shear and moment diagrams directly below the free-body diagram of the beam.

EXAMPLE 6-1

Draw the shear and moment diagrams for the beam shown in Fig. 6–4a.

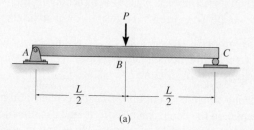

(a)

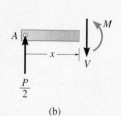

(b)

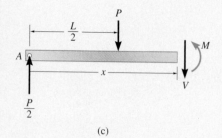

(c)

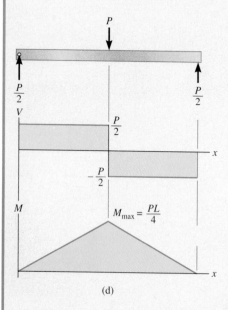

(d)

Fig. 6–4

SOLUTION

Support Reactions. The support reactions have been determined, Fig. 6–4d.

Shear and Moment Functions. The beam is sectioned at an arbitrary distance x from the support A, extending within region AB, and the free-body diagram of the left segment is shown in Fig. 6–4b. The unknowns **V** and **M** are indicated acting in the *positive sense* on the right-hand face of the segment according to the established sign convention. Applying the equilibrium equations yields

$$+\uparrow \Sigma F_y = 0; \qquad V = \frac{P}{2} \qquad (1)$$

$$\zeta + \Sigma M = 0; \qquad M = \frac{P}{2}x \qquad (2)$$

A free-body diagram for a left segment of the beam extending a distance x within region BC is shown in Fig. 6–4c. As always, **V** and **M** are shown acting in the positive sense. Hence,

$$+\uparrow \Sigma F_y = 0; \qquad \frac{P}{2} - P - V = 0$$

$$V = -\frac{P}{2} \qquad (3)$$

$$\zeta + \Sigma M = 0; \qquad M + P\left(x - \frac{L}{2}\right) - \frac{P}{2}x = 0$$

$$M = \frac{P}{2}(L - x) \qquad (4)$$

The shear diagram represents a plot of Eqs. 1 and 3, and the moment diagram represents a plot of Eqs. 2 and 4, Fig. 6–4d. These equations can be checked in part by noting that $dV/dx = -w$ and $dM/dx = V$ in each case. (These relationships are developed in the next section as Eqs. 6–1 and 6–2.)

EXAMPLE 6-2

Draw the shear and moment diagrams for the beam shown in Fig. 6–5a.

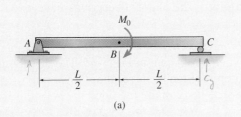

$\Sigma F_y = A_y + \dfrac{M_0}{L}$

$A_y = -\dfrac{M_0}{L}$

$M_A = -M_0 + L C_y = 0$

$C_y = \dfrac{M_0}{L}$

(a)

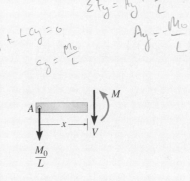

(b)

SOLUTION

Support Reactions.   The support reactions have been determined in Fig. 6–5d.

Shear and Moment Functions.   This problem is similar to the previous example, where two x coordinates must be used to express the shear and moment in the beam throughout its length. For the segment within region AB, Fig. 6–5b, we have

$+\uparrow \Sigma F_y = 0;$ $V = -\dfrac{M_0}{L}$ $-\dfrac{M_0}{L} - V = 0$

$\curvearrowleft + \Sigma M = 0;$ $M = -\dfrac{M_0}{L}x$

And for the segment within region BC, Fig. 6–5c,

$+\uparrow \Sigma F_y = 0;$ $V = -\dfrac{M_0}{L}$

$\curvearrowleft + \Sigma M = 0;$ $M = M_0 - \dfrac{M_0}{L}x$

$M = M_0\left(1 - \dfrac{x}{L}\right)$

$M_A = -M_0 + M(L-x) = 0$

$M(L-x) = M_0$

$M = \dfrac{M_0}{L-x}$

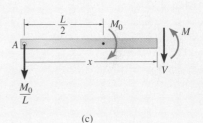

(c)

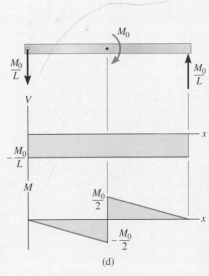

(d)

Fig. 6–5

Shear and Moment Diagrams.   When the above functions are plotted, the shear and moment diagrams shown in Fig. 6–5d are obtained. In this case, notice that the shear is constant over the entire length of the beam; i.e., it is not affected by the couple moment $\mathbf{M}_0$ acting at the center of the beam. Just as a force creates a jump in the shear diagram, Example 6–1, a couple moment creates a jump in the moment diagram.

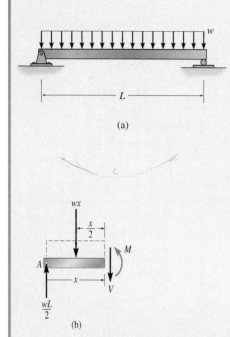

(a)

(b)

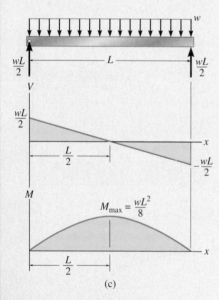

(c)

Fig. 6–6

EXAMPLE 6–3

Draw the shear and moment diagrams for the beam shown in Fig. 6–6a.

SOLUTION

Support Reactions.   The support reactions have been computed in Fig. 6–6c.

Shear and Moment Functions.   A free-body diagram of the left segment of the beam is shown in Fig. 6–6b. The distributed loading on this segment is represented by its resultant force only *after* the segment is isolated as a free-body diagram. Since the segment has a length x, the *magnitude* of the *resultant force* is wx. This force acts through the centroid of the area comprising the distributed loading, a distance of $x/2$ from the right end. Applying the two equations of equilibrium yields

$$+\uparrow \Sigma F_y = 0; \qquad \frac{wL}{2} - wx - V = 0$$

$$V = w\left(\frac{L}{2} - x\right) \qquad (1)$$

$$\zeta^+ \Sigma M = 0; \qquad -\left(\frac{wL}{2}\right)x + (wx)\left(\frac{x}{2}\right) + M = 0$$

$$M = \frac{w}{2}(Lx - x^2) \qquad (2)$$

These results for V and M can be checked by noting that $dV/dx = -w$. This is indeed correct, since positive w acts downward. Also, notice that $dM/dx = V$.

Shear and Moment Diagrams.   The shear and moment diagrams shown in Fig. 6–6c are obtained by plotting Eqs. 1 and 2. The point of *zero shear* can be found from Eq. 1:

$$V = w\left(\frac{L}{2} - x\right) = 0$$

$$x = \frac{L}{2}$$

From the moment diagram, this value of x happens to represent the point on the beam where the *maximum moment* occurs, since by Eq. 6–2, the slope $V = 0 = dM/dx$. From Eq. 2, we have

$$M_{max} = \frac{w}{2}\left[L\left(\frac{L}{2}\right) - \left(\frac{L}{2}\right)^2\right]$$

$$= \frac{wL^2}{8}$$

EXAMPLE 6–4

Draw the shear and moment diagrams for the beam shown in Fig. 6–7a.

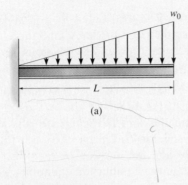

(a)

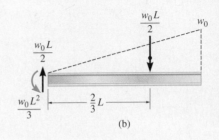

(b)

SOLUTION

Support Reactions.   The distributed load is replaced by its resultant force and the reactions have been determined as shown in Fig. 6–7b.

Shear and Moment Functions.   A free-body diagram of a beam segment of length x is shown in Fig. 6–7c. Note that the intensity of the triangular load at the section is found by proportion, that is, $w/x = w_0/L$ or $w = w_0x/L$. With the load intensity known, the resultant of the distributed loading is determined from the area under the diagram, Fig. 6–7c. Thus,

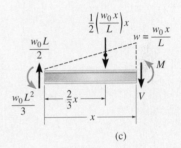

(c)

$$+\uparrow \Sigma F_y = 0; \qquad \frac{w_0L}{2} - \frac{1}{2}\left(\frac{w_0x}{L}\right)x - V = 0$$

$$V = \frac{w_0}{2L}(L^2 - x^2) \qquad (1)$$

$$\overset{\curvearrowright}{+}\Sigma M = 0; \quad \frac{w_0L^2}{3} - \frac{w_0L}{2}(x) + \frac{1}{2}\left(\frac{w_0x}{L}\right)x\left(\frac{1}{3}x\right) + M = 0$$

$$M = \frac{w_0}{6L}(-2L^3 + 3L^2x - x^3) \qquad (2)$$

These results can be checked by applying Eqs. 6–1 and 6–2, that is,

$$w = -\frac{dV}{dx} = -\frac{w_0}{2L}(0 - 2x) = \frac{w_0x}{L} \qquad \text{OK}$$

$$V = \frac{dM}{dx} = \frac{w_0}{6L}(-0 + 3L^2 - 3x^2) = \frac{w_0}{2L}(L^2 - x^2) \qquad \text{OK}$$

Shear and Moment Diagrams.   The graphs of Eqs. 1 and 2 are shown in Fig. 6–7d.

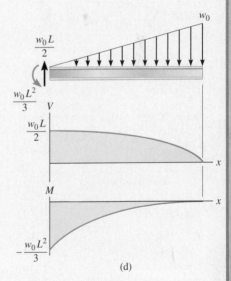

(d)

Fig. 6–7

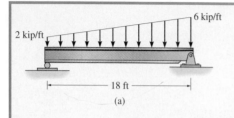

(a)

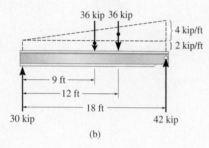

(b)

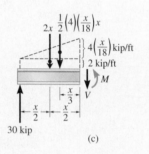

(c)

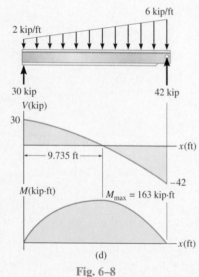

(d)

Fig. 6–8

EXAMPLE 6–5

Draw the shear and moment diagrams for the beam shown in Fig. 6–8a.

SOLUTION

Support Reactions. The distributed load is divided into triangular and rectangular component loadings and these loadings are then replaced by their resultant forces. The reactions have been determined as shown on the beam's free-body diagram, Fig. 6–8b.

Shear and Moment Functions. A free-body diagram of the left segment is shown in Fig. 6–8c. As above, the trapezoidal loading is replaced by rectangular and triangular distributions. Note that the intensity of the triangular load at the section is found by proportion. The resultant force and the location of each distributed loading are also shown. Applying the equilibrium equations, we have

$$+\uparrow \Sigma F_y = 0; \quad 30 \text{ kip} - (2 \text{ kip/ft})x - \frac{1}{2}(4 \text{ kip/ft})\left(\frac{x}{18 \text{ ft}}\right)x - V = 0$$

$$V = \left(30 - 2x - \frac{x^2}{9}\right) \text{kip} \qquad (1)$$

$$\zeta^+ \Sigma M = 0;$$

$$-30 \text{ kip}(x) + (2 \text{ kip/ft})\, x\left(\frac{x}{2}\right) + \frac{1}{2}(4 \text{ kip/ft})\left(\frac{x}{18 \text{ ft}}\right)x\left(\frac{x}{3}\right) + M = 0$$

$$M = \left(30x - x^2 - \frac{x^3}{27}\right) \text{kip} \cdot \text{ft} \qquad (2)$$

Equation 2 may be checked by noting that $dM/dx = V$, that is, Eq. 1. Also, $w = -dV/dx = 2 + \frac{2}{9}x$. This equation checks, since when $x = 0$, $w = 2$ kip/ft, and when $x = 18$ ft, $w = 6$ kip/ft, Fig. 6–8a.

Shear and Moment Diagrams. Equations 1 and 2 are plotted in Fig. 6–8d. Since the point of maximum moment occurs when $dM/dx = V = 0$, then, from Eq. 1,

$$V = 0 = 30 - 2x - \frac{x^2}{9}$$

Choosing the positive root,

$$x = 9.735 \text{ ft}$$

Thus, from Eq. 2,

$$M_{max} = 30(9.735) - (9.735)^2 - \frac{(9.735)^3}{27}$$

$$= 163 \text{ kip} \cdot \text{ft}$$

EXAMPLE 6-6

Draw the shear and moment diagrams for the beam shown in Fig. 6–9a.

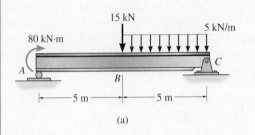

(a)

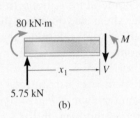

(b)

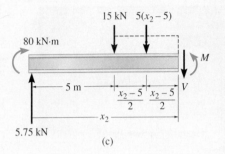

(c)

SOLUTION

Support Reactions. The reactions at the supports have been determined and are shown on the free-body diagram of the beam, Fig. 6–9d.

Shear and Moment Functions. Since there is a discontinuity of distributed load and also a concentrated load at the beam's center, two regions of x must be considered in order to describe the shear and moment functions for the entire beam.

$0 \le x_1 < 5$ m, Fig. 6–9b:

$+\uparrow \Sigma F_y = 0;$ $\qquad$ $5.75 \text{ kN} - V = 0$

$\qquad\qquad\qquad$ $V = 5.75 \text{ kN}$ $\qquad\qquad$ (1)

$\zeta + \Sigma M = 0;$ $\qquad$ $-80 \text{ kN} \cdot \text{m} - 5.75 \text{ kN } x_1 + M = 0$

$\qquad\qquad\qquad$ $M = (5.75x_1 + 80) \text{ kN} \cdot \text{m}$ $\qquad$ (2)

5 m $< x_2 \le 10$ m, Fig. 6–9c:

$+\uparrow \Sigma F_y = 0;$ $\quad$ $5.75 \text{ kN} - 15 \text{ kN} - 5 \text{ kN/m}(x_2 - 5 \text{ m}) - V = 0$

$\qquad\qquad\qquad$ $V = (15.75 - 5x_2) \text{ kN}$ $\qquad$ (3)

$\zeta + \Sigma M = 0;$ $\qquad$ $-80 \text{ kN} \cdot \text{m} - 5.75 \text{ kN } x_2 + 15 \text{ kN}(x_2 - 5 \text{ m})$

$\qquad\qquad$ $+ 5 \text{ kN/m}(x_2 - 5 \text{ m})\left(\dfrac{x_2 - 5 \text{ m}}{2}\right) + M = 0$

$\qquad$ $M = (-2.5x_2^2 + 15.75x_2 + 92.5) \text{ kN} \cdot \text{m}$ $\quad$ (4)

These results can be checked in part by noting that by applying $w = -dV/dx$ and $V = dM/dx$. Also, when $x_1 = 0$, Eqs. 1 and 2 give $V = 5.75$ kN and $M = 80$ kN $\cdot$ m; when $x_2 = 10$ m, Eqs. 3 and 4 give $V = -34.25$ kN and $M = 0$. These values check with the support reactions shown on the free-body diagram, Fig. 6–9d.

Shear and Moment Diagrams. Equations 1 through 4 are plotted in Fig. 6–9d.

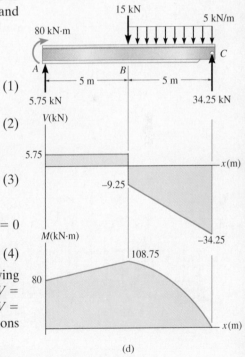

(d)

Fig. 6–9

6.2   GRAPHICAL METHOD FOR CONSTRUCTING SHEAR AND MOMENT DIAGRAMS

In cases where a beam is subjected to *several* different loadings, determining V and M as functions of x and then plotting these equations can become quite tedious. In this section a simpler method for constructing the shear and moment diagrams is discussed—a method based on two differential relations that exist among distributed load, shear, and moment.

Regions of Distributed Load.   For purposes of generality, consider the beam shown in Fig. 6–10a, which is subjected to an arbitrary loading. A free-body diagram for a small segment Δx of the beam is shown in Fig. 6–10b. Since this segment has been chosen at a position x where there is no concentrated force or couple moment, the results to be obtained will *not* apply at these points of concentrated loading.

Notice that all the loadings shown on the segment act in their positive directions according to the established sign convention, Fig. 6–3. Also, both the internal resultant shear and moment, acting on the right face of the segment, must be changed by a small finite amount in order to keep the segment in equilibrium. The distributed load has been replaced by a resultant force $w(x)\,\Delta x$ that acts at a fractional distance $k(\Delta x)$ from the right end, where $0 < k < 1$ [for example, if $w(x)$ is *uniform*, $k = \frac{1}{2}$]. Applying the two equations of equilibrium to the segment, we have

As shown, failure of this table occurred at the brace support on its right side. The bending moment diagram for the table loading would indicate this to be the point of maximum internal moment.

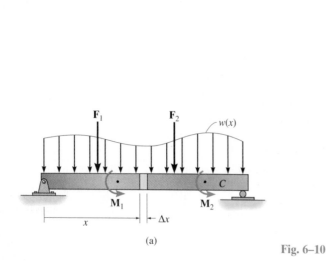

Free-body diagram of segment Δx

Cross-sectional area of segment

(a)

(b)

Fig. 6–10

$+\uparrow \Sigma F_y = 0;$ $\qquad V - w(x)\,\Delta x - (V + \Delta V) = 0$

$$\Delta V = -w(x)\,\Delta x$$

$\zeta + \Sigma M_O = 0;$ $\qquad -V\,\Delta x - M + w(x)\,\Delta x[k(\Delta x)] + (M + \Delta M) = 0$

$$\Delta M = V\,\Delta x - w(x)\,k(\Delta x)^2$$

Dividing by Δx and taking the limit as $\Delta x \to 0$, the above two equations become

$$\frac{dV}{dx} = -w(x)$$

$$\begin{matrix} \text{slope of} & = & -\text{distributed} \\ \text{shear diagram} & & \text{load intensity} \\ \text{at each point} & & \text{at each point} \end{matrix}$$

(6–1)

$$\frac{dM}{dx} = V$$

$$\begin{matrix} \text{slope of} & = & \text{shear} \\ \text{moment diagram} & & \text{at each} \\ \text{at each point} & & \text{point} \end{matrix}$$

(6–2)

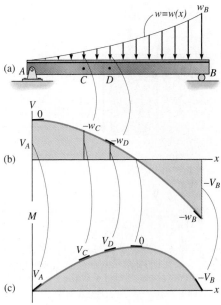

(a)

(b)

(c)

Fig. 6–11

These two equations provide a convenient means for quickly obtaining the shear and moment diagrams for a beam. Equation 6–1 states that at a point the *slope* of the shear diagram equals the negative of the intensity of the distributed loading. For example, consider the beam in Fig. 6–11*a*. The distributed loading is positive and increases from zero to w_B. Therefore, the shear diagram will be a curve that has a *negative slope*, increasing from zero to $-w_B$. Specific slopes $w_A = 0, -w_C,$ $-w_D,$ and $-w_B$ are shown in Fig. 6–11*b*.

In a similar manner, Eq. 6–2 states that at a point the *slope* of the moment diagram is equal to the shear. Notice that the shear diagram in Fig. 6–11*b* starts at $+V_A$, decreases to zero, and then becomes negative and decreases to $-V_B$. The moment diagram will then have an initial slope of $+V_A$ which decreases to zero, then the slope becomes negative and decreases to $-V_B$. Specific slopes $V_A, V_C, V_D, 0,$ and $-V_B$ are shown in Fig. 6–11*c*.

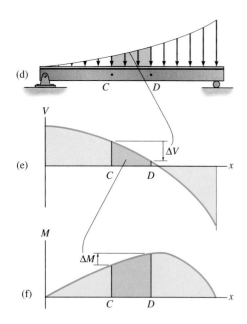

(d)

(e)

(f)

Fig. 6–11

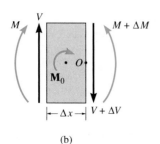

(a)

(b)

Fig. 6–12

Equations 6–1 and 6–2 may also be rewritten in the form $dV = -w(x)\,dx$ and $dM = V\,dx$. Noting that $w(x)\,dx$ and $V\,dx$ represent differential areas under the distributed loading and shear diagram, respectively, we can integrate these areas between any two points C and D on the beam, Fig. 6–11d and write

$$\Delta V = -\int w(x)\,dx$$

$$\text{change in} = \text{\quad-area under}$$
$$\text{shear} \quad \text{distributed loading}$$

(6-3)

$$\Delta M = \int V(x)\,dx$$

$$\text{change in} = \text{\quad area under}$$
$$\text{moment} \quad \text{shear diagram}$$

(6-4)

Equation 6–3 states that the *change in shear* between points C and D is equal to the (negative) *area* under the distributed-loading curve between these two points, Fig. 6–11d. Similarly, from Eq. 6–4, the change in moment between C and D, Fig. 6–11f, is equal to the area under the shear diagram within the region from C to D.

As stated previously, the above equations do not apply at points where a concentrated force or couple moment acts.

Regions of Concentrated Force and Moment.   A free-body diagram of a small segment of the beam in Fig. 6–10a taken from under one of the forces is shown in Fig. 6–12a. Here it can be seen that force equilibrium requires

$$+\uparrow \Sigma F_y = 0; \qquad V - F - (V + \Delta V) = 0$$
$$\Delta V = -F \qquad (6\text{–}5)$$

Thus, when $\mathbf{F}$ acts *downward* on the beam, ΔV is *negative* so the shear will "jump" *downward*. Likewise, if $\mathbf{F}$ acts *upward*, the jump (ΔV) will be *upward*.

From Fig. 6–12b, moment equilibrium requires the change in moment to be

$$\zeta + \Sigma M_O = 0; \qquad M + \Delta M - M_0 - V\,\Delta x - M = 0$$

Letting $\Delta x \to 0$, we get

$$\Delta M = M_0 \qquad (6\text{–}6)$$

In this case, if $\mathbf{M}_0$ is applied *clockwise*, ΔM is *positive* so the moment diagram will "jump" *upward*. Likewise, when $\mathbf{M}_0$ acts *counterclockwise*, the jump (ΔM) will be *downward*.

Table 6–1 illustrates application of Eqs. 6–1, 6–2, 6–5, and 6–6 to some common loading cases. None of these results should be memorized; rather, each should be *carefully studied* so that you become fully aware of how the shear and moment diagrams can be constructed on the basis of knowing the *variation of the slope* from the load and shear diagrams, respectively. It would be well worth the time and effort to self-test your understanding of these concepts by covering over the shear and moment diagram columns in the table and then trying to reconstruct these diagrams on the basis of knowing the loading.

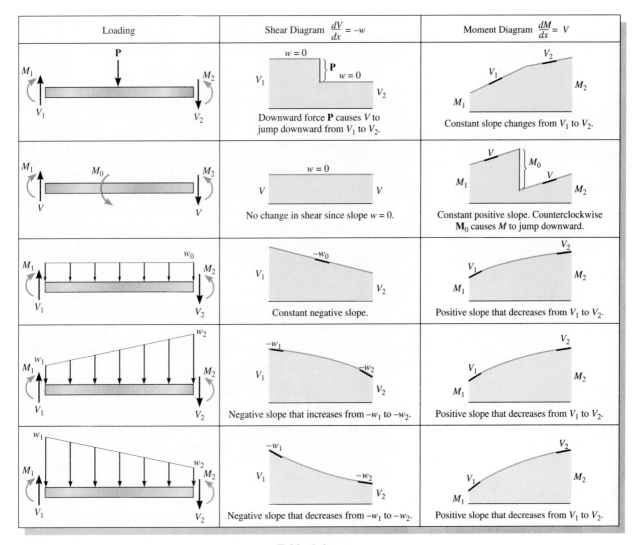

Table 6–1

PROCEDURE FOR ANALYSIS

The following procedure provides a method for constructing the shear and moment diagrams for a beam based on the relations among distributed load, shear, and moment.

Support Reactions.

• Determine the support reactions and resolve the forces acting on the beam into components that are perpendicular and parallel to the beam's axis.

Shear Diagram.

• Establish the V and x axes and plot the known values of the shear at the two *ends* of the beam.

• Since $dV/dx = -w$, the *slope* of the *shear diagram* at any point is equal to the (negative) intensity of the *distributed loading* at the point. Note that w is positive when it acts downward.

• If a numerical value of the shear is to be determined at a point, one can find this value either by using the method of sections and the equation of force equilibrium, or by using $\Delta V = -\int w(x)\,dx$, which states that the *change in the shear* between any two points is equal to the (*negative*) *area under the load diagram* between the two points.

• Since $w(x)$ must be *integrated* to obtain ΔV, then if $w(x)$ is a curve of degree n, $V(x)$ will be a curve of degree $n + 1$; for example, if $w(x)$ is uniform, $V(x)$ will be linear.

Moment Diagram.

• Establish the M and x axes and plot the known values of the moment at the *ends* of the beam.

• Since $dM/dx = V$, the *slope* of the moment diagram at any point is equal to the *shear* at the point.

• At the point where the shear is zero, $dM/dx = 0$, and therefore this would be a point of maximum or minimum moment.

• If a numerical value of the moment is to be determined at the point, one can find this value either by using the method of sections and the equation of moment equilibrium, or by using $\Delta M = \int V(x)\,dx$, which states that the *change in moment* between any two points is equal to the *area under the shear diagram* between the two points.

• Since $V(x)$ must be *integrated* to obtain ΔM, then if $V(x)$ is a curve of degree n, $M(x)$ will be a curve of degree $n + 1$; for example, if $V(x)$ is linear, $M(x)$ will be parabolic.

EXAMPLE 6–7

Draw the shear and moment diagrams for the beam in Fig. 6–13a.

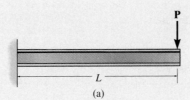

(a)

SOLUTION

Support Reactions.   The reactions are shown on a free-body diagram, Fig. 6–13b.

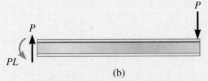

(b)

Shear Diagram.   According to the sign convention, Fig. 6–3, at $x = 0, V = +P$ and at $x = L, V = +P$. These points are plotted in Fig. 6–13b. Since $w = 0$, Fig. 6–13a, the *slope* of the shear diagram will be zero ($dV/dx = -w = 0$) at all points, and therefore a horizontal straight line connects the end points.

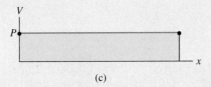

(c)

Moment Diagram.   At $x = 0, M = -PL$ and at $x = L, M = 0$, Fig. 6–13d. The shear diagram indicates that the shear is constant positive and therefore the *slope* of the moment diagram will be *constant positive*, $dM/dx = V = +P$ at all points. Hence, the end points are connected by a straight positive sloped line as shown in Fig. 6–13d.

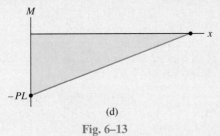

(d)

Fig. 6–13

EXAMPLE 6–8

Draw the shear and moment diagrams for the beam shown in Fig. 6–14*a*.

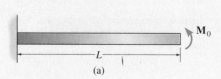

(a)

SOLUTION

Support Reactions.   The reaction at the fixed support is shown on the free-body diagram, Fig. 6–14*b*.

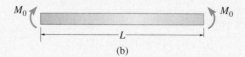

(b)

Shear Diagram.   The shear $V = 0$ at each end is plotted first, Fig. 6–14*c*. Since no distributed load exists on the beam the shear diagram will have zero *slope*, at all points. Therefore, a horizontal line connects the end points, which indicates that the shear is zero throughout the beam.

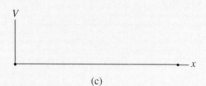

(c)

Moment Diagram.   The moment M_0 at the beam's end points is plotted first, Fig. 6–14*d*. From the shear diagram the *slope* of the moment diagram will be zero since $V = 0$. Therefore, a horizontal line connects the end points as shown.

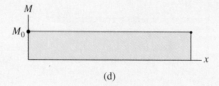

(d)

Fig. 6–14

EXAMPLE 6-9

Draw the shear and moment diagrams for the beam shown in Fig. 6–15a.

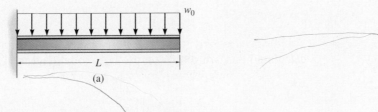

(a)

SOLUTION

Support Reactions.   The reactions at the fixed support are shown on the free-body diagram, Fig. 6–15b.

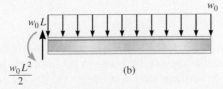

(b)

Shear Diagram.   The shear at each end point, is plotted first, Fig. 6–15c. The distributed loading on the beam is constant positive, and so the *slope* of the shear diagram will be constant negative ($dV/dx = -w_0$). This requires a straight negative sloped line connects the end points.

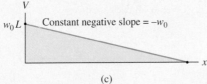

(c)

Moment Diagram.   The moment at each end point is plotted first, Fig. 6–15d. The shear diagram indicates that V is positive and decreases from w_0L to zero, and so the moment diagram must start with a positive slope of w_0L and decrease to zero. Specifically, since the shear diagram is a straight sloping line, the moment diagram will be *parabolic*, having a decreasing slope as shown in the figure.

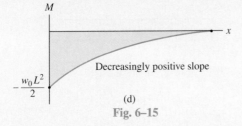

(d)

Fig. 6–15

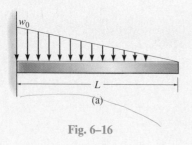

Fig. 6–16

EXAMPLE 6–10

Draw the shear and moment diagrams for the beam shown in Fig. 6–16a.

SOLUTION

Support Reactions. The reactions at the fixed support have been calculated and are shown on the free-body diagram, Fig. 6–16b.

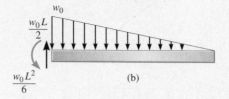

(b)

Shear Diagram. The shear at each end point is plotted first, Fig. 6–16c. The distributed loading on the beam is positive yet decreasing. Therefore the *slope* of the shear diagram will be *negatively decreasing*. At $x = 0$ the slope begins at $-w_0$ and goes to zero at $x = L$. Since the loading is *linear,* the shear diagram is a *parabola* having a negatively decreasing slope.

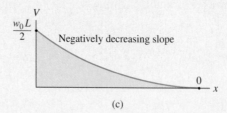

(c)

Moment Diagram. The moment at each end is plotted first, Fig. 6–16d. From the shear diagram V is positive but decreases from $w_0L/2$ at $x = 0$ to zero at $x = L$. The curve of the moment diagram having this slope behavior is a *cubic* function of x, as shown in the figure.

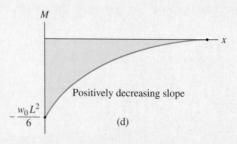

(d)

EXAMPLE 6-11

Draw the shear and moment diagrams for the beam in Fig. 6–17a.

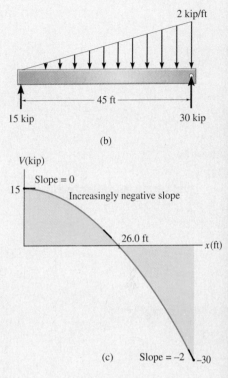

2 kip/ft

45 ft

15 kip   30 kip

(b)

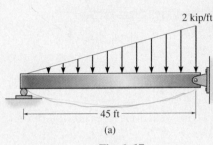

2 kip/ft

45 ft

(a)

Fig. 6–17

SOLUTION

Support Reactions.   The reactions have been determined and are shown on the free-body diagram, Fig. 6–17b.

Shear Diagram.   The end points $x = 0$, $V = +15$, and $x = 45$, $V = -30$ are plotted first, Fig. 6–17c. From the behavior of the distributed load, the *slope* of the shear diagram will vary from zero at $x = 0$ to -2 at $x = 45$. As a result, the shear diagram is a parabola having the shape shown.

The point of zero shear can be found by using the method of sections for a beam segment of length x, Fig. 6–17e. We require that $V = 0$, so that

$$+\uparrow \Sigma F_y = 0; \quad 15 \text{ kip} - \frac{1}{2}\left[2 \text{ kip/ft}\left(\frac{x}{45 \text{ ft}}\right)\right]x = 0; \quad x = 26.0 \text{ ft}$$

Moment Diagram.   The end points $x = 0$, $M = 0$ and $x = 45$, $M = 0$ are plotted first, Fig. 6–17d. From the behavior of the shear diagram, the slope of the moment diagram will begin at $+15$, then it becomes *decreasingly positive* until it reaches zero at 26.0 ft. It then becomes *increasingly negative* reaching -30 at $x = 45$ft. Here the moment diagram is a cubic function of x. Why?

Notice that the maximum moment is at $x = 26.0$, since $dM/dx = V = 0$ at this point. From the free-body diagram in Fig. 6–17e we have

$$\zeta + \Sigma M = 0;$$

$$-15 \text{ kip}(26.0 \text{ ft}) + \frac{1}{2}\left[2 \text{ kip/ft}\left(\frac{26.0 \text{ ft}}{45 \text{ ft}}\right)\right](26.0 \text{ ft})\left(\frac{26.0 \text{ ft}}{3}\right) + M = 0$$

$$M = 260 \text{ kip} \cdot \text{ft}$$

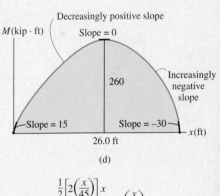

V(kip)

Slope = 0

15

Increasingly negative slope

26.0 ft

x(ft)

(c)   Slope = -2 -30

Decreasingly positive slope

M(kip · ft)   Slope = 0

260

Increasingly negative slope

Slope = 15   Slope = -30

x(ft)

26.0 ft

(d)

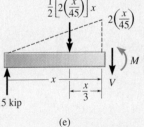

$\frac{1}{2}\left[2\left(\frac{x}{45}\right)\right]x$ $2\left(\frac{x}{45}\right)$

M

x $\frac{x}{3}$ V

15 kip

(e)

EXAMPLE 6–12

Draw the shear and moment diagrams for the beam shown in Fig. 6–18a.

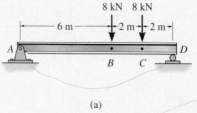

(a)

SOLUTION

Support Reactions. The reactions are indicated on the free-body diagram, Fig. 6–18b.

Shear Diagram. At $x = 0$, $V_A = +4.8$ kN, and at $x = 10$, $V_D = -11.2$ kN, Fig. 6–18c. At intermediate points between each force the *slope* of the shear diagram will be zero. Why? Hence the shear retains its value of +4.8 up to point B. At B the shear is *discontinuous*, since there is a *concentrated force* of 8 kN there. The value of the shear just to the right of B can be found by sectioning the beam at this point, Fig. 6–18e, where for equilibrium $V = -3.2$ kN. Use the method of sections and show that the diagram "jumps" again at C, as shown, then closes to the value of −11.2 kN at D.

It should be noted that based on Eq. 6–5, $\Delta V = -F$, the shear diagram can also be constructed by "following the load" on the free-body diagram. Beginning at A the 4.8-kN force acts upward, so $V_A = +4.8$ kN. No distributed load acts between A and B, so the shear remains constant ($dV/dx = 0$). At B the 8-kN force is down, so the shear jumps down 8 kN, from +4.8 kN to −3.2 kN. Again the shear is constant from B to C (no distributed load), then at C it jumps down another 8 kN to −11.2 kN. Finally, with no distributed load between C and D, it ends at −11.2 kN.

Moment Diagram. The moment at each end of the beam is zero, Fig. 6–18d. The *slope* of the moment diagram from A to B is constant at +4.8. Why? The value of the moment at B can be determined by using statics, Fig. 6–18c, or by finding the area under the shear diagram between A and B, that is, $\Delta M_{AB} = (4.8 \text{ kN})(6 \text{ m}) = 28.8 \text{ kN} \cdot \text{m}$. Since $M_A = 0$, then $M_B = M_A + \Delta M_{AB} = 0 + 28.8 \text{ kN} \cdot \text{m} = 28.8 \text{ kN} \cdot \text{m}$. From point B, the slope of the moment diagram is −3.2 until point C is reached. Again, the value of the moment can be obtained by statics or by finding the area under the shear diagram from B to C, that is, $\Delta M_{BC} = (-3.2 \text{ kN})(2 \text{ m}) = -6.4 \text{ kN} \cdot \text{m}$, so that $M_C = 28.8 \text{ kN} \cdot \text{m} - 6.4 \text{ kN} \cdot \text{m} = 22.4 \text{ kN} \cdot \text{m}$. Continuing in this manner, verify that closure occurs at D.

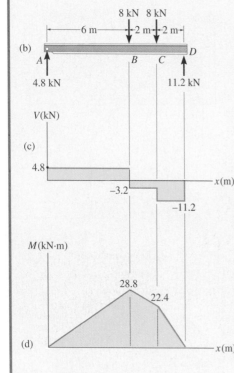

Fig. 6–18

EXAMPLE 6–13

Draw the shear and moment diagrams for the overhanging beam shown in Fig. 6–19a.

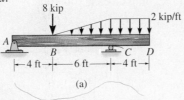

(a)

SOLUTION

Support Reactions. The free-body diagram with the calculated support reactions is shown in Fig. 6–19b.

Shear Diagram. As usual we start by plotting the end shears $V_A = +4.40$ kip and $V_D = 0$, Fig. 6–19c. The shear diagram will have *zero* slope from A to B. It then jumps down 8 kip to -3.60 kip. It then has a slope that is *increasingly negative*. The shear at C can be determined from the area under the load diagram, $V_C = V_B + \Delta V_{BC} = -3.60$ kip $-(1/2)(6$ ft$)$ $(2$ kip/ft$) = -9.60$ kip. It then jumps up 17.6 kip to 8 kip. Finally, from C to D, the slope of the shear diagram will be *constant yet negative*, until the shear reaches zero at D.

Moment Diagram. The end moments $M_A = 0$ and $M_D = 0$ are plotted first, Fig. 6–19d. Study the diagram and note how the slopes and therefore the various curves are established from the shear diagram using $dM/dx = V$. Verify the numerical values for the peaks using the method of sections and statics or by computing the appropriate areas under the shear diagram to find the change in moment between two points. In particular, the point of zero moment can be determined by establishing M as a function of x, where, for convenience, x extends *from* point B into region BC, Fig. 6–19e. Hence,

$$\zeta + \Sigma M = 0;$$

$$-4.40 \text{ kip}(4 \text{ ft} + x) + 8 \text{ kip}(x) + \frac{1}{2}\left(\frac{2 \text{ kip/ft}}{6 \text{ ft}}\right)x(x)\left(\frac{x}{3}\right) + M = 0$$

$$M = \left(-\frac{1}{18}x^3 - 3.60x + 17.6\right) \text{ kip} \cdot \text{ft} = 0$$

$$x = 3.94 \text{ ft}$$

Reviewing these diagrams, we see that because of the integration process for region AB the load is zero, shear is constant, and moment is linear; for region BC the load is linear, shear is parabolic, and moment is cubic; and for region CD the load is constant, the shear is linear, and the moment is parabolic. It is recommended that Examples 6–1 through 6–6 also be solved using this method.

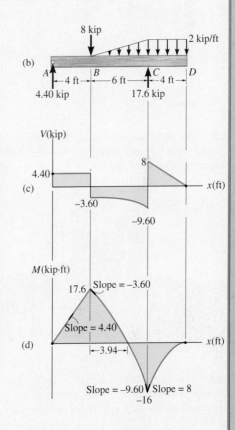

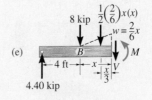

Fig. 6–19

PROBLEMS

6–1. Draw the shear and moment diagrams for the shaft. The bearings at *A* and *B* exert only vertical reactions on the shaft.

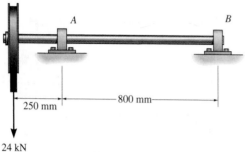

Prob. 6–1

6–2. The shaft is subjected to the loadings caused by belts passing over the two pulleys. Draw the shear and moment diagrams. The bearings at *A* and *B* exert only vertical reactions on the shaft.

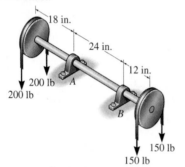

Prob. 6–2

6–3. The three traffic lights each have a mass of 10 kg, and the cantilevered pipe *AB* has a mass of 1.5 kg/m. Draw the shear and moment diagrams for the pipe. Neglect the mass of the sign.

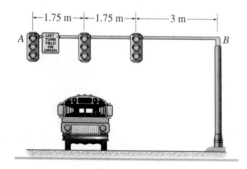

Prob. 6–3

***6–4.** The steam pipe *P* rests on the roller guide *CD* to allow for its thermal expansion. Draw the shear and moment diagrams for the beam *AB* if the pipe weighs 800 lb. The bearing supports at *C* and *D* exert only vertical forces on the beam.

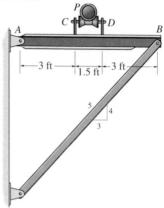

Prob. 6–4

6–5. A reinforced concrete pier is used to support the stringers for a bridge deck. Draw the shear and moment diagrams for the pier when it is subjected to the stringer loads shown. Assume the columns at *A* and *B* exert only vertical reactions on the pier.

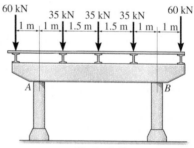

Prob. 6–5

6–6. Draw the shear and moment diagrams for the shaft. The bearings at *A* and *B* exert only vertical reactions on the shaft. Also, express the shear and moment in the shaft as a function of *x* within the region 125 mm $< x <$ 725 mm.

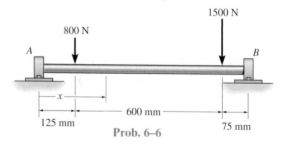

Prob. 6–6

6–7. Draw the shear and moment diagrams for the rod segment *ABC*. The end *A* is subjected to a force of 5 kN. *Hint:* The reactions at pin *D* must be replaced by equivalent loadings at *C* on the axis of the rod. The journal bearing at *B* exerts only a vertical force on the rod.

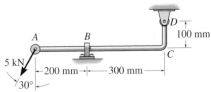

Prob. 6–7

*6–8. Draw the shear and moment diagrams for the shaft and determine the shear and moment throughout the shaft as a function of *x*. The bearings at *A* and *B* exert only vertical reactions on the shaft.

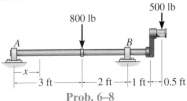

Prob. 6–8

6–9. Draw the shear and moment diagrams for the pipe. The end screw is subjected to a horizontal force of 5 kN. *Hint:* The reactions at the pin *C* must be replaced by equivalent loadings at point *B* on the axis of the pipe.

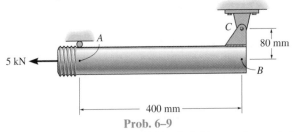

Prob. 6–9

6–10. The walking beam of an oil-field pumping unit is subjected to the draw force of 800 lb. Determine the force **P** in the pitman arm and the reactions at the pin *C*. Then draw the shear and moment diagrams for portion *AB* of the beam. *Hint:* The reactions at *C* must be replaced by equivalent loadings at point *C'* on the axis of the beam.

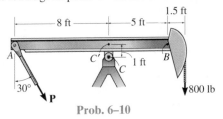

Prob. 6–10

6–11. The engine crane is used to support the engine, which has a weight of 1200 lb. Draw the shear and moment diagrams of the boom *ABC* when it is in the horizontal position shown.

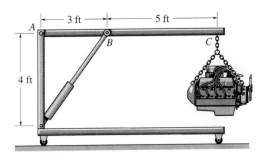

Prob. 6–11

*6–12. Draw the shear and moment diagrams for the compound beam. It is supported by a smooth plate at *A* which slides within the groove and so it cannot support a vertical force, although it can support a moment and axial load.

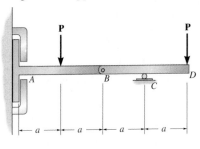

Prob. 6–12

6–13. The overhanging beam has been fabricated with a projected arm *BD* on it. Draw the shear and moment diagrams for the beam *ABC* if it supports a load of 800 lb. *Hint:* The loading in the supporting strut *DE* must be replaced by equivalent loads at point *B* on the axis of the beam.

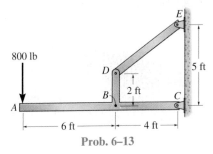

Prob. 6–13

6–14. Draw the shear and moment diagrams for the beam.

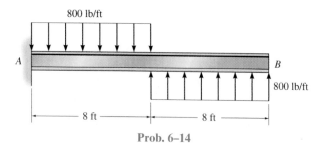

800 lb/ft

A

B

800 lb/ft

8 ft

8 ft

Prob. 6–14

6–15. The beam is subjected to the uniformly distributed moment m (moment / length). Draw the shear and moment diagrams for the beam.

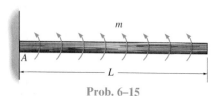

m

A

L

Prob. 6–15

*6–16. The beam is subjected to the uniformly distributed moment m (moment / length). Draw the shear and moment diagrams for the beam.

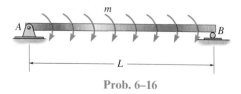

m

A

B

L

Prob. 6–16

6–17. Draw the shear and moment diagrams for the beam. The bearings at A and B only exert vertical reactions on the beam.

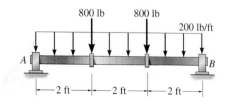

800 lb   800 lb

200 lb/ft

A

B

2 ft   2 ft   2 ft

Prob. 6–17

6–18. Draw the shear and moment diagrams for the beam, and determine the shear and moment throughout the beam as functions of x.

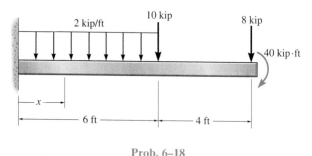

2 kip/ft   10 kip   8 kip

40 kip·ft

x

6 ft   4 ft

Prob. 6–18

6–19. Draw the shear and moment diagrams for the compound beam.

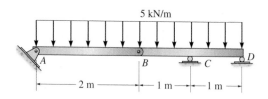

5 kN/m

A   B   C   D

2 m   1 m   1 m

Prob. 6–19

*6–20. Draw the shear and moment diagrams for the beam. It is supported by a smooth plate at A which slides within the groove and so it cannot support a vertical force, although it can support a moment and axial load.

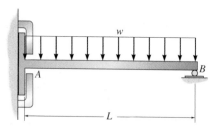

w

A   B

L

Prob. 6–20

6–21. Draw the shear and moment diagrams for the beam, and determine the shear and moment throughout the beam as functions of x.

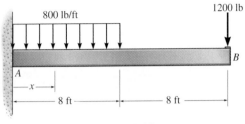

Prob. 6–21

6–22. Draw the shear and moment diagrams for the beam, and determine the shear and moment throughout the beam as functions of x.

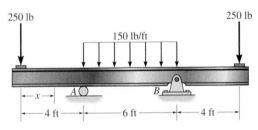

Prob. 6–22

6–23. Determine the equilibrium torque T acting on the propeller, and then draw the shear and moment diagrams for the propeller.

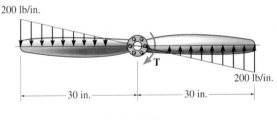

Prob. 6–23

*6–24. Draw the shear and moment diagrams for the beam.

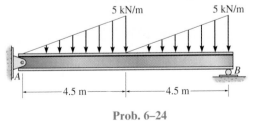

Prob. 6–24

6–25. Draw the shear and moment diagrams for the compound beam.

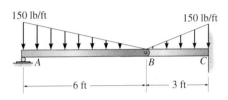

Prob. 6–25

6–26. The dead-weight loading along the centerline of the airplane wing is shown. If the wing is fixed to the fuselage at A, determine the reactions at A, and then draw the shear and moment diagram for the wing.

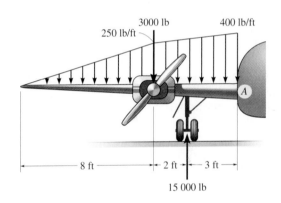

Prob. 6–26

6–27. Draw the shear and moment diagrams for the beam.

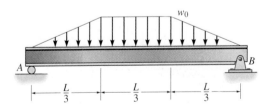

Prob. 6–27

*6–28. The boom *ABC* has a weight of 30 lb/ft and is used to lift the load of 2000 lb. Draw the shear and moment diagrams of the boom when it is in the horizontal position shown.

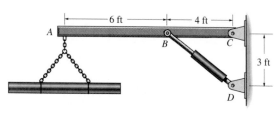

Prob. 6–28

6–29. The T-beam is subjected to the loading shown. Draw the shear and moment diagrams.

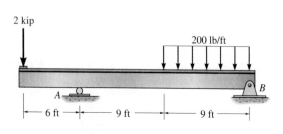

Prob. 6–29

6–30. Draw the shear and moment diagrams for the beam.

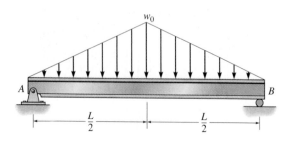

Prob. 6–30

6–31. Draw the shear and moment diagrams for the beam, and determine the shear and moment in the beam as functions of *x*.

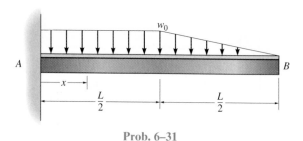

Prob. 6–31

*6–32. Draw the shear and moment diagrams for the wood beam, and determine the shear and moment throughout the beam as functions of *x*.

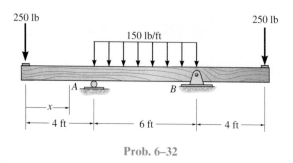

Prob. 6–32

6–33. The ski supports the 180-lb weight of the man. If the snow loading on its bottom surface is trapezoidal as shown, determine the intensity *w*, and then draw the shear and moment diagrams for the ski.

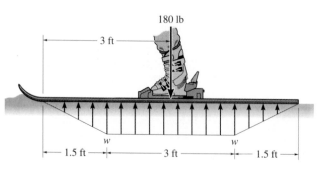

Prob. 6–33

6–34. The smooth pin is supported by two leaves A and B and subjected to a compressive load of 0.4 kN/m caused by bar C. Determine the intensity of the distributed load w_0 of the leaves on the pin and draw the shear and moment diagrams for the pin.

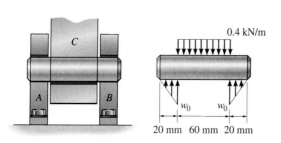

Prob. 6–34

6–35. Draw the shear and moment diagrams for the beam. The two segments are joined together at B.

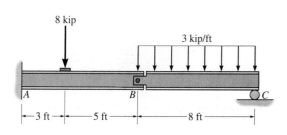

Prob. 6–35

*6–36. Draw the shear and moment diagrams for the beam.

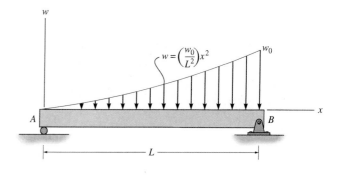

Prob. 6–36

6–37. The truck is to be used to transport the concrete column. If the column has a uniform weight of w (force/length), determine the equal placement a of the supports from the ends so that the absolute maximum bending moment in the column is as small as possible. Also, draw the shear and moment diagrams for the column.

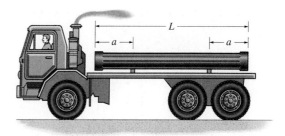

Prob. 6–37

6.3   BENDING DEFORMATION OF A STRAIGHT MEMBER

In this section we will discuss the deformations that occur when a straight prismatic beam, made of a homogeneous material, is subjected to bending. The discussion will be limited to beams having a cross-sectional area that is symmetrical with respect to an axis, and the bending moment is applied about an axis perpendicular to this axis of symmetry as shown in Fig. 6–20. The behavior of members that have unsymmetrical cross sections, or are made from several different materials, is based on similar observations and will be discussed separately in later sections of this chapter.

By using a highly deformable material such as rubber, we can physically illustrate what happens when a straight prismatic member is subjected to a bending moment. Consider, for example, the undeformed bar in Fig. 6–21*a*, which has a square cross section and is marked with longitudinal and transverse grid lines. When a bending moment is applied, it tends to distort these lines into the pattern shown in Fig. 6–21*b*. Here it can be seen that the longitudinal lines become *curved* and the vertical transverse lines *remain straight* and yet undergo a *rotation*.

The behavior of any deformable bar subjected to a bending moment causes the material within the bottom portion of the bar to stretch and the material within the top portion to compress. Consequently, between these two regions there must be a surface, called the *neutral surface*, in which longitudinal fibers of the material will not undergo a change in length, Fig. 6–20.

Axis of
symmetry   *y*

M

z

x

Neutral
surface

Longitudinal
axis

Fig. 6–20

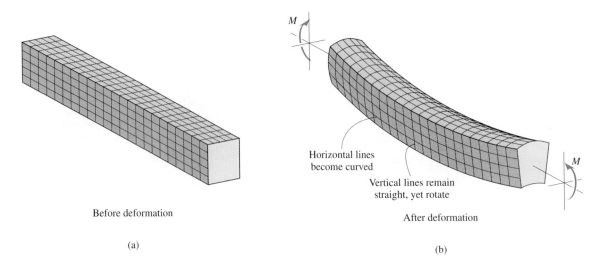

Before deformation

(a)

Horizontal lines
become curved

Vertical lines remain
straight, yet rotate

After deformation

(b)

M

M

Fig. 6–21

From these observations we will make the following three assumptions regarding the way the stress deforms the material. First, the *longitudinal axis x*, which lies within the neutral surface, Fig. 6–22a, does *not* experience any *change in length*. Rather the moment will tend to deform the beam so that this line *becomes a curve* that lies in the $x-y$ plane of symmetry, Fig. 6–22b. Second, all **cross sections** of the beam **remain plane** and perpendicular to the longitudinal axis during the deformation. And third, any *deformation* of the *cross section* within its own plane, as noticed in Fig. 6–21b, will be *neglected*. In particular, the z axis, lying in the plane of the cross section and about which the cross section rotates, is called the *neutral axis*, Fig. 6–22b. Its location will be determined in the next section.

In order to show how this distortion will strain the material, we will isolate a segment of the beam that is located a distance x along the beam's length and has an undeformed thickness Δx, Fig. 6–22a. This element, taken from the beam, is shown in profile view in the

Note the distortion of the lines due to bending of this rubber bar. The top line stretches, the bottom line compresses, and the center line remains the same length. Furthermore the vertical lines rotate and yet remain straight.

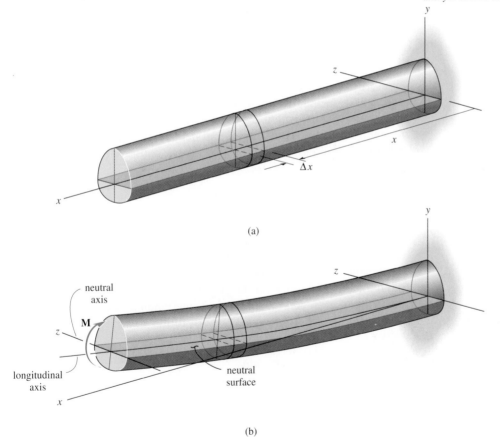

(a)

(b)

Fig. 6–22

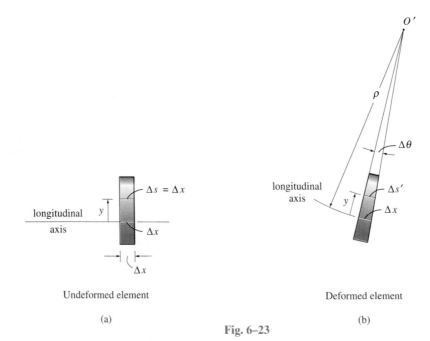

Undeformed element

(a)

Deformed element

(b)

Fig. 6–23

undeformed and deformed positions in Fig. 6–23. Notice that any line segment Δx, located on the neutral surface, does not change its length, whereas any line segment Δs, located at the arbitrary distance y above the neutral surface, will contract and become $\Delta s'$ after deformation. By definition, the normal strain along Δs is determined from Eq. 2–2, namely,

$$\epsilon = \lim_{\Delta s \to 0} \frac{\Delta s' - \Delta s}{\Delta s}$$

We will now represent this strain in terms of the location y of the segment and the radius of curvature ρ of the longitudinal axis of the element. Before deformation, $\Delta s = \Delta x$, Fig. 6–23a. After deformation Δx has a radius of curvature ρ, with center of curvature at point O', Fig. 6–23b. Since $\Delta \theta$ defines the angle between the cross-sectional sides of the element, $\Delta x = \Delta s = \rho \, \Delta \theta$. In the same manner, the deformed length of Δs becomes $\Delta s' = (\rho - y) \, \Delta \theta$. Substituting into the above equation, we get

$$\epsilon = \lim_{\Delta \theta \to 0} \frac{(\rho - y) \, \Delta \theta - \rho \, \Delta \theta}{\rho \, \Delta \theta}$$

or

$$\epsilon = -\frac{y}{\rho} \qquad (6\text{–}7)$$

This important result indicates that the longitudinal normal strain of any element within the beam depends on its location y on the cross

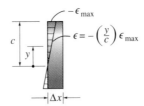

Normal strain distribution

Fig. 6–24

section and the radius of curvature of the beam's longitudinal axis at the point. In other words, for any specific cross section, the ***longitudinal normal strain will vary linearly*** with y from the neutral axis. A contraction $(-\epsilon)$ will occur in fibers located above the neutral axis $(+y)$, whereas elongation $(+\epsilon)$ will occur in fibers located below the axis $(-y)$. This variation in strain over the cross section is shown in Fig. 6–24. Here the maximum strain occurs at the outermost fiber, located a distance c from the neutral axis. Using Eq. 6–7, since $\epsilon_{max} = c/\rho$, then by division,

$$\frac{\epsilon}{\epsilon_{max}} = \frac{-y/\rho}{c/\rho}$$

So that

$$\epsilon = -\left(\frac{y}{c}\right)\epsilon_{max} \qquad (6\text{–}8)$$

This normal strain depends only on the assumptions made with regards to the *deformation*. Provided only a moment is applied to the beam, then it is reasonable to further assume that this moment causes a *normal stress only* in the longitudinal or x direction. All the other components of normal and shear stress are zero, since the beam's surface is free of any other load. It is this uniaxial state of stress that causes the material to have the longitudinal normal strain component ϵ_x, ($\sigma_x = E\epsilon_x$), defined by Eq. 6–8. Furthermore, by Poisson's ratio, there must *also* be associated strain components $\epsilon_y = -\nu\epsilon_x$ and $\epsilon_z = -\nu\epsilon_x$, which deform the plane of the cross-sectional area, although here we have neglected these deformations. Such deformations will, however, cause the *cross-sectional dimensions* to become smaller below the neutral axis and larger above the neutral axis. For example, if the beam has a square cross section, it will actually deform as shown in Fig. 6–25.

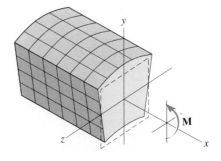

Fig. 6–25

6.4   THE FLEXURE FORMULA

In this section we will develop an equation that relates the longitudinal stress distribution in a beam to the internal resultant bending moment acting on the beam's cross section. To do this we will assume that the material behaves in a linear-elastic manner so that Hooke's law applies, that is, $\sigma = E\epsilon$. A **linear variation of normal strain**, Fig. 6–26a, must then be the consequence of a **linear variation in normal stress**, Fig. 6–26b. Hence, like the normal strain variation, σ will vary from zero at the member's neutral axis to a maximum value, σ_{max}, a distance c farthest from the neutral axis. Because of the proportionality of triangles, Fig. 6–26b, or by using Hooke's law, $\sigma = E\epsilon$, and Eq. 6–8, we can write

Normal strain variation
(profile view)

(a)

$$\sigma = -\left(\frac{y}{c}\right)\sigma_{max} \qquad (6\text{–}9)$$

Bending stress variation
(profile view)

(b)

Fig. 6–26

This equation represents the stress distribution over the cross-sectional area. The sign convention established here is significant. For positive $\mathbf{M}$, which acts in the $+z$ direction, positive values of y give negative values for σ, that is, a compressive stress since it acts in the negative x direction. Similarly, negative y values will give positive or tensile values for σ. If a volume element of material is selected at a specific point on the cross section, only these tensile or compressive normal stresses will act on it. For example, the element located at $+y$ is shown in Fig. 6–26c.

We can locate the position of the neutral axis on the cross section by satisfying the condition that the *resultant force* produced by the stress distribution over the cross-sectional area must be equal to *zero*. Noting that the force $dF = \sigma \, dA$ acts on the arbitrary element dA in Fig. 6–26c, we require

This wood specimen failed in bending due to its fibers being crushed at its top and torn apart at its bottom.

$$F_R = \Sigma F_x; \qquad 0 = \int_A dF = \int_A \sigma \, dA$$

$$= \int_A -\left(\frac{y}{c}\right)\sigma_{max} \, dA$$

$$= \frac{-\sigma_{max}}{c} \int_A y \, dA$$

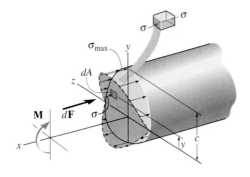

Bending stress variation

(c)

Since σ_{max}/c is not equal to zero, then

$$\int_A y \, dA = 0 \tag{6–10}$$

In other words, the first moment of the member's cross-sectional area about the neutral axis must be zero. This condition can only be satisfied if the *neutral axis* is also the horizontal *centroidal axis* for the cross section.* Consequently, once the centroid for the member's cross-sectional area is determined, the location of the neutral axis is known.

We can determine the stress in the beam from the requirement that the resultant internal moment M must be equal to the moment produced by the stress distribution about the neutral axis. The moment of dF in Fig. 6–26c about the neutral axis is $dM = y \, dF$. This moment is *positive* since, by the right-hand rule, the thumb is directed along the positive z axis when the fingers are curled with the sense of rotation caused by dM. Since $dF = \sigma \, dA$, using Eq. 6–9, we have for the entire cross section,

$$(M_R)_z = \Sigma M_z; \quad M = \int_A y \, dF = \int_A y \, (\sigma \, dA) = \int_A y \left(\frac{y}{c} \sigma_{max} \right) dA$$

or

$$M = \frac{\sigma_{max}}{c} \int_A y^2 \, dA \tag{6–11}$$

---

*Recall that the location $\bar{y}$ for the centroid of the cross-sectional area is defined from the equation $\bar{y} = \int y \, dA / \int dA$. If $\int y \, dA = 0$, then $\bar{y} = 0$, and so the centroid lies on the reference (neutral) axis. See Appendix A.

Here the integral represents the *moment of inertia* of the cross-sectional area, computed about the neutral axis. We symbolize its value as *I*. Hence, Eq. 6–11 can be solved for σ_{max} and written in general form as

$$\sigma_{max} = \frac{Mc}{I} \qquad (6\text{–}12)$$

Here

σ_{max} = the maximum normal stress in the member, which occurs at a point on the cross-sectional area *farthest away* from the neutral axis

M = the resultant internal moment, determined from the method of sections and the equations of equilibrium, and computed about the neutral axis of the cross section

I = the moment of inertia of the cross-sectional area computed about the neutral axis

c = the perpendicular distance from the neutral axis to a point farthest away from the neutral axis, where σ_{max} acts

Since $\sigma_{max}/c = -\sigma/y$, Eq. 6–9, the normal stress at the intermediate distance y can be determined from an equation similar to Eq. 6–12. We have

$$\sigma = -\frac{My}{I} \qquad (6\text{–}13)$$

Note that the negative sign is necessary since it agrees with the established *x*, *y*, *z* axes. By the right-hand rule, *M* is positive along the +*z* axis, *y* is positive upward, and σ therefore must be negative (compressive) since it acts in the negative *x* direction, Fig. 6–26c.

Either of the above two equations is often referred to as the **flexure formula**. It is used to determine the normal stress in a straight member, having a cross section that is symmetrical with respect to an axis, and the moment is applied perpendicular to this axis. Although we have assumed that the member is prismatic, we can in most cases of engineering design also use the flexure formula to determine the normal stress in members that have a *slight taper*. For example, using a mathematical analysis based on the theory of elasticity, a member having a rectangular cross section and a length that is tapered 15° will have an actual maximum normal stress that is about 5.4% *less* than that calculated using the flexure formula.

IMPORTANT POINTS

- The cross section of a straight beam *remains plane* when the beam deforms due to bending. This causes tensile stress on one side of the beam and compressive stress on the other side. The *neutral axis* is subjected to *zero stress.*

- Due to the deformation, the *longitudinal strain* varies *linearly* from zero at the neutral axis to a maximum at the outer fibers of the beam. Provided the material is homogeneous and Hooke's law applies, the *stress* also varies in a *linear* fashion over the cross section.

- For linear-elastic material the neutral axis passes through the *centroid* of the cross-sectional area. This conclusion is based on the fact that the resultant normal force acting on the cross section must be zero.

- The flexure formula is based on the requirement that the resultant moment on the cross section is equal to the moment produced by the linear normal stress distribution about the neutral axis.

PROCEDURE FOR ANALYSIS

In order to apply the flexure formula, the following procedure is suggested.

Internal Moment.

- Section the member at the point where the bending or normal stress is to be determined, and obtain the internal moment M at the section. The centroidal or neutral axis for the cross section must be known, since M *must* be computed about this axis.

- If the absolute maximum bending stress is to be determined, then draw the moment diagram in order to determine the maximum moment in the beam.

Section Property.

- Determine the moment of inertia of the cross-sectional area about the neutral axis. Methods used for its computation are discussed in Appendix A, and a table listing values of I for several common shapes is given on the inside front cover.

Normal Stress.

- Specify the distance y, measured perpendicular to the neutral axis to the point where the normal stress is to be determined. Then apply the equation $\sigma = -My/I$, or if the maximum bending stress is to be calculated, use $\sigma_{max} = Mc/I$. When substituting the data, make sure the units are consistent.

- The stress acts in a direction such that the force it creates at the point contributes a moment about the neutral axis that is in the same direction as the internal moment **M**, Fig. 6–26c. In this manner the stress distribution acting over the entire cross section can be sketched, or a volume element of the material can be isolated and used to represent graphically the normal stress acting at the point.

EXAMPLE 6–14

A beam has a rectangular cross section and is subjected to the stress distribution shown in Fig. 6–27a. Determine the internal moment **M** at the section caused by the stress distribution (*a*) using the flexure formula, (*b*) by finding the resultant of the stress distribution using basic principles.

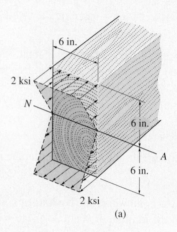

Fig. 6–27

SOLUTION

Part (a). The flexure formula is $\sigma_{max} = Mc/I$. From Fig. 6–27a, $c = 6$ in. and $\sigma_{max} = 2$ ksi. The neutral axis is defined as line *NA*, because the stress is zero along this line. Since the cross section has a rectangular shape, the moment of inertia for the area about *NA* is determined from the formula for a rectangle given on the inside front cover; i.e.,

$$I = \frac{1}{12}bh^3 = \frac{1}{12}(6 \text{ in.})(12 \text{ in.})^3 = 864 \text{ in}^4$$

Therefore,

$$\sigma_{max} = \frac{Mc}{I}; \qquad 2 \text{ kip/in}^2 = \frac{M(6 \text{ in.})}{864 \text{ in}^4}$$

$$M = 288 \text{ kip} \cdot \text{in.} = 24 \text{ kip} \cdot \text{ft} \qquad \textit{Ans.}$$

Part (b).   First we will show that the resultant force of the stress distribution is zero. As shown in Fig. 6–27b, the stress acting on the arbitrary element strip $dA = (6\text{ in.})\,dy$, located y from the neutral axis, is

$$\sigma = \left(\frac{-y}{6\text{ in.}}\right)(2\text{ kip/in}^2)$$

The force created by this stress is $dF = \sigma\,dA$, and thus, for the entire cross section,

$$F_R = \int_A \sigma\,dA = \int_{-6\text{ in.}}^{6\text{ in.}} \left[\left(\frac{-y}{6\text{ in.}}\right)(2\text{ kip/in}^2)\right](6\text{ in.})\,dy$$

$$= (-1\text{ kip/in}^2)\,y^2\,\Big|_{-6\text{ in.}}^{+6\text{ in.}} = 0$$

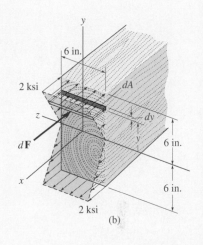

(b)

The resultant moment of the stress distribution about the neutral axis (z axis) must equal M. Since the magnitude of the moment of $d\mathbf{F}$ about this axis is $dM = y\,dF$, and $d\mathbf{M}$ is *always positive*, Fig. 6–27b, then for the entire area,

$$M = \int_A y\,dF = \int_{-6\text{ in.}}^{6\text{ in.}} y\left[\left(\frac{y}{6\text{ in.}}\right)(2\text{ kip/in}^2)\right](6\text{ in.})\,dy$$

$$= \left(\frac{2}{3}\text{ kip/in}^2\right)y^3\,\Big|_{-6\text{ in.}}^{+6\text{ in.}}$$

$$= 288\text{ kip}\cdot\text{in.} = 24\text{ kip}\cdot\text{ft} \qquad\qquad \textit{Ans.}$$

The above result can *also* be determined without the need for integration. The resultant force for each of the two *triangular* stress distributions in Fig. 6–27c is graphically equivalent to the *volume* contained within each stress distribution. Thus, each volume is

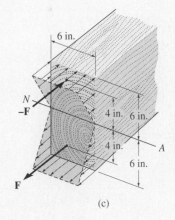

(c)

$$F = \frac{1}{2}(6\text{ in.})(2\text{ kip/in}^2)(6\text{ in.}) = 36\text{ kip}$$

These forces, which form a couple, act in the same direction as the stresses within each distribution, Fig. 6–27c. Furthermore, they act through the *centroid* of each volume, i.e., $\frac{1}{3}(6\text{ in.}) = 2$ in. from the top and bottom of the beam. Hence the distance between them is 8 in. as shown. The moment of the couple is therefore

$$M = 36\text{ kip (8 in.)} = 288\text{ kip}\cdot\text{in.} = 24\text{ kip}\cdot\text{ft} \qquad \textit{Ans.}$$

EXAMPLE 6–15

The simply supported beam in Fig. 6–28a has the cross-sectional area shown in Fig. 6–28b. Determine the absolute maximum bending stress in the beam and draw the stress distribution over the cross section at this location.

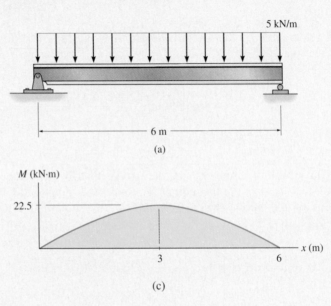

5 kN/m

6 m

(a)

M (kN·m)

22.5

3   6

x (m)

(c)

Fig. 6–28

SOLUTION

Maximum Internal Moment. The maximum internal moment in the beam, $M = 22.5$ kN · m, occurs at the center as shown on the bending moment diagram, Fig. 6–28c. See Example 6–3.

Section Property. By reasons of symmetry, the centroid C and thus the neutral axis pass through the midheight of the beam, Fig. 6–28b. The area is subdivided into the three parts shown, and the moment of inertia of each part is computed about the neutral axis using the parallel-axis theorem. (See Eq. A–5 of Appendix A.) Choosing to work in meters, we have

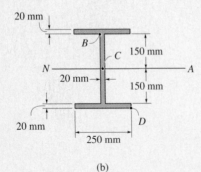

20 mm

B

C 150 mm

N ———————— A

20 mm

150 mm

20 mm

D

250 mm

(b)

$$I = \Sigma(\bar{I} + Ad^2)$$

$$= 2\left[\frac{1}{12}(0.25 \text{ m})(0.020 \text{ m})^3 + (0.25 \text{ m})(0.020 \text{ m})(0.160 \text{ m})^2\right]$$

$$+ \left[\frac{1}{12}(0.020 \text{ m})(0.300 \text{ m})^3\right]$$

$$= 301.3(10^{-6}) \text{ m}^4$$

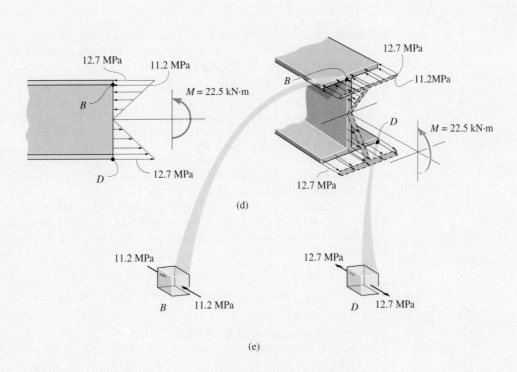

(d)

(e)

Bending Stress.   Applying the flexure formula, with $c = 170$ mm, the absolute maximum bending stress is

$$\sigma_{\max} = \frac{Mc}{I}; \quad \sigma_{\max} = \frac{22.5 \text{ kN} \cdot \text{m}(0.170 \text{ m})}{301.3(10^{-6}) \text{ m}^4} = 12.7 \text{ MPa} \qquad \textit{Ans.}$$

Two-and-three-dimensional views of the stress distribution are shown in Fig. 6–28d. Notice how the stress at each point on the cross section develops a force that contributes a moment $d\mathbf{M}$ about the neutral axis such that it has the same direction as $\mathbf{M}$. Specifically, at point B, $y_B = 150$ mm, and so

$$\sigma_B = \frac{My_B}{I}; \quad \sigma_B = \frac{22.5 \text{ kN} \cdot \text{m}(0.150 \text{ m})}{301.3(10^{-6}) \text{ m}^4} = 11.2 \text{ MPa}$$

The normal stress acting on elements of material located at points B and D is shown in Fig. 6–28e.

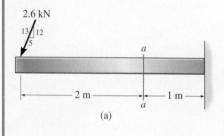

(a)

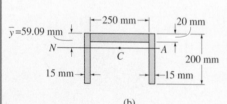

(b)

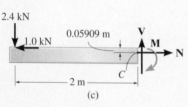

(c)

Fig. 6–29

EXAMPLE 6–16

The beam shown in Fig. 6–29a has a cross-sectional area in the shape of a channel, Fig. 6–29b. Determine the maximum bending stress that occurs in the beam at section a–a.

SOLUTION

Internal Moment. Here the beam's support reactions do not have to be determined. Instead, by the method of sections, the segment to the left of section a–a can be used, Fig. 6–29c. In particular, note that the resultant internal axial force **N** passes through the centroid of the cross section. Also, realize that *the resultant internal moment must be computed about the beam's neutral axis* at section a–a.

To find the location of the neutral axis, the cross-sectional area is subdivided into three composite parts as shown in Fig. 6–29b. Since the neutral axis passes through the centroid, then using Eq. A–2 of Appendix A, we have

$$\bar{y} = \frac{\Sigma \tilde{y} A}{\Sigma A} = \frac{2[0.100 \text{ m}](0.200 \text{ m})(0.015 \text{ m}) + [0.010 \text{ m}](0.02 \text{ m})(0.250 \text{ m})}{2(0.200 \text{ m})(0.015 \text{ m}) + 0.020 \text{ m}(0.250 \text{ m})}$$

$$= 0.05909 \text{ m} = 59.09 \text{ mm}$$

This dimension is shown in Fig. 6–29c.

Applying the moment equation of equilibrium about the neutral axis, we have

$$\zeta + \Sigma M_{NA} = 0; \quad 2.4 \text{ kN}(2 \text{ m}) + 1.0 \text{ kN}(0.05909 \text{ m}) - M = 0$$

$$M = 4.859 \text{ kN} \cdot \text{m}$$

Section Property. The moment of inertia about the neutral axis is determined using the parallel-axis theorem applied to each of the three composite parts of the cross-sectional area. Working in meters, we have

$$I = \left[\frac{1}{12}(0.250 \text{ m})(0.020 \text{ m})^3 + (0.250 \text{ m})(0.020 \text{ m})(0.05909 \text{ m} - 0.010 \text{ m})^2 \right]$$

$$+ 2 \left[\frac{1}{12}(0.015 \text{ m})(0.200 \text{ m})^3 + (0.015 \text{ m})(0.200 \text{ m})(0.100 \text{ m} - 0.05909 \text{ m})^2 \right]$$

$$= 42.26(10^{-6}) \text{ m}^4$$

Maximum Bending Stress. The maximum bending stress occurs at points farthest away from the neutral axis. This is at the bottom of the beam, $c = 0.200 \text{ m} - 0.05909 \text{ m} = 0.1409 \text{ m}$. Thus,

$$\sigma_{max} = \frac{Mc}{I} = \frac{4.859 \text{ kN} \cdot \text{m}(0.1409 \text{ m})}{42.26(10^{-6}) \text{ m}^4} = 16.2 \text{ MPa} \quad \textit{Ans.}$$

Show that at the top of the beam the bending stress is $\sigma' = 6.79 \text{ MPa}$. Note that in addition to this effect of bending, the normal force of $N = 1 \text{ kN}$ and shear force $V = 2.4 \text{ kN}$ will also contribute additional stress on the cross section. The superposition of all these effects will be discussed in a later chapter.

EXAMPLE 6-17

The member having a rectangular cross section, Fig. 6–30a, is designed to resist a moment of 40 N · m. In order to increase its strength and rigidity, it is proposed that two small ribs be added at its bottom, Fig. 6–30b. Determine the maximum normal stress in the member for both cases.

SOLUTION

Without Ribs.   Clearly the neutral axis is at the center of the cross section, Fig. 6-30a, so $\bar{y} = c = 15$ mm $= 0.015$ m. Thus,

$$I = \frac{1}{12}bh^3 = \frac{1}{12}(0.06 \text{ m})(0.03 \text{ m})^3 = 0.135(10^{-6}) \text{ m}^4$$

Therefore the maximum normal stress is

$$\sigma_{max} = \frac{Mc}{I} = \frac{(40 \text{ N} \cdot \text{m})(0.015 \text{ m})}{0.135(10^{-6}) \text{ m}^4} = 4.44 \text{ MPa} \qquad \textit{Ans.}$$

With Ribs.   From Fig. 6–30b, segmenting the area into the large main rectangle and the bottom two rectangles (ribs), the location $\bar{y}$ of the centroid and the neutral axis is determined as follows:

$$\bar{y} = \frac{\Sigma \tilde{y}A}{\Sigma A}$$

$$= \frac{[0.015 \text{ m}](0.030 \text{ m})(0.060 \text{ m}) + 2[0.0325 \text{ m}](0.005 \text{ m})(0.010 \text{ m})}{(0.03 \text{ m})(0.060 \text{ m}) + 2(0.005 \text{ m})(0.010 \text{ m})}$$

$$= 0.01592 \text{ m}$$

This value does not represent c. Instead

$$c = 0.035 \text{ m} - 0.01592 \text{ m} = 0.01908 \text{ m}$$

Using the parallel-axis theorem, the moment of inertia about the neutral axis is

$$I = \left[\frac{1}{12}(0.060 \text{ m})(0.030 \text{ m})^3 + (0.060 \text{ m})(0.030 \text{ m})(0.01592 \text{ m} - 0.015 \text{ m})^2 \right]$$

$$+ 2\left[\frac{1}{12}(0.010 \text{ m})(0.005 \text{ m})^3 + (0.010 \text{ m})(0.005 \text{ m})(0.0325 \text{ m} - 0.01592 \text{ m})^2 \right]$$

$$= 0.1642(10^{-6}) \text{ m}^4$$

Therefore, the maximum normal stress is

$$\sigma_{max} = \frac{Mc}{I} = \frac{40 \text{ N} \cdot \text{m}(0.01908 \text{ m})}{0.1642 (10^{-6}) \text{ m}^4} = 4.65 \text{ MPa} \qquad \textit{Ans.}$$

This surprising result indicates that the addition of the ribs to the cross section will *increase* the normal stress rather than decrease it, and for this reason they should be omitted.

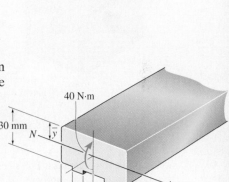

(a)

(b)

Fig. 6–30

PROBLEMS

6–38. The beam is subjected to a moment M. Determine the percentage of this moment that is resisted by the stresses acting on both the top and bottom boards, A and B, of the beam.

6–39. Determine the moment M that should be applied to the beam in order to create a compressive stress at point D of $\sigma_D = 30$ MPa. Also sketch the stress distribution acting over the cross section and compute the maximum stress developed in the beam.

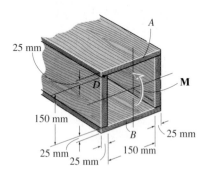

Probs. 6–38/6–39

*6–40. A member has the triangular cross section shown. Determine the largest internal moment M that can be applied to the cross section without exceeding allowable tensile and compressive stresses of $(\sigma_{allow})_t = 22$ ksi and $(\sigma_{allow})_c = 15$ ksi, respectively.

6–41. A member has the triangular cross section shown. If a moment of $M = 800$ lb · ft is applied to the cross section, determine the maximum tensile and compressive bending stresses in the member. Also, sketch a three-dimensional view of the stress distribution acting over the cross section.

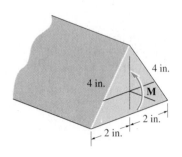

Probs. 6–40/6–41

6–42. Two considerations have been proposed for the design of a beam. Determine which one will support a moment of $M = 150$ kN · m with the least amount of bending stress. What is that stress? By what percentage is it more effective?

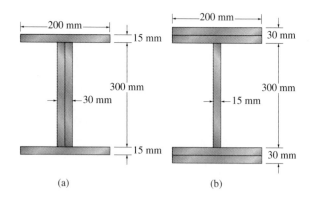

Prob. 6–42

6–43. Determine the moment M that will produce a maximum stress of 10 ksi on the cross section.

*6–44. Determine the maximum tensile and compressive bending stress in the beam if it is subjected to a moment of $M = 4$ kip · ft.

6–45. Determine the resultant force the bending stresses produce on the horizontal top flange plate AB of the beam if $M = 4$ kip · ft.

6–46. Determine the resultant force the bending stresses produce on the web CD of the beam if $M = 4$ kip · ft.

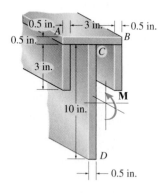

Probs. 6–43/6–44/6–45/6–46

6–47. The aluminum machine part is subjected to a moment of $M = 75$ N · m. Determine the bending stress created at points B and C on the cross section. Sketch the results on a volume element located at each of these points.

*6–48. The aluminum machine part is subjected to a moment of $M = 75$ N · m. Determine the maximum tensile and compressive bending stresses in the part.

6–51. The channel strut is used as a guide rail for a trolley. If the maximum moment in the strut is $M = 30$ N · m, determine the bending stress at points A, B, and C.

*6–52. The channel strut is used as a guide rail for a trolley. If the allowable bending stress for the material is $\sigma_{\text{allow}} = 175$ MPa, determine the maximum bending moment the strut will resist.

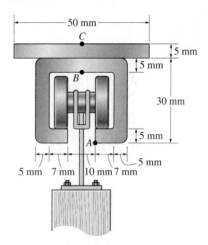

Probs. 6–51/6–52

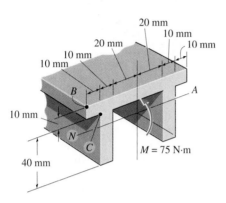

Probs. 6–47/6–48

6–49. The beam is made from three boards nailed together as shown. If the moment acting on the cross section is $M = 1$ kip · ft, determine the maximum bending stress in the beam. Sketch a three-dimensional view of the stress distribution acting over the cross section.

6–50. Determine the resultant force the bending stresses produce on the top board A of the beam if $M = 1$ kip · ft.

6–53. A beam is constructed from four pieces of wood, glued together as shown. If the moment acting on the cross section is $M = 450$ N · m, determine the resultant force the bending stress produces on the top board A and on the side board B.

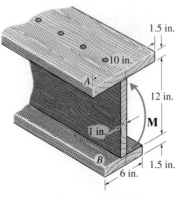

Probs. 6–49/6–50

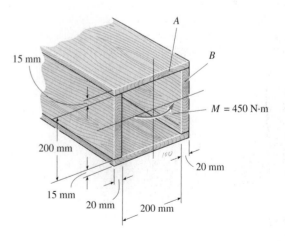

Prob. 6–53

6–54. The control lever is used on a riding lawn mower. Determine the maximum bending stress in the lever at section *a–a* if a force of 20 lb is applied to the handle. The lever is supported by a pin at *A* and a wire at *B*. Section *a–a* is square, 0.25 in. by 0.25 in.

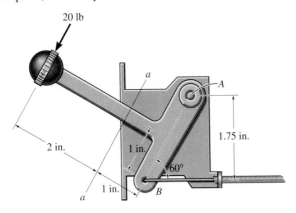

Prob. 6–54

6–55. The beam is subjected to a moment of 15 kip · ft. Determine the resultant force the stress produces on the top flange *A* and bottom flange *B*. Also compute the maximum stress developed in the beam.

*6–56. The beam is subjected to a moment of 15 kip · ft. Determine the percentage of this moment that is resisted by the web *D* of the beam.

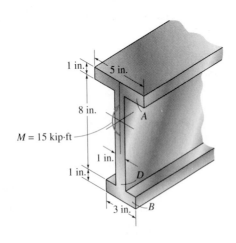

Probs. 6–55/6–56

6–57. The beam is subjected to a moment of $M = 30$ lb · ft. Determine the bending stress acting at points *A* and *B*. Also, sketch a three-dimensional view of the stress distribution acting over the entire cross-sectional area.

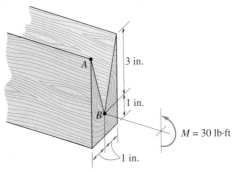

Prob. 6–57

6–58. Two considerations have been proposed for the design of a beam. Determine which one will support a moment of $M = 150$ kN · m with the least amount of bending stress. What is that stress? By what percentage is it more effective?

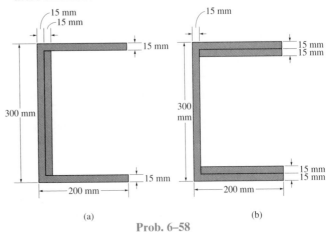

Prob. 6–58

6–59. If the shaft has a diameter of 50 mm, determine the absolute maximum bending stress in the shaft.

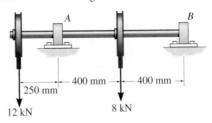

Prob. 6–59

*6–60. If the beam has a square cross section of 9 in. on each side, determine the absolute maximum bending stress in the beam.

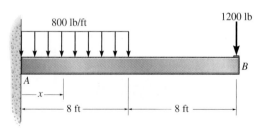

800 lb/ft

1200 lb

A

B

x

8 ft

8 ft

Prob. 6–60

6–61. If the beam has a rectangular cross section with a width of 8 in. and a height of 16 in., determine the absolute maximum bending stress in the beam.

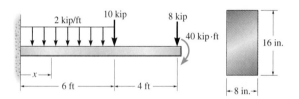

2 kip/ft   10 kip   8 kip

40 kip·ft

16 in.

x

6 ft

4 ft

8 in.

Prob. 6–61

6–62. If the beam has a cross section as shown, determine the absolute maximum bending stress in the beam.

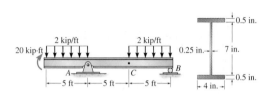

20 kip·ft   2 kip/ft   2 kip/ft

0.5 in.

0.25 in.   7 in.

A   *C*   *B*

0.5 in.

5 ft   5 ft   5 ft

4 in.

Prob. 6–62

6–63. The axle of the freight car is subjected to wheel loadings of 20 kip. If it is supported by two journal bearings at *C* and *D*, determine the maximum bending stress developed at the center of the axle, where the diameter is 5.5 in.

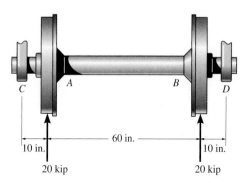

C   *A*   *B*   *D*

60 in.

10 in.   10 in.

20 kip   20 kip

Prob. 6–63

*6–64. If the beam in Prob. 6–27 has a rectangular cross section with a width *b* and a height *h*, determine the absolute maximum bending stress in the beam.

6–65. If the compound beam in Prob. 6-19 has a square cross section, determine its dimension *a* if the allowable bending stress is $\sigma_{\text{allow}} = 150$ MPa.

6–66. If the compound beam in Prob. 6-25 has a square cross section, determine its required height and width, *a*, if the allowable bending stress for the material is $\sigma_{\text{allow}} = 1.50$ ksi.

6–67. Determine the absolute maximum bending stress in the beam. The cross section of the beam is as shown.

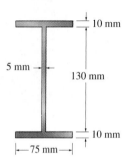

10 mm

5 mm

130 mm

10 mm

75 mm

Prob. 6–67

*6–68.   Determine the absolute maximum bending stress in the T-beam in Prob. 6–29. The cross section of the beam is as shown.

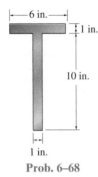

Prob. 6–68

6–69.   If the beam in Prob. 6–17 has a rectangular cross section, determine its required height h and width b if $h = 2b$. The allowable bending stress for the material is $\sigma_{allow} = 24$ ksi.

6–70.   If the beam in Prob. 6–17 is to be a pipe having an outer diameter of 3 in., determine its inner diameter if the allowable bending stress for the material is $\sigma_{allow} = 24$ ksi.

6–71.   Determine the absolute maximum bending stress in the 30-mm-diameter shaft which is subjected to the concentrated forces. The sleeve bearings at A and B support only vertical forces.

*6–72.   Determine the smallest allowable diameter of the shaft which is subjected to the concentrated forces. The sleeve bearings at A and B support only vertical forces, and the allowable bending stress is $\sigma_{allow} = 160$ MPa.

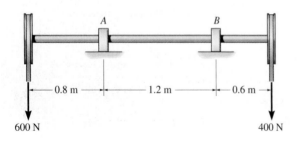

Probs. 6–71/6–72

6–73.   The beam has a rectangular cross section as shown. Determine the largest load P that can be supported on its overhanging ends so that the bending stress in the beam does not exceed $\sigma_{max} = 10$ MPa.

6–74.   The beam has the rectangular cross section shown. If $P = 1.5$ kN, determine the maximum bending stress in the beam. Sketch the stress distribution acting over the cross section.

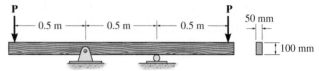

Probs. 6–73/6–74

6–75.   Determine the absolute maximum bending stress in the 1.5-in.-diameter shaft which is subjected to the concentrated forces. The sleeve bearings at A and B support only vertical forces.

*6–76.   Determine the smallest allowable diameter of the shaft which is subjected to the concentrated forces. The sleeve bearings at A and B support only vertical forces, and the allowable bending stress is $\sigma_{allow} = 22$ ksi.

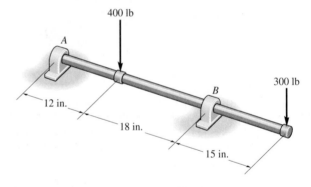

Probs. 6–75/6–76

6–77.   The beam is subjected to the loading shown. Determine its required cross-sectional dimension a, if the allowable bending stress for the material is $\sigma_{allow} = 150$ MPa.

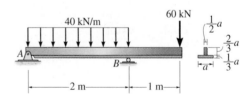

Probs. 6–76/6–77

6–78. Determine the magnitude of the maximum load **P** that can be applied to the beam if the beam is made of a material having an allowable bending stress of $(\sigma_{allow})_c$ = 12 ksi in compression and $(\sigma_{allow})_t$ = 22 ksi in tension.

6–79. Determine the magnitude of the maximum load **P** that can be applied to the beam if the beam is made of a material having an allowable bending stress of $(\sigma_{allow})_c$ = 16 ksi in compression and $(\sigma_{allow})_t$ = 18 ksi in tension.

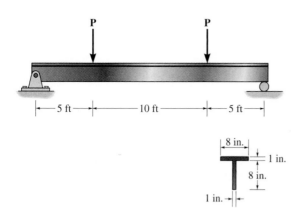

Probs. 6–78/6–79

***6–80.** The two solid steel rods are bolted together along their length and support the loading shown. Assume the support at A is a pin and B is a roller. Determine the required diameter d of each of the rods if the allowable bending stress is σ_{allow} = 130 MPa.

6–81. Solve Prob. 6–80 if the rods are rotated 90° so that both rods rest on the supports at A (pin) and B (roller).

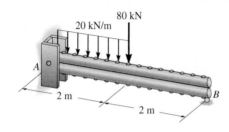

Probs. 6–80/6–81

6–82. The strut *CD* on the utility pole supports the cable having a weight of 600 lb. Determine the absolute maximum bending stress in the strut if A, B, and C are assumed to be pinned.

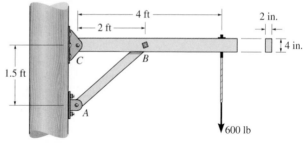

Prob. 6–82

6–83. The beam is subjected to the load **P** at its center. Determine the placement a of the supports so that the absolute maximum bending stress in the beam is as large as possible. What is this stress?

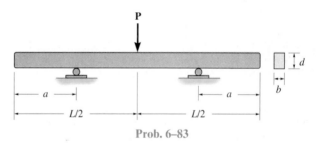

Prob. 6–83

***6–84.** Determine the absolute maximum bending stress in the tubular shaft if d_i = 160 mm and d_o = 200 mm.

6–85. The tubular shaft is to have a cross section such that its inner diameter and outer diameter are related by $d_i = 0.8d_o$. Determine these required dimensions if the allowable bending stress is σ_{allow} = 155 MPa.

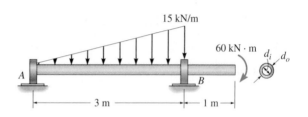

Probs. 6–84/6–85

6–86. The pin is used to connect the three links together. Due to wear, the load is distributed over the top and bottom of the pin as shown on the free-body diagram. If the diameter of the pin is 0.40 in., determine the maximum bending stress on the cross-sectional area at the center section a–a. For the solution it is first necessary to determine the load intensities w_1 and w_2.

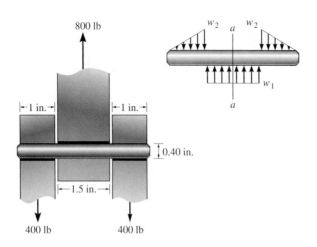

Prob. 6–86

6–87. The man has a mass of 78 kg and stands motionless at the end of the diving board. If the board has the cross section shown, determine the maximum normal strain developed in the board. The modulus of elasticity for the material is $E = 125$ GPa. Assume A is a pin and B is a roller.

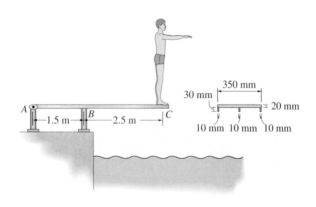

Prob. 6–87

*6–88. The rod is supported by smooth journal bearings at A and B that only exert vertical reactions on the shaft. If $d = 90$ mm, determine the absolute maximum bending stress in the beam, and sketch the stress distribution acting over the cross section.

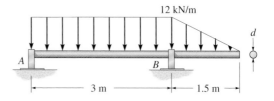

Prob. 6–88

6–89. The rod is supported by smooth journal bearings at A and B that only exert vertical reactions on the shaft. Determine its smallest diameter d if the allowable bending stress is $\sigma_{\text{allow}} = 180$ MPa.

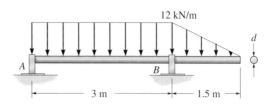

Prob. 6–89

6–90. The wooden beam has a rectangular cross section for which $b = 60$ mm. Determine the absolute maximum bending stress in the beam and sketch the stress distribution acting over the cross section.

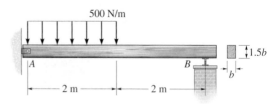

Prob. 6–90

6–91. The wing spar *ABD* of a light plane is made from 2014–T6 aluminum and has a cross-sectional area of 1.27 in., a depth of 3 in., and a moment of inertia about its neutral axis of 2.68 in⁴. Determine the absolute maximum bending stress in the spar if the anticipated loading is to be as shown. Assume *A*, *B*, and *C* are pins. Connection is made along the central longitudinal axis of the spar.

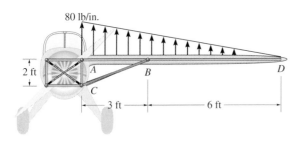

Prob. 6–91

*6–92. The boat has a weight of 2300 lb and a center of gravity at *G*. If it rests on the trailer at the smooth contact *A* and can be considered pinned at *B*, determine the absolute maximum bending stress developed in the main strut of the trailer. Consider the strut to be a box-beam, having the dimensions shown, and pinned at *C*.

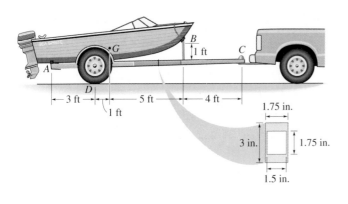

Prob. 6–92

6–93. Determine the maximum allowable uniform load w that can be supported by the fiberglass beam if $b = 125$ mm and the allowable bending stress for the material is $\sigma_{allow} = 13$ MPa.

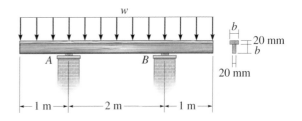

Prob. 6–93

6–94. The wood beam is subjected to the uniform load of $w = 200$ lb/ft. If the allowable bending stress for the material is $\sigma_{allow} = 1.40$ ksi, determine the required dimension b of its cross section. Assume the support at *A* is a pin and *B* is a roller.

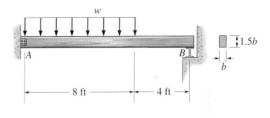

Prob. 6–94

6–95. The cantilevered beam has a thickness of 4 in. and a variable depth that can be described by the function $y = 2[(x + 2)/4]^{0.2}$, where x is in inches. Determine the maximum bending stress in the beam at its center.

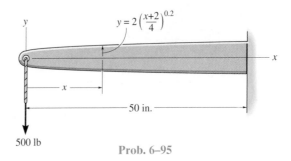

Prob. 6–95

6.5   Unsymmetric Bending

When developing the flexure formula we imposed a condition that the cross-sectional area be *symmetric* about an axis perpendicular to the neutral axis; furthermore, the resultant internal moment **M** acts along the neutral axis. Such is the case for the "T" or channel sections shown in Fig. 6–31. These conditions, however, are unnecessary, and in this section we will show that the flexure formula can also be applied either to a beam having a cross-sectional area of any shape or to a beam having a resultant internal moment that acts in any direction.

Moment Applied Along Principal Axis.   Consider the beam's cross section to have the unsymmetrical shape shown in Fig. 6–32a. As in Sec. 6.4, the right-handed x, y, z coordinate system is established such that the origin is located at the centroid C of the cross section, and the resultant internal moment **M** acts along the $+z$ axis. We require the stress distribution acting over the entire cross-sectional area to have a zero force resultant, the resultant internal moment about the y axis to be zero, and the resultant internal moment about the z axis to equal **M**.* These three conditions can be expressed mathematically by considering the force acting on the differential element dA located at $(0, y, z)$, Fig. 6–32a. This force is $dF = \sigma\, dA$, and therefore we have

$$F_R = \Sigma F_x; \qquad\qquad 0 = \int_A \sigma\, dA \qquad\qquad (6\text{–}14)$$

$$(M_R)_y = \Sigma M_y; \qquad\qquad 0 = \int_A z\sigma\, dA \qquad\qquad (6\text{–}15)$$

$$(M_R)_z = \Sigma M_z; \qquad\qquad M = \int_A -y\sigma\, dA \qquad\qquad (6\text{–}16)$$

---

*The condition that moments about the y axis be equal to zero was not considered in Sec. 6.4, since the bending-stress distribution was *symmetric* with respect to the y axis and such a distribution of stress automatically produces zero moment about the y axis. See Fig. 6–26c.

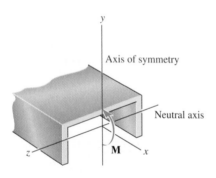

Axis of symmetry

Neutral axis

M

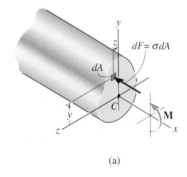

Axis of symmetry

Neutral axis

M

Figs. 6–31

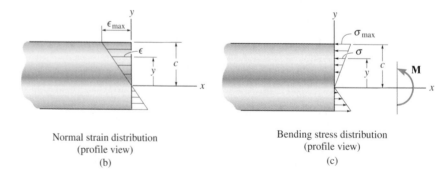

(a)

Normal strain distribution
(profile view)
(b)

Bending stress distribution
(profile view)
(c)

Figs. 6–32

As shown in Sec. 6.4, Eq. 6–14 is satisfied since the z axis passes through the *centroid* of the cross-sectional area. Also, since the z axis represents the *neutral axis* for the cross section, the normal strain will vary linearly from zero at the neutral axis, to a maximum at a point located the largest y coordinate distance, $y = c$, from the neutral axis, Fig. 6–32b. Provided the material behaves in a linear-elastic manner, the normal-stress distribution over the cross-sectional area is *also* linear, so that $\sigma = -(y/c)\sigma_{max}$, Fig. 6–32c. When this equation is substituted into Eq. 6–16 and integrated, it leads to the flexure formula $\sigma_{max} = Mc/I$. When it is substituted into Eq. 6–15, we get

$$0 = \frac{-\sigma_{max}}{c} \int_A yz \, dA$$

which requires

$$\int_A yz \, dA = 0$$

This integral is called the **product of inertia** for the area. As indicated in Appendix A, it will indeed be zero provided the y and z axes are chosen as **principal axes of inertia** for the area. For an arbitrarily shaped area, the orientation of the principal axes can always be determined, using either the inertia transformation equations or Mohr's circle of inertia as explained in Appendix A, Secs. A–4 and A–5. If the area has an axis of symmetry, however, the **principal axes** can easily be established **since they will always be oriented along the axis of symmetry and perpendicular to it.**

In summary, then, Eqs. 6–14 through 6–16 will *always* be satisfied regardless of the direction of the applied moment **M**. For example, consider the members shown in Fig. 6–33. In each of these cases, y and z define the principal axes of inertia for the cross section having the origin located at the area's centroid. In Fig. 6–33a and 6–33b, the principal axes are located by symmetry, and in Fig. 6–33c and 6–33d their orientation is determined using the methods of Appendix A. Since **M** is applied about one of the principal axes (z axis), the stress distribution is determined from the flexure formula, $\sigma = My/I_z$, and is shown for each case.

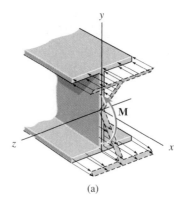

(a)

Figs. 6–33

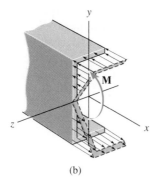

(b)

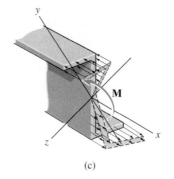

(c)

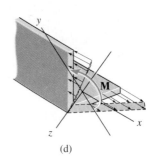

(d)

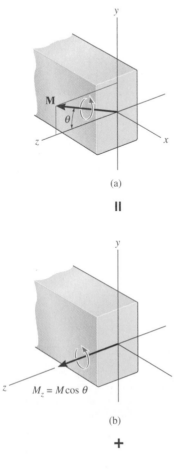

(a)

‖

$M_z = M \cos \theta$

(b)

+

$M_y = M \sin \theta$

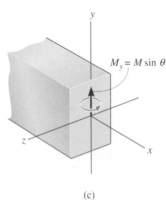

(c)

Figs. 6–34

Moment Arbitrarily Applied. Sometimes a member may be loaded such that the resultant internal moment does not act about one of the principal axes of the cross section. When this occurs, the moment should first be resolved into components directed along the principal axes. The flexure formula can then be used to determine the normal stress caused by each moment component. Finally, using the principle of superposition, the resultant normal stress at the point can be determined.

To show this, consider the beam to have a rectangular cross section and to be subjected to the moment **M**, Fig. 6–34a. Here **M** makes an angle θ with the *principal z* axis. We will assume θ is positive when it is directed from the $+z$ axis toward the $+y$ axis, as shown. Resolving **M** into components along the z and y axes, we have $M_z = M \cos \theta$ and $M_y = M \sin \theta$, respectively. Each of these components is shown separately on the cross section in Fig. 6–34b and 6–34c. The normal-stress distributions that produce **M** and its components **M**$_z$ and **M**$_y$ are shown in Fig. 6–34d, 6–34e, and 6–34f, respectively. Here it is assumed that $(\sigma_x)_{max} > (\sigma'_x)_{max}$. By inspection, the maximum tensile and compressive stresses $[(\sigma_x)_{max} + (\sigma'_x)_{max}]$ occur at two opposite corners of the cross section, Fig. 6–34d.

Applying the flexure formula to each moment component in Fig. 6–34b and 6–34c, we can express the resultant normal stress at any point on the cross section, Fig. 6–34d, in general terms as

$$\sigma = -\frac{M_z y}{I_z} + \frac{M_y z}{I_y} \qquad (6\text{–}17)$$

where

σ = the normal stress at the point

y, z = the coordinates of the point measured from x, y, z axes having their origin at the centroid of the cross-sectional area and forming a right-handed coordinate system. The x axis is directed outward from the cross-section and the y and z axes represent respectively the principal axes of minimum and maximum moment of inertia for the area

M_y, M_z = the resultant internal moment components directed along the principal y and z axes. They are positive if directed along the $+y$ and $+z$ axes, otherwise they are negative. Or, stated another way, $M_y = M \sin\theta$ and $M_z = M \cos\theta$, where θ is measured positive from the $+z$ axis toward the $+y$ axis

I_y, I_z = the *principal moments of inertia* computed about the y and z axes, respectively. See Appendix A

As noted previously, it is *very important* that the x, y, z axes form a right-handed system and that the proper algebraic signs be assigned to the moment components and the coordinates when applying this equation. The resulting stress will be *tensile* if it is *positive* and *compressive* if it is *negative*.

Orientation of the Neutral Axis.   The angle α of the neutral axis in Fig. 6–34d can be determined by applying Eq. 6–17 with $\sigma = 0$, since by definition no normal stress acts on the neutral axis. We have

$$y = \frac{M_y I_z}{M_z I_y} z$$

Since $M_z = M \cos \theta$ and $M_y = M \sin \theta$, then

$$y = \left(\frac{I_z}{I_y} \tan \theta \right) z \qquad (6\text{–}18)$$

This is the equation of the line that defines neutral axis for the cross section. Since the slope of this line is $\tan \alpha = y/z$, then

$$\boxed{\tan \alpha = \frac{I_z}{I_y} \tan \theta} \qquad (6\text{–}19)$$

Here it can be seen that for *unsymmetrical bending* the angle θ, defining the direction of the moment **M**, Fig. 6–34a, is *not equal* to α, the angle defining the inclination of the neutral axis, Fig. 6–34d, unless $I_z = I_y$. Instead, if as in Fig. 6–34a the y axis is chosen as the principal axis for the *minimum* moment of inertia, and the z axis is chosen as the principal axis for the *maximum* moment of inertia, so that $I_y < I_z$, then from Eq. 6–19 we can conclude that the angle α, which is measured positive from the $+z$ axis toward the $+y$ axis, will lie *between* the line of action of **M** and the y axis, i.e., $\theta \le \alpha \le 90°$.

IMPORTANT POINTS

- The flexure formula can be applied only when bending occurs about axes that represent the *principal axes of inertia* for the cross section. These axes have their origin at the centroid and are orientated along an axis of symmetry, if there is one, and perpendicular to it.

- If the moment is applied about some arbitrary axis, then the moment must be resolved into components along each of the principal axes, and the stress at a point is determined by superposition of the stress caused by each of the moment components.

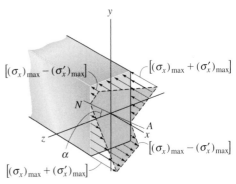

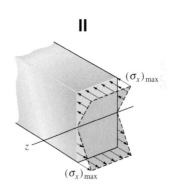

(d)

$\parallel$

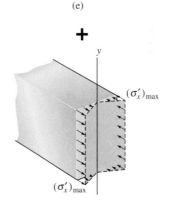

(e)

$+$

(f)

EXAMPLE 6–18

The rectangular cross section shown in Fig. 6–35a is subjected to a bending moment of $M = 12$ kN · m. Determine the normal stress developed at each corner of the section, and specify the orientation of the neutral axis.

SOLUTION

Internal Moment Components. By inspection it is seen that the y and z axes represent the principal axes of inertia since they are axes of symmetry for the cross section. As required we have established the z *axis* as the principal axis for *maximum* moment of inertia. The moment is resolved into its y and z components, where

$$M_y = -\frac{4}{5}(12 \text{ kN} \cdot \text{m}) = -9.60 \text{ kN} \cdot \text{m}$$

$$M_z = \frac{3}{5}(12 \text{ kN} \cdot \text{m}) = 7.20 \text{ kN} \cdot \text{m}$$

Section Properties. The moments of inertia about the y and z axes are

$$I_y = \frac{1}{12}(0.4 \text{ m})(0.2 \text{ m})^3 = 0.2667(10^{-3}) \text{ m}^4$$

$$I_z = \frac{1}{12}(0.2 \text{ m})(0.4 \text{ m})^3 = 1.067(10^{-3}) \text{ m}^4$$

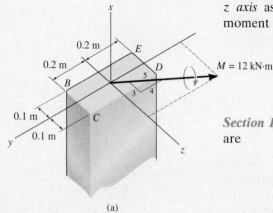

(a)

Figs. 6–35

Bending Stress. Thus,

$$\sigma = -\frac{M_z y}{I_z} + \frac{M_y z}{I_y}$$

$$\sigma_B = -\frac{7.20(10^3) \text{ N} \cdot \text{m}(0.2 \text{ m})}{1.067(10^{-3}) \text{ m}^4} + \frac{-9.60(10^3) \text{ N} \cdot \text{m}(-0.1 \text{ m})}{0.2667(10^{-3}) \text{ m}^4} = 2.25 \text{ MPa} \qquad \textit{Ans.}$$

$$\sigma_C = -\frac{7.20(10^3) \text{ N} \cdot \text{m}(0.2 \text{ m})}{1.067(10^{-3}) \text{ m}^4} + \frac{-9.60(10^3) \text{ N} \cdot \text{m}(0.1 \text{ m})}{0.2667(10^{-3}) \text{ m}^4} = -4.95 \text{ MPa} \qquad \textit{Ans.}$$

$$\sigma_D = -\frac{7.20(10^3) \text{ N} \cdot \text{m}(-0.2 \text{ m})}{1.067(10^{-3}) \text{ m}^4} + \frac{-9.60(10^3) \text{ N} \cdot \text{m}(0.1 \text{ m})}{0.2667(10^{-3}) \text{ m}^4} = -2.25 \text{ MPa} \qquad \textit{Ans.}$$

$$\sigma_E = -\frac{7.20(10^3) \text{ N} \cdot \text{m}(-0.2 \text{ m})}{1.067(10^{-3}) \text{ m}^4} + \frac{-9.60(10^3) \text{ N} \cdot \text{m}(-0.1 \text{ m})}{0.2667(10^{-3}) \text{ m}^4} = 4.95 \text{ MPa} \qquad \textit{Ans.}$$

The resultant normal-stress distribution has been sketched using these values, Fig. 6–35b. Since superposition applies, the distribution is linear as shown.

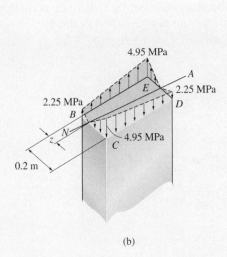

(b)

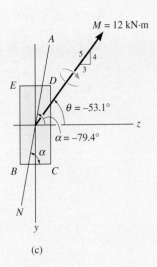

(c)

Orientation of Neutral Axis.   The location z of the neutral axis (NA), Fig. 6–35b, can be established by proportion. Along the edge BC, we require

$$\frac{2.25 \text{ MPa}}{z} = \frac{4.95 \text{ MPa}}{(0.2 \text{ m} - z)}.$$

$$0.450 - 2.25z = 4.95z$$

$$z = 0.0625 \text{ m}$$

In the same manner this is also the distance from D to the neutral axis in Fig. 6–35b.

We can also establish the orientation of the NA using Eq. 6–19, which is used to specify the angle α that the axis makes with the z or *maximum* principal axis. According to our sign convention, θ must be measured from the $+z$ axis toward the $+y$ axis. By comparison, in Fig. 6–35c, $\theta = -\tan^{-1}\frac{4}{3} = -53.1°$ (or $\theta = +306.9°$). Thus,

$$\tan \alpha = \frac{I_z}{I_y} \tan \theta$$

$$\tan \alpha = \frac{1.067(10^{-3}) \text{ m}^4}{0.2667(10^{-3}) \text{ m}^4} \tan(-53.1°)$$

$$\alpha = -79.4° \hspace{4em} \textit{Ans.}$$

This result is shown in Fig. 6–35c. Using the value of z calculated above, verify, using the geometry of the cross section, that one obtains the same answer.

EXAMPLE 6-19

A T-beam is subjected to the bending moment of 15 kN · m as shown in Fig. 6–36a. Determine the maximum normal stress in the beam and the orientation of the neutral axis.

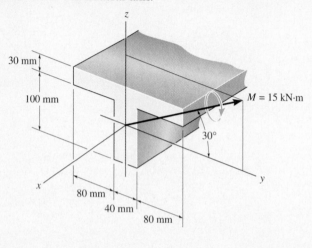

(a)

SOLUTION

Internal Moment Components.   The y and z axes are principal axes of inertia. Why? From Fig. 6–36a, both moment components are positive. We have

$$M_y = (15 \text{ kN} \cdot \text{m}) \cos 30° = 12.99 \text{ kN} \cdot \text{m}$$
$$M_z = (15 \text{ kN} \cdot \text{m}) \sin 30° = 7.50 \text{ kN} \cdot \text{m}$$

Section Properties.   With reference to Fig. 6–36b, working in units of meters, we have

$$\bar{z} = \frac{\Sigma \tilde{z} A}{\Sigma A} = \frac{[0.05 \text{ m}](0.100 \text{ m})(0.04 \text{ m}) + [0.115 \text{ m}](0.03 \text{ m})(0.200 \text{ m})}{(0.100 \text{ m})(0.04 \text{ m}) + (0.03 \text{ m})(0.200 \text{ m})}$$

$$= 0.0890 \text{ m}$$

Using the parallel-axis theorem of Appendix A, $I = \bar{I} + Ad^2$, the principal moments of inertia are thus

$$I_z = \frac{1}{12}(0.100 \text{ m})(0.04 \text{ m})^3 + \frac{1}{12}(0.03 \text{ m})(0.200 \text{ m})^3 = 20.53(10^{-6}) \text{m}^4$$

$$I_y = \left[\frac{1}{12}(0.04 \text{ m})(0.100 \text{ m})^3 + (0.100 \text{ m})(0.04 \text{ m})(0.0890 \text{ m} - 0.05 \text{ m})^2 \right]$$

$$+ \left[\frac{1}{12}(0.200 \text{ m})(0.03 \text{ m})^3 + (0.200 \text{ m})(0.03 \text{ m})(0.115 \text{ m} - 0.0890 \text{ m})^2 \right]$$

$$= 13.92(10^{-6}) \text{ m}^4$$

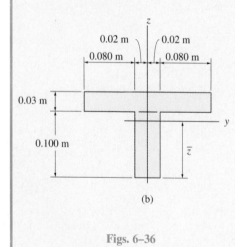

(b)

Figs. 6–36

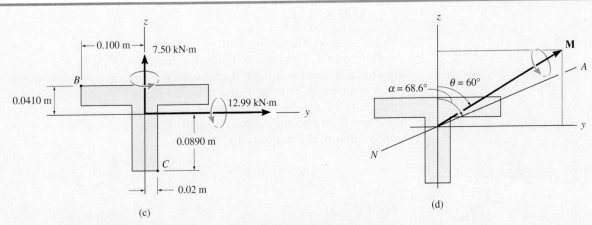

(c)

(d)

Maximum Bending Stress.   The moment components are shown in Fig. 6–36c. By inspection, the largest *tensile* stress occurs at point *B*, since by superposition both moment components create a tensile stress there. Likewise, the greatest *compressive* stress occurs at point *C*. Thus,

$$\sigma = -\frac{M_z y}{I_z} + \frac{M_y z}{I_y}$$

$$\sigma_B = -\frac{7.50 \text{ kN} \cdot \text{m} \,(-0.100 \text{ m})}{20.53(10^{-6}) \text{ m}^4} + \frac{12.99 \text{ kN} \cdot \text{m}(0.0410 \text{ m})}{13.92(10^{-6}) \text{ m}^4}$$

$$= 74.8 \text{ MPa}$$

$$\sigma_C = -\frac{7.50 \text{ kN} \cdot \text{m} \,(0.020 \text{ m})}{20.53(10^{-6}) \text{ m}^4} + \frac{12.99 \text{ kN} \cdot \text{m} \,(-0.0890 \text{ m})}{13.92(10^{-6}) \text{ m}^4}$$

$$= -90.4 \text{ MPa} \qquad\qquad\qquad \textit{Ans.}$$

By comparison, the largest normal stress is therefore compressive and occurs at point *C*.

Orientation of Neutral Axis.   When applying Eq. 6–19 it is important to be sure the angles α and θ are defined correctly. As previously stated, y must represent the axis for *minimum* principal moment of inertia, and z must represent the axis for *maximum* principal moment of inertia. These axes are properly positioned here since $I_y < I_z$. Using this setup, θ and α are measured positive from the $+z$ axis toward the $+y$ axis. Hence, from Fig. 6–36a, $\theta = +60°$. Thus,

$$\tan \alpha = \left(\frac{20.53(10^{-6}) \text{ m}^4}{13.92(10^{-6}) \text{ m}^4} \right) \tan 60°$$

$$\alpha = 68.6° \qquad\qquad\qquad \textit{Ans.}$$

The neutral axis is shown in Fig. 6–36d. As expected, it lies between the y axis and the line of action of **M**.

EXAMPLE 6-20

The Z-section shown in Fig. 6–37a is subjected to the bending moment of $M = 20$ kN · m. Using the methods of Appendix A (see Example A–4 or A–5), the principal axes y and z are oriented as shown, such that they represent the minimum and maximum principal moments of inertia, $I_y = 0.960(10^{-3})$ m^4 and $I_z = 7.54(10^{-3})$ m^4, respectively. Determine the normal stress at point P and the orientation of the neutral axis.

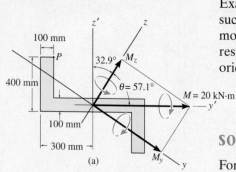

(a)

SOLUTION

For use of Eq. 6–19, it is important that the z axis be the principal axis for the *maximum* moment of inertia, which it is because most of the area is located furthest from this axis.

Internal Moment Components.   From Fig. 6–37a,

$$M_y = 20 \text{ kN} \cdot \text{m} \sin 57.1° = 16.79 \text{ kN} \cdot \text{m}$$
$$M_z = 20 \text{ kN} \cdot \text{m} \cos 57.1° = 10.86 \text{ kN} \cdot \text{m}$$

Bending Stress.   The y and z coordinates of point P must be determined first. Note that the y', z' coordinates of P are $(-0.2$ m, 0.35 m). Using the colored and shaded triangles from the construction shown in Fig. 6–37b, we have

$$y_P = -0.35 \sin 32.9° - 0.2 \cos 32.9° = -0.3580 \text{ m}$$
$$z_P = 0.35 \cos 32.9° - 0.2 \sin 32.9° = 0.1852 \text{ m}$$

Applying Eq. 6–17, we have

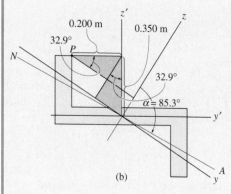

(b)

Figs. 6–37

$$\sigma_P = -\frac{M_z y_P}{I_z} + \frac{M_y z_P}{I_y}$$

$$= -\frac{(10.86 \text{ kN} \cdot \text{m})(-0.3580 \text{ m})}{7.54(10^{-3}) \text{ m}^4} + \frac{(16.79 \text{ kN} \cdot \text{m})(0.1852 \text{ m})}{0.960(10^{-3}) \text{ m}^4}$$

$$= 3.76 \text{ MPa} \qquad\qquad\qquad\qquad\qquad\qquad\qquad \textit{Ans.}$$

Orientation of Neutral Axis.   The angle $\theta = 57.1°$ is shown in Fig. 6–37a. Thus,

$$\tan \alpha = \left[\frac{7.54(10^{-3}) \text{ m}^4}{0.960(10^{-3}) \text{ m}^4} \right] \tan 57.1°$$

$$\alpha = 85.3° \qquad\qquad\qquad\qquad\qquad\qquad\qquad \textit{Ans.}$$

The neutral axis is located as shown in Fig. 6–37b.

PROBLEMS

*6–96. The member has a square cross section and is subjected to a resultant moment of $M = 850$ N · m as shown. Determine the bending stress at each corner and sketch the stress distribution produced by **M**. Set $\theta = 45°$.

6–97. The member has a square cross section and is subjected to a resultant moment of $M = 850$ N · m as shown. Determine the bending stress at each corner and sketch the stress distribution produced by **M**. Set $\theta = 30°$.

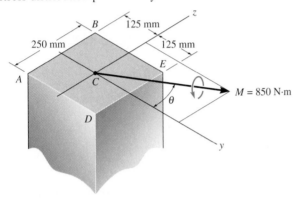

Probs. 6–96/6–97

6–98. Determine the maximum magnitude of the bending moment **M** so that the bending stress in the member does not exceed 24 ksi. The location $\bar{y}$ of the centroid C must be determined.

6–99. The moment acting on the cross section of the T-beam has a magnitude of $M = 15$ kip · ft and is directed as shown. Determine the bending stress at points A and B. The location $\bar{y}$ of the centroid C must be determined.

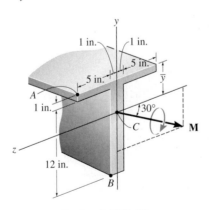

Probs. 6–98/6–99

*6–100. If the internal moment acting on the cross section of the strut has a magnitude of $M = 800$ N · m and is directed as shown, determine the bending stress at points A and B. The location $\bar{z}$ of the centroid C of the strut's cross-sectional area must be determined. Also, specify the orientation of the neutral axis.

6–101. The resultant moment acting on the cross section of the aluminum strut has a magnitude of $M = 800$ N · m and is directed as shown. Determine the maximum bending stress in the strut. The location $\bar{y}$ of the centroid C of the strut's cross-sectional area must be determined. Also, specify the orientation of the neutral axis.

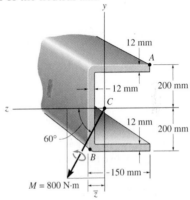

Probs. 6–100/6–101

6–102. The cantilevered wide-flange steel beam is subjected to the concentrated force **P** at its end. Determine the largest magnitude of this force so that the bending stress developed at section A does not exceed $\sigma_{allow} = 180$ MPa.

6–103. The cantilevered wide-flange steel beam is subjected to the concentrated force of $P = 600$ N at its end. Determine the maximum bending stress developed in the beam at section A.

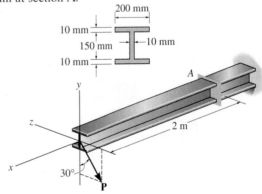

Probs. 6–102/6–103

*6–104. The steel beam has the cross-sectional area shown. Determine the largest intensity of distributed load w that it can support so that the bending stress does not exceed $\sigma_{max} = 22$ ksi.

6–105. The steel beam has the cross-sectional area shown. If $w = 5$ kip/ft, determine the absolute maximum bending stress in the beam.

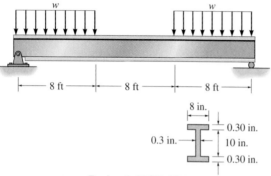

Probs. 6–104/6–105

6–106. The 30-mm-diameter shaft is subjected to the vertical and horizontal loadings of two pulleys as shown. It is supported on two journal bearings at A and B which offer no resistance to axial loading. Furthermore, the coupling to the motor at C can be assumed not to offer any support to the shaft. Determine the maximum bending stress developed in the shaft.

6–107. The shaft is subjected to the vertical and horizontal loadings of two pulleys D and E as shown. It is supported on two journal bearings at A and B which offer no resistance to axial loading. Furthermore, the coupling to the motor at C can be assumed not to offer any support to the shaft. Determine the required diameter d of the shaft if the allowable bending stress for the material is $\sigma_{allow} = 180$ MPa.

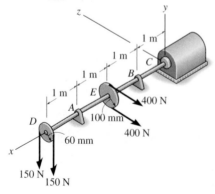

Probs. 6–106/6–107

*6–108. The 65-mm-diameter steel shaft is subjected to the two loads that act in the directions shown. If the journal bearings at A and B do not exert an axial force on the shaft, determine the absolute maximum bending stress developed in the shaft.

6–109. The steel shaft is subjected to the two loads that act in the directions shown. If the journal bearings at A and B do not exert an axial force on the shaft, determine the required diameter of the shaft if the allowable bending stress is $\sigma_{allow} = 180$ MPa.

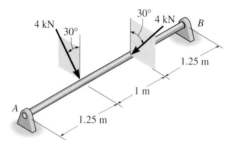

Probs. 6–108/6–109

6–110. Consider the general case of a prismatic beam subjected to bending moment components $\mathbf{M}_y$ and $\mathbf{M}_z$, as shown, where the x, y, z axes pass through the centroid of the cross section. If the material is linear-elastic, the normal stress in the beam is a linear function of position such that $\sigma = a + by + cz$. Using the equilibrium conditions $0 = \int_A \sigma \, dA$, $M_y = \int_A z\sigma \, dA$, $M_z = \int_A -y\sigma \, dA$, determine the constants a, b, and c, and show that the normal stress can be determined from the equation $\sigma = \left[-(M_z I_y + M_y I_{yz})y + (M_y I_z + M_z I_{yz})z \right] / (I_y I_z - I_{yz}^2)$, where the moments and products of inertia are defined in Appendix A.

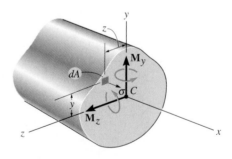

Prob. 6–110

6–111. Using the techniques outlined in Appendix A, Example A–5 or Example A–6, the Z section has principal moments of inertia of $I_{y'} = 0.060(10^{-3})$ m^4 and $I_{z'} = 0.471(10^{-3})$ m^4, computed about the principal axes of inertia y' and z', respectively. If the section is subjected to a moment of $M = 250$ N · m directed horizontally as shown, determine the bending stress produced at point A. Solve the problem using Eq. 6–17.

*6–112. Solve Prob. 6–111 using the equation developed in Prob. 6–110.

6–113. Using the techniques outlined in Appendix A, Example A–5 or Example A–6, the Z section has principal moments of inertia of $I_{y'} = 0.060(10^{-3})$ m^4 and $I_{z'} = 0.471(10^{-3})$ m^4, computed about the principal axes of inertia y' and z', respectively. If the section is subjected to a moment of $M = 250$ N · m directed horizontally as shown, determine the bending stress produced at point B. Solve the problem using Eq. 6–17.

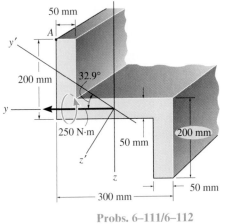

Probs. 6–111/6–112

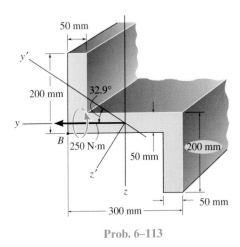

Prob. 6–113

*6.6 COMPOSITE BEAMS

Beams constructed of two or more different materials are referred to as *composite beams*. Examples include those made of wood with straps of steel used at the bottom or top, Fig. 6–38a, or more commonly, concrete beams reinforced with steel rods, Fig. 6–38b. Engineers purposely design beams in this manner in order to develop a more efficient means for carrying applied loads. For example, it has been shown in Sec. 3.3 that concrete is excellent in resisting compressive stress but is very poor in resisting tensile stress. Hence, the steel reinforcing rods shown in Fig. 6–38b have been placed in the tension zone of the beam's cross section so that they resist the tensile stresses that result from the moment **M**.

Since the flexure formula was developed for beams whose material is homogeneous, this formula cannot be applied directly to determine the normal stress in a composite beam. In this section, however, we will develop a method for modifying or "transforming" the beam's cross section into one made of a single material. Once this has been done, the flexure formula can then be used for the stress analysis.

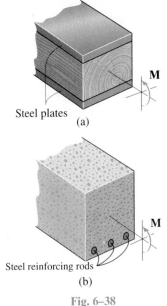

Fig. 6–38

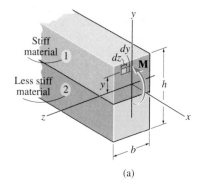

(a)

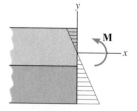

Normal strain variation
(profile view)
(b)

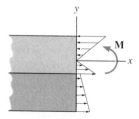

Bending stress variation
(profile view)
(c)

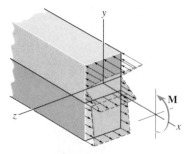

Bending stress variation
(d)

Fig. 6–39

To explain how to apply the *transformed-section method*, consider the composite beam to be made of two materials, 1 and 2, which have the cross-sectional areas shown in Fig. 6–39a. If a bending moment is applied to this beam, then, like one that is homogeneous, the total cross-sectional area will *remain plane* after bending, and hence the normal strains will vary linearly from zero at the neutral axis to a maximum in the material located farthest from this axis, Fig. 6–39b. Provided the material has linear-elastic behavior, Hooke's law applies, and at any point the normal stress in material 1 is determined from $\sigma = E_1 \epsilon$. Likewise, for material 2 the stress distribution is found from $\sigma = E_2 \epsilon$. Obviously, if material 1 is *stiffer* than material 2, e.g., steel versus rubber, most of the load will be carried by material 1, since $E_1 > E_2$. Assuming this to be the case, the stress distribution will look like that shown in Fig. 6–39c or 6–39d. In particular, notice the jump in stress that occurs at the juncture of the materials. Here the *strain* is the *same*, but since the modulus of elasticity or stiffness for the materials suddenly changes, so does the stress. Location of the neutral axis, and determination of the maximum stress in the beam, using this stress distribution, can be based on a trial-and-error procedure. This requires satisfying the conditions that the stress distribution produces a zero resultant force on the cross section and the moment of the stress distribution about the neutral axis must be equal to **M**.

A simpler way to satisfy these two conditions, however, is to transform the beam into one made of a *single material*. For example, if the beam is thought to consist entirely of the less stiff material 2, then the cross section would have to look like that shown in Fig. 6–39e. Here the height h of the beam remains the *same*, since the strain distribution shown in Fig. 6–39b must be preserved. However, the upper portion of the beam must be widened in order to carry a load *equivalent* to that carried by the stiffer material 1 in Fig. 6–39d. The necessary width can be determined by considering the force $d\mathbf{F}$ acting on an area $dA = dz\, dy$ of the beam in Fig. 6–39a. It is $dF = \sigma\, dA = (E_1 \epsilon) dz\, dy$. On the other hand, if the width of a *corresponding element* of height dy in Fig. 6–39e is $n\, dz$, then $dF' = \sigma'\, dA' = (E_2 \epsilon) n\, dz\, dy$. Equating these forces, so that they produce the same moment about the z axis, we have

$$E_1 \epsilon\, dz\, dy = E_2 \epsilon n\, dz\, dy$$

or

$$n = \frac{E_1}{E_2} \qquad (6\text{–}20)$$

This dimensionless number n is called the ***transformation factor***. It indicates that the cross section, having a width b on the original beam, Fig. 6–39a, must be increased in width to $b_2 = nb$ in the region where material 1 is being transformed into material 2, Fig. 6–39e. In a similar manner, if the less stiff material 2 is transformed into the stiffer material 1, the cross section will look like that shown in Fig. 6–39f. Here the width of material 2 has been changed to $b_1 = n'b$, where $n' = E_2/E_1$. Note that in this case the transformation factor n' must be *less than one* since $E_1 > E_2$. In other words, we need less of the stiffer material to support a given moment.

Once the beam has been transformed into one having a *single material*, the normal-stress distribution over the transformed cross section will be linear as shown in Fig. 6–39g or 6–39h. Consequently, the centroid (neutral axis) and moment of inertia for the transformed area can be determined and the flexure formula applied in the usual manner to determine the stress at each point on the transformed beam. Realize that the stress in the transformed beam is equivalent to the stress in the *same material* of the actual beam. For the *transformed material*, however, the stress found on the transformed section has to be multiplied by the transformation factor n (or n'), since the area of the transformed material, $dA' = n\, dz\, dy$, is n times the area of actual material $dA = dz\, dy$. That is,

$$dF = \sigma\, dA = \sigma'\, dA'$$
$$\sigma\, dz\, dy = \sigma'n\, dz\, dy$$
$$\sigma = n\sigma' \qquad (6\text{–}21)$$

Examples 6–21 and 6–22 numerically illustrate application of the transformed section method.

IMPORTANT POINTS

- *Composite beams* are made from different materials in order to efficiently carry a load. Application of the flexure formula requires the material to be homogeneous, and so the cross section of the beam must be transformed into a single material if this formula is to be used to compute the bending stress.

- The *transformation factor* is a ratio of the moduli of the different materials that make up the beam. Used as a multiplier, it converts the dimensions of the cross section of the composite beam into a beam made from a single material so that this beam has the same strength as the composite beam. Stiff material will thus be replaced by more of the softer material and vice versa.

- Once the stress in the transformed section is determined, it must be multiplied by the transformation factor to obtain the stress in the actual beam.

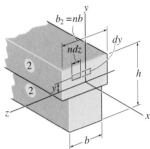

Beam transformed to material ②
(e)

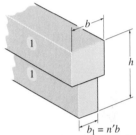

Beam transformed to material ①
(f)

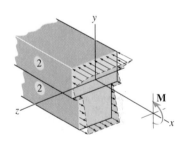

Bending-stress variation for beam transformed to material ②
(g)

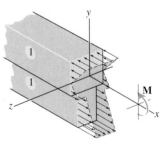

Bending-stress variation for beam transformed to material ①
(h)

EXAMPLE 6-21

A composite beam is made of wood and reinforced with a steel strap located on its bottom side. It has the cross-sectional area shown in Fig. 6–40a. If the beam is subjected to a bending moment of $M = 2$ kN · m, determine the normal stress at points B and C. Take $E_w = 12$ GPa and $E_{st} = 200$ GPa.

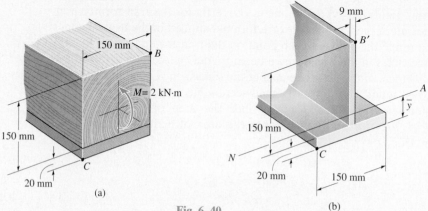

Fig. 6–40

SOLUTION

Section Properties.   Although the choice is arbitrary, here we will transform the section into one made entirely of steel. Since steel has a greater stiffness than wood ($E_{st} > E_w$), the width of the wood must be *reduced* to an equivalent width for steel. Hence n must be less than one. For this to be the case, $n = E_w/E_{st}$, so that

$$b_{st} = nb_w = \frac{12 \text{ GPa}}{200 \text{ GPa}} (150 \text{ mm}) = 9 \text{ mm}$$

The transformed section is shown in Fig. 6–40b.

The location of the centroid (neutral axis), computed from a reference axis located at the *bottom* of the section, is

$$\bar{y} = \frac{\Sigma \tilde{y}A}{\Sigma A} = \frac{[0.01 \text{ m}](0.02 \text{ m})(0.150 \text{ m}) + [0.095 \text{ m}](0.009 \text{ m})(0.150 \text{ m})}{0.02 \text{ m}(0.150 \text{ m}) + 0.009 \text{ m}(0.150 \text{ m})} = 0.03638 \text{ m}$$

The moment of inertia about the neutral axis is therefore

$$I_{NA} = \left[\frac{1}{12} (0.150 \text{ m})(0.02 \text{ m})^3 + (0.150 \text{ m})(0.02 \text{ m})(0.03638 \text{ m} - 0.01 \text{ m})^2 \right]$$

$$+ \left[\frac{1}{12} (0.009 \text{ m})(0.150 \text{ m})^3 + (0.009 \text{ m})(0.150 \text{ m})(0.095 \text{ m} - 0.03638 \text{ m})^2 \right]$$

$$= 9.358(10^{-6}) \text{ m}^4$$

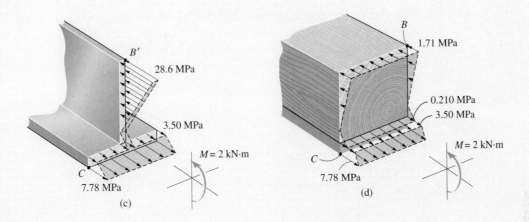

(c)

(d)

Normal Stress. Applying the flexure formula, the normal stress at B' and C is

$$\sigma_{B'} = \frac{2 \text{ kN} \cdot \text{m}(0.170 \text{ m} - 0.03638 \text{ m})}{9.358(10^{-6}) \text{ m}^4} = 28.6 \text{ MPa}$$

$$\sigma_C = \frac{2 \text{ kN} \cdot \text{m}(0.03638 \text{ m})}{9.358(10^{-6}) \text{ m}^4} = 7.78 \text{ MPa} \qquad \textit{Ans.}$$

The normal-stress distribution on the transformed (all steel) section is shown in Fig. 6–40c.

The normal stress in the wood, located at B in Fig. 6–40a, is determined from Eq. 6–21; that is,

$$\sigma_B = n\sigma_{B'} = \frac{12 \text{ GPa}}{200 \text{ GPa}}(28.56 \text{ MPa}) = 1.71 \text{ MPa} \qquad \textit{Ans.}$$

Using these concepts, show that the normal stress in the steel and the wood at the point where they are in contact is $\sigma_{st} = 3.50$ MPa and $\sigma_w = 0.210$ MPa, respectively. The normal-stress distribution in the actual beam is shown in Fig. 6–40d.

EXAMPLE 6–22

In order to reinforce the steel beam, an oak board is placed between its flanges as shown in Fig. 6–41a. If the allowable normal stress for the steel is $(\sigma_{\text{allow}})_{st} = 24$ ksi, and for the wood $(\sigma_{\text{allow}})_w = 3$ ksi, determine the maximum bending moment the beam can support, with and without the wood reinforcement. $E_{st} = 29(10^3)$ ksi, $E_w = 1.60(10^3)$ ksi. The moment of inertia of the steel beam is $I_z = 20.3$ in⁴, and its cross-sectional area is $A = 8.79$ in².

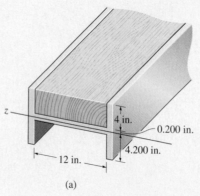

(a)

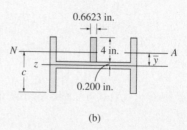

(b)

Fig. 6–41

SOLUTION

Without Board.   Here the neutral axis coincides with the z axis. Direct application of the flexure formula to the steel beam yields

$$(\sigma_{\text{allow}})_{st} = \frac{Mc}{I_z}$$

$$24 \text{ kip/in}^2 = \frac{M(4.200 \text{ in.})}{20.3 \text{ in}^4}$$

$$M = 116 \text{ kip} \cdot \text{in.} \qquad Ans.$$

With Board.   Since now we have a composite beam, we must transform the section to a single material. It will be easier to transform the wood to an equivalent amount of steel. To do this, $n = E_w/E_{st}$. Why? Thus, the width of an equivalent amount of steel is

$$b_{st} = nb_w = \frac{1.60(10^3) \text{ ksi}}{29(10^3) \text{ ksi}}(12 \text{ in.}) = 0.662 \text{ in.}$$

The transformed section is shown in Fig. 6–41b. The neutral axis is at

$$\bar{y} = \frac{\Sigma \tilde{y} A}{\Sigma A} = \frac{[0](8.79 \text{ in}^2) + [2.20 \text{ in.}](4 \text{ in.})(0.662 \text{ in.})}{8.79 \text{ in}^2 + 4(0.662 \text{ in}^2)}$$

$$= 0.5093 \text{ in.}$$

And the moment of inertia about the neutral axis is

$$I = [20.3 \text{ in}^4 + (8.79 \text{ in}^2)(0.5093 \text{ in.})^2] +$$

$$\left[\frac{1}{12}(0.662 \text{ in.})(4 \text{ in.})^3 + (0.662 \text{ in.})(4 \text{ in.})(2.200 \text{ in.} - 0.5093 \text{ in.})^2 \right]$$

$$= 33.68 \text{ in}^4$$

The maximum normal stress in the steel will occur at the *bottom* of the beam, Fig. 6–41b. Here $c = 4.200 \text{ in.} + 0.5093 \text{ in.} = 4.7093 \text{ in.}$ The maximum moment based on the allowable stress for the steel is therefore

$$(\sigma_{\text{allow}})_{st} = \frac{Mc}{I}$$

$$24 \text{ kip/in}^2 = \frac{M(4.7093 \text{ in.})}{33.68 \text{ in}^4}$$

$$M = 172 \text{ kip} \cdot \text{in.}$$

The maximum normal stress in the wood occurs at the top of the beam, Fig. 6–41b. Here $c' = 4.20 \text{ in.} - 0.5093 \text{ in.} = 3.6907 \text{ in.}$ Since $\sigma_w = n\sigma_{st}$, the maximum moment based on the allowable stress for the wood is

$$(\sigma_{\text{allow}})_w = n\frac{M'c'}{I}$$

$$3 \text{ kip/in}^2 = \left[\frac{1.60(10^3) \text{ ksi}}{29(10^3) \text{ ksi}} \right] \frac{M'(3.6907 \text{ in.})}{33.68 \text{ in}^4}$$

$$M' = 496 \text{ kip} \cdot \text{in.}$$

By comparison, the maximum moment is limited by the allowable stress in the steel. Thus,

$$M = 172 \text{ kip} \cdot \text{in.} \qquad \textit{Ans.}$$

Note also that by using the board as reinforcement, one provides an additional 48% moment capacity for the beam.

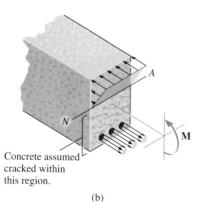

(a)

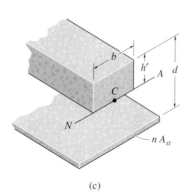

Concrete assumed
cracked within
this region.

(b)

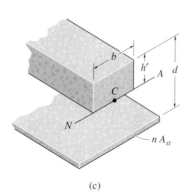

(c)

Fig. 6–42

*6.7   REINFORCED CONCRETE BEAMS

All beams subjected to pure bending must resist both tensile and compressive stresses. Concrete, however, is very susceptible to cracking when it is in tension, and therefore by itself would not be suitable for resisting a bending moment.* In order to circumvent this shortcoming, engineers place steel reinforcing rods within a concrete beam at a location where the concrete is in tension, Fig. 6–42a. To be most effective, these rods are located farthest from the beam's neutral axis, so that the moment created by the forces developed in the rods is greatest about the neutral axis. On the other hand, the rods are also required to have some concrete coverage so as to protect them from corrosion or loss of strength in the event of a fire. In actual reinforced concrete design, the ability of concrete to support any tensile loading is neglected since the possible cracking of concrete is unpredictable. As a result, the normal stress distribution acting on the cross-sectional area of a reinforced concrete beam is assumed to look like that shown in Fig. 6–42b.

The stress analysis requires locating the neutral axis and determining the maximum stress in the steel and concrete. To do this, the area of steel A_{st} is first transformed into an equivalent area of concrete using the transformation factor $n = E_{st}/E_{conc}$. This ratio, which gives $n > 1$, is chosen since a "greater" amount of concrete is needed to replace the steel. The transformed area is nA_{st} and the transformed section looks like that shown in Fig. 6–42c. Here d represents the distance from the top of the beam to the (transformed) steel, b is the beam's width, and h' is the yet unknown distance from the top of the beam to the neutral axis. We can obtain h' using the fact that the centroid C of the cross-sectional area of the transformed section lies on the neutral axis, Fig. 6–42c. With reference to the neutral axis, therefore, the moment of the two areas, $\Sigma \tilde{y}A$, must be zero, since $\bar{y} = \Sigma \tilde{y}A/\Sigma A = 0$. Thus,

$$bh'\left(\frac{h'}{2}\right) - nA_{st}(d - h') = 0$$

$$\frac{b}{2}h'^2 + nA_{st}h' - nA_{st}d = 0$$

Once h' is obtained from this quadratic equation, the solution proceeds in the usual manner for obtaining the stress in the beam.

---

*Inspection of its particular stress–strain diagram in Fig. 3–11 reveals that concrete can be 12.5 times stronger in compression than in tension.

EXAMPLE 6-23

The reinforced concrete beam has the cross-sectional area shown in Fig. 6–43a. If it is subjected to a bending moment of $M = 60$ kip · ft, determine the normal stress in each of the steel reinforcing rods and the maximum normal stress in the concrete. Take $E_{st} = 29(10^3)$ ksi and $E_{conc} = 3.6(10^3)$ ksi.

SOLUTION

Since the beam is made from concrete, in the following analysis we will neglect its strength in supporting a tensile stress.

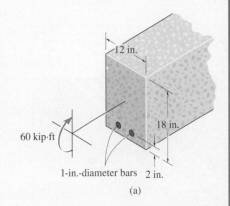

(a)

Section Properties. The total area of steel, $A_{st} = 2[\pi(0.5 \text{ in.})^2] = 1.571$ in^2 will be transformed into an equivalent area of concrete, Fig. 6–43b. Here

$$A' = nA_{st} = \frac{29(10^3) \text{ ksi}}{3.6(10^3) \text{ ksi}}(1.571 \text{ in}^2) = 12.65 \text{ in}^2$$

We require the centroid to lie on the neutral axis. Thus $\Sigma \tilde{y}A = 0$, or

$$12 \text{ in.}(h')\frac{h'}{2} - 12.65 \text{ in}^2 (16 \text{ in.} - h') = 0$$

$$h'^2 + 2.11h' - 33.7 = 0$$

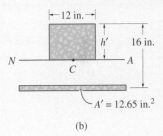

(b)

Solving for the positive root,

$$h' = 4.85 \text{ in.}$$

Using this value for h', the moment of inertia of the transformed section, computed about the neutral axis, is

$$I = \left[\frac{1}{12}(12 \text{ in.})(4.85 \text{ in.})^3 + 12 \text{ in.}(4.85 \text{ in.})\left(\frac{4.85 \text{ in.}}{2}\right)^2\right] + 12.65 \text{ in}^2(16 \text{ in.} - 4.85 \text{ in.})^2$$

$$= 2029 \text{ in}^4$$

Normal Stress. Applying the flexure formula to the transformed section, the maximum normal stress in the concrete is

$$(\sigma_{conc})_{max} = \frac{[60 \text{ kip} \cdot \text{ft}(12 \text{ in./ft})](4.85 \text{ in.})}{2029 \text{ in}^4} = 1.72 \text{ ksi} \quad \textbf{Ans.}$$

The normal stress resisted by the "concrete" strip, which replaced the steel, is

$$\sigma'_{conc} = \frac{[60 \text{ kip} \cdot \text{ft}(12 \text{ in./ft})](16 \text{ in.} - 4.85 \text{ in.})}{2029 \text{ in}^4} = 3.96 \text{ ksi}$$

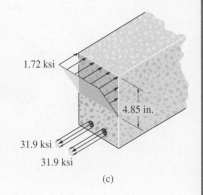

(c)

Fig. 6–43

The normal stress in each of the two reinforcing rods is therefore

$$\sigma_{st} = n\sigma'_{conc} = \left(\frac{29(10^3) \text{ ksi}}{3.6(10^3) \text{ ksi}}\right)3.96 \text{ ksi} = 31.9 \text{ ksi} \quad \textbf{Ans.}$$

The normal-stress distribution is shown graphically in Fig. 6–43c.

*6.8   CURVED BEAMS

The flexure formula applies to a prismatic member that is *straight*, since it was shown that for a straight member the normal strain varies linearly from the neutral axis. If the member is *curved*, however, this assumption becomes inaccurate, and so we must develop another equation that describes the stress distribution. In this section we will consider the analysis of a *curved beam*, that is, a member that has a curved axis and is subjected to bending. Typical examples include hooks and chain links. In all cases, the members are not slender, but rather have a sharp curve, and their cross-sectional dimensions are large compared with their radius of curvature.

The analysis to be considered assumes that the cross-sectional area is constant and has an axis of symmetry that is perpendicular to the direction of the applied moment **M**, Fig. 6–44a. Also, the material is homogeneous and isotropic, and it behaves in a linear-elastic manner when the load is applied. Like the case of a straight beam, we will, for a curved beam, also assume that the *cross sections* of the member *remain plane* after the moment is applied. Furthermore, any distortion of the cross section within its own plane will be neglected.

To perform the analysis, three radii, extending from the center of curvature O' of the member, are identified in Fig. 6–44a. They are as follows: $\bar{r}$ references the known location of the *centroid* for the cross-sectional area, R references the yet unspecified location of the *neutral axis*, and r locates the *arbitrary point* or area element dA on the cross section. Notice that the neutral axis lies within the cross section, since the moment **M** creates compression in the beam's top fibers and tension in its bottom fibers, and by definition, the neutral axis is a line of zero stress and strain.

The bending stress in this crane hook can be estimated using the curved-beam formula.

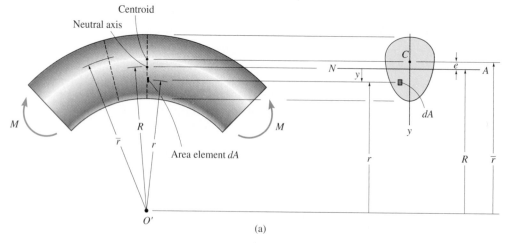

(a)

Fig. 6–44

If we isolate a differential segment of the beam, Fig. 6–44b, the stress tends to deform the material such that each cross section will rotate through an angle $\delta\theta/2$. The normal strain ϵ in the arbitrary strip of material located at r will now be determined. This strip has an original length $r\,d\theta$, Fig. 6–44b. Due to the rotations $\delta\theta/2$, however, the strip's total change in length is $\delta\theta(R - r)$. Consequently,

$$\epsilon = \frac{\delta\theta(R - r)}{r\,d\theta}$$

Defining $k = \delta\theta/d\theta$, which is constant for any particular element, we have

$$\epsilon = k\left(\frac{R - r}{r}\right)$$

Unlike the case of straight beams, here it can be seen that the **normal strain** is a nonlinear function of r, in fact it varies in a **hyperbolic fashion**. This occurs even though the cross section of the beam remains plane after deformation. Since the moment causes the material to behave elastically, Hooke's law applies, and therefore the stress as a function of position is

$$\sigma = Ek\left(\frac{R - r}{r}\right) \tag{6–22}$$

This variation is also hyperbolic, and since it has now been established, we can determine the location of the neutral axis and relate the stress distribution to the resultant internal moment M.

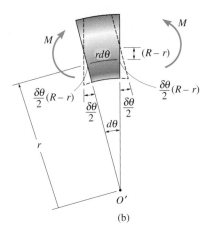

(b)

To obtain the location R of the neutral axis, we require the resultant internal force caused by the stress distribution acting over the cross section to be equal to zero; i.e.,

$$F_R = \Sigma F_x; \qquad\qquad \int_A \sigma \, dA = 0$$

$$\int_A Ek \left(\frac{R - r}{r} \right) dA = 0$$

Since Ek and R are constants, we have

$$R \int_A \frac{dA}{r} - \int_A dA = 0$$

Solving for R yields

$$R = \frac{A}{\displaystyle\int_A \frac{dA}{r}} \qquad\qquad (6\text{–}23)$$

Here

$R =$ the location of the neutral axis, specified from the center of curvature O' of the member

$A =$ the cross-sectional area of the member

$r =$ the arbitrary position of the area element dA on the cross section, specified from the center of curvature O' of the member

The integral in Eq. 6–23 may be evaluated for various cross-sectional geometries. The results for some common cross sections are listed in Table 6–2.

Shape	Area	$\int_A \frac{dA}{r}$
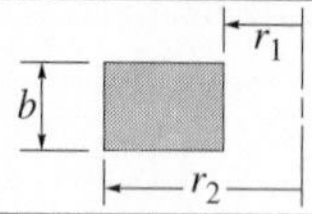	$b(r_2 - r_1)$	$b \ln \frac{r_2}{r_1}$
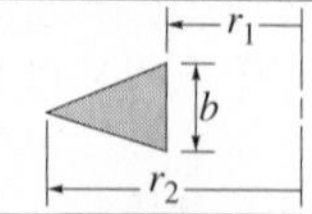	$\frac{b}{2}(r_2 - r_1)$	$\frac{b \, r_2}{(r_2 - r_1)} \left(\ln \frac{r_2}{r_1} \right) - b$
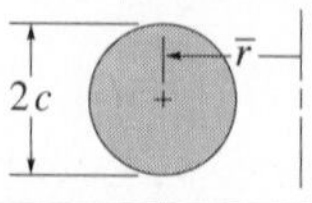	πc^2	$2\pi \left(\bar{r} - \sqrt{\bar{r}^2 - c^2} \right)$
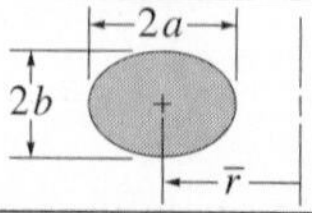	πab	$\frac{2\pi b}{a} \left(\bar{r} - \sqrt{\bar{r}^2 - a^2} \right)$

Table 6–2

In order to relate the stress distribution to the resultant bending moment, we require the resultant internal moment to be equal to the moment of the stress distribution computed about the neutral axis. From Fig. 6–44a, the stress σ, acting on the area element dA and located a distance y from the neutral axis, creates a force of $dF = \sigma\, dA$ on the element and a moment about the neutral axis of $dM = y\,(\sigma\, dA)$. This moment is positive, since by the right-hand rule it is directed in the same direction as **M**. For the entire cross section, we require $M = \int y\sigma\, dA$.

Since $y = R - r$, and σ is defined by Eq. 6–22, we have

$$M = \int_A (R - r)Ek\left(\frac{R - r}{r}\right) dA$$

Expanding, realizing that Ek and R are constants, gives

$$M = Ek\left(R^2 \int_A \frac{dA}{r} - 2R \int_A dA + \int_A r\, dA\right)$$

The first integral is equivalent to A/R as determined from Eq. 6–23, and the second integral is simply the cross-sectional area A. Realizing that the location of the centroid of the cross section is determined from $\bar{r} = \int r\, dA/A$, the third integral can be replaced by $\bar{r}A$. Thus, we can write

$$M = EkA(\bar{r} - R)$$

Solving for Ek in Eq. 6–22, substituting into the above equation, and solving for σ, we have

$$\sigma = \frac{M(R - r)}{Ar(\bar{r} - R)} \tag{6–24}$$

Here

σ = the normal stress in the member

M = the internal moment, determined from the method of sections and the equations of equilibrium and computed about the neutral axis for the cross section. This moment is *positive* if it tends to increase the member's radius of curvature, i.e., it tends to straighten out the member

A = the cross-sectional area of the member

R = the distance measured from the center of curvature to the neutral axis, determined from Eq. 6–23

$\bar{r}$ = the distance measured from the center of curvature to the centroid of the cross-sectional area

r = the distance measured from the center of curvature to the point where the stress σ is to be determined

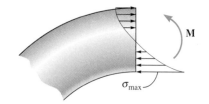

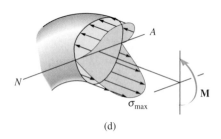

Bending stress variation
(profile view)

(c)

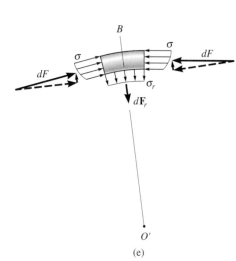

(d)

(e)

Fig. 6–44

From Fig. 6–44a, $y = R - r$ or $r = R - y$. Also, the constant and usually very small distance $e = \bar{r} - R$. When these results are substituted into Eq. 6–24, we can also write

$$\sigma = \frac{My}{Ae(R - y)}$$

(6–25)

These last two equations represent two forms of the so-called *curved-beam formula*, which like the flexure formula can be used to determine the normal-stress distribution in a curved member. This distribution is, as previously stated, hyperbolic; an example is shown in Fig. 6–44c and 6–44d. Since the stress acts in the direction of the circumference of the beam, it is sometimes called ***circumferential stress***. It should be realized, however, that due to the curvature of the beam, the circumferential stress will create a corresponding component of ***radial stress***, so called since this component acts in the radial direction. To show how it is developed, consider the free-body diagram shown in Fig. 6–44e, which is a top segment of the differential element in Fig. 6–44b. Here the radial stress σ_r is necessary since it creates the force dF_r that is required to balance the components of circumferential forces dF, which act along the line $O'B$.

Sometimes the radial stresses within curved members may become significant, especially if the member is constructed from thin plates and has, for example, the shape of an I-section. In this case the radial stress can become as large as the circumferential stress, and consequently the member must be designed to resist both stresses. For most cases, however, these stresses can be neglected, especially if the member's cross section is a *solid section*. Here the curved-beam formula gives results that are in very close agreement with those determined either by experiment or by a mathematical analysis based on the theory of elasticity.

The curved-beam formula is normally used when the curvature of the member is very pronounced, as in the case of hooks or rings. However, if the radius of curvature is greater than five times the depth of the member, the *flexure formula* can normally be used to determine the stress. Specifically, for rectangular sections for which this ratio equals 5, the maximum normal stress, when determined by the flexure formula, will be about 7% *less* than its value when determined by the curved-beam formula. This error is further reduced when the radius of curvature-to-depth ratio is more than 5.*

---

*See, for example, Boresi, A. P., et al., *Advanced Mechanics of Materials*, 3rd ed., p. 333, 1978, John Wiley & Sons, New York.

IMPORTANT POINTS

• The *curved-beam formula* should be used to determine the circumferential stress in a beam when the radius of curvature is less than five times the depth of the beam.

• Due to the curvature of the beam, the normal strain in the beam *does not* vary linearly with depth as in the case of a straight beam. As a result, the neutral axis does not pass through the centroid of the cross section.

• The radial stress component caused by bending can generally be neglected, especially if the cross section is a solid section and not made from thin plates.

PROCEDURE FOR ANALYSIS

In order to apply the curved-beam formula the following procedure is suggested.

Section Properties.

• Determine the cross-sectional area A and the location of the centroid, $\bar{r}$, measured from the center of curvature.

• Compute the location of the neutral axis, R, using Eq. 6–23 or Table 6–2. If the cross-sectional area consists of n "composite" parts, compute $\int dA/r$ for *each part*. Then, from Eq. 6–23, for the entire section, $R = \Sigma A / \Sigma \left(\int dA/r \right)$. In all cases, $R < \bar{r}$.

Normal Stress.

• The normal stress located at a point r away from the center of curvature is determined from Eq. 6–24. If the distance y to the point is measured from the neutral axis, then compute $e = \bar{r} - R$ and use Eq. 6–25.

• Since $\bar{r} - R$ generally produces a very *small number*, it is best to calculate $\bar{r}$ and R with sufficient accuracy so that the subtraction leads to a number e having at least three significant figures.

• If the stress is positive it will be tensile, whereas if it is negative it will be compressive.

• The stress distribution over the entire cross section can be graphed, or a volume element of the material can be isolated and used to represent the stress acting at the point on the cross section where it has been calculated.

EXAMPLE 6-24

A steel bar having a rectangular cross section is shaped into a circular arc as shown in Fig. 6–45a. If the allowable normal stress is $\sigma_{allow} = 20$ ksi, determine the maximum bending moment M that can be applied to the bar. What would this moment be if the bar was straight?

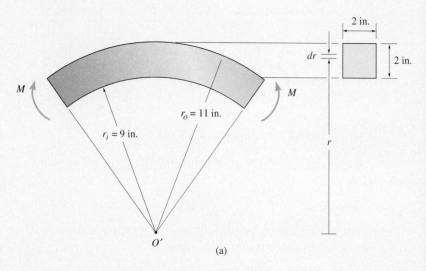

(a)

Fig. 6–45

SOLUTION

Internal Moment. Since M tends to increase the bar's radius of curvature, it is positive.

Section Properties. The location of the neutral axis is determined using Eq. 6–23. From Fig. 6–45a, we have

$$\int_A \frac{dA}{r} = \int_{9\,in.}^{11\,in.} \frac{(2\ in.)\ dr}{r} = (2\ in.) \ln r \Big|_{9\,in.}^{11\,in.} = 0.40134\ in.$$

This same result can of course be obtained directly from Table 6–2. Thus,

$$R = \frac{A}{\int_A \frac{dA}{r}} = \frac{(2\ in.)(2\ in.)}{0.40134\ in.} = 9.9666\ in.$$

It should be noted that throughout the above calculations R must be determined to several significant figures to ensure that $(\bar{r} - R)$ is accurate to at least three significant figures.

It is unknown if the normal stress reaches its maximum at the top or at the bottom of the bar, and so we must compute the moment M for each case separately. Since the normal stress at the bar's top is compressive, $\sigma = -20$ ksi,

$$\sigma = \frac{M(R - r_o)}{Ar_o(\bar{r} - R)}$$

$$-20 \text{ kip/in}^2 = \frac{M(9.9666 \text{ in.} - 11 \text{ in.})}{(2 \text{ in.})(2 \text{ in.})(11 \text{ in.})(10 \text{ in.} - 9.9666 \text{ in.})}$$

$$M = 28.5 \text{ kip} \cdot \text{in.}$$

Likewise, at the bottom of the bar the normal stress will be tensile, so $\sigma = +20$ ksi. Therefore,

$$\sigma = \frac{M(R - r_i)}{Ar_i(\bar{r} - R)}$$

$$20 \text{ kip/in}^2 = \frac{M(9.9666 \text{ in.} - 9 \text{ in.})}{(2 \text{ in.})(2 \text{ in.})(9 \text{ in.})(10 \text{ in.} - 9.9666 \text{ in.})}$$

$$M = 24.9 \text{ kip} \cdot \text{in.} \qquad \textit{Ans.}$$

By comparison, the maximum moment that can be applied is 24.9 kip · in. and so maximum normal stress occurs at the bottom of the bar. The compressive stress at the top of the bar is then

$$\sigma = \frac{24.9 \text{ kip} \cdot \text{in.}(9.9666 \text{ in.} - 11 \text{ in.})}{(2 \text{ in.})(2 \text{ in.})(11 \text{ in.})(10 \text{ in.} - 9.9666 \text{ in.})}$$

$$= -17.5 \text{ ksi}$$

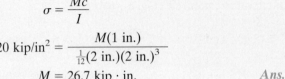

(b)

The stress distribution is shown in Fig. 6–45b.

If the bar was straight, then

$$\sigma = \frac{Mc}{I}$$

$$20 \text{ kip/in}^2 = \frac{M(1 \text{ in.})}{\frac{1}{12}(2 \text{ in.})(2 \text{ in.})^3}$$

$$M = 26.7 \text{ kip} \cdot \text{in.} \qquad \textit{Ans.}$$

This represents an error of about 7% from the more exact value determined above.

EXAMPLE 6–25

The curved bar has a cross-sectional area shown in Fig. 6–46a. If it is subjected to bending moments of 4 kN · m, determine the maximum normal stress developed in the bar.

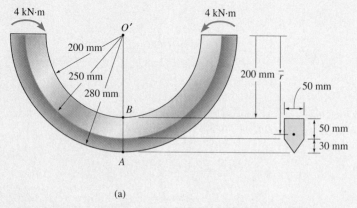

(a)

Fig. 6–46

SOLUTION

Internal Moment.   Each section of the bar is subjected to the same resultant internal moment of 4 kN · m. Since this moment tends to decrease the bar's radius of curvature, it is negative. Thus $M = -4$ kN ·m.

Section Properties.   Here we will consider the cross section to be composed of a rectangle and triangle. The total cross-sectional area is

$$\Sigma A = (0.05 \text{ m})^2 + \frac{1}{2}(0.05 \text{ m})(0.03 \text{ m}) = 3.250(10^{-3}) \text{ m}^2$$

The location of the centroid is determined with reference to the center of curvature, point O', Fig. 6–46a.

$$\bar{r} = \frac{\Sigma \tilde{r} A}{\Sigma A}$$

$$= \frac{[0.225 \text{ m}](0.05 \text{ m})(0.05 \text{ m}) + [0.260 \text{ m}]\frac{1}{2}(0.050 \text{ m})(0.030 \text{ m})}{3.250(10^{-3}) \text{ m}^2}$$

$$= 0.23308 \text{ m}$$

We can compute $\int_A dA/r$ for each part using Table 6–2. For the rectangle,

$$\int_A \frac{dA}{r} = 0.05 \text{ m}\left(\ln \frac{0.250 \text{ m}}{0.200 \text{ m}}\right) = 0.011157 \text{ m}$$

And for the triangle,

$$\int_A \frac{dA}{r} = \frac{(0.05 \text{ m})(0.280 \text{ m})}{(0.280 \text{ m} - 0.250 \text{ m})}\left(\ln \frac{0.280 \text{ m}}{0.250 \text{ m}}\right) - 0.05 \text{ m} = 0.0028867 \text{ m}$$

Thus the location of the neutral axis is determined from

$$R = \frac{\Sigma A}{\Sigma \int_A dA/r} = \frac{3.250(10^{-3}) \text{ m}^2}{0.011157 \text{ m} + 0.0028867 \text{ m}} = 0.23142 \text{ m}$$

Note that $R < \bar{r}$ as expected. Also, the calculations were performed with sufficient accuracy so that $(\bar{r} - R) = 0.23308 \text{ m} - 0.23142 \text{ m} = 0.00166 \text{ m}$ is now accurate to three significant figures.

Normal Stress. The maximum normal stress occurs either at A or B. Applying the curved-beam formula to calculate the normal stress at B, $r_B = 0.200 \text{ m}$, we have

$$\sigma_B = \frac{M(R - r_B)}{Ar_B(\bar{r} - R)} = \frac{(-4 \text{ kN} \cdot \text{m})(0.23142 \text{ m} - 0.200 \text{ m})}{3.2500(10^{-3}) \text{ m}^2(0.200 \text{ m})(0.00166 \text{ m})}$$

$$= -116 \text{ MPa}$$

At point A, $r_A = 0.280 \text{ m}$ and the normal stress is

$$\sigma_A = \frac{M(R - r_A)}{Ar_A(\bar{r} - R)} = \frac{(-4 \text{ kN} \cdot \text{m})(0.23142 \text{ m} - 0.280 \text{ m})}{3.2500(10^{-3}) \text{ m}^2(0.280 \text{ m})(0.00166 \text{ m})}$$

$$= 129 \text{ MPa} \qquad\qquad Ans.$$

By comparison, the maximum normal stress is at A. A two-dimensional representation of the stress distribution is shown in Fig. 6–46b.

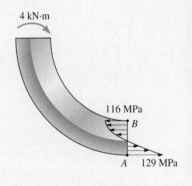

(b)

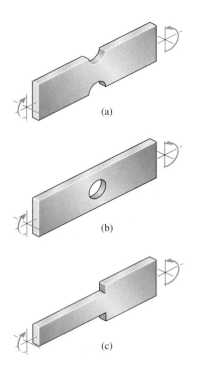

(a)

(b)

(c)

Fig. 6–47

6.9 STRESS CONCENTRATIONS

The flexure formula, $\sigma_{max} = Mc/I$, can be used to determine the stress distribution within regions of a member where the cross-sectional area is constant or tapers slightly. If the cross section suddenly changes, however, the normal-stress and strain distributions at the section become *nonlinear* and can be obtained only through experiment or, in some cases, by a mathematical analysis using the theory of elasticity. Common discontinuities include members having notches on their surfaces, Fig. 6–47a, holes for passage of fasteners or other items, Fig. 6–47b, or abrupt changes in the outer dimensions of the member's cross section, Fig. 6–47c. The *maximum* normal stress at each of these discontinuities occurs at the section taken through the *smallest* cross-sectional area.

For design, it is generally important to know the maximum normal stress developed at these sections, not the actual stress distribution itself. As in the previous cases of axially loaded bars and torsionally loaded shafts, we can obtain the maximum normal stress due to bending using a stress-concentration factor K. For example, Fig. 6–48 gives values of K for a flat bar that has a change in cross section using shoulder fillets. To use this graph simply find the geometric ratios w/h and r/h

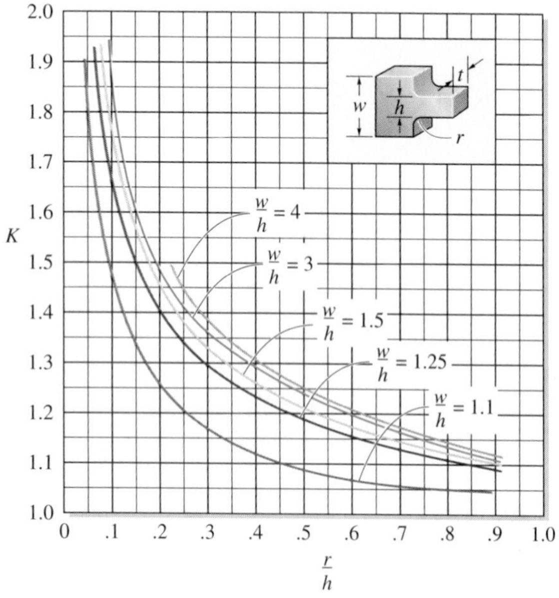

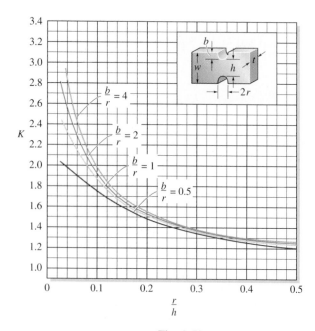

Fig. 6–48

Fig. 6–50

and then find the corresponding value of K for a particular geometry. Once K is obtained, the maximum bending stress is determined using

$$\sigma_{max} = K\frac{Mc}{I}$$ (6–26)

Fig. 6–49

Here the flexure formula is applied to the *smaller* cross-sectional area, since σ_{max} occurs at the base of the fillet, Fig. 6–49. In the same manner, Fig. 6–50 can be used if the discontinuity consists of circular grooves or notches.

Like axial load and torsion, stress concentration for bending should always be considered when designing members made of brittle materials or those that are subjected to fatigue or cyclic loadings. Realize also that stress-concentration factors apply only when the material is subjected to *elastic behavior*. If the applied moment causes yielding of the material, as is the case with ductile materials, the stress becomes redistributed throughout the member, and the maximum stress that results will be *lower* than that determined using stress-concentration factors. This phenomenon is discussed further in the next section.

Stress concentrations caused by bending occur at the sharp corners of this window lintel and are responsible for the cracks at the corners.

IMPORTANT POINTS

- Stress concentrations in members subjected to bending occur at points of cross-sectional change, such as notches and holes, because here the stress and strain become nonlinear. The more severe the change, the larger the stress concentration.

- For design or analysis, it is not necessary to know the exact stress distribution around the cross-sectional change. Instead, the maximum normal stress occurs at the *smallest* cross-sectional area. It is possible to obtain this stress using a stress concentration factor, K, that has been determined through experiment and is only a function of the geometry of the member.

- Normally, the stress concentration in a ductile material subjected to a static moment will not have to be considered in design; however, if the material is *brittle*, or subjected to *fatigue* loading, then stress concentrations become important.

EXAMPLE 6–26

The transition in the cross-sectional area of the steel bar is achieved using shoulder fillets as shown in Fig. 6–51*a*. If the bar is subjected to a bending moment of 5 kN · m, determine the maximum normal stress developed in the steel. The yield stress is $\sigma_Y = 500$ MPa.

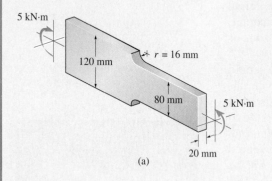

(a)

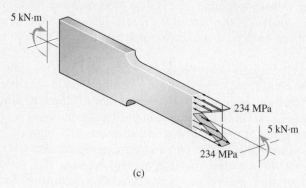

(c)

Fig. 6–51

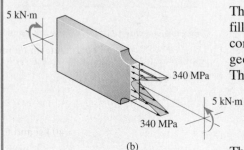

(b)

SOLUTION

The moment creates the largest stress in the bar at the base of the fillet, where the cross-sectional area is smallest. The stress-concentration factor can be determined by using Fig. 6–48. From the geometry of the bar, we have $r = 16$ mm, $h = 80$ mm, $w = 120$ mm. Thus,

$$\frac{r}{h} = \frac{16 \text{ mm}}{80 \text{ mm}} = 0.2 \qquad \frac{w}{h} = \frac{120 \text{ mm}}{80 \text{ mm}} = 1.5$$

These values give $K = 1.45$.

Applying Eq. 6–26, we have

$$\sigma_{max} = K\frac{Mc}{I} = (1.45)\frac{(5 \text{ kN} \cdot \text{m})(0.04 \text{ m})}{[\frac{1}{12}(0.020 \text{ m})(0.08 \text{ m})^3]} = 340 \text{ MPa}$$

This result indicates that the steel remains elastic since the stress is below the yield stress (500 MPa).

The normal-stress distribution is nonlinear and is shown in Fig. 6–51*b*. Realize, however, that by Saint-Venant's principle, Sec. 4.1, these localized stresses smooth out and become linear when one moves (approximately) a distance of 80 mm or more to the right of the transition. In this case, the flexure formula gives $\sigma_{max} = 234$ MPa, Fig. 6–51*c*. Also note that the choice of a larger-radius fillet will significantly reduce σ_{max}, since as r increases in Fig. 6–48, K will decrease.

PROBLEMS

6–114. The composite beam is made of 6061-T6 aluminum (A) and C83400 red brass (B). Determine the dimension h of the brass strip so that the neutral axis of the beam is located at the seam of the two metals. What maximum moment will this beam support if the allowable bending stress for the aluminum is $(\sigma_{allow})_{al} = 128$ MPa and for the brass $(\sigma_{allow})_{br} = 35$ MPa?

6–115. The composite beam is made of 6061-T6 aluminum (A) and C83400 red brass (B). If the height $h = 40$ mm, determine the maximum moment that can be applied to the beam if the allowable bending stress for the aluminum is $(\sigma_{allow})_{al} = 128$ MPa and for the brass $(\sigma_{allow})_{br} = 35$ MPa.

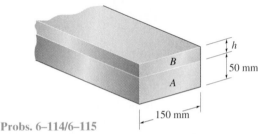

150 mm

Probs. 6–114/6–115

*6–116. The composite beam is made of steel (A) bonded to brass (B) and has the cross section shown. If it is subjected to a moment of $M = 6.5$ kN · m, determine the maximum bending stress in the brass and steel. Also, what is the stress in each material at the seam where they are bonded together? $E_{br} = 100$ GPa, $E_{st} = 200$ GPa.

6–117. The composite beam is made of steel (A) bonded to brass (B) and has the cross section shown. If the allowable bending stress for the steel is $(\sigma_{allow})_{st} = 180$ MPa, and for the brass $(\sigma_{allow})_{br} = 60$ MPa, determine the maximum moment M that can be applied to the beam. $E_{br} = 100$ GPa, $E_{st} = 200$ GPa.

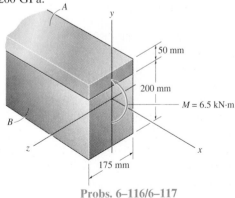

Probs. 6–116/6–117

6–118. The composite beam consists of a Douglas fir wood core and three A-36 steel plates. If the allowable bending stress for the wood is $(\sigma_{allow})_w = 20$ MPa and for the steel $(\sigma_{allow})_{st} = 130$ MPa, determine the maximum moment that can be applied to the beam.

6–119. Solve Prob. 6–118 if the moment is applied about the y axis instead of the z axis as shown.

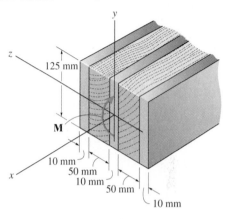

Probs. 6–118/6–119

*6–120. The top plate is made of 2014-T6 aluminum and is used to reinforce a Kevlar 49 plastic beam. Determine the maximum stress in the aluminum and in the Kevlar if the beam is subjected to a moment of $M = 900$ lb · ft.

6–121. The top plate made of 2014-T6 aluminum is used to reinforce a Kevlar 49 plastic beam. If the allowable bending stress for the aluminum is $(\sigma_{allow})_{al} = 40$ ksi and for the Kevlar $(\sigma_{allow})_k = 8$ ksi, determine the maximum moment M that can be applied to the beam.

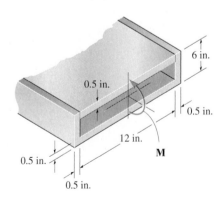

Probs. 6–120/6–121

6–122. A white spruce beam is reinforced with A-36 steel straps at its top and bottom as shown. Determine the maximum stress developed in the wood and steel if the beam is subjected to a bending moment of $M = 8$ kip · ft. Sketch the stress distribution acting over the cross section.

6–123. A white spruce beam is reinforced with A-36 steel straps at its top and bottom as shown. Determine the bending moment M it can support if $(\sigma_{allow})_{st} = 22$ ksi and $(\sigma_{allow})_w = 2.0$ ksi.

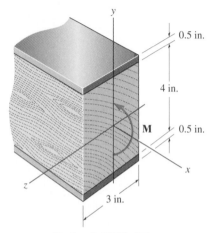

Probs. 6–122/6–123

*6–124. The Douglas fir beam is reinforced with A-36 steel straps at its center and sides. Determine the maximum stress developed in the wood and steel if the beam is subjected to a bending moment of $M_z = 7.50$ kip · ft. Sketch the stress distribution acting over the cross section.

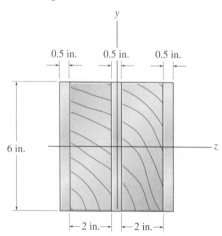

Prob. 6–124

6–125. The reinforced concrete beam is made using two steel reinforcing rods. If the allowable tensile stress for the steel is $(\sigma_{st})_{allow} = 40$ ksi, and the allowable compressive stress for the concrete is $(\sigma_{conc})_{allow} = 3$ ksi, determine the maximum moment M that can be applied to the section. Assume the concrete cannot support a tensile stress. $E_{st} = 29(10^3)$ ksi, $E_{conc} = 3.8(10^3)$ ksi.

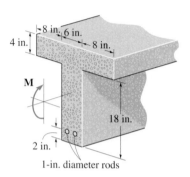

Prob. 6–125

6–126. The composite beam is made of A-36 steel (A) bonded to C83400 red brass (B) and has the cross section shown. If it is subjected to a moment of $M = 6.5$ kN · m, determine the maximum stress in the brass and steel. Also, what is the stress in each material at the seam where they are bonded together?

6–127. The composite beam is made of A-36 steel (A) bonded to C83400 red brass (B) and has the cross section shown. If the allowable bending stress for the steel is $(\sigma_{allow})_{st} = 180$ MPa and for the brass $(\sigma_{allow})_{br} = 60$ MPa, determine the maximum moment M that can be applied to the beam.

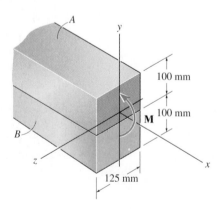

Probs. 6–126/6–127

*6–128. The curved beam is subjected to a moment of $M = 40$ lb · ft. Determine the maximum bending stress in the beam. Also, sketch a two-dimensional view of the stress distribution acting on section a–a.

6–129. The curved beam is made from a material having an allowable bending stress of $\sigma_{allow} = 24$ ksi. Determine the maximum moment M that can be applied to the beam.

*6–132. The circular spring clamp produces a compressive force of 3 N on the plates. Determine the maximum bending stress produced in the spring at A. The spring has a rectangular cross section as shown.

6–133. Determine the maximum compressive force the spring clamp can exert on the plates if the allowable bending stress for the clamp is $\sigma_{allow} = 4$ MPa.

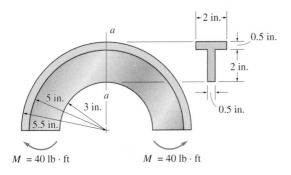

Probs. 6–128/6–129

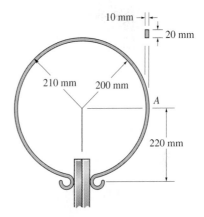

Probs. 6–132/6–133

6–130. The curved beam is subjected to a bending moment of $M = 900$ N · m as shown. Determine the stress at points A and B, and show the stress on a volume element located at each of these points.

6–131. The curved beam is subjected to a bending moment of $M = 900$ N · m. Determine the stress at point C.

6–134. The steel rod has a circular cross section. If it is gripped at its ends and a couple moment of $M = 12$ lb · in. is developed at each grip, determine the stress acting at points A and B and at the centroid C.

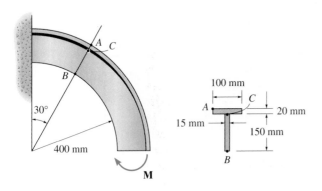

Probs. 6–130/6–131

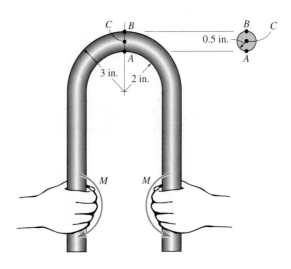

Prob. 6–134

6–135. The ceiling-suspended C-arm is used to support the x-ray camera used in medical diagnoses. If the camera has a mass of 150 kg, with center of mass at G, determine the maximum bending stress at section A.

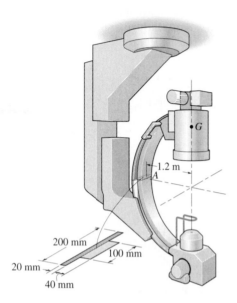

Prob. 6–135

*6–136. While in flight, the curved rib on the jet plane is subjected to an anticipated moment of $M = 16$ N · m at the section. Determine the maximum bending stress in the rib at this section, and sketch a two-dimensional view of the stress distribution.

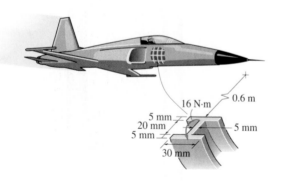

Prob. 6–136

6–137. For the curved beam in Fig. 6–44a, show that when the radius of curvature approaches infinity, the curved-beam formula, Eq. 6–24, reduces to the flexure formula, Eq. 6–13.

6–138. The member has an elliptical cross section. If it is subjected to a moment of $M = 100$ N · m, determine the stress at points A and B. Is the stress at point A', which is located on the member near the wall, the same as that at A? Explain.

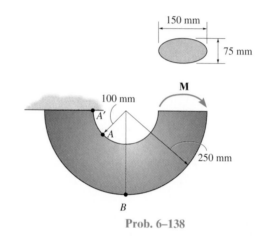

Prob. 6–138

6–139. The member has an elliptical cross section. If the allowable bending stress is $\sigma_{allow} = 125$ MPa, determine the maximum moment M that can be applied to the member.

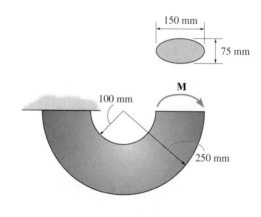

Prob. 6–139

*6–140. The bar is subjected to a moment of $M = 40 \text{ N} \cdot \text{m}$. Determine the smallest radius r of the fillets so that an allowable bending stress of $\sigma_{\text{allow}} = 124$ MPa is not exceeded.

6–141. The bar is subjected to a moment of $M = 17.5 \text{ N} \cdot \text{m}$. If $r = 5$ mm, determine the maximum bending stress in the material.

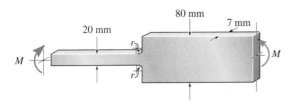

Probs. 6–140/6–141

6–142. The bar is subjected to a moment of $M = 20 \text{ N} \cdot \text{m}$. Determine the maximum bending stress in the bar and sketch, approximately, how the stress varies over the critical section.

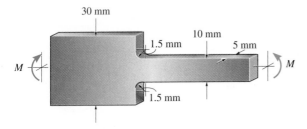

Prob. 6–142

6–143. The allowable bending stress for the bar is $\sigma_{\text{allow}} = 175$ MPa. Determine the maximum moment M that can be applied to the bar.

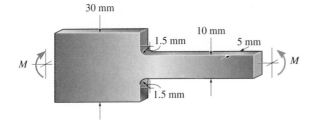

Prob. 6–143

*6–144. Determine the length L of the center portion of the bar so that the maximum bending stress at sections A, B, and C is the same. The bar has a thickness of 10 mm.

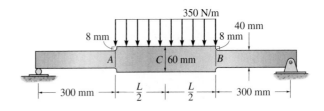

Prob. 6–144

6–145. The bar has a thickness of 0.5 in. and is subjected to a moment of 60 lb · ft. Determine the maximum bending stress in the bar.

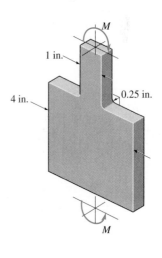

Prob. 6–145

6–146. The symmetric notched plate is subjected to bending. If the radius of each notch is $r = 10$ mm and the applied moment is $M = 2$ kN $\cdot$ m, determine the maximum bending stress in the plate.

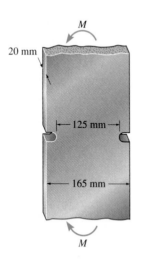

Prob. 6–146

6–147. The bar has a thickness of 0.5 in. and is made of a material having an allowable bending stress of $\sigma_{allow} = 20$ ksi. Determine the maximum moment M that can be applied.

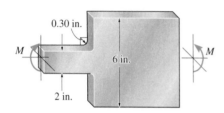

Prob. 6–147

*6–148. The bar has a thickness of 0.5 in. and is subjected to a moment of 600 lb $\cdot$ ft. Determine the maximum bending stress in the bar.

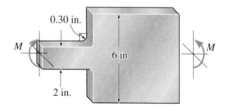

Prob. 6–148

6–149. The bar is subjected to four couple moments. If $M = 180$ lb $\cdot$ ft and $M' = 70$ lb $\cdot$ ft, determine the maximum bending stress developed in the bar.

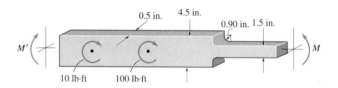

Prob. 6–149

6–150. The bar is subjected to a moment of $M = 153$ N $\cdot$ m. Determine the smallest radius r of the fillets so that an allowable bending stress of $\sigma_{allow} = 120$ MPa is not exceeded.

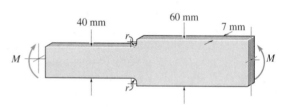

Prob. 6–150

*6.10   INELASTIC BENDING

The equations for determining the normal stress due to bending that have previously been developed are valid only if the material behaves in a linear-elastic manner. If the applied moment causes the material to *yield*, a plastic analysis must then be used to determine the stress distribution. For both elastic and plastic cases, however, realize that for bending of straight members three conditions must be met.

Linear Normal-Strain Distribution.   Based on geometric consider-ations, it was shown in Sec. 6.3 that the normal strains that develop in the material always vary *linearly* from zero at the neutral axis of the cross section to a maximum at the farthest point from the neutral axis.

Resultant Force Equals Zero.   Since there is only a resultant internal moment acting on the cross section, the resultant force caused by the stress distribution must be equal to zero. Since σ creates a force on the area dA of $dF = \sigma\, dA$, Fig. 6–52, then for the entire cross-sectional area A, we have

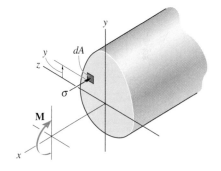

$$F_R = \Sigma F_x; \qquad \int_A \sigma\, dA = 0 \qquad (6\text{–}27)$$

This equation provides a means for obtaining the *location of the neutral axis*.

Resultant Moment.   The resultant moment at the section must be equivalent to the moment caused by the stress distribution about the neutral axis. Since the moment of the force $dF = \sigma\, dA$ about the neutral axis is $dM = y(\sigma\, dA)$, then summing the results over the entire cross section, Fig. 6–52, we have,

Fig. 6–52

$$(M_R)_z = \Sigma M_z; \qquad M = \int_A y\,(\sigma\, dA) \qquad (6\text{–}28)$$

These conditions of geometry and loading will now be used to show how to determine the stress distribution in a beam when it is subjected to a resultant internal moment that causes yielding of the material. Throughout the discussion we will assume that the material has a stress–strain diagram that is the *same* in tension as it is in compression. For simplicity, we will begin by considering the beam to have a cross-sectional area with two axes of symmetry; in this case, a rectangle of height h and width b, as shown in Fig. 6–53a. Three cases of loading that are of special interest will be considered.

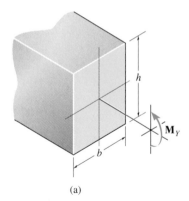

(a)

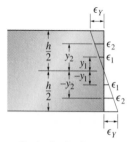

Strain distribution
(profile view)

(b)

σ

σ_Y

σ_2

σ_1

$\epsilon_1\ \epsilon_2\ \epsilon_Y$ ϵ

Stress–strain diagram
(elastic region)

(c)

Fig. 6–53

Maximum Elastic Moment.   Assume that the applied moment $M = M_Y$ is just sufficient to produce yielding strains in the top and bottom fibers of the beam as shown in Fig. 6–53b. Since the strain distribution is linear, we can determine the corresponding stress distribution by using the stress–strain diagram, Fig. 6–53c. Here it is seen that the yield strain ϵ_Y causes the yield stress σ_Y, and the intermediate strains ϵ_1 and ϵ_2 cause stresses σ_1 and σ_2, respectively. When these stresses, and others like them, are plotted at the measured points $y = h/2$, $y = y_1$, $y = y_2$, etc., the stress distribution in Fig. 6–53d or 6–53e results. The linearity of the stress is, of course, a consequence of Hooke's law.

Now that the stress distribution has been established, we can check to see if Eq. 6–27 is satisfied. To do so we will first calculate the resultant force for each of the two portions of the stress distribution in Fig. 6–53e. Geometrically this is equivalent to finding the *volumes* under the two triangular blocks. As shown, the top cross section of the member is subjected to compression and the bottom portion is subjected to tension. We have

$$T = C = \frac{1}{2}\left(\frac{h}{2}\ \sigma_Y\right)b = \frac{1}{4}\ bh\sigma_Y$$

Since $\mathbf{T}$ is equal but opposite to $\mathbf{C}$, Eq. 6–27 is satisfied and indeed the neutral axis passes through the centroid of the cross-sectional area.

The maximum elastic moment M_Y is determined from Eq. 6–28, which states that M_Y is equivalent to the moment of the stress distribution about the neutral axis. To apply this equation geometrically, we must determine the moments created by $\mathbf{T}$ and $\mathbf{C}$ in Fig. 6–53e about the neutral axis. Since each of the forces acts through the centroid of the volume of its associated triangular stress block, we have

$$M_Y = C\left(\frac{2}{3}\right)\frac{h}{2} + T\left(\frac{2}{3}\right)\frac{h}{2} = 2\left(\frac{1}{4}\ bh\ \sigma_Y\right)\left(\frac{2}{3}\right)\frac{h}{2}$$

$$= \frac{1}{6}\ bh^2\sigma_Y \tag{6–29}$$

This same result can of course be obtained in a more direct manner by using the flexure formula, that is, $\sigma_Y = M_Y(h/2)/[bh^3/12]$, or $M_Y = bh^2\sigma_Y/6$.

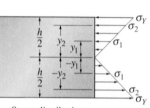

Stress distribution
(profile view)

(d)

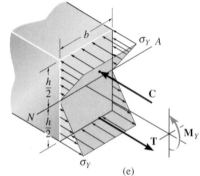

(e)

Plastic Moment. Some materials, such as steel, tend to exhibit elastic-perfectly plastic behavior when the stress in the material exceeds σ_Y. Consider, for example, the member in Fig. 6–54a. If the internal moment $M > M_Y$, the material at the top and bottom of the beam will begin to yield, causing a redistribution of stress over the cross section until the required internal moment M is developed. If the normal-strain distribution so produced is as shown in Fig. 6–54b, the corresponding normal-stress distribution is determined from the stress–strain diagram in the same manner as in the elastic case. Using the stress–strain diagram for the material shown in Fig. 6–54c, the strains ϵ_1, ϵ_Y, ϵ_2, ϵ_3 correspond to stresses σ_1, σ_Y, σ_Y, σ_Y, respectively. When these and other stresses are plotted on the cross section, we obtain the stress distribution shown in Fig. 6–54d or 6–54e. Here the compression and tension stress "blocks" each consist of component rectangular and triangular blocks. Their volumes are

$$T_1 = C_1 = \frac{1}{2} \, y_Y \sigma_Y b$$

$$T_2 = C_2 = \left(\frac{h}{2} - y_Y \right) \sigma_Y b$$

Because of the symmetry, Eq. 6–27 is satisfied and the neutral axis passes through the centroid of the cross section as shown. The applied moment M can be related to the yield stress σ_Y using Eq. 6–28. From Fig. 6–54e, we require

$$M = T_1 \left(\frac{2}{3} y_Y \right) + C_1 \left(\frac{2}{3} y_Y \right) + T_2 \left[y_Y + \frac{1}{2} \left(\frac{h}{2} - y_Y \right) \right]$$

$$+ C_2 \left[y_Y + \frac{1}{2} \left(\frac{h}{2} - y_Y \right) \right]$$

$$= 2 \left(\frac{1}{2} y_Y \sigma_Y b \right) \left(\frac{2}{3} y_Y \right) + 2 \left[\left(\frac{h}{2} - y_Y \right) \sigma_Y b \right] \left[\frac{1}{2} \left(\frac{h}{2} + y_Y \right) \right]$$

$$= \frac{1}{4} bh^2 \sigma_Y \left(1 - \frac{4}{3} \frac{y_Y^2}{h^2} \right)$$

Or using Eq. 6–29,

$$M = \frac{3}{2} M_Y \left(1 - \frac{4}{3} \frac{y_Y^2}{h^2} \right) \qquad (6\text{–}30)$$

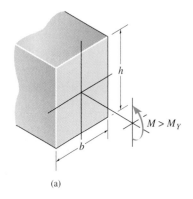

(a)

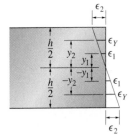

Strain distribution
(profile view)

(b)

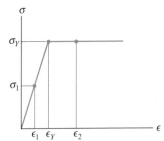

Stress–strain diagram
(elastic-plastic region)

(c)

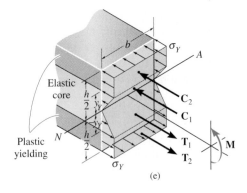

(e)

Fig. 6–54

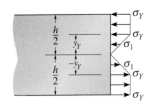

Stress distribution
(profile view)

(d)

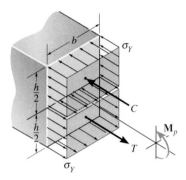

Plastic moment

(f)

Fig. 6–54

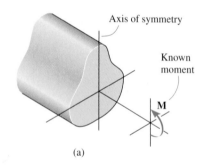

(a)

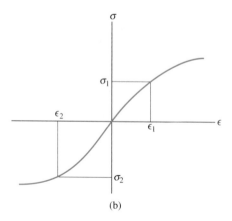

(b)

Fig. 6–55

Inspection of Fig. 6–54e reveals that **M** produces two zones of plastic yielding and an elastic core in the member. The boundary between them is located a distance $\pm y_Y$ from the neutral axis. As **M** increases in magnitude, y_Y approaches zero. This would render the material entirely plastic and the stress distribution would then look like that shown in Fig. 6–54f. From Eq. 6–30 with $y_Y = 0$, or by finding the moments of the stress "blocks" around the neutral axis, we can write this limiting value as

$$M_p = \frac{1}{4} bh^2 \sigma_Y \tag{6–31}$$

Using Eq. 6–29, we have

$$M_p = \frac{3}{2} M_Y \tag{6–32}$$

This moment is referred to as the **plastic moment**. Its value is unique only for the rectangular section shown in Fig. 6–54f, since the analysis depends on the geometry of the cross section.

Beams used in steel buildings are sometimes designed to resist a plastic moment. When this is the case, codes usually list a design property for a beam called the shape factor. The **shape factor** is defined as the ratio

$$k = \frac{M_p}{M_Y} \tag{6–33}$$

This value specifies the additional moment capacity that a beam can support beyond its maximum elastic moment. For example, from Eq. 6–32, a beam having a rectangular cross section has a shape factor of $k = 1.5$. We may therefore conclude that this section will support 50% more bending moment than its maximum elastic moment when it becomes fully plastic.

Ultimate Moment. Consider now the more general case of a beam having a cross section that is symmetrical only with respect to the vertical axis, while the moment is applied about the horizontal axis, Fig. 6–55a. We will assume that the material exhibits strain hardening and that its stress–strain diagrams for tension and compression are different, Fig. 6–55b.

If the moment **M** produces yielding of the beam, difficulty arises in finding *both* the location of the neutral axis and the maximum strain that is produced in the beam. This is because the cross section is unsymmetrical about the horizontal axis and the stress–strain behavior of the material is unsymmetrical in tension and compression. To solve this problem, a trial-and-error procedure requires the following steps:

1. For a given moment **M**, *assume* the location of the neutral axis and the slope of the "linear" strain distribution, Fig. 6–55c.
2. Graphically establish the stress distribution on the member's cross section using the σ–ϵ curve to plot values of stress corresponding to values of strain. The resulting stress distribution, Fig. 6–55d, will then have the same shape as the σ–ϵ curve.

3. Determine the volumes enclosed by the tensile and compressive stress "blocks." (As an approximation, this may require dividing each block into composite regions.) Equation 6–27 requires the volumes of these blocks to be *equal*, since they represent the resultant tensile force **T** and resultant compressive force **C** on the section, Fig. 6–55e. If these forces are unequal, an adjustment as to the *location* of the neutral axis must be made (point of *zero strain*) and the process repeated until Eq. 6–27 ($T = C$) is satisfied.

4. Once $T = C$, the moments produced by **T** and **C** can be computed about the neutral axis. Here the moment arms for **T** and **C** are measured from the neutral axis to the *centroids of the volumes* defined by the stress distributions, Fig. 6–55e. Equation 6–28 requires $M = Ty' + Cy''$. If this equation is not satisfied, the *slope* of the *strain distribution* must be adjusted and the computations for T and C and the moment must be repeated until close agreement is obtained.

This calculation process is obviously very tedious, and fortunately it does not occur very often in engineering practice. Most beams are symmetric about two axes, and they are constructed from materials that are assumed to have similar tension-and-compression stress–strain diagrams. Whenever this occurs, the neutral axis will pass through the centroid of the cross-section, and the process of relating the stress distribution to the resultant moment is thereby simplified.

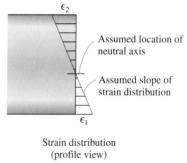

Strain distribution
(profile view)

(c)

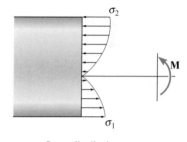

Stress distribution
(profile view)

(d)

IMPORTANT POINTS

• The *normal strain distribution* over the cross section of a beam is based only on geometric considerations and has been found to always remain *linear*, regardless of the applied load. The normal stress distribution, however, must be determined from the material behavior, or stress/strain diagram once the strain distribution is established.

• The *location of the neutral axis* is determined from the condition that the *resultant force* on the cross section is *zero*.

• The resultant internal moment on the cross section must be equal to the moment of the stress distribution about the neutral axis.

• Perfectly plastic behavior assumes the normal stress distribution is *constant* over the cross section, and the beam will continue to bend, with no increase in moment. This moment is called the *plastic moment*.

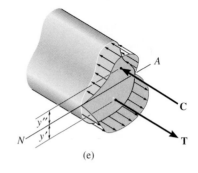

(e)

Fig. 6–55

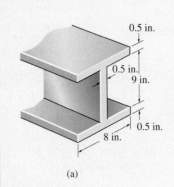

0.5 in.

0.5 in.
9 in.

0.5 in.

8 in.

(a)

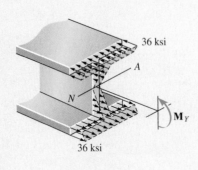

36 ksi

A

N

$\mathbf{M}_Y$

36 ksi

(b)

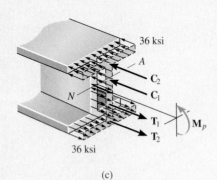

36 ksi

A

C_2
C_1

N

T_1
T_2

$\mathbf{M}_p$

36 ksi

(c)

Fig. 6–56

EXAMPLE 6–27

The steel wide-flange beam has the dimensions shown in Fig. 6–56a. If it is made of an elastic perfectly plastic material having a tensile and compressive yield stress of $\sigma_Y = 36$ ksi, determine the shape factor for the beam.

SOLUTION

In order to determine the shape factor, it is first necessary to compute the maximum elastic moment M_Y and the plastic moment M_p.

Maximum Elastic Moment. The normal-stress distribution for the maximum elastic moment is shown in Fig. 6–56b. The moment of inertia about the neutral axis is

$$I = \left[\frac{1}{12} (0.5 \text{ in.})(9 \text{ in.})^3 \right] +$$
$$2 \left[\frac{1}{12} (8 \text{ in.})(0.5 \text{ in.})^3 + 8 \text{ in.}(0.5 \text{ in.})(4.75 \text{ in.})^2 \right] = 211.0 \text{ in}^4$$

Applying the flexure formula, we have

$$\sigma_{max} = \frac{Mc}{I}; \qquad 36 \text{ kip/in}^2 = \frac{M_Y (5 \text{ in.})}{211.0 \text{ in}^4}$$
$$M_Y = 1519.5 \text{ kip} \cdot \text{in.}$$

Plastic Moment. The plastic moment causes the steel over the entire cross section of the beam to yield, so that the normal-stress distribution looks like that shown in Fig. 6–56c. Due to symmetry of the cross-sectional area and since the tension and compression stress–strain diagrams are the same, the neutral axis passes through the centroid of the cross section. In order to determine the plastic moment, the stress distribution is divided into four composite rectangular "blocks," and the force produced by each "block" is equal to the volume of the block. Therefore, we have

$$C_1 = T_1 = 36 \text{ kip/in}^2 (0.5 \text{ in.})(4.5 \text{ in.}) = 81 \text{ kip}$$
$$C_2 = T_2 = 36 \text{ kip/in}^2 (0.5 \text{ in.})(8 \text{ in.}) = 144 \text{ kip}$$

These forces act through the *centroid* of the volume for each block. Computing the moments of these forces about the neutral axis, we obtain the plastic moment:

$$M_p = 2 [(2.25 \text{ in.})(81 \text{ kip})] + 2 [(4.75 \text{ in.})(144 \text{ kip})] = 1732.5 \text{ kip} \cdot \text{in.}$$

Shape Factor. Applying Eq. 6–33 gives

$$k = \frac{M_p}{M_Y} = \frac{1732.5 \text{ kip} \cdot \text{in.}}{1519.5 \text{ kip} \cdot \text{in.}} = 1.14 \qquad \qquad Ans.$$

This value indicates that a wide-flange beam provides a very efficient section for resisting an *elastic moment*. Most of the moment is developed in the flanges, i.e., in the top and bottom segments, whereas the web or vertical segment contributes very little. In this particular case, only 14% additional moment can be supported by the beam beyond that which can be supported elastically.

EXAMPLE 6-28

A T-beam has the dimensions shown in Fig. 6–57a. If it is made of an elastic perfectly plastic material having a tensile and compressive yield stress of $\sigma_Y = 250$ MPa, determine the plastic moment that can be resisted by the beam.

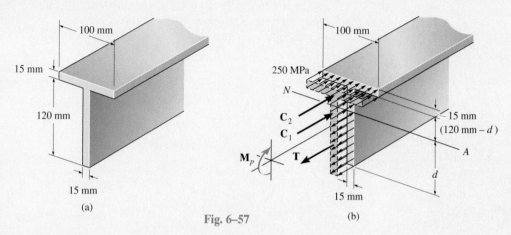

Fig. 6–57

SOLUTION

The "plastic" stress distribution acting over the beam's cross-sectional area is shown in Fig. 6–57b. In this case the cross section is not symmetric with respect to a horizontal axis, and consequently, the neutral axis will *not* pass through the centroid of the cross section. To determine the *location* of the neutral axis, d, we require the stress distribution to produce a zero resultant force on the cross section. Assuming that $d \leq 120$ mm, we have

$$\int_A \sigma \, dA = 0; \qquad T - C_1 - C_2 = 0$$

$$250 \text{ MPa}(0.015 \text{ m})(d) - 250 \text{ MPa}(0.015 \text{ m})(0.120 \text{ m} - d)$$
$$- 250 \text{ MPa}(0.015 \text{ m})(0.100 \text{ m}) = 0$$

$$d = 0.110 \text{ m} < 0.120 \text{ m} \qquad \text{OK}$$

Using this result, the forces acting on each segment are +

$$T = 250 \text{ MN/m}^2(0.015 \text{ m})(0.110 \text{ m}) = 412.5 \text{ kN}$$
$$C_1 = 250 \text{ MN/m}^2(0.015 \text{ m})(0.010 \text{ m}) = 37.5 \text{ kN}$$
$$C_2 = 250 \text{ MN/m}^2(0.015 \text{ m})(0.100 \text{ m}) = 375 \text{ kN}$$

Hence the resultant plastic moment about the neutral axis is

$$M_p = 412.5 \text{ kN}\left(\frac{0.110 \text{ m}}{2}\right) + 37.5 \text{ kN}\left(\frac{0.01 \text{ m}}{2}\right) + 375 \text{ kN}\left(0.01 \text{ m} + \frac{0.015 \text{ m}}{2}\right)$$

$$M_p = 29.4 \text{ kN} \cdot \text{m} \qquad\qquad\qquad\qquad\qquad\qquad \textit{Ans.}$$

EXAMPLE 6–29

The beam in Fig. 6–58a is made of an alloy of titanium that has a stress–strain diagram that can in part be approximated by two straight lines. If the material behavior is the *same* in both tension and compression, determine the bending moment that can be applied to the beam that will cause the material at the top and bottom of the beam to be subjected to a strain of 0.050 in./in.

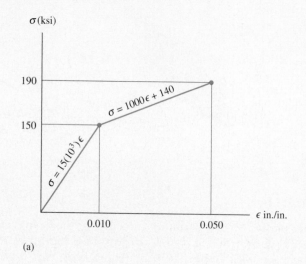

(a)

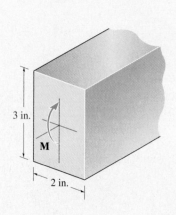

SOLUTION I

By inspection of the stress–strain diagram, the material is said to exhibit "elastic-plastic behavior with strain hardening." Since the cross section is symmetric and the tension–compression σ–ϵ diagrams are the same, the neutral axis must pass through the centroid of the cross section. The strain distribution, which is always linear, is shown in Fig. 6–58b. In particular, the point where maximum elastic strain (0.010 in./in.) occurs has been determined by proportion, such that 0.05/1.5 in. = 0.010/y or $y = 0.3$ in.

The corresponding normal-stress distribution acting over the cross section is shown in Fig. 6–58c. The moment produced by this distribution can be calculated by finding the "volume" of the stress blocks. To do so we will subdivide this distribution into two triangular blocks and a rectangular block in both the tension and compression regions, Fig. 6–58d. Since the beam is 2 in. wide, the resultants and their locations are determined as follows:

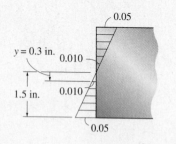

Strain distribution

(b)

Fig. 6–58

$$T_1 = C_1 = \frac{1}{2}(1.2 \text{ in.})(40 \text{ kip/in}^2)(2 \text{ in.}) = 48 \text{ kip}$$

$$y_1 = 0.3 \text{ in.} + \frac{2}{3}(1.2 \text{ in.}) = 1.10 \text{ in.}$$

$$T_2 = C_2 = (1.2 \text{ in.})(150 \text{ kip/in}^2)(2 \text{ in.}) = 360 \text{ kip}$$

$$y_2 = 0.3 \text{ in.} + \frac{1}{2}(1.2 \text{ in.}) = 0.90 \text{ in.}$$

$$T_3 = C_3 = \frac{1}{2}(0.3 \text{ in.})(150 \text{ ksi})(2 \text{ in.}) = 45 \text{ kip}$$

$$y_3 = \frac{2}{3}(0.3 \text{ in.}) = 0.2 \text{ in.}$$

The moment produced by this normal-stress distribution about the neutral axis is therefore

$$M = 2\,[48 \text{ kip}\,(1.10 \text{ in.}) + 360 \text{ kip}\,(0.90 \text{ in.}) + 45 \text{ kip}\,(0.2 \text{ in.})]$$
$$= 772 \text{ kip} \cdot \text{in.} \qquad\qquad\qquad \textit{Ans.}$$

SOLUTION II

Rather than using the above semigraphical technique, it is also possible to compute the moment analytically. To do this we must express the stress distribution in Fig. 6–58c as a function of position y along the beam. Note that $\sigma = f(\epsilon)$ has been given in Fig. 6–58a. Also, from Fig. 6–58b, the normal strain can be determined as a function of position y by proportional triangles; i.e.,

$$\epsilon = \frac{0.05}{1.5}y \qquad 0 \le y \le 1.5 \text{ in.}$$

Substituting this into the $\sigma\text{–}\epsilon$ functions shown in Fig. 6–58a gives

$$\sigma = 500y \qquad\qquad 0 \le y \le 0.3 \text{ in.} \qquad (1)$$
$$\sigma = 33.33y + 140 \qquad 0.3 \text{ in.} \le y \le 1.5 \text{ in.} \qquad (2)$$

From Fig. 6–58e, the moment caused by σ acting on the area strip $dA = 2\,dy$ is

$$dM = y\,(\sigma\,dA) = y\sigma\,(2\,dy)$$

Using Eqs. 1 and 2, the moment for the entire cross section is thus

$$M = 2\left[2\int_0^{0.3} 500y^2\,dy + 2\int_{0.3}^{1.5}(33.3y^2 + 140y)\,dy\right]$$

$$= 772 \text{ kip} \cdot \text{in.} \qquad\qquad\qquad \textit{Ans.}$$

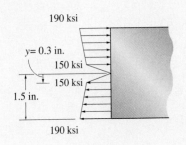

190 ksi

$y = 0.3$ in.

150 ksi

150 ksi

1.5 in.

190 ksi

Stress distribution

(c)

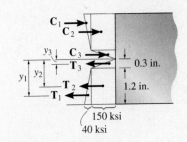

C_1
C_2
y_3 C_3
T_3 0.3 in.
y_1 y_2
T_2 1.2 in.
T_1

150 ksi
40 ksi

(d)

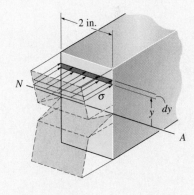

2 in.

N σ y dy

A

(e)

*6.11   RESIDUAL STRESS

If a beam is loaded such that it causes the material to yield, then removal of the load will cause **residual stress** to be developed in the beam. Since residual stresses are often important when considering fatigue and other types of mechanical behavior, we will discuss a method used for their computation when a member is subjected to bending.

Like the case for torsion, we can calculate the residual-stress distribution by using the principles of superposition and elastic recovery. To explain how this is done, consider the beam shown in Fig. 6–59a, which has a rectangular cross section and is made of an elastic perfectly plastic material having the same stress–strain diagram in tension as in compression, Fig. 6–59b. Application of the plastic moment M_p causes a stress distribution in the member to be idealized as shown in Fig. 6–59c. From Eq. 6–31, this moment is

$$M_p = \frac{1}{4} bh^2 \sigma_Y$$

If M_p causes the material at the top and bottom of the beam to be strained to ϵ_1 ($>> \epsilon_Y$), as shown by point B on the σ–ϵ curve in Fig. 6–59b, then a release of this moment will cause this material to recover some of this strain elastically by following the dashed path BC. Since this recovery is elastic, we can superimpose on the stress distribution in Fig. 6–59c a linear stress distribution caused by applying the plastic moment in the opposite direction, Fig. 6–59d. Here the maximum stress,

(a)

Fig. 6–59

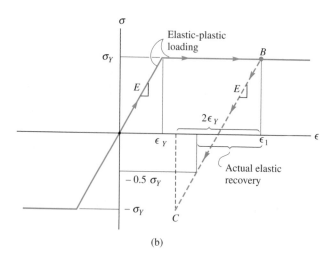

(b)

which is called the **modulus of rupture** for bending, σ_r, can be determined from the flexure formula when the beam is loaded with the plastic moment. We have

$$\sigma_{max} = \frac{Mc}{I}; \qquad \sigma_r = \frac{M_p(\frac{1}{2}h)}{(\frac{1}{12}bh^3)} = \frac{(\frac{1}{4}bh^2\sigma_Y)(\frac{1}{2}h)}{(\frac{1}{12}bh^3)}$$

$$= 1.5\sigma_Y$$

Note that reversed application of the plastic moment using a linear stress distribution is possible here, since *elastic recovery* of the material at the top and bottom of the beam can have a *maximum recovery strain* of $2\epsilon_Y$ as shown in Fig. 6–59b. This would correspond to a maximum stress of $2\sigma_Y$ at the top and bottom of the beam, which is greater than the *required* stress of $1.5\sigma_Y$ as calculated above, Fig. 6–59d.

The superposition of the plastic moment, Fig. 6–59c, and its removal, Fig. 6–59d, gives the residual-stress distribution shown in Fig. 6–59e. As an exercise, use the component triangular "blocks" that represent this stress distribution and show that it results in a zero-force and zero-moment resultant on the member as required.

The following example numerically illustrates application of these principles.

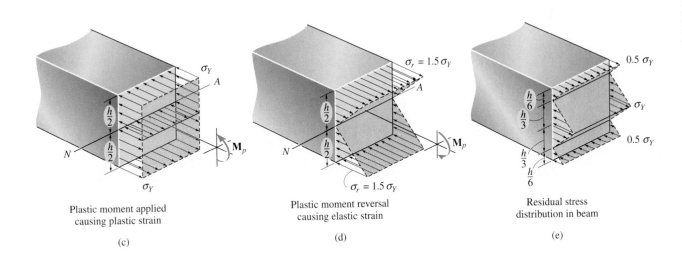

Plastic moment applied
causing plastic strain

(c)

Plastic moment reversal
causing elastic strain

(d)

Residual stress
distribution in beam

(e)

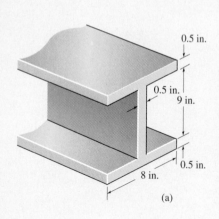

0.5 in.

0.5 in.

9 in.

0.5 in.

8 in.

(a)

EXAMPLE 6–30

The steel wide-flange beam shown in Fig. 6–60a is subjected to a fully plastic moment of $\mathbf{M}_p$. If this moment is removed, determine the residual-stress distribution in the beam. The material is elastic perfectly plastic and has a yield stress of $\sigma_Y = 36$ ksi.

SOLUTION

The normal-stress distribution in the beam caused by $\mathbf{M}_p$ is shown in Fig. 6–60b. When $\mathbf{M}_p$ is removed, the material responds elastically. Removal of $\mathbf{M}_p$ requires applying $\mathbf{M}_p$ in its reverse direction and therefore leads to an assumed elastic stress distribution as shown in Fig. 6–60c. The modulus of rupture σ_r is computed from the flexure formula. Using $M_p = 1732.5$ kip · in. and $I = 211.0$ in^4 from Example 6–27, we have

$$\sigma_{max} = \frac{Mc}{I}; \quad \sigma_r = \frac{1732.5 \text{ kip} \cdot \text{in.}(5 \text{ in.})}{211.0 \text{ in}^4} = 41.1 \text{ ksi}$$

As expected, $\sigma_r < 2\sigma_Y$.

Superposition of the stresses gives the residual-stress distribution shown in Fig. 6–60d. Note that the point of zero normal stress was determined by proportion; i.e., from Fig. 6–60b and 6–60c, we require that

$$\frac{41.1 \text{ ksi}}{5 \text{ in.}} = \frac{36 \text{ ksi}}{y}$$

$$y = 4.38 \text{ in.}$$

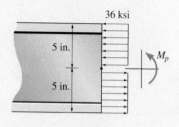

Plastic moment applied
(profile view)

(b)

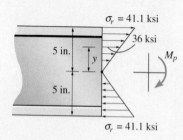

Plastic moment reversed
(profile view)

(c)

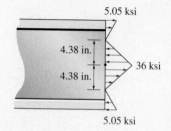

Residual stress distribution

(d)

Fig. 6–60

PROBLEMS

6–151. The channel strut is made of an elastic-perfectly plastic material for which $\sigma_Y = 250$ MPa. Determine the maximum elastic moment and the plastic moment that can be applied to the cross section.

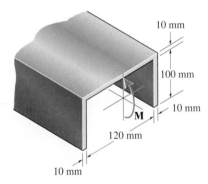

10 mm

100 mm

10 mm

M

120 mm

10 mm

Prob. 6–151

*6–152. A rectangular A-36 steel bar has a width of 1 in. and height of 3 in. Determine the moment applied about the horizontal axis that will cause half the bar to yield.

6–153. Determine the shape factor of the beam's cross section.

6–154. The box beam is made of an elastic-perfectly plastic material for which $\sigma_Y = 250$ MPa. Determine the residual stress in the top and bottom of the beam after the plastic moment $\mathbf{M}_p$ is applied and then released.

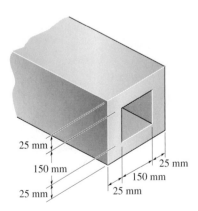

25 mm

150 mm

25 mm

25 mm

150 mm

25 mm

Probs. 6–153/6–154

6–155. Determine the shape factor for the cross section of the H-beam.

*6–156. The H-beam is made of an elastic-plastic material for which $\sigma_Y = 250$ MPa. Determine the residual stress in the top and bottom of the beam after the plastic moment $\mathbf{M}_p$ is applied and then released.

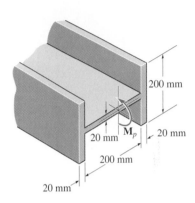

200 mm

20 mm $\mathbf{M}_p$ 20 mm

200 mm

20 mm

Probs. 6–155/6–156

6–157. The T-beam is made of an elastic-perfectly plastic material. Determine the maximum elastic moment and the plastic moment that can be applied to the cross section. $\sigma_Y = 36$ ksi.

6–158. Determine the plastic section modulus and the shape factor of its cross section.

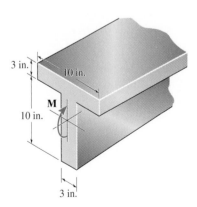

3 in.

10 in.

M

10 in.

3 in.

Probs. 6–157/6–158

6–159. Determine the plastic moment $\mathbf{M}_p$ that can be supported by a beam having the cross section shown. $\sigma_Y = 30$ ksi.

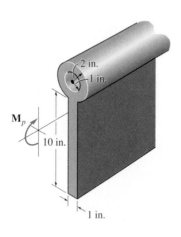

Prob. 6–159

6–162. Determine the plastic section modulus and the shape factor of the cross section.

6–163. The beam is made of elastic-perfectly plastic material. Determine the maximum elastic moment and the plastic moment that can be applied to the cross section. Take $a = 2$ in. and $\sigma_Y = 36$ ksi.

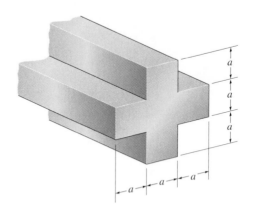

Probs. 6–162/6–163

*6–160.** Determine the plastic section modulus and the shape factor for the member having the box cross section.

6–161. The box beam is made of elastic-perfectly plastic material. Determine the maximum elastic moment and the plastic moment that can be applied to the cross section. Take $a = 100$ mm and $\sigma_Y = 250$ MPa.

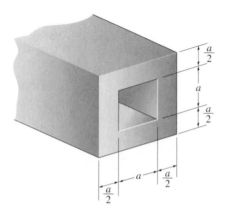

Probs. 6–160/6–161

*6–164.** Determine the plastic section modulus and the shape factor for the member having the tubular cross section.

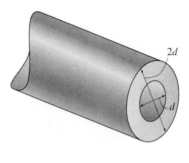

Prob. 6–164

6–165. Determine the plastic section modulus and the shape factor for the member.

6–166. The member is made of elastic-perfectly plastic material for which $\sigma_Y = 230$ MPa. Determine the maximum elastic moment and the plastic moment that can be applied to the cross section. Take $b = 50$ mm and $h = 80$ mm.

Probs. 6–165/6–166

6–167. The beam is made from an elastic-plastic material for which $\sigma_Y = 200$ MPa. Determine the magnitude of force **P** that causes this moment to be (*a*) the largest elastic moment and (*b*) the largest plastic moment.

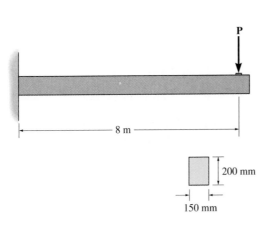

Prob. 6–167

*6–168. The box beam is made from an elastic-plastic material for which $\sigma_Y = 25$ ksi. Determine the intensity of the distributed load w_0 that will cause this moment to be (*a*) the largest elastic moment and (*b*) the largest plastic moment.

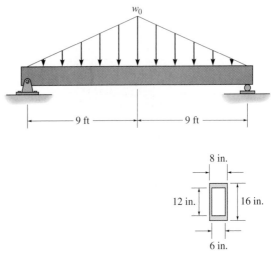

Prob. 6–168

6–169. The beam is made of a polyester that has the stress–strain curve shown. If the curve can be represented by the equation $\sigma = [20\tan^{-1}(15\epsilon)]$ ksi, where $\tan^{-1}(15\epsilon)$ is in radians, determine the magnitude of the force **P** that can be applied to the beam without causing the maximum strain in its fibers at the critical section to exceed $\epsilon_{max} = 0.003$ in./in.

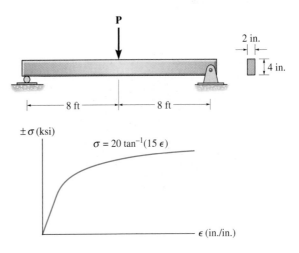

Prob. 6–169

6–170. The stress–strain diagram for a titanium alloy can be approximated by the two straight lines. If a strut made of this material is subjected to bending, determine the moment resisted by the strut if the maximum stress reaches a value of (a) σ_A and (b) σ_B.

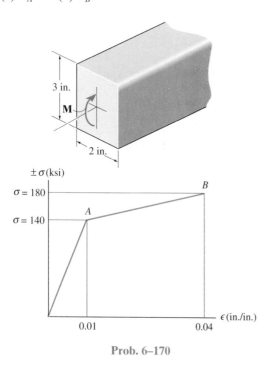

Prob. 6–170

6–171. A material has a stress–strain diagram such that within the elastic range the tensile or compressive stress can be related to the tensile or compressive strain by the equation $\sigma^n = K\epsilon$, where K and n are constants. If the material is subjected to a bending moment M, derive an expression between the maximum stress in the material and the moment. The cross section has a moment of inertia of I about its neutral axis.

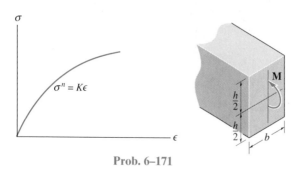

Prob. 6–171

*6–172. The plexiglass bar has a stress–strain curve that can be approximated by the straight-line segments shown. Determine the largest moment M that can be applied to the bar before it fails.

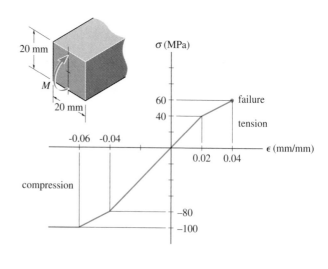

Prob. 6–172

6–173. Determine the residual stress at the top and bottom of an A-36 steel bar having a circular cross section that has been unloaded from a fully plastic moment. The radius of the bar is 4 in.

REVIEW PROBLEMS

6–174. Draw the shear and moment diagrams for the beam. *Hint:* The 20-kip load must be replaced by equivalent loadings at point *C* on the axis of the beam.

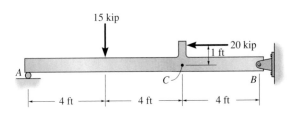

Prob. 6–174

6–175. The member has a square cross section. If it is made of an elastic-perfectly plastic material, determine the shape factor and the plastic section modulus *Z*.

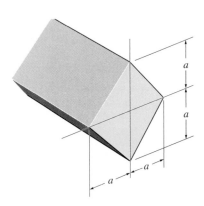

Prob. 6–175

*6–176. The beam is made of an elastic plastic material for which $\sigma_Y = 250$ MPa. Determine the residual stress in the beam at its top and bottom after the plastic moment $\mathbf{M}_p$ is applied and then released.

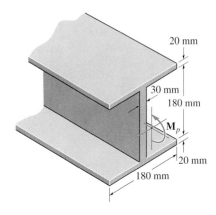

Prob. 6–176

6–177. The beam is made from three boards nailed together as shown. Determine the maximum tensile and compressive stresses in the beam.

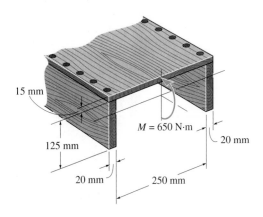

Prob. 6–177

6–178. The beam is constructed from four pieces of wood, glued together as shown. If the internal bending moment is $M = 80$ kip · ft, determine the maximum bending stress in the beam. Sketch a three-dimensional view of the stress distribution acting over the cross section.

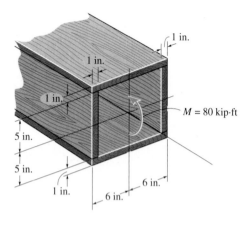

Prob. 6–178

*6–180. Draw the shear and moment diagrams for the beam.

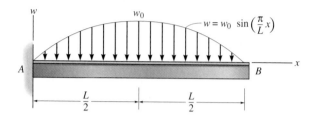

Prob. 6–180

6–179. Draw the shear and moment diagrams for the shaft if it is subjected to the vertical loadings of the belt, gear, and flywheel. The bearings at A and B exert only vertical reactions on the shaft.

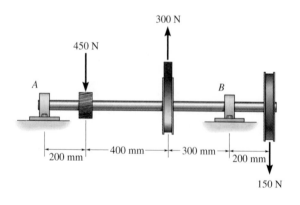

Prob. 6–179

6–181. The strut has a square cross section a by a and is subjected to the bending moment $\mathbf{M}$ applied at an angle θ as shown. Determine the maximum bending stress in terms of a, M, and θ. What angle θ will give the largest bending stress in the strut? Specify the orientation of the neutral axis for this case.

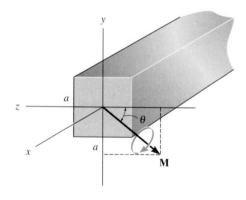

Prob. 6–181

6–182. For the section, $I_z = 114(10^{-6})\,\text{m}^4$, $31.7(10^{-6})\,\text{m}^4$, $I_{yz} = 15.1(10^{-6})\,\text{m}^4$. Using the techniques outlined in Appendix A, the member's cross-sectional area has principal moments of inertia of $I_{y'} = 29(10^{-6})\,\text{m}^4$ and $I_{z'} = 117(10^{-6})\,\text{m}^4$, computed about the principal axes of inertia y' and z', respectively. If the section is subjected to a moment of $M = 2\,\text{kN}\cdot\text{m}$ directed as shown, determine the stress produced at point A, (a) using Eq. 6–11 and (b) using the equation developed in Prob. 6–110.

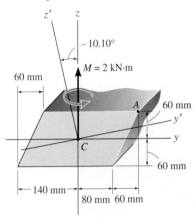

Prob. 6–182

6–183. A shaft is made of a polymer having a parabolic cross section. If it resists an internal moment of $M = 125\,\text{N}\cdot\text{m}$, determine the maximum bending stress developed in the material (a) using the flexure formula and (b) using integration. Sketch a three-dimensional view of the stress distribution acting over the cross-sectional area. *Hint:* The moment of inertia is determined using Eq. A–3 of Appendix A.

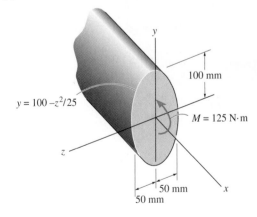

Prob. 6–183

*6–184. Determine the maximum bending stress in the handle of the cable cutter at section a–a. A force of 45 lb is applied to the handles. The cross-sectional area is shown in the figure.

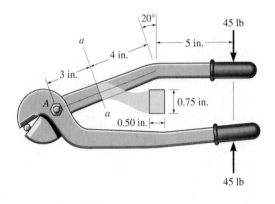

Prob. 6–184

Railroad ties act as beams that support very large transverse shear loadings. As a result, wooden ties tend to split at their ends, where the shear loads are the largest.

7 TRANSVERSE SHEAR

CHAPTER OBJECTIVES

In this chapter we will develop a method for finding the shear stress in a beam having a prismatic cross section and made from homogeneous material that behaves in a linear-elastic manner. The method of analysis to be developed will be somewhat limited to special cases of cross-sectional geometry. Although this is the case, it has many wide-range applications in engineering design and analysis. The concept of shear flow, along with shear stress, will be discussed for beams and thin-walled members. The chapter ends with a discussion of the shear center.

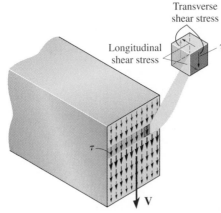

Fig. 7–1

7.1 SHEAR IN STRAIGHT MEMBERS

It was shown in Sec. 6.1 that beams generally support both shear and moment loadings. The shear V is the result of a transverse shear-stress distribution that acts over the beam's cross section, Fig. 7–1. Due to the complementary property of shear, notice that associated longitudinal shear stresses will also act along longitudinal planes of the beam. For example, a typical element removed from the interior point on the cross section is subjected to both transverse and longitudinal shear stress as shown in Fig. 7–1.

Boards not bonded together
(a)

Boards bonded together
(b)

Fig. 7–2

Shear connectors are "tack welded" to this corrugated metal floor liner so that when the concrete floor is poured, the connectors will prevent the concrete slab from slipping on the liner surface. The two materials will thus act as a composite slab.

It is possible to physically illustrate why shear stress develops on the longitudinal planes of a beam by considering the beam to be made from three boards, Fig. 7–2a. If the top and bottom surfaces of each board are smooth, and the boards are not bonded together, then application of the load **P** will cause the boards to *slide* relative to one another, and so the beam will deflect as shown. On the other hand, if the boards are bonded together, then the longitudinal shear stresses between the boards will prevent the relative sliding of the boards, and consequently the beam will act as a single unit, Fig. 7–2b.

As a result of the shear stress, shear strains will be developed and these will tend to distort the cross section in a rather complex manner. To show this consider a bar made of a highly deformable material and marked with horizontal and vertical grid lines, Fig. 7–3a. When a shear **V** is applied, it tends to deform these lines into the pattern shown in Fig. 7–3b. This nonuniform shear-strain distribution over the cross section will cause the cross section to *warp*, that is, *not* to remain plane.

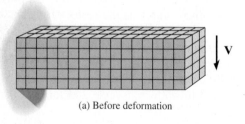

(a) Before deformation

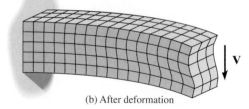

(b) After deformation

Fig. 7–3

Recall that in the development of the flexure formula, we assumed that cross sections must *remain plane* and perpendicular to the longitudinal axis of the beam after deformation. Although this is *violated* when the beam is subjected to *both* bending and shear, we can generally assume the cross-sectional warping described above is small enough so that it can be neglected. This assumption is particularly true for the most common case of a *slender beam*; that is, one that has a small depth compared with its length.

In the previous chapters we developed the axial load, torsion, and flexure formulas by first determining the strain distribution, based on assumptions regarding the deformation of the cross section. In the case of transverse shear, however, the shear-strain distribution throughout the depth of a beam *cannot* be easily expressed mathematically. For example, it is not uniform or linear for rectangular cross sections as we have shown. Therefore, the foregoing analysis of shear stress will be developed in a manner different from that used to study the previous loadings. Specifically, we will develop a formula for shear stress *indirectly*; that is, using the the flexure formula and the relationship between moment and shear ($V = dM/dx$).

7.2 THE SHEAR FORMULA

Development of a relationship between the shear-stress distribution, acting over the cross section of a beam, and the resultant shear force at the section is based on a study of the *longitudinal shear stress* and the results of Eq. 6–2, $V = dM/dx$. To show how this relationship is established, we will consider the *horizontal force equilibrium* of a portion of the element taken from the beam in Fig. 7–4a and shown in Fig. 7–4b. A free-body diagram of the *element* that shows *only* the normal-stress distribution acting on it is shown in Fig. 7–4c. This distribution is caused by the bending moments M and $M + dM$. We have excluded the effects of V, $V + dV$, and $w(x)$ on the free-body diagram since these loadings are vertical and will therefore not be involved in a horizontal force summation. The element in Fig. 7–4c will indeed satisfy $\Sigma F_x = 0$ since the stress distribution on each side of the element forms only a couple moment and therefore a zero force resultant.

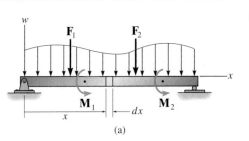

(a)

(b)

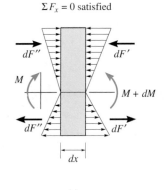

(c)

Fig. 7–4

Now consider the shaded top *segment* of the element that has been sectioned at y' from the neutral axis, Fig. 7–4b. This segment has a width t at the section, and the cross-sectional sides each have an area A'. Because the resultant moments on each side of the element differ by dM, it can be seen in Fig. 7–4d that $\Sigma F_x = 0$ will not be satisfied *unless* a longitudinal shear stress τ acts over the bottom face of the segment. In the following analysis, we will assume this shear stress is *constant* across the width t of the bottom face. It acts on the area $t\, dx$. Applying the equation of horizontal force equilibrium, and using the flexure formula, Eq. 6–13, we have

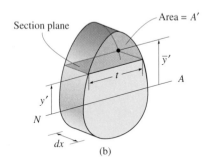

Section plane   Area = A'

(b)

$$\overset{+}{\leftarrow} \Sigma F_x = 0; \qquad \int_{A'} \sigma'\, dA - \int_{A'} \sigma\, dA - \tau(t\, dx) = 0$$

$$\int_{A'} \left(\frac{M + dM}{I} \right) y\, dA - \int_{A'} \left(\frac{M}{I} \right) y\, dA - \tau(t\, dx) = 0$$

$$\left(\frac{dM}{I} \right) \int_{A'} y\, dA = \tau(t\, dx) \qquad (7\text{–}1)$$

Solving for τ, we get

$$\tau = \frac{1}{It} \left(\frac{dM}{dx} \right) \int_{A'} y\, dA$$

This equation can be simplified by noting that $V = dM/dx$ (Eq. 6–2). Also, the integral represents the first moment of the area A' about the neutral axis. We will denote it by the symbol Q. Since the location of the centroid of the area A' is determined from $\overline{y}' = \int_{A'} y\, dA/A'$, we can also write

$$Q = \int_{A'} y\, dA = \overline{y}'A' \qquad (7\text{–}2)$$

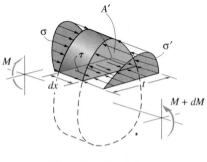

A'

M

dx

$M + dM$

Three-dimensional view

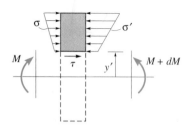

σ

σ'

M

τ

y'

$M + dM$

(d)

Profile view

Fig. 7–4

The final result is therefore

$$\tau = \frac{VQ}{It}$$
(7–3)

Here

τ = the shear stress in the member at the point located a distance y' from the neutral axis, Fig. 7–4b. This stress is assumed to be constant and therefore *averaged* across the width t of the member, Fig. 7–4d

V = the internal resultant shear force, determined from the method of sections and the equations of equilibrium

I = the moment of inertia of the *entire* cross-sectional area computed about the neutral axis

t = the width of the member's cross-sectional area, measured at the point where τ is to be determined

$Q = \int_{A'} y\, dA' = \bar{y}'A'$, where A' is the top (or bottom) portion of the member's cross-sectional area, defined from the section where t is measured, and $\bar{y}'$ is the distance to the centroid of A', measured from the neutral axis

The above equation is referred to as the *shear formula*. Although in the derivation we considered only the shear stresses acting on the beam's longitudinal plane, the formula applies as well for finding the transverse shear stress on the beam's cross-sectional area. This, of course, is because the transverse and longitudinal shear stresses are complementary and numerically equal.

Since Eq. 7–3 was derived indirectly from the flexure formula, it is necessary that the material behave in a linear-elastic manner and have a modulus of elasticity that is the *same* in tension as it is in compression. The shear stress in composite members, that is, those having cross sections made of different materials, can also be obtained using the shear formula. To do so it is necessary to compute Q and I from the *transformed section* of the member as discussed in Sec. 6.6. The thickness t in the formula, however, remains the actual width t of the cross section at the point where τ is to be calculated.

7.3   SHEAR STRESSES IN BEAMS

In order to develop some insight as to the method of applying the shear formula and also discuss some of its limitations, we will now study the shear-stress distributions in a few common types of beam cross sections. Numerical applications of the shear formula will then be given in the examples that follow.

Rectangular Cross Section.   Consider the beam to have a rectangular cross section of width b and height h as shown in Fig. 7–5a. The distribution of the shear stress throughout the cross section can be determined by computing the shear stress at an *arbitrary height y* from the neutral axis, Fig. 7–5b, and then plotting this function. Here the dark color shaded area A' will be used for computing τ.* Hence

$$Q = \bar{y}'A' = \left[y + \frac{1}{2}\left(\frac{h}{2} - y \right) \right]\left(\frac{h}{2} - y \right)b$$

$$= \frac{1}{2}\left(\frac{h^2}{4} - y^2 \right)b$$

Typical shear failure of this wooden beam occurred at the support and through the approximate center of its cross section.

Applying the shear formula, we have

$$\tau = \frac{VQ}{It} = \frac{V(\frac{1}{2})[(h^2/4) - y^2]b}{(\frac{1}{12}bh^3)b}$$

or

$$\tau = \frac{6V}{bh^3}\left(\frac{h^2}{4} - y^2 \right) \qquad (7\text{–}4)$$

This result indicates that the shear-stress distribution over the cross section is ***parabolic***. As shown in Fig. 7–5c, the intensity varies from zero at the top and bottom, $y = \pm h/2$, to a maximum value at the neutral axis, $y = 0$. Specifically, since the area of the cross section is $A = bh$, then at $y = 0$ we have, from Eq. 7–4,

$$\tau_{max} = 1.5\frac{V}{A} \qquad (7\text{–}5)$$

---

*The area below y can also be used [$A' = b(h/2 + y)$], but doing so involves a bit more algebraic manipulation.

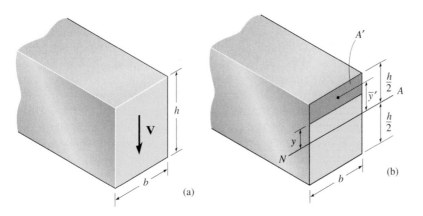

(a)

(b)

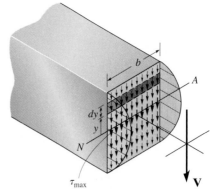

Shear–stress distribution

(c)

This same value for τ_{max} can be obtained directly from the shear formula, $\tau = VQ/It$, by realizing that τ_{max} occurs where Q is *largest*, since V, I, and t are *constant*. By inspection, Q will be a maximum when the area above (or below) the neutral axis is considered; that is, $A' = bh/2$ and $\bar{y}' = h/4$. Thus,

$$\tau_{max} = \frac{VQ}{It} = \frac{V(h/4)(bh/2)}{[\frac{1}{12}bh^3]b} = 1.5\frac{V}{A}$$

By comparison, τ_{max} is 50% greater than the *average* shear stress determined from Eq. 1–7; that is, $\tau_{avg} = V/A$.

It is important to remember that for every τ acting on the cross-sectional area in Fig. 7–5c, there is a corresponding τ acting in the longitudinal direction along the beam. For example, if the beam is sectioned by a longitudinal plane through its neutral axis, then as noted above, the *maximum shear stress* acts on this plane, Fig. 7–5d. It is this stress that can cause a timber beam to fail as shown in Fig. 7–6. Here horizontal splitting of the wood starts to occur through the neutral axis at the beam's ends, since there the vertical reactions subject the beam to large shear stress and wood has a low resistance to shear along its grains, which are oriented in the longitudinal direction.

It is instructive to show that when the shear-stress distribution, Eq. 7–4, is integrated over the cross section it yields the resultant shear V. To do this, a differential strip of area $dA = b\,dy$ is chosen, Fig. 7–5c, and since τ acts uniformly over this strip, we have

$$\int_A \tau\,dA = \int_{-h/2}^{h/2} \frac{6V}{bh^3}\left(\frac{h^2}{4} - y^2\right)b\,dy$$

$$= \frac{6V}{h^3}\left[\frac{h^2}{4}y - \frac{1}{3}y^3\right]_{-h/2}^{h/2}$$

$$= \frac{6V}{h^3}\left[\frac{h^2}{4}\left(\frac{h}{2} + \frac{h}{2}\right) - \frac{1}{3}\left(\frac{h^3}{8} + \frac{h^3}{8}\right)\right] = V$$

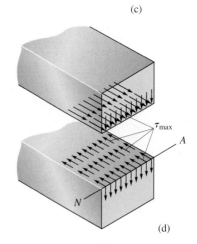

(d)

Fig. 7–5

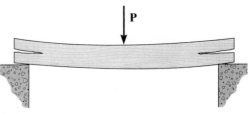

Fig. 7–6

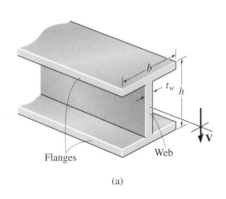

Flanges   Web

(a)

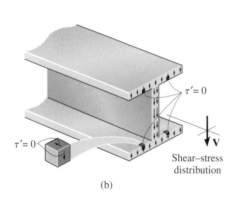

$\tau'=0$

$\tau'=0$

Shear–stress
distribution

(b)

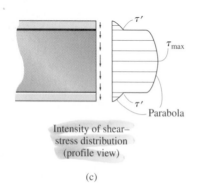

τ'

τ_{max}

τ'

Parabola

Intensity of shear–
stress distribution
(profile view)

(c)

Fig. 7–7

Wide-Flange Beam. A *wide-flange beam* consists of two (wide) "flanges" and a "web" as shown in Fig. 7–7a. Using an analysis similar to that just given we can determine the shear-stress distribution acting over the cross section. The results are depicted graphically in Fig. 7–7b and 7–7c. Like the rectangular cross section, the shear stress varies *parabolically* over the beam's depth, since the cross section can be treated like the rectangular section, first having the width of the top flange, b, then the thickness of the web, t_w, and again the width of the bottom flange, b. In particular, notice that the shear stress will vary *only slightly* throughout the web, and also, a *jump* in shear stress occurs at the flange–web junction since the cross-sectional thickness changes at this point, or in other words, t in the shear formula changes. By comparison, the web will carry significantly more of the shear force than the flanges. This will be illustrated numerically in Example 7–2.

Limitations on the Use of the Shear Formula. One of the major assumptions used in the development of the shear formula is that the shear stress is *uniformly* distributed over the *width* t at the section where the shear stress is determined. In other words, the *average* shear stress is computed across the width. We can test the accuracy of this assumption by comparing it with a more exact mathematical analysis based on the theory of elasticity. In this regard, if the beam's cross section is rectangular, the *actual* shear-stress distribution across the neutral axis as calculated from the theory of elasticity varies as shown in Fig. 7–8. The maximum value, τ'_{max}, occurs at the *edges* of the cross section, and its magnitude depends on the ratio b/h (width/depth). For sections having a $b/h = 0.5$, τ'_{max} is only about 3% greater than the shear stress calculated from the shear formula, Fig. 7–8a. However, for *flat sections*, say $b/h = 2$, τ'_{max} is about 40% greater than τ_{max}, Fig. 7–8b. The error becomes even greater as the section becomes flatter, or as the b/h ratio increases. Errors of this magnitude are certainly intolerable if one uses the shear formula to determine the shear stress in the *flange* of a wide-flange beam, as discussed above.

It should also be pointed out that the shear formula will not give accurate results when used to determine the shear stress at the flange–web junction of a wide-flange beam, since this is a point of sudden cross-sectional change and therefore a *stress concentration* occurs here. Furthermore, the inner regions of the flanges are free boundaries, Fig. 7–7b, and as a result the shear stress on these boundaries must be zero. If the shear formula is applied to determine the shear stress at these boundaries, however, one obtains a value of τ' that is *not* equal to zero, Fig. 7–7c. Fortunately, these limitations for applying the shear formula to the flanges of a wide-flange beam are not important in engineering practice. Most often engineers must only calculate the *average maximum shear stress*, which occurs at the neutral axis, where the b/h (width/depth) ratio is *very small*, and therefore the calculated result is very close to the *actual* maximum shear stress as explained above.

Another important limitation on the use of the shear formula can be illustrated with reference to Fig. 7–9a, which shows a beam having a cross section with an irregular or nonrectangular boundary. If we apply the shear formula to determine the (average) shear stress τ along the line AB, it will be directed as shown in Fig. 7–9b. Consider now an element of material taken from the boundary point B, such that one of its faces is located on the outer surface of the beam, Fig. 7–9c. Here the calculated shear stress τ on the front face of the element is resolved into components, τ' and τ''. By inspection, the component τ' must be equal to zero since its corresponding longitudinal component τ', acting on the stress-free boundary surface, must be zero. To satisfy this boundary condition therefore, the shear stress acting on the element at the boundary must be directed tangent to the boundary. The shear-stress distribution across line AB would then be directed as shown in Fig. 7–9d. Due to the greatest inclination of the shear stresses at A and B, the maximum shear stress will occur at these points. Specific values for this shear stress must be obtained using the principles of the theory of elasticity. Note, however, that we can apply the shear formula to obtain the shear stress acting across each of the colored lines in Fig. 7–9a. Here these lines intersect the tangents to the boundary of the cross section at *right angles*, and as shown in Fig. 7–9e, the transverse shear stress is vertical and constant along each line.

To summarize the above points, the shear formula does not give accurate results when applied to members having cross sections that are *short or flat*, or at points where the cross section suddenly changes. Nor should it be applied across a section that intersects the boundary of the member at an angle other than 90°. Instead, for these cases the shear stress should be determined using more advanced methods based on the theory of elasticity.

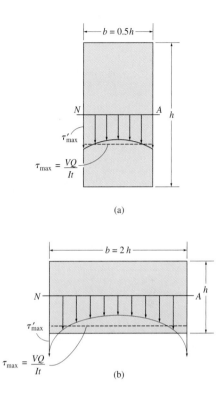

(a)

(b)

Fig. 7–8

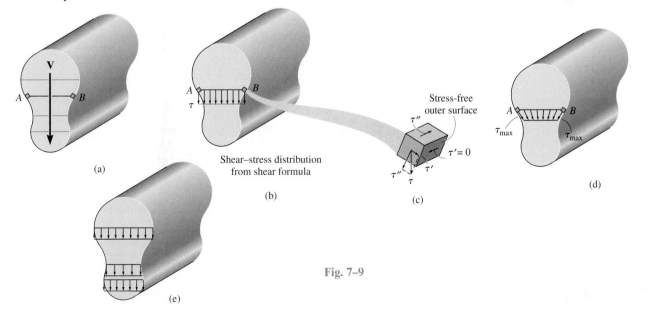

(a)

Shear–stress distribution
from shear formula

(b)

Stress-free
outer surface

(c)

(d)

(e)

Fig. 7–9

IMPORTANT POINTS

• Shear forces in beams cause *nonlinear shear-strain* distributions over the cross section, causing it to *warp*.

• Due to the complementary property of shear stress, the shear stress developed in a beam acts on both the cross section and on longitudinal planes.

• The *shear formula* was derived by considering horizontal force equilibrium of the longitudinal shear stress and bending-stress distributions acting on a portion of a differential segment of the beam.

• The shear formula is to be used on straight prismatic members made of homogeneous material that has linear-elastic behavior. Also, the internal resultant shear force must be directed along an axis of symmetry for the cross-sectional area.

• For a beam having a rectangular cross section, the *shear stress varies parabolically* with depth. The maximum shear stress is along the neutral axis.

• The shear formula should not be used to determine the shear stress on cross sections that are short or flat, or at points of sudden cross-sectional changes, or at a point on an inclined boundary.

PROCEDURE FOR ANALYSIS

In order to apply the shear formula, the following procedure is suggested.

Internal Shear.

• Section the member perpendicular to its axis at the point where the shear stress is to be determined, and obtain the internal shear **V** at the section.

Section Properties.

• Determine the location of the neutral axis, and determine the moment of inertia I of the *entire cross-sectional area* about the neutral axis.

• Pass an imaginary horizontal section through the point where the shear stress is to be determined. Measure the width t of the area at this section.

• The portion of the area lying either above or below this section is A'. Determine Q either by integration, $Q = \int_{A'} y \, dA'$, or by using $Q = \bar{y}' A'$. Here $\bar{y}'$ is the distance to the centroid of A', measured from the neutral axis. It may be helpful to realize that A' is the portion of the member's cross-sectional area that is being "held onto the member" by the longitudinal shear stresses, Fig. 7–4d.

Shear Stress.

• Using a consistent set of units, substitute the data into the shear formula and compute the shear stress τ.

• It is suggested that the proper direction of the transverse shear stress τ be established on a volume element of material located at the point where it is computed. This can be done by realizing that τ acts on the cross section in the same direction as **V**. From this, corresponding shear stresses acting on the other three planes of the element can then be established.

EXAMPLE 7-1

The beam shown in Fig. 7–10a is made of wood and is subjected to a resultant internal vertical shear force of $V = 3$ kip. (a) Determine the shear stress in the beam at point P, and (b) compute the maximum shear stress in the beam.

SOLUTION

Part (a)

Section Properties.   The moment of inertia of the cross-sectional area computed about the neutral axis is

$$I = \frac{1}{12}bh^3 = \frac{1}{12}(4 \text{ in.})(5 \text{ in.})^3 = 41.7 \text{ in}^4$$

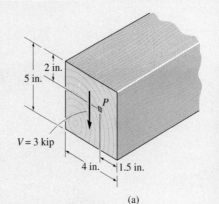

(a)

A horizontal section line is drawn through point P and the partial area A' is shown shaded in Fig. 7–10b. Hence

$$Q = \bar{y}'A' = \left[0.5 \text{ in.} + \frac{1}{2}(2 \text{ in.})\right](2 \text{ in.})(4 \text{ in.}) = 12 \text{ in}^3$$

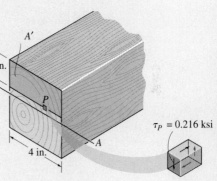

(b)

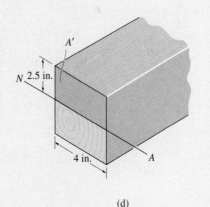

(c)

$\tau_P = 0.216$ ksi

Shear Stress.   The shear force at the section is $V = 3$ kip. Applying the shear formula, we have

$$\tau_P = \frac{VQ}{It} = \frac{(3 \text{ kip})(12 \text{ in}^3)}{(41.7 \text{ in}^4)(4 \text{ in.})} = 0.216 \text{ ksi} \qquad Ans.$$

Since τ_P contributes to V, it acts downward at P on the cross section. Consequently, a volume element of the material at this point would have shear stresses acting on it as shown in Fig. 7–10c.

Part (b)

Section Properties.   Maximum shear stress occurs at the neutral axis, since t is constant throughout the cross section and Q is largest for this case. For the dark shaded area A' in Fig. 7–10d, we have

$$Q = \bar{y}'A' = \left[\frac{2.5 \text{ in.}}{2}\right](4 \text{ in.})(2.5 \text{ in.}) = 12.5 \text{ in}^3$$

Shear Stress.   Applying the shear formula yields

$$\tau_{max} = \frac{VQ}{It} = \frac{(3 \text{ kip})(12.5 \text{ in}^3)}{(41.7 \text{ in}^4)(4 \text{ in.})} = 0.225 \text{ ksi} \qquad Ans.$$

Note that this is equivalent to

$$\tau_{max} = 1.5\frac{V}{A} = 1.5\frac{3 \text{ kip}}{(4 \text{ in.})(5 \text{ in.})} = 0.225 \text{ ksi} \qquad Ans.$$

(d)

Fig. 7–10

EXAMPLE 7-2

A steel wide-flange beam has the dimensions shown in Fig. 7–11a. If it is subjected to a shear of $V = 80$ kN, (a) plot the shear-stress distribution acting over the beam's cross-sectional area, and (b) determine the shear force resisted by the web.

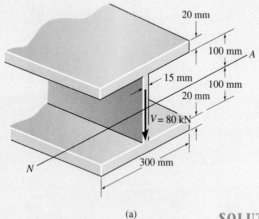

(a)

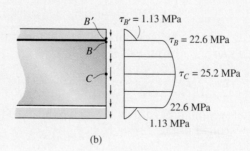

(b)

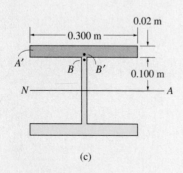

(c)

Fig. 7–11

SOLUTION

Part (a).   The shear-stress distribution will be parabolic and varies in the manner shown in Fig. 7–11b. Due to symmetry, only the shear stresses at points B', B, and C have to be computed. To show how these values are obtained, we must first determine the moment of inertia of the cross-sectional area about the neutral axis. Working in meters, we have

$$I = \left[\frac{1}{12}(0.015 \text{ m})(0.200 \text{ m})^3 \right]$$
$$+ 2\left[\frac{1}{12}(0.300 \text{ m})(0.02 \text{ m})^3 + (0.300 \text{ m})(0.02 \text{ m})(0.110 \text{ m})^2 \right]$$
$$= 155.6(10^{-6}) \text{ m}^4$$

For point B', $t_{B'} = 0.300$ m, and A' is the dark shaded area shown in Fig. 7–11c. Thus,

$$Q_{B'} = \bar{y}'A' = [0.110 \text{ m}](0.300 \text{ m})(0.02 \text{ m}) = 0.660(10^{-3}) \text{ m}^3$$

so that

$$\tau_{B'} = \frac{VQ_{B'}}{It_{B'}} = \frac{80 \text{ kN}(0.660(10^{-3}) \text{ m}^3)}{155.6(10^{-6}) \text{ m}^4(0.300 \text{ m})} = 1.13 \text{ MPa}$$

For point B, $t_B = 0.015$ m and $Q_B = Q_{B'}$, Fig. 7–11c. Hence

$$\tau_B = \frac{VQ_B}{It_B} = \frac{80 \text{ kN}(0.660(10^{-3}) \text{ m}^3)}{155.6(10^{-6}) \text{ m}^4(0.015 \text{ m})} = 22.6 \text{ MPa}$$

Note from the discussion of "Limitations on the Use of the Shear Formula" that the calculated value for both $\tau_{B'}$ and τ_B will actually be very misleading. Why?

For point C, $t_C = 0.015$ m and A' is the dark shaded area shown in Fig. 7–11d. Considering this area to be composed of two rectangles, we have

$$Q_C = \Sigma \bar{y}'A' = [0.110 \text{ m}](0.300 \text{ m})(0.02 \text{ m})$$
$$+ [0.05 \text{ m}](0.015 \text{ m})(0.100 \text{ m})$$
$$= 0.735(10^{-3}) \text{ m}^3$$

Thus,

$$\tau_C = \tau_{max} = \frac{VQ_C}{It_C} = \frac{80 \text{ kN}[0.735(10^{-3}) \text{ m}^3]}{155.6(10^{-6}) \text{ m}^4(0.015 \text{ m})} = 25.2 \text{ MPa}$$

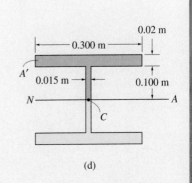

(d)

Part (b).   The shear force in the web will be determined by first formulating the shear stress at the *arbitrary* location y within the web, Fig. 7–11e. Using units of meters, we have

$$I = 155.6(10^{-6}) \text{ m}^4$$
$$t = 0.015 \text{ m}$$
$$A' = (0.300 \text{ m})(0.02 \text{ m}) + (0.015 \text{ m})(0.1 \text{ m} - y)$$

$$Q = \Sigma \bar{y}'A' = (0.11 \text{ m})(0.300 \text{ m})(0.02 \text{ m})$$

$$+[y + \tfrac{1}{2}(0.1 \text{ m} - y)](0.015 \text{ m})(0.1 \text{ m} - y)$$

$$= (0.735 - 7.50 \, y^2)(10^{-3}) \text{ m}^3$$

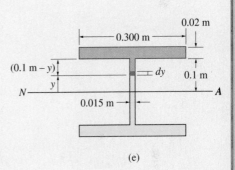

(e)

so that

$$\tau = \frac{VQ}{It} = \frac{80 \text{ kN}(0.735 - 7.50 \, y^2)(10^{-3}) \text{ m}^3}{(155.6(10^{-6}) \text{ m}^4)(0.015 \text{ m})}$$

$$= (25.192 - 257.07 \, y^2) \text{ MPa}$$

This stress acts on the area strip $dA = 0.015 \, dy$ shown in Fig. 7–11e, and therefore the shear force resisted by the web is

$$V_w = \int_{A_w} \tau \, dA = \int_{-0.1 \text{ m}}^{0.1 \text{ m}} (25.192 - 257.07 \, y^2)(10^6)(0.015 \text{ m}) \, dy$$

$$V_w = 73.0 \text{ kN} \qquad\qquad \textit{Ans.}$$

Note that by comparison, the web supports 91% of the total shear (80 kN), whereas the flanges support the remaining 9%. Try solving this problem by finding the force in one of the flanges (3.496 kN) using the same method. Then $V_w = V - 2V_f = 80 \text{ kN} - 2(3.496 \text{kN}) = 73.0 \text{ kN}$.

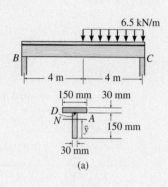

(a)

EXAMPLE 7–3

The beam shown in Fig. 7–12a is made from two boards. Determine the maximum shear stress in the glue necessary to hold the boards together along the seam where they are joined. The supports at B and C exert only vertical reactions on the beam.

SOLUTION

Internal Shear. The support reactions and the shear diagram for the beam are shown in Fig. 7–12b. It is seen that the maximum shear in the beam is 19.5 kN.

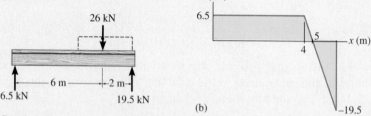

(b)

Section Properties. The centroid and therefore the neutral axis will be determined from the reference axis placed at the bottom of the cross-sectional area, Fig. 7–12a. Working in units of meters, we have

$$\bar{y} = \frac{\Sigma \bar{y} A}{\Sigma A} = \frac{[0.075 \text{ m}](0.150 \text{ m})(0.030 \text{ m}) + [0.165 \text{ m}](0.030 \text{ m})(0.150 \text{ m})}{(0.150 \text{ m})(0.030 \text{ m}) + (0.030 \text{ m})(0.150 \text{ m})} = 0.120 \text{ m}$$

The moment of inertia, computed about the neutral axis, Fig. 7–12a, is therefore

$$I = \left[\frac{1}{12}(0.030 \text{ m})(0.150 \text{ m})^3 + (0.150 \text{ m})(0.030 \text{ m})(0.120 \text{ m} - 0.075 \text{ m})^2 \right]$$
$$+ \left[\frac{1}{12}(0.150 \text{ m})(0.030 \text{ m})^3 + (0.030 \text{ m})(0.150 \text{ m})(0.165 \text{ m} - 0.120 \text{ m})^2 \right] = 27.0(10^{-6}) \text{ m}^4$$

The top board (flange) is being held onto the bottom board (web) by the glue, which is applied over the thickness $t = 0.03$ m. Consequently A' is defined as the area of the top board, Fig. 7–12a. We have

$$Q = \bar{y}'A' = [0.180 \text{ m} - 0.015 \text{ m} - 0.120 \text{ m}](0.03 \text{ m})(0.150 \text{ m})$$
$$= 0.2025(10^{-3}) \text{ m}^3$$

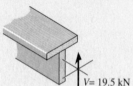

Plane containing glue

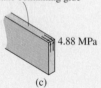

4.88 MPa

(c)

Fig. 7–12

Shear Stress. Using the above data and applying the shear formula yields

$$\tau_{max} = \frac{VQ}{It} = \frac{19.5 \text{ kN}(0.2025(10^{-3}) \text{ m}^3)}{27.0(10^{-6}) \text{ m}^4(0.030 \text{ m})} = 4.88 \text{ MPa} \qquad Ans.$$

The shear stress acting at the top of the bottom board is shown in Fig. 7–12c. Note that it is the glue's resistance to this lateral or *horizontal shear* stress that is necessary to hold the boards from slipping at the support C.

PROBLEMS

7–1. The beam is fabricated from three steel plates, and it is subjected to a shear force of $V = 150$ kN. Determine the shear stress at points A and C where the plates are joined. Show $\bar{y} = 0.080196$ m from the bottom and $I_{NA} = 4.8646(10^{-6})$ m^4.

7–2. The beam is fabricated from three steel plates, and it is subjected to a shear force of $V = 150$ kN. Determine the shear stress at point B where the plates are joined. Show $\bar{y} = 0.080196$ m from the bottom and $I_{NA} = 4.8646(10^{-6})$ m^4.

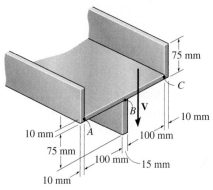

10 mm
75 mm
100 mm
15 mm
10 mm
75 mm
10 mm
100 mm
B **V**
A
C

Probs. 7–1/7–2

7–3. If the T-beam is subjected to a vertical shear of $V = 12$ kip, determine the maximum shear stress in the beam. Also, compute the shear-stress jump at the flange–web junction AB. Sketch the variation of the shear-stress intensity over the entire cross section.

*7–4. If the T-beam is subjected to a vertical shear of $V = 12$ kip, determine the vertical shear force resisted by the flange.

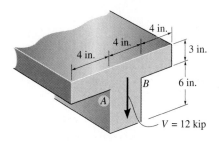

4 in.
4 in.
4 in.
3 in.
B
6 in.
A
$V = 12$ kip

Probs. 7–3/7–4

7–5. If the T-beam is subjected to a vertical shear of $V = 10$ kip, determine the maximum shear stress in the beam. Also, compute the shear-stress jump at the flange–web junction AB. Sketch the variation of the shear-stress intensity over the entire cross section. Show that $I_{NA} = 532.04$ in^4.

7–6. If the T-beam is subjected to a vertical shear of $V = 10$ kip, determine the vertical shear force resisted by the flange. Show that $I_{NA} = 532.04$ in^4.

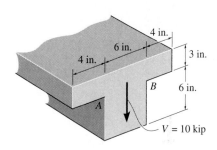

4 in.
6 in.
4 in.
3 in.
B 6 in.
A
$V = 10$ kip

Probs. 7–5/7–6

7–7. Determine the shear stress at point B on the web of the cantilevered strut at section a–a.

*7–8. Determine the maximum shear stress acting at section a–a of the cantilevered strut.

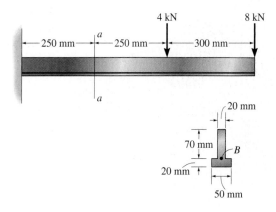

4 kN   8 kN
250 mm a 250 mm   300 mm
a
20 mm
70 mm
B
20 mm
50 mm

Probs. 7–7/7–8

7–9. Sketch the intensity of the shear-stress distribution acting over the beam's cross-sectional area, and determine the resultant shear force acting on the segment AB. The shear acting at the section is $V = 35$ kip. Show that $I_{NA} = 872.49$ in^4.

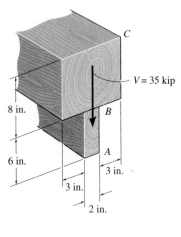

Prob. 7–9

7–10. Determine the largest shear force V that the member can sustain if the allowable shear stress is $\tau_{allow} = 8$ ksi.

7–11. If the applied shear force $V = 18$ kip, determine the maximum shear stress in the member.

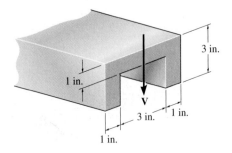

Probs. 7–10/7–11

*7–12. Determine the maximum shear stress in the strut if it is subjected to a shear force of $V = 20$ kN.

7–13. Determine the maximum shear force V that the strut can support if the allowable shear stress for the material is $\tau_{allow} = 40$ MPa.

7–14. Plot the intensity of the shear stress distributed over the cross section of the strut if it is subjected to a shear force of $V = 15$ kN.

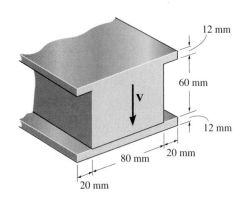

Probs. 7–12/7–13/7–14

7–15. Determine the maximum shear stress in the shaft, which has a circular cross section of radius r and is subjected to the shear force $\mathbf{V}$. State the answer in terms of the area A of the cross section.

Prob. 7–15

*7–16. Plot the shear-stress distribution over the cross section of a rod that has a radius c. By what factor is the maximum shear stress greater than the average shear stress acting over the cross section?

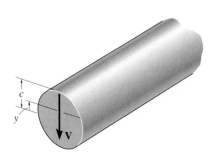

Prob. 7–16

7–17. Determine the largest end forces P that the member can support if the allowable shear stress is $\tau_{allow} = 10$ ksi. The supports at A and B only exert vertical reactions on the beam.

7–18. If the force $P = 800$ lb, determine the maximum shear stress in the beam at the critical section. The supports at A and B only exert vertical reactions on the beam.

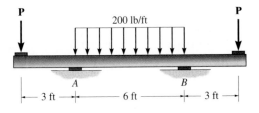

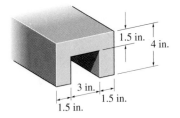

Probs. 7–17/7–18

7–19. The T-beam is subjected to the loading shown. Determine the maximum transverse shear stress in the beam at the critical section.

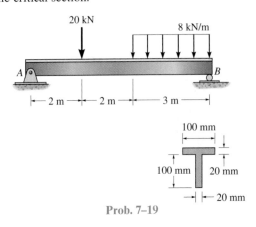

Prob. 7–19

*7–20. The supports at A and B exert vertical reactions on the wood beam. If the distributed load $w = 4$ kip/ft, determine the maximum shear stress in the beam at section a–a.

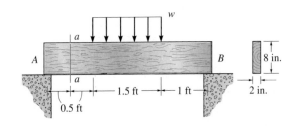

Prob. 7–20

7–21. The supports at A and B exert vertical reactions on the wood beam. If the allowable shear stress is $\tau_{allow} = 400$ psi, determine the intensity w of the largest distributed load that can be applied to the beam.

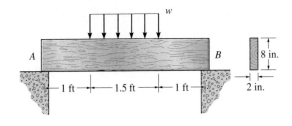

Prob. 7–21

7–22. Railroad ties must be designed to resist large shear loadings. If the tie is subjected to the 30-kip rail loadings and the gravel bed exerts a distributed reaction as shown, determine the intensity w for equilibrium, and find the maximum shear stress in the tie.

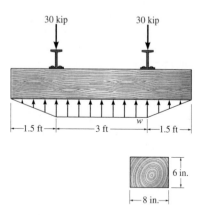

Prob. 7–22

7–23. The beam is made from three plastic pieces glued together at the seams A and B. If it is subjected to the loading shown, determine the shear stress developed in the glued joints at the critical section. The supports at C and D exert only vertical reactions on the beam.

***7–24.** The beam is made from three plastic pieces glued together at the seams A and B. If it is subjected to the loading shown, determine the vertical shear force resisted by the top flange of the beam at the critical section. The supports at C and D exert only vertical reactions on the beam.

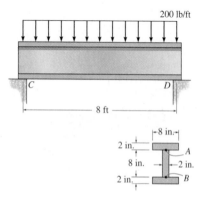

Probs. 7–23/7–24

7–25. The beam is made from three boards glued together at the seams A and B. If it is subjected to the loading shown, determine the maximum vertical shear force resisted by the top flange of the beam. The supports at C and D exert only vertical reactions on the beam.

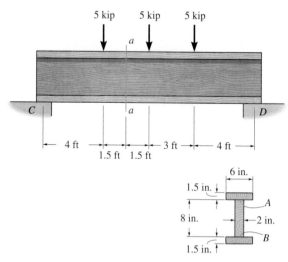

Prob. 7–25

7–26. Determine the maximum shear stress acting in the fiberglass beam at the critical section.

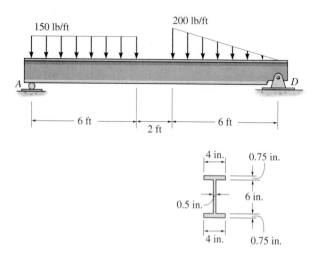

Prob. 7–26

7–27. Determine the variation of the shear stress over the cross section of a hollow rivet. What is the maximum shear stress in the rivet? Also, show that if $r_i \to r_o$, then $\tau_{max} = 2(V/A)$.

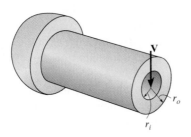

Prob. 7–27

*7–28. The beam has a square cross section and is subjected to the shear force **V**. Sketch the shear-stress distribution over the cross section and specify the maximum shear stress. Also, from the neutral axis, locate where a crack along the member will first start to appear due to shear.

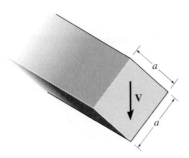

Prob. 7–28

7–29. The beam has a rectangular cross section and is subjected to a load **P** that is just large enough to develop a fully plastic moment $M_p = PL$ at the fixed support. If the material is elastic plastic, then at a distance $x < L$ the moment $M = Px$ creates a region of plastic yielding with an associated elastic core having a height $2y'$. This situation has been described by Eq. 6–30 and the moment **M** is distributed over the cross section as shown in Fig. 6–54e. Prove that the maximum shear stress developed in the beam is given by $\tau_{max} = \frac{3}{2}(P/A')$, where $A' = 2\,y'b$, the cross-sectional area of the elastic core.

7–30. The beam in Fig. 6–54f is subjected to a fully plastic moment M_p. Prove that the longitudinal and transverse shear stresses in the beam are zero. *Hint:* Consider an element of the beam as shown in Fig. 7–4d.

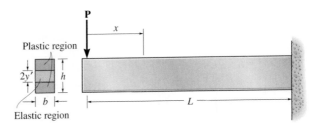

Probs. 7–29/7–30

■7–31. Write a computer program that can be used to determine the maximum shear stress in the beam that has the cross section shown and is subjected to a specific constant distributed load w and concentrated force **P**. Show an application of the program using the values $L = 4$ m, $a = 2$ m, $P = 1.5$ kN, $d_1 = 0$, $d_2 = 2$ m, $w = 400$ N/m, $t_1 = 15$ mm, $t_2 = 20$ mm, $b = 50$ mm, and $h = 150$ mm.

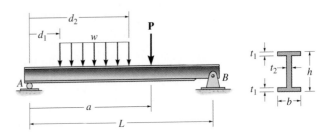

Prob. 7–31

Fig. 7–13

7.4 SHEAR FLOW IN BUILT-UP MEMBERS

Occasionally in engineering practice members are "built up" from several composite parts in order to achieve a greater resistance to loads. Some examples are shown in Fig. 7–13. If the loads cause the members to bend, fasteners such as nails, bolts, welding material, or glue may be needed to keep the component parts from sliding relative to one another, Fig. 7–2. In order to design these fasteners it is necessary to know the shear force that must be resisted by the fastener along the member's *length*. This loading, when measured as a force per unit length, is referred to as the ***shear flow q.***[*]

The magnitude of the shear flow along any longitudinal section of a beam can be obtained using a development similar to that for finding the shear stress in the beam. To show this, we will consider finding the shear flow along the juncture where the composite part in Fig. 7–14a is connected to the flange of the beam. As shown in Fig. 7–14b, three horizontal forces must act on this part. Two of these forces, F and $F + dF$, are developed by normal stresses caused by the moments M and $M + dM$, respectively. The third force, which for equilibrium equals dF, acts at the juncture and is to be supported by the fastener. Realizing that dF is the result of dM, then, as in the case of the shear formula, Eq. 7–1, we have

$$dF = \frac{dM}{I} \int_{A'} y \, dA'$$

The integral represents Q, that is, the moment of the light colored area A' in Fig. 7–14b about the neutral axis for the cross section. Since the segment has a length dx, the shear flow, or force per unit length along the beam, is $q = dF/dx$. Hence dividing both sides by dx and noting that $V = dM/dx$, Eq. 6–2, we can write

$$q = \frac{VQ}{I} \qquad (7\text{–}6)$$

Here

q = the shear flow, measured as a force per unit length along the beam
V = the internal resultant shear force, determined from the method of sections and the equations of equilibrium
I = the moment of inertia of the *entire* cross-sectional area computed about the neutral axis
$Q = \int_{A'} y \, dA' = \bar{y}'A'$, where A' is the cross-sectional area of the segment that is connected to the beam at the juncture where the shear flow is to be calculated, and $\bar{y}'$ is the distance from the neutral axis to the centroid of A'

---

[*]The use of the word "flow" in this terminology will become meaningful as it pertains to the discussion in Sec. 7.5.

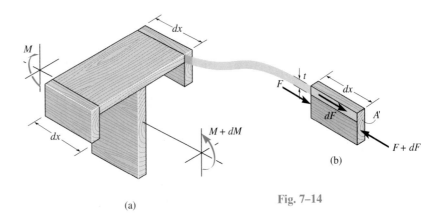

(a)

(b)

Fig. 7–14

Application of this equation follows the same "procedure for analysis" as outlined in Sec. 7.3 for the shear formula. In this regard it is very important to correctly identify the proper value for Q when determining the shear flow at a particular junction on the cross section. A few examples should serve to illustrate how this is done. Consider the beam cross sections shown in Fig. 7–15. The shaded composite parts are connected to the beam by fasteners. At the planes of connection, the necessary shear flow q is determined by using a value of Q computed from A' and $\bar{y}'$ indicated in each figure. Notice that this value of q will be resisted by a *single* fastener in Fig. 7–15a and 7–15b, by *two* fasteners in Fig. 7–15c, and by *three* fasteners in Fig. 7–15d. In other words, the fastener in Figs. 7–15a and 7–15b supports the calculated value of q and in Figs. 7–15c and 7–15d, each fastener supports $q/2$ and $q/3$, respectively.

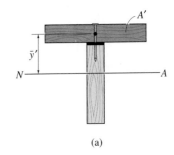

(a)

IMPORTANT POINTS

- *Shear flow* is a measure of the force per unit length along a longitudinal axis of a beam. This value is found from the shear formula and is used to determine the shear force developed in fasteners and glue that holds the various segments of a beam together.

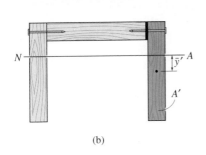

(b)

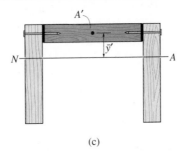

(c)

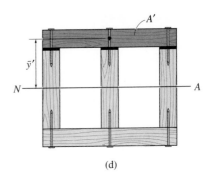

(d)

Fig. 7–15

EXAMPLE 7-4

The beam is constructed from four boards glued together as shown in Fig. 7–16a. If it is subjected to a shear of $V = 850$ kN, determine the shear flow at B and C that must be resisted by the glue.

SOLUTION

Section Properties.   The neutral axis (centroid) will be located from the bottom of the beam, Fig. 7–16a. Working in units of meters, we have

$$\bar{y} = \frac{\Sigma \bar{y}A}{\Sigma A} = \frac{2[0.15 \text{ m}](0.3 \text{ m})(0.01 \text{ m}) + [0.205 \text{ m}](0.125 \text{ m})(0.01 \text{ m}) + [0.305 \text{ m}](0.250 \text{ m})(0.01 \text{ m})}{2(0.3 \text{ m})(0.01 \text{ m}) + 0.125 \text{ m}(0.01 \text{ m}) + 0.250 \text{ m}(0.01 \text{ m})}$$

$$= 0.1968 \text{ m}$$

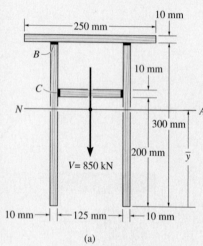

The moment of inertia computed about the neutral axis is thus

$$I = 2\left[\frac{1}{12}(0.01 \text{ m})(0.3 \text{ m})^3 + (0.01 \text{ m})(0.3 \text{ m})(0.1968 \text{ m} - 0.150 \text{ m})^2\right]$$

$$+ \left[\frac{1}{12}(0.125 \text{ m})(0.01 \text{ m})^3 + (0.125 \text{ m})(0.01 \text{ m})(0.205 \text{ m} - 0.1968 \text{ m})^2\right]$$

$$+ \left[\frac{1}{12}(0.250 \text{ m})(0.01 \text{ m})^3 + (0.250 \text{ m})(0.01 \text{ m})(0.305 \text{ m} - 0.1968 \text{ m})^2\right]$$

$$= 87.52(10^{-6}) \text{ m}^4$$

Since the glue at B and B' holds the top board to the beam, Fig. 7–16b, we have

$$Q_B = \bar{y}'_B A'_B = [0.305 \text{ m} - 0.1968 \text{ m}](0.250 \text{ m})(0.01 \text{ m})$$

$$= 0.270(10^{-3}) \text{ m}^3$$

Likewise, the glue at C and C' holds the inner board to the beam, Fig. 7–16b, and so

$$Q_C = \bar{y}'_C A'_C = [0.205 \text{ m} - 0.1968 \text{ m}](0.125 \text{ m})(0.01 \text{ m})$$

$$= 0.01025(10^{-3}) \text{ m}^3$$

Shear Flow.   For B and B' we have

$$q'_B = \frac{VQ_B}{I} = \frac{850 \text{ kN}(0.270(10^{-3}) \text{ m}^3)}{87.52(10^{-6}) \text{ m}^4} = 2.62 \text{ MN/m}$$

And for C and C',

$$q'_C = \frac{VQ_C}{I} = \frac{850 \text{ kN}(0.01025(10^{-3}) \text{ m}^3)}{87.52(10^{-6}) \text{ m}^4} = 0.0995 \text{ MN/m}$$

Since *two seams* are used to secure each board, the glue per meter length of beam at each seam must be strong enough to resist *one-half* of each calculated value of q'. Thus,

$$q_B = 1.31 \text{ MN/m} \qquad \text{and} \qquad q_C = 0.0498 \text{ MN/m} \qquad Ans.$$

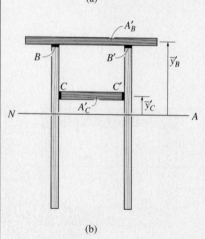

(b)

Fig. 7–16

EXAMPLE 7-5

A box beam is to be constructed from four boards nailed together as shown in Fig. 7–17a. If each nail can support a shear force of 30 lb, determine the maximum spacing s of nails at B and at C so that the beam will support the vertical force of 80 lb.

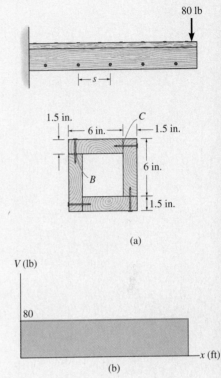

(a)

SOLUTION

Internal Shear. If the beam is sectioned at an *arbitrary point* along its length, the internal shear required for equilibrium is always $V = 80$ lb, and so the shear diagram is shown in Fig. 7–17b.

Section Properties. The moment of inertia of the cross-sectional area about the neutral axis can be determined by considering a 7.5-in. × 7.5-in. square minus a 4.5-in. × 4.5-in. square.

$$I = \frac{1}{12}(7.5 \text{ in.})(7.5 \text{ in.})^3 - \frac{1}{12}(4.5 \text{ in.})(4.5 \text{ in.})^3 = 229.5 \text{ in}^4$$

The shear flow at B is determined using Q_B found from the darker shaded area shown in Fig. 7–17c. It is this "symmetric" portion of the beam that is to be "held" onto the rest of the beam by nails on the left side and by the fibers of the board on the right side. Thus,

$$Q_B = \bar{y}'A' = [3 \text{ in.}](7.5 \text{ in.})(1.5 \text{ in.}) = 33.75 \text{ in}^3$$

Likewise, the shear flow at C can be determined using the "symmetric" shaded area shown in Fig. 7–17d. We have

$$Q_C = \bar{y}'A' = [3 \text{ in.}](4.5 \text{ in.})(1.5 \text{ in.}) = 20.25 \text{ in}^3$$

Shear Flow.

$$q_B = \frac{VQ_B}{I} = \frac{80 \text{ lb}(33.75 \text{ in}^3)}{229.5 \text{ in}^4} = 11.76 \text{ lb/in.}$$

$$q_C = \frac{VQ_C}{I} = \frac{80 \text{ lb}(20.25 \text{ in}^3)}{229.5 \text{ in}^4} = 7.059 \text{ lb/in.}$$

These values represent the shear force per unit length of the beam that must be resisted by the nails at B and the fibers at B', Fig. 7–17c, and the nails at C and the fibers at C', Fig. 7–17d, respectively. Since in each case the shear flow is resisted at *two* surfaces and each nail can resist 30 lb, for B the spacing is

$$s_B = \frac{30 \text{ lb}}{(11.76/2) \text{ lb/in.}} = 5.10 \text{ in.} \qquad \text{Use } s_B = 5 \text{ in.} \qquad \textit{Ans.}$$

And for C,

$$s_C = \frac{30 \text{ lb}}{(7.059/2) \text{ lb/in.}} = 8.50 \text{ in.} \qquad \text{Use } s_C = 8.5 \text{ in.} \qquad \textit{Ans.}$$

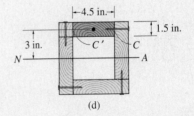

V (lb)

(b)

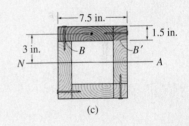

(c)

(d)

Fig. 7–17

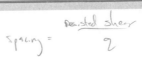

EXAMPLE 7-6

Nails having a total shear strength of 40 lb are used in a beam that can be constructed either as in Case I or as in Case II, Fig. 7–18. If the nails are spaced at 9 in., determine the largest vertical shear that can be supported in each case so that the fasteners will not fail.

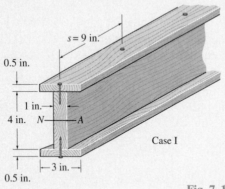

Fig. 7–18

SOLUTION

Since the geometry is the same in both cases, the moment of inertia about the neutral axis is

$$I = \frac{1}{12}(3 \text{ in.})(5 \text{ in.})^3 - 2\left[\frac{1}{12}(1 \text{ in.})(4 \text{ in.})^3\right] = 20.58 \text{ in}^4$$

Case I.   For this design a single row of nails holds the top or bottom flange onto the web. For one of these flanges,

$$Q = \bar{y}'A' = [2.25 \text{ in.}](3 \text{ in.}(0.5 \text{ in.})) = 3.375 \text{ in}^3$$

so that

$$q = \frac{VQ}{I}$$

$$\frac{40 \text{ lb}}{9 \text{ in.}} = \frac{V(3.375 \text{ in}^3)}{20.58 \text{ in}^4}$$

$$V = 27.1 \text{ lb} \qquad\qquad\qquad \textit{Ans.}$$

Case II.   Here a single row of nails holds one of the side boards onto the web. Thus,

$$Q = \bar{y}'A' = [2.25 \text{ in.}](1 \text{ in.}(0.5 \text{ in.})) = 1.125 \text{ in}^3$$

$$q = \frac{VQ}{I}$$

$$\frac{40 \text{ lb}}{9 \text{ in.}} = \frac{V(1.125 \text{ in}^3)}{20.58 \text{ in}^4}$$

$$V = 81.3 \text{ lb} \qquad\qquad\qquad \textit{Ans.}$$

PROBLEMS

*7–32. The beam is constructed from three boards. Determine the maximum shear V that it can sustain if the allowable shear stress for the wood is $\tau_{allow} = 400$ psi. What is the required spacing s of the nails if each nail can resist a shear force of 400 lb?

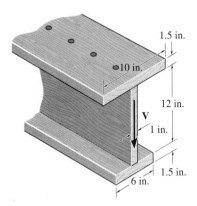

Prob. 7–32

7–33. The beam is constructed from two boards fastened together at the top and bottom with two rows of nails spaced every 6 in. If an internal shear force of $V = 600$ lb is applied to the boards, determine the shear force resisted by each nail.

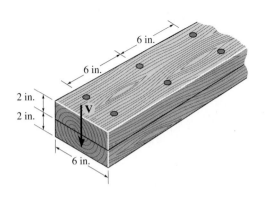

Prob. 7–33

7–34. A beam is constructed from five boards bolted together as shown. Determine the maximum shear force developed in each bolt if the bolts are spaced $s = 250$ mm apart and the applied shear is $V = 35$ kN.

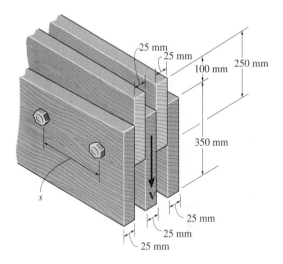

Prob. 7–34

7–35. The box beam is made from four pieces of plastic that are glued together as shown. If the shear $V = 2$ kip, determine the shear stress resisted by the seam at each of the glued joints.

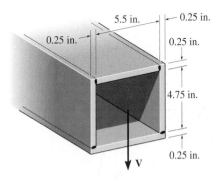

Prob. 7–35

*7–36. A beam is constructed from three boards bolted together as shown. Determine the shear force developed in each bolt if the bolts are spaced $s = 250$ mm apart and the applied shear is $V = 35$ kN.

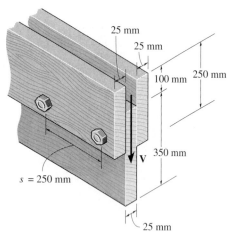

Prob. 7–36

7–37. The beam is fabricated from two equivalent structural tees and two plates. Each plate has a height of 6 in. and a thickness of 0.5 in. If a shear of $V = 50$ kip is applied to the cross section, determine the maximum spacing of the bolts. Each bolt can resist a shear force of 15 kip.

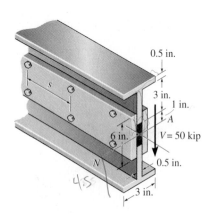

Prob. 7–37

7–38. The beam is fabricated from two equivalent structural tees and two plates. Each plate has a height of 6 in. and a thickness of 0.5 in. If the bolts are spaced at $s = 8$ in., determine the maximum shear force V that can be applied to the cross section. Each bolt can resist a shear force of 15 kip.

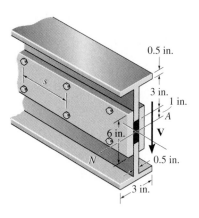

Prob. 7–38

7–39. The beam is made from three polystyrene strips that are glued together as shown. If the glue has a shear strength of 80 kPa, determine the maximum load P that can be applied without causing the glue to lose its bond.

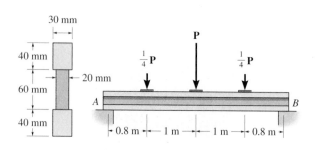

Prob. 7–39

*7–40. The beam is fabricated from two equivalent channels and two plates. Each plate has a height of 6 in. and a thickness of 0.5 in. If a shear of $V = 50$ kip is applied to the cross section, determine the maximum spacing of the bolts. Each bolt can resist a shear force of 15 kip.

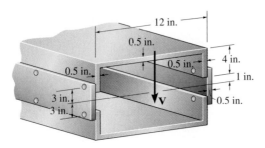

12 in.

0.5 in.

0.5 in.

0.5 in.

4 in.

1 in.

0.5 in.

3 in.

3 in.

V

Prob. 7–40

7–41. The strut is constructed from three pieces of plastic that are glued together as shown. If the allowable shear stress for the plastic is $\tau_{allow} = 800$ psi and each glue joint can withstand 250 lb/in., determine the largest allowable distributed loading w that can be applied to the strut.

7–42. The strut is constructed from three pieces of plastic that are glued together as shown. If the distributed load $w = 200$ lb/ft, determine the shear stress that must be resisted by each glue joint.

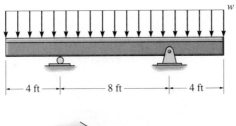

w

4 ft   8 ft   4 ft

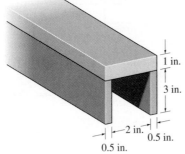

1 in.

3 in.

2 in.

0.5 in.

0.5 in.

Probs. 7–41/7–42

7–43. The beam is subjected to the loading shown, where $P = 7$ kN. Determine the average shear stress developed in the nails within region AB of the beam. The nails are located on each side of the beam and are spaced 100 mm apart. Each nail has a diameter of 5 mm.

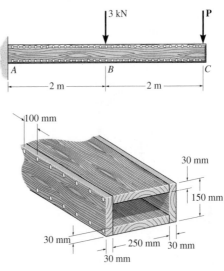

3 kN

P

A   B   C

2 m   2 m

100 mm

30 mm

150 mm

30 mm

250 mm   30 mm

30 mm

Prob. 7–43

*7–44. The beam is constructed from four boards which are nailed together. If the nails are on both sides of the beam and each can resist a shear of 3 kN, determine the maximum load P that can be applied to the end of the beam.

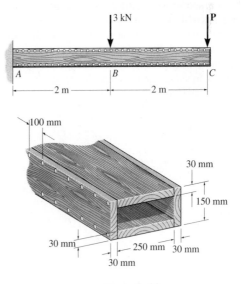

3 kN

P

A   B   C

2 m   2 m

100 mm

30 mm

150 mm

30 mm

250 mm   30 mm

30 mm

Prob. 7–44

7–45. The double T-beam is fabricated by welding the three plates together as shown. Determine the shear stress in the weld necessary to support a shear force of $V = 80$ kN.

7–46. The double T-beam is fabricated by welding the three plates together as shown. If the weld can resist a shear stress $\tau_{allow} = 90$ MPa, determine the maximum shear V that can be applied to the beam.

■**7–49.** The timber T-beam is subjected to a load consisting of n concentrated forces, P_n. If the allowable shear V_{nail} for each of the nails is known, write a computer program that will specify the nail spacing between each load. Show an application of the program using the values $L = 15$ ft, $a_1 = 4$ ft, $P_1 = 600$ lb, $a_2 = 8$ ft, $P_2 = 1500$ lb, $b_1 = 1.5$ in., $h_1 = 10$ in., $b_2 = 8$ in., $h_2 = 1$ in., and $V_{nail} = 200$ lb.

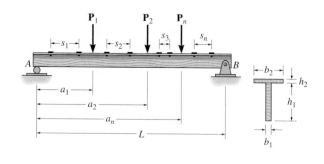

Prob. 7–49

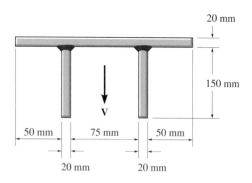

Probs. 7–45/7–46

7–47. The double-web girder is constructed from two plywood sheets that are secured to wood members at its top and bottom. If each fastener can support 600 lb in single shear, determine the required spacing s of the fasteners needed to support the loading $P = 3000$ lb. Assume A is pinned and B is a roller.

*__7–48.__ The double-web girder is constructed from two plywood sheets that are secured to wood members at its top and bottom. The allowable bending stress for the wood is $\sigma_{allow} = 2$ ksi and the allowable shear stress is $\tau_{allow} = 800$ psi. If the fasteners are spaced $s = 6$ in. and each fastener can support 600 lb in single shear, determine the maximum load P that can be applied to the beam.

■**7–50.** A built-up timber beam is made from n boards, each having a rectangular cross section. Write a computer program that can be used to determine the maximum shear stress in the beam when it is subjected to any shear V. Show an application of the program using a cross section that is in the form of a "T" and a box.

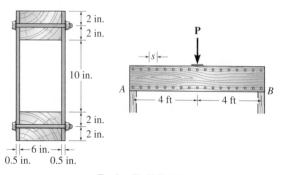

Probs. 7–47/7–48

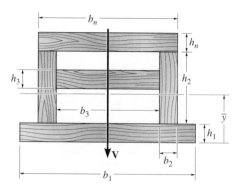

Prob. 7–50

7.5 SHEAR FLOW IN THIN-WALLED MEMBERS

In the previous section we developed the shear-flow equation, $q = VQ/I$, and showed how it can be used to determine the shear flow acting along any longitudinal plane of a member. In this section we will show how to apply this equation to find the shear-flow *distribution* throughout a member's cross-sectional area. Here we will assume that the member has *thin walls*, that is, the wall thickness is small compared with the height or width of the member. As will be shown in the next section, this analysis has important applications in structural and mechanical design.

Before we determine the shear-flow distribution on a cross section, we will first show how the shear flow is related to the shear stress. To do this, consider the segment dx of the wide-flange beam in Fig. 7–19a. A free-body diagram of a portion of the flange is shown in Fig. 7–19b. The force dF is developed along the shaded longitudinal section in order to balance the normal forces F and $F + dF$ created by the moments M and $M + dM$, respectively. Since the segment has a length dx, then the shear flow or force per length along the section is $q = dF/dx$. Because the flange wall is *thin*, the shear stress τ will not vary much over the thickness t of the section; and so we will assume that it is *constant*. Hence, $dF = \tau dA = \tau(t\ dx) = q\ dx$, or

$$q = \tau t \qquad (7\text{–}7)$$

This same result can also be determined by comparing the shear-flow equation, $q = VQ/I$, with the shear formula, $\tau = VQ/It$.

Like the shear stress, the shear flow acts on both longitudinal and transverse planes. For example, if the corner element at point B in Fig. 7–19b is removed, Fig. 7–19c, the shear flow acts as shown on the side face of the element. Although it exists, we will *neglect* the vertical transverse component of shear flow, because as shown in Fig. 7–19d, this component, like the shear stress, is approximately zero throughout the thickness of the element. This is because the walls are assumed to be thin and the top and bottom surfaces of the element are free of stress. To summarize, then, only the shear-flow component that acts *parallel* to the walls of the member will be considered.

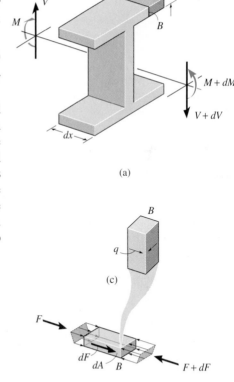

(a)

(c)

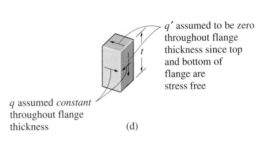

q' assumed to be zero throughout flange thickness since top and bottom of flange are stress free

q assumed *constant* throughout flange thickness

(d)

(b)

Fig. 7–19

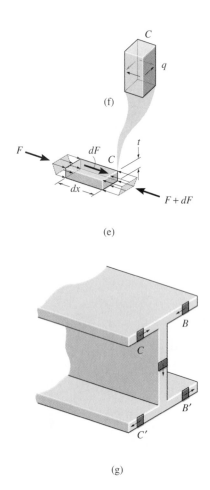

(f)

(e)

(g)

Fig. 7–19

By a similar analysis, isolation of the left-hand segment on the top flange, Fig. 7–19e, will establish the correct direction of the shear flow on the corner element C of the segment, Fig. 7–19f. Using this method, show that the shear flow at corresponding points B' and C' on the bottom flange is directed as shown in Fig. 7–19g.

This example illustrates how the *direction* of the shear flow can be established at any point on a beam's cross section. Using the shear-flow formula, $q = VQ/I$, we will now show how to determine the distribution of the shear flow throughout the cross section. It is to be expected that this formula will give reasonable results for the shear flow since, as stated in Sec. 7.3, the accuracy of this equation improves for members having thin rectangular cross sections. For any application, however, the shear force **V** must act along an axis of symmetry or principal centroidal axis of inertia for the cross section.

We will begin by finding the distribution of shear flow along the top right flange of the wide-flange beam in Fig. 7–20a. To do this, consider the shear flow q, acting on the colored element, located an arbitrary distance x from the centerline of the cross section, Fig. 7–20b. It is determined using Eq. 7–6 with $Q = \bar{y}'A' = [d/2](b/2 - x)t$. Thus,

$$q = \frac{VQ}{I} = \frac{V[d/2]((b/2) - x)t}{I} = \frac{Vtd}{2I}\left(\frac{b}{2} - x\right) \qquad (7\text{–}8)$$

By inspection, this distribution is *linear*, varying from $q = 0$ at $x = b/2$ to $(q_{max})_f = Vtdb/4I$ at $x = 0$. (The limitation of $x = 0$ is possible here since the member is assumed to have "thin walls" and so the thickness of the web is neglected.) Due to symmetry, a similar analysis yields the same distribution of shear flow for the other flanges, so that the results are as shown in Fig. 7–20d.

The total force developed in the left and right portions of a flange can be determined by integration. Since the force on the colored element in Fig. 7–20b is $dF = q\,dx$, then

$$F_f = \int q\,dx = \int_0^{b/2} \frac{Vtd}{2I}\left(\frac{b}{2} - x\right)dx = \frac{Vtdb^2}{16I}$$

Also, we can determine this result by finding the area under the triangle in Fig. 7–20d since q is a distribution of force/length. Hence,

$$F_f = \frac{1}{2}(q_{max})_f\left(\frac{b}{2}\right) = \frac{Vtdb^2}{16I}$$

All four of these flange forces are shown in Fig. 7–20e, and we can see from the direction of these forces that horizontal force equilibrium of the cross section is maintained.

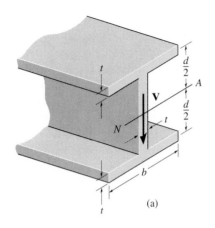

(a)

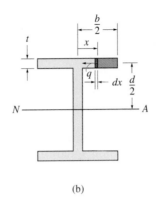

(b)

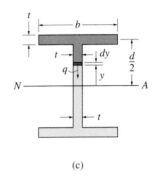

(c)

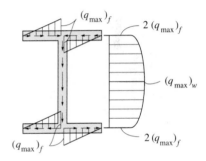

Shear-flow distribution

(d)

A similar analysis can be performed for the web, Fig. 7–20c. Here we have $Q = \Sigma \bar{y}'A' = [d/2](bt) + [y + (1/2)(d/2 - y)]t(d/2 - y) = btd/2 + (t/2)(d^2/4 - y^2)$, so that

$$q = \frac{VQ}{I} = \frac{Vt}{I}\left[\frac{db}{2} + \frac{1}{2}\left(\frac{d^2}{4} - y^2\right)\right] \qquad (7\text{–}9)$$

For the web, the shear flow varies in a *parabolic* manner, from $q = 2(q_{max})_f = Vtdb/2I$ at $y = d/2$ to a maximum of $q = (q_{max})_w = (Vtd/I)(b/2 + d/8)$ at $y = 0$, Fig. 7–20d.

In order to determine the force in the web, F_w, we must integrate Eq. 7–9, that is,

$$F_w = \int q \, dy = \int_{-d/2}^{d/2} \frac{Vt}{I}\left[\frac{db}{2} + \frac{1}{2}\left(\frac{d^2}{4} - y^2\right)\right]dy$$

$$= \frac{Vt}{I}\left[\frac{db}{2}y + \frac{1}{2}\left(\frac{d^2}{4}y - \frac{1}{3}y^3\right)\right]\Bigg|_{-d/2}^{d/2}$$

$$= \frac{Vtd^2}{4I}\left(2b + \frac{1}{3}d\right)$$

Simplification is possible by noting that the moment of inertia for the cross-sectional area is

$$I = 2\left[\frac{1}{12}bt^3 + bt\left(\frac{d}{2}\right)^2\right] + \frac{1}{12}td^3$$

Neglecting the first term, since the thickness of each flange is small, we get

$$I = \frac{td^2}{4}\left(2b + \frac{1}{3}d\right)$$

Substituting into the above equation, we see that $F_w = V$, which is to be expected, Fig. 7–20e.

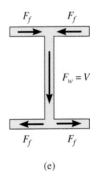

(e)

Fig. 7–20

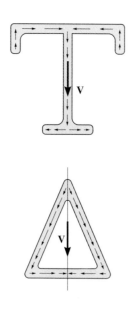

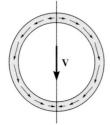

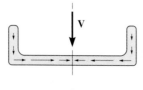

Shear flow q

Fig. 7–21

From the foregoing analysis three important points should be observed. First, the value of q changes over the cross section, since Q will be different for each area segment A' for which it is determined. In particular, q will vary *linearly* along segments (flanges) that are *perpendicular* to the direction of **V**, and *parabolically* along segments (web) that are *inclined or parallel* to **V**. Second, q will *always act parallel to the walls* of the member, since the section on which q is calculated is taken perpendicular to the walls. And third, the *directional sense* of q is such that the shear appears to "*flow*" through the cross section, *inward* at the beam's top flange, "combining" and then "flowing" *downward* through the web, since it must contribute to the shear force **V**, and then separating and "flowing" *outward* at the bottom flange. If one is able to "visualize" this "flow" it will provide an easy means for establishing not only the direction of q, but *also* the corresponding direction of τ. Other examples of how q is directed along the segments of thin-walled members are shown in Fig. 7–21. In all cases, symmetry prevails about an axis that is collinear with **V**, and as a result, q "flows" in a direction such that it will provide the necessary vertical force components equivalent to **V** and yet also satisfy the horizontal force equilibrium requirements for the cross section.

IMPORTANT POINTS

- If a member is made from segments having thin walls, only the shear flow *parallel* to the walls of the member is important.

- The shear flow varies *linearly* along segments that are *perpendicular* to the direction of the shear **V**.

- The shear flow varies *parabolically* along segments that are *inclined* or *parallel* to the direction of the shear **V**.

- On the cross section, the shear "flows" along the segments so that it contributes to the shear **V** yet satisfies horizontal and vertical force equilibrium.

EXAMPLE 7-7

The thin-walled box beam in Fig. 7–22a is subjected to a shear of 10 kip. Determine the variation of the shear flow throughout the cross section.

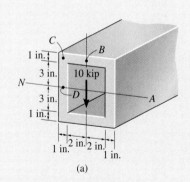

(a)

SOLUTION

By symmetry, the neutral axis passes through the center of the cross section. The moment of inertia is

$$I = \frac{1}{12}(6 \text{ in.})(8 \text{ in.})^3 - \frac{1}{12}(4 \text{ in.})(6 \text{ in.})^3 = 184 \text{ in}^4$$

Only the shear flow at points B, C, and D has to be determined. For point B, the area $A' \approx 0$, Fig. 7–22b, since it can be thought of as being located entirely at point B. Alternatively, A' can also represent the *entire* cross-sectional area, in which case $Q_B = \bar{y}'A' = 0$ since $\bar{y}' = 0$. Because $Q_B = 0$, then

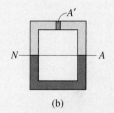

(b)

$$q_B = 0$$

For point C, the area A' is shown dark shaded in Fig. 7–22c. Here we have used the mean dimensions since point C is on the centerline of each segment. We have

$$Q_C = \bar{y}'A' = (3.5 \text{ in.})(5 \text{ in.})(1 \text{ in.}) = 17.5 \text{ in}^3$$

Thus,

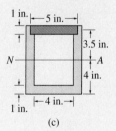

(c)

$$q_C = \frac{VQ_C}{I} = \frac{10 \text{ kip}(17.5 \text{ in}^3/2)}{184 \text{ in}^4} = 0.951 \text{ kip/in.}$$

The shear flow at D is computed using the three dark-shaded rectangles shown in Fig. 7–22d. We have

$$Q_D = \Sigma\bar{y}'A' = 2[2 \text{ in.}](1 \text{ in.})(4 \text{ in.}) + [3.5 \text{ in.}](4 \text{ in.})(1 \text{ in.}) = 30 \text{ in}^3$$

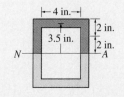

(d)

So that

$$q_D = \frac{VQ_D}{I} = \frac{10 \text{ kip}(30 \text{ in}^3/2)}{184 \text{ in}^4} = 1.63 \text{ kip/in.}$$

Using these results, and the symmetry of the cross section, the shear-flow distribution is plotted in Fig. 7–22e. As expected the distribution is linear along the horizontal segments (perpendicular to **V**) and parabolic along the vertical segments (parallel to **V**).

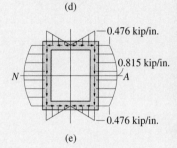

(e)

Fig. 7–22

*7.6   SHEAR CENTER

In the previous section it was assumed that the internal shear **V** was applied along a principal centroidal axis of inertia that *also* represents an *axis of symmetry* for the cross section. In this section we will consider the effect of applying the shear along a principal centroidal axis that is *not* an axis of symmetry. As before, only thin-walled members will be analyzed, so the dimensions to the centerline of the walls of the members will be used. A typical example of this case is the channel section shown in Fig. 7–23*a*, which is cantilevered from a fixed support and is subjected to the force **P**. If this force is applied along the once vertical, unsymmetrical axis that passes through the *centroid C* of the cross-sectional area, the channel will not only bend downward, *it will also twist clockwise as shown.*

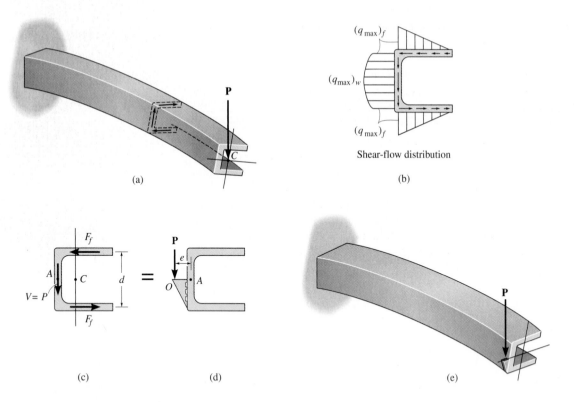

Shear-flow distribution

Fig. 7–23

In order to understand why the member twists, it is necessary to study the shear-flow distribution along the channel's flanges and web, Fig. 7–23b. When this distribution is integrated over the flange and web areas, it will give resultant forces of F_f in each flange and a force of $V = P$ in the web, Fig. 7–23c. If the moments of these forces are summed about point A, it can be seen that the couple or torque created by the flange forces is responsible for causing the member to twist. The actual twist is clockwise when viewed from the front of the beam as shown in Fig. 7–23a, since *reactive* internal "equilibrium" forces F_f cause the twisting. In order to *prevent* this twisting it is therefore necessary to apply **P** at a point O located a distance e from the web of the channel, Fig. 7–23d. We require $\Sigma M_A = F_f d = Pe$, or

$$e = \frac{F_f d}{P}$$

Using the method discussed in Sec. 7.5, F_f can be evaluated in terms of $P (= V)$ and the dimensions of the flanges and web. Once this is done, then P will cancel upon substitution into the above equation, and it becomes possible to express e simply as a function of the cross-sectional geometry and *not* as a function of P or its location along the length of the beam (see Example 7–9). The point O so located is called the **shear center** or **flexural center**. When **P** is applied at the shear center, the **beam will bend without twisting** as shown in Fig. 7–23e. Design handbooks often list the location of this point for a variety of beams having thin-walled cross sections that are commonly used in practice.

When doing this analysis it should be noted that **the shear center will always lie on an axis of symmetry** of a member's cross-sectional area. For example, if the channel in Fig. 7–23a is rotated 90° and **P** is applied at A, Fig. 7–24a, no twisting will occur since the shear flow in the web and flanges for this case is *symmetrical*, and therefore the force resultants in these elements will create zero moments about A, Fig. 7–24b. Obviously, if a member has a cross section with *two* axes of symmetry, as in the case of a wide-flange beam, the shear center will then coincide with the intersection of these axes (the centroid).

Demonstration of how a cantilever beam deflects when loaded through the centroid (above) and through the shear center (below).

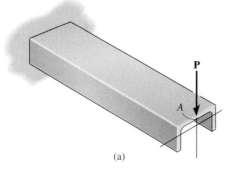

(a)

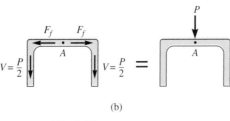

(b)

Fig. 7–24

IMPORTANT POINTS

- The *shear center* is the point through which a force can be applied which will cause a beam to bend and yet not twist.

- The shear center will always lie on an axis of symmetry of the cross section.

- The location of the shear center is only a function of the geometry of the cross section and does not depend upon the applied loading.

PROCEDURE FOR ANALYSIS

The location of the shear center for a thin-walled member for which the internal shear is in the *same direction* as a principal centroidal axis for the cross section may be determined by using the following procedure.

Shear-Flow Resultants.

- Determine the direction as to how the shear "flows" through the various segments of the cross section, and sketch the force resultants on each segment of the cross section. (For example, see Fig. 7–23c.) Since the shear center is determined by taking the moments of these force resultants about a point (A), choose this point at a location that eliminates the moments of as many force resultants as possible.

- The magnitudes of the force resultants that create a moment about A must be calculated. For any segment this is done by determining the shear flow q at an arbitrary point on the segment and then integrating q along the segment's length. Note that **V** will create a *linear* variation of shear flow in segments that are *perpendicular* to **V**, and a *parabolic* variation of shear flow in segments that are *parallel or inclined* to **V**.

Shear Center.

- Sum the moments of the shear-flow resultants about point A and set this moment equal to the moment of **V** about A. By solving this equation one is able to determine the moment-arm distance e, which locates the line of action of **V** from A.

- If an *axis of symmetry* for the cross section exists, the shear center lies at the point where this axis intersects the line of action of **V**. If, however, no axes of symmetry exist, rotate the cross section 90° and repeat the process to obtain another line of action for **V**. The shear center then lies at the point of intersection of the two 90° lines.

EXAMPLE 7-9

Determine the location of the shear center for the thin-walled channel section having the dimensions shown in Fig. 7–25a.

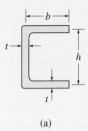

(a)

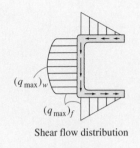

$(q_{max})_w$

$(q_{max})_f$

Shear flow distribution

(b)

SOLUTION

Shear-Flow Resultants.   A vertical downward shear **V** applied to the section causes the shear to flow through the flanges and web as shown in Fig. 7–25b. This causes force resultants F_f and V in the flanges and web as shown in Fig. 7–25c. We will take moments about point A so that only the force F_f on the lower flange has to be determined.

The cross-sectional area can be divided into three component rectangles—a web and two flanges. Since each component is assumed to be thin, the moment of inertia of the area about the neutral axis is

$$I = \frac{1}{12}th^3 + 2\left[bt\left(\frac{h}{2}\right)^2\right] = \frac{th^2}{2}\left(\frac{h}{6} + b\right)$$

From Fig. 7–25d, q at the arbitrary position x is

$$q = \frac{VQ}{I} = \frac{V(h/2)[b - x]t}{(th^2/2)[(h/6) + b]} = \frac{V(b - x)}{h[(h/6) + b]}$$

Hence, the force F_f is

$$F_f = \int_0^b q \, dx = \frac{V}{h[(h/6) + b]} \int_0^b (b - x) \, dx = \frac{Vb^2}{2h[(h/6) + b]}$$

This same result can also be determined by first finding $(q_{max})_f$, Fig. 7–25b, then determining the triangular area $\frac{1}{2}b(q_{max})_f = F_f$.

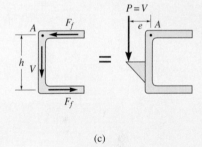

$P = V$

e A

(c)

Shear Center.   Summing moments about point A, Fig. 7–25c, we require

$$Ve = F_f h = \frac{Vb^2h}{2h[(h/6) + b]}$$

Thus

$$e = \frac{b^2}{[(h/3) + 2b]} \qquad Ans.$$

As stated previously, e depends only on the geometry of the cross section.

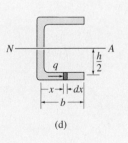

N ——————— A

q

$\frac{h}{2}$

x dx

b

(d)

Fig. 7–25

EXAMPLE 7–10

Determine the location of the shear center for the angle having equal legs, Fig. 7–26a. Also, calculate the internal shear force resultant in each leg.

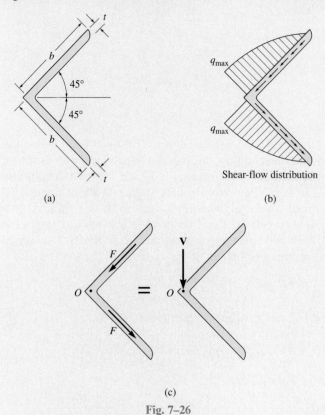

(a)

(b)

Shear-flow distribution

(c)

Fig. 7–26

SOLUTION

When a vertical downward shear $\mathbf{V}$ is applied at the section, the shear flow and shear-flow resultants are directed as shown in Fig. 7–26b and 7–26c, respectively. Note that the force F in each leg must be equal, since for equilibrium the sum of their horizontal components must be equal to zero. Also, the lines of action of both forces intersect point O; therefore, this point *must be the shear center* since the sum of the moments of these forces and $\mathbf{V}$ about O is zero, Fig. 7–26c.

The magnitude of $\mathbf{F}$ can be determined by first finding the shear flow at the arbitrary location s along the top leg, Fig. 7–26d. Here

$$Q = \bar{y}'A' = \frac{1}{\sqrt{2}}\left((b-s) + \frac{s}{2}\right)ts = \frac{1}{\sqrt{2}}\left(b - \frac{s}{2}\right)st$$

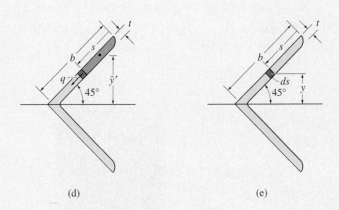

(d)   (e)

The moment of inertia of the angle, computed about the neutral axis, must be determined from "first principles," since the legs are inclined with respect to the neutral axis. For the area element $dA = t\,ds$, Fig. 7–26e, we have

$$I = \int_A y^2\, dA = 2\int_0^b \left[\frac{1}{\sqrt{2}}(b-s)\right]^2 t\, ds = t\left(b^2 s - bs^2 + \frac{1}{3}s^3\right)\Big|_0^b = \frac{tb^3}{3}$$

Thus, the shear flow is

$$q = \frac{VQ}{I} = \frac{V}{(tb^3/3)}\left[\frac{1}{\sqrt{2}}\left(b - \frac{s}{2}\right)st\right]$$

$$= \frac{3V}{\sqrt{2}b^3}\, s\left(b - \frac{s}{2}\right)$$

The variation of q is parabolic, and it reaches a maximum value when $s = b$ as shown in Fig. 7–26b. The force F is therefore

$$F = \int_0^b q\, ds = \frac{3V}{\sqrt{2}b^3}\int_0^b s\left(b - \frac{s}{2}\right) ds$$

$$= \frac{3V}{\sqrt{2}b^3}\left(b\frac{s^2}{2} - \frac{1}{6}s^3\right)\Big|_0^b$$

$$= \frac{1}{\sqrt{2}}V \qquad\qquad\qquad \textit{Ans.}$$

This result can be easily verified since the sum of the vertical components of the force F in each leg must equal V and, as stated previously, the sum of the horizontal components equals zero.

PROBLEMS

7–51. The assembly is subjected to a vertical shear of $V = 7$ kip. Determine the shear flow at points A and B and the maximum shear flow in the cross section.

7–53. A shear force of $V = 18$ kN is applied to the symmetric box girder. Determine the shear flow at A and B.

7–54. A shear force of $V = 18$ kN is applied to the box girder. Determine the shear flow at C.

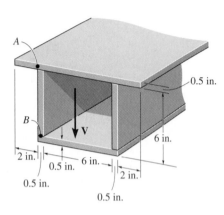

Prob. 7–51

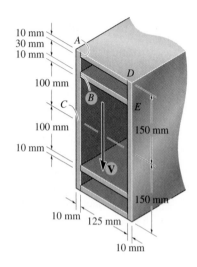

Probs. 7–53/7–54

*7–52. The beam supports a vertical shear of $V = 7$ kip. Determine the resultant force developed in segment AB of the beam.

7–55. The box girder is subjected to a shear of $V = 15$ kN. Determine (*a*) the shear flow developed at point B and (*b*) the maximum shear flow in the girder's web AB.

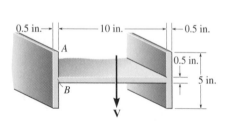

Prob. 7–52

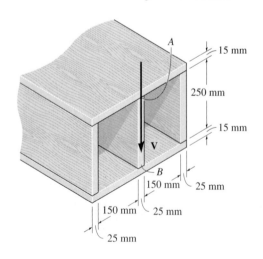

Prob. 7–55

*7–56. The beam is made from three thin plates welded together as shown. If it is subjected to a shear of $V = 48$ kN, determine the shear flow at points A and B. Also, calculate the maximum shear stress in the beam.

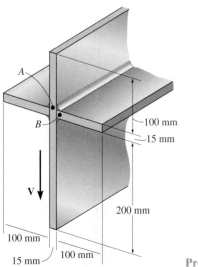

Prob. 7–56

7–57. The pipe is subjected to a shear force of $V = 8$ kip. Determine the shear flow in the pipe at points A and B.

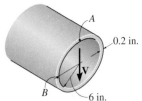

Prob. 7–57

7–58. The square tube is subjected to a shear force of $V = 50$ kN. Determine the maximum shear flow in the tube. The tube is 10 mm thick.

7–59. The square tube is subjected to a shear force of $V = 50$ kN. Determine the shear flow in the tube at point A. The tube is 10 mm thick.

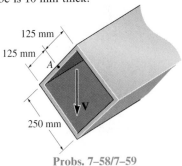

Probs. 7–58/7–59

*7–60. The member is subjected to a shear force of $V = 10$ kip. Sketch the shear-flow distribution along the vertical plate AB. Indicate numerical values of all peaks.

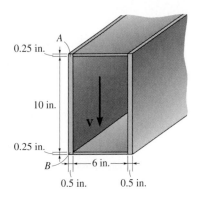

Prob. 7–60

7–61. Determine the location e of the shear center, point O, for the thin-walled member having the cross section shown. The member segments have the same thickness t.

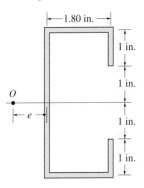

Prob. 7–61

7–62. Determine the location e of the shear center, point O, for the thin-walled member having the cross section shown. The member segments have the same thickness t.

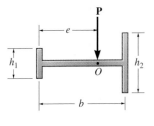

Prob. 7–62

7–63. Determine the location e of the shear center, point O, for the thin-walled member having the cross section shown. The member segments have the same thickness t.

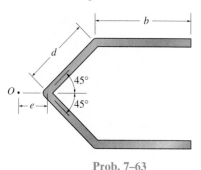

Prob. 7–63

7–66. Determine the location e of the shear center, point O, for the thin-walled member having the cross-sectional area shown, where $b_2 > b_1$. The member segments have the same thickness t.

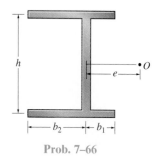

Prob. 7–66

*7–64. Determine the location e of the shear center, point O, for the thin-walled member having the cross section shown. The member segments have the same thickness t.

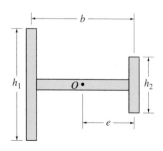

Prob. 7–64

7–67. Determine the location e of the shear center, point O, for the thin-walled member having the cross section shown. The member segments have the same thickness t.

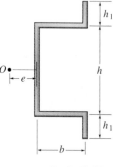

Prob. 7–67

7–65. Determine the location e of the shear center, point O, for the thin-walled member having the cross section shown. The member segments have the same thickness t.

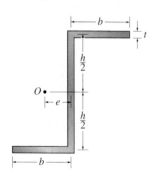

Prob. 7–65

*7–68. Determine the location e of the shear center, point O, for the thin-walled member having the cross section shown. The member segments have the same thickness t.

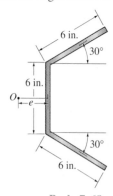

Prob. 7–68

7–69. Determine the location e of the shear center, point O, for the thin-walled member having the cross section shown. The member segments have the same thickness t.

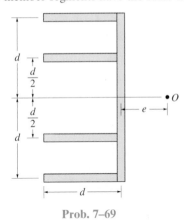

Prob. 7–69

7–70. Determine the location e of the shear center, point O, for the thin-walled member having a slit along its side. Each element has a constant thickness t.

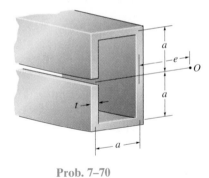

Prob. 7–70

7–71. Determine the location e of the shear center, point O, for the thin-walled member having the cross section shown.

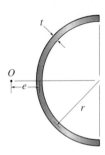

Prob. 7–71

*7–72. Determine the location e of the shear center, point O, for the thin-walled member having the cross section shown.

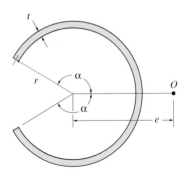

Prob. 7–72

7–73. Determine the location e of the shear center, point O, for the tube having a slit along its length.

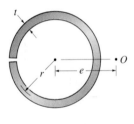

Prob. 7–73

7–74. Determine the location e of the shear center, point O, for the thin-walled tube having the cross section shown. The member segments have the same thickness t.

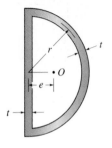

Prob. 7–74

REVIEW PROBLEMS

7–75. Determine the maximum shear stress acting at section *a–a* in the beam.

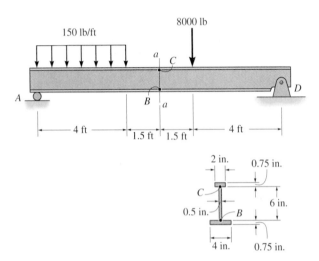

Prob. 7–75

7–77. The beam is constructed from four boards glued together at their seams. If the glue can withstand 75 lb/in., what is the maximum vertical shear *V* that the beam can support?

7–78. Solve Prob. 7–77 if the beam is rotated 90° from the position shown.

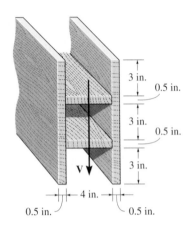

Probs. 7–77/7–78

*7–76. The beam is subjected to a shear of $V = 25$ kN. Determine the shear stress at points *A* and *B* and compute the maximum shear stress in the beam. Assume the gap at *C* is closed so that the center plate is fixed to the top plate.

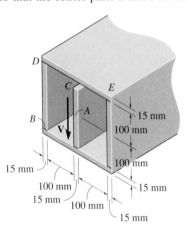

Prob. 7–76

7–79. The member is subjected to a shear force of $V = 2$ kN. Determine the shear flow at points *A*, *B*, and *C*. The thickness of each thin-walled segment is 15 mm.

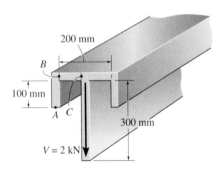

Prob. 7–79

*7–80.   Determine the location e of the shear center, point O, for the beam having the cross section shown. The thickness is t.

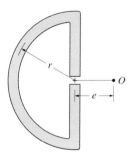

Prob. 7–80

7–81.   If the beam is subjected to a shear of $V = 15$ kN, determine the web's shear stress at A and B. Indicate the shear-stress components on a volume element located at these points.

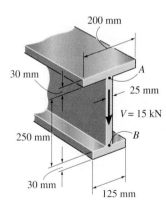

Prob. 7–81

7–82.   A steel plate having a thickness of 0.25 in. is formed into the thin-walled section shown. If it is subjected to a shear force of $V = 250$ lb, determine the shear stress at points A and C. Indicate the results on volume elements located at these points.

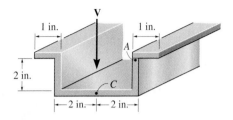

Prob. 7–82

7–83.   A steel plate having a thickness of 0.25 in. is formed into the thin-walled section shown. If it is subjected to a shear force of $V = 250$ lb, determine the shear stress at point B.

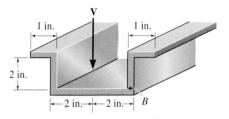

Prob. 7–83

*7–84.   A steel plate having a thickness of 0.25 in. is formed into the thin-walled section shown. If it is subjected to a shear force of $V = 250$ lb, determine the maximum shear stress in the plate.

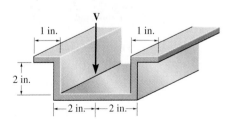

Prob. 7–84

The offset column supporting this sign is subjected to the combined loadings of normal force, shear force, bending moment, and torsion.

8 COMBINED LOADINGS

CHAPTER OBJECTIVES

This chapter serves as a review of the stress analysis that has been developed in the previous chapters regarding axial load, torsion, bending, and shear. We will discuss the solution of problems where several of these internal loads occur simultaneously on a member's cross section. Before doing this, however, the chapter begins with an analysis of stress developed in thin-walled pressure vessels.

8.1 THIN-WALLED PRESSURE VESSELS

Cylindrical or spherical vessels are commonly used in industry to serve as boilers or tanks. When under pressure, the material of which they are made is subjected to a loading from all directions. Although this is the case, the vessel can be analyzed in a simpler manner provided it has a thin wall. In general, "***thin wall***" refers to a vessel having an inner-radius-to-wall-thickness ratio of 10 or more ($r/t \geq 10$). Specifically, when $r/t = 10$ the results of a thin-wall analysis will predict a stress that is approximately 4% *less* than the actual maximum stress in the vessel. For larger r/t ratios this error will be even smaller.

When the vessel wall is "thin," the stress distribution throughout its thickness will not vary significantly, and so we will assume that it is *uniform* or *constant*. Using this assumption, we will now analyze the state of stress in thin-walled cylindrical and spherical pressure vessels. In both cases, the pressure in the vessel is understood to be the *gauge pressure*, since it measures the pressure *above* atmospheric pressure, which is assumed to exist both inside and outside the vessel's wall.

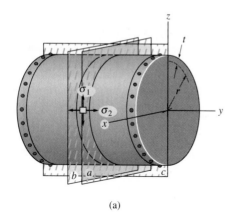

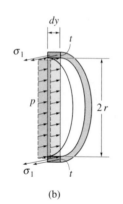

(a)

Cylindrical Vessels. Consider the cylindrical vessel having a wall thickness t and inner radius r as shown in Fig. 8–1a. A gauge pressure p is developed within the vessel by a contained gas or fluid, which is assumed to have negligible weight. Due to the uniformity of this loading, an element of the vessel that is sufficiently removed from the ends and oriented, as shown, is subjected to normal stresses σ_1 in the **circumferential or hoop direction** and σ_2 in the **longitudinal or axial direction**. Both of these stress components exert tension on the material. We wish to determine the magnitude of each of these components in terms of the vessel's geometry and the internal pressure. To do this requires using the method of sections and applying the equations of force equilibrium.

For the hoop stress, consider the vessel to be sectioned by planes a, b, and c. A free-body diagram of the back segment along with the contained gas or fluid is shown in Fig. 8–1b. Here only the loadings in the x direction are shown. These loadings are developed by the uniform hoop stress σ_1, acting throughout the vessel's wall, and the pressure acting on the vertical face of the sectioned gas or fluid. For equilibrium in the x direction, we require

$$\Sigma F_x = 0; \qquad 2[\sigma_1(t\,dy)] - p(2r\,dy) = 0$$

$$\boxed{\sigma_1 = \frac{pr}{t}} \qquad (8\text{–}1)$$

In order to obtain the longitudinal stress σ_2, we will consider the left portion of section b of the cylinder, Fig. 8–1a. As shown in Fig. 8–1c, σ_2 acts uniformly throughout the wall, and p acts on the section of gas or fluid. Since the mean radius is approximately equal to the vessel's inner radius, equilibrium in the y direction requires

$$\Sigma F_y = 0; \qquad \sigma_2(2\pi r t) - p(\pi r^2) = 0$$

$$\boxed{\sigma_2 = \frac{pr}{2t}} \qquad (8\text{–}2)$$

In the above equations,

$\sigma_1, \sigma_2 =$ the normal stress in the hoop and longitudinal directions, respectively. Each is assumed to be *constant* throughout the wall of the cylinder, and each subjects the material to tension

$p =$ the internal gauge pressure developed by the contained gas or fluid

$r =$ the inner radius of the cylinder

$t =$ the thickness of the wall ($r/t \geq 10$)

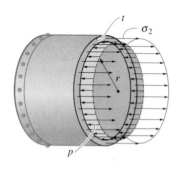

(b)

(c)

Fig. 8–1

Comparing Eqs. 8–1 and 8–2, it should be noted that the hoop or circumferential stress is twice as large as the longitudinal or axial stress. Consequently, when fabricating cylindrical pressure vessels from rolled-formed plates, the longitudinal joints must be designed to carry twice as much stress as the circumferential joints.

Spherical Vessels. We can analyze a spherical pressure vessel in a similar manner. For example, consider the vessel to have a wall thickness t and inner radius r and to be subjected to an internal gauge pressure p, Fig. 8–2a. If the vessel is sectioned in half using section a, the resulting free-body diagram is shown in Fig. 8–2b. Like the cylinder, equilibrium in the y direction requires

$$\Sigma F_y = 0; \qquad \sigma_2(2\pi rt) - p(\pi r^2) = 0$$

$$\sigma_2 = \frac{pr}{2t} \tag{8–3}$$

By comparison, this is the *same result* as that obtained for the longitudinal stress in the cylindrical pressure vessel. Furthermore, from the analysis, this stress will be the same *regardless* of the orientation of the hemispheric free-body diagram. Consequently, an element of the material is subjected to the state of stress shown in Fig. 8–2a.

The above analysis indicates that an element of material taken from either a cylindrical or a spherical pressure vessel is subjected to **biaxial stress**, i.e., normal stress existing in only two directions. Actually, material of the vessel is also subjected to a **radial stress**, σ_3, which acts along a radial line. This stress has a maximum value equal to the pressure p at the interior wall and decreases through the wall to zero at the exterior surface of the vessel, since the gauge pressure there is zero. For thin-walled vessels, however, we will *ignore* the radial-stress component, since our limiting assumption of $r/t = 10$ results in σ_2 and σ_1 being, respectively, 5 and 10 times *higher* than the maximum radial stress, $(\sigma_3)_{max} = p$. Lastly, realize that the above formulas should be used only for vessels subjected to an internal gauge pressure. If the vessel is subjected to an external pressure, the compressive stress developed within the thin wall may cause the vessel to become unstable, and collapse may occur by buckling.

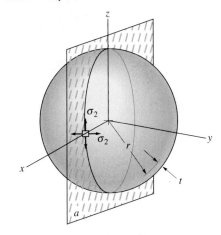

Shown is the barrel of a shotgun which was clogged with debris just before firing. Gas pressure from the charge increased the circumferential stress within the barrel enough to cause the rupture.

(a)

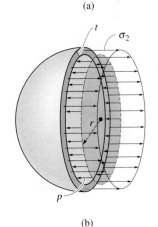

(b)

Fig. 8–2

EXAMPLE 8-1

A cylindrical pressure vessel has an inner diameter of 4 ft and a thickness of $\frac{1}{2}$ in. Determine the maximum internal pressure it can sustain so that neither its circumferential nor its longitudinal stress component exceeds 20 ksi. Under the same conditions, what is the maximum internal pressure that a similar-size spherical vessel can sustain?

SOLUTION

Cylindrical Pressure Vessel. The maximum stress occurs in the circumferential direction. From Eq. 8–1 we have

$$\sigma_1 = \frac{pr}{t}; \qquad 20 \text{ kip/in}^2 = \frac{p(24 \text{ in.})}{\frac{1}{2} \text{ in.}}$$

$$p = 417 \text{ psi} \qquad\qquad Ans.$$

Note that when this pressure is reached, from Eq. 8–2, the stress in the longitudinal direction will be $\sigma_2 = \frac{1}{2}(20 \text{ ksi}) = 10 \text{ ksi}$. Furthermore, the *maximum stress* in the *radial direction* occurs on the material at the inner wall of the vessel and is $(\sigma_3)_{\max} = p = 417$ psi. This value is 48 times smaller than the circumferential stress (20 ksi), and as stated earlier, its effects will be neglected.

Spherical Vessel. Here the maximum stress occurs in any two perpendicular directions on an element of the vessel, Fig. 8–2a. From Eq. 8–3, we have

$$\sigma_2 = \frac{pr}{2t}; \qquad 20 \text{ kip/in}^2 = \frac{p(24 \text{ in.})}{2(\frac{1}{2} \text{ in.})}$$

$$p = 833 \text{ psi} \qquad\qquad Ans.$$

Although it is more difficult to fabricate, the spherical pressure vessel will carry twice as much internal pressure as a cylindrical vessel.

PROBLEMS

8–1. A spherical gas tank has an inner radius of $r = 1.5$ m. If it is subjected to an internal pressure of $p = 300$ kPa, determine its required thickness if the maximum normal stress is not to exceed 12 MPa.

8–2. The open-ended polyvinyl chloride pipe has an inner diameter of 4 in. and thickness of 0.2 in. If it carries flowing water at 60 psi pressure, determine the state of stress in the walls of the pipe.

8–3. If the flow of water within the pipe in Prob. 8–2 is stopped due to the closing of a valve, determine the state of stress in the walls of the pipe. Neglect the weight of the water. Assume the supports only exert vertical forces on the pipe.

Probs. 8–2/8–3

*8–4.** The open-ended pipe has a wall thickness of 2 mm and an internal diameter of 40 mm. Calculate the pressure that ice exerted on the interior wall of the pipe to cause it to burst in the manner shown. The maximum stress that the material can support at freezing temperatures is $\sigma_{max} = 360$ MPa. Show the stress acting on a small element of material just before the pipe fails.

Prob. 8–4

8–5. Two hemispheres having an inner radius of 2 ft and wall thickness of 0.25 in. are fitted together, and the inside gauge pressure is reduced to -10 psi. If the coefficient of static friction is $\mu_s = 0.5$ between the hemispheres, determine (a) the torque T needed to initiate the rotation of the top hemisphere relative to the bottom one, (b) the vertical force needed to pull the top hemisphere off the bottom one, and (c) the horizontal force needed to slide the top hemisphere off the bottom one.

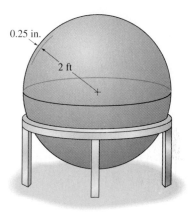

Prob. 8–5

8–6. The 304 stainless steel band initially fits snugly around the smooth rigid cylinder. If the band is then subjected to a nonlinear temperature drop of $\Delta T = 20 \sin^2 \theta$ °F, where θ is in radians, determine the circumferential stress in the band.

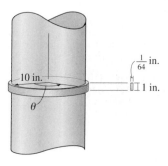

Prob. 8–6

8–7. The A-36-steel band is 2 in. wide and is secured around the smooth rigid cylinder. If the bolts are tightened so that the tension in them is 400 lb, determine the normal stress in the band, the pressure exerted on the cylinder, and the distance half the band stretches.

8–9. A wood pipe having an inner diameter of 3 ft is bound together using steel hoops having a cross-sectional area of 0.2 in². If the allowable stress for the hoops is $\sigma_{\text{allow}} = 12$ ksi, determine their maximum spacing s along the section of pipe so that the pipe can resist an internal gauge pressure of 4 psi. Assume each hoop supports the pressure loading acting along the length s of the pipe.

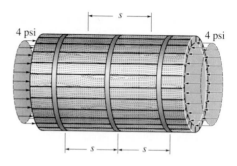

Prob. 8–9

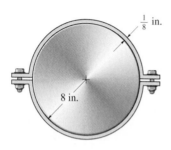

Prob. 8–7

8–10. An A-36-steel hoop has an inner diameter of 23.99 in., thickness of 0.25 in., and width of 1 in. If it and the 24-in.-diameter rigid cylinder have a temperature of 65°F, determine the temperature to which the hoop should be heated in order for it to just slip over the cylinder. What is the pressure the hoop exerts on the cylinder, and the tensile stress in the ring when it cools back down to 65°F?

*8–8. A pressure-vessel head is fabricated by gluing the circular plate to the end of the vessel as shown. If the vessel sustains an internal pressure of 450 kPa, determine the average shear stress in the glue and the state of stress in the wall of the vessel.

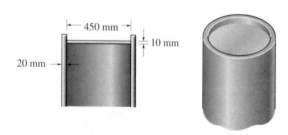

Prob. 8–8

Prob. 8–10

8–11. The inner ring A has an inner radius r_1 and outer radius r_2. Before heating, the outer ring B has an inner radius r_3 and an outer radius r_4 and $r_2 > r_3$. If the outer ring is heated and then fitted over the inner ring, determine the pressure between the two rings when ring B reaches the temperature of the inner ring. The material has a modulus of elasticity of E and a coefficient of thermal expansion of α.

8–13. A boiler is constructed of 8-mm steel plates that are fastened together at their ends using a butt joint consisting of two 8-mm cover plates and rivets having a diameter of 10 mm and spaced 50 mm apart as shown. If the steam pressure in the boiler is 1.35 MPa, determine (*a*) the circumferential stress in the boiler's plate apart from the seam, (*b*) the circumferential stress in the outer cover plate along the rivet line *a–a*, and (*c*) the shear stress in the rivets.

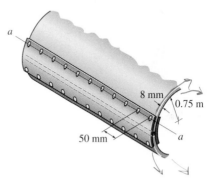

Prob. 8–13

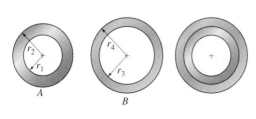

A

B

Prob. 8–11

8–14. A closed-ended pressure vessel is fabricated by cross winding glass filaments over a mandrel, so that the wall thickness t of the vessel is composed entirely of filament and an epoxy binder as shown. Consider a segment of the vessel of width w and wrapped at an angle θ. If the vessel is subjected to an internal pressure p, show that the force in the segment is $F_\theta = \sigma_0 wt$, where σ_0 is the stress in the filaments. Also, show that the stresses in the hoop and longitudinal directions are $\sigma_h = \sigma_0 \sin^2\theta$ and $\sigma_l = \sigma_0 \cos^2\theta$, respectively. At what angle θ (optimum winding angle) would the filaments have to be wound so that the hoop and longitudinal stresses are equivalent?

*8–12. The ring, having the dimensions shown, is placed over a flexible membrane which is pumped up with a pressure p. Determine the change in the internal radius of the ring after this pressure is applied. The modulus of elasticity for the ring is E.

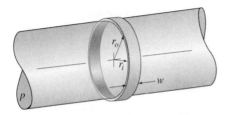

Prob. 8–12

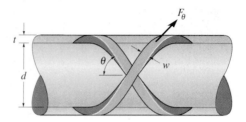

Prob. 8–14

8.2   STATE OF STRESS CAUSED BY COMBINED LOADINGS

In previous chapters we developed methods for determining the stress distributions in a member subjected to either an internal axial force, a shear force, a bending moment, or a torsional moment. Most often, however, the cross section of a member is subjected to several of these types of loadings *simultaneously*, and as a result, the method of superposition, if it applies, can be used to determine the *resultant* stress distribution caused by the loads. For application, the stress distribution due to *each loading* is first determined, and then these distributions are superimposed to determine the resultant stress distribution. As stated in Sec. 4.3, the principle of superposition can be used for this purpose provided a *linear relationship* exists between the *stress* and the *loads*. Also, the geometry of the member should *not* undergo *significant change* when the loads are applied. This is necessary in order to ensure that the stress produced by one load is not related to the stress produced by any other load. The discussion will be confined to meet these two criteria.

This chimney is subjected to the combined loading of wind and weight. It is important to investigate the tensile stress in the chimney since masonry is weak in tension.

PROCEDURE FOR ANALYSIS

The following procedure provides a general means for establishing the normal and shear stress components at a point in a member when the member is subjected to several different types of loadings simultaneously. It is assumed that the material is homogeneous and behaves in a linear-elastic manner. Also, Saint-Venant's principle requires that the point where the stress is to be determined is far removed from any discontinuities in the cross section or points of applied load.

Internal Loading.

• Section the member perpendicular to its axis at the point where the stress is to be determined and obtain the resultant internal normal and shear force components and the bending and torsional moment components.

• The force components should act through the *centroid* of the cross section, and the moment components should be computed about *centroidal axes*, which represent the principal axes of inertia for the cross section.

Average Normal Stress.

• Compute the stress component associated with *each* internal loading. For each case, represent the effect either as a distribution of stress acting over the entire cross-sectional area, or show the stress on an element of the material located at a specified point on the cross section.

NORMAL FORCE. The internal normal force is developed by a uniform normal-stress distribution determined from $\sigma = P/A$.

SHEAR FORCE. The internal shear force in a member that is subjected to bending is developed by a shear-stress distribution determined from the shear formula, $\tau = VQ/It$. Special care, however, must be exercised when applying this equation, as noted in Sec. 7.3.

BENDING MOMENT. For *straight members* the internal bending moment is developed by a normal-stress distribution that varies linearly from zero at the neutral axis to a maximum at the outer boundary of the member. The stress distribution is determined from the flexure formula, $\sigma = -My/I$. If the member is *curved*, the stress distribution is nonlinear and is determined from $\sigma = My/[Ae(R - y)]$.

TORSIONAL MOMENT. For circular shafts and tubes the internal torsional moment is developed by a shear-stress distribution that varies linearly from the central axis of the shaft to a maximum at the shaft's outer boundary. The shear-stress distribution is determined from the torsional formula, $\tau = T\rho/J$. If the member is a closed thin-walled tube, use $\tau = T/2A_m t$.

THIN-WALLED PRESSURE VESSELS. If the vessel is a thin-walled cylinder, the internal pressure p will cause a biaxial state of stress in the material such that the hoop or circumferential stress component is $\sigma_1 = pr/t$ and the longitudinal stress component is $\sigma_2 = pr/2t$.. If the vessel is a thin-walled sphere, then the biaxial state of stress is represented by two equivalent components, each having a magnitude of $\sigma_2 = pr/2t$.

Superposition.

• Once the normal and shear stress components for each loading have been calculated, use the principle of superposition and determine the resultant normal and shear stress components.

• Represent the results on an element of material located at the point, or show the results as a distribution of stress acting over the member's cross-sectional area.

Problems in this section, which involve combined loadings, serve as a basic *review* of the application of many of the important stress equations mentioned above. A thorough understanding of how these equations are applied, as indicated in the previous chapters, is necessary if one is to successfully solve the problems at the end of this section. The following examples should be carefully studied before proceeding to solve the problems.

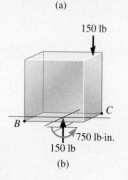

(a)

150 lb

B

750 lb·in.

150 lb

(b)

Fig. 8–3

EXAMPLE 8–2

A force of 150 lb is applied to the edge of the member shown in Fig. 8–3a. Neglect the weight of the member and determine the state of stress at points B and C.

SOLUTION

Internal Loadings.   The member is sectioned through B and C. For equilibrium at the section there must be an axial force of 150 lb acting through the centroid and a bending moment of 750 lb · in. about the centroidal or principal axis, Fig. 8–3b.

Stress Components.

NORMAL FORCE.   The uniform normal-stress distribution due to the normal force is shown in Fig. 8–3c. Here

$$\sigma = \frac{P}{A} = \frac{150 \text{ lb}}{(10 \text{ in.})(4 \text{ in.})} = 3.75 \text{ psi}$$

BENDING MOMENT.   The normal-stress distribution due to the bending moment is shown in Fig. 8–3d. The maximum stress is

$$\sigma_{max} = \frac{Mc}{I} = \frac{750 \text{ lb} \cdot \text{in.}(5 \text{ in.})}{\left[\frac{1}{12}(4 \text{ in.})(10 \text{ in.})^3\right]} = 11.25 \text{ psi}$$

Superposition.   If the above normal-stress distributions are added algebraically, the resultant stress distribution is shown in Fig. 8–3e. Although it is not needed here, the location of the line of zero stress can be determined by proportional triangles; i.e.,

$$\frac{7.5 \text{ psi}}{x} = \frac{15 \text{ psi}}{(10 \text{ in.} - x)}$$

$$x = 3.33 \text{ in.}$$

Elements of material at B and C are subjected only to normal or *uniaxial* stress as shown in Fig. 8–3f and 8–3g. Hence,

$$\sigma_B = 7.5 \text{ psi} \quad \text{(tension)} \qquad \qquad Ans.$$
$$\sigma_C = 15 \text{ psi} \quad \text{(compression)} \qquad Ans.$$

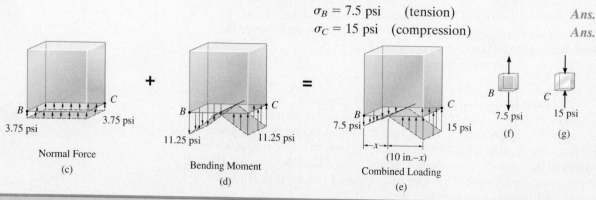

3.75 psi

3.75 psi

Normal Force

(c)

+

11.25 psi

11.25 psi

Bending Moment

(d)

=

7.5 psi

15 psi

Combined Loading

(e)

B

7.5 psi

(f)

C

15 psi

(g)

EXAMPLE 8–3

The tank in Fig. 8–4a has an inner radius of 24 in. and a thickness of 0.5 in. It is filled to the top with water having a specific weight of $\gamma_w = 62.4$ lb/ft^3. If it is made of steel having a specific weight of $\gamma_{st} = 490$ lb/ft^3, determine the state of stress at point A. The tank is open at the top.

$t = 0.5$ in.

$r = 24$ in.

3 ft

A

(a)

SOLUTION

Internal Loadings. The free-body diagram of the section of both the tank and the water above point A is shown in Fig. 8–4b. Notice that the weight of the water is supported by the water surface just *below* the section, *not* by the walls of the tank. In the vertical direction, the walls simply hold up the weight of the tank. This weight is

$$W_{st} = \gamma_{st} V_{st} = (490 \text{ lb/ft}^3)\left[\pi\left(\frac{24.5}{12}\text{ ft}\right)^2 - \pi\left(\frac{24}{12}\text{ ft}\right)^2\right](3 \text{ ft}) = 777.7 \text{ lb}$$

The stress in the circumferential direction is developed by the water pressure at level A. To obtain this pressure we must use *Pascal's law*, which states that the pressure at a point located a depth z in the water is $p = \gamma_w z$. Consequently, the pressure on the tank at level A is

$$p = \gamma_w z = (62.4 \text{ lb/ft}^3)(3 \text{ ft}) = 187.2 \text{ lb/ft}^2 = 1.30 \text{ psi}$$

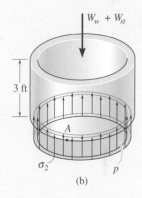

$W_w + W_{st}$

3 ft

A

σ_2

p

(b)

Stress Components.

CIRCUMFERENTIAL STRESS.   *Applying Eq. 8–1, using the inner radius* $r = 24$ in., we have

$$\sigma_1 = \frac{pr}{t} = \frac{1.30 \text{ lb/in}^2 (24 \text{ in.})}{(0.5 \text{ in.})} = 62.4 \text{ psi} \qquad Ans.$$

LONGITUDINAL STRESS.   Since the weight of the tank is supported uniformly by the walls, we have

$$\sigma_2 = \frac{W_{st}}{A_{st}} = \frac{777.7 \text{ lb}}{\pi[(24.5 \text{ in.})^2 - (24 \text{ in.})^2]} = 10.2 \text{ psi} \qquad Ans.$$

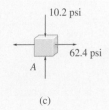

10.2 psi

62.4 psi

A

(c)

Fig. 8–4

Note that Eq. 8–2, $\sigma_2 = pr/2t$, does *not apply* here, since the tank is open at the top and therefore, as stated previously, the water cannot develop a loading on the walls in the longitudinal direction.

Point A is therefore subjected to the biaxial stress shown in Fig. 8–4c.

EXAMPLE 8-4

The member shown in Fig. 8–5a has a rectangular cross section.
Determine the state of stress that the loading produces at point C.

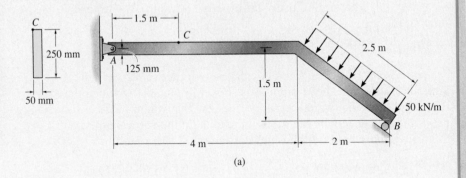

(a)

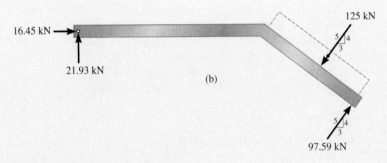

(b)

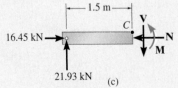

(c)

Fig. 8–5

SOLUTION

Internal Loadings. The support reactions on the member have been
determined and are shown in Fig. 8–5b. If the left segment AC of the
member is considered, Fig. 8–5c, the resultant internal loadings at the
section consist of a normal force, a shear force, and a bending
moment. Solving,

$$N = 16.45 \text{ kN} \qquad V = 21.93 \text{ kN} \qquad M = 32.89 \text{ kN} \cdot \text{m}$$

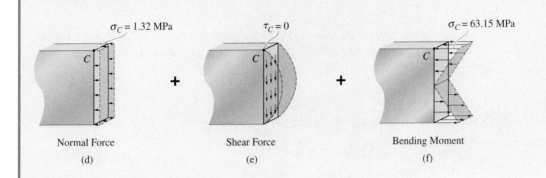

Normal Force   Shear Force   Bending Moment
   (d)   (e)   (f)

Stress Components.

NORMAL FORCE.   The uniform normal-stress distribution acting over the cross section is produced by the normal force, Fig. 8–5d. At point C,

$$\sigma_C = \frac{P}{A} = \frac{16.45 \text{ kN}}{(0.050 \text{ m})(0.250 \text{ m})} = 1.32 \text{ MPa}$$

SHEAR FORCE.   Here the area $A' = 0$, since point C is located at the top of the member. Thus $Q = \bar{y}'A' = 0$ and for C, Fig. 8–5e, the shear stress

$$\tau_C = 0$$

BENDING MOMENT.   Point C is located at $y = c = 125$ mm from the neutral axis, so the normal stress at C, Fig. 8–5f, is

$$\sigma_C = \frac{Mc}{I} = \frac{(32.89 \text{ kN} \cdot \text{m})(0.125 \text{ m})}{[\frac{1}{12}(0.050 \text{ m})(0.250)^3]} = 63.15 \text{ MPa}$$

Superposition.   The shear stress is zero. Adding the normal stresses determined above gives a compressive stress at C having a value of

$$\sigma_C = 1.32 \text{ MPa} + 63.15 \text{ MPa} = 64.5 \text{ MPa} \qquad \textit{Ans.}$$

64.5 MPa

(g)

This result, acting on an element at C, is shown in Fig. 8–5g.

EXAMPLE 8–5

The solid rod shown in Fig. 8–6a has a radius of 0.75 in. If it is subjected to the loading shown, determine the state of stress at point A.

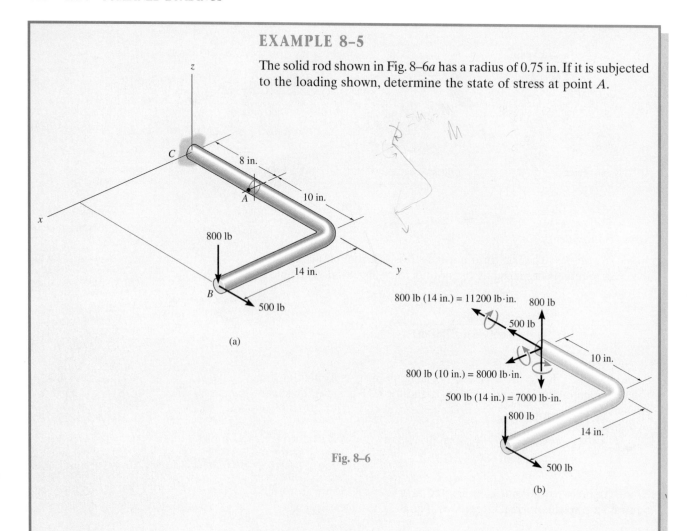

(a)

(b)

Fig. 8–6

SOLUTION

Internal Loadings. The rod is sectioned through point A. Using the free-body diagram of segment AB, Fig. 8–6b, the resultant internal loadings can be determined from the six equations of equilibrium. Verify these results. The normal force (500 lb) and shear force (800 lb) must act through the centroid of the cross section and the bending-moment components (8000 lb · in. and 7000 lb · in.) are applied about centroidal (principal) axes. In order to better "visualize" the stress distributions due to each of these loadings, we will consider the *equal but opposite resultants* acting on AC, Fig. 8–6c.

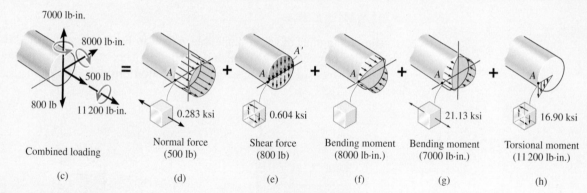

Combined loading   Normal force   Shear force   Bending moment   Bending moment   Torsional moment

(c)   (d)   (e)   (f)   (g)   (h)

Stress Components.

NORMAL FORCE.   The normal-stress distribution is shown in Fig. 8–6*d*. For point *A*, we have

$$\sigma_A = \frac{P}{A} = \frac{500 \text{ lb}}{\pi(0.75 \text{ in.})^2} = 283 \text{ psi} = 0.283 \text{ ksi}$$

SHEAR FORCE.   The shear-stress distribution is shown in Fig. 8–6*e*. For point *A*, *Q* is determined from the shaded *semicircular* area. Using the table on the inside front cover, we have

$$Q = \bar{y}'A' = \frac{4(0.75 \text{ in.})}{3\pi}\left[\frac{1}{2}\pi\left(0.75 \text{ in.}\right)^2\right] = 0.2813 \text{ in}^3$$

so that

$$\tau_A = \frac{VQ}{It} = \frac{800 \text{ lb}(0.2813 \text{ in}^3)}{[\frac{1}{4}\pi(0.75 \text{ in.})^4]2(0.75 \text{ in.})} = 604 \text{ psi} = 0.604 \text{ ksi}$$

BENDING MOMENTS.   For the 8000-lb · in. component, point *A* lies on the neutral axis, Fig. 8–6*f*, so the normal stress is

$$\sigma_A = 0$$

For the 7000-lb · in. moment, *c* = 0.75 in., so the normal stress at point *A*, Fig. 8–6*g*, is

$$\sigma_A = \frac{Mc}{I} = \frac{7000 \text{ lb} \cdot \text{in.}(0.75 \text{ in.})}{[\frac{1}{4}\pi(0.75 \text{ in.})^4]} = 21\,126 \text{ psi} = 21.13 \text{ ksi}$$

TORSIONAL MOMENT.   At point *A*, $\rho_A = c = 0.75$ in., Fig. 8–6*h*. Thus the shear stress is

$$\tau_A = \frac{Tc}{J} = \frac{11\,200 \text{ lb} \cdot \text{in.}(0.75 \text{ in.})}{[\frac{1}{2}\pi(0.75 \text{ in.})^4]} = 16\,901 \text{ psi} = 16.90 \text{ ksi}$$

Superposition.   When the above results are superimposed, it is seen that an element of material at *A* is subjected to both normal and shear stress components, Fig. 8–6*i*.

EXAMPLE 8–6

The rectangular block of negligible weight in Fig. 8–7a is subjected to a vertical force of 40 kN, which is applied to its corner. Determine the normal-stress distribution acting on a section through $ABCD$.

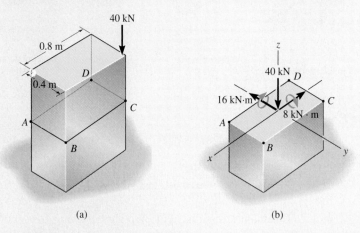

(a)   (b)

Fig. 8–7

SOLUTION

Internal Loadings.   If we consider the equilibrium of the bottom segment of the block, Fig. 8–7b, it is seen that the 40-kN force must act through the centroid of the cross section and *two* bending-moment components must also act about the centroidal or principal axes of inertia for the section. Verify these results.

Stress Components.

NORMAL FORCE.   The uniform normal-stress distribution is shown in Fig. 8–7c. We have

$$\sigma = \frac{P}{A} = \frac{40 \text{ kN}}{(0.8 \text{ m})(0.4 \text{ m})} = 125 \text{ kPa}$$

BENDING MOMENTS.   The normal-stress distribution for the 8-kN · m moment is shown in Fig. 8–7d. The maximum stress is

$$\sigma_{\max} = \frac{M_x c_y}{I_x} = \frac{8 \text{ kN} \cdot \text{m}(0.2 \text{ m})}{\left[\frac{1}{12}(0.8 \text{ m})(0.4 \text{ m})^3\right]} = 375 \text{ kPa}$$

Likewise, for the 16-kN · m moment, Fig. 8–7*e*, the maximum normal stress is

$$\sigma_{max} = \frac{M_y c_x}{I_y} = \frac{16 \text{ kN} \cdot \text{m}(0.4 \text{ m})}{\left[\frac{1}{12}(0.4 \text{ m})(0.8 \text{ m})^3\right]} = 375 \text{ kPa}$$

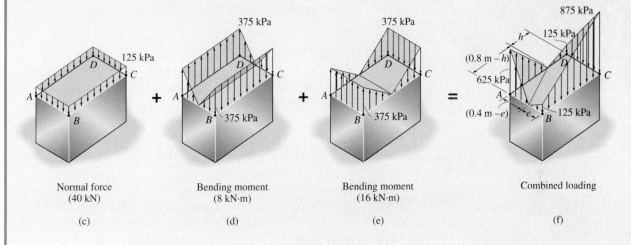

Normal force
(40 kN)

(c)

Bending moment
(8 kN·m)

(d)

Bending moment
(16 kN·m)

(e)

Combined loading

(f)

Superposition. The normal stress at each corner point can be determined by algebraic addition. Assuming that tensile stress is positive, we have

$$\sigma_A = -125 \text{ kPa} + 375 \text{ kPa} + 375 \text{ kPa} = 625 \text{ kPa}$$
$$\sigma_B = -125 \text{ kPa} - 375 \text{ kPa} + 375 \text{ kPa} = -125 \text{ kPa}$$
$$\sigma_C = -125 \text{ kPa} - 375 \text{ kPa} - 375 \text{ kPa} = -875 \text{ kPa}$$
$$\sigma_D = -125 \text{ kPa} + 375 \text{ kPa} - 375 \text{ kPa} = -125 \text{ kPa}$$

Since the stress distributions due to bending moment are linear, the resultant stress distribution is also linear and therefore looks like that shown in Fig. 8–7*f*. The line of zero stress can be located along each side by proportional triangles. From the figure we require

$$\frac{(0.4 \text{ m} - e)}{625 \text{ kPa}} = \frac{e}{125 \text{ kPa}}$$
$$e = 0.0667 \text{ m}$$

and

$$\frac{(0.8 \text{ m} - h)}{625 \text{ kPa}} = \frac{h}{125 \text{ kPa}}$$
$$h = 0.133 \text{ m}$$

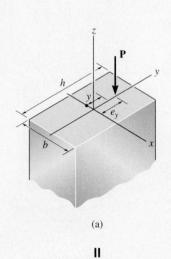

(a)

$$\parallel$$

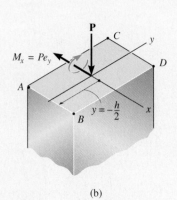

(b)

Fig. 8-8

EXAMPLE 8-7

A rectangular block has a negligible weight and is subjected to a vertical force **P**, Fig. 8–8a. (a) Determine the range of values for the eccentricity e_y of the load along the y axis so that it does not cause any tensile stress in the block. (b) Specify the region on the cross section where **P** may be applied without causing a tensile stress in the block.

SOLUTION

Part (a). When **P** is moved to the centroid of the cross section, Fig. 8–8b, it is necessary to add a couple moment $M_x = Pe_y$ in order to maintain a statically equivalent loading. The combined normal stress at any coordinate location y on the cross section caused by these two loadings is

$$\sigma = -\frac{P}{A} - \frac{(Pe_y)y}{I_x} = -\frac{P}{A}\left(1 + \frac{Ae_yy}{I_x}\right)$$

Here the negative sign indicates compressive stress. For positive e_y, Fig. 8–8a, the *smallest* compressive stress will occur along edge AB, where $y = -h/2$, Fig. 8–8b. (By inspection, **P** causes compression there, but $\mathbf{M}_x$ causes tension.) Hence

$$\sigma_{min} = -\frac{P}{A}\left(1 - \frac{Ae_yh}{2I_x}\right)$$

This stress will remain negative, i.e., compressive, provided the term in parentheses is positive; i.e.,

$$1 > \frac{Ae_yh}{2I_x}$$

Since $A = bh$ and $I_x = \frac{1}{12}bh^3$, then

$$1 > \frac{6e_y}{h}$$

or

$$e_y < \frac{1}{6}h$$

Ans.

In other words, if $-\frac{1}{6}h \le e_y \le \frac{1}{6}h$, the stress in the block along edge AB or CD will be zero or remain *compressive*. This is sometimes referred to as the "*middle-third rule.*" It is very important to keep this rule in mind when loading columns or arches having a rectangular cross section and made of material such as stone or concrete, which can support little or no tensile stress.

Part (b). We can extend the above analysis in two directions by assuming that **P** acts in the positive quadrant of the $x-y$ plane, Fig. 8–8c. The equivalent static loading when **P** acts at the centroid is shown in Fig. 8–8d. At any coordinate point x,y on the cross section, the combined normal stress due to both normal and bending loadings is

$$\sigma = -\frac{P}{A} - \frac{Pe_y y}{I_x} - \frac{Pe_x x}{I_y}$$

$$= -\frac{P}{A}\left(1 + \frac{Ae_y y}{I_x} + \frac{Ae_x x}{I_y}\right)$$

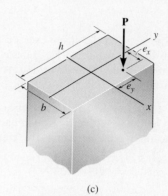

(c)

By inspection, Fig. 8–8d, the moments both create tensile stress at point A and the normal force creates a compressive stress there. Hence, the smallest compressive stress will occur at point A, for which $x = -b/2$ and $y = -h/2$. Thus,

$$\sigma_A = -\frac{P}{A}\left(1 - \frac{Ae_y h}{2I_x} - \frac{Ae_x b}{2I_y}\right)$$

As before, the normal stress remains negative or compressive at point A, provided the terms in the parentheses remain positive; i.e.,

$$0 < \left(1 - \frac{Ae_y h}{2I_x} - \frac{Ae_x b}{2I_y}\right)$$

Substituting $A = bh$, $I_x = \frac{1}{12}bh^3$, $I_y = \frac{1}{12}hb^3$ yields

$$0 < 1 - \frac{6e_y}{h} - \frac{6e_x}{b} \qquad \textit{Ans.}$$

Hence, regardless of the magnitude of **P**, if it is applied at any point within the boundary of line GH shown in Fig. 8–8e, the normal stress at point A will remain compressive. In a similar manner, the normal stress at the other corners of the cross section will be compressive if **P** is confined within the boundaries of lines EG, FE, and HF. The shaded parallelogram so defined is referred to as the *core* or *kern* of the section. From the "middle-third rule" of part (a), the diagonals of the parallelogram will have lengths of $b/3$ and $h/3$.

$\|$

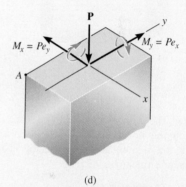

(d)

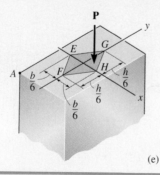

(e)

Here is an example of where combined axial and bending stress can occur.

PROBLEMS

8–15. The screw of the clamp exerts a compressive force of 500 lb on the wood blocks. Determine the maximum normal stress developed along section a–a. The cross section there is rectangular, 0.75 in. by 0.50 in.

*8–16. The screw of the clamp exerts a compressive force of 500 lb on the wood blocks. Sketch the stress distribution along section a–a of the clamp. The cross section there is rectangular, 0.75 in. by 0.50 in.

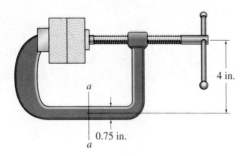

Probs. 8–15/8–16

8–17. The joint is subjected to a force of 250 lb as shown. Sketch the normal-stress distribution acting over section a–a if the member has a rectangular cross section of width 0.5 in. and thickness 0.75 in.

8–18. The joint is subjected to a force of 250 lb as shown. Determine the state of stress at points A and B, and sketch the results on differential elements located at these points. The member has a rectangular cross-sectional area of width 0.5 in. and thickness 0.75 in.

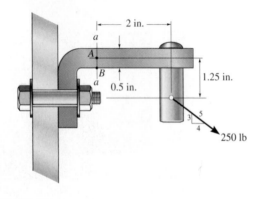

Probs. 8–17/8–18

8–19. The offset link supports the loading of $P = 30$ kN. Determine its required width w if the allowable normal stress is $\sigma_{\text{allow}} = 73$ MPa. The link has a thickness of 40 mm.

*8–20. The offset link has a width of $w = 200$ mm and a thickness of 40 mm. If the allowable normal stress is $\sigma_{\text{allow}} = 75$ MPa, determine the maximum load P that can be applied to the cables.

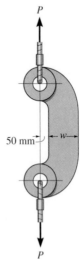

Probs. 8–19/ 8–20

8–21. Determine the maximum and minimum normal stress in the bracket at section a when the load is applied at $x = 0$.

8–22. Determine the maximum and minimum normal stress in the bracket at section a when the load is applied at $x = 50$ mm.

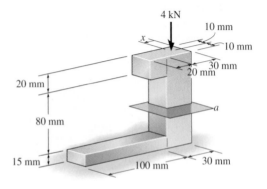

Probs. 8–21/8–22

8–23. The bent link is subjected to the cable load of P = 500 N. Determine its required diameter *d* if the allowable normal stress for the material is σ_{allow} = 175 MPa. Consider the critical section to be at *A*.

***8–24.** The bent link has a diameter of *d* = 15 mm and is made of a material having an allowable normal stress of σ_{allow} = 175 MPa. Determine the maximum load *P* it will safely support. Consider the critical section to be at *A*.

8–27. The stepped support is subjected to the bearing load of 50 kN. Determine the maximum and minimum compressive stress in the material.

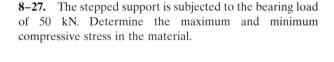
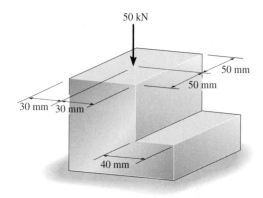

Prob. 8–27

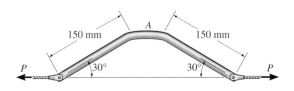

Probs. 8–23/8–24

8–25. The bar has a diameter of 40 mm. If it is subjected to a force of 800 N as shown, determine the stress components that act at point *A* and show the results on a volume element located at this point.

8–26. Solve Prob. 8–25 for point *B*.

***8–28.** The pin support is made from a steel rod and has a diameter of 20 mm. Determine the stress components at points *A* and *B* and represent the results on a volume element located at each of these points.

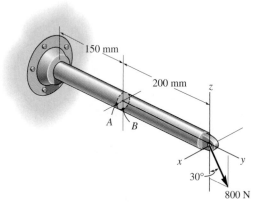

Probs. 8–25/8–26

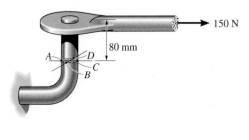

Prob. 8–28

8–29. Since concrete can support little or no tension, this problem can be avoided by using wires or rods to *prestress* the concrete once it is formed. Consider the simply supported beam shown, which has a rectangular cross section of 18 in. by 12 in. If concrete has a specific weight of 150 lb/ft³, determine the required tension in rod AB, which runs through the beam so that no tensile stress is developed in the concrete at its center section a–a. Neglect the size of the rod and any deflection of the beam.

8–30. Solve Prob. 8–29 if the rod has a diameter of 0.5 in. Use the transformed area method discussed in Sec. 6–6. $E_{st} = 29(10^3)$ ksi, $E_c = 3.60(10^3)$ ksi.

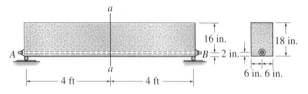

Probs. 8–29/8–30

8–31. The chimney is subjected to the uniform wind loading of $w = 150$ lb/ft and has a weight of 2200 lb/ft. If the mortar between the bricks cannot support a tensile stress, determine if the chimney is safe. Take $d = 6$ ft. The thickness of the brick wall is 1 ft.

*8–32. The chimney is subjected to the uniform wind pressure of $p = 25$ lb/ft². It is to be constructed with 1-ft-thick brick walls. If the bricks and mortar have a specific weight of 145 lb/ft³, determine the smallest outer diameter d of the chimney so that no tensile stress is developed in the material. The wind loading can be approximated by $w = pd$.

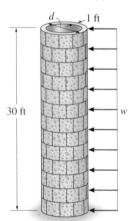

Probs. 8–31/8–32

8–33. The block is subjected to the two axial loads shown. Determine the normal stress developed at points A and B. Neglect the weight of the block.

8–34. The block is subjected to the two axial loads shown. Sketch the normal stress distribution acting over the cross section at section a–a. Neglect the weight of the block.

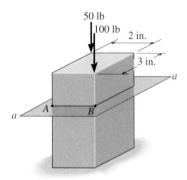

Probs. 8–33/8–34

8–35. A bar having a square cross section of 30 mm by 30 mm is 2 m long and is held upward. If it has a mass of 5 kg/m, determine the largest angle θ, measured from the vertical, at which it can be supported before it is subjected to a tensile stress near the grip.

*8–36. Solve Prob. 8–35 if the bar has a circular cross section of 30-mm diameter.

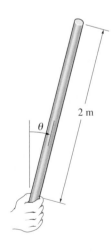

Probs. 8–35/8–36

8–37. The frame supports the distributed load shown. Determine the state of stress acting at point D. Show the results on a differential element located at this point.

8–38. The frame supports the distributed load shown. Determine the state of stress acting at point E. Show the results on a differential element located at this point.

*8–40. Determine the state of stress at point A when the beam is subjected to the cable force of 4 kN. Indicate the result as a differential volume element.

8–41. Determine the state of stress at point B when the beam is subjected to the cable force of 4 kN. Indicate the result as a differential volume element.

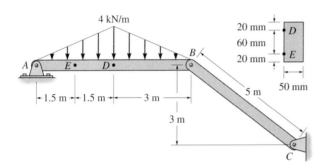

Probs. 8–37/ 8–38

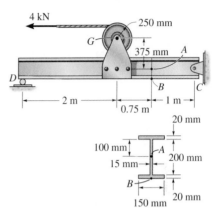

Probs. 8–40/8–41

8–39. The cylinder of negligible weight rests on a smooth floor. Determine the eccentric distance e_y at which the load can be placed so that the normal stress at point A is zero.

8–42. The sign is subjected to the uniform wind loading. Determine the stress components at points A and B on the 100-mm-diameter supporting post. Show the results on a volume element located at each of these points.

8–43. The sign is subjected to the uniform wind loading. Determine the stress components at points C and D on the 100-mm-diameter supporting post. Show the results on a volume element located at each of these points..

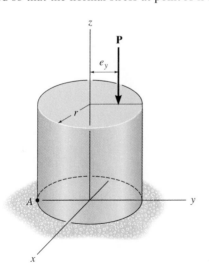

Prob. 8–39

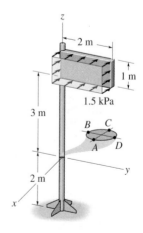

Probs. 8–42/8–43

*8–44. The solid rod is subjected to the loading shown. Determine the state of stress developed in the material at point A, and show the results on a differential volume element at this point.

8–45. The solid rod is subjected to the loading shown. Determine the state of stress at point B, and show the results on a differential volume element located at this point.

8–46. The solid rod is subjected to the loading shown. Determine the state of stress at point C, and show the results on a differential volume element located at this point.

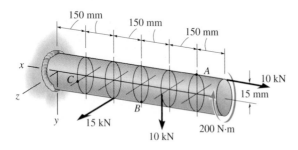

Probs. 8–44/8–45/8–46

8–47. The bent shaft is fixed in the wall at A. If a force $\mathbf{F}$ is applied at B, determine the stress components at points D and E. Show the results on a differential element located at each of these points. Take $F = 12$ lb and $\theta = 90°$.

*8–48. The bent shaft is fixed in the wall at A. If a force $\mathbf{F}$ is applied at B, determine the stress components at points D and E. Show the results on a volume element located at each of these points. Take $F = 12$ lb and $\theta = 45°$.

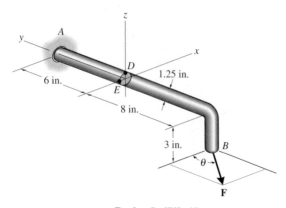

Probs. 8–47/8–48

8–49. The caster wheel supports a reactive load of 180 N. Determine the state of stress at points A and B on one of the two supporting leaves. Show the results on a differential volume element located at each point.

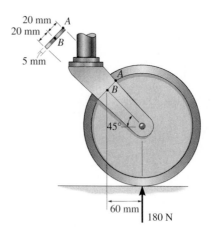

Prob. 8–49

8–50. The crane boom is subjected to the load of 500 lb. Determine the state of stress at points A and B. Show the results on a differential volume element located at each of these points.

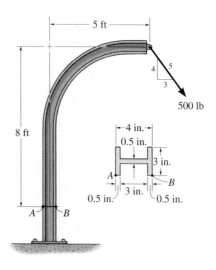

Prob. 8–50

8–51. The beam supports the loading shown. Determine the state of stress at points E and F at section a–a, and represent the results on a differential volume element located at each of these points.

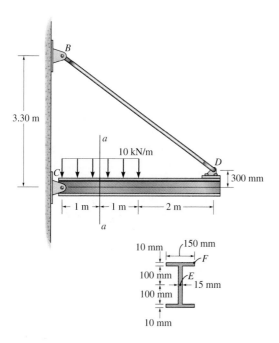

Prob. 8–51

***8–52.** The symmetrically loaded spreader bar is used to lift the 2000-lb tank. Determine the state of stress at points A and B, and indicate the results on a differential volume elements.

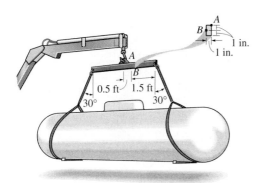

Prob. 8–52

8–53. The tine ABC of the fork lift is subjected to a uniform distributed loading as shown. If it is pin connected at C and roller supported at B, determine the state of stress at points D and E, and show the results on differential elements. The tine is 3 in. wide and 0.5 in. thick.

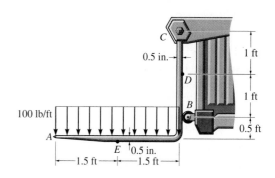

Prob. 8–53

8–54. The 1-in.-diameter rod is subjected to the loads shown. Determine the state of stress at point A, and show the results on a differential element located at this point.

8–55. The 1-in.-diameter rod is subjected to the loads shown. Determine the state of stress at point B, and show the results on a differential element located at this point.

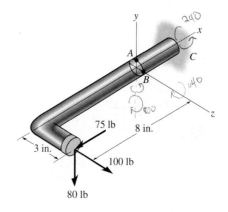

Probs. 8–54/8–55

*8–56. The member is subjected to the combination of distributed and concentrated loadings shown. Determine the state of stress at point A, and show the results on a differential volume element located at this point. *Hint:* Use Table 5-1.

8–57. The member is subjected to the combination of distributed and concentrated loadings shown. Determine the state of stress at point B, and show the results on a differential volume element located at this point. *Hint:* Use Table 5-1.

8–58. The member is subjected to the combination of distributed and concentrated loadings shown. Determine the state of stress at point C, and show the results on a differential volume element located at this point. *Hint:* Use Table 5-1.

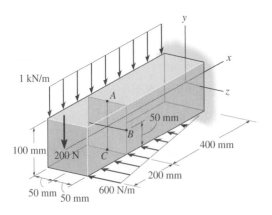

Probs. 8–56/8–57/8–58

8–59. A post having the dimensions shown is subjected to the bearing load **P**. Specify the region to which this load can be applied without causing tensile stress to be developed at points A, B, C, and D.

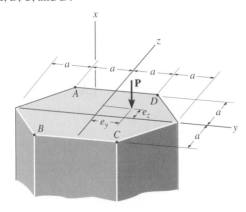

Prob. 8–59

*8–60. The masonry pier is subjected to the 800-kN load. Determine the equation of the line $y = f(x)$ along which the load can be placed without causing a tensile stress in the pier. Neglect the weight of the pier.

8–61. The masonry pier is subjected to the 800-kN load. If $x = 0.25$ m and $y = 0.5$ m, determine the normal stress at each corner A, B, C, D (not shown) and plot the stress distribution over the cross section. Neglect the weight of the pier.

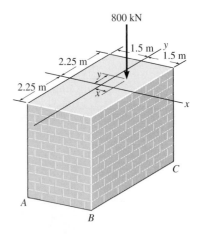

Probs. 8–60/8–61

8–62. The eye is subjected to the force of 50 lb. Determine the maximum tensile and compressive stresses at section a–a. The cross section is circular and has a diameter of 0.25 in. Use the curved-beam formula to compute the bending stress.

8–63. Solve Prob. 8–62 if the cross section is square, having dimensions of 0.25 in. by 0.25 in.

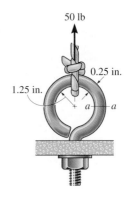

Probs. 8–62/8–63

*8–64. The C-clamp applies a compressive stress on the cylindrical block of 80 psi. Determine the maximum normal stress developed in the clamp.

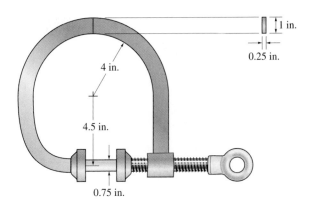

1 in.

0.25 in.

4 in.

4.5 in.

0.75 in.

Prob. 8–64

8–65. The hook is used to lift the force of 600 lb. Determine the maximum tensile and compressive stresses at section *a–a*. The cross section is circular and has a diameter of 1 in. Use the curved-beam formula to compute the bending stress.

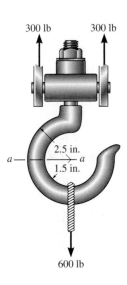

300 lb   300 lb

2.5 in.
a —————————— *a*
1.5 in.

600 lb

Prob. 8–65

8–66. The support is subjected to the compressive load **P**. Determine the absolute maximum and minimum normal stress acting in the material.

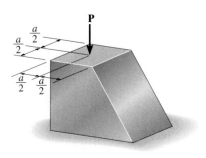

$\frac{a}{2}$ $\frac{a}{2}$

$\frac{a}{2}$ $\frac{a}{2}$

P

Prob. 8–66

8–67. The support is subjected to the compressive load **P**. Determine the maximum and minimum normal stress acting in the material.

P

r

Prob. 8–67

REVIEW PROBLEMS

*8–68.   Air pressure in the cylinder is increased by exerting forces $P = 2$ kN on the two pistons, each having a radius of 45 mm. If the cylinder has a wall thickness of 2 mm, determine the state of stress in the wall of the cylinder.

8–69.   Determine the maximum force P that can be exerted on each of the two pistons so that the circumferential stress component in the cylinder does not exceed 3 MPa. Each piston has a radius of 45 mm and the cylinder has a wall thickness of 2 mm.

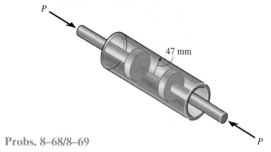

47 mm

Probs. 8–68/8–69

8–70.   The cap on the cylindrical tank is bolted to the tank along the flanges. The tank has an inner diameter of 1.5 m and a wall thickness of 18 mm. If the largest normal stress is not to exceed 150 MPa, determine the maximum pressure the tank can sustain. Also, compute the number of bolts required to attach the cap to the tank if each bolt has a diameter of 20 mm. The allowable stress for the bolts is $(\sigma_{allow})_b = 180$ MPa.

8–71.   The cap on the cylindrical tank is bolted to the tank along the flanges. The tank has an inner diameter of 1.5 m and a wall thickness of 18 mm. If the pressure in the tank is $p = 1.20$ MPa, determine the force in the 16 bolts that are used to attach the cap to the tank. Also, specify the state of stress in the wall of the tank.

Probs. 8–70/8–71

*8–72.   The crowbar is used to pull out the nail at A. If a force of 8 lb is required, determine the stress components in the bar at points D and E. Show the results on a differential volume element located at each of these points. The bar has a circular cross section with a diameter of 0.5 in. No slipping occurs at B.

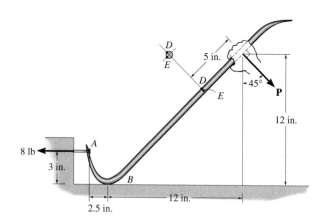

Prob. 8–72

8–73.   The steel bracket is used to connect the ends of two cables. If the applied force $P = 500$ lb, determine the maximum normal stress in the bracket. The bracket has a thickness of 0.5 in. and a width of 0.75 in.

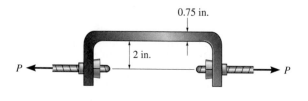

Prob. 8–73

8–74. The wall hanger has a thickness of 0.25 in. and is used to support the vertical reactions of the beam that is loaded as shown. If the load is transferred uniformly to each strap of the hanger, determine the state of stress at points C and D on the strap at A. Assume the vertical reaction $\mathbf{F}$ at this end acts in the center and on the edge of the bracket as shown.

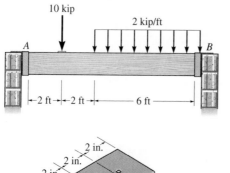

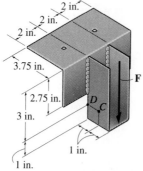

Prob. 8–74

8–75. The clamp is made from members AB and AC, which are pin connected at A. If the compressive force at C and B is 180 N, determine the state of stress at point F, and indicate the results on a differential volume element. The screw DE is subjected only to a tensile force along its axis.

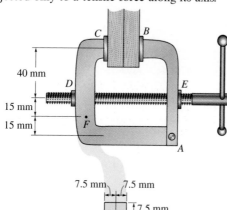

Prob. 8–75

***8–76.** The clamp is made from members AB and AC, which are pin-connected at A. If the compressive force at C and B is 180 N, determine the state of stress at point G, and indicate the results on a differential volume element. The screw DE is subjected only to a tensile force along its axis.

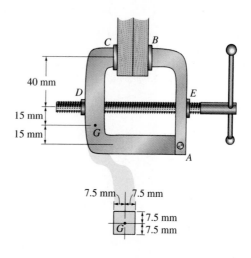

Prob. 8–76

8–77. The wide-flange beam is subjected to the loading shown. Determine the state of stress at points A and B, and show the results on a differential volume element located at each of these points.

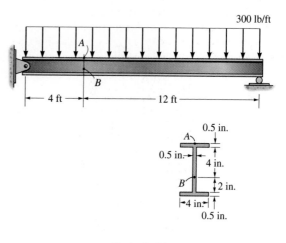

Prob. 8–77

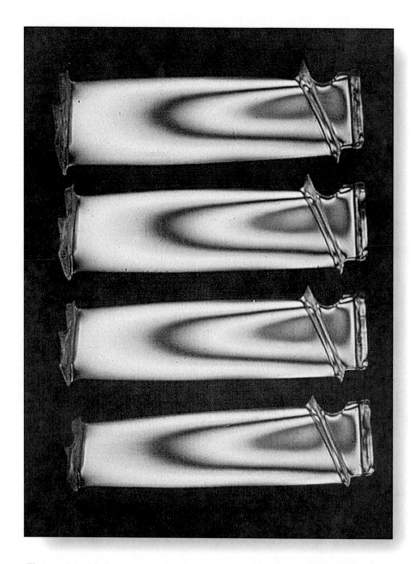

These turbine blades are subjected to a complex pattern of stress, which is illustrated by the shaded bands that appear on the blades when they are made of transparent material and viewed through polarized light. For proper design, engineers must be able to determine where and in what direction the maximum stress occurs. (Courtesy of Measurements Group, Inc., Raleigh, North Carolina 27611, USA.)

9 STRESS TRANSFORMATION

CHAPTER OBJECTIVES

In this chapter we will show how to transform the stress components that are associated with a particular coordinate system into components associated with a coordinate system having a different orientation. Once the necessary transformation equations are established, we will then be able to obtain the maximum normal and maximum shear stress at a point and find the orientation of elements upon which they act. Plane-stress transformation will be discussed in the first part of the chapter, since this condition is most common in engineering practice. At the end of the chapter we will discuss a method for finding the absolute maximum shear stress at a point when the material is subjected to both plane and three-dimensional states of stress.

9.1 PLANE-STRESS TRANSFORMATION

It was shown in Sec. 1.3 that the general state of stress at a point is characterized by *six* independent normal and shear stress components, which act on the faces of an element of material located at the point, Fig. 9–1a. This state of stress, however, is not often encountered in engineering practice. Instead, engineers frequently make approximations or simplifications of the loadings on a body in order that the stress produced in a structural member or mechanical element can be analyzed in a *single plane*. When this is the case, the material is said to be subjected to *plane stress*, Fig. 9–1b. For example, if there is no load on the surface of a body, then the normal and shear stress components will be zero on the face of an element that lies on the surface. Consequently, the corresponding stress components on the opposite face will also be zero, and so the material at the point will be subjected to plane stress.

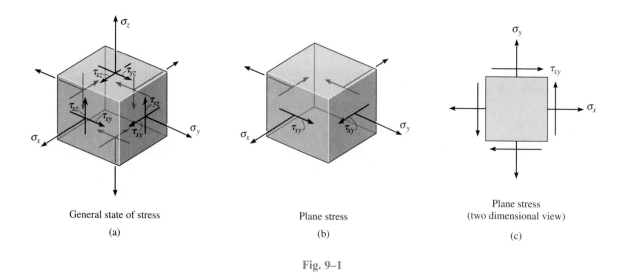

General state of stress

(a)

Plane stress

(b)

Plane stress
(two dimensional view)

(c)

Fig. 9–1

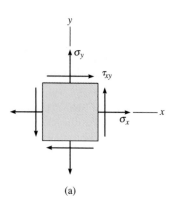

(a)

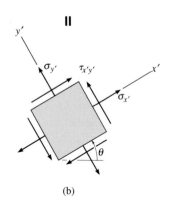

(b)

Fig. 9–2

The general state of *plane stress* at a point is therefore represented by a combination of two normal-stress components, σ_x, σ_y, and one shear-stress component, τ_{xy}, which act on four faces of the element. For convenience, in this text we will view this state of stress in the $x-y$ plane, Fig. 9–1c. Realize that if the state of stress at a point is defined by the three stress components shown on the element in Fig. 9–2a, then an element having a different orientation, such as in Fig. 9–2b, will be subjected to three different stress components. In other words, **the state of plane stress at the point is uniquely represented by three components acting on an element that has a specific orientation at the point**.

In this section, by using numerical examples, we will show how to *transform* the stress components from one orientation of an element to an element having a different orientation. That is, if the state of stress is defined by the components σ_x, σ_y, τ_{xy}, oriented along the x, y axes, Fig. 9–2a, we will show how to obtain the components $\sigma_{x'}$, $\sigma_{y'}$, $\tau_{x'y'}$, oriented along the x', y' axes, Fig. 9–2b, so that they represent the *same* state of stress at the point. This is like knowing two force components, say, $\mathbf{F}_x$ and $\mathbf{F}_y$, directed along the x, y axes, that produce a resultant force $\mathbf{F}_R$, and then trying to find the force components $\mathbf{F}_{x'}$ and $\mathbf{F}_{y'}$, directed along the x', y' axes, so they produce the *same* resultant. The transformation of stress components, however, is more difficult than that of force components, since for *stress*, the transformation must account for the magnitude and direction of each stress component *and* the orientation of the area upon which each component acts. For force, the transformation must account only for the force component's magnitude and direction.

PROCEDURE FOR ANALYSIS

If the state of stress at a point is known for a given orientation of an element of material, Fig. 9–3a, then the state of stress for some other orientation, Fig. 9–3b, can be determined using the following procedure.

- To determine the normal and shear stress components $\sigma_{x'}$, $\tau_{x'y'}$ acting on the x' face of the element, Fig. 9–3b, section the element in Fig. 9–3a as shown in Fig. 9–3c. If it is assumed the sectioned area is ΔA, then the adjacent areas of the segment will be $\Delta A \sin \theta$ and $\Delta A \cos \theta$.

- Draw the free-body diagram of the segment, which requires showing the *forces* that act on the element. This is done by multiplying the stress components on each face by the area upon which they act.

- Apply the equations of force equilibrium in the x' and y' directions to obtain the two unknown stress components $\sigma_{x'}$ and $\tau_{x'y'}$.

- If $\sigma_{y'}$, acting on the $+y'$ face of the element in Fig. 9–3b, is to be determined, then it is necessary to consider a segment of the element as shown in Fig. 9–3d and follow the same procedure just described. Here, however, the shear stress $\tau_{x'y'}$ does not have to be determined if it was previously calculated since it is complementary, that is, it has the same magnitude on each of the four faces of the element, Fig. 9–3b.

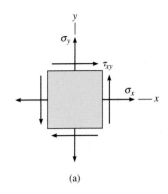

(a)

||

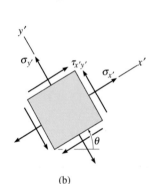

(b)

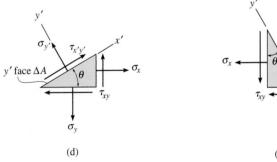

(d)   (c)

Fig. 9–3

EXAMPLE 9-1

The state of plane stress at a point on the surface of the airplane fuselage is represented on the element oriented as shown in Fig. 9–4a. Represent the state of stress at the point on an element that is oriented 30° clockwise from the position shown.

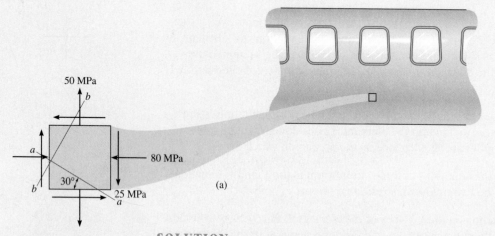

(a)

SOLUTION

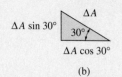

(b)

The element is sectioned by the line a–a in Fig. 9–4a, the bottom segment is removed, and assuming the sectioned (inclined) plane has an area ΔA, the horizontal and vertical planes have the areas shown in Fig. 9–4b. The free-body diagram of the segment is shown in Fig. 9–4c. Applying the equations of force equilibrium in the x' and y' directions to avoid a simultaneous solution for the two unknowns $\sigma_{x'}$ and $\tau_{x'y'}$, we have

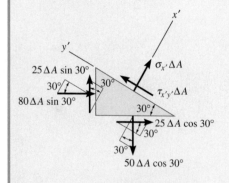

(c)

Fig. 9–4

$$+\nearrow \Sigma F_{x'} = 0; \quad \sigma_{x'} \Delta A - (50 \, \Delta A \cos 30°) \cos 30°$$
$$+(25 \, \Delta A \cos 30°) \sin 30° + (80 \, \Delta A \sin 30°) \sin 30°$$
$$+(25 \, \Delta A \sin 30°) \cos 30° = 0$$
$$\sigma_{x'} = -4.15 \text{ MPa} \qquad \textit{Ans.}$$

$$+\nwarrow \Sigma F_{y'} = 0; \quad \tau_{x'y'} \Delta A - (50 \, \Delta A \cos 30°) \sin 30°$$
$$-(25 \, \Delta A \cos 30°) \cos 30° - (80 \, \Delta A \sin 30°) \cos 30°$$
$$+(25 \, \Delta A \sin 30°) \sin 30° = 0$$
$$\tau_{x'y'} = 68.8 \text{ MPa} \qquad \textit{Ans.}$$

Since $\sigma_{x'}$ is negative, it acts in the opposite direction of that shown in Fig. 9–4c. The results are shown on the *top* of the element in Fig. 9–4d, since this surface is the one considered in Fig. 9–4c.

We must now repeat the procedure to obtain the stress on the *perpendicular* plane *b–b*. Sectioning the element in Fig. 9–4*a* along *b–b* results in a segment having sides with areas shown in Fig. 9–4*e*. Orientating the $+x'$ axis outward, perpendicular to the sectioned face, the associated free-body diagram is shown in Fig. 9–4*f*. Thus,

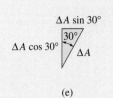

(e)

$$+\searrow \Sigma F_{x'} = 0; \quad \sigma_{x'} \, \Delta A - (25 \, \Delta A \cos 30°) \sin 30°$$
$$+ (80 \, \Delta A \cos 30°) \cos 30° - (25 \, \Delta A \sin 30°) \cos 30°$$
$$- (50 \, \Delta A \sin 30°) \sin 30° = 0$$
$$\sigma_{x'} = -25.8 \text{ MPa} \qquad \textbf{\textit{Ans.}}$$

$$+\nearrow \Sigma F_{y'} = 0; \quad -\tau_{x'y'} \, \Delta A + (25 \, \Delta A \cos 30°) \cos 30°$$
$$+ (80 \, \Delta A \cos 30°) \sin 30° - (25 \, \Delta A \sin 30°) \sin 30°$$
$$+ (50 \, \Delta A \sin 30°) \cos 30° = 0$$
$$\tau_{x'y'} = 68.8 \text{ MPa} \qquad \textbf{\textit{Ans.}}$$

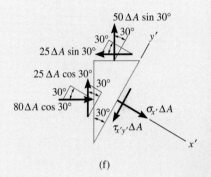

(f)

Since $\sigma_{x'}$ is a negative quantity, it acts opposite to its direction shown in Fig. 9–4*f*. The stress components are shown acting on the *right side* of the element in Fig. 9–4*d*.

From this analysis we may therefore conclude that the state of stress at the point can be represented by choosing an element oriented as shown in Fig. 9–4*a*, or by choosing one oriented as shown in Fig. 9–4*d*. In other words, the states of stress are equivalent.

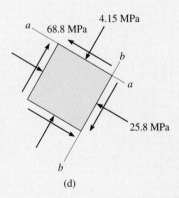

(d)

9.2   GENERAL EQUATIONS OF PLANE-STRESS TRANSFORMATION

The method of transforming the normal and shear stress components from the x, y to the x', y' coordinate axes, as discussed in the previous section, will now be developed in a general manner and expressed as a set of stress-transformation equations.

Sign Convention.   Before the transformation equations are derived, we must first establish a sign convention for the stress components. Here we will adopt the same one used in Sec. 1.3. Briefly stated, once the x, y or x', y' axes have been established, a normal or shear stress component is *positive* provided it acts in the *positive* coordinate direction on the *positive* face of the element, or it acts in the *negative* coordinate direction on the *negative* face of the element, Fig. 9–5a. For example, σ_x is positive since it acts to the right on the right-hand vertical face, and it acts to the left ($-x$ direction) on the left-hand vertical face. The shear stress in Fig. 9–5a is shown acting in the positive direction on all four faces of the element. On the right-hand face, τ_{xy} acts upward ($+y$ direction); on the bottom face, τ_{xy} acts to the left ($-x$ direction), and so on.

All the stress components shown in Fig. 9–5a maintain equilibrium of the element, and because of this, knowing the direction of τ_{xy} on one face of the element defines its direction on the other three faces. Hence, the above sign convention can also be remembered by simply noting that *positive normal stress acts outward from all faces and positive shear stress acts upward on the right-hand face of the element.*

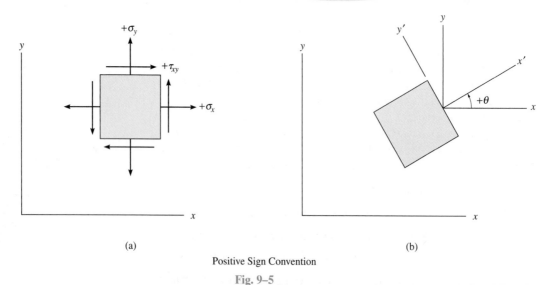

(a)   (b)

Positive Sign Convention

Fig. 9–5

Given the state of plane stress shown in Fig. 9–5a, the orientation of the inclined plane on which the normal and shear stress components are to be determined will be defined using the angle θ. To show this angle properly, it is first necessary to establish a positive x' axis, *directed outward*, perpendicular or normal to the plane, and an associated y' axis, directed along the plane, Fig. 9–5b. Notice that the unprimed and primed sets of axes both form right-handed coordinate systems; that is, the positive z (or z') axis is established by the right-hand rule. Curling the fingers from x (or x') toward y (or y') gives the direction for the positive z (or z') axis that points outward. The *angle θ* is measured from the positive x to the positive x' axis. It is *positive* provided it follows the curl of the right-hand fingers, i.e., counterclockwise as shown in Fig. 9–5b.

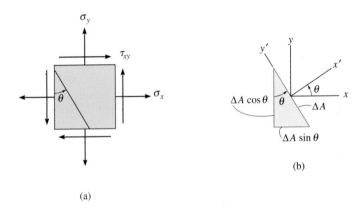

(a)

(b)

Fig. 9–6

Normal and Shear Stress Components.   Using the established sign convention, the element in Fig. 9–6a is sectioned along the inclined plane and the segment shown in Fig. 9–6b is isolated. Assuming the sectioned area is ΔA, then the horizontal and vertical faces of the segment have an area of $\Delta A \sin \theta$ and $\Delta A \cos \theta$, respectively.

The resulting *free-body diagram* of the segment is shown in Fig. 9–6c. Applying the equations of force equilibrium to determine the unknown normal and shear stress components $\sigma_{x'}$ and $\tau_{x'y'}$, we obtain

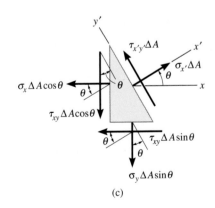

$$+\nearrow \Sigma F_{x'} = 0; \quad \sigma_{x'}\, \Delta A - (\tau_{xy}\, \Delta A \sin \theta) \cos \theta - (\sigma_y\, \Delta A \sin \theta) \sin \theta$$
$$- (\tau_{xy}\, \Delta A \cos \theta) \sin \theta - (\sigma_x\, \Delta A \cos \theta) \cos \theta = 0$$

$$\sigma_{x'} = \sigma_x \cos^2 \theta + \sigma_y \sin^2 \theta + \tau_{xy}(2 \sin \theta \cos \theta)$$

$$+\nwarrow \Sigma F_{y'} = 0; \quad \tau_{x'y'}\, \Delta A + (\tau_{xy}\, \Delta A \sin \theta) \sin \theta - (\sigma_y\, \Delta A \sin \theta) \cos \theta$$
$$- (\tau_{xy}\, \Delta A \cos \theta) \cos \theta + (\sigma_x\, \Delta A \cos \theta) \sin \theta = 0$$

$$\tau_{x'y'} = (\sigma_y - \sigma_x) \sin \theta \cos \theta + \tau_{xy}(\cos^2 \theta - \sin^2 \theta)$$

(c)

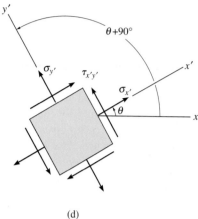

Fig. 9–6   (d)

These two equations may be simplified by using the trigonometric identities $\sin 2\theta = 2 \sin \theta \cos \theta$, $\sin^2 \theta = (1 - \cos 2\theta)/2$, and $\cos^2 \theta = (1 + \cos 2\theta)/2$, in which case,

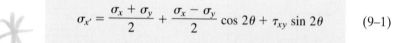

$$\sigma_{x'} = \frac{\sigma_x + \sigma_y}{2} + \frac{\sigma_x - \sigma_y}{2} \cos 2\theta + \tau_{xy} \sin 2\theta \qquad (9\text{–}1)$$

$$\tau_{x'y'} = -\frac{\sigma_x - \sigma_y}{2} \sin 2\theta + \tau_{xy} \cos 2\theta \qquad (9\text{–}2)$$

If the normal stress acting in the y' direction is needed, it can be obtained by simply substituting $(\theta = \theta + 90°)$ for θ into Eq. 9–1, Fig. 9–6d. This yields

$$\sigma_{y'} = \frac{\sigma_x + \sigma_y}{2} - \frac{\sigma_x - \sigma_y}{2} \cos 2\theta - \tau_{xy} \sin 2\theta \qquad (9\text{–}3)$$

If $\sigma_{y'}$ is calculated as a positive quantity, this indicates that it acts in the positive y' direction as shown in Fig. 9–6d.

PROCEDURE FOR ANALYSIS

To apply the stress transformation Eqs. 9–1 and 9–2, it is simply necessary to substitute in the known data for σ_x, σ_y, τ_{xy}, and θ in accordance with the established sign convention, Fig. 9–5. If $\sigma_{x'}$ and $\tau_{x'y'}$ are calculated as positive quantities, then these stresses act in the positive direction of the x' and y' axes.

For convenience these equations can easily be programmed on a pocket calculator.

EXAMPLE 9–2

The state of plane stress at a point is represented by the element shown in Fig. 9–7a. Determine the state of stress at the point on another element oriented 30° clockwise from the position shown.

SOLUTION

This problem was solved in Example 9–1 using basic principles. Here we will apply Eqs. 9–1 and 9–2. From the established sign convention, Fig. 9–5, it is seen that

$$\sigma_x = -80 \text{ MPa} \qquad \sigma_y = 50 \text{ MPa} \qquad \tau_{xy} = -25 \text{ MPa}$$

Plane CD. To obtain the stress components on plane *CD*, Fig. 9–7b, the positive x' axis is directed outward, perpendicular to *CD*, and the associated y' axis is directed along *CD*. The angle measured from the x to the x' axis is $\theta = -30°$ (clockwise). Applying Eqs. 9–1 and 9–2 yields

$$\sigma_{x'} = \frac{\sigma_x + \sigma_y}{2} + \frac{\sigma_x - \sigma_y}{2} \cos 2\theta + \tau_{xy} \sin 2\theta$$

$$= \frac{-80 + 50}{2} + \frac{-80 - 50}{2} \cos 2(-30°) + (-25) \sin 2(-30°)$$

$$= -25.8 \text{ MPa} \qquad\qquad\qquad\qquad\qquad\qquad \textit{Ans.}$$

$$\tau_{x'y'} = -\frac{\sigma_x - \sigma_y}{2} \sin 2\theta + \tau_{xy} \cos 2\theta$$

$$= -\frac{-80 - 50}{2} \sin 2(-30°) + (-25) \cos 2(-30°)$$

$$= -68.8 \text{ MPa} \qquad\qquad\qquad\qquad\qquad\qquad \textit{Ans.}$$

The negative signs indicate that $\sigma_{x'}$ and $\tau_{x'y'}$ act in the negative x' and y' directions, respectively. The results are shown acting on the element in Fig. 9–7d.

Plane BC. In a similar manner, the stress components acting on face *BC*, Fig. 9–7c, are obtained using $\theta = 60°$. Applying Eqs. 9–1 and 9–2,* we get

$$\sigma_{x'} = \frac{-80 + 50}{2} + \frac{-80 - 50}{2} \cos 2(60°) + (-25) \sin 2(60°)$$

$$= -4.15 \text{ MPa} \qquad\qquad\qquad\qquad\qquad\qquad \textit{Ans.}$$

$$\tau_{x'y'} = -\frac{-80 - 50}{2} \sin 2(60°) + (-25) \cos 2(60°)$$

$$= 68.8 \text{ MPa} \qquad\qquad\qquad\qquad\qquad\qquad \textit{Ans.}$$

Here $\tau_{x'y'}$ has been computed twice in order to provide a check. The negative sign for $\sigma_{x'}$ indicates that this stress acts in the negative x' direction, Fig. 9–7c. The results are shown on the element in Fig. 9–7d.

---

*Alternatively, we could apply Eq. 9–3 with $\theta = -30°$ rather than Eq. 9–1.

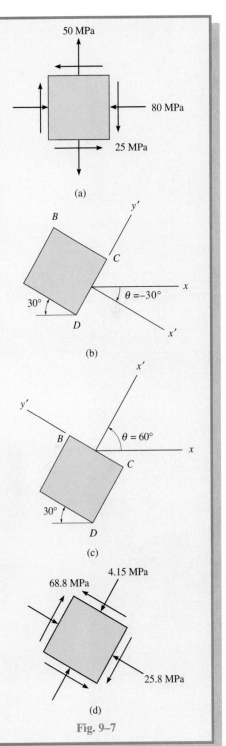

(a)

(b)

(c)

(d)

Fig. 9–7

9.3 PRINCIPAL STRESSES AND MAXIMUM IN-PLANE SHEAR STRESS

From Eqs. 9–1 and 9–2, it can be seen that $\sigma_{x'}$ and $\tau_{x'y'}$ depend on the angle of inclination θ of the planes on which these stresses act. In engineering practice it is often important to determine the orientation of the planes that causes the normal stress to be a maximum and a minimum and the orientation of the planes that causes the shear stress to be a maximum. In this section each of these problems will be considered.

In-Plane Principal Stresses. To determine the maximum and minimum *normal stress* we must differentiate Eq. 9–1 with respect to θ and set the result equal to zero. This gives

why set equal to zero?

$$\frac{d\sigma_{x'}}{d\theta} = -\frac{\sigma_x - \sigma_y}{2} (2 \sin 2\theta) + 2\tau_{xy} \cos 2\theta = 0$$

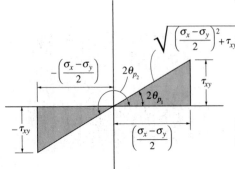

Fig. 9–8

Solving this equation we obtain the orientation $\theta = \theta_p$ of the planes of maximum and minimum normal stress.

$$\tan 2\theta_p = \frac{\tau_{xy}}{(\sigma_x - \sigma_y)/2} \tag{9–4}$$

The solution has two roots, θ_{p_1}, and θ_{p_2}. Specifically, the values of $2\theta_{p_1}$ and $2\theta_{p_2}$ are 180° apart, so θ_{p_1} and θ_{p_2} will be 90° apart.

The values of θ_{p_1} and θ_{p_2} must be substituted into Eq. 9–1 if we are to obtain the required normal stresses. We can obtain the necessary sine and cosine of $2\theta_{p_1}$ and $2\theta_{p_2}$ from the shaded triangles shown in Fig. 9–8. The construction of these triangles is based on Eq. 9–4, assuming that τ_{xy} and $(\sigma_x - \sigma_y)$ are both positive or both negative quantities. We have

for θ_{p_1}, $$\sin 2\theta_{p_1} = \tau_{xy} \bigg/ \sqrt{\left(\frac{\sigma_x - \sigma_y}{2}\right)^2 + \tau_{xy}^2}$$

$$\cos 2\theta_{p_1} = \left(\frac{\sigma_x - \sigma_y}{2}\right) \bigg/ \sqrt{\left(\frac{\sigma_x - \sigma_y}{2}\right)^2 + \tau_{xy}^2}$$

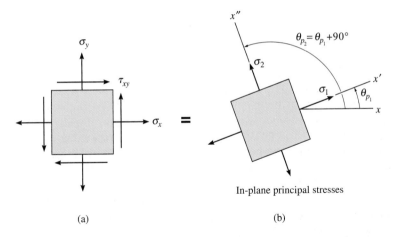

In-plane principal stresses

(a)   (b)

Fig. 9–9

for θ_{p2},
$$\sin 2\theta_{p2} = -\tau_{xy} \Big/ \sqrt{\left(\frac{\sigma_x - \sigma_y}{2}\right)^2 + \tau_{xy}^2}$$

$$\cos 2\theta_{p2} = -\left(\frac{\sigma_x - \sigma_y}{2}\right) \Big/ \sqrt{\left(\frac{\sigma_x - \sigma_y}{2}\right)^2 + \tau_{xy}^2}$$

The cracks in this concrete beam were caused by tension stress, even though the beam was subjected to both an internal moment and shear. The stress transformation equations can be used to predict the direction of the cracks, and the principal normal stresses that caused it.

If either of these two sets of trigonometric relations are substituted into Eq. 9–1 and simplified, we obtain

$$\sigma_{1,2} = \frac{\sigma_x + \sigma_y}{2} \pm \sqrt{\left(\frac{\sigma_x - \sigma_y}{2}\right)^2 + \tau_{xy}^2} \qquad (9–5)$$

Depending upon the sign chosen, this result gives the maximum or minimum in-plane normal stress acting at a point, where $\sigma_1 \geq \sigma_2$. This particular set of values are called the in-plane **principal stresses**, and the corresponding planes on which they act are called the **principal planes** of stress, Fig. 9–9b. Furthermore, if the trigonometric relations for θ_{p1} and θ_{p2} are substituted into Eq. 9–2, it can be seen that $\tau_{x'y'} = 0$; that is, **no shear stress acts on the principal planes**.

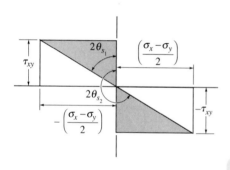

Fig. 9–10

Maximum In-Plane Shear Stress.   The orientation of an element that is subjected to maximum shear stress on its faces can be determined by taking the derivative of Eq. 9–2 with respect to θ and setting the result equal to zero. This gives

$$\tan 2\theta_s = \frac{-(\sigma_x - \sigma_y)/2}{\tau_{xy}} \tag{9–6}$$

The two roots of this equation, θ_{s_1} and θ_{s_2}, can be determined from the shaded triangles shown in Fig. 9–10. By comparison with Fig. 9–8, each root of $2\theta_s$ is 90° from $2\theta_p$. Thus, the roots θ_s and θ_p are 45° apart, and as a result *the planes for maximum shear stress can be determined by orienting an element 45° from the position of an element that defines the planes of principal stress.*

Using either one of the roots θ_{s_1} or θ_{s_2}, the maximum shear stress can be found by taking the trigonometric values of $\sin 2\theta_s$ and $\cos 2\theta_s$ from Fig. 9–10 and substituting them into Eq. 9–2. The result is

$$\tau_{\substack{\text{max} \\ \text{in-plane}}} = \sqrt{\left(\frac{\sigma_x - \sigma_y}{2}\right)^2 + \tau_{xy}^2} \tag{9–7}$$

The value of $\tau_{\substack{\text{max} \\ \text{in-plane}}}$ as calculated by Eq. 9–7 is referred to as the **maximum in-plane shear stress** because it acts on the element in the $x-y$ plane.

Substituting the values for $\sin 2\theta_s$ and $\cos 2\theta_s$ into Eq. 9–1, we see that there is also a normal stress on the planes of maximum in-plane shear stress. We get

$$\sigma_{\text{avg}} = \frac{\sigma_x + \sigma_y}{2} \tag{9–8}$$

Like the stress-transformation equations, it may be convenient to program the above equations so they can be used on a pocket calculator.

IMPORTANT POINTS

• The *principal stresses* represent the maximum and minimum normal stress at the point.

• When the state of stress is represented by the principal stresses, *no shear stress* will act on the element.

• The state of stress at the point can also be represented in terms of the *maximum in-plane shear stress*. In this case an *average normal stress* will also act on the element.

• The element representing the maximum in-plane shear stress with the associated average normal stresses is *oriented* 45° from the element representing the principal stresses.

EXAMPLE 9-3

When the torsional loading T is applied to the bar in Fig. 9–11a, it produces a state of pure shear stress in the material. Determine a) the maximum in-plane shear stress and the associated average normal stress, and b) the principal stress.

SOLUTION

From the established sign convention,

$$\sigma_x = 0 \qquad \sigma_y = 0 \qquad \tau_{xy} = -\tau$$

Maximum In-Plane Shear Stress. Applying Eqs. 9–7 and 9–8, we have

$$\tau_{\substack{\max \\ \text{in-plane}}} = \sqrt{\left(\frac{\sigma_x - \sigma_y}{2}\right)^2 + \tau_{xy}^2} = \sqrt{(0)^2 + (-\tau)^2} = \pm\tau \qquad \textit{Ans.}$$

$$\sigma_{\text{avg}} = \frac{\sigma_x + \sigma_y}{2} = \frac{0 + 0}{2} = 0 \qquad \textit{Ans.}$$

Thus, as expected, the maximum in-plane shear stress is represented by the element in Fig. 9– 11a.

Through experiment it has been found that materials that are *ductile* will *fail* due to *shear stress*. As a result, if a torque is applied to a bar made from mild steel, the maximum in-plane shear stress will cause it to fail as shown in the adjacent photo.

Principal Stress. Applying Eqs. 9–4 and 9–5 yields

$$\tan 2\theta_p = \frac{\tau_{xy}}{(\sigma_x - \sigma_y)/2} = \frac{-\tau}{(0 - 0)/2} \ , \sigma_{p_2} = 45°, \sigma_{p_1} = 135°$$

$$\sigma_{1,2} = \frac{\sigma_x + \sigma_y}{2} \pm \sqrt{\left(\frac{\sigma_x - \sigma_y}{2}\right)^2 + \tau_{xy}^2} = 0 \pm \sqrt{(0)^2 + \tau^2} = \pm\tau \qquad \textit{Ans.}$$

If we now apply Eq. 9–1 with $\theta_{p_2} = 45°$, then

$$\sigma_{x'} = \frac{\sigma_x + \sigma_y}{2} + \frac{\sigma_x - \sigma_y}{2}\cos 2\theta + \tau_{xy}\sin 2\theta = 0 + 0 + (-\tau)\sin 90° = -\tau$$

Thus, $\sigma_2 = -\tau$ acts at $\theta_{p_2} = 45°$ as shown in Fig. 9–11b, and $\sigma_1 = \tau$ acts on the other face, $\theta_{p_1} = 135°$.

Materials that are *brittle* fail due to *normal stress*. That is why when a brittle material, such as cast iron, is subjected to torsion it will fail in tension at a 45° inclination as seen in the adjacent photo.

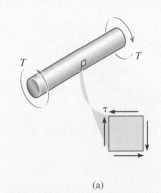

(a)

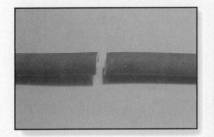

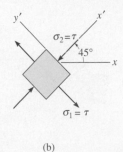

(b)

Fig. 9–11

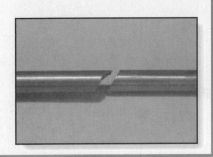

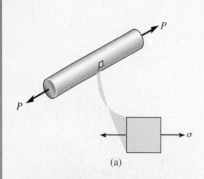

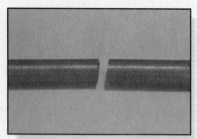

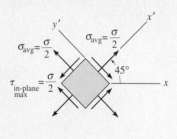

(a)

(b)

Fig. 9–12

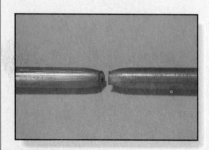

EXAMPLE 9–4

When the axial loading P is applied to the bar in Fig. 9–12a, it produces a tensile stress in the material. Determine a) the principal stress and b) the maximum in-plane shear stress and associated average normal stress.

SOLUTION

From the established sign convention,

$$\sigma_x = \sigma \qquad \sigma_y = 0 \qquad \tau_{xy} = 0$$

Principal Stress. By observation, the element orientated as shown in Fig. 9–12a illustrates a condition of principal stress since no shear stress acts on this element. This can also be shown by direct substitution of the above values into Eqs. 9–4 and 9–5. Thus,

$$\sigma_1 = \sigma \qquad \sigma_2 = 0 \qquad \qquad Ans.$$

Since experiments have shown that normal stress causes brittle materials to fail, then if the bar is made from *brittle material*, such as cast iron, it will cause failure as shown in the adjacent photo.

Maximum In-Plane Shear Stress. Applying Eqs. 9–6, 9–7, and 9–8, we have

$$\tan 2\theta_s = \frac{-(\sigma_x - \sigma_y)/2}{\tau_{xy}} = \frac{-(\sigma - 0)/2}{0}; \theta_{s_1} = 45°, \theta_{s_2} = 135°$$

$$\tau_{\substack{\text{max} \\ \text{in-plane}}} = \sqrt{\left(\frac{\sigma_x - \sigma_y}{2}\right)^2 + \tau_{xy}^2} = \sqrt{\left(\frac{\sigma - 0}{2}\right)^2 + (0)^2} = \pm\frac{\sigma}{2} \quad Ans.$$

$$\sigma_{\text{avg}} = \frac{\sigma_x + \sigma_y}{2} = \frac{\sigma + 0}{2} = \frac{\sigma}{2} \qquad \qquad Ans.$$

To determine the proper orientation of the element, apply Eq. 9–2.

$$\tau_{x'y'} = -\frac{\sigma_x - \sigma_y}{2} \sin 2\theta + \tau_{xy} \cos 2\theta = -\frac{\sigma - 0}{2} \sin 90° + 0 = -\frac{\sigma}{2}$$

This negative shear stress acts on the x' face, in the negative y' direction as shown in Fig. 9–12b.

If the bar is made from a *ductile material* such as mild steel then shear stress will cause it to fail when it is subjected to *tension*. This can be noted in the adjacent photo, where within the region of necking, shear stress has caused "slipping" along the steel's crystalline boundaries, resulting in a plane of failure that has formed a *cone* around the bar oriented at approximately 45° as calculated above.

EXAMPLE 9-5

The state of plane stress at a point on a body is shown on the element in Fig. 9–13a. Represent this stress state in terms of the principal stresses.

SOLUTION

From the established sign convention, we have

$$\sigma_x = -20 \text{ MPa} \qquad \sigma_y = 90 \text{ MPa} \qquad \tau_{xy} = 60 \text{ MPa}$$

Orientation of Element. Applying Eq. 9–4, we have

$$\tan 2\theta_p = \frac{\tau_{xy}}{(\sigma_x - \sigma_y)/2} = \frac{60}{(-20 - 90)/2}$$

Solving, and referring to this root as θ_{p_2}, as will be shown below, yields

$$2\theta_{p_2} = -47.49° \qquad \theta_{p_2} = -23.7°$$

Since the difference between $2\theta_{p_1}$ and $2\theta_{p_2}$ is 180°, we have

$$2\theta_{p_1} = 180° + 2\theta_{p_2} = 132.51° \qquad \theta_{p_1} = 66.3°$$

Recall that θ is measured positive *counterclockwise* from the x axis to the outward normal (x' axis) on the face of the element, and so the results are shown in Fig. 9–13b.

Principal Stresses. We have

$$\sigma_{1,2} = \frac{\sigma_x + \sigma_y}{2} \pm \sqrt{\left(\frac{\sigma_x - \sigma_y}{2}\right)^2 + \tau_{xy}^2}$$

$$= \frac{-20 + 90}{2} \pm \sqrt{\left(\frac{-20 - 90}{2}\right)^2 + (60)^2}$$

$$= 35.0 \pm 81.4$$

$$\sigma_1 = 116 \text{ MPa} \qquad\qquad\qquad \textit{Ans.}$$

$$\sigma_2 = -46.4 \text{ MPa} \qquad\qquad\qquad \textit{Ans.}$$

The principal plane on which each normal stress acts can be determined by applying Eq. 9–1 with, say, $\theta = \theta_{p_2} = -23.7°$. We have

$$\sigma_{x'} = \frac{\sigma_x + \sigma_y}{2} + \frac{\sigma_x - \sigma_y}{2} \cos 2\theta + \tau_{xy} \sin 2\theta$$

$$= \frac{-20 + 90}{2} + \frac{-20 - 90}{2} \cos 2(-23.7°) + 60 \sin 2(-23.7°)$$

$$= -46.4 \text{ MPa}$$

Hence, $\sigma_2 = -46.4$ MPa acts on the plane defined by $\theta_{p_2} = -23.7°$, whereas $\sigma_1 = 116$ MPa acts on the plane defined by $\theta_{p_1} = 66.3°$. The results are shown on the element in Fig. 9–13c. Recall that no shear stress acts on this element.

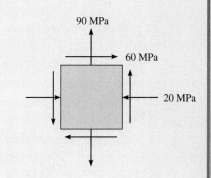

(a)

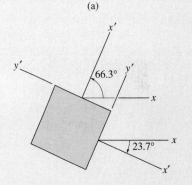

(b)

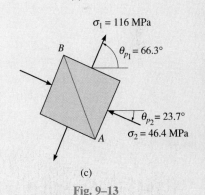

(c)

Fig. 9–13

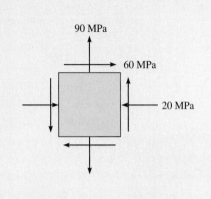

90 MPa

60 MPa

20 MPa

(a)

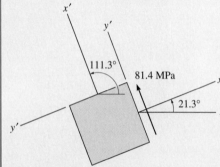

x'

y'

111.3°

81.4 MPa

x'

21.3°

x

y'

(b)

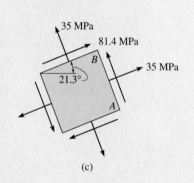

35 MPa

81.4 MPa

B

35 MPa

21.3°

A

(c)

Fig. 9–14

EXAMPLE 9–6

The state of plane stress at a point on a body is represented on the element shown in Fig. 9–14a. Represent this stress state in terms of the maximum in-plane shear stress and associated average normal stress.

SOLUTION

Orientation of Element. Since $\sigma_x = -20$ MPa, $\sigma_y = 90$ MPa, and $\tau_{xy} = 60$ MPa, applying Eq 9–6, we have

$$\tan 2\theta_s = \frac{-(\sigma_x - \sigma_y)/2}{\tau_{xy}} = \frac{-(-20 - 90)/2}{60}$$

$$2\theta_{s_2} = 42.5° \qquad \theta_{s_2} = 21.3°$$

$$2\theta_{s_1} = 180° + 2\theta_{s_2} \qquad \theta_{s_1} = 111.3°$$

Note that these angles shown in Fig. 9–14b are 45° away from the principal planes of stress, which was determined in Example 9–5.

Maximum In-Plane Shear Stress. Applying Eq. 9–7,

$$\tau_{\substack{\max \\ \text{in-plane}}} = \sqrt{\left(\frac{\sigma_x - \sigma_y}{2}\right)^2 + \tau_{xy}^2} = \sqrt{\left(\frac{-20 - 90}{2}\right)^2 + (60)^2}$$

$$= 81.4 \text{ MPa} \qquad\qquad Ans.$$

The proper direction of $\tau_{\substack{\max \\ \text{in-plane}}}$ on the element can be determined by considering $\theta = \theta_{s_2} = 21.3°$, and applying Eq. 9–2. We have

$$\tau_{x'y'} = -\left(\frac{\sigma_x - \sigma_y}{2}\right) \sin 2\theta + \tau_{xy} \cos 2\theta$$

$$= -\left(\frac{-20 - 90}{2}\right) \sin 2(21.3°) + 60 \cos 2(21.3°)$$

$$= 81.4 \text{ MPa}$$

Thus, $\tau_{\substack{\max \\ \text{in-plane}}} = \tau_{x'y'}$ acts in the *positive y'* direction on this face ($\theta = 21.3°$), Fig. 9–14b. The shear stresses on the other three faces are directed as shown in Fig. 9–14c.

Average Normal Stress. Besides the maximum shear stress, as calculated above, the element is also subjected to an average normal stress determined from Eq. 9–8; that is,

$$\sigma_{avg} = \frac{\sigma_x + \sigma_y}{2} = \frac{-20 + 90}{2} = 35 \text{ MPa} \qquad\qquad Ans.$$

This is a tensile stress. The results are shown in Fig. 9–14c.

PROBLEMS

9–1. Prove that the sum of the normal stresses $\sigma_x + \sigma_y = \sigma_{x'} + \sigma_{y'}$ is constant.

9–2. The state of stress at a point in a member is shown on the element. Determine the stress components acting on the inclined plane AB. Solve the problem using the method of equilibrium described in Sec. 9.1.

9–6. The state of stress at a point in a member is shown on the element. Determine the stress components acting on the inclined plane AB. Solve the problem using the method of equilibrium described in Sec. 9.1.

9–7. Solve Prob. 9–6 using the stress-transformation equations developed in Sec. 9.2.

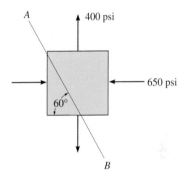

Probs. 9–6/ 9–7

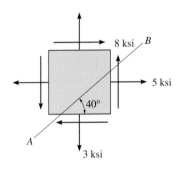

Probs. 9–2/ 9–3

9–3. Solve Prob. 9–2 using the stress-transformation equations developed in Sec. 9.2.

*9–4. The state of stress at a point in a member is shown on the element. Determine the stress components acting on the inclined plane AB. Solve the problem using the method of equilibrium described in Sec. 9.1.

9–5. Solve Prob. 9–4 using the stress-transformation equations developed in Sec. 9.2. Show the result on a sketch.

*9–8. Determine the equivalent state of stress on an element if the element is oriented 60° clockwise from the element shown.

9–9. Determine the equivalent state of stress on an element if the element is oriented 30° counterclockwise from the element shown.

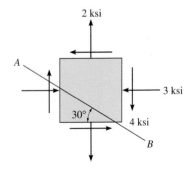

Probs. 9–4/9–5

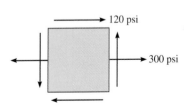

Probs. 9–8/9–9

9–10. The state of stress at a point is shown on the element. Determine (*a*) the principal stresses and (*b*) the maximum in-plane shear stress and average normal stress at the point. Specify the orientation of the element in each case.

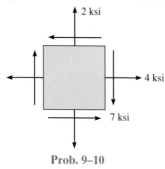

Prob. 9–10

9–11. Determine the equivalent state of stress on an element if it is oriented 50° counterclockwise from the element shown. Use the stress-transformation equations.

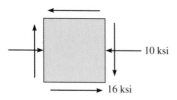

Prob. 9–11

***9–12.** The state of stress at a point is shown on the element. Determine (*a*) the principal stresses and (*b*) the maximum in-plane shear stress and average normal stress at the point. Specify the orientation of the element in each case.

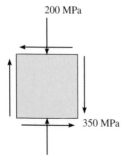

Prob. 9–12

9–13. The state of stress at a point is shown on the element. Determine (*a*) the principal stresses and (*b*) the maximum in-plane shear stress and average normal stress at the point. Specify the orientation of the element in each case.

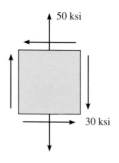

Prob. 9–13

9–14. The state of stress at a point on the upper surface of the airplane wing is shown on the element. Determine (*a*) the principal stresses and (*b*) the maximum in-plane shear stress and average normal stress at the point. Specify the orientation of the element in each case.

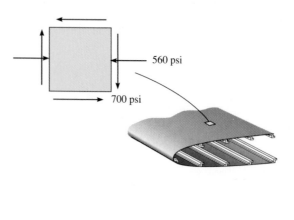

Prob. 9–14

9–15. The steel bar has a thickness of 0.5 in. and is subjected to the edge loading shown. Determine the principal stresses developed in the bar.

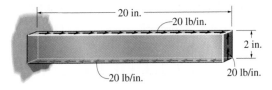

Prob. 9–15

*9–16. The steel plate has a thickness of 10 mm and is subjected to the edge loading shown. Determine the maximum in-plane shear stress and the average normal stress developed in the steel.

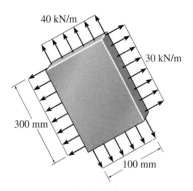

Prob. 9–16

■9–17. Consider the general case of plane stress as shown. Write a computer program that can be used to determine the normal and shear stresses, $\sigma_{x'}$ and $\tau_{x'y'}$, on the plane of an element oriented at an angle θ from the horizontal. Also, compute the principal stresses, the maximum in-plane shear stress, the average normal stress, and the element's orientation.

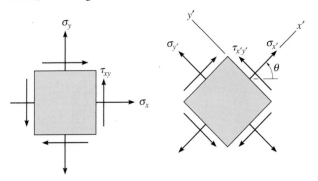

Prob. 9–17

9–18. A point on a thin plate is subjected to the two successive states of stress shown. Determine the resultant state of stress represented on the element oriented as shown on the right.

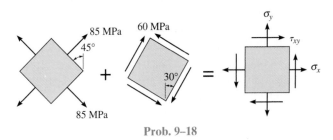

Prob. 9–18

9–19. The stress acting on two planes at a point is indicated. Determine the shear stress on plane a–a and the principal stresses at the point.

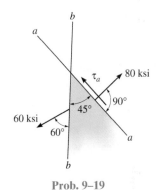

Prob. 9–19

*9–20. The stress acting on two planes at a point is indicated. Determine the normal stress σ_b and the principal stresses at the point.

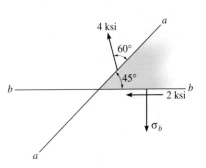

Prob. 9–20

The following problems involve material covered in Chapter 8.

9–21. The wooden beam is subjected to a load of 12 kN. If grains of wood in the beam at point A make an angle of $25°$ with the horizontal as shown, determine the normal and shear stresses that act perpendicular and parallel to the grains due to the loading.

9–22. The wooden beam is subjected to a load of 12 kN. Determine the principal stresses at point A and specify the orientation of the element.

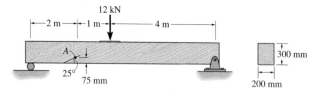

Probs. 9–21/9–22

9–23. The clamp bears down on the smooth surface at E by tightening the bolt. If the tensile force in the bolt is 40 kN, determine the principal stresses at points A and B and show the results on elements located at each of these points. The cross-sectional area at A and B is shown in the adjacent figure.

***9–24.** Solve Prob. 9–23 for points C and D.

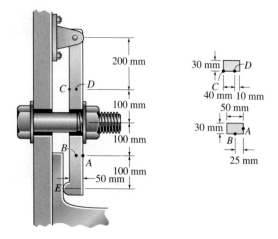

Probs. 9–23/9–24

9–25. The square steel plate has a thickness of 10 mm and is subjected to the edge loading shown. Determine the maximum in-plane shear stress and the average normal stress developed in the steel.

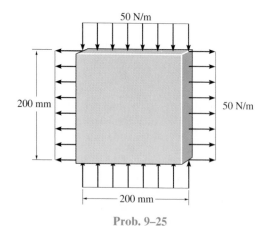

Prob. 9–25

9–26. The T-beam is subjected to the distributed loading that is applied along its centerline. Determine the principal stresses at points A and B and show the results on elements located at each of these points.

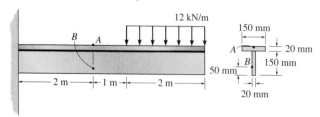

Prob. 9–26

9–27. The bent rod has a diameter of 15 mm and is subjected to the force of 600 N. Determine the principal stresses and the maximum in-plane shear stress that are developed at point A and point B. Show the results on properly oriented elements located at these points.

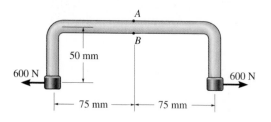

Prob. 9–27

*9–28. The beam has a rectangular cross section and is subjected to the loading shown. Determine the principal stresses and the maximum in-plane shear stress that are developed at point A and point B. Show the results on properly oriented elements located at these points.

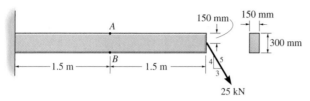

Prob. 9–28

9–29. The beam has a rectangular cross section and is subjected to the loadings shown. Determine the principal stresses and the maximum in-plane shear stress that are developed at point A and point B. These points are just to the left of the 2000-lb load. Show the results on properly oriented elements located at these points.

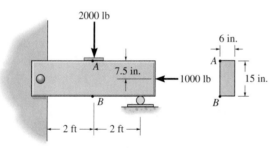

Prob. 9–29

9–30. The wide-flange beam is subjected to the loading shown. Determine the principal stress in the beam at point A and at point B. These points are located at the top and bottom of the web, respectively. Although it is not very accurate, use the shear formula to compute the shear stress.

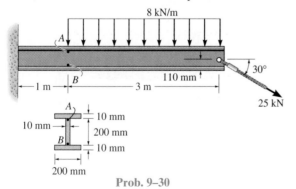

Prob. 9–30

9–31. The shaft has a diameter d and is subjected to the loadings shown. Determine the principal stresses and the maximum in-plane shear stress that is developed anywhere on the surface of the shaft.

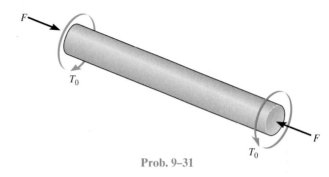

Prob. 9–31

*9–32. The shaft has a diameter d and is subjected to the loadings shown. Determine the principal stresses and the maximum in-plane shear stress that is developed at point A. The bearings only support vertical reactions.

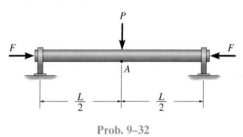

Prob. 9–32

9–33. A rod has a circular cross section with a diameter of 2 in. It is subjected to a torque of 12 kip · in. and a bending moment **M**. The greater principal stress at the point of maximum flexural stress is 15 ksi. Determine the magnitude of the bending moment.

Prob. 9–33

9–34. The nose wheel of the plane is subjected to a design load of 12 kN. Determine the principal stresses acting on the aluminum wheel support at point A.

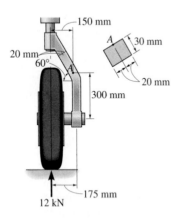

Prob. 9–34

9–35. The internal loadings at a section of the beam are shown. Determine the principal stresses at point A. Also compute the maximum in-plane shear stress at this point.

*9–36. Solve Prob. 9–35 for point B.

9–37. Solve Prob. 9–35 for point C, located in the center on the bottom of the web.

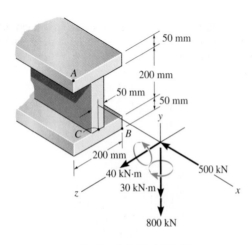

Probs. 9–35/9–36/9–37

9–38. The internal loadings at a section of the beam consist of an axial force of 6 kip, a shear force of 12 kip, and a moment of 5 kip · ft. Determine the principal stresses at point A. Also compute the maximum in-plane shear stress at this point.

9–39. Solve Prob. 9–38 for point B.

*9–40. Solve Prob. 9–38 for point C.

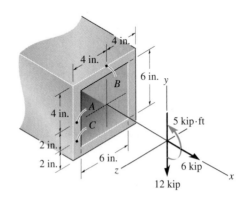

Probs. 9–38/9–39/9–40

9–41. The beam has a rectangular cross section and is subjected to the loadings shown. Determine the principal stresses that are developed at point A and point B, which are located just to the left of the 20-kN load. Show the results on elements located at these points.

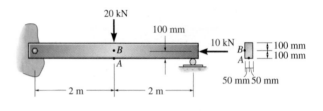

Prob. 9–41

9–42. The solid shaft is subjected to a torque, bending moment, and shear force as shown. Determine the principal stresses acting at point A.

9–43. Solve Prob. 9-42 for point B.

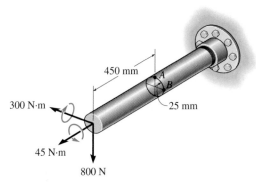

450 mm

300 N·m

25 mm

45 N·m

800 N

Probs. 9–42/9–43

*9–44. The internal loadings at a cross section through the 6-in.-diameter drive shaft of a turbine consist of an axial force of 2500 lb, a bending moment of 800 lb · ft, and a torsional moment of 1500 lb · ft. Determine the principal stresses at point A. Also calculate the maximum in-plane shear stress at this point.

9–45. The internal loadings at a cross section through the 6-in.-diameter drive shaft of a turbine consist of an axial force of 2500 lb, a bending moment of 800 lb · ft, and a torsional moment of 1500 lb · ft. Determine the principal stresses at point B. Also calculate the maximum in-plane shear stress at this point.

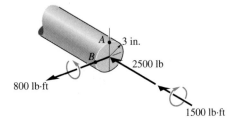

3 in.

2500 lb

800 lb·ft

1500 lb·ft

Probs. 9–44/9–45

9–46. The box beam is subjected to the loading shown. Determine the principal stresses in the beam at points A and B.

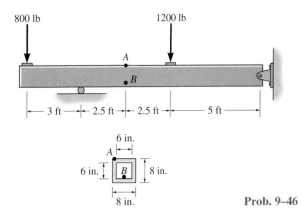

800 lb   1200 lb

A

$\cdot B$

3 ft — 2.5 ft — 2.5 ft — 5 ft

6 in.

A

6 in. B 8 in.

8 in.

Prob. 9–46

9–47. The beam has a rectangular cross section and is subjected to the loads shown. Write a computer program that can be used to determine the principal stresses at points A, B, C, and D. Show an application of the program using the values $h = 12$ in., $b = 8$ in., $N_x = 400$ lb, $V_y = 300$ lb, $V_z = 0$, $M_y = 0$, and $M_z = -150$ lb · ft.

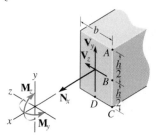

Prob. 9–47

*9–48. The member has a rectangular cross section and is subjected to the loading shown. Write a computer program that can be used to determine the principal stresses at points A, B, and C. Show an application of the program using the values $b = 150$ mm, $h = 200$ mm, $P = 1.5$ kN, $x = 75$ mm, $z = -50$ mm, $V_x = 300$ N, and $V_z = 600$ N.

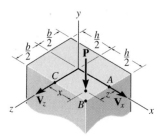

Prob. 9–48

9.4 MOHR'S CIRCLE—PLANE STRESS

In this section we will show that the equations for plane stress transformation have a graphical solution that is often convenient to use and easy to remember. Furthermore, this approach will allow us to "visualize" how the normal and shear stress components $\sigma_{x'}$ and $\tau_{x'y'}$ vary as the plane on which they act is oriented in different directions, Fig. 9–15a.

Equations 9–1 and 9–2 can be rewritten in the form

$$\sigma_{x'} - \left(\frac{\sigma_x + \sigma_y}{2}\right) = \left(\frac{\sigma_x - \sigma_y}{2}\right) \cos 2\theta + \tau_{xy} \sin 2\theta \tag{9–9}$$

$$\tau_{x'y'} = -\left(\frac{\sigma_x - \sigma_y}{2}\right) \sin 2\theta + \tau_{xy} \cos 2\theta \tag{9–10}$$

The parameter θ can be *eliminated* by squaring each equation and adding the equations together. The result is

$$\left[\sigma_{x'} - \left(\frac{\sigma_x + \sigma_y}{2}\right)\right]^2 + \tau_{x'y'}^2 = \left(\frac{\sigma_x - \sigma_y}{2}\right)^2 + \tau_{xy}^2$$

For a specific problem, σ_x, σ_y, τ_{xy} are *known constants*. Thus the above equation can be written in a more compact form as

$$(\sigma_{x'} - \sigma_{avg})^2 + \tau_{x'y'}^2 = R^2 \tag{9–11}$$

where

$$\sigma_{avg} = \frac{\sigma_x + \sigma_y}{2}$$

$$R = \sqrt{\left(\frac{\sigma_x - \sigma_y}{2}\right)^2 + \tau_{xy}^2} \tag{9–12}$$

If we establish coordinate axes, σ *positive to the right* and τ *positive downward*, and then plot Eq. 9–11, it will be seen that this equation represents a *circle* having a radius R and center on the σ axis at point $C(\sigma_{avg}, 0)$, Fig. 9–15b. This circle is called *Mohr's circle*, because it was developed by the German engineer Otto Mohr.

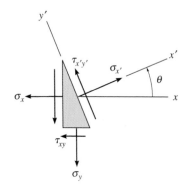

(a)

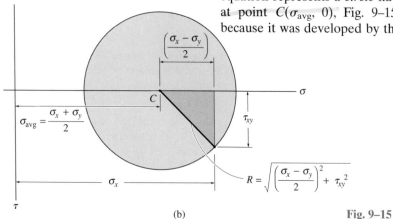

(b)

Fig. 9–15

To draw Mohr's circle it is necessary to first establish the σ and τ axis, Fig. 9–16c. Since the stress components σ_x, σ_y, τ_{xy} are known, the center of the circle can then be plotted, $C(\sigma_{avg}, 0)$. To obtain the radius, we need to know at least one point on the circle. Consider the case when the x' axis is coincident with the x axis as shown in Fig. 9–16a. Then $\theta = 0°$ and $\sigma_{x'} = \sigma_x$, $\tau_{x'y'} = \tau_{xy}$. We will refer to this as the "reference point" A and plot its coordinates $A(\sigma_x, \tau_{xy})$, Fig. 9–16c. Applying the Pythagorean theorem to the shaded triangle, the radius R can now be determined, which checks with Eq. 9–12. With points C and A known, the circle can be drawn as shown.

Now consider rotating the x' axis 90° counterclockwise, Fig. 9–16b. Then $\sigma_{x'} = \sigma_y$, $\tau_{x'y'} = -\tau_{xy}$. These values are the coordinates of point $G(\sigma_y, -\tau_{xy})$ on the circle, Fig. 9–16c. Hence, the radial line CG is 180° counterclockwise from the "reference line" CA. In other words, a rotation θ of the x' axis on the element will correspond to a rotation 2θ on the circle in the *same direction*.*

Once established, Mohr's circle can be used to determine the principal stresses, the maximum in-plane shear stress and associated average normal stress, or the stress on any arbitrary plane. The method for doing this is explained in the following procedure for analysis.

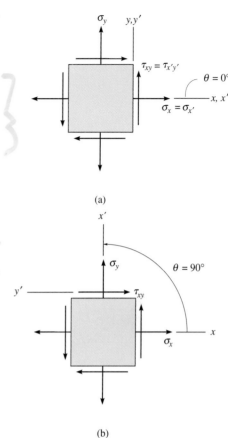

(a)

(b)

_____

*If instead the τ axis is constructed *positive upwards*, then the angle 2θ on the circle would be measured in the *opposite direction* to the orientation θ of the plane.

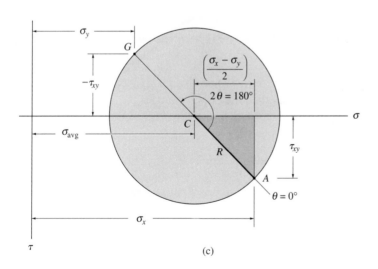

(c)

Fig. 9–16

PROCEDURE FOR ANALYSIS

The following steps are required to draw and use Mohr's circle.

Construction of the Circle.

- Establish a coordinate system such that the abscissa represents the normal stress σ, with *positive to the right*, and the ordinate represents the shear stress τ, with *positive downward*, Fig. 9–17a.*

- Using the positive sign convention for σ_x, σ_y, τ_{xy}, as shown in Fig. 9–17b, plot the center of the circle C, which is located on the σ axis at a distance $\sigma_{\text{avg}} = (\sigma_x + \sigma_y)/2$ from the origin, Fig. 9–17a.

- Plot the "reference point" A having coordinates $A(\sigma_x, \tau_{xy})$. This point represents the normal and shear stress components on the element's right-hand vertical face, and since the x' axis coincides with the x axis, this represents $\theta = 0°$, Fig. 9–17b.

- Connect point A with the center C of the circle and determine CA by trigonometry. This distance represents the radius R of the circle, Fig. 9–17a.

- Once R has been determined, sketch the circle.

Principal Stress.

- The principal stresses σ_1 and σ_2 ($\sigma_1 \geq \sigma_2$) are represented by the two points B and D where the circle intersects the σ axis, i.e., where $\tau = 0$, Fig. 9–17a.

- These stresses act on planes defined by angles θ_{p_1} and θ_{p_2}, Fig. 9–17c. They are represented on the circle by angles $2\theta_{p_1}$ (shown) and $2\theta_{p_2}$ (not shown) and are measured *from* the radial reference line CA *to* lines CB and CD, respectively.

- Using trigonometry, only one of these angles needs to be calculated from the circle, since θ_{p_1} and θ_{p_2} are 90° apart. Remember that the direction of rotation $2\theta_p$ on the circle (here counterclockwise) represents the *same* direction of rotation θ_p from the reference axis ($+x$) to the principal plane ($+x'$), Fig. 9–17c.*

Maximum In-Plane Shear Stress.

• The average normal stress and maximum in-plane shear stress components are determined from the circle as the coordinates of either point E or F, Fig. 9–17a.

• In this case the angles θ_{s_1} and θ_{s_2} give the orientation of the planes that contain these components, Fig. 9–17d. The angle $2\theta_{s_1}$ is shown in Fig. 9–17a and can be determined using trigonometry. Here the rotation is clockwise, and so θ_{s_1} must be clockwise on the element, Fig. 9–17d.*

Stresses on Arbitrary Plane.

• The normal and shear stress components $\sigma_{x'}$ and $\tau_{x'y'}$ acting on a specified plane defined by the angle θ, Fig. 9–17e, can be obtained from the circle using trigonometry to determine the coordinates of point P, Fig. 9–17a.

• To locate P, the known angle θ for the plane (in this case counterclockwise), Fig. 9–17e, must be measured on the circle in the *same direction* 2θ (counterclockwise), *from* the radial reference line CA to the radial line CP, Fig. 9–17a.*

———————
* If instead the τ axis is constructed *positive upwards*, then the angle 2θ on the circle would be measured in the *opposite direction* to the orientation θ of the plane.

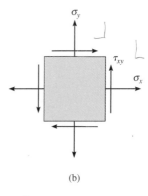

(b)

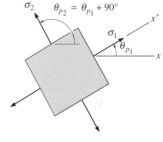

(c)

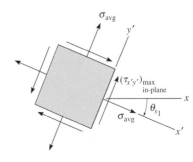

(d)

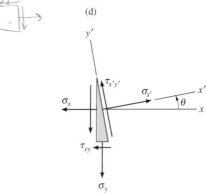

(e)

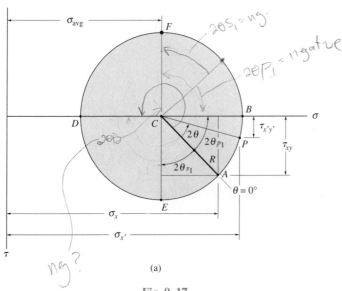

(a)

Fig. 9–17

EXAMPLE 9-7

The axial loading P produces the state of stress in the material as shown in Fig. 9–18 *a*. Draw Mohr's circle for this case.

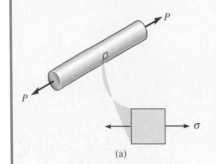

(a)

SOLUTION

Construction of the Circle.   From Fig. 9–18a,

$$\sigma_x = \sigma \quad \sigma_y = 0 \quad \tau_{xy} = 0$$

The σ and τ axes are established in Fig. 9–18b. The center of the circle C is on the σ axis at

$$\sigma_{avg} = \frac{\sigma_x + \sigma_y}{2} = \frac{\sigma + 0}{2} = \frac{\sigma}{2}$$

From the right-hand face of the element, Fig. 9–18*a*, the reference point for $\theta = 0°$ has coordinates $A(\sigma, 0)$. Hence the radius of the circle CA is $R = \sigma/2$, Fig. 9–18*b*.

Stresses.   Note that the principal stresses are at points A and D.

$$\sigma_1 = \sigma \qquad \sigma_2 = 0$$

The element in Fig. 9–18*a* represents this principal state of stress.

The maximum in-plane shear stress and associated average normal stress is identified on the circle as point E or F, Fig. 9–18*b*. At E we have

$$\tau_{\substack{max \\ \text{in-plane}}} = \frac{\sigma}{2}$$

$$\sigma_{avg} = \frac{\sigma}{2}$$

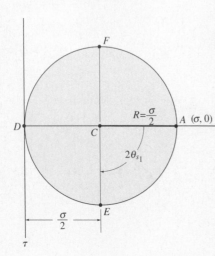

(b)

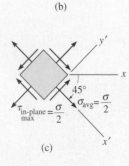

(c)

Fig. 9–18

By observation, the clockwise angle $2\theta_{s_1} = 90°$. Therefore $\theta_{s_1} = 45°$, so that the x' axis is oriented 45° clockwise from the x axis; Fig. 9–18c. Since E has positive coordinates, then σ_{avg} and $\tau_{\substack{max \\ \text{in-plane}}}$ act in the positive x' and y' directions, respectively.

EXAMPLE 9-8

The torsional loading T produces the state of stress in the shaft as shown in Fig. 9–19*a*. Draw Mohr's circle for this case.

SOLUTION

Construction of the Circle. From Fig. 9–19*a*,

$$\sigma_x = 0 \qquad \sigma_y = 0 \qquad \tau_{xy} = -\tau$$

The σ and τ axes are established in Fig. 9–19*b*. The center of the circle C is on the σ axis at

$$\sigma_{avg} = \frac{\sigma_x + \sigma_y}{2} = \frac{0 + 0}{2} = 0$$

From the right-hand face of the element, Fig. 9–19*a*, the reference point for $\theta = 0°$ has coordinates $A(0, -\tau)$, Fig. 9–19*b*. Hence the radius CA is $R = \tau$.

Stresses. Here point A represents a point of average normal stress and maximum in-plane shear stress, Fig. 9–19*b*. Thus,

$$\tau_{\substack{\text{max} \\ \text{in-plane}}} = -\tau$$

$$\sigma_{avg} = 0$$

The principal stresses are identified as points B and D on the circle. Thus,

$$\sigma_1 = \tau$$

$$\sigma_2 = -\tau$$

The clockwise angle from CA to CB is $2\theta_{p_1} = 90°$, so that $\theta_{p_1} = 45°$. This clockwise angle defines the direction of σ_1 (or the x' axis). The results are shown in Fig. 9–19*c*.

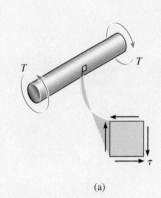

(a)

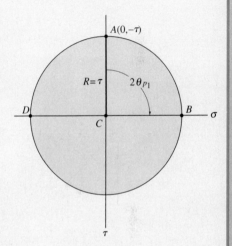

(b)

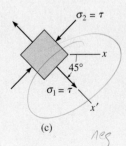

(c)

Fig. 9–19

EXAMPLE 9-9

Due to the applied loading, the element at point A on the solid cylinder in Fig. 9–20a is subjected to the state of stress shown. Determine the principal stresses acting at this point.

SOLUTION

Construction of the Circle.   From Fig. 9–20a,

$$\sigma_x = -12 \text{ ksi} \qquad \sigma_y = 0 \qquad \tau_{xy} = -6 \text{ ksi}$$

The center of the circle is at

$$\sigma_{avg} = \frac{-12 + 0}{2} = -6 \text{ ksi}$$

The initial point $A(-12, -6)$ and the center $C(-6, 0)$ are plotted in Fig. 9-20b. The circle is constructed having a radius of

$$R = \sqrt{(12 - 6)^2 + (6)^2} = 8.49 \text{ ksi}$$

Principal Stresses.   The principal stresses are indicated by the coordinates of points B and D. We have, for $\sigma_1 > \sigma_2$,

$$\sigma_1 = 8.49 - 6 = 2.49 \text{ ksi} \qquad\qquad \textit{Ans.}$$

$$\sigma_2 = -6 - 8.49 = -14.5 \text{ ksi} \qquad\qquad \textit{Ans.}$$

The orientation of the element can be determined by calculating the *counterclockwise* angle $2\theta_{p_2}$ in Fig. 9–20b, which defines the direction θ_{p_2} of σ_2 and its associated principal plane. We have

$$2\theta_{p_2} = \tan^{-1} \frac{6}{(12 - 6)} = 45.0°$$

$$\theta_{p_2} = 22.5°$$

The element is oriented such that the x' axis or σ_2 is directed 22.5° *counterclockwise* from the horizontal (x axis) as shown in Fig. 9–20c.

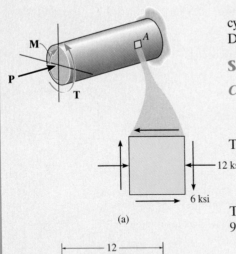

(a)

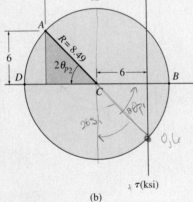

(b)

Fig. 9–20

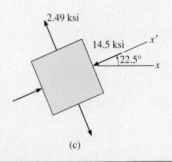

(c)

EXAMPLE 9–10

The state of plane stress at a point is shown on the element in Fig. 9–21a. Determine the maximum in-plane shear stresses and the orientation of the element upon which they act.

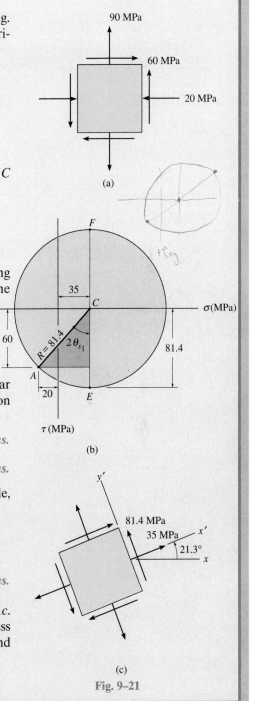

SOLUTION

Construction of the Circle.   From the problem data,

$$\sigma_x = -20 \text{ MPa} \qquad \sigma_y = 90 \text{ MPa} \qquad \tau_{xy} = 60 \text{ MPa}$$

The σ, τ axes are established in Fig. 9–21b. The center of the circle C is located on the σ axis, at the point

$$\sigma_{\text{avg}} = \frac{-20 + 90}{2} = 35 \text{ MPa}$$

Point C and the reference point A $(-20, 60)$ are plotted. Applying the Pythagorean theorem to the shaded triangle to determine the circle's radius CA, we have

$$R = \sqrt{(60)^2 + (55)^2} = 81.4 \text{ MPa}$$

Maximum In-Plane Shear Stress.   The maximum in-plane shear stress and the average normal stress are identified by point E or F on the circle. In particular, the coordinates of point $E(35, 81.4)$ give

$$\tau_{\max}_{\text{in-plane}} = 81.4 \text{ MPa} \qquad \textit{Ans.}$$

$$\sigma_{\text{avg}} = 35 \text{ MPa} \qquad \textit{Ans.}$$

The *counterclockwise* angle θ_{s_1} can be found from the circle, identified as $2\theta_{s_1}$. We have

$$2\theta_{s_1} = \tan^{-1}\left(\frac{20 + 35}{60}\right) = 42.5°$$

$$\theta_{s_1} = 21.3° \qquad \textit{Ans.}$$

This *counterclockwise* angle defines the direction of the x' axis, Fig. 9–21c. Since point E has *positive* coordinates, then the average normal stress and the maximum in-plane shear stress both act in the *positive* x' and y' directions as shown.

Fig. 9–21

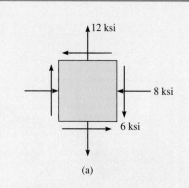

(a)

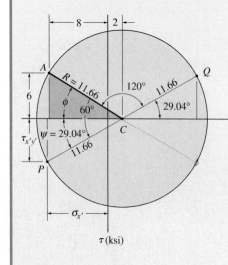

(b)

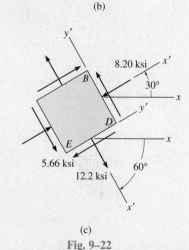

(c)

Fig. 9–22

EXAMPLE 9–11

The state of plane stress at a point is shown on the element in Fig. 9–22a. Represent this state of stress on an element oriented 30° counterclockwise from the position shown.

SOLUTION

Construction of the Circle.   From the problem data,

$$\sigma_x = -8 \text{ ksi} \qquad \sigma_y = 12 \text{ ksi} \qquad \tau_{xy} = -6 \text{ ksi}$$

The σ and τ axes are established in Fig. 9–22b. The center of the circle C is on the σ axis at

$$\sigma_{avg} = \frac{-8 + 12}{2} = 2 \text{ ksi}$$

The initial point for $\theta = 0°$ has coordinates $A(-8, -6)$. Hence from the shaded triangle the radius CA is

$$R = \sqrt{(10)^2 + (6)^2} = 11.66$$

Stresses on 30° Element.   Since the element is to be rotated 30° *counterclockwise*, we must construct a radial line CP, $2(30°) = 60°$ *counterclockwise*, measured from CA ($\theta = 0°$), Fig. 9–22b. The coordinates of point P ($\sigma_{x'}$, $\tau_{x'y'}$) must now be obtained. From the geometry of the circle,

$$\phi = \tan^{-1}\frac{6}{10} = 30.96° \qquad \psi = 60° - 30.96° = 29.04°$$

$$\sigma_{x'} = 2 - 11.66 \cos 29.04° = -8.20 \text{ ksi} \qquad\qquad Ans.$$
$$\tau_{x'y'} = 11.66 \sin 29.04° = 5.66 \text{ ksi} \qquad\qquad Ans.$$

These two stress components act on face BD of the element shown in Fig. 9–22c since the x' axis for this face is oriented 30° counterclockwise from the x axis.

The stress components acting on the adjacent face DE of the element, which is 60° *clockwise* from the positive x axis, Fig. 9–22c, are represented by the coordinates of point Q on the circle. This point lies on the radial line CQ, which is 180° from CP. The coordinates of point Q are

$$\sigma_{x'} = 2 + 11.66 \cos 29.04° = 12.2 \text{ ksi} \qquad\qquad Ans.$$
$$\tau_{x'y'} = -(11.66 \sin 29.04) = -5.66 \text{ ksi} \qquad (check)$$

Note that here $\tau_{x'y'}$ acts in the $-y'$ direction.

9.5   STRESS IN SHAFTS DUE TO AXIAL LOAD AND TORSION

Occasionally circular shafts are subjected to the combined effects of both an axial load and torsion. Provided the material remains linear elastic, and is only subjected to small deformations, then we can use the principle of superposition to obtain the resultant stress in the shaft due to both of these loadings. The principal stresses can then be determined using either the stress transformation equations or Mohr's circle.

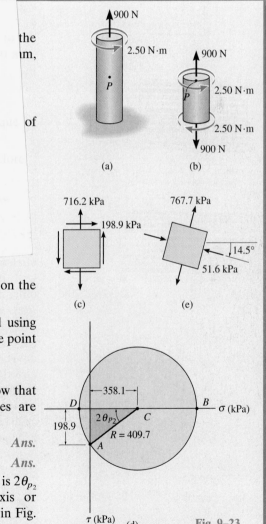

(a)   (b)

(c)   (e)

τ (kPa)   (d)   Fig. 9–23

EXAMPLE 9-

An axial force o   the
shaft as shown i   im,
determine the pr

SOLUTION

Internal Loading.   of
2.50 N·m and the

Stress Component

$$\tau = \frac{Tc}{J} = \frac{2.50 \text{ N} \cdot}{\frac{\pi}{2}(0}$$

$$\sigma = \frac{P}{A} = \frac{900 \text{ N}}{\pi(0.02 \text{ m}}$$

The state of stress defined by these two components is shown on the element at P in Fig. 9–23c.

Principal Stresses. The principal stresses can be determined using Mohr's circle, Fig. 9–23d. Here the center of the circle C is at the point

$$\sigma_{\text{avg}} = \frac{0 + 716.2}{2} = 358.1 \text{ kPa}$$

Plotting C (358.1, 0) and the reference point A (0, 198.9), show that the radius of the circle is $R = 409.7$. The principal stresses are represented by points B and D. Therefore,

$$\sigma_1 = 358.1 + 409.7 = 767.8 \text{ kPa} \qquad Ans.$$

$$\sigma_2 = 358.1 - 409.7 = -51.6 \text{ kPa} \qquad Ans.$$

The clockwise angle $2\theta_{p_2}$ can be determined from the circle. It is $2\theta_{p_2} = 29.1°$. The element is orientated such that the x' axis or σ_2 is directed clockwise $\theta_{p_1} = 14.5°$ with the x axis as shown in Fig. 9–23e.

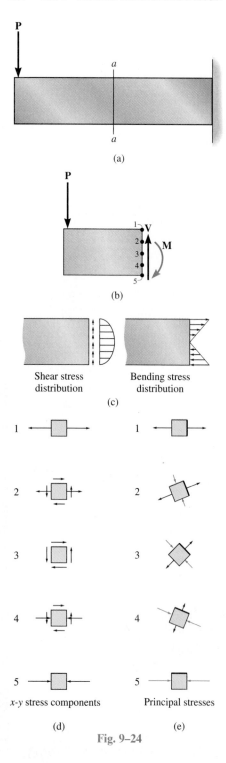

(a)

(b)

Shear stress
distribution

Bending stress
distribution

(c)

1

2

3

4

5

x-y stress components

1

2

3

4

5

Principal stresses

(d)

(e)

Fig. 9–24

9.6   STRESS VARIATIONS THROUGHOUT A PRISMATIC BEAM

Since beams resist both internal shear and moment loadings, the stress analysis of a beam requires application of the shear and flexure formulas. Here we will discuss the general results obtained when these equations are applied to various points in a cantilevered beam that has a rectangular cross section and supports a load **P** at its end, Fig. 9–24a.

In general, at an arbitrary section a–a along the beam's axis, Fig. 9–24b, the internal shear **V** and moment **M** are developed from a *parabolic* shear-stress distribution, and a *linear* normal-stress distribution, Fig. 9–24c. As a result, the stresses acting on elements located at points 1 through 5 along the section will be as shown in Fig. 9–24d. Note that elements 1 and 5 are subjected only to the maximum normal stress, whereas element 3, which is on the neutral axis, is subjected only to the maximum shear stress. The intermediate elements 2 and 4 resist *both* normal and shear stress.

In each case the state of stress can be transformed into *principal stresses*, using either the stress-transformation equations or Mohr's circle. The results are shown in Fig. 9–24e. Here each successive element, 1 through 5, undergoes a counterclockwise orientation. Specifically, relative to element 1, considered to be at the 0° position, element 3 is oriented at 45° and element 5 is oriented at 90°. Also, the *maximum tensile stress* acting on the vertical faces of element 1 becomes smaller on the corresponding faces of each of the successive elements, until it is zero on the horizontal faces of element 5. In a similar manner, the *maximum compressive stress* on the vertical faces of element 5 reduces to zero on the horizontal faces of element 1.

If this analysis is extended to many vertical sections along the beam other than a–a, a profile of the results can be represented by curves called *stress trajectories*. Each of these curves indicates the *direction* of a principal stress having a constant magnitude. Some of these trajectories are shown for the cantilevered beam in Fig. 9–25. Here the solid lines represent the direction of the tensile principal stresses and the dashed lines represent the direction of the compressive principal stresses. As expected, the lines intersect the neutral axis at 45° angles, and the solid and dashed lines always intersect at 90°. Why? Knowing the direction of these lines can help engineers decide where to reinforce a beam so that it does not crack or become unstable.

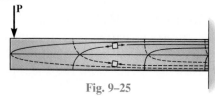

Fig. 9–25

EXAMPLE 9-13

The beam shown in Fig. 9–26a is subjected to the distributed loading of $w = 120$ kN/m. Determine the principal stresses in the beam at point P, which lies at the top of the web. Neglect the size of the fillets and stress concentrations at this point. $I = 67.4(10^{-6})$m^4.

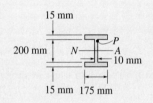

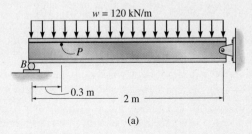

(a)

SOLUTION

Internal Loadings. The support reaction on the beam at B is determined, and equilibrium of the sectioned beam shown in Fig. 9–26b yields

$$V = 84 \text{ kN} \qquad M = 30.6 \text{ kN} \cdot \text{m}$$

Stress Components. At point P,

$$\sigma = \frac{-My}{I} = \frac{30.6(10^3) \text{ N} \cdot \text{m} \ (0.100 \text{ m})}{67.4(10^{-6}) \text{ m}^4} = -45.4 \text{ MPa} \qquad \textit{Ans.}$$

$$\tau = \frac{VQ}{It} = \frac{84(10^3) \text{ N} \ [(0.1075 \text{ m})(0.175 \text{ m})(0.015 \text{ m})]}{67.4(10^{-6}) \text{ m}^4(0.010 \text{ m})}$$

$$= 35.2 \text{ MPa} \qquad \textit{Ans.}$$

These results are shown in Fig. 9–26c.

Principal Stresses. Using Mohr's circle the principal stresses at P can be determined. As shown in Fig. 9–26d, the center of the circle is at $(-45.4 + 0)/2 = -22.7$, and point A has coordinates of $A(-45.4, -35.2)$. Show that the radius is $R = 41.9$, and therefore

$$\sigma_1 = (41.9 - 22.7) = 19.2 \text{ MPa}$$
$$\sigma_2 = -(22.7 + 41.9) = -64.6 \text{ MPa}$$

The counterclockwise angle $2\theta_{p_2} = 57.2°$, so that

$$\theta_{p_2} = 28.6°$$

These results are shown in Fig. 9–26e.

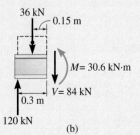

(b)

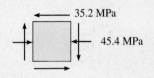

(c)

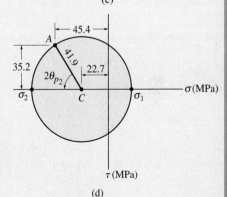

(d)

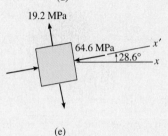

(e)

Fig. 9–26

PROBLEMS

9–49. Solve Prob. 9–3 using Mohr's circle.

9–50. Solve Prob. 9–5 using Mohr's circle.

9–51. Solve Prob. 9–6 using Mohr's circle.

*9–52. Solve Prob. 9–8 using Mohr's circle.

9–53. Solve Prob. 9–10 using Mohr's circle.

9–54. Solve Prob. 9–11 using Mohr's circle.

9–55. Solve Prob. 9–13 using Mohr's circle.

*9–56. Solve Prob. 9–12 using Mohr's circle.

9–57. Solve Prob. 9–14 using Mohr's circle.

9–58. Solve Prob. 9–16 using Mohr's circle.

9–59. Determine the equivalent state of stress if an element is oriented 60° clockwise from the element shown.

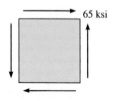

Prob. 9–59

*9–60. Determine the equivalent state of stress if an element is oriented 60° counterclockwise from the element shown.

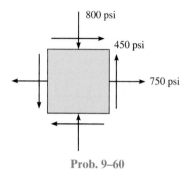

800 psi

450 psi

750 psi

Prob. 9–60

9–61. Determine the equivalent state of stress if an element is oriented 30° clockwise from the element shown.

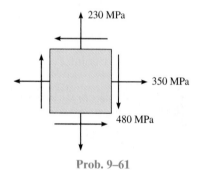

230 MPa

350 MPa

480 MPa

Prob. 9–61

9–62. Determine the equivalent state of stress if an element is oriented 25° counterclockwise from the element shown.

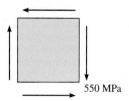

550 MPa

Prob. 9–62

9–63. Determine (*a*) the principal stresses and (*b*) the maximum in-plane shear stress and average normal stress. Specify the orientation of the element in each case.

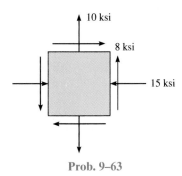

Prob. 9–63

9–66. Draw Mohr's circle that describes each of the following states of stress.

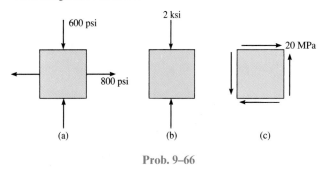

(a)   (b)   (c)

Prob. 9–66

***9–64.** Determine (*a*) the principal stresses and (*b*) the maximum in-plane shear stress and average normal stress. Specify the orientation of the element in each case.

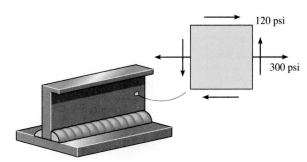

Prob. 9–64

9–65. Determine (*a*) the principal stress and (*b*) the maximum in-plane shear stress and average normal stress. Specify the orientation of the element in each case.

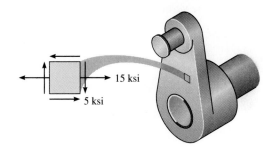

Prob. 9–65

9–67. A point on a thin plate is subjected to two successive states of stress as shown. Determine the resulting state of stress with reference to an element oriented as shown at the right.

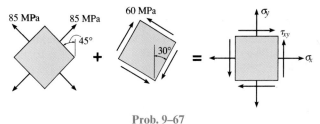

Prob. 9–67

***9–68.** Draw Mohr's circle that describes each of the following states of stress.

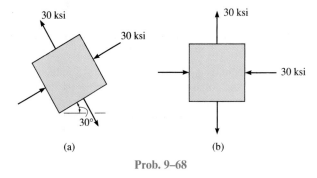

(a)   (b)

Prob. 9–68

9–69. Mohr's circle for the state of stress in Fig. 9–15*a* is shown in Fig. 9–15*b*. Show that finding the coordinates of point $P(\sigma_{x'}, \tau_{x'y'})$ on the circle gives the same value as the stress-transformation Eqs. 9–1 and 9–2.

The following problems involve material covered in Chapter 8.

9–70. The cantilevered rectangular bar is subjected to the force of 5 kip. Determine the principal stresses at point A.

9–71. Solve Prob. 9–70 for the principal stresses at point B.

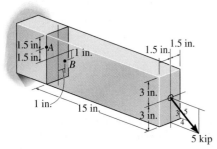

Probs. 9–70/9–71

***9–72.** The grains of wood in the board make an angle of 20° with the horizontal as shown. Determine the normal and shear stresses that act perpendicular and parallel to the grains if the board is subjected to an axial load of 250 N.

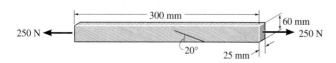

Prob. 9–72

9–73. The bent rod has a diameter of 15 mm and is subjected to the force of 600 N. Determine the principal stresses and the maximum in-plane shear stress that are developed at point A and point B. Show the results on elements located at these points.

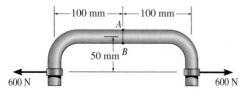

Prob. 9–73

9–74. The beam has a rectangular cross section and is subjected to the loadings shown. Determine the principal stresses and the maximum in-plane shear stress that is developed at point A and point B. Show the results on elements located at these points.

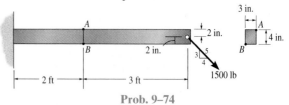

Prob. 9–74

9–75. A spherical pressure vessel has an inner radius of 5 ft and a wall thickness of 0.5 in. Draw Mohr's circle for the state of stress at a point on the vessel and explain the significance of the result. The vessel is subjected to an internal pressure of 80 psi.

***9–76.** A rod has a circular cross section with a diameter of 2 in. It is subjected to a torque of 12 kip · in. and a bending moment **M**. If the greatest principal stress at the point of maximum flexural stress is 15 ksi, determine the magnitude of the bending moment.

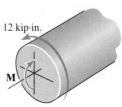

Prob. 9–76

9–77. The beam is subjected to the two forces shown. Determine the principal stresses at point A.

9–78. The beam is subjected to the two forces shown. Determine the principal stresses at point B, which is located at the bottom of the vertical segment of the cross section.

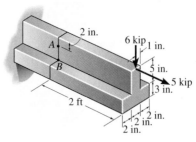

Probs. 9–77/9–78

9–79. The pedal crank for a bicycle has the cross section shown. If it is fixed to the gear at *B* and does not rotate while subjected to a force of 75 lb, determine the principal stresses in the material on the cross section at point *C*.

9–82. The pipe has an inner radius of 25 mm and an outer radius of 27 mm. If it is subjected to an internal pressure of 8 MPa and a torsional moment of 500 N · m, determine the principal stresses and the maximum in-plane shear stress at point *A*, which lies on the pipe's outer surface.

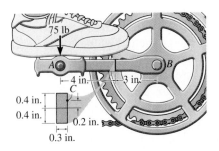

Prob. 9–79

Prob. 9–82

*9–80. The frame supports the distributed loading of 200 N/m. Determine the normal and shear stresses at point *D* that act perpendicular and parallel, respectively, to the grains. The grains at this point make an angle of 30° with the horizontal as shown.

9–81. The frame supports the distributed loading of 200 N/m. Determine the normal and shear stresses at point *E* that act perpendicular and parallel, respectively, to the grains. The grains at this point make an angle of 60° with the horizontal as shown.

9–83. The cylindrical pressure vessel has an inner radius of 1.25 m and a wall thickness of 15 mm. It is made from steel plates that are welded along the 45° seam. Determine the normal and shear stress components along this seam if the vessel is subjected to an internal pressure of 8 MPa.

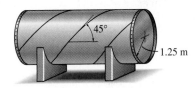

Prob. 9–83

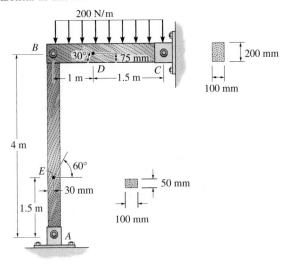

Probs. 9–80/9–81

*9–84. The thin-walled pipe has an inner diameter of 0.5 in. and a thickness of 0.025 in. If it is subjected to an internal pressure of 500 psi and the axial tension and torsional loadings shown, determine the principal stresses at a point on the surface of the pipe.

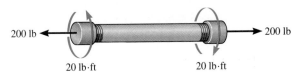

Prob. 9–84

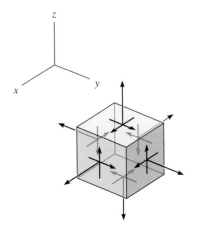

(a)

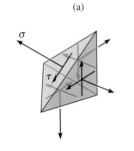

(b)

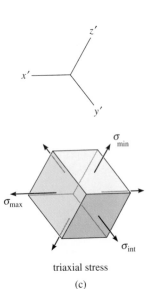

triaxial stress

(c)

9.7   ABSOLUTE MAXIMUM SHEAR STRESS

When a point in a body is subjected to a general three-dimensional state of stress, an element of material has a normal-stress and two shear-stress components acting on each of its faces, Fig. 9–27a. Like the case of plane stress, it is possible to develop stress-transformation equations that can be used to determine the normal and shear stress components σ and τ acting on *any* skewed plane of the element, Fig. 9–27b. Furthermore, at the point it is also possible to determine the unique orientation of an element having only principal stresses acting on its faces. As shown in Fig. 9–27c, these principal stresses are assumed to have magnitudes of maximum, intermediate, and minimum intensity, i.e., $\sigma_{max} \geq \sigma_{int} \geq \sigma_{min}$.

A discussion of the transformation of stress in three dimensions is beyond the scope of this text; however, it is discussed in books related to the theory of elasticity. For our purposes, we will assume that the principal orientation of the element and the principal stresses are known, Fig. 9–27c. This is a condition known as ***triaxial stress***. If we view this element in two dimensions, that is, in the y'–z', x'–z', and x'–y' planes, Fig. 9–28a, 9–28b, and 9–28c, we can then use Mohr's circle to determine the *maximum in-plane shear stress* for each case. For example, the diameter of Mohr's circle extends between the principal stresses σ_{int} and σ_{min} for the case shown in Fig. 9–28a. From this circle, Fig. 9–28d, the maximum in-plane shear stress is $(\tau_{y'z'})_{max} = (\sigma_{int} - \sigma_{min})/2$, and the associated average normal stress is $(\sigma_{int} + \sigma_{min})/2$. As shown in Fig. 9–28e, the element having these stress components on it must be oriented 45° from the position of the element shown in Fig. 9–28a. Mohr's circles for the elements in Fig. 9–28b and 9–28c have also been constructed in Fig. 9–28d. The corresponding elements having a 45° orientation and subjected to maximum in-plane shear and average normal stress components are shown in Fig. 9–28f and 9–28g, respectively.

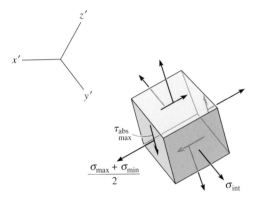

Fig. 9–27

(d)

Comparing the three circles in Fig. 9–28d, it is seen that the **absolute maximum shear stress**, $\tau_{\substack{abs \\ max}}$, is defined by the circle having the largest radius, which occurs for the element shown in Fig. 9–28b. In other words, the element in Fig. 9–28f is oriented by a rotation of 45° about the y' axis from the element in Fig. 9–27b. Notice that this condition can also be *determined directly* by simply choosing the maximum and minimum principal stresses from Fig. 9–27c, in which case the absolute maximum shear stress will be

$$\tau_{\substack{abs \\ max}} = \frac{\sigma_{max} - \sigma_{min}}{2} \qquad (9\text{–}13)$$

And the associated average normal stress will be

$$\sigma_{avg} = \frac{\sigma_{max} + \sigma_{min}}{2} \qquad (9\text{–}14)$$

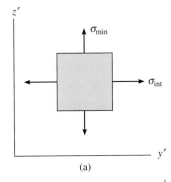

(a)

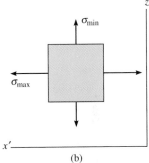

(b)

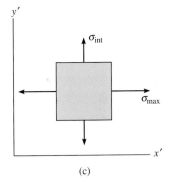

(c)

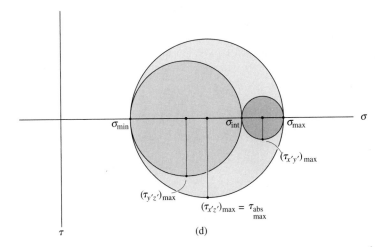

(d)

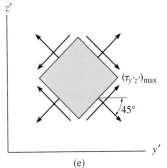

(e)

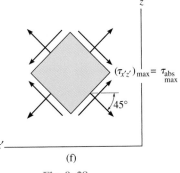

(f)

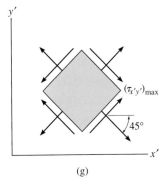

(g)

Fig. 9–28

The analysis considered only the stress components acting on elements located in positions found from rotations about the x', y', or z' axis. If we had used the three-dimensional stress-transformation equations of the theory of elasticity to obtain values of the normal and shear stress components acting on any arbitrary skewed plane at the point, as in Fig. 9–27b, it could be shown that regardless of the orientation of the plane, specific values of the shear stress τ on the plane will *always be less* than the absolute maximum shear stress found from Eq. 9–13. Furthermore, the normal stress σ acting on any plane will have a value lying between the maximum and minimum principal stresses, that is, $\sigma_{max} \geq \sigma \geq \sigma_{min}$.

Plane Stress.   The above results have an important implication for the case of plane stress, particularly when the in-plane principal stresses have the *same sign*, i.e., they are both tensile or both compressive. For example, consider the material to be subjected to plane stress such that the in-plane principal stresses are represented as σ_{max} and σ_{int}, in the x' and y' directions, respectively; while the out-of-plane principal stress in the z' direction is $\sigma_{min} = 0$, Fig. 9–29a. Mohr's circles that describe this state of stress for element orientations about each of the three coordinate axes are shown in Fig. 9–29b. Here it is seen that although the maximum in-plane shear stress is $(\tau_{x'y'})_{max} = (\sigma_{max} - \sigma_{int})/2$, this value is *not* the absolute maximum shear stress to which the material is subjected. Instead, from Eq. 9–13 or Fig. 9–29b,

$$\tau_{abs \atop max} = (\tau_{x'z'})_{max} = \frac{\sigma_{max} - 0}{2} = \frac{\sigma_{max}}{2} \qquad (9\text{--}15)$$

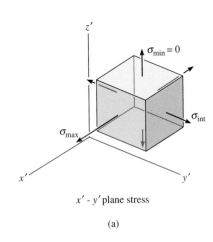

x' - y' plane stress

(a)

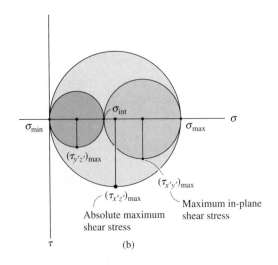

Absolute maximum
shear stress

Maximum in-plane
shear stress

(b)

Fig. 9–29

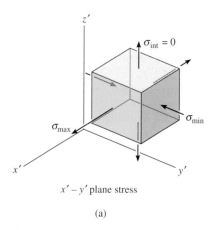

$x' - y'$ plane stress

(a)

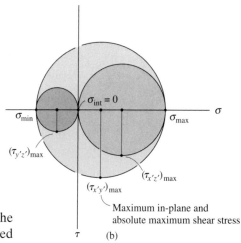

Maximum in-plane and absolute maximum shear stress

(b)

Fig. 9–30

In the case where one of the in-plane principal stresses has the *opposite sign* of that of the other, then these stresses will be represented as σ_{max} and σ_{min}, and the out-of-plane principal stress $\sigma_{int} = 0$, Fig. 9–30a. Mohr's circles that describe this state of stress for element orientations about each coordinate axis are shown in Fig. 9–30b. Clearly, in this case

$$\tau_{abs} = (\tau_{x'y'})_{max} = \frac{\sigma_{max} - \sigma_{min}}{2} \qquad (9\text{–}16)$$

Calculation of the absolute maximum shear stress as indicated here is important when designing members made of a ductile material, since the strength of the material depends on its ability to resist shear stress. This situation will be discussed further in Sec. 10.7.

IMPORTANT POINTS

• The general three-dimensional state of stress at a point can be represented by an element oriented so that only three principal stresses act on it.

• From this orientation, the orientation of the element representing the absolute maximum shear stress can be obtained by rotating the element 45° about the axis defining the direction of σ_{int}.

• If the in-plane principal stresses both have the *same sign*, the *absolute maximum shear stress* will occur *out of the plane* and has a value of $\tau_{abs} = \sigma_{max}/2$.

• If the in-plane principal stresses are of *opposite signs*, then the *absolute maximum shear stress equals the maximum in-plane shear stress*; that is, $\tau_{abs} = (\sigma_{max} - \sigma_{min})/2$.

EXAMPLE 9–14

Due to the applied loading, the element at the point on the frame in Fig. 9–31a is subjected to the state of plane stress shown. Determine the principal stresses and the absolute maximum shear stress at the point.

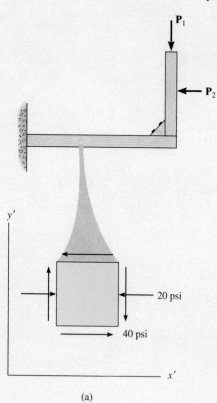

(a)

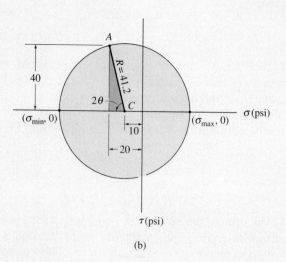

(b)

Fig. 9–31

SOLUTION

Principal Stresses. The in-plane principal stresses can be determined from Mohr's circle. The center of the circle is on the σ axis at $\sigma_{avg} = (-20 + 0)/2 = -10$ psi. Plotting the controlling point $A(-20, -40)$, the circle can be drawn as shown in Fig. 9–31b. The radius is

$$R = \sqrt{(20 - 10)^2 + (40)^2} = 41.2 \text{ psi}$$

The principal stresses are at the points where the circle intersects the σ axis; i.e.,

$$\sigma_{max} = -10 + 41.2 = 31.2 \text{ psi}$$
$$\sigma_{min} = -10 - 41.2 = -51.2 \text{ psi}$$

From the circle, the *counterclockwise* angle 2θ, measured from CA to the $-\sigma$ axis, is

$$2\theta = \tan^{-1}\left(\frac{40}{(20-10)}\right) = 76.0°$$

Thus,

$$\theta = 38.0°$$

This *counterclockwise* rotation defines the direction of the x' axis or σ_{min} and its associated principal plane, Fig. 9–31c. Since there is no principal stress on the element in the z direction, we have

$$\sigma_{max} = 31.2 \text{ psi} \qquad \sigma_{int} = 0 \qquad \sigma_{min} = -51.2 \text{ psi} \qquad \textit{Ans.}$$

Absolute Maximum Shear Stress.   Applying Eqs. 9–13 and 9–14, we have

$$\tau_{\substack{abs \\ max}} = \frac{\sigma_{max} - \sigma_{min}}{2} = \frac{31.2 - (-51.2)}{2} = 41.2 \text{ psi} \qquad \textit{Ans.}$$

$$\sigma_{avg} = \frac{\sigma_{max} + \sigma_{min}}{2} = \frac{31.2 - 51.2}{2} = -10 \text{ psi}$$

These same results can also be obtained by drawing Mohr's circle for each orientation of an element about the x', y', and z' axes, Fig. 9–31d. Since σ_{max} and σ_{min} are of *opposite signs*, then the absolute maximum shear stress equals the maximum in-plane shear stress. This results from a 45° rotation of the element in Fig. 9–31c about the z' axis, so that the properly oriented element is shown in Fig. 9–31e.

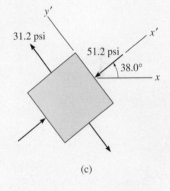

(c)

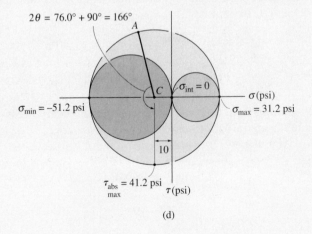

(d)

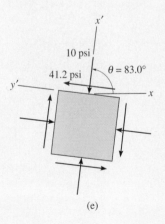

(e)

EXAMPLE 9-15

The point on the surface of the cylindrical pressure vessel in Fig. 9–32a is subjected to the state of plane stress. Determine the absolute maximum shear stress at this point.

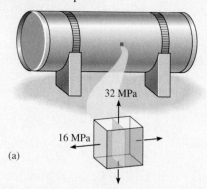

32 MPa

16 MPa

(a)

SOLUTION

The principal stresses are $\sigma_{max} = 32$ MPa, $\sigma_{int} = 16$ MPa, and $\sigma_{min} = 0$. If these stresses are plotted along the σ axis, the three Mohr's circles can be constructed that describe the stress state viewed in each of the three perpendicular planes, Fig. 9–32b. The largest circle has a radius of 16 MPa and describes the state of stress in the plane containing $\sigma_{max} = 32$ MPa and $\sigma_{min} = 0$, shown shaded in Fig. 9–32a. An orientation of an element 45° within this plane yields the state of absolute maximum shear stress and the associated average normal stress, namely,

$$\tau_{abs \atop max} = 16 \text{ MPa} \qquad \textit{Ans.}$$

$$\sigma_{avg} = 16 \text{ MPa}$$

These same results can be obtained from direct application of Eqs. 9–13 and 9–14; that is,

$$\tau_{abs \atop max} = \frac{\sigma_{max} - \sigma_{min}}{2} = \frac{32 - 0}{2} = 16 \text{ MPa} \qquad \textit{Ans.}$$

$$\sigma_{avg} = \frac{\sigma_{max} + \sigma_{min}}{2} = \frac{32 + 0}{2} = 16 \text{ MPa}$$

By comparison, the maximum in-plane shear stress can be determined from the Mohr's circle drawn between $\sigma_{max} = 32$ MPa and $\sigma_{int} = 16$ MPa, Fig. 9–32b. This gives a value of

$$\tau_{max \atop in\text{-}plane} = \frac{32 - 16}{2} = 8 \text{ MPa}$$

$$\sigma_{avg} = 16 + \frac{32 - 16}{2} = 24 \text{ MPa}$$

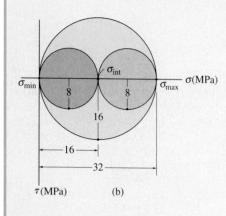

σ_{min} σ_{int} σ_{max} $\sigma(\text{MPa})$

8   8

16

16

32

$\tau(\text{MPa})$ (b)

Fig. 9–32

PROBLEMS

9-85. Draw the three Mohr's circles that describe each of the following states of stress.

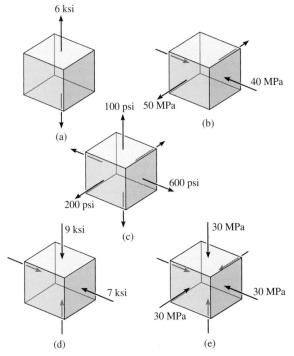

(a)

(b)

(c)

(d)

(e)

Prob. 9-85

9-86. The principal stresses acting at a point in a body are shown. Draw the three Mohr's circles that describe this state of stress, and find the maximum in-plane shear stresses and associated average normal stresses for the x–y, y–z, and x–z planes. For each case, show the results on the element oriented in the appropriate direction.

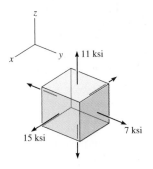

Prob. 9-86

9-87 The stress at a point is shown on the element. Determine the principal stresses and the absolute maximum shear stress.

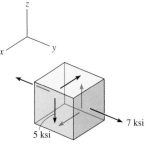

Prob. 9-87

***9-88.** The stress at a point is shown on the element. Determine the principal stresses and the absolute maximum shear stress.

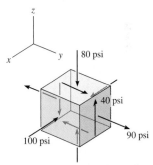

Prob. 9-88

9-89 The principal stresses acting at a point in a body are shown. Draw the three Mohr's circles that describe this state of stress and find the maximum in-plane shear stresses and associated average normal stresses for the x–y, y–z, and x–z planes. For each case, show the results on the element oriented in the appropriate direction.

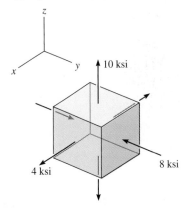

Prob. 9-89

The following problems involve material covered in Chapter 8.

9–90. The solid cylinder having a radius r is placed in a sealed container and subjected to a pressure p. Determine the stress components acting at point A located on the center line of the cylinder. Draw Mohr's circles for the element at this point.

Prob. 9–90

9–91. Determine the principal stresses and the absolute maximum shear stress at point A on the frame. The cross-sectional area at this point is shown.

*9–92. Solve Prob. 9–91 for point B.

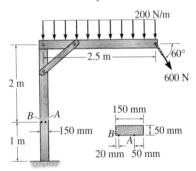

Probs. 9–91/9–92

9–93. The bolt is fixed to its support at C. If a force of 18 lb is applied to the wrench to tighten it, determine the principal stresses and the absolute maximum shear stress developed in the bolt shank at point A. Represent the results on an element located at this point. The shank has a diameter of 0.25 in.

9–94. Solve Prob. 9–93 for point B.

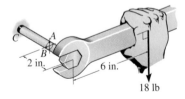

Probs. 9–93/9–94

REVIEW PROBLEMS

9–95. The internal loadings on a cross section through the 6-in.-diameter drive shaft of a turbine consist of an axial force of 2500 lb, a bending moment of 800 lb · ft. and a torsional moment of 1500 lb · ft. Determine the principal stresses at point A. Also compute the maximum in-plane shear stress at this point.

*9–96. The internal loadings at a cross section through the 6-in.-diameter drive shaft of a turbine consist of an axial force of 2500 lb, a bending moment of 800 lb · ft, and a torsional moment of 1500 lb · ft. Determine the principal stresses at point B. Also compute the maximum in-plane shear stress at this point.

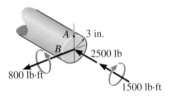

Probs. 9–95/9–96

9–97. The state of stress at a point is shown on the element. Determine (a) the principal stresses and (b) the maximum in-plane shear stress and the average normal stress at the point. Specify the orientation of the element in each case.

9–98. Draw the three Mohr's circles for the state of stress shown, and determine the absolute maximum shear stress.

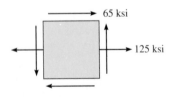

Probs. 9–97/9–98

9–99. A bar has a circular cross section with a diameter of 1 in. It is subjected to a torque and a bending moment. At the point of maximum bending stress the principal tensile stresses are 20 ksi and -10 ksi. Determine the torque and the bending moment.

*9–100. The box beam is subjected to the loading shown. Determine the principal stresses in the beam at points A and B.

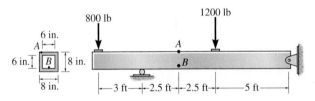

Prob. 9–100

9–101. Determine the equivalent state of stress if an element is oriented 40° clockwise from the element shown. Use Mohr's circle.

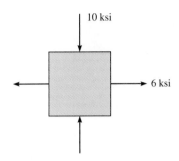

Prob. 9–101

9–102. The square steel plate has a thickness of 10 mm and is subjected to the edge loading shown. Determine the maximum in-plane shear stress and the average normal stress developed in the steel.

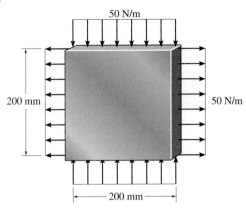

Prob. 9–102

9–103. The wooden strut is subjected to the loading shown. If grains of wood in the strut at point C make an angle of 60° with the horizontal as shown, determine the normal and shear stresses that act perpendicular and parallel to the grains, respectively, due to the loading. The strut is supported by a bolt (pin) at B and smooth support at A.

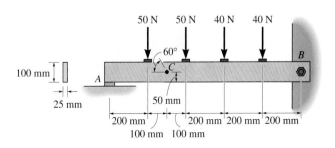

Prob. 9–103

*9–104. The state of stress at a point is shown on the element. Determine (*a*) the principal stresses and (*b*) the maximum in-plane shear stress and average normal stress at the point. Specify the orientation of the element in each case.

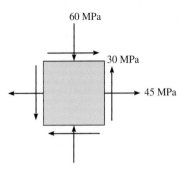

Prob. 9–104

Complex stresses developed within this airplane wing are analyzed from strain gauge data. (Courtesy of Measurements Group, Inc., Raleigh, North Carolina, 27611, USA.)

10 STRAIN TRANSFORMATION

CHAPTER OBJECTIVES

The transformation of strain at a point is similar to the transformation of stress, and as a result the methods of Chapter 9 will be applied in this chapter. Here we will also discuss various ways for measuring strain and develop some important material-property relationships, including a generalized form of Hooke's law. At the end of the chapter, a few of the theories used to predict the failure of a material will be discussed.

10.1 PLANE STRAIN

As outlined in Sec. 2.2, the general state of strain at a point in a body is represented by a combination of three components of normal strain, ϵ_x, ϵ_y, ϵ_z, and three components of shear strain γ_{xy}, γ_{xz}, γ_{yz}. These six components tend to deform each face of an element of the material, and like stress, the normal and shear strain *components* at the point will vary according to the orientation of the element. The *strain* components at a point are often determined by using strain gauges, which measure these components in *specified directions*. For both analysis and design, however, engineers must sometimes transform this data in order to obtain the strain components in other directions.

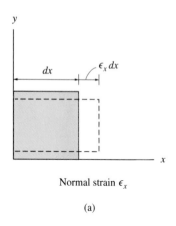

Normal strain ϵ_x

(a)

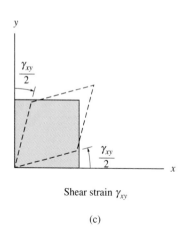

Shear strain γ_{xy}

(c)

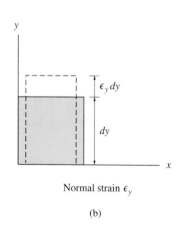

Normal strain ϵ_y

(b)

Fig. 10–1

The rubber specimen is constrained between the two fixed supports, and so it will undergo plane strain when loads are applied to it in the horizontal plane.

To understand how this is done, we will first confine our attention to a study of **plane strain**. Specifically, we will not consider the effects of the components ϵ_z, γ_{xz}, and γ_{yz}. In general, then, a plane-strained element is subjected to two components of normal strain, ϵ_x, ϵ_y, and one component of shear strain, γ_{xy}. The deformations of an element caused by each of these strains are shown graphically in Fig. 10–1. Note that the normal strains are produced by *changes in length* of the element in the *x* and *y* directions, and the shear strain is produced by the *relative rotation* of two adjacent sides of the element.

Although plane strain and plane stress each have three components lying in the same plane, realize that plane stress *does not* necessarily cause plane strain or vice versa. The reason for this has to do with the Poisson effect discussed in Sec. 3.6. For example, if the element in Fig. 10–2 is subjected to plane stress σ_x and σ_y, not only are normal strains ϵ_x and ϵ_y produced, but there is *also* an associated normal strain, ϵ_z. This is obviously *not* a case of plane strain. In general, then, unless $\nu = 0$, the Poisson effect will *prevent* the simultaneous occurrence of plane strain and plane stress. It should also be pointed out that since shear stress and shear strain are *not* affected by Poisson's ratio, a condition of $\tau_{xz} = \tau_{yz} = 0$ requires $\gamma_{xz} = \gamma_{yz} = 0$.

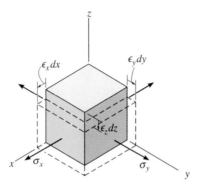

Plane stress, σ_x, σ_y, does not cause plane strain in the x–y plane since $\epsilon_z \neq 0$

Fig. 10–2

10.2   GENERAL EQUATIONS OF PLANE-STRAIN TRANSFORMATION

It is important in plane-strain analysis to establish transformation equations that can be used to determine the x', y' components of normal and shear strain at a point, provided the x, y components of strain are known. Essentially this problem is one of geometry and requires relating the deformations and rotations of differential line segments, which represent the sides of differential elements that are parallel to each set of axes.

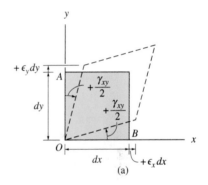

(a)

Sign Convention.   Before the strain-transformation equations can be developed, we must first establish a sign convention for the strains. This convention is the same as that established in Sec. 2.2 and will be restated here for the condition of plane strain. With reference to the differential element shown in Fig. 10–3*a*, *normal strains* ϵ_x and ϵ_y are *positive* if they cause *elongation* along the x and y axes, respectively, and the *shear strain* γ_{xy} is *positive* if the interior angle *AOB becomes smaller* than 90°. This sign convention also follows the corresponding one used for plane stress, Fig. 9–5*a*, that is, positive σ_x, σ_y, τ_{xy} will cause the element to *deform* in the positive ϵ_x, ϵ_y, γ_{xy} directions, respectively.

The problem here will be to determine at a point the normal and shear strains $\epsilon_{x'}$, $\epsilon_{y'}$, $\gamma_{x'y'}$, measured relative to the x', y' axes, if we know ϵ_x, ϵ_y, γ_{xy}, measured relative to the x, y axes. If the angle between the x and x' axes is θ, then, like the case of plane stress, θ will be *positive* provided it follows the curl of the right-hand fingers, i.e., counterclockwise, as shown in Fig. 10–3*b*.

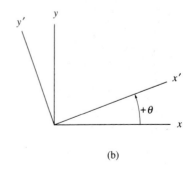

(b)

Positive sign convention

Fig. 10–3

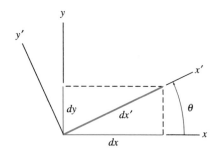

Before deformation

(a)

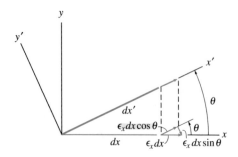

Normal strain ϵ_x

(b)

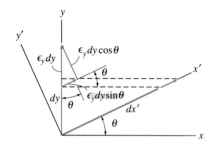

Normal strain ϵ_y

(c)

Fig. 10–4

Normal and Shear Strains. In order to develop the strain-transformation equation for determining $\epsilon_{x'}$, we must determine the elongation of a line segment dx' that lies along the x' axis and is subjected to strain components ϵ_x, ϵ_y, γ_{xy}. As shown in Fig. 10–4a, the components of the line dx' along the x and y axes are

$$dx = dx' \cos \theta$$
$$dy = dx' \sin \theta \tag{10-1}$$

When the positive normal strain ϵ_x occurs, Fig. 10–4b, the line dx is elongated $\epsilon_x \, dx$, which causes line dx' to elongate $\epsilon_x \, dx \cos \theta$. Likewise, when ϵ_y occurs, Fig. 10–4c, line dy elongates $\epsilon_y \, dy$, which causes line dx' to elongate $\epsilon_y \, dy \sin \theta$. Lastly, assuming that dx remains fixed in position, the shear strain γ_{xy}, which is the change in angle between dx and dy, causes the top of line dy to be displaced $\gamma_{xy} \, dy$ to the right, as shown in Fig. 10–4d. This causes dx' to elongate $\gamma_{xy} \, dy \cos \theta$. If all three of these elongations are added together, the resultant elongation of dx' is then

$$\delta x' = \epsilon_x \, dx \cos \theta + \epsilon_y \, dy \sin \theta + \gamma_{xy} \, dy \cos \theta$$

From Eq. 2–2, the normal strain along the line dx' is $\epsilon_{x'} = \delta x'/dx'$. Using Eq. 10–1, we therefore have

$$\epsilon_{x'} = \epsilon_x \cos^2 \theta + \epsilon_y \sin^2 \theta + \gamma_{xy} \sin \theta \cos \theta \tag{10-2}$$

The strain-transformation equation for determining $\gamma_{x'y'}$ can be developed by considering the amount of rotation each of the line segments dx' and dy' undergo when subjected to the strain components ϵ_x, ϵ_y, γ_{xy}. First we will consider the rotation of dx', which is defined by the counterclockwise angle α shown in Fig. 10–4e. It can be determined from the displacement $\delta y'$ using $\alpha = \delta y'/dx'$. To obtain $\delta y'$, consider the following three displacement components acting in the y' direction: one from ϵ_x, giving $-\epsilon_x \, dx \sin \theta$, Fig. 10–4b; another from ϵ_y, giving $\epsilon_y \, dy \cos \theta$, Fig. 10–4c; and the last from γ_{xy}, giving $-\gamma_{xy} \, dy \sin \theta$, Fig. 10–4d. Thus, $\delta y'$, as caused by all three strain components, is

$$\delta y' = -\epsilon_x \, dx \sin \theta + \epsilon_y \, dy \cos \theta - \gamma_{xy} \, dy \sin \theta$$

Using Eq. 10–1, with $\alpha = \delta y'/dx'$, we have

$$\alpha = (-\epsilon_x + \epsilon_y) \sin \theta \cos \theta - \gamma_{xy} \sin^2 \theta \tag{10-3}$$

As shown in Fig. 10–4e, the line dy' rotates by an amount β. We can determine this angle by a similar analysis, or by simply substituting $\theta + 90°$ for θ into Eq. 10–3. Using the identities $\sin (\theta + 90°) = \cos \theta$, $\cos(\theta + 90°) = -\sin \theta$, we have

$$\beta = (-\epsilon_x + \epsilon_y) \sin (\theta + 90°) \cos (\theta + 90°) - \gamma_{xy} \sin^2 (\theta + 90°)$$

$$= -(-\epsilon_x + \epsilon_y) \cos \theta \sin \theta - \gamma_{xy} \cos^2 \theta$$

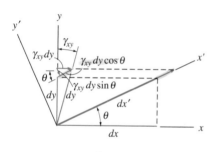

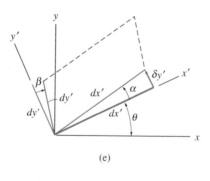

(e)

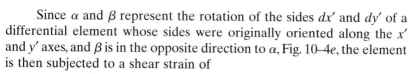

Shear strain γ_{xy}

(d)

Fig. 10–4

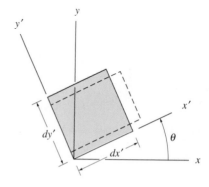

Positive normal strain, $\epsilon_{x'}$

(a)

Since α and β represent the rotation of the sides dx' and dy' of a differential element whose sides were originally oriented along the x' and y' axes, and β is in the opposite direction to α, Fig. 10–4e, the element is then subjected to a shear strain of

$$\gamma_{x'y'} = \alpha - \beta = -2(\epsilon_x - \epsilon_y)\sin\theta\cos\theta + \gamma_{xy}(\cos^2\theta - \sin^2\theta) \quad (10\text{–}4)$$

Using the trigonometric identities $\sin 2\theta = 2\sin\theta\cos\theta$, $\cos^2\theta = (1 + \cos 2\theta)/2$, and $\sin^2\theta + \cos^2\theta = 1$, we can rewrite Eqs. 10–2 and 10–4 in the final form

$$\epsilon_{x'} = \frac{\epsilon_x + \epsilon_y}{2} + \frac{\epsilon_x - \epsilon_y}{2}\cos 2\theta + \frac{\gamma_{xy}}{2}\sin 2\theta \quad (10\text{–}5)$$

$$\frac{\gamma_{x'y'}}{2} = -\left(\frac{\epsilon_x - \epsilon_y}{2}\right)\sin 2\theta + \frac{\gamma_{xy}}{2}\cos 2\theta \quad (10\text{–}6)$$

These strain-transformation equations give the normal strain $\epsilon_{x'}$ in the x' direction and the shear strain $\gamma_{x'y'}$ of an element oriented at an angle θ, as shown in Fig. 10–5. According to the established sign convention, if $\epsilon_{x'}$ is *positive*, the element *elongates* in the positive x' direction, Fig. 10–5a, and if $\gamma_{x'y'}$ is positive, the element deforms as shown in Fig. 10–5b. Note that these deformations occur as if positive normal stress $\sigma_{x'}$ and positive shear stress $\tau_{x'y'}$ act on the element.

If the normal strain in the y' direction is required, it can be obtained from Eq. 10–5 by simply substituting $(\theta + 90°)$ for θ. The result is

$$\epsilon_{y'} = \frac{\epsilon_x + \epsilon_y}{2} - \frac{\epsilon_x - \epsilon_y}{2}\cos 2\theta - \frac{\gamma_{xy}}{2}\sin 2\theta \quad (10\text{–}7)$$

The similarity between the above three equations and those for plane-stress transformation, Eqs. 9–1, 9–2, and 9–3, should be noted. By comparison, σ_x, σ_y, $\sigma_{x'}$, $\sigma_{y'}$ correspond to ϵ_x, ϵ_y, $\epsilon_{x'}$, $\epsilon_{y'}$; and τ_{xy}, $\tau_{x'y'}$ correspond to $\gamma_{xy}/2$, $\gamma_{x'y'}/2$.

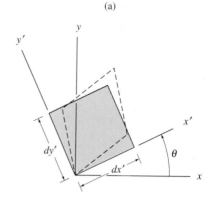

Positive shear strain, $\gamma_{x'y'}$

(b)

Fig. 10–5

Principal Strains. Like stress, the orientation of an element at a point can be determined such that the element's deformation is represented by normal strains, with *no* shear strain. When this occurs the normal strains are referred to as *principal strains*, and if the material is isotropic, the axes along which these strains occur coincide with the axes that define the planes of principal stress.

From Eqs. 9–4 and 9–5, and the correspondence between stress and strain mentioned above, the direction of the axes and the two values of the principal strains ϵ_1 and ϵ_2 are determined from

$$\tan 2\theta_p = \frac{\gamma_{xy}}{\epsilon_x - \epsilon_y} \qquad (10\text{–}8)$$

$$\epsilon_{1,2} = \frac{\epsilon_x + \epsilon_y}{2} \pm \sqrt{\left(\frac{\epsilon_x - \epsilon_y}{2}\right)^2 + \left(\frac{\gamma_{xy}}{2}\right)^2} \qquad (10\text{–}9)$$

Maximum In-Plane Shear Strain. Using Eqs. 9–6, 9–7, and 9–8, the direction of the axis, and the maximum in-plane shear strain and associated average normal strain are determined from the following equations:

$$\tan 2\theta_s = -\left(\frac{\epsilon_x - \epsilon_y}{\gamma_{xy}}\right) \qquad (10\text{–}10)$$

$$\frac{\gamma_{\substack{\max \\ \text{in-plane}}}}{2} = \sqrt{\left(\frac{\epsilon_x - \epsilon_y}{2}\right)^2 + \left(\frac{\gamma_{xy}}{2}\right)^2} \qquad (10\text{–}11)$$

$$\epsilon_{\text{avg}} = \frac{\epsilon_x + \epsilon_y}{2} \qquad (10\text{–}12)$$

Complex stresses are often developed at the joints where vessels are connected together. The stresses are determined by making measurements of strain.

IMPORTANT POINTS

- Due to the Poisson effect, the state of plane strain is not a state of plane stress, and vice versa.

- A point on a body is subjected to plane stress when the surface of the body is stress free. Plane-strain analysis may be used within the plane of the stresses to analyze the results from the gauges. Remember, though, there is a normal strain that is perpendicular to the gauges.

- When the state of strain is represented by the principal strains, no shear strain will act on the element.

- The state of strain at the point can also be represented in terms of the maximum in-plane shear strain. In this case an average normal strain will also act on the element.

- The element representing the maximum in-plane shear strain and its associated average normal strains is 45° from the element representing the principal strains.

EXAMPLE 10-1

A differential element of material at a point is subjected to a state of plane strain $\epsilon_x = 500(10^{-6})$, $\epsilon_y = -300(10^{-6})$, $\gamma_{xy} = 200(10^{-6})$, which tends to distort the element as shown in Fig. 10–6a. Determine the equivalent strains acting on an element oriented at the point, *clockwise* 30° from the original position.

SOLUTION

The strain-transformation Eqs. 10–5 and 10–6 will be used to solve the problem. Since θ is *positive counterclockwise*, then for this problem $\theta = -30°$. Thus,

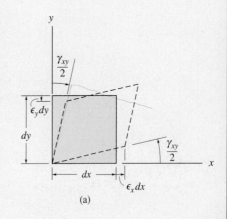

$$\epsilon_{x'} = \frac{\epsilon_x + \epsilon_y}{2} + \frac{\epsilon_x - \epsilon_y}{2} \cos 2\theta + \frac{\gamma_{xy}}{2} \sin 2\theta$$

$$= \left[\frac{500 + (-300)}{2}\right](10^{-6}) + \left[\frac{500 - (-300)}{2}\right](10^{-6}) \cos(2(-30°))$$

$$+ \left[\frac{200(10^{-6})}{2}\right] \sin(2(-30°))$$

$$\epsilon_{x'} = 213(10^{-6}) \hspace{4cm} \textit{Ans.}$$

(a)

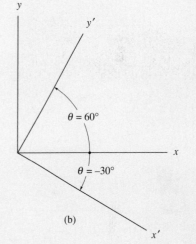

$$\frac{\gamma_{x'y'}}{2} = -\left(\frac{\epsilon_x - \epsilon_y}{2}\right) \sin 2\theta + \frac{\gamma_{xy}}{2} \cos 2\theta$$

$$= -\left[\frac{500 - (-300)}{2}\right](10^{-6}) \sin(2(-30°)) + \frac{200(10^{-6})}{2} \cos(2(-30°))$$

$$\gamma_{x'y'} = 793(10^{-6}) \hspace{3cm} \textit{Ans.}$$

(b)

The strain in the y' direction can be obtained from Eq. 10–7 with $\theta = -30°$. However, we can also obtain $\epsilon_{y'}$ using Eq. 10–5 with $\theta = 60°$ ($\theta = -30° + 90°$), Fig. 10–6b. We have with ϵ_y replacing $\epsilon_{x'}$,

$$\epsilon_{y'} = \frac{\epsilon_x + \epsilon_y}{2} + \frac{\epsilon_x - \epsilon_y}{2} \cos 2\theta + \frac{\gamma_{xy}}{2} \sin 2\theta$$

$$= \left[\frac{500 + (-300)}{2}\right](10^{-6}) + \left[\frac{500 - (-300)}{2}\right](10^{-6}) \cos(2(60°))$$

$$+ \frac{200(10^{-6})}{2} \sin(2(60°))$$

$$\epsilon_{y'} = -13.4(10^{-6}) \hspace{3cm} \textit{Ans.}$$

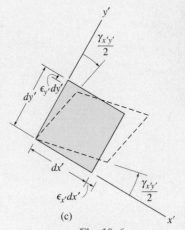

(c)

These results tend to distort the element as shown in Fig. 10–6c.

Fig. 10–6

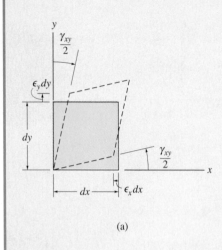

(a)

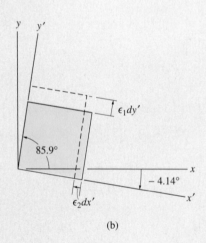

(b)

Fig. 10–7

EXAMPLE 10-2

A differential element of material at a point is subjected to a state of plane strain defined by $\epsilon_x = -350(10^{-6})$, $\epsilon_y = 200(10^{-6})$, $\gamma_{xy} = 80(10^{-6})$, which tends to distort the element as shown in Fig. 10–7a. Determine the principal strains at the point and the associated orientation of the element.

SOLUTION

Orientation of the Element.   From Eq. 10–8 we have

$$\tan 2\theta_p = \frac{\gamma_{xy}}{\epsilon_x - \epsilon_y}$$

$$= \frac{80(10^{-6})}{(-350 - 200)(10^{-6})}$$

Thus, $2\theta_p = -8.28°$ and $-8.28° + 180° = 172°$, so that

$$\theta_p = -4.14° \text{ and } 85.9° \qquad\qquad Ans.$$

Each of these angles is measured *positive counterclockwise*, from the x axis to the outward normals on each face of the element, Fig. 10–7b.

Principal Strains.   The principal strains are determined from Eq. 10–9. We have

$$\epsilon_{1,2} = \frac{\epsilon_x + \epsilon_y}{2} \pm \sqrt{\left(\frac{\epsilon_x - \epsilon_y}{2}\right)^2 + \left(\frac{\gamma_{xy}}{2}\right)^2}$$

$$= \frac{(-350 + 200)(10^{-6})}{2} \pm \left[\sqrt{\left(\frac{-350 - 200}{2}\right)^2 + \left(\frac{80}{2}\right)^2}\right](10^{-6})$$

$$= -75.0(10^{-6}) \pm 277.9(10^{-6})$$

$$\epsilon_1 = 203(10^{-6}) \qquad \epsilon_2 = -353(10^{-6}) \qquad\qquad Ans.$$

We can determine which of these two strains deforms the element in the x' direction by applying Eq. 10–5 with $\theta = -4.14°$. Thus,

$$\epsilon_{x'} = \frac{\epsilon_x + \epsilon_y}{2} + \frac{\epsilon_x - \epsilon_y}{2}\cos 2\theta + \frac{\gamma_{xy}}{2}\sin 2\theta$$

$$= \left(\frac{-350 + 200}{2}\right)(10^{-6}) + \left(\frac{-350 - 200}{2}\right)(10^{-6})\cos 2(-4.14°)$$

$$+ \frac{80(10^{-6})}{2}\sin 2(-4.14°)$$

$$\epsilon_{x'} = -353(10^{-6})$$

Hence $\epsilon_{x'} = \epsilon_2$. When subjected to the principal strains, the element is distorted as shown in Fig. 10–7b.

EXAMPLE 10-3

A differential element of material at a point is subjected to a state of plane strain defined by $\epsilon_x = -350(10^{-6})$, $\epsilon_y = 200(10^{-6})$, $\gamma_{xy} = 80(10^{-6})$, which tends to distort the element as shown in Fig. 10–8a. Determine the maximum in-plane shear strain at the point and the associated orientation of the element.

SOLUTION

Orientation of the Element. From Eq. 10–10 we have

$$\tan 2\theta_s = -\left(\frac{\epsilon_x - \epsilon_y}{\gamma_{xy}}\right) = -\frac{(-350 - 200)(10^{-6})}{80(10^{-6})}$$

Thus, $2\theta_s = 81.72°$ and $81.72° + 180° = 261.72°$, so that

$$\theta_s = 40.9° \text{ and } 130.9°$$

Note that this orientation is $45°$ from that shown in Fig. 10–7b in Example 10–2 as expected.

Maximum In-Plane Shear Strain. Applying Eq. 10–11 gives

$$\frac{\gamma_{\substack{max \\ in\text{-}plane}}}{2} = \sqrt{\left(\frac{\epsilon_x - \epsilon_y}{2}\right)^2 + \left(\frac{\gamma_{xy}}{2}\right)^2}$$

$$= \left[\sqrt{\left(\frac{-350 - 200}{2}\right)^2 + \left(\frac{80}{2}\right)^2}\right](10^{-6})$$

$$\gamma_{\substack{max \\ in\text{-}plane}} = 556(10^{-6}) \qquad \textit{Ans.}$$

The proper sign of $\gamma_{\substack{max \\ in\text{-}plane}}$ can be obtained by applying Eq. 10–6 with $\theta_s = 40.9°$. We have

$$\frac{\gamma_{x'y'}}{2} = -\frac{\epsilon_x - \epsilon_y}{2}\sin 2\theta + \frac{\gamma_{xy}}{2}\cos 2\theta$$

$$= -\left(\frac{-350 - 200}{2}\right)(10^{-6})\sin 2(40.9°) + \frac{80(10^{-6})}{2}\cos 2(40.9°)$$

$$\gamma_{x'y'} = 556(10^{-6})$$

Thus, $\gamma_{\substack{max \\ in\text{-}plane}}$ tends to distort the element so that the right angle between dx' and dy' is decreased (positive sign convention), Fig. 10–8b.

Also, there are associated average normal strains imposed on the element that are determined from Eq. 10–12:

$$\epsilon_{avg} = \frac{\epsilon_x + \epsilon_y}{2} = \frac{-350 + 200}{2}(10^{-6}) = -75(10^{-6})$$

These strains tend to cause the element to contract, Fig. 10–8b.

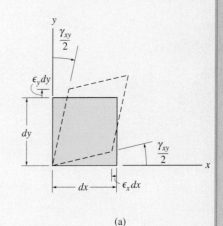

(a)

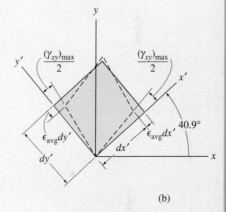

(b)

Fig. 10–8

*10.3 MOHR'S CIRCLE—PLANE STRAIN

Since the equations of plane-strain transformation are mathematically similar to the equations of plane-stress transformation, we can also solve problems involving the transformation of strain using Mohr's circle. This approach has the advantage of making it possible to see graphically how the normal and shear strain components at a point vary from one orientation of the element to the next.

Like the case for stress, the parameter θ in Eqs. 10–5 and 10–6 can be eliminated and the result rewritten in the form

$$(\epsilon_x - \epsilon_{avg})^2 + \left(\frac{\gamma_{xy}}{2}\right)^2 = R^2 \qquad (10\text{–}13)$$

where

$$\epsilon_{avg} = \frac{\epsilon_x + \epsilon_y}{2}$$

$$R = \sqrt{\left(\frac{\epsilon_x - \epsilon_y}{2}\right)^2 + \left(\frac{\gamma_{xy}}{2}\right)^2}$$

Equation 10–13 represents the equation of Mohr's circle for strain. It has a center on the ϵ axis at point $C(\epsilon_{avg}, 0)$ and a radius R.

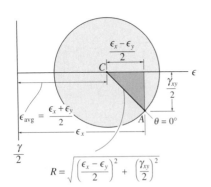

Fig. 10–9

PROCEDURE FOR ANALYSIS

The procedure for drawing Mohr's circle for strain follows the same one established for stress.

Construction of the Circle.

• Establish a coordinate system such that the abscissa represents the normal strain ϵ, with *positive to the right*, and the ordinate represents *half* the value of the shear strain, $\gamma/2$, with *positive downward*, Fig. 10–9.

• Using the positive sign convention for ϵ_x, ϵ_y, γ_{xy}, as shown in Fig. 10–3, determine the center of the circle C, which is located on the ϵ axis at a distance $\epsilon_{avg} = (\epsilon_x + \epsilon_y)/2$ from the origin, Fig. 10–9.

• Plot the reference point A having coordinates $A(\epsilon_x, \gamma_{xy}/2)$. This point represents the case for which the x' axis coincides with the x axis. Hence $\theta = 0°$, Fig. 10–9.

• Connect point A with the center C of the circle and from the shaded triangle determine the radius R of the circle, Fig. 10–9.

• Once R has been determined, sketch the circle.

Principal Strains.

- The principal strains ϵ_1 and ϵ_2 are determined from the circle as the coordinates of points B and D, that is where $\gamma/2 = 0$, Fig. 10–10a.

- The orientation of the plane on which ϵ_1 acts can be determined from the circle by calculating $2\theta_{p_1}$ using trigonometry. Here this angle is measured counterclockwise *from* the radial reference line CA to line CB, Fig. 10–10a. Remember that the *rotation* of θ_{p_1} must be in this *same direction*, from the element's reference axis x to the x' axis, Fig. 10–10b.*

- When ϵ_1 and ϵ_2 are indicated as being positive as in Fig. 10–10a, the element in Fig. 10–10b will elongate in the x' and y' directions as shown by the dashed outline.

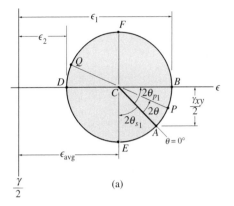

(a)

Maximum In-Plane Shear Strain.

- The average normal strain and half the maximum in-plane shear strain are determined from the circle as the coordinates of points E and F, Fig. 10–10a.

- The orientation of the plane on which $\frac{\gamma_{max}}{\text{in-plane}}$ and ϵ_{avg} act can be determined from the circle by calculating $2\theta_{s_1}$ using trigonometry. Here this angle is measured clockwise *from* the radial reference line CA to line CE, Fig. 10–10a. Remember that the *rotation* of θ_{s_1} must be in this *same direction*, from the element's reference axis x to the x' axis, Fig. 10–10c.*

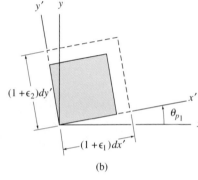

(b)

Strains on Arbitrary Plane.

- The normal and shear strain components $\epsilon_{x'}$ and $\gamma_{x'y'}$ for a plane specified at an angle θ, Fig. 10–10d, can be obtained from the circle using trigonometry to determine the coordinates of point P, Fig. 10–10a.

- To locate P, the known angle θ of the x' axis is measured on the circle as 2θ. This measurement is made *from* the radial reference line CA to the radial line CP. Remember that measurements for 2θ on the circle must be in the same direction as θ for the x' axis.*

- If the value of $\epsilon_{y'}$ is required, it can be determined by calculating the ϵ coordinate of point Q in Fig. 10–10a. The line CQ lies 180° away from CP and thus represents a rotation of 90° of the x' axis.

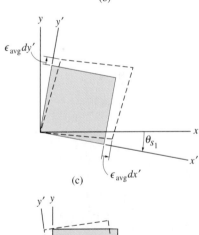

(c)

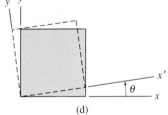

(d)

*If instead the $\gamma/2$ axis is constructed *positive upwards*, then the angle 2θ on the circle would be measured in the *opposite direction* to the orientation θ of the plane.

Fig. 10–10

EXAMPLE 10–4

The state of plane strain at a point is represented by the components $\epsilon_x = 250(10^{-6})$, $\epsilon_y = -150(10^{-6})$, and $\gamma_{xy} = 120(10^{-6})$. Determine the principal strains and the orientation of the element.

SOLUTION

Construction of the Circle. The ϵ and $\gamma/2$ axes are established in Fig. 10–11a. Remember that the *positive* $\gamma/2$ axis must be directed *downward* so that *counterclockwise* rotations of the element correspond to *counterclockwise* rotation around the circle, and vice versa. The center of the circle C is located on the ϵ axis at

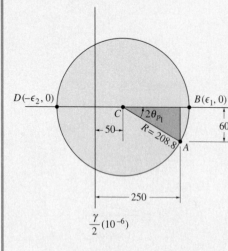

(a)

$$\epsilon_{\text{avg}} = \frac{250 + (-150)}{2}(10^{-6}) = 50(10^{-6})$$

Since $\gamma_{xy}/2 = 60(10^{-6})$, the reference point A ($\theta = 0°$) has coordinates $A(250(10^{-6}), 60(10^{-6}))$. From the shaded triangle in Fig. 10–11a, the radius of the circle is CA; that is,

$$R = \left[\sqrt{(250 - 50)^2 + (60)^2}\right](10^{-6}) = 208.8(10^{-6})$$

Principal Strains. The ϵ coordinates of points B and D represent the principal strains. They are

$$\epsilon_1 = (50 + 208.8)(10^{-6}) = 259(10^{-6}) \qquad \textit{Ans.}$$

$$\epsilon_2 = (50 - 208.8)(10^{-6}) = -159(10^{-6}) \qquad \textit{Ans.}$$

The direction of the positive principal strain ϵ_1 is defined by the *counterclockwise* angle $2\theta_{p_1}$, measured from the radial reference line CA to the line CB. We have

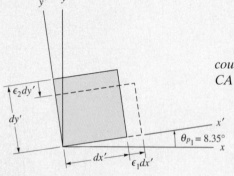

(b)

Fig. 10–11

$$\tan 2\theta_{p_1} = \frac{60}{(250 - 50)}$$

$$\theta_{p_1} = 8.35° \qquad \textit{Ans.}$$

Hence the side dx' of the element is oriented *counterclockwise* 8.35° as shown in Fig. 10–11b. This also defines the direction of ϵ_1. The deformation of the element is also shown in the figure.

EXAMPLE 10–5

The state of plane strain at a point is represented by the components $\epsilon_x = 250(10^{-6})$, $\epsilon_y = -150(10^{-6})$, and $\gamma_{xy} = 120(10^{-6})$. Determine the maximum in-plane shear strains and the orientation of an element.

SOLUTION

The circle has been established in the previous example and is shown in Fig. 10–12a.

Maximum In-Plane Shear Strain.   Half the maximum in-plane shear strain and average normal strain are represented by the coordinates of point E or F on the circle. From the coordinates of point E,

$$\frac{(\gamma_{x'y'})^{\text{max}}_{\text{in-plane}}}{2} = 208.8(10^{-6})$$

$$(\gamma_{x'y'})^{\text{max}}_{\text{in-plane}} = 418(10^{-6}) \qquad Ans.$$

$$\epsilon_{\text{avg}} = 50(10^{-6})$$

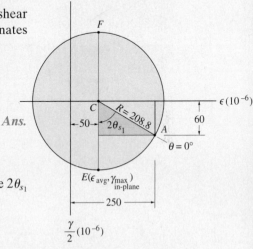

(a)

Fig. 10–12

To orient the element, we can determine the clockwise angle $2\theta_{s_1}$ from the circle.

$$2\theta_{s_1} = 90° - 2(8.35°)$$
$$\theta_{s_1} = 36.6° \qquad Ans.$$

This angle is shown in Fig. 10–12b. Since the shear strain defined from point E on the circle has a positive value and the average normal strain is also positive, corresponding positive shear stress and positive average normal stress deform the element into the dashed shape shown in the figure.

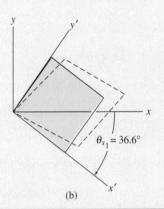

(b)

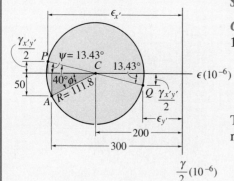

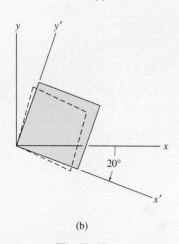

(b)

Fig. 10–13

EXAMPLE 10–6

The state of plane strain at a point is represented on an element having components $\epsilon_x = -300(10^{-6})$, $\epsilon_y = -100(10^{-6})$, and $\gamma_{xy} = 100(10^{-6})$. Determine the state of strain on an element oriented 20° clockwise from this reported position.

SOLUTION

Construction of the Circle.   The ϵ and $\gamma/2$ axes are established in Fig. 10–13a. The center of the circle is on the ϵ axis at

$$\epsilon_{avg} = \left(\frac{-300 - 100}{2}\right)(10^{-6}) = -200(10^{-6})$$

The reference point A has coordinates $A(-300(10^{-6}), 50(10^{-6}))$. The radius CA determined from the shaded triangle is therefore

$$R = \left[\sqrt{(300 - 200)^2 + (50)^2}\right](10^{-6}) = 111.8(10^{-6})$$

Strains on Inclined Element.   Since the element is to be oriented 20° *clockwise*, we must establish a radial line CP, $2(20°) = 40°$ *clockwise*, measured from CA ($\theta = 0°$), Fig. 10–13a. The coordinates of point P ($\epsilon_{x'}, -\gamma_{x'y'}/2$) are obtained from the geometry of the circle. Note that

$$\phi = \tan^{-1}\left(\frac{50}{(300 - 200)}\right) = 26.57°, \qquad \psi = 40° - 26.57° = 13.43°$$

Thus,

$$\epsilon_{x'} = -(200 + 111.8 \cos 13.43°)(10^{-6})$$
$$= -309(10^{-6}) \hspace{3cm} \textit{Ans.}$$

$$\frac{\gamma_{x'y'}}{2} = -(111.8 \sin 13.43°)(10^{-6})$$

$$\gamma_{x'y'} = -52.0(10^{-6}) \hspace{3cm} \textit{Ans.}$$

The normal strain $\epsilon_{y'}$ can be determined from the ϵ coordinate of point Q on the circle, Fig. 10–13a. Why?

$$\epsilon_{y'} = -(200 - 111.8 \cos 13.43°)(10^{-6}) = -91.3(10^{-6})$$
$$\textit{Ans.}$$

As a result of these strains, the element deforms relative to x', y' axes as shown in Fig. 10–13b.

PROBLEMS

10–1. Prove that the sum of the normal strains in perpendicular directions is constant.

10–2. The state of strain at the point on the bracket has components $\epsilon_x = -200(10^{-6})$, $\epsilon_y = -650(10^{-6})$, $\gamma_{xy} = -175(10^{-6})$. Use the strain-transformation equations to determine the equivalent in-plane strains on an element oriented at an angle of $\theta = 20°$ counterclockwise from the original position. Sketch the deformed element due to these strains within the x–y plane.

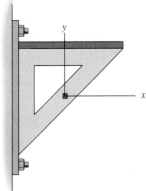

Prob. 10–2

10–3. A differential element on the bracket is subjected to plane strain that has the following components: $\epsilon_x = 150(10^{-6})$, $\epsilon_y = 200(10^{-6})$, $\gamma_{xy} = -700(10^{-6})$. Use the strain-transformation equations and determine the equivalent in-plane strains on an element oriented at an angle of $\theta = 60°$ counterclockwise from the original position. Sketch the deformed element within the x–y plane due to these strains.

*10–4. Solve Prob. 10-3 for an element oriented $\theta = 30°$ clockwise.

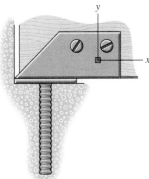

Probs. 10–3/10–4

10–5. Due to the load **P**, the state of strain at the point on the bracket has components of $\epsilon_x = 500(10^{-6})$, $\epsilon_y = 350(10^{-6})$, and $\gamma_{xy} = -430(10^{-6})$. Use the strain-transformation equations to determine the equivalent in-plane strains on an element oriented at an angle of $\theta = 30°$ clockwise from the original position. Sketch the deformed element due to these strains with in the x–y plane.

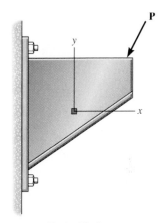

Prob. 10–5

10–6. The state of strain at the point on the boom of the hydraulic engine crane has components of $\epsilon_x = 250(10^{-6})$, $\epsilon_y = 300(10^{-6})$, and $\gamma_{xy} = -180(10^{-6})$. Use the strain-transformation equations to determine (a) the in-plane principal strains and (b) the maximum in-plane shear strain and average normal strain. In each case, specify the orientation of the element and show how the strains deform the element within the x–y plane.

Prob. 10–6

10–7. The state of strain at the point on the gear tooth has components $\epsilon_x = 850(10^{-6})$, $\epsilon_y = 480(10^{-6})$, $\gamma_{xy} = 650(10^{-6})$. Use the strain-transformation equations to determine (a) the in-plane principal strains and (b) the maximum in-plane shear strain and average normal strain. In each case, specify the orientation of the element and show how the strains deform the element within the x–y plane.

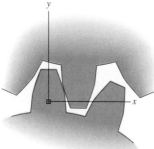

Prob. 10–7

*10–8.** The state of strain at the point on the fan blade has components of $\epsilon_x = 250(10^{-6})$, $\epsilon_y = -450(10^{-6})$, and $\gamma_{xy} = -825(10^{-6})$. Use the strain transformation equations to determine (a) the in-plane principal strains and (b) the maximum in-plane shear strain and average normal strain. In each case specify the orientation of the element and show how the strains deform the element within the x–y plane.

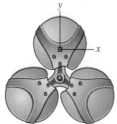

Prob. 10–8

10–9. The state of strain at the point on the spanner wrench has components of $\epsilon_x = 260(10^{-6})$, $\epsilon_y = 320(10^{-6})$, and $\gamma_{xy} = 180(10^{-6})$. Use the strain-transformation equations to determine (a) the in-plane principal strains and (b) the maximum in-plane shear strain and average normal strain. In each case, specify the orientation of the element and show how the strains deform the element within the x–y plane.

Prob. 10–9

10–10. The state of strain at the point on the wrench has components $\epsilon_x = 120(10^{-6})$, $\epsilon_y = -180(10^{-6})$, $\gamma_{xy} = 150(10^{-6})$. Use the strain-transformation equations to determine (a) the in-plane principal strains and (b) the maximum in-plane shear strain and average normal strain. In each case, specify the orientation of the element and show how the strains deform the element within the x–y plane.

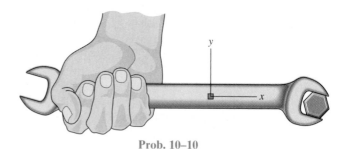

Prob. 10–10

■**10–11.** Consider the general case of plane strain where ϵ_x, ϵ_y, and γ_{xy} are known. Write a computer program that can be used to determine the normal and shear strains, $\epsilon_{x'}$ and $\gamma_{x'y'}$, on the plane of an element oriented θ from the horizontal. Also, compute the principal strains and the element's orientation, and the maximum in-plane shear strain, the average normal strain, and the element's orientation.

*10–12.** Solve Prob. 10–2 using Mohr's circle.

10–13. Solve Prob. 10–3 using Mohr's circle.

10–14. Solve Prob. 10–4 using Mohr's circle.

10–15. Solve Prob. 10–5 using Mohr's circle.

*10–16.** Solve Prob. 10–8 using Mohr's circle.

10–17. Solve Prob. 10–6 using Mohr's circle.

10–18. Solve Prob. 10–7 using Mohr's circle.

10–19. Solve Prob. 10–9 using Mohr's circle.

*10.4   ABSOLUTE MAXIMUM SHEAR STRAIN

In Sec. 9.7 it was pointed out that in three dimensions the state of stress at a point can be represented by an element oriented in a specific direction, such that the element is subjected only to *principal stresses* having maximum, intermediate, and minimum values, σ_{max}, σ_{int}, and σ_{min}. These stresses subject the material to associated *principal strains* ϵ_{max}, ϵ_{int}, and ϵ_{min}. Also, if the material is both homogeneous and isotropic, the element will *not* be subjected to shear strains since the shear stress on the principal planes is zero.

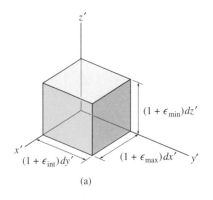

(a)

Assume that the three principal strains cause elongations along the x', y', z' axes as shown in Fig. 10–14a. If we view the element in two dimensions, that is, in the $x'-y'$, $x'-z'$, and $y'-z'$ planes, Fig. 10–14b, 10–14c, and 10–14d, we can then use Mohr's circle to determine the *maximum in-plane shear strain* for each case. For example, from the view of the element in the $x'-y'$ plane, Fig. 10–14b, the diameter of Mohr's circle extends between ϵ_{max} and ϵ_{int}, Fig. 10–14e. This circle gives the normal and shear strain components on each element oriented about the z' axis. Likewise, Mohr's circles for each element oriented about the y' and x' axes are also shown in Fig. 10–14e.

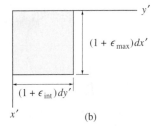

(b)

From these three circles it is seen that the ***absolute maximum shear strain*** is determined from the circle having the largest radius. It occurs on the element oriented $45°$ about the y' axis from the element shown in its original position, Fig. 10–14a or 10–14c. For this condition,

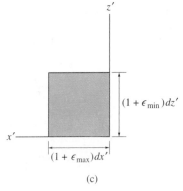

(c)

$$\gamma_{\substack{abs \\ max}} = \epsilon_{max} - \epsilon_{min} \qquad (10\text{--}14)$$

and

$$\epsilon_{avg} = \frac{\epsilon_{max} + \epsilon_{min}}{2} \qquad (10\text{--}15)$$

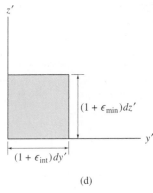

(d)

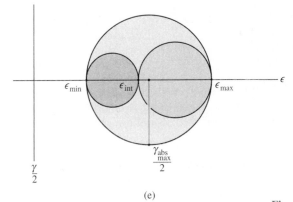

(e)

Fig. 10–14

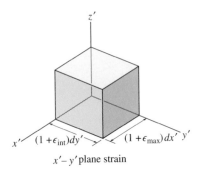

$x' - y'$ plane strain

(a)

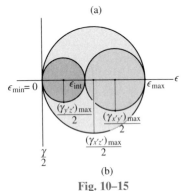

(b)

Fig. 10–15

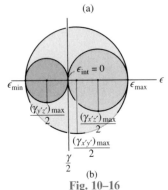

$x' - y'$ plane strain

(a)

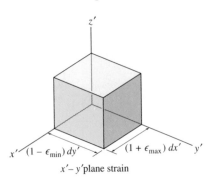

(b)

Fig. 10–16

Plane Strain. As in the case of plane stress, the previous analysis has an important implication when the material is subjected to *plane strain*, especially when the *principal strains* have the *same sign*, i.e., both cause elongation or both cause contraction. For example, if the principal in-plane strains are ϵ_{max} and ϵ_{int}, while the out-of-plane principal strain is $\epsilon_{min} = 0$, Fig. 10–15a, then the three Mohr's circles describing the normal and shear strain components for elements oriented about the x', y', and z' axes are shown in Fig. 10–15b. By inspection, the largest circle has a radius $R = (\gamma_{x'z'})_{max}/2$. Hence,

$$\gamma_{\substack{abs \\ max}} = (\gamma_{x'z'})_{max} = \epsilon_{max}$$

This value represents the *absolute maximum shear strain* for the material. Note that it is *larger* than the maximum in-plane shear strain, which is $(\gamma_{x'y'})_{max} = \epsilon_{max} - \epsilon_{int}$.

On the other hand, if one of the in-plane principal strains is of *opposite sign* to the other in-plane principal strain, then ϵ_{max} causes elongation, ϵ_{min} causes contraction, and the out-of-plane principal strain is $\epsilon_{int} = 0$, Fig. 10–16a. Mohr's circles, which describe the strains on each element's orientation about the x', y', z' axes, are shown in Fig. 10–16b. In this case,

$$\gamma_{\substack{abs \\ max}} = (\gamma_{x'y'})_{max} = \epsilon_{max} - \epsilon_{min}$$

We may therefore summarize the above two points as follows. If the in-plane principal strains both have the *same sign*, the *absolute maximum shear strain* will occur *out of plane* and has a value of $\gamma_{\substack{abs \\ max}} = \epsilon_{max}$. However, if the in-plane principal strains are of *opposite signs*, then the absolute maximum shear strain *equals* the maximum in-plane shear strain.

IMPORTANT POINTS

- The general three-dimensional state of strain at a point can be represented by an element oriented so that only three principal strains act on it.

- From this orientation, the orientation of the element representing the absolute maximum shear strain can be obtained by rotating the element 45° about the axis defining the direction of ϵ_{int}.

- The absolute maximum shear strain will be *larger* than the maximum in-plane shear strain whenever the in-plane principal strains have the *same sign*. When this occurs the absolute maximum shear strain will act out of the plane.

EXAMPLE 10-7

The state of plane strain at a point is represented by the strain components $\epsilon_x = -400(10^{-6})$, $\epsilon_y = 200(10^{-6})$, $\gamma_{xy} = 150(10^{-6})$. Determine the maximum in-plane shear strain and the absolute maximum shear strain.

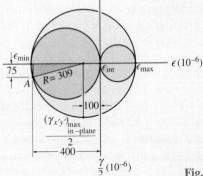

Fig. 10–17

SOLUTION

Maximum In-Plane Strain.   We will solve this problem using Mohr's circle. From the strain components, the center of the circle is on the ϵ axis at

$$\epsilon_{avg} = \frac{-400 + 200}{2}(10^{-6}) = -100(10^{-6})$$

Since $\gamma_{xy}/2 = 75(10^{-6})$, the reference point has coordinates $A(-400(10^{-6}), 75(10^{-6}))$. As shown in Fig. 10–17, the radius of the circle is therefore

$$R = \left[\sqrt{(400 - 100)^2 + (75)^2} \right](10^{-6}) = 309(10^{-6})$$

Computing the in-plane principal strains, we have

$$\epsilon_{max} = (-100 + 309)(10^{-6}) = 209(10^{-6})$$

$$\epsilon_{min} = (-100 - 309)(10^{-6}) = -409(10^{-6})$$

From the circle, the maximum in-plane shear strain is

$$\gamma_{\substack{max \\ \text{in-plane}}} = \epsilon_{max} - \epsilon_{min} = [209 - (-409)](10^{-6}) = 618(10^{-6}) \qquad \textit{Ans.}$$

Absolute Maximum Shear Strain.   From the above results, we have $\epsilon_{max} = 209(10^{-6})$, $\epsilon_{int} = 0$, $\epsilon_{min} = -409(10^{-6})$. The three Mohr's circles, plotted for element orientations about each of the x', y', z' axes, are also shown in Fig. 10–17. It is seen that since the *principal in-plane strains have opposite signs*, the maximum in-plane shear strain is *also* the absolute maximum shear strain; i.e.,

$$\gamma_{\substack{abs \\ max}} = 618(10^{-6}) \qquad \textit{Ans.}$$

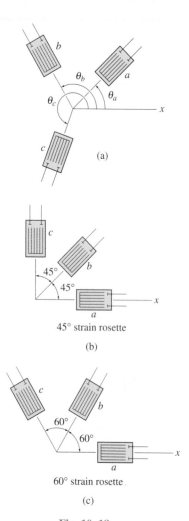

(a)

45° strain rosette

(b)

60° strain rosette

(c)

Fig. 10–18

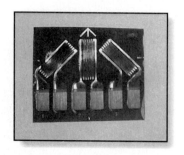

Typical electrical resistance 45° strain rosette.

10.5 STRAIN ROSETTES

It was mentioned in Sec. 3.1 that the normal strain in a tension-test specimen can be measured using an **electrical-resistance strain gauge**, which consists of a wire grid or piece of metal foil bonded to the specimen. However, for a general loading on a body, the *normal strains* at a point on its free surface are often determined using a cluster of three electrical-resistance strain gauges, arranged in a specified pattern. This pattern is referred to as a **strain rosette**, and once the readings on the three gauges are made, the data can then be used to specify the state of strain at the point. It should be noted, however, that these strains are measured *only* in the plane of the gauges, and since the body is stress-free on its surface, the gauges may be subjected to *plane stress* but *not* plane strain. In this regard, the normal line to the free surface is a principal axis of strain, and so the principal normal strain along this axis is *not* measured by the strain rosette. What is important here is that the out-of-plane displacement caused by this principal strain will *not* affect the in-plane measurements of the gauges.

In the general case, the axes of the three gauges are arranged at the angles θ_a, θ_b, θ_c as shown in Fig. 10–18a. If the readings ϵ_a, ϵ_b, ϵ_c are taken, we can determine the strain components ϵ_x, ϵ_y, γ_{xy} at the point by applying the strain-transformation equation, Eq. 10–2, for each gauge. We have

$$\epsilon_a = \epsilon_x \cos^2 \theta_a + \epsilon_y \sin^2 \theta_a + \gamma_{xy} \sin \theta_a \cos \theta_a$$
$$\epsilon_b = \epsilon_x \cos^2 \theta_b + \epsilon_y \sin^2 \theta_b + \gamma_{xy} \sin \theta_b \cos \theta_b \qquad (10\text{–}16)$$
$$\epsilon_c = \epsilon_x \cos^2 \theta_c + \epsilon_y \sin^2 \theta_c + \gamma_{xy} \sin \theta_c \cos \theta_c$$

The values of ϵ_x, ϵ_y, γ_{xy} are determined by solving these three equations simultaneously.

Strain rosettes are often arranged in 45° or 60° patterns. In the case of the 45° or "rectangular" strain rosette shown in Fig. 10–18b, $\theta_a = 0°$, $\theta_b = 45°$, $\theta_c = 90°$, so that Eq. 10–16 gives

$$\epsilon_x = \epsilon_a$$
$$\epsilon_y = \epsilon_c$$
$$\gamma_{xy} = 2\epsilon_b - (\epsilon_a + \epsilon_c)$$

And for the 60° strain rosette in Fig. 10–18c, $\theta_a = 0°$, $\theta_b = 60°$, $\theta_c = 120°$. Here Eq. 10–16 gives

$$\epsilon_x = \epsilon_a$$
$$\epsilon_y = \frac{1}{3}(2\epsilon_b + 2\epsilon_c - \epsilon_a) \qquad (10\text{–}17)$$
$$\gamma_{xy} = \frac{2}{\sqrt{3}}(\epsilon_b - \epsilon_c)$$

Once ϵ_x, ϵ_y, γ_{xy} are determined, the transformation equations of Sec. 10.2 or Mohr's circle can then be used to determine the principal in-plane strains and the maximum in-plane shear strain at the point.

EXAMPLE 10-8

The state of strain at point A on the bracket in Fig. 10–19a is measured using the strain rosette shown in Fig. 10–19b. Due to the loadings, the readings from the gauges give $\epsilon_a = 60(10^{-6})$, $\epsilon_b = 135(10^{-6})$, and $\epsilon_c = 264(10^{-6})$. Determine the in-plane principal strains at the point and the directions in which they act.

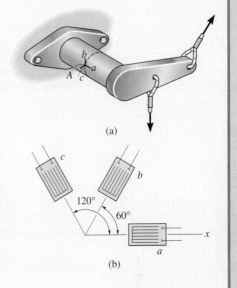

(a)

SOLUTION

We will use Eq. 10–16 for the solution. Establishing an x axis as shown in Fig. 10–19b and measuring the angles counterclockwise from the $+x$ axis to the center-lines of each gauge, we have $\theta_a = 0°$, $\theta_b = 60°$, and $\theta_c = 120°$. Substituting these results, along with the problem data, into Eq. 10–16 gives

$$60(10^{-6}) = \epsilon_x \cos^2 0° + \epsilon_y \sin^2 0° + \gamma_{xy} \sin 0° \cos 0°$$

$$= \epsilon_x \qquad (1)$$

$$135(10^{-6}) = \epsilon_x \cos^2 60° + \epsilon_y \sin^2 60° + \gamma_{xy} \sin 60° \cos 60°$$

$$= 0.25\epsilon_x + 0.75\epsilon_y + 0.433\gamma_{xy} \qquad (2)$$

$$264(10^{-6}) = \epsilon_x \cos^2 120° + \epsilon_y \sin^2 120° + \gamma_{xy} \sin 120° \cos 120°$$

$$= 0.25\epsilon_x + 0.75\epsilon_y - 0.433\gamma_{xy} \qquad (3)$$

Using Eq. (1) and solving Eqs. (2) and (3) simultaneously, we get

$$\epsilon_x = 60(10^{-6}) \qquad \epsilon_y = 246(10^{-6}) \qquad \gamma_{xy} = -149(10^{-6})$$

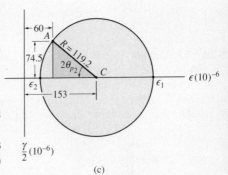

(b)

These same results can also be obtained in a more direct manner from Eq. 10–17.

The in-plane principal strains can be determined using Mohr's circle. The reference point on the circle is at A $(60(10^{-6}), -74.5(10^{-6}))$ and the center of the circle, C, is on the ϵ axis at $\epsilon_{avg} = 153(10^{-6})$, Fig. 10–19c. From the shaded triangle, the radius is

$$R = \left[\sqrt{(153 - 60)^2 + (74.5)^2} \right](10^{-6}) = 119.2(10^{-6})$$

The in-plane principal strains are thus

$$\epsilon_1 = 153(10^{-6}) + 119.2(10^{-6}) = 272(10^{-6}) \qquad Ans.$$

$$\epsilon_2 = 153(10^{-6}) - 119.2(10^{-6}) = 33.8(10^{-6}) \qquad Ans.$$

$$2\theta_{p_2} = \tan^{-1} \frac{74.5}{(153 - 60)} = 38.7°$$

$$\theta_{p_2} = 19.3° \qquad Ans.$$

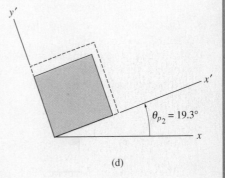

The deformed element is shown in the dashed position in Fig. 10–19d. Realize that, due to the Poisson effect, the element is *also* subjected to an out-of-plane strain, i.e., in the z direction, although this value does not influence the calculated results.

(c)

(d)

Fig. 10–19

PROBLEMS

*10–20. The strain at a point has components of $\epsilon_x = -480(10^{-6})$, $\epsilon_y = 650(10^{-6})$, $\gamma_{xy} = 780(10^{-6})$, and $\epsilon_z = 0$ Determine (a) the principal strains, (b) the maximum shear strain in the x–y plane, and (c) the absolute maximum shear strain.

10–21. The strain at a point has components of $\epsilon_x = -480(10^{-6})$, $\epsilon_y = 300(10^{-6})$, $\gamma_{xy} = -650(10^{-6})$, and $\epsilon_z = 0$. Determine (a) the principal strains, (b) the maximum shear strain in the x–y plane, and (c) the absolute maximum shear strain.

10–22. The strain at a point has components of $\epsilon_x = 130(10^{-6})$, $\epsilon_y = 280(10^{-6})$, and $\gamma_{xy} = 75(10^{-6})$ and $\epsilon_z = 0$. Determine (a) the principal strains at the point, (b) the maximum in-plane shear strain in the x–y plane, and (c) the absolute maximum shear strain.

10–23. The strain at a point has components of $\epsilon_x = -520(10^{-6})$, $\epsilon_y = -350(10^{-6})$, $\gamma_{xy} = 720(10^{-6})$, and $\epsilon_z = 0$. Determine (a) the principal strains at the point, (b) the maximum shear strain in the x−y plane, and (c) the absolute maximum shear strain.

*10–24. The strain at a point has components of $\epsilon_x = 350(10^{-6})$, $\epsilon_y = -460(10^{-6})$, $\gamma_{xy} = -560(10^{-6})$, and $\epsilon_z = 0$. Determine (a) the principal strains at the point, (b) the maximum shear strain in the x−y plane, and (c) the absolute maximum shear strain.

10–25. The strain at a point has components of $\epsilon_x = 450(10^{-6})$, $\epsilon_y = 825(10^{-6})$, $\gamma_{xy} = 275(10^{-6})$, and $\epsilon_z = 0$. Determine (a) the principal strains at the point, (b) the maximum shear strain in the x–y plane, and (c) the absolute maximum shear strain.

10–26. The state of strain at a point has components of $\epsilon_x = -400(10^{-6})$, $\epsilon_y = -200(10^{-6})$, $\gamma_{xy} = -250(10^{-6})$, and $\epsilon_z = 0$. Determine (a) the principal strains at the point, (b) the maximum shear strain in the x–y plane, and (c) the absolute maximum shear strain.

10–27. The strain at point A on the bracket has components $\epsilon_x = 300(10^{-6})$, $\epsilon_y = 550(10^{-6})$, $\gamma_{xy} = -650(10^{-6})$, $\epsilon_z = 0$. Determine (a) the principal strains at A, (b) the maximum shear strain in the x−y plane, and (c) the absolute maximum shear strain.

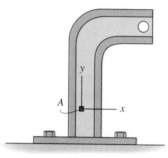

Prob. 10–27

*10–28. The 45° strain rosette is mounted on a steel shaft. The following readings are obtained from each gauge: $\epsilon_a = 300(10^{-6})$, $\epsilon_b = 180(10^{-6})$, and $\epsilon_c = -250(10^{-6})$. Determine the in-plane principal strains and their orientation.

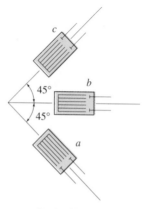

Prob. 10–28

10–29. The strain at point A on the leg of the angle has components $\epsilon_x = -140(10^{-6})$, $\epsilon_y = 180(10^{-6})$, $\gamma_{xy} = -125(10^{-6})$, $\epsilon_z = 0$. Determine (a) the principal strains at A, (b) the maximum shear strain in the $x-y$ plane, and (c) the absolute maximum shear strain.

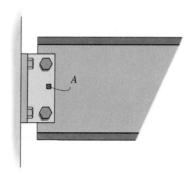

Prob. 10–29

10–30. The 45° strain rosette is mounted on the link of the backhoe. The following readings are obtained from each gauge: $\epsilon_a = 650(10^{-6})$, $\epsilon_b = -300(10^{-6})$, $\epsilon_c = 480(10^{-6})$. Determine (a) the in-plane principal strains and (b) the maximum in-plane shear strain and associated average normal strain.

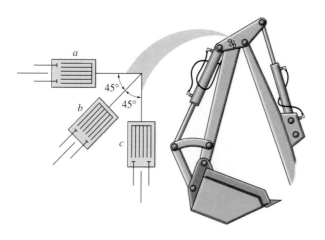

Prob. 10–30

10–31. The 60° strain rosette is mounted on the surface of an aluminum plate. The following readings are obtained from each gauge: $\epsilon_a = 950(10^{-6})$, $\epsilon_b = 380(10^{-6})$, $\epsilon_c = -220(10^{-6})$. Determine the in-plane principal strains and their orientation.

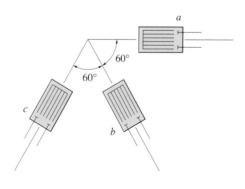

Prob. 10–31

■*10–32. Consider the general orientation of three strain gauges at a point as shown. Write a computer program that can be used to determine the principal in-plane strains and the maximum in-plane shear strain at the point. Show an application of the program using the values and $\theta_a = 40°$, $\epsilon_a = 160(10^{-6})$, $\theta_b = 125°$, $\epsilon_b = 100(10^{-6})$, $\theta_c = 220°$, and $\epsilon_c = 80(10^{-6})$.

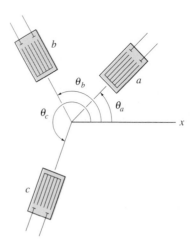

Prob. 10–32

10.6   MATERIAL-PROPERTY RELATIONSHIPS

Now that the general principles of multiaxial stress and strain have been presented, we will use these principles to develop some important relationships involving the material's properties. To do so we will assume that the material is homogeneous and isotropic and behaves in a linear-elastic manner.

Generalized Hooke's Law.   If the material at a point is subjected to a state of triaxial stress, σ_x, σ_y, σ_z, Fig. 10–20a, associated normal strains ϵ_x, ϵ_y, ϵ_z are developed in the material. The stresses can be related to the strains by using the principle of superposition, Poisson's ratio, $\epsilon_{lat} = -\nu\epsilon_{long}$, and Hooke's law, as it applies in the uniaxial direction, $\epsilon = \sigma/E$. To show how this is done we will first consider the normal strain of the element in the x direction, caused by separate application of each normal stress. When σ_x is applied, Fig. 10–20b, the element elongates in the x direction and the strain ϵ'_x in this direction is

$$\epsilon'_x = \frac{\sigma_x}{E}$$

Application of σ_y causes the element to contract with a strain ϵ''_x in the x direction, Fig. 10–20c. Here

$$\epsilon''_x = -\nu\frac{\sigma_y}{E}$$

Likewise, application of σ_z, Fig. 10–20d, causes a contraction in the x direction such that

$$\epsilon'''_x = -\nu\frac{\sigma_z}{E}$$

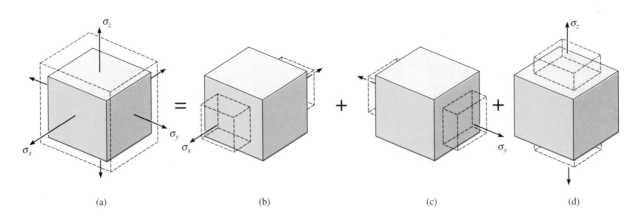

(a)   (b)   (c)   (d)

Fig. 10–20

When these three normal strains are superimposed, the normal strain ϵ_x is determined for the state of stress in Fig. 10–20a. Similar equations can be developed for the normal strains in the y and z directions. The final results can be written as

$$\epsilon_x = \frac{1}{E}\left[\sigma_x - \nu(\sigma_y + \sigma_z)\right]$$

$$\epsilon_y = \frac{1}{E}\left[\sigma_y - \nu(\sigma_x + \sigma_z)\right] \qquad (10\text{–}18)$$

$$\epsilon_z = \frac{1}{E}\left[\sigma_z - \nu(\sigma_x + \sigma_y)\right]$$

$\epsilon_x = \frac{1}{29(10^3)}\left[P - 0.3(P+P)\right]$
$.4P$

These three equations express Hooke's law in a general form for a triaxial state of stress. As noted in the derivation, they are valid only if the principle of superposition applies, which requires a *linear-elastic* response of the material and application of strains that do not severely alter the shape of the material—i.e., small deformations are required. When applying these equations, note that tensile stresses are considered positive quantities, and compressive stresses are negative. If a resulting normal strain is *positive*, it indicates that the material *elongates*, whereas a *negative* normal strain indicates the material *contracts*.

Since the material is isotropic, the element in Fig. 10–20a will *remain a rectangular block* when subjected to the normal stresses; i.e., *no shear strains* will be produced in the material. If we now apply a shear stress τ_{xy} to the element, Fig. 10–21a, experimental observations indicate that the material will deform *only* due to a shear strain γ_{xy}; that is, τ_{xy} will not cause other strains in the material. Likewise, τ_{yz} and τ_{xz} will only cause shear strains γ_{yz} and γ_{xz}, respectively. Hooke's law for shear stress and shear strain can therefore be written as

$$\gamma_{xy} = \frac{1}{G}\tau_{xy} \qquad \gamma_{yz} = \frac{1}{G}\tau_{yz} \qquad \gamma_{xz} = \frac{1}{G}\tau_{xz} \qquad (10\text{–}19)$$

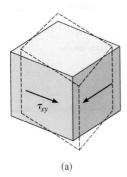

(a)

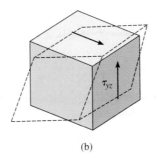

(b)

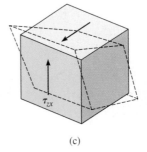

(c)

Fig. 10–21

Relationship Involving E, ν, and G.   In Sec. 3.7 we stated that the modulus of elasticity E is related to the shear modulus G by Eq. 3–11, namely,

$$G = \frac{E}{2(1 + \nu)} \qquad (10\text{–}20)$$

One way to derive this relationship is to consider an element of the material to be subjected to pure shear ($\sigma_x = \sigma_y = \sigma_z = 0$), Fig. 10–22a. Applying Eq. 9–5 to obtain the principal stresses yields $\sigma_{max} = \tau_{xy}$ and $\sigma_{min} = -\tau_{xy}$. From Eq. 9–4, the element must be oriented $\theta_{p_1} = 45°$ counterclockwise from the x axis in order to define the direction of the plane on which σ_{max} acts, Fig. 10–22b. If the three principal stresses $\sigma_{max} = \tau_{xy}$, $\sigma_{int} = 0$, and $\sigma_{min} = -\tau_{xy}$ are substituted into the first of Eq. 10–18, the principal strain ϵ_{max} can be related to the shear stress τ_{xy}. The result is

$$\epsilon_{max} = \frac{\tau_{xy}}{E}(1 + \nu) \qquad (10\text{–}21)$$

This strain, which deforms the element along the x' axis, can also be related to the shear strain γ_{xy} using the strain transformation equations or Mohr's circle for strain. To do this, first note that since $\sigma_x = \sigma_y = \sigma_z = 0$, then from Eq. 10–18 $\epsilon_x = \epsilon_y = 0$. Substituting these results into the transformation Eq. 10–9, we get

$$\epsilon_1 = \epsilon_{max} = \frac{\gamma_{xy}}{2}$$

By Hooke's law, $\gamma_{xy} = \tau_{xy}/G$, so that $\epsilon_{max} = \tau_{xy}/2G$. Substituting into Eq. 10–21 and rearranging terms gives the final result, namely, Eq. 10–20.

Dilatation and Bulk Modulus.   When an elastic material is subjected to normal stress, its volume will change. In order to compute this change, consider a volume element which is subjected to the principal stresses σ_x, σ_y, σ_z. The sides of the element are originally dx, dy, dz, Fig. 10–23a; however, after application of the stress they become $(1 + \epsilon_x)\, dx$, $(1 + \epsilon_y)\, dy$, $(1 + \epsilon_z)\, dz$, respectively, Fig. 10–23b. The change in volume of the element is therefore

$$\delta V = (1 + \epsilon_x)(1 + \epsilon_y)(1 + \epsilon_z)\, dx\, dy\, dz - dx\, dy\, dz$$

Neglecting the products of the strains since the strains are very small, we have

$$\delta V = (\epsilon_x + \epsilon_y + \epsilon_z)\, dx\, dy\, dz$$

The change in volume per unit volume is called the "volumetric strain" or the **dilatation** e. It can be written as

$$e = \frac{\delta V}{dV} = \epsilon_x + \epsilon_y + \epsilon_z \qquad (10\text{–}22)$$

By comparison, the shear strains will *not* change the volume of the element, rather they will only change its rectangular shape.

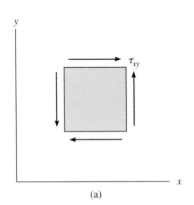

(a)

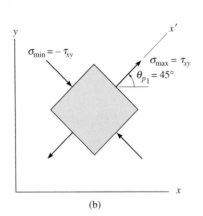

(b)

Fig. 10–22

If we use the generalized Hooke's law, as defined by Eq. 10–18, we can write the dilatation in terms of the applied stress. We have

$$e = \frac{1 - 2\nu}{E}(\sigma_x + \sigma_y + \sigma_z) \qquad (10\text{–}23)$$

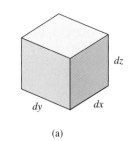

(a)

When a volume element of material is subjected to the uniform pressure p of a liquid, the pressure on the body is the same in all directions and is always normal to any surface on which it acts. Shear stresses are *not present*, since the shear resistance of a liquid is zero. This state of "hydrostatic" loading requires the normal stresses to be equal in any and all directions, and therefore an element of the body is subjected to principal stresses $\sigma_x = \sigma_y = \sigma_z = -p$, Fig. 10–24. Substituting into Eq. 10–23 and rearranging terms yields

$$\frac{p}{e} = -\frac{E}{3(1 - 2\nu)} \qquad (10\text{–}24)$$

The term on the right consists *only* of the material's properties E and ν. It is equal to the ratio of the uniform normal stress p to the dilatation or "volumetric strain." Since this ratio is *similar* to the ratio of linear-elastic stress to strain, which defines E, i.e., $\sigma/\epsilon = E$, the terms on the right are called the *volume modulus of elasticity* or the **bulk modulus**. It has the same units as stress and will be symbolized by the letter k; that is,

$$k = \frac{E}{3(1 - 2\nu)} \qquad (10\text{–}25)$$

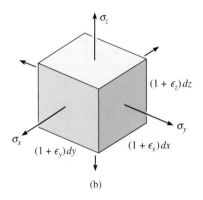

Fig. 10–23

Note that for most metals $\nu \approx \frac{1}{3}$ so $k \approx E$. If a material existed that did not change its volume then $\delta V = 0$, and so k would have to be infinite. From Eq. 10–25 the theoretical *maximum* value for Poisson's ratio is therefore $\nu = 0.5$. Also, during yielding, no actual volume change of the material is observed, and so $\nu = 0.5$ is used when plastic yielding occurs.

IMPORTANT POINTS

- When a homogeneous and isotropic material is subjected to a state of triaxial stress, the strain in one of the stress directions is influenced by the strains produced by *all* the stresses. This is the result of the Poisson effect, and results in the form of a generalized Hooke's law.

- A shear stress applied to homogeneous and isotropic material will only produce shear strain in the same plane.

- The material constants, E, G, and ν, are related mathematically.

- *Dilatation*, or *volumetric strain*, is caused only by normal strain, not shear strain.

- The *bulk modulus* is a measure of the stiffness of a volume of material. This material property provides an upper limit to Poisson's ratio of $\nu = 0.5$, which remains at this value while plastic yielding occurs.

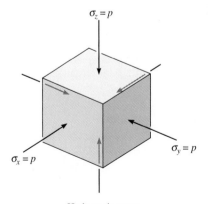

Hydrostatic stress

Fig. 10–24

EXAMPLE 10-9

The bracket in Example 10–8, Fig. 10–25a, is made of steel for which $E_{st} = 200$ GPa, $\nu_{st} = 0.3$. Determine the principal stresses at point A.

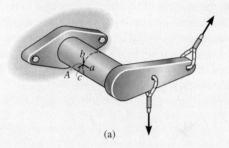

(a)

Fig. 10–25

SOLUTION I

From Example 10–8 the principal strains have been determined as

$$\epsilon_1 = 272(10^{-6})$$
$$\epsilon_2 = 33.8(10^{-6})$$

Since point A is on the *surface* of the bracket for which there is no loading, the stress on the surface is zero, and so point A is subjected to plane stress. Applying Hooke's law with $\sigma_3 = 0$, we have

$$\epsilon_1 = \frac{\sigma_1}{E} - \frac{\nu}{E}\sigma_2; \qquad 272(10^{-6}) = \frac{\sigma_1}{200(10^9)} - \frac{0.3}{200(10^9)}\sigma_2$$

$$54.4(10^6) = \sigma_1 - 0.3\sigma_2 \qquad (1)$$

$$\epsilon_2 = \frac{\sigma_2}{E} - \frac{\nu}{E}\sigma_1; \qquad 33.8(10^{-6}) = \frac{\sigma_2}{200(10^9)} - \frac{0.3}{200(10^9)}\sigma_1$$

$$6.76(10^6) = \sigma_2 - 0.3\sigma_1 \qquad (2)$$

Solving Eqs. (1) and (2) simultaneously yields

$$\sigma_1 = 62.0 \text{ MPa} \qquad\qquad\qquad Ans.$$
$$\sigma_2 = 25.4 \text{ MPa} \qquad\qquad\qquad Ans.$$

SOLUTION II

It is also possible to solve the problem using the given state of strain,

$$\epsilon_x = 60(10^{-6}) \qquad \epsilon_y = 246(10^{-6}) \qquad \gamma_{xy} = -149(10^{-6})$$

as specified in Example 10–8. Applying Hooke's law in the $x-y$ plane, we have

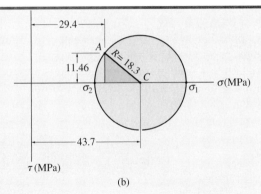

(b)

$$\epsilon_x = \frac{\sigma_x}{E} - \frac{\nu}{E}\sigma_y; \qquad 60(10^{-6}) = \frac{\sigma_x}{200(10^9)\ \text{Pa}} - \frac{0.3\sigma_y}{200(10^9)\ \text{Pa}}$$

$$\epsilon_y = \frac{\sigma_y}{E} - \frac{\nu}{E}\sigma_x; \qquad 246(10^{-6}) = \frac{\sigma_y}{200(10^9)\ \text{Pa}} - \frac{0.3\sigma_x}{200(10^9)\ \text{Pa}}$$

$$\sigma_x = 29.4\ \text{MPa} \qquad \sigma_y = 58.0\ \text{MPa}$$

The shear stress is determined using Hooke's law for shear. First, however, we must calculate G.

$$G = \frac{E}{2(1 + \nu)} = \frac{200\ \text{GPa}}{2(1 + 0.3)} = 76.9\ \text{GPa}$$

Thus,

$$\tau_{xy} = G\gamma_{xy}; \qquad \tau_{xy} = 76.9(10^9)[-149(10^{-6})] = -11.46\ \text{MPa}$$

The Mohr's circle for this state of plane stress has a reference point $A(29.4\ \text{MPa}, -11.46\ \text{MPa})$ and center at $\sigma_{\text{avg}} = 43.7\ \text{MPa}$, Fig. 10–25b. The radius is determined from the shaded triangle.

$$R = \sqrt{(43.7 - 29.4)^2 + (11.46)^2} = 18.3\ \text{MPa}$$

Therefore,

$$\sigma_1 = 43.7\ \text{MPa} + 18.3\ \text{MPa} = 62.0\ \text{MPa} \qquad \textit{Ans.}$$
$$\sigma_2 = 43.7\ \text{MPa} - 18.3\ \text{MPa} = 25.4\ \text{MPa} \qquad \textit{Ans.}$$

Note that each of these solutions is valid provided the material is both linear elastic and isotropic, since then the principal planes of stress and strain coincide.

EXAMPLE 10-10

The copper bar in Fig. 10–26 is subjected to a uniform loading along its edges as shown. If it has a length $a = 300$ mm, width $b = 50$ mm, and thickness $t = 20$ mm before the load is applied, determine its new length, width, and thickness after application of the load. Take $E_{cu} = 120$ GPa, $\nu_{cu} = 0.34$.

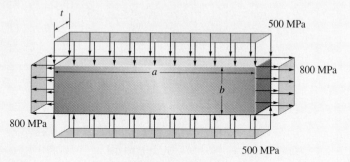

Fig. 10–26

SOLUTION

By inspection, the bar is subjected to a state of plane stress. From the loading we have

$$\sigma_x = 800 \text{ MPa} \qquad \sigma_y = -500 \text{ MPa} \qquad \tau_{xy} = 0 \qquad \sigma_z = 0$$

The associated normal strains are determined from the generalized Hooke's law, Eq. 10–18; that is,

$$\epsilon_x = \frac{\sigma_x}{E} - \frac{\nu}{E}(\sigma_y + \sigma_z)$$

$$= \frac{800 \text{ MPa}}{120(10^3) \text{ MPa}} - \frac{0.34}{120(10^3) \text{ MPa}}(-500 \text{ MPa}) = 0.00808$$

$$\epsilon_y = \frac{\sigma_y}{E} - \frac{\nu}{E}(\sigma_x + \sigma_z)$$

$$= \frac{-500 \text{ MPa}}{120(10^3) \text{ MPa}} - \frac{0.34}{120(10^3) \text{ MPa}}(800 \text{ MPa} + 0) = -0.00643$$

$$\epsilon_z = \frac{\sigma_z}{E} - \frac{\nu}{E}(\sigma_x + \sigma_y)$$

$$= 0 - \frac{0.34}{120(10^3) \text{ MPa}}(800 \text{ MPa} - 500 \text{ MPa}) = -0.000850$$

The new bar length, width, and thickness are therefore

$$a' = 300 \text{ mm} + 0.00808(300 \text{ mm}) = 302.4 \text{ mm} \qquad \textit{Ans.}$$

$$b' = 50 \text{ mm} + (-0.00643)(50 \text{ mm}) = 49.68 \text{ mm} \qquad \textit{Ans.}$$

$$t' = 20 \text{ mm} + (-0.000850)(20 \text{ mm}) = 19.98 \text{ mm} \qquad \textit{Ans.}$$

EXAMPLE 10-11

If the rectangular block shown in Fig. 10–27 is subjected to a uniform pressure of $p = 20$ psi, determine the dilatation and the change in length of each side. Take $E = 600$ psi, $\nu = 0.45$.

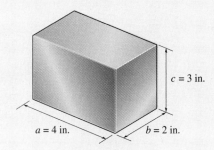

$c = 3$ in.

$a = 4$ in. $\quad b = 2$ in.

Fig. 10–27

SOLUTION

Dilatation. The dilatation can be determined using Eq. 10–23 with $\sigma_x = \sigma_y = \sigma_z = -20$ psi. We have

$$e = \frac{1 - 2\nu}{E}(\sigma_x + \sigma_y + \sigma_z)$$

$$= \frac{1 - 2(0.45)}{600 \text{ psi}}[3(-20 \text{ psi})]$$

$$= -0.01 \text{ in}^3/\text{in}^3 \qquad\qquad \textit{Ans.}$$

Change in Length. The normal strain on each side can be determined from Hooke's law, Eq. 10–18; that is,

$$\epsilon = \frac{1}{E}[\sigma_x - \nu(\sigma_y + \sigma_z)]$$

$$= \frac{1}{600 \text{ psi}}[-20 \text{ psi} - (0.45)(-20 \text{ psi} - 20 \text{ psi})] = -0.00333 \text{ in./in.}$$

Thus the change in length of each side is

$$\delta a = -0.00333(4 \text{ in.}) = -0.0133 \text{ in.} \qquad \textit{Ans.}$$
$$\delta b = -0.00333(2 \text{ in.}) = -0.00667 \text{ in.} \qquad \textit{Ans.}$$
$$\delta c = -0.00333(3 \text{ in.}) = -0.0100 \text{ in.} \qquad \textit{Ans.}$$

The negative signs indicate that each dimension is decreased.

$$\sigma_1 = \frac{Pr}{t} \qquad P = \frac{F}{A} \qquad \epsilon = \frac{\delta}{L_0} \qquad \frac{Pr}{t}$$
$$\sigma = EE = \frac{Pr}{t}$$

PROBLEMS

10–33. For the case of plane stress, show that Hooke's law can be written as

$$\sigma_x = \frac{E}{(1 - \nu^2)}(\epsilon_x + \nu\epsilon_y), \qquad \sigma_y = \frac{E}{(1 - \nu^2)}(\epsilon_y + \nu\epsilon_x)$$

10–34. Use Hooke's law, Eq. 10-18, to develop the strain-transformation equations, Eqs. 10-5 and 10-6, from the stress-transformation equations, Eqs. 9-1 and 9-2.

10–35. Determine the bulk modulus for gray cast iron if $E_{fe} = 14(10^3)$ ksi and $\nu_{fe} = 0.20$.

*__10–36.__ Determine the bulk modulus for hard rubber if $E_r = 0.68(10^3)$ ksi and $\nu_r = 0.43$.

10–37. The polyvinyl chloride bar is subjected to an axial force of 900 lb. If it has the original dimensions shown, determine the value of Poisson's ratio if the angle θ decreases by $\Delta\theta = 0.01°$ after the load is applied. $E_{pvc} = 800(10^3)$ psi.

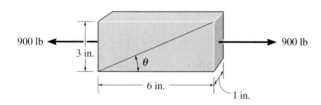

900 lb ← 900 lb
3 in.
θ
6 in.
1 in.

Prob. 10–37

10–38. The rod is made of aluminum 2014-T6. If it is subjected to the tensile load of 700 N and has a diameter of 20 mm, determine the principal strains at a point on the surface of the rod.

700 N   700 N

Prob. 10–38

10–39. The strain gauge is placed on the surface of a thin-walled steel boiler as shown. If it is 0.5 in. long, determine the pressure in the boiler when the gauge elongates $0.2(10^{-3})$ in. The boiler has a thickness of 0.5 in. and inner diameter of 60 in. Also, determine the maximum x, y in-plane shear strain in the material. $E_{st} = 29(10^3)$ ksi, $\nu_{st} = 0.3$.

y
x
0.5 in.
60 in.

Prob. 10–39

*__10–40.__ The shaft has a radius of 15 mm and is made of L2 tool steel. Determine the strains in the x' and y' directions if a torque $T = 2$ kN · m is applied to the shaft.

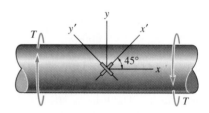

T
y
y'
x'
45°
x
T

Prob. 10–40

10–41. Determine the principal strains that occur at a point on a steel member where the principal stresses are $\sigma_{max} = 18$ ksi, $\sigma_{int} = 15$ ksi, $\sigma_{min} = -28$ ksi. $E_{st} = 29(10^3)$ ksi and $\nu_{st} = 0.3$.

10–42. A bar of plastic having a diameter of 0.5 in. is loaded in a tension machine, and it is determined that $\epsilon_x = 530(10^{-6})$ when the load is 80 lb. Determine the modulus of elasticity, E_p, and the dilatation, e_p, of the plastic $\nu_p = 0.26$.

10–43. A rod has a radius of 10 mm. If it is subjected to an axial load of 15 N such that the axial strain in the rod is $\epsilon_x = 2.75(10^{-6})$, determine the modulus of elasticity E and the change in its diameter. $\nu = 0.23$.

*__10–44.__ From experiment, the principal strains in a plane at a point on a steel shell are $\epsilon_1 = 350(10^{-6})$ and $\epsilon_2 = -250(10^{-6})$. If $E_{st} = 200$ GPa and $\nu_{st} = 0.3$, determine the principal plane stresses in this plane.

10–45. The principal plane stresses and associated strains in a plane at a point are $\sigma_1 = 40$ ksi, $\sigma_2 = 25$ ksi, $\epsilon_1 = 1.15(10^{-3})$, and $\epsilon_2 = 0.450(10^{-3})$. If this is a case of plane stress, determine the modulus of elasticity and Poisson's ratio.

10–46. The spherical pressure vessel has an inner diameter of 2 m and a thickness of 10 mm. A strain gauge having a length of 20 mm is attached to it, and it is observed to increase in length by 0.012 mm when the vessel is pressurized. Determine the pressure causing this deformation, and find the maximum in-plane shear stress, and the absolute maximum shear stress at a point on the outer surface of the vessel. The material is steel, for which $E_{st} = 200$ GPa and $\nu_{st} = 0.3$.

Prob. 10–46

10–47. The principal strains in a plane, measured experimentally at a point on the aluminum fuselage of a jet aircraft, are $\epsilon_1 = 630(10^{-6})$ and $\epsilon_2 = 350(10^{-6})$. If this is a case of plane stress, determine the associated principal stresses at the point in the same plane. $E_{al} = 10(10^3)$ ksi and $\nu_{al} = 0.33$.

*__10–48.__ A single strain gauge, placed in the vertical plane on the outer surface and at an angle of 60° to the axis of the pipe, gives a reading at point A of $\epsilon_A = -250(10^{-6})$. Determine the vertical force P if the pipe has an outer diameter of 1 in. and an inner diameter of 0.6 in. The pipe is made of C86100 bronze.

10–49. A single strain gauge, placed in the vertical plane on the outer surface and at an angle of 60° to the axis of the pipe, gives a reading at point A of $\epsilon_A = -250(10^{-6})$. Determine the principal strains in the pipe at point A. The pipe has an outer diameter of 1 in. and an inner diameter of 0.6 in. and is made of C86100 bronze.

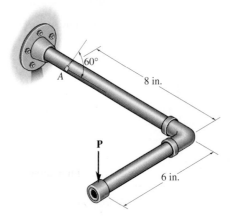

Probs. 10–48/10–49

10–50. Air is pumped into the steel thin-walled pressure vessel at C. If the ends of the vessel are closed using two pistons connected by a rod AB, determine the increase in the diameter of the pressure vessel when the internal gauge pressure is 5 MPa. Also, what is the tensile stress in rod AB if it has a diameter of 100 mm? The inner radius of the vessel is 400 mm, and its thickness is 10 mm. $E_{st} = 200$ GPa and $\nu_{st} = 0.3$.

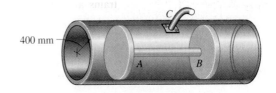

Prob. 10–50

10–51. The strain gauge is placed on the surface of a thin-walled boiler as shown. If it is 0.2 in. long, determine the pressure in the boiler when the gauge elongates $0.08(10^{-3})$ in. The boiler has a thickness of 0.5 in. and inner diameter of 30 in., and it is made of 304 stainless steel. Also, determine the maximum x, y in-plane shear strain and the absolute maximum shear strain in the material.

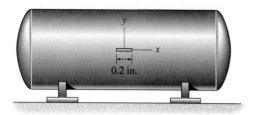

0.2 in.

Prob. 10–51

***10–52.** The shaft has a radius of 15 mm and is made of L2 tool steel. Determine the strains in the x' and y' directions if a torque of $T = 2$ kN · m is applied to the shaft.

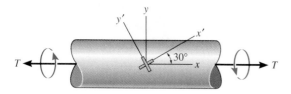

30°

Prob. 10–52

10–53. The shaft has a radius of 15 mm and is made of L2 tool steel. Determine the torque T in the shaft if the two strain gauges, attached to the surface of the shaft, report strains of $\epsilon_{x'} = -45(10^{-6})$ and $\epsilon_{y'} = 45(10^{-6})$. Also, compute the strains acting in the x and y directions.

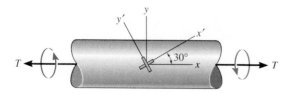

30°

Prob. 10–53

10–54. Determine the change in volume of the tapered plate when it is subjected to the axial load **P**. The material has a thickness t, a modulus of elasticity E, and Poisson's ratio is v.

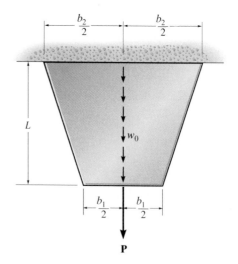

$\frac{b_2}{2}$ $\frac{b_2}{2}$

L

w_0

$\frac{b_1}{2}$ $\frac{b_1}{2}$

P

Prob. 10–54

10–55. A thin-walled cylindrical pressure vessel has an inner radius r, thickness t, and length L. If it is subjected to an internal pressure p, show that the increase in its inner radius is $\delta r = pr^2(2 - v)/2Et$ and the increase in its length is $\delta L = pLr(1 - 2v)/2Et$. Using these results, show that the change in internal volume becomes $\delta V = \pi r^2(1 + \epsilon_1)^2(1 + \epsilon_2)L - \pi r^2 L$. Since ϵ_1 and ϵ_2 are small quantities, show further that the change in volume per unit volume, called *volumetric strain*, can be written as $\delta V/V = (pr/2Et)(5 - 4v)$.

***10–56.** The A-36 steel pipe is subjected to the axial loading of 60 kN. Determine the change in volume of the material after the load is applied.

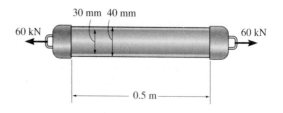

30 mm 40 mm

60 kN   60 kN

0.5 m

Prob. 10–56

10–57. A soft material is placed within the confines of a rigid cylinder which rests on a rigid support. Determine the factor by which the apparent modulus of elasticity will be increased from not being confined when a load is applied. Take $\nu = 0.3$ for the material.

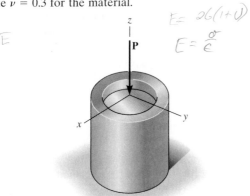

$E = 2G(1+\nu)$

$E = \dfrac{\sigma}{\epsilon}$

Prob. 10–57

10–58. A thin-walled spherical pressure vessel having an inner radius r and thickness t is subjected to an internal pressure p. Show that the increase in volume within the vessel is $\delta V = (2p\pi r^4/Et)(1-\nu)$. Use a small-strain analysis.

10–59. The thin-walled cylindrical pressure vessel of inner radius r and thickness t is subjected to an internal pressure p. If the material constants are E and ν, determine the strains in the circumferential and longitudinal directions. Using these results, compute the increase in both the diameter and the length of a steel pressure vessel filled with air and having an internal gauge pressure of 20 MPa. The vessel is 2 m long and has an inner radius of 0.4 m and a thickness of 10 mm. $E_{st} = 200$ GPa, and $\nu_{st} = 0.3$.

*10–60. Estimate the increase in volume of the tank in Prob. 10–59. *Suggestion:* Use the results of Prob. 10–55 as a check.

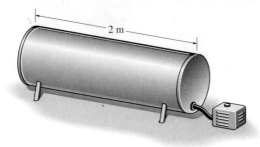

Probs. 10–59/10–60

10–61. The smooth rigid-body cavity is filled with liquid 6061-T6 aluminum. When cooled it is 0.012 in. from the top of the cavity. If the top of the cavity is covered and the temperature is increased by 200°F, determine the stress components σ_x, σ_y, and σ_z in the aluminum. *Hint:* Use Eq. 10–18 with an additional strain term of $\alpha\Delta T$ (Eq. 4–4).

10–62. The smooth rigid-body cavity is filled with liquid 6061-T6 aluminum. When cooled it is 0.012 in. from the top of the cavity. If the top of the cavity is not covered and the temperature is increased by 200°F, determine the strain components ϵ_x, ϵ_y, and ϵ_z in the aluminum. *Hint:* Use Eq. 10–18 with an additional strain term of $\alpha\Delta T$ (Eq. 4–4).

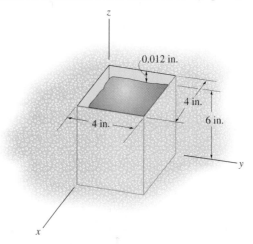

Probs. 10–61/10–62

10–63. The block is fitted between the fixed supports. If the glued joint can resist a maximum shear stress of $\tau_{allow} = 2$ ksi, determine the temperature rise that will cause the joint to fail. Take $E = 10\,(10^3)$ ksi, $\nu = 0.2$, and $\alpha = 6.0\,(10^{-6})/°F$. *Hint:* Use Eq. 10–18 with an additional strain term of $\alpha\Delta T$ (Eq. 4–4).

Prob. 10–63

*10.7   THEORIES OF FAILURE

When an engineer is faced with the problem of design using a specific material, it becomes important to place an upper *limit* on the state of stress that defines the material's failure. If the material is *ductile*, failure is usually specified by the initiation of *yielding*, whereas if the material is *brittle*, it is specified by *fracture*. These modes of failure are readily defined if the member is subjected to a uniaxial state of stress, as in the case of simple tension; however, if the member is subjected to biaxial or triaxial stress, the criterion for failure becomes more difficult to establish.

In this section we will discuss four theories that are often used in engineering practice to predict the failure of a material subjected to a *multiaxial* state of stress. These theories, and others like them, are also used to determine the allowable stresses reported in many design codes. No single theory of failure, however, can be applied to a specific material at *all times*, because a material may behave in either a ductile or brittle manner depending on the temperature, rate of loading, chemical environment, or the way the material is shaped or formed. When using a particular theory of failure, it is first necessary to calculate the normal and shear stress components at points where they are the largest in the member. This may be done by using the fundamentals of mechanics of materials and applying stress-concentration factors where applicable, or in complex situations, the largest stress components can be found by using either a mathematical analysis based on the theory of elasticity or by an appropriate experimental technique. In any case, once this state of stress is established, the *principal stresses* at these critical points are then determined, since each of the following theories is based on knowing the principal stress.

Ductile Materials

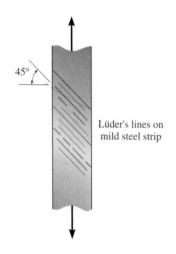

45°

Lüder's lines on mild steel strip

Fig. 10–28

Maximum-Shear-Stress Theory.   The most common cause of *yielding of a ductile material* such as steel is *slipping*, which occurs along the contact planes of randomly ordered crystals that make up the material. This *slipping* is due to *shear stress*, and if we make a specimen into a highly polished thin strip and subject it to a simple tension test we can see how it causes the material to *yield*, Fig. 10–28. The edges of the planes of slipping as they appear on the surface of the strip are referred to as *Lüder's lines*. These lines clearly indicate the slip planes in the strip, which occur at approximately 45° with the axis of the strip.

Consider now an element of the material taken from a tension specimen, which is subjected only to the yield stress σ_Y, Fig. 10–29a. The maximum shear stress can be determined by drawing Mohr's circle for the element, Fig. 10–29b. The results indicate that

$$\tau_{\text{max}} = \frac{\sigma_Y}{2} \qquad (10\text{–}26)$$

Furthermore, this shear stress acts on planes that are 45° from the planes of principal stress, Fig. 10–29c, and these planes *coincide* with the direction of the Lüder lines shown on the specimen, indicating that indeed failure occurs by shear.

Using this idea, that ductile materials fail by shear, Henri Tresca in 1868 proposed the ***maximum-shear-stress theory*** or ***Tresca yield criterion***. This theory can be used to predict the failure stress of a ductile material subjected to any type of loading. The maximum-shear-stress theory states that yielding of the material begins when the absolute maximum shear stress in the material reaches the shear stress that causes the same material to yield when it is subjected *only* to axial tension. To avoid failure, therefore, the maximum-shear-stress theory requires $\tau_{abs\ max}$ in the material to be less than or equal to $\sigma_Y/2$, where σ_Y is determined from a simple tension test.

For application we will express the absolute maximum shear stress in terms of the *principal stresses*. The procedure for doing this was discussed in Sec. 9.7 with reference to a condition of *plane stress*, that is, where the out-of-plane principal stress is zero. If the two in-plane principal stresses have the *same sign*, i.e., they are both tensile or both compressive, then failure will occur *out of the plane*, and from Eq. 9–15,

$$\tau_{abs\ max} = \frac{\sigma_{max}}{2}$$

On the other hand, if the in-plane principal stresses are of *opposite signs*, then failure occurs in the plane, and from Eq. 9–16,

$$\tau_{abs\ max} = \frac{\sigma_{max} - \sigma_{min}}{2}$$

Using these equations and Eq. 10–26, the maximum-shear-stress theory for *plane stress* can be expressed for any two in-plane principal stresses as σ_1 and σ_2 by the following criteria:

$$\left.\begin{array}{l} |\sigma_1| = \sigma_Y \\ |\sigma_2| = \sigma_Y \end{array}\right\} \quad \sigma_1, \sigma_2 \text{ have same signs}$$

$$|\sigma_1 - \sigma_2| = \sigma_Y\} \quad \sigma_1, \sigma_2 \text{ have opposite signs}$$

(10–27)

A graph of these equations is given in Fig. 10–30. Clearly, if any point of the material is subjected to plane stress, and its in-plane principal stresses are represented by a coordinate (σ_1, σ_2) plotted *on the boundary* or *outside* the shaded hexagonal area shown in this figure, the material will yield at the point and failure is said to occur.

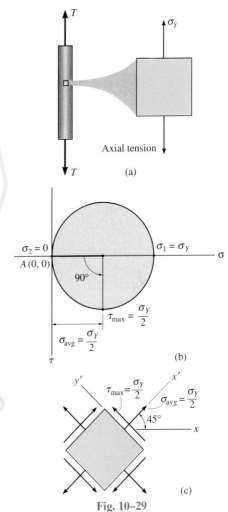

Axial tension

(a)

(b)

(c)

Fig. 10–29

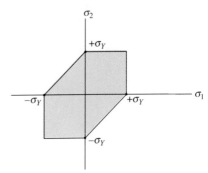

Maximum-shear-stress theory

Fig. 10–30

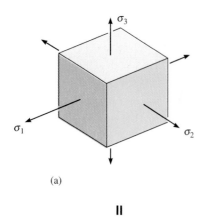

(a)

$=$

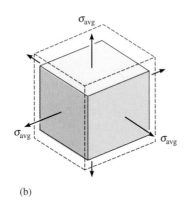

(b)

$+$

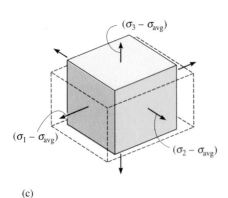

(c)

Fig. 10–31

Maximum-Distortion-Energy Theory. It was stated in Sec. 3.5 that a material, when deformed by an external loading, tends to store energy *internally* throughout its volume. The energy per unit volume of material is called the **strain-energy density**, and if the material is subjected to a uniaxial stress, σ, the strain-energy density, defined by Eq. 3–6, can be written as

$$u = \frac{1}{2}\sigma\epsilon \tag{10–28}$$

It is possible to formulate a failure criterion based on the distortions caused by strain energy. Before doing this, however, we need to determine the strain-energy density in a volume element of material subjected to the three principal stresses σ_1, σ_2, and σ_3, Fig. 10–31a. Here, each principal stress contributes a portion of the total strain-energy density, so that

$$u = \frac{1}{2}\sigma_1\epsilon_1 + \frac{1}{2}\sigma_2\epsilon_2 + \frac{1}{2}\sigma_3\epsilon_3$$

If the material behaves in a linear-elastic manner, then Hooke's law applies. Therefore, substituting Eq. 10–18 into the above equation and simplifying, we get

$$u = \frac{1}{2E}[\sigma_1^2 + \sigma_2^2 + \sigma_3^2 - 2\nu(\sigma_1\sigma_2 + \sigma_1\sigma_3 + \sigma_3\sigma_2)] \tag{10–29}$$

This strain-energy density can be considered as the sum of two parts, one part representing the energy needed to cause a *volume change* of the element with no change in shape, and the other part representing the energy needed to *distort* the element. Specifically, the energy stored in the element as a result of its volume being changed is caused by application of the average principal stress, $\sigma_{avg} = (\sigma_1 + \sigma_2 + \sigma_3)/3$, since this stress causes equal principal strains in the material, Fig. 10–31b. The remaining portion of the stress, $(\sigma_1 - \sigma_{avg})$, $(\sigma_2 - \sigma_{avg})$, $(\sigma_3 - \sigma_{avg})$, causes the energy of distortion, Fig. 10–31c.

Experimental evidence has shown that materials do not yield when subjected to a uniform (hydrostatic) stress, such as σ_{avg} discussed above. As a result, in 1904, M. Huber proposed that yielding in a ductile material occurs when the *distortion energy* per unit volume of the material equals or exceeds the distortion energy per unit volume of the same material when it is subjected to yielding in a simple tension test. This theory is called the **maximum-distortion-energy theory**, and since it was later redefined independently by R. von Mises and H. Hencky, it sometimes also bears their names.

To obtain the distortion energy per unit volume, we will substitute the stresses $(\sigma_1 - \sigma_{avg})$, $(\sigma_2 - \sigma_{avg})$, and $(\sigma_3 - \sigma_{avg})$ for σ_1, σ_2, and σ_3, respectively, into Eq. 10–29, realizing that $\sigma_{avg} = (\sigma_1 + \sigma_2 + \sigma_3)/3$. Expanding and simplifying, we obtain

$$u_d = \frac{1+\nu}{6E}[(\sigma_1 - \sigma_2)^2 + (\sigma_2 - \sigma_3)^2 + (\sigma_3 - \sigma_1)^2]$$

In the case of *plane stress*, $\sigma_3 = 0$, and this equation reduces to

$$u_d = \frac{1 + \nu}{3E}(\sigma_1{}^2 - \sigma_1\sigma_2 + \sigma_2{}^2)$$

For a *uniaxial* tension test, $\sigma_1 = \sigma_Y$, $\sigma_2 = \sigma_3 = 0$, and so

$$(u_d)_Y = \frac{1 + \nu}{3E}\sigma_Y{}^2$$

Since the maximum-distortion-energy theory requires $u_d = (u_d)_Y$, then for the case of plane or biaxial stress, we have

$$\boxed{\sigma_1{}^2 - \sigma_1\sigma_2 + \sigma_2{}^2 = \sigma_Y{}^2} \tag{10–30}$$

This equation represents an elliptical curve, Fig. 10–32. Thus, if a point in the material is stressed such that the stress coordinate (σ_1, σ_2) is plotted on the boundary or outside the shaded area, the material is said to fail.

A comparison of the above two failure criteria is shown in Fig. 10–33. Note that both theories give the same results when the principal stresses are equal, i.e., from Eqs. 10–27 and 10–30, $\sigma_1 = \sigma_2 = \sigma_Y$, or when one of the principal stresses is zero and the other has a magnitude of σ_Y. On the other hand, if the material is subjected to pure shear, τ, then the theories have the largest discrepancy in predicting failure. The stress coordinates of these points on the curves have been determined by considering the element shown in Fig. 10–34a. From the associated Mohr's circle for this state of stress, Fig. 10–34b, we obtain principal stresses $\sigma_1 = \tau$ and $\sigma_2 = -\tau$. Applying Eqs. 10–27 and 10–30, the maximum-shear-stress theory and maximum-distortion-energy theory yield $\sigma_1 = \sigma_Y/2$ and $\sigma_1 = \sigma_Y/\sqrt{3}$, respectively, Fig. 10–33.

Actual torsion tests, used to develop a condition of pure shear in a ductile specimen, have shown that the maximum-distortion-energy theory gives more accurate results for pure-shear failure than the maximum-shear-stress theory. In fact, since $(\sigma_Y/\sqrt{3})/(\sigma_Y/2) = 1.15$, the shear stress for yielding of the material, as given by the maximum-distortion-energy theory, is 15% more accurate than that given by the maximum-shear-stress theory.

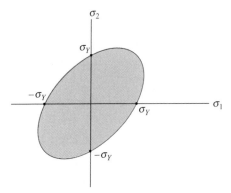

Maximum-distortion-energy theory

Fig. 10–32

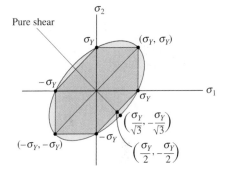

Fig. 10–33

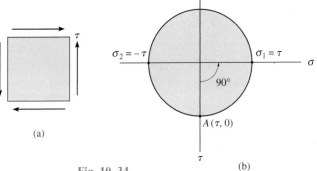

Fig. 10–34

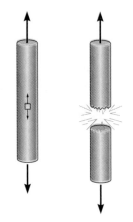

Failure of a brittle material
in tension

(a)

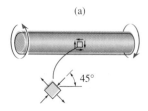

45°

45°

Failure of a brittle material
in torsion

(b)

Fig. 10–35

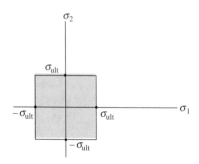

Maximum-normal-stress theory

Fig. 10–36

Brittle Materials

Maximum-Normal-Stress Theory. It was previously stated that brittle materials, such as gray cast iron, tend to fail suddenly by *fracture* with no apparent yielding. In a *tension test*, the fracture occurs when the normal stress reaches the ultimate stress σ_{ult}, Fig. 10–35a. Also, in a *torsion test*, brittle fracture occurs due to a *maximum tensile stress* since the plane of fracture for an element is at 45° to the shear direction, Fig. 10–35b. The fracture surface is therefore helical as shown.* Experiments have shown further that during torsion the material's strength is somewhat *unaffected* by the presence of the associated principal compressive stress being at right angles to the principal tensile stress. Consequently, the tensile stress needed to fracture a specimen during a torsion test is approximately the same as that needed to fracture a specimen in simple tension. Because of this, the **maximum-normal-stress theory** states that a brittle material will fail when the maximum principal stress σ_1 in the material reaches a limiting value that is equal to the ultimate normal stress the material can sustain when it is subjected to simple tension.

If the material is subjected to *plane stress*, we require that

$$\boxed{\begin{aligned} |\sigma_1| &= \sigma_{ult} \\ |\sigma_2| &= \sigma_{ult} \end{aligned}}$$ (10–31)

These equations are shown graphically in Fig. 10–36. Here it is seen that if the stress coordinate (σ_1, σ_2) at a point in the material falls on the boundary or outside the shaded area, the material is said to fracture. This theory is generally credited to W. Rankine, who proposed it in the mid-1800s. Experimentally it has been found to be in close agreement with the behavior of brittle materials that have stress–strain diagrams that are *similar* in both tension and compression.

Mohr's Failure Criterion. In some brittle materials the tension and compression properties are *different*. When this occurs a criterion based on the use of Mohr's circle may be used to predict failure of the material. This method was developed by Otto Mohr and is sometimes referred to as **Mohr's failure criterion**. To apply it, one first performs *three tests* on the material. A uniaxial tensile test and uniaxial compressive test are used to determine the ultimate tensile and compressive stresses $(\sigma_{ult})_t$ and $(\sigma_{ult})_c$, respectively. Also a torsion test is performed to determine the ultimate shear stress τ_{ult} of the material. Mohr's circle for each of these stress conditions is then plotted as shown in Fig. 10–37. Circle *A* represents the stress condition $\sigma_1 = \sigma_2 = 0, \sigma_3 = -(\sigma_{ult})_c$; circle *B* represents the stress

---

*A stick of blackboard chalk fails in this way when its ends are twisted with the fingers.

conditions $\sigma_1 = (\sigma_{\text{ult}})_t$, $\sigma_2 = \sigma_3 = 0$; and circle C represents the pure-shear-stress condition caused by τ_{ult}. These three circles are contained in a "failure envelope" indicated by the extrapolated colored curve that is drawn tangent to all three circles. If a plane-stress condition at a point is represented by a circle that is contained within the envelope, the material is said not to fail. If, however, the circle has a point of tangency with the envelope, or if it extends beyond the envelope's boundary, then failure is said to occur.

We may also represent this criterion on a graph of principal stresses σ_1 and σ_2 ($\sigma_3 = 0$). This is shown in Fig. 10–38. Here failure occurs when the absolute value of either one of the principal stresses reaches a value equal to or greater than $(\sigma_{\text{ult}})_t$ or $(\sigma_{\text{ult}})_c$ or in general, if the state of stress at a point is defined by the stress coordinate (σ_1, σ_2), which is plotted on the boundary or outside the shaded area.

Either of the above two criteria can be used in practice to predict the failure of a brittle material. However, it should be realized that their usefulness is quite limited. A tensile fracture occurs very suddenly, and its initiation generally depends on stress concentrations developed at microscopic imperfections of the material such as inclusions or voids, surface indentations, and small cracks. Since each of these irregularities varies from specimen to specimen, it becomes difficult to specify failure on the basis of a single test. On the other hand, cracks and other irregularities tend to close up when the specimen is compressed, and therefore they do not form points of failure as they would when the specimen is subjected to tension.

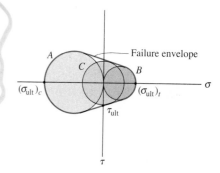

Fig. 10–37

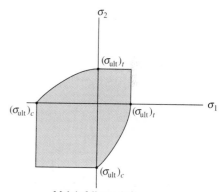

Mohr's failure criteria

Fig. 10–38

IMPORTANT POINTS

- If a material is *ductile*, failure is specified by the initiation of *yielding*, whereas if it is *brittle*, it is specified by *fracture*.

- *Ductile failure* can be defined when *slipping* occurs between the crystals that compose the material. This slipping is due to *shear stress* and the *maximum-shear-stress theory* is based on this idea.

- *Strain energy* is stored in a material when it is subjected to normal stress. The *maximum-distortion-energy theory* depends on the *strain energy* that *distorts* the material, and not the part that increases its volume.

- The fracture of a *brittle material* is caused only by the *maximum tensile stress* in the material, and not the compressive stress. This is the basis of the *maximum-normal-stress theory*, and it is applicable if the stress–strain diagram is *similar* in tension and compression.

- If a *brittle material* has a stress–strain diagram that is *different* in tension and compression, then *Mohr's failure criterion* may be used to predict failure.

- Due to material imperfections, *tensile fracture* of a brittle material is *difficult to predict*, and so theories of failure for brittle materials should be used with caution.

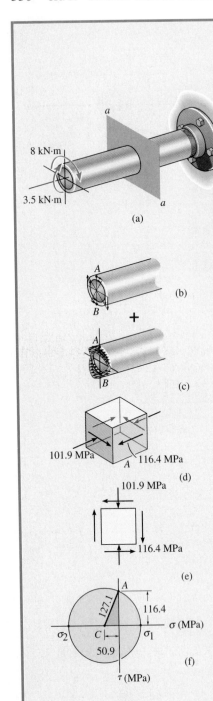

Fig. 10–39

EXAMPLE 10–12

The steel pipe shown in Fig. 10–39a has an inner diameter of 60 mm and an outer diameter of 80 mm. If it is subjected to a torsional moment of 8 kN · m and a bending moment of 3.5 kN · m, determine if these loadings cause failure as defined by the maximum-distortion-energy theory. The yield stress for the steel found from a tension test is $\sigma_Y = 250$ MPa.

SOLUTION

To solve this problem we must investigate a point on the pipe that is subjected to a state of maximum critical stress. Both the torsional and bending moments are uniform throughout the pipe's length. At the arbitrary section a–a, Fig. 10–39a, these loadings produce the stress distributions shown in Fig. 10–39b and 10–39c. By inspection, points A and B are subjected to the same state of critical stress. Here we will investigate the state of stress at A. Thus,

$$\tau_A = \frac{Tc}{J} = \frac{(8000 \text{ N} \cdot \text{m})(0.04 \text{ m})}{(\pi/2)[(0.04 \text{ m})^4 - (0.03 \text{ m})^4]} = 116.4 \text{ MPa}$$

$$\sigma_A = \frac{Mc}{I} = \frac{(3500 \text{ N} \cdot \text{m})(0.04 \text{ m})}{(\pi/4)[(0.04 \text{ m})^4 - (0.03 \text{ m})^4]} = 101.9 \text{ MPa}$$

These results are shown on a three-dimensional view of an element of material at point A, Fig. 10–39d, and also, since the material is subjected to plane stress, it is shown in two dimensions, Fig. 10–39e.

Mohr's circle for this state of plane stress has a center located at

$$\sigma_{\text{avg}} = \frac{0 - 101.9}{2} = -50.9 \text{ MPa}$$

The reference point A(0, −116.4 MPa) is plotted and the circle is constructed, Fig. 10–39f. Here the radius has been calculated from the shaded triangle to be R = 127.1 and so the in-plane principal stresses are

$$\sigma_1 = -50.9 + 127.1 = 76.2 \text{ MPa}$$
$$\sigma_2 = -50.9 - 127.1 = -178.0 \text{ MPa}$$

Using Eq. 10–30, we require

$$(\sigma_1^2 - \sigma_1\sigma_2 + \sigma_2^2) \leq \sigma_Y^2$$
$$[(76.2)^2 - (76.2)(-178.0) + (-178.0)^2] \stackrel{?}{\leq} (250)^2 \quad 51\,100 < 62\,500 \quad \text{OK}$$

Since the criterion has been met, the material within the pipe will *not* yield ("fail") according to the maximum-distortion-energy theory.

EXAMPLE 10-13

The solid cast-iron shaft shown in Fig. 10–40a is subjected to a torque of $T = 400$ lb · ft. Determine its smallest radius so that it does not fail according to the maximum-normal-stress theory. A specimen of cast iron, tested in tension, has an ultimate stress of $(\sigma_{\text{ult}})_t = 20$ ksi.

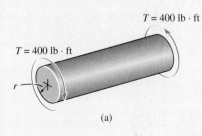

(a)

Fig. 10–40

(b)

SOLUTION

The maximum or critical stress occurs at a point located on the surface of the shaft. Assuming the shaft to have a radius r, the shear stress is

$$\tau_{\text{max}} = \frac{Tc}{J} = \frac{(400 \text{ lb} \cdot \text{ft})(12 \text{ in./ft})r}{(\pi/2)r^4} = \frac{3055.8 \text{ lb} \cdot \text{in.}}{r^3}$$

Mohr's circle for this state of stress (pure shear) is shown in Fig. 10–40b. Since $R = \tau_{\text{max}}$, then

$$\sigma_1 = -\sigma_2 = \tau_{\text{max}} = \frac{3055.8 \text{ lb} \cdot \text{in.}}{r^3}$$

The maximum-normal-stress theory, Eq. 10–31, requires

$$|\sigma_1| \leq \sigma_{\text{ult}}$$

$$\frac{3055.8 \text{ lb} \cdot \text{in.}}{r^3} \leq 20 \, 000 \text{ lb/in}^2$$

Thus, the smallest radius of the shaft is determined from

$$\frac{3055.8 \text{ lb} \cdot \text{in.}}{r^3} = 20 \, 000 \text{ lb/in}^2$$

$$r = 0.535 \text{ in.} \qquad \qquad Ans.$$

EXAMPLE 10-14

The solid shaft shown in Fig. 10–41a has a radius of 0.5 in. and is made of steel having a yield stress of $\sigma_Y = 36$ ksi. Determine if the loadings cause the shaft to fail according to the maximum-shear-stress theory and the maximum-distortion-energy theory.

SOLUTION

The state of stress in the shaft is caused by both the axial force and the torque. Since maximum shear stress caused by the torque occurs in the material at the outer surface, we have

$$\sigma_x = \frac{P}{A} = \frac{15 \text{ kip}}{\pi(0.5 \text{ in.})^2} = 19.10 \text{ ksi}$$

$$\tau_{xy} = \frac{Tc}{J} = \frac{3.25 \text{ kip} \cdot \text{in.}(0.5 \text{ in.})}{\frac{\pi}{2}(0.5 \text{ in.})^4} = 16.55 \text{ ksi}$$

The stress components are shown acting on an element of material at point A in Fig. 10–41b. Rather than using Mohr's circle, the principal stresses can also be obtained using the stress-transformation equations, Eq. 9–5.

$$\sigma_{1,2} = \frac{\sigma_x + \sigma_y}{2} \pm \sqrt{\left(\frac{\sigma_x - \sigma_y}{2}\right)^2 + \tau_{xy}{}^2}$$

$$= \frac{-19.10 + 0}{2} \pm \sqrt{\left(\frac{-19.10 - 0}{2}\right)^2 + (16.55)^2}$$

$$= -9.55 \pm 19.11$$

$$\sigma_1 = 9.56 \text{ ksi}$$

$$\sigma_2 = -28.66 \text{ ksi}$$

Maximum-Shear-Stress Theory. Since the principal stresses have *opposite signs*, then from Sec. 9.7, the absolute maximum shear stress will occur in the plane, and therefore, applying the second of Eq. 10–27, we have

$$|\sigma_1 - \sigma_2| \leq \sigma_Y$$

$$|9.56 - (-28.66)| \overset{?}{\leq} 36$$

$$38.2 > 36$$

Thus, shear failure of the material will occur according to this theory.

Maximum-Distortion-Energy Theory. Applying Eq. 10–30, we have

$$(\sigma_1{}^2 - \sigma_1\sigma_2 + \sigma_2{}^2) \leq \sigma_Y$$

$$[(9.56)^2 - (9.56)(-28.66) + (-28.66)^2] \overset{?}{\leq} (36)^2$$

$$1187 \leq 1296$$

Using this theory, failure will not occur.

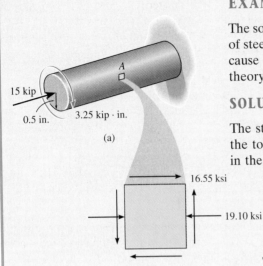

15 kip

0.5 in.   3.25 kip · in.

(a)

16.55 ksi

19.10 ksi

Fig. 10–41

PROBLEMS

*10–64. A material is subjected to plane stress. Express the maximum-shear-stress theory of failure in terms of σ_x, σ_y, and τ_{xy}. Assume that the principal stresses are of different algebraic signs.

10–65. A material is subjected to plane stress. Express the maximum-distortion-energy theory of failure in terms of σ_x, σ_y, and τ_{xy}.

10–66. The state of plane stress at a critical point in a steel machine bracket is shown. If the yield stress for steel is $\sigma_Y = 36$ ksi, determine if yielding occurs using the maximum-distortion-energy theory.

10–67. Solve Prob. 10–66 using the maximum-shear-stress theory.

10–70. The yield stress for heat-treated beryllium copper is $\sigma_Y = 130$ ksi. If this material is subjected to plane stress and elastic failure occurs when one principal stress is 145 ksi, what is the smallest magnitude of the other principal stress? Use the maximum-distortion-energy theory.

10–71. The yield stress for a plastic material is $\sigma_Y = 110$ MPa. If this material is subjected to plane stress and elastic failure occurs when one principal stress is 120 MPa, what is the smallest magnitude of the other principal stress? Use the maximum-distortion-energy theory.

*10–72. Solve Prob. 10–71 using the maximum-shear-stress theory. Both principal stresses have opposite signs.

10–73. The element is subjected to the stresses shown. If $\sigma_Y = 36$ ksi, determine the factor of safety for the loading based on the maximum-shear-stress theory.

10–74. Solve Prob. 10–73 using the maximum-distortion-energy theory.

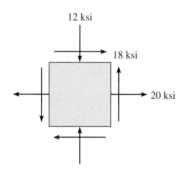

Probs. 10–66/10–67

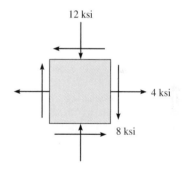

Probs. 10–73/10–74

*10–68. The yield stress for a zirconium-magnesium alloy is $\sigma_Y = 15.3$ ksi. If a machine part is made of this material and a critical point in the material is subjected to in-plane principal stresses σ_1 and $\sigma_2 = -0.5\sigma_1$, determine the magnitude of σ_1 that will cause yielding according to the maximum-shear-stress theory.

10–69. Solve Prob. 10–68 using the maximum-distortion-energy theory.

10–75. A bar with a circular cross-sectional area is made of SAE 1045 carbon steel having a yield stress of $\sigma_Y = 150$ ksi. If the bar is subjected to a torque of 30 kip · in. and a bending moment of 56 kip · in., determine the required diameter of the bar according to the maximum-distortion-energy theory. Use a factor of safety of 2 with respect to yielding.

*10–76.   A bar with a square cross-sectional area is made of a material having a yield stress of $\sigma_Y = 120$ ksi. If the bar is subjected to a bending moment of 75 kip · in., determine the required size of the bar according to the maximum-distortion-energy theory. Use a factor of safety of 1.5 with respect to yielding.

10–77.   Solve Prob. 10–76 using the maximum-shear-stress theory.

10–78.   The principal plane stresses acting on a differential element are shown. If the material is machine steel having a yield stress of $\sigma_Y = 700$ MPa, determine the factor of safety with respect to yielding if the maximum-shear-stress theory is considered.

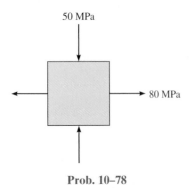

Prob. 10–78

10–79.   The state of stress acting at a critical point on a machine element is shown in the figure. Determine the smallest yield stress for a steel that might be selected for the part, based on the maximum-shear-stress theory.

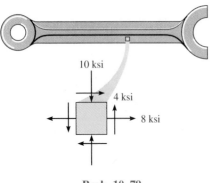

Prob. 10–79

*10–80.   Derive an expression for an equivalent torque T_e that, if applied alone to a solid shaft with a circular cross section, would cause the same energy of distortion as the combination of an applied bending moment M and torque T.

10–81.   If a shaft is made of nickel for which $\sigma_Y = 65$ ksi, determine the maximum torsional stress required to cause yielding using (a) the maximum-shear-stress theory and (b) the maximum-distortion-energy theory.

10–82.   Derive an expression for an equivalent bending moment M_e that, if applied alone to a solid bar with a circular cross section, would cause the same maximum shear stress as the combination of an applied moment M and torque T. Assume that the principal stresses are of opposite algebraic signs.

10–83.   The state of stress acting at a critical point on the seat frame of an automobile during a crash is shown in the figure. Determine the smallest yield stress for a steel that can be selected for the member, based on the maximum-shear-stress theory.

*10–84.   Solve Prob. 10–83 using the maximum-distortion-energy theory.

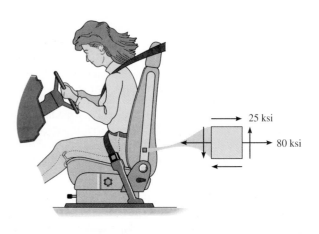

Probs. 10–83/10–84

10-85. If a machine part is made of titanium (Ti-6A1-4V) and a critical point in the material is subjected to plane stress, such that the principal stresses are σ_1 and $\sigma_2 = 0.5\sigma_1$, determine the magnitude of σ_1 in MPa that will cause yielding according to (a) the maximum-shear-stress theory, and (b) the maximum-distortion-energy theory.

10-86. An aluminum alloy 6061-T6 is to be used for a drive shaft such that it transmits 50 hp at 1800 rev/min. Using a factor of safety of F.S. = 2, with respect to yielding, determine the smallest-diameter shaft that can be selected based on the maximum-distortion-energy theory.

10-87. Solve Prob. 10-86 using the maximum-shear-stress theory.

*10-88. The element is subjected to the stresses shown. If $\sigma_Y = 50$ ksi, determine the factor of safety for this loading based on (a) the maximum-shear-stress theory and (b) the maximum-distortion-energy theory.

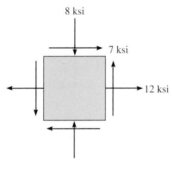

Prob. 10-88

10-89. An aluminum alloy 6061-T6 is to be used for a solid drive shaft such that it transmits 40 hp at 2400 rev/min. Using a factor of safety of F.S. = 2, with respect to yielding, determine the smallest-diameter shaft that can be selected based on the maximum-shear-stress theory.

10-90. Solve Prob. 10-89 using the maximum-distortion-energy theory.

10-91. The principal stresses acting at a point on a thin-walled cylindrical pressure vessel are $\sigma_1 = pr/t$, $\sigma_2 = pr/2t$, and $\sigma_3 = 0$. If the yield stress is σ_Y, determine the maximum value of p based on (a) the maximum-shear-stress theory and (b) the maximum-distortion-energy theory.

*10-92. The state of stress acting at a critical point on a wrench is shown in the figure. Determine the smallest yield stress for steel that might be selected for the part, based on the maximum-distortion-energy theory.

10-93. The state of stress acting at a critical point on a wrench is shown in the figure. Determine the smallest yield stress for steel that might be selected for the part, based on the maximum-shear-stress theory.

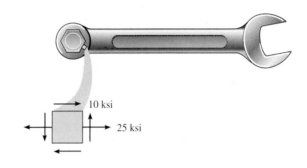

Probs. 10-92/10-93

10-94. The internal loadings at a critical section along the steel drive shaft of a ship are calculated to be a torque of 2650 lb · ft, a bending moment of 2800 lb · ft, and an axial thrust of 3700 lb. If the yield points for tension and shear are $\sigma_Y = 100$ ksi and $\tau_Y = 50$ ksi, respectively, determine the required diameter of the shaft using the maximum-shear-stress theory.

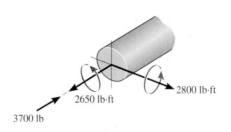

Prob. 10-94

10–95. The cast iron cylinder having a diameter of 100 mm is subjected to a torque of 600 N · m and an axial compressive force of 15 kN. Determine if it fails according to the maximum-normal-stress theory. The ultimate stress of the cast iron is $\sigma_{ult} = 170$ MPa.

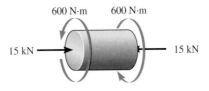

600 N·m   600 N·m

15 kN   15 kN

Prob. 10–95

*10–96.** The internal loadings at a critical section along the steel drive shaft of a ship are calculated to be a torque of 2650 lb · ft, a bending moment of 2800 lb · ft, and an axial thrust of 3700 lb. If the yield points for tension and shear are $\sigma_Y = 100$ ksi and $\tau_Y = 50$ ksi, respectively, determine the required diameter of the shaft using the maximum distortion-energy theory.

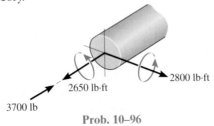

2800 lb·ft

2650 lb·ft

3700 lb

Prob. 10–96

10–97. The short concrete cylinder having a diameter of 50 mm is subjected to a torque of 500 N · m and an axial compressive force of 2 kN. Determine if it fails according to the maximum-normal-stress theory. The ultimate stress of the concrete is $\sigma_{ult} = 28$ MPa.

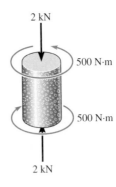

2 kN

500 N·m

500 N·m

2 kN

Prob. 10–97

REVIEW PROBLEMS

10–98. The strain at point A on the shell has components $\epsilon_x = 250(10^{-6})$, $\epsilon_y = 400(10^{-6})$, $\gamma_{xy} = 275(10^{-6})$, $\epsilon_z = 0$. Determine (a) the principal strains at A, (b) the maximum shear strain in the $x-y$ plane, and (c) the absolute maximum shear strain.

y

A

x

Prob. 10–98

10–99. If a solid shaft having a diameter d is subjected to a torque **T** and moment **M**, show that by the maximum-shear-stress theory the maximum allowable shear stress is $\tau_{allow} = (16/\pi d^3) \sqrt{M^2 + T^2}$. Assume the principal stresses to be of opposite algebraic signs.

*10–100.** Derive an expression for an equivalent bending moment M_e that if applied alone to a solid bar with a circular cross section would cause the same energy of distortion as the combination of an applied bending moment M and torque T.

10–101. The state of strain at a point on the arm has components of $\epsilon_x = 250(10^{-6})$, $\epsilon_y = -450(10^{-6})$, $\gamma_{xy} = -825(10^{-6})$. Use the strain-transformation equations to determine (a) the in-plane principal strains and (b) the maximum in-plane shear strain and average normal strain. In each case specify the orientation of the element and show how the strains deform the element within the x–y plane.

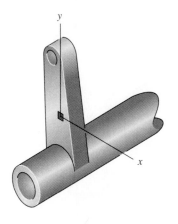

Prob. 10–101

10–102. The aluminum beam has the rectangular cross section shown. If it is subjected to a bending moment of $M = 60$ kip · in., determine the increase in the 2-in. dimension at the top of the beam and the decrease in this dimension at the bottom. $E_{al} = 10(10^3)$ ksi, $\nu_{al} = 0.3$.

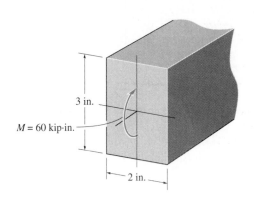

Prob. 10–102

10–103. Determine the bulk modulus for each of the following materials: (a) rubber, $E_r = 0.4$ ksi, $\nu_r = 0.48$, and (b) glass, $E_g = 8(10^3)$ ksi, $\nu_g = 0.24$.

*10–104. A thin-walled spherical pressure vessel has an inner radius r, thickness t, and is subjected to an internal pressure p. If the material constants are E and ν, determine the strain in the circumferential direction in terms of the stated parameters.

10–105. The 60° strain rosette is mounted on a beam. The following readings are obtained for each gauge: $\epsilon_a = 600(10^{-6})$, $\epsilon_b = -700(10^{-6})$, and $\epsilon_c = 350(10^{-6})$. Determine (a) the in-plane principal strains and (b) the maximum in-plane shear strain and average normal strain. In each case show the deformed element due to these strains.

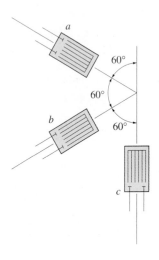

Prob. 10–105

Beams are important structural members that are used to support roof and floor loadings.

11 DESIGN OF BEAMS AND SHAFTS

CHAPTER OBJECTIVES

In this chapter, we will discuss how to design a beam so that it is able to resist both bending and shear loads. Specifically, methods used for designing prismatic beams and determining the shape of fully stressed beams will be developed. At the end of the chapter, we will consider the design of shafts based on the resistance of both bending and torsional moments.

11.1 BASIS FOR BEAM DESIGN

Beams are structural members designed to support loadings applied perpendicular to their longitudinal axes. Because of these loadings, beams develop an internal shear force and bending moment that, in general, vary from point to point along the axis of the beam. Some beams may also be subjected to an internal axial force; however, the effects of this force are often neglected in design, since the axial stress is generally much smaller than the stresses developed by shear and bending. A beam that is chosen to resist both shear and bending stresses is said to be designed on the *basis of strength*. To design a beam in this way requires the use of the shear and flexure formulas developed in Chapters 6 and 7. Application of these formulas, however, is limited to beams made of a homogeneous material that has linear-elastic behavior.

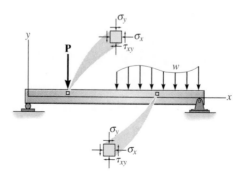

Fig. 11–1

Beams normally support a large shear load at their supports. For this reason, a metal stiffener is often connected to the web of the beam, as shown here, to prevent localized deformation of the beam.

The stress analysis of a beam generally neglects the effects caused by external distributed loadings and concentrated forces applied to the beam. As shown in Fig. 11–1, these loadings will create additional stresses in the beam *directly under the load*. Notably, a compressive stress σ_y will be developed, in addition to the bending stress σ_x and shear stress τ_{xy} discussed previously. Using advanced methods of analysis, as treated in the theory of elasticity, it can be shown, however, that the stress σ_y diminishes rapidly throughout the beam's depth, and for *most* beam span-to-depth ratios used in engineering practice, the maximum value of σ_y generally represents only a small percentage compared to the bending stress σ_x, that is, $\sigma_x \gg \sigma_y$. Furthermore, the direct application of concentrated loads is generally avoided in beam design. Instead, *bearing plates* are used to spread these loads more evenly onto the surface of the beam.

Although beams are designed mainly for strength, they must also be braced properly along their sides so that they do not buckle or suddenly become unstable. Furthermore, in some cases beams must be designed to resist a limited amount of *deflection*, as when they support ceilings made of brittle materials such as plaster. Methods for finding beam deflections will be discussed in Chapter 12, and limitations placed on beam buckling are often discussed in codes on structural or mechanical design.

11.2   PRISMATIC BEAM DESIGN

In order to design a beam on the basis of *strength*, it is required that the actual bending stress and shear stress in the beam do not exceed allowable bending and shear stress for the material as defined by structural or mechanical codes. If the suspended span of the beam is relatively long, so that the internal moments become large, the engineer will first consider a design based upon bending and then check the shear strength. A bending design requires a determination of the beam's **section modulus**, which is the ratio of I and c, that is, $S = I/c$. Using the flexure formula, $\sigma = Mc/I$, we have

$$S_{\text{req'd}} = \frac{M}{\sigma_{\text{allow}}} \tag{11-1}$$

Here M is determined from the beam's moment diagram, and the allowable bending stress, σ_{allow}, is specified in a design code. In many cases the beam's unknown weight will be small and can be neglected in comparison with the loads the beam must carry. However, if the additional moment caused by the weight is to be included in the design, a selection for S is made so that it slightly *exceeds* $S_{\text{req'd}}$.

Once $S_{\text{req'd}}$ is known, if the beam has a simple cross-sectional shape, such as a square, a circle, or a rectangle of known width-to-height proportions, its *dimensions* can be determined directly from $S_{\text{req'd}}$, since by definition $S_{\text{req'd}} = I/c$. However, if the cross section is made from several elements, such as a wide-flange section, then an infinite number of web and flange dimensions can be determined that satisfy the value of $S_{\text{req'd}}$. In practice, however, engineers choose a particular beam meeting the requirement that $S > S_{\text{req'd}}$ from a handbook that lists the standard shapes available from manufacturers. Often several beams that have the same section modulus can be selected from these tables. If deflections are not restricted, usually the beam having the smallest cross-sectional area is chosen, since it is made of less material and is therefore both lighter and more economical than the others.

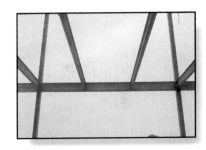

The two floor beams are connected to the girder, which transmits the load to the columns of this building frame. For a force analysis, the connections can be considered to act as pins.

The above discussion assumes that the material's allowable bending stress is the *same* for both tension and compression. If this is the case, then a beam having a cross section that is *symmetric* with respect to the neutral axis should be chosen. However, if the allowable tensile and compressive bending stresses are *not* the same, then the choice of an unsymmetric cross section may be more efficient. Under these circumstances the beam must be designed to resist *both* the largest positive and the largest negative moment in the span.

Once the beam has been selected, the shear formula $\tau_{\text{allow}} \geq VQ/It$ can then be used to check that the allowable shear stress is not exceeded. Often this requirement will not present a problem. However, if the beam is "short" and supports large concentrated loads, the shear-stress limitation may dictate the size of the beam. This limitation is particularly important in the design of wood beams, because wood tends to split along its grain due to shear (see Fig. 7–6).

Fabricated Beams.   Since beams are often made of steel or wood, we will now discuss some of the tabulated properties of beams made from these materials.

Steel Sections.   Most manufactured steel beams are produced by rolling a hot ingot of steel until the desired shape is formed. These so-called *rolled shapes* have properties that are tabulated in the American Institute of Steel Construction (AISC) manual. A representative listing for wide-flange beams taken from this manual is given in Appendix B. As noted in this appendix, the wide-flange shapes are designated by their depth and weight per unit length; for example, W18 × 46 indicates a wide-flange cross section (W) having a depth of 18 in. and a weight of 46 lb/ft, Fig. 11–2. For any given section, the weight per length, dimensions, cross-sectional area, moment of inertia, and section modulus are reported. Also included is the radius of gyration r, which is a geometric property related to the section's buckling strength. This will be discussed in Chapter 13. Appendix B and the *AISC Manual* also list data on other members such as channels and angles.

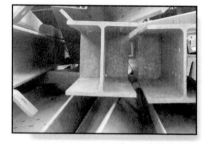

Typical profile view of a steel wide-flange beam.

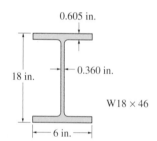

Fig. 11–2

Wood Sections.   Most beams made of wood have rectangular cross sections because such beams are easy to manufacture and handle. Manuals, such as that of the National Forest Products Association, list the dimensions of lumber often used in the design of wood beams. Often, both the nominal and actual dimensions are reported. Lumber is identified by its *nominal* dimensions, such as 2 × 4 (2 in. by 4 in.); however, its actual or "dressed" dimensions are smaller, being 1.5 in. by 3.5 in. The reduction in the dimensions occurs due to the requirement of obtaining smooth surfaces from lumber that is rough sawn. Obviously, the *actual dimensions* must be used whenever stress calculations are performed on wood beams.

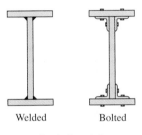

Welded   Bolted

Steel plate girders

Fig. 11–3

Built-up Sections. A *built-up section* is constructed from two or more parts joined together to form a single unit. As indicated by Eq. 11–1, the capacity of the beam to resist a moment will vary directly with its section modulus S, and since $S = I/c$, then S is *increased* if I is *increased*. In order to increase I, *most of the material* should be placed as far *away* from the neutral axis as practical. This, of course, is what makes a deep wide-flange beam so efficient in resisting a moment. For very large loads, however, an available rolled-steel section may not have a section modulus great enough to support a given moment. Rather than using several available beams to support the load, engineers will usually "build up" a beam made from plates and angles. A deep I-shaped section having this form is called a *plate girder*. For example, the steel plate girder in Fig. 11–3 has two flange plates that are either welded or, using angles, bolted to the web plate.

Wood beams are also "built up," usually in the form of a box beam section, Fig. 11–4a. They may be made having plywood webs and larger boards for the flanges. For very large spans, *glulam beams* are used. These members are made from several boards glue-laminated together to form a single unit, Fig. 11–4b.

Just as in the case of rolled sections or beams made from a single piece, the design of built-up sections requires that the bending and shear stresses be checked. In addition, the shear stress in the fasteners, such as weld, glue, nails, etc., must be checked to be certain the beam acts as a single unit. The principles for doing this were outlined in Sec. 7.4.

Wooden box beam

(a)

Glulam beam

(b)

Fig. 11–4

IMPORTANT POINTS

- Beams support loadings that are applied perpendicular to their axes. If they are designed on the basis of strength, they must resist allowable shear and bending stresses.

- The maximum bending stress in the beam is assumed to be much greater than the localized stresses caused by the application of loadings on the surface of the beam.

PROCEDURE FOR ANALYSIS

Based on the previous discussion, the following procedure provides a rational method for the design of a beam on the basis of strength.

Shear and Moment Diagrams.

• Determine the maximum shear and moment in the beam. Often this is done by constructing the beam's shear and moment diagrams.

• For built-up beams, shear and moment diagrams are useful for identifying *regions* where the shear and moment are excessively large and may require additional structural reinforcement or fasteners.

Average Normal Stress.

• If the beam is relatively long, it is designed by finding its section modulus using the flexure formula, $S_{req'd} = M_{max}/\sigma_{allow}$.

• Once $S_{req'd}$ is determined, the cross-sectional dimensions for simple shapes can then be computed, since $S_{req'd} = I/c$.

• If rolled-steel sections are to be used, several possible values of S may be selected from the tables in Appendix B. Of these, choose the one having the smallest cross-sectional area, since this beam has the least weight and is therefore the most economical.

• Make sure that the selected section modulus, S, is *slightly greater* than $S_{req'd}$, so that the additional moment created by the beam's weight is considered.

Shear Stress.

• Normally beams that are short and carry large loads, especially those made of wood, are first designed to resist shear and then later checked against the allowable-bending-stress requirements.

• Using the shear formula, check to see that the allowable shear stress is not exceeded; that is, use $\tau_{allow} \geq V_{max} Q/It$.

• If the beam has a solid *rectangular* cross section, the shear formula becomes $\tau_{allow} \geq 1.5(V_{max}/A)$, Eq. 7–5, and if the cross section is a *wide flange*, it is generally appropriate to assume that the shear stress is *constant* over the cross-sectional area of the beam's web so that $\tau_{allow} \geq V_{max}/A_{web}$, where A_{web} is determined from the product of the beam's depth and the web's thickness. (See Sec. 7.3.)

Adequacy of Fasteners.

• The adequacy of fasteners used on built-up beams depends upon the shear stress the fasteners can resist. Specifically, the required spacing of nails or bolts of a particular size is determined from the allowable shear flow, $q_{allow} = VQ/I$, calculated at points on the cross section where the fasteners are located. (See Sec. 7.4.)

EXAMPLE 11-1

A beam is to be made of steel that has an allowable bending stress of $\sigma_{\text{allow}} = 24$ ksi and an allowable shear stress of $\tau_{\text{allow}} = 14.5$ ksi. Select an appropriate W shape that will carry the loading shown in Fig. 11–5a.

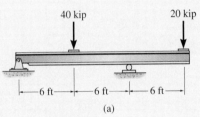

SOLUTION

Shear and Moment Diagrams. The support reactions have been calculated, and the shear and moment diagrams are shown in Fig. 11–5b. From these diagrams, $V_{\text{max}} = 30$ kip and $M_{\text{max}} = 120$ kip · ft.

Bending Stress. The required section modulus for the beam is determined from the flexure formula,

$$S_{\text{req'd}} = \frac{M_{\text{max}}}{\sigma_{\text{allow}}} = \frac{120 \text{ kip} \cdot \text{ft}(12 \text{ in./ft})}{24 \text{ kip/in}^2} = 60 \text{ in}^3$$

Using the table in Appendix B, the following beams are adequate:

$$
\begin{array}{ll}
\text{W18} \times 40 & S = 68.4 \text{ in}^3 \\
\text{W16} \times 45 & S = 72.7 \text{ in}^3 \\
\text{W14} \times 43 & S = 62.7 \text{ in}^3 \\
\text{W12} \times 50 & S = 64.7 \text{ in}^3 \\
\text{W10} \times 54 & S = 60.0 \text{ in}^3 \\
\text{W8} \times 67 & S = 60.4 \text{ in}^3
\end{array}
$$

The beam having the least weight per foot is chosen, i.e.,

$$\text{W18} \times 40$$

The *actual* maximum moment M_{max}, which includes the weight of the beam, can be computed and the adequacy of the selected beam can be checked. In comparison with the applied loads, however, the beam's weight, $(0.040 \text{ kip/ft})(18 \text{ ft}) = 0.720$ kip, will only *slightly increase* $S_{\text{req'd}}$. In spite of this,

$$S_{\text{req'd}} = 60 \text{ in}^3 < 68.4 \text{ in}^3 \qquad \text{OK}$$

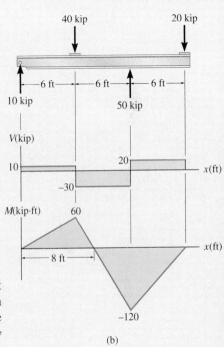

Fig. 11–5

Shear Stress. Since the beam is a *wide-flange section*, the *average shear stress* within the web will be considered. Here the web is assumed to extend from the very top to the very bottom of the beam. From Appendix B, for a W18 × 40, $d = 17.90$ in, $t_w = 0.315$ in. Thus,

$$\tau_{\text{avg}} = \frac{V_{\text{max}}}{A_w} = \frac{30 \text{ kip}}{(17.90 \text{ in.})(0.315 \text{ in.})} = 5.32 \text{ ksi} < 14.5 \text{ ksi} \qquad \text{OK}$$

Use a W18 × 40. *Ans.*

EXAMPLE 11–2

The wooden T-beam shown in Fig. 11–6a is made from two 200 mm × 30 mm boards. If the allowable bending stress is $\sigma_{allow} = 12$ MPa and the allowable shear stress is $\tau_{allow} = 0.8$ MPa, determine if the beam can safely support the loading shown. Also, specify the maximum spacing of nails needed to hold the two boards together if each nail can safely resist 1.50 kN in shear.

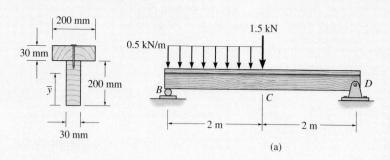

(a)

SOLUTION

Shear and Moment Diagrams.   The reactions on the beam are shown, and the shear and moment diagrams are drawn in Fig. 11–6b. Here $V_{max} = 1.5$ kN, $M_{max} = 2$ kN · m.

Bending Stress.   The neutral axis (centroid) will be located from the bottom of the beam. Working in units of meters, we have

$$\bar{y} = \frac{\Sigma \tilde{y} A}{\Sigma A}$$

$$= \frac{(0.1 \text{ m})(0.03 \text{ m})(0.2 \text{ m}) + 0.215 \text{ m}(0.03 \text{ m})(0.2 \text{ m})}{0.03 \text{ m}(0.2 \text{ m}) + 0.03 \text{ m}(0.2 \text{ m})} = 0.1575 \text{ m}$$

Thus,

$$I = \left[\frac{1}{12}(0.03 \text{ m})(0.2 \text{ m})^3 + (0.03 \text{ m})(0.2 \text{ m})(0.1575 \text{ m} - 0.1 \text{ m})^2 \right]$$

$$+ \left[\frac{1}{12}(0.2 \text{ m})(0.03 \text{ m})^3 + (0.03 \text{ m})(0.2 \text{ m})(0.215 \text{ m} - 0.1575 \text{ m})^2 \right]$$

$$= 60.125(10^{-6}) \text{ m}^4$$

Since $c = 0.1575$ m (not 0.230 m − 0.1575 m = 0.0725 m), we require

$$\sigma_{allow} \geq \frac{M_{max} c}{I}$$

$$12(10^3) \text{ kPa} \geq \frac{2 \text{ kN} \cdot \text{m}(0.1575 \text{ m})}{60.125(10^{-6}) \text{ m}^4} = 5.24(10^3) \text{ kPa}   \quad \text{OK}$$

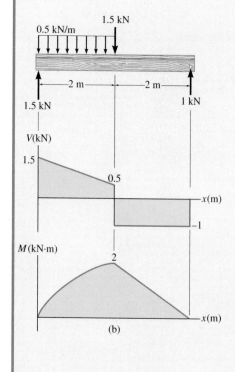

Fig. 11–6

Shear Stress.   Maximum shear stress in the beam depends upon the magnitude of Q and t. It occurs at the neutral axis, since Q is a maximum there and the neutral axis is in the web, where the thickness $t = 0.03$ m is smallest for the cross section. For simplicity, we will use the rectangular area below the neutral axis to calculate Q, rather than a two-part composite area above this axis, Fig. 11–6c. We have

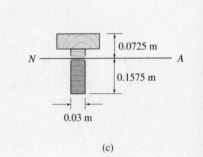

(c)

$$Q = \bar{y}'A' = \left(\frac{0.1575 \text{ m}}{2}\right)[(0.1575 \text{ m})(0.03 \text{ m})] = 0.372(10^{-3}) \text{ m}^3$$

So that

$$\tau_{allow} \geq \frac{V_{max}Q}{It}$$

$$800 \text{ kPa} \geq \frac{1.5 \text{ kN}[0.372(10^{-3})] \text{ m}^3}{60.125(10^{-6}) \text{ m}^4(0.03 \text{ m})} = 309 \text{ kPa} \qquad \text{OK}$$

Nail Spacing.   From the shear diagram it is seen that the shear varies over the entire span. Since the nail spacing depends on the magnitude of shear in the beam, for simplicity (and to be conservative), we will design the spacing on the basis of $V = 1.5$ kN for region BC and $V = 1$ kN for region CD. Since the nails join the flange to the web, Fig. 11–6d, we have

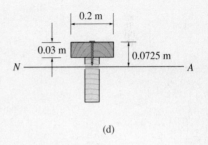

(d)

$$Q = \bar{y}'A' = (0.0725 \text{ m} - 0.015 \text{ m})[(0.2 \text{ m})(0.03 \text{ m})] = 0.345(10^{-3}) \text{ m}^3$$

The shear flow for each region is therefore

$$q_{BC} = \frac{V_{BC}Q}{I} = \frac{1.5 \text{ kN}[0.345(10^{-3})] \text{ m}^3}{60.125(10^{-6}) \text{ m}^4} = 8.61 \text{ kN/m}$$

$$q_{CD} = \frac{V_{CD}Q}{I} = \frac{1 \text{ kN}[0.345(10^{-3})] \text{ m}^3}{60.125(10^{-6}) \text{ m}^4} = 5.74 \text{ kN/m}$$

One nail can resist 1.50 kN in shear, so the spacing becomes

$$s_{BC} = \frac{1.50 \text{ kN}}{8.61 \text{ kN/m}} = 0.174 \text{ m}$$

$$s_{CD} = \frac{1.50 \text{ kN}}{5.74 \text{ kN/m}} = 0.261 \text{ m}$$

For ease of measuring, use

$$s_{BC} = 150 \text{ mm} \qquad\qquad \text{Ans.}$$

$$s_{CD} = 250 \text{ mm} \qquad\qquad \text{Ans.}$$

EXAMPLE 11-3

The laminated wooden beam shown in Fig. 11–7a supports a uniform distributed loading of 12 kN/m. If the beam is to have a height-to-width ratio of 1.5, determine its smallest width. The allowable bending stress is σ_{allow} = 9 MPa and the allowable shear stress is τ_{allow} = 0.6 MPa. Neglect the weight of the beam.

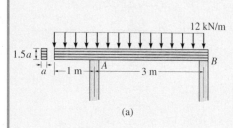

SOLUTION

Shear and Moment Diagrams. The support reactions at A and B have been calculated and the shear and moment diagrams are shown in Fig. 11–7b. Here V_{max} = 20 kN, M_{max} = 10.67 kN · m.

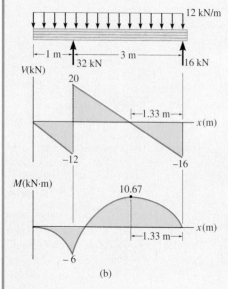

(b)

Fig. 11–7

Bending Stress. Applying the flexure formula yields

$$S_{req'd} = \frac{M_{max}}{\sigma_{allow}} = \frac{10.67 \text{ kN} \cdot \text{m}}{9(10^3) \text{ kN/m}^2} = 0.00119 \text{ m}^3$$

Assuming that the width is a, then the height is h = 1.5a, Fig. 11–7a. Thus,

$$S_{req'd} = \frac{I}{c} = \frac{\frac{1}{12}(a)(1.5a)^3}{(0.75a)} = 0.00119 \text{ m}^3$$

$$a^3 = 0.003160 \text{ m}^3$$

$$a = 0.147 \text{ m}$$

Shear Stress. Applying the shear formula for rectangular sections (which is a special case of τ_{max} = VQ/It), we have

$$\tau_{max} = 1.5 \frac{V_{max}}{A} = (1.5) \frac{20 \text{ kN}}{(0.147 \text{ m})(1.5)(0.147 \text{ m})}$$

$$= 0.929 \text{ MPa} > 0.6 \text{ MPa}$$

EQUATION

Since the shear criterion fails, the beam must be redesigned on the basis of shear.

$$\tau_{allow} = \frac{3}{2} \frac{V_{max}}{A}$$

$$600 \text{ kN/m}^2 = \frac{3}{2} \frac{20 \text{ kN}}{(a)(1.5a)}$$

$$a = 0.183 \text{ m} = 183 \text{ mm} \qquad \textit{Ans.}$$

This larger section will also adequately resist the normal stress.

PROBLEMS

11–1. The wooden beam has a rectangular cross section and is used to support a load of 1200 lb. If the allowable bending stress is $\sigma_{\text{allow}} = 2$ ksi and the allowable shear stress is $\tau_{\text{allow}} = 750$ psi, determine the height h of the cross section to the nearest $\frac{1}{4}$ in. if it is to be rectangular and have a width of $b = 3$ in. Assume the supports at A and B only exert vertical reactions on the beam.

11–2. Solve Prob. 11–1 if the cross section has an unknown width but is to be square, i.e., $h = b$.

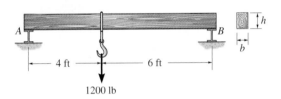

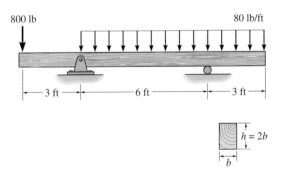

Probs. 11–1/11–2

11–3. The beam is made of Douglas fir having an allowable bending stress of $\sigma_{\text{allow}} = 1.1$ ksi and an allowable shear stress of $\tau_{\text{allow}} = 0.70$ ksi. Determine the width b of the beam if the height $h = 2b$.

Prob. 11–3

*11–4. Select the lightest-weight steel wide-flange beam from Appendix B that will safely support the machine loading shown. The allowable bending stress is $\sigma_{\text{allow}} = 24$ ksi and the allowable shear stress is $\tau_{\text{allow}} = 14$ ksi.

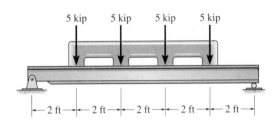

Prob. 11–4

11–5. Select the lightest-weight steel wide-flange beam from Appendix B that will safely support the loading shown, where $w = 6$ kip/ft and $P = 5$ kip. The allowable bending stress is $\sigma_{\text{allow}} = 24$ ksi, and the allowable shear stress is $\tau_{\text{allow}} = 14$ ksi.

11–6. Select the lightest-weight steel wide-flange beam having the shortest height from Appendix B that will safely support the loading shown, where $w = 0$ and $P = 10$ kip. The allowable bending stress is $\sigma_{\text{allow}} = 24$ ksi, and the allowable shear stress is $\tau_{\text{allow}} = 14$ ksi.

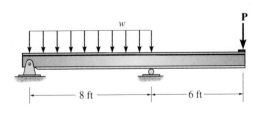

Probs. 11–5/11–6

11–7. The spreader beam AB is used to lift slowly the 3000-lb pipe that is centrally located on the straps at C and D. If the beam is a W 12×45, determine if it can safely support the load. The allowable bending stress is $\sigma_{allow} = 22$ ksi and the allowable shear stress is $\tau_{allow} = 12$ ksi.

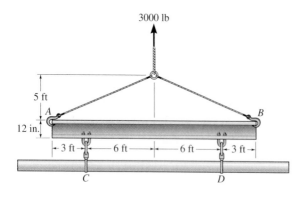

Prob. 11–7

***11–8.** Select the lightest-weight steel structural wide-flange beam with the shortest depth from Appendix B that will safely support the loading shown. The allowable bending stress is $\sigma_{allow} = 24$ ksi and the allowable shear stress is $\tau_{allow} = 14$ ksi.

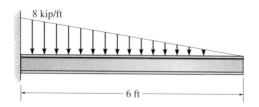

Prob. 11–8

11–9. The wooden beam has a rectangular cross section and is used to support a chain hoist that carries a load of 12 kip. If the allowable bending stress is $\sigma_{allow} = 26$ ksi and the allowable shear stress is $\tau_{allow} = 12$ ksi, determine the height of the cross section to the nearest $\frac{1}{4}$ in. if it is to be rectangular and have a width of $b = 3$ in.

11–10. The wooden beam has a rectangular cross section and is used to support a chain hoist that carries a load of 12 kip. If the allowable bending stress is $\sigma_{allow} = 26$ ksi and the allowable shear stress is $\tau_{allow} = 12$ ksi, determine the height of the cross section to the nearest $\frac{1}{4}$ in. if it has an unknown width but is to be square, i.e., $h = b$.

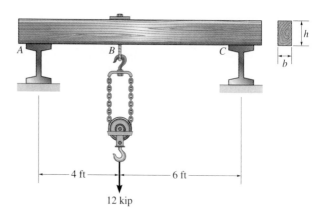

Probs. 11–9/11–10

11–11. The beam is made of a ceramic material having an allowable bending stress of $\sigma_{allow} = 735$ psi and an allowable shear stress of $\tau_{allow} = 400$ psi. Determine the width b of the beam if the height $h = 2b$.

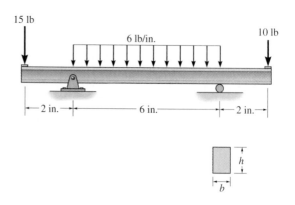

Prob. 11–11

*11–12.   The simply supported beam is composed of two W12 × 22 sections built up as shown. Determine the maximum uniform loading w the beam will support if the allowable bending stress is σ_{allow} = 22 ksi and the allowable shear stress is τ_{allow} = 14 ksi.

11–13.   The simply supported beam is composed of two W12 × 22 sections built up as shown. Determine if the beam will safely support a loading of w = 2 kip/ft. The allowable bending stress is σ_{allow} = 22 ksi and the allowable shear stress is τ_{allow} = 14 ksi.

11–15.   Draw the shear and moment diagrams for the W12 × 14 beam and check if the beam will safely support the loading. The allowable bending stress is σ_{allow} = 22 ksi and the allowable shear stress is τ_{allow} = 12 ksi.

*11–16.   Select the lightest-weight steel wide-flange beam from Appendix B that will safely support the loading shown. The allowable bending stress is σ_{allow} = 22 ksi and the allowable shear stress is τ_{allow} = 12 ksi.

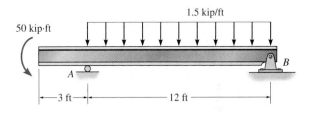

Prob. 11–15/11–16

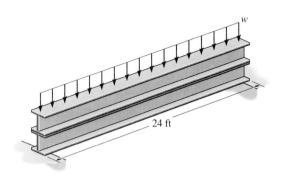

Probs. 11–12/11–13

11–14.   Select the lightest-weight steel wide-flange beam from Appendix B that will safely support the loading shown. The allowable bending stress is σ_{allow} = 24 ksi and the allowable shear stress is τ_{allow} = 14 ksi.

11–17.   Determine the smallest diameter rod that will safely support the loading shown. The allowable bending stress is σ_{allow} = 167 MPa and the allowable shear stress is τ_{allow} = 97 MPa.

11–18.   The pipe has an outer diameter of 15 mm. Determine the smallest inner diameter so that it will safely support the loading shown. The allowable bending stress is σ_{allow} = 167 MPa and the allowable shear stress is τ_{allow} = 97 MPa.

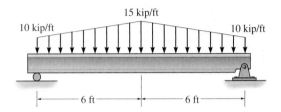

Prob. 11–14

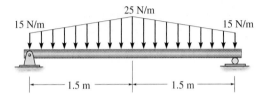

Probs. 11–17/11–18

11–19. Two acetyl plastic members are to be glued together and used to support the loading shown. If the allowable bending stress for the plastic is $\sigma_{allow} = 13$ ksi and the allowable shear stress is $\tau_{allow} = 4$ ksi, determine the greatest load P that can be supported and specify the required shear stress capacity of the glue.

11–21. Determine the minimum width b of the beam to the nearest $\frac{1}{4}$ in. that will safely support the loading of $P = 8$ kip. The allowable bending stress is $\sigma_{allow} = 24$ ksi and the allowable shear stress is $\tau_{allow} = 15$ ksi.

11–22. Solve Prob. 11–21 if $P = 10$ kip.

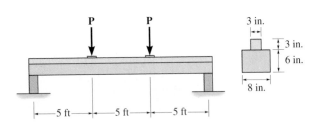

Prob. 11–19

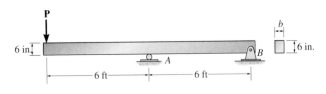

Probs. 11–21/11–22

*11–20. The timber beam has a rectangular cross section. If the width of the beam is 6 in., determine its height h so that it simultaneously reaches its allowable bending stress of $\sigma_{allow} = 1.50$ ksi and an allowable shear stress of $\tau_{allow} = 50$ psi. Also, what is the maximum load P that the beam can then support?

11–23. Draw the shear and moment diagrams for the shaft, and determine its required diameter to the nearest $\frac{1}{4}$ in. if $\sigma_{allow} = 7$ ksi and $\tau_{allow} = 3$ ksi. The bearings at A and D exert only vertical reactions on the shaft. The loading is applied to the pulleys at B, C, and E. Take $P = 110$ lb.

*11–24. Draw the shear and moment diagrams for the shaft, and determine its required diameter to the nearest $\frac{1}{4}$ in. if $\sigma_{allow} = 7$ ksi and $\tau_{allow} = 3$ ksi. The bearings at A and D exert only vertical reactions on the shaft. The loading is applied to the pulleys at B, C, and E. Take $P = 80$ lb.

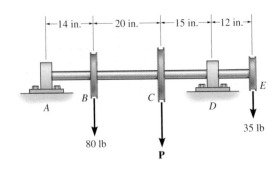

Probs. 11–23/24

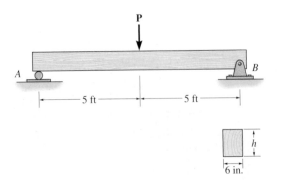

Prob. 11–20

11–25. The brick wall exerts a uniform distributed load of 1.20 kip/ft on the beam. If the allowable bending stress is $\sigma_{\text{allow}} = 22$ ksi, determine the required width b of the flange to the nearest $\frac{1}{4}$ in.

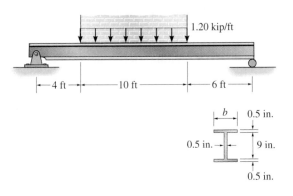

Prob. 11–25

11–26. The brick wall exerts a uniform distributed load of 1.20 kip/ft on the beam. If the allowable bending stress is $\sigma_{\text{allow}} = 22$ ksi   and   the   allowable   shear   stress   is $\tau_{\text{allow}} = 12$ ksi, select the lightest wide-flange section with the shortest depth from Appendix B that will safely support the load.

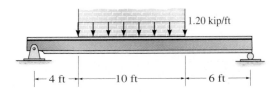

Prob. 11–26

11–27. The simply supported joist is used in the construction of a floor for a building. In order to keep the floor low with respect to the sill beams C and D, the ends of the joists are notched as shown. If the allowable shear stress for the wood is $\tau_{\text{allow}} = 350$ psi and the allowable bending stress is $\sigma_{\text{allow}} = 1500$ psi, determine the height h that will cause the beam to reach both allowable stresses at the same time. Also, what load P causes this to happen? Neglect the stress concentration at the notch.

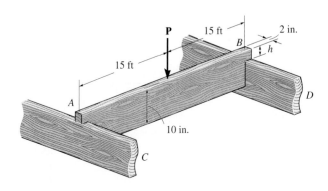

Prob. 11–27

***11–28.** The joist AB used in housing construction is to be made from 8-in. by 1.5-in. Southern-pine boards. If the design loading on each board is placed as shown, determine the largest room width L that the boards can span. The allowable bending stress for the wood is $\sigma_{\text{allow}} = 2$ ksi and the allowable shear stress is $\tau_{\text{allow}} = 180$ psi. Assume that the beam is simply supported from the walls at A and B.

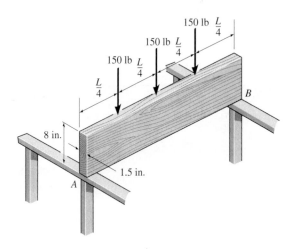

Prob. 11–28

*11.3   FULLY STRESSED BEAMS

In the previous section, we developed a method for determining the dimensions of the cross section of a *prismatic beam* so that it resists the maximum moment, M_{max}, within its span. Since the moment in the beam generally *varies* over the beam's length, the choice of a prismatic beam is usually inefficient since it is never fully stressed at points along the beam where $M < M_{max}$. In order to refine the design so as to reduce a beam's weight, engineers sometimes choose a beam having a *variable* cross-sectional area, such that at each cross section along the beam, the bending stress reaches its maximum allowable value. Beams having a variable cross-sectional area are called *nonprismatic beams*. They are often used in machines since they can be readily formed by casting. Examples are shown in Fig. 11–8a. In structures such beams may be "haunched" at their ends as shown in Fig. 11–8b. Also, beams may be "built up" or fabricated in a shop using plates. An example is a girder made from a rolled-shaped prismatic beam and having cover plates welded to it in the region where the moment is a maximum, Fig. 11–8c.

The stress analysis of a nonprismatic beam is generally very difficult to perform and is beyond the scope of this text. Most often these shapes are analyzed by using experimental methods or the theory of elasticity. The results obtained from such an analysis, however, do indicate that the assumptions used in the derivation of the flexure formula are approximately correct for predicting the bending stresses in nonprismatic sections, provided the taper or slope of the upper or lower boundary of the beam is not too severe. On the other hand, the shear formula cannot be used for nonprismatic beam design, since the results obtained from it are very misleading.

Although caution is advised when applying the flexure formula to nonprismatic beam design, we will show here, in principle, how this formula can be used as an approximate means for obtaining the beam's general shape. In this regard, the *size* of the cross section of a nonprismatic beam that supports a given loading can be determined using the flexure formula written as

$$S = \frac{M}{\sigma_{allow}}$$

If we express the internal moment M in terms of its position x along the beam, then since σ_{allow} is a known constant, the section modulus S or the beam's dimensions become a function of x. A beam designed in this manner is called a **fully stressed beam**. Although *only* bending stresses have been considered in approximating its final shape, attention must also be given to ensure that the beam will resist shear, especially at points where concentrated loads are applied. As a result, the ideal shape of the beam cannot be entirely determined from the flexure formula alone.

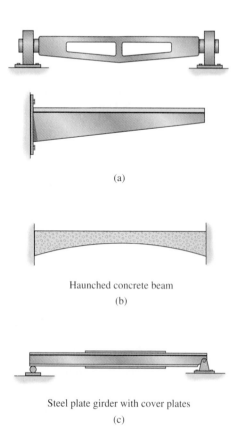

(a)

Haunched concrete beam

(b)

Steel plate girder with cover plates

(c)

Fig. 11–8

The leaf spring supporting this train car represents a nonprismatic beam.

EXAMPLE 11-4

Determine the shape of a fully stressed, simply supported beam that supports a concentrated force at its center, Fig. 11–9a. The beam has a rectangular cross section of constant width b, and the allowable stress is σ_{allow}.

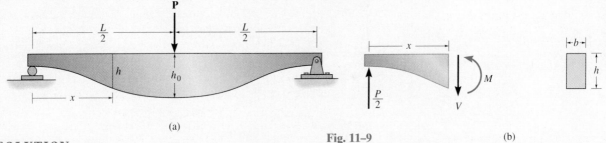

(a)

Fig. 11–9

(b)

SOLUTION

The internal moment in the beam, Fig. 11–9b, expressed as a function of position, $0 \le x < L/2$, is

$$M = \frac{P}{2}x$$

Hence the required section modulus is

$$S = \frac{M}{\sigma_{\text{allow}}} = \frac{P}{2\sigma_{\text{allow}}}x$$

Since $S = I/c$, then for a cross-sectional area h by b we have

$$\frac{I}{c} = \frac{\frac{1}{12}bh^3}{h/2} = \frac{P}{2\sigma_{\text{allow}}}x$$

$$h^2 = \frac{3P}{\sigma_{\text{allow}}b}x$$

If $h = h_0$ at $x = L/2$, then

$$h_0{}^2 = \frac{3PL}{2\sigma_{\text{allow}}b}$$

so that

$$h^2 = \left(\frac{2h_0{}^2}{L}\right)x \qquad\qquad \textit{Ans.}$$

By inspection, the depth h must therefore vary in a *parabolic* manner with the distance x. In practice this *shape* is the basis for the design of leaf springs used to support the rear-end axles of most heavy trucks. Note that although this result indicates that $h = 0$ at $x = 0$, it is necessary that the beam resist shear stress at the supports, and so practically speaking, it must be required that $h > 0$ at the supports, Fig. 11–9a.

EXAMPLE 11–5

The cantilevered beam shown in Fig. 11–10a is formed into a trapezoidal shape having a depth h_0 at A and a depth $3h_0$ at B. If it supports a load **P** at its end, determine the absolute maximum normal stress in the beam. The beam has a rectangular cross section of constant width b.

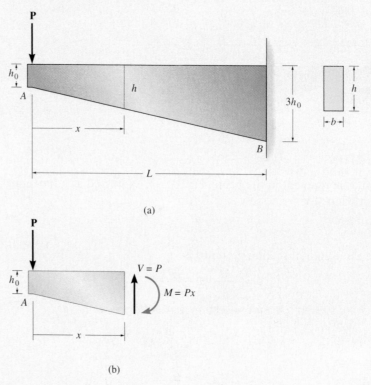

(a)

(b)

Fig. 11–10

SOLUTION

At any cross section, the maximum normal stress occurs at the top and bottom surface of the beam. However, since $\sigma_{max} = M/S$ and the section modulus S increases as x increases, the absolute maximum normal stress does *not* necessarily occur at the wall B, where the moment is maximum. Using the flexure formula, we can express the maximum normal stress at an arbitrary section in terms of its position x, Fig. 11–10b. Here the internal moment has a magnitude of $M = Px$. Since the slope of the bottom of the beam is $2h_0/L$, Fig. 11–10a, the depth of the beam at position x is

$$h = \frac{2h_0}{L}x + h_0 = \frac{h_0}{L}(2x + L)$$

Applying the flexure formula, we have

$$\sigma = \frac{Mc}{I} = \frac{Px(h/2)}{(\frac{1}{12} bh^3)} = \frac{6PL^2x}{bh_0^2(2x + L)^2} \qquad (1)$$

To determine the position x where the absolute maximum normal stress occurs, we must take the derivative of σ with respect to x and set it equal to zero. This gives

$$\frac{d\sigma}{dx} = \left(\frac{6PL^2}{bh_0^2}\right) \frac{1(2x + L)^2 - x(2)(2x + L)(2)}{(2x + L)^4} = 0$$

Thus,

$$4x^2 + 4xL + L^2 - 8x^2 - 4xL = 0$$
$$L^2 - 4x^2 = 0$$
$$x = \frac{1}{2}L$$

Substituting into Eq. 1 and simplifying, the absolute maximum normal stress is therefore

$$\sigma_{\substack{abs \\ max}} = \frac{3}{4} \frac{PL}{bh_0^2} \qquad \qquad Ans.$$

Note that at the wall, B, the maximum normal stress is

$$(\sigma_{max})_B = \frac{Mc}{I} = \frac{PL(1.5h_0)}{[\frac{1}{12} b(3h_0)^3]} = \frac{2}{3} \frac{PL}{bh_0^2}$$

which is 11.1% smaller than $\sigma_{\substack{abs \\ max}}$.

It should be recalled that the flexure formula was derived on the basis of assuming the beam to be *prismatic*. Since this is not the case here, some error is to be expected in this analysis and that of Example 11–4. A more exact mathematical analysis, using the theory of elasticity, reveals that application of the flexure formula as in the above example gives only small errors in the normal stress if the angle of beam taper is small. For example, if this angle is 15°, the stress calculated above is about 5% greater than that calculated by the more exact analysis. It may also be worth noting that the calculation of $(\sigma_{max})_B$ was done only for illustrative purposes, since, by Saint-Venant's principle, the actual stress distribution at the support (wall) is highly irregular.

*11.4 SHAFT DESIGN

Shafts that have circular cross sections are frequently used in many types of mechanical equipment and machinery. As a result, they are often subjected to cyclic or fatigue stress, which is caused by the combined bending and torsional loads they must transmit or resist. In addition to these loadings, stress concentrations may exist on a shaft due to keys, couplings, and sudden transitions in its cross-sectional area (Sec. 5.8). In order to design a shaft properly, it is therefore necessary to take all of these effects into account.

In this section we will discuss some of the important aspects of the design of uniform shafts required to transmit power. These shafts are often subjected to loads applied to attached pulleys and gears, such as the one shown in Fig. 11–11*a*. Since the loads can be applied to the shaft at various angles, the internal bending and torsional moments at any cross section can be determined by first replacing the loads by their statically equivalent counterparts and then resolving these loads into components in two perpendicular planes, Fig. 11–11*b*. The bending-moment diagrams for the loads *in each plane* can then be drawn, and the resultant internal moment at any section along the shaft is then determined by vector addition, $M = \sqrt{M_x^2 + M_z^2}$, Fig. 11–11*c*. In addition to the moment, segments of the shaft are also subjected to different internal torques, Fig. 11–11*b*. Here we see that, for equilibrium, the torque developed at one gear must balance the torque developed at the other gear. To account for the general variation of torque along the shaft, a ***torque diagram*** may also be drawn, Fig. 11–11*d*.

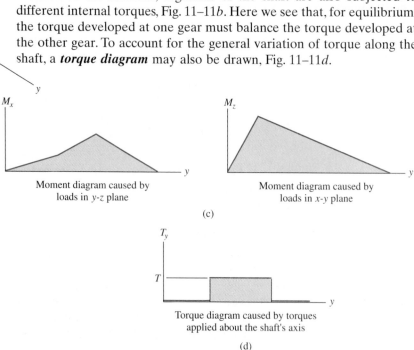

(a)

(b)

Moment diagram caused by
loads in *y-z* plane

Moment diagram caused by
loads in *x-y* plane

(c)

Torque diagram caused by torques
applied about the shaft's axis

(d)

Fig. 11–11

Once the moment and torque diagrams have been established, it is then possible to investigate certain critical sections along the shaft where the *combination* of a resultant moment **M** and a torque **T** creates the worst stress situation. In this regard, the moment of inertia of the shaft is the *same* about *any* diametrical axis, and since this axis represents a *principal axis of inertia* for the cross section, we can apply the flexure formula using the *resultant moment* to obtain the maximum bending stress. As shown in Fig. 11–11e, this stress will occur on two elements, *C* and *D*, each located on the outer boundary of the shaft. If a torque **T** is also resisted at this section, then a maximum shear stress is also developed on these elements, Fig. 11–11f. Furthermore, the external forces will also create shear stress in the shaft determined from $\tau = VQ/It$; however, this stress will generally contribute a much smaller stress distribution on the cross section compared with that developed by bending and torsion. In some cases, it must be investigated, but for simplicity, we will neglect its effect in the following analysis. In general, then, the critical element *D* (or *C*) on the shaft is subjected to *plane stress* as shown in Fig. 11–11g, where

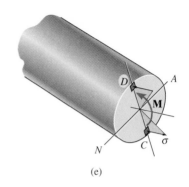

(e)

$$\sigma = \frac{Mc}{I} \quad \text{and} \quad \tau = \frac{Tc}{J}$$

If the allowable normal or shear stress for the material is known, the size of the shaft is then based on the use of these equations and selection of an appropriate theory of failure. For example, if the material is known to be ductile, then the maximum-shear-stress theory may be appropriate. As stated in Sec. 10.7, this theory requires the allowable shear stress, which is determined from the results of a simple tension test, to be equal to the maximum shear stress in the element. Using the stress-transformation equation, Eq. 9–7, for the stress state in Fig. 11–11g, we have

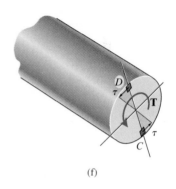

(f)

$$\tau_{\text{allow}} = \sqrt{\left(\frac{\sigma}{2}\right)^2 + \tau^2}$$

$$= \sqrt{\left(\frac{Mc}{2I}\right)^2 + \left(\frac{Tc}{J}\right)^2}$$

Since $I = \pi c^4/4$ and $J = \pi c^4/2$, this equation becomes

$$\tau_{\text{allow}} = \frac{2}{\pi c^3} \sqrt{M^2 + T^2}$$

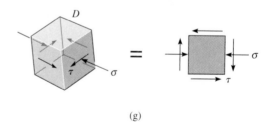

(g)

Solving for the radius of the shaft, we get

$$c = \left(\frac{2}{\pi \tau_{\text{allow}}} \sqrt{M^2 + T^2}\right)^{1/3} \qquad (11\text{–}2)$$

Application of any other theory of failure will, of course, lead to a different formulation for *c*. However, in all cases it may be necessary to apply this formulation at various "critical sections" along the shaft in order to determine the particular combination of *M* and *T* that gives the largest value for *c*.

The following example illustrates the procedure numerically.

EXAMPLE 11–6

The shaft in Fig. 11–12a is supported by smooth journal bearings at A and B. Due to the transmission of power to and from the shaft, the belts on the pulleys are subjected to the tensions shown. Determine the smallest diameter of the shaft using the maximum-shear-stress theory, with $\tau_{allow} = 50$ MPa.

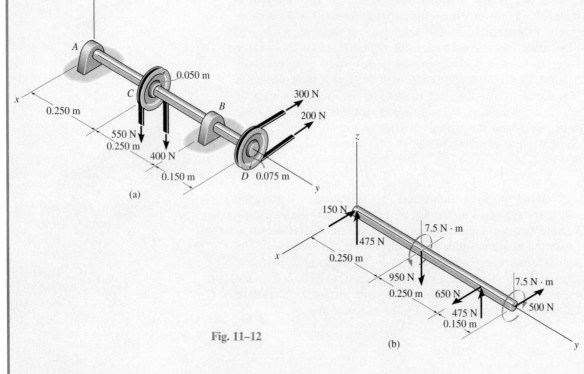

Fig. 11–12

SOLUTION

The support reactions have been calculated and are shown on the free-body diagram of the shaft, Fig. 11–12b. Bending-moment diagrams for M_x and M_z are shown in Figs. 11–12c and 11–12d, respectively. The torque diagram is shown in Fig. 11–12e. By inspection, critical points for bending moment occur either at C or B. Also, just to the right of C *and* at B the torsional moment is 7.5 N · m. At C, the resultant moment is

$$M_C = \sqrt{(118.75 \text{ N} \cdot \text{m})^2 + (37.5 \text{ N} \cdot \text{m})^2} = 124.5 \text{ N} \cdot \text{m}$$

whereas at B it is smaller, namely

$$M_B = 75 \text{ N} \cdot \text{m}$$

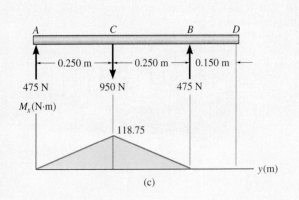

(c)

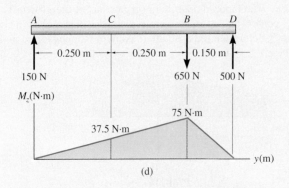

(d)

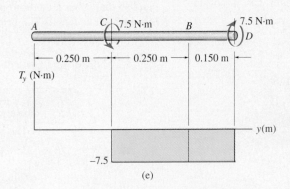

(e)

Since the design is based on the maximum-shear-stress theory, Eq. 11–2 applies. The radical $\sqrt{M^2 + T^2}$ will be the largest at a section just to the right of C. We have

$$c = \left(\frac{2}{\pi\tau_{allow}} \sqrt{M^2 + T^2} \right)^{1/3}$$

$$= \left(\frac{2}{\pi(50)(10^6) \text{ N/m}^2} \sqrt{(124.5 \text{ N} \cdot \text{m})^2 + (7.5 \text{ N} \cdot \text{m})^2} \right)^{1/3}$$

$$= 0.0117 \text{ m}$$

Thus, the smallest allowable diameter is

$$d = 2(0.0117 \text{ m}) = 23.3 \text{ mm} \qquad \textbf{\textit{Ans.}}$$

PROBLEMS

11–29. The tapered beam supports a concentrated force **P** at its center. If it is made from a plate that has a constant width b, determine the absolute maximum bending stress in the beam.

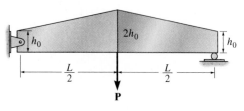

Prob. 11–29

11–30. Determine the variation in the width w as a function of x for the cantilevered beam that supports a concentrated force **P** at its end so that it has a maximum bending stress σ_{allow} throughout its length. The beam has a constant thickness t.

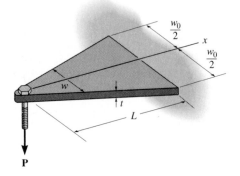

Prob. 11–30

11–31. The tapered cantilevered beam supports the concentrated force **P** at its end. Determine the absolute maximum bending stress in the beam.

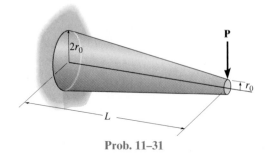

Prob. 11–31

*11–32. Determine the variation of the radius r of the cantilevered beam that supports the uniform distributed load so that it has a constant maximum bending stress σ_{max} throughout its length.

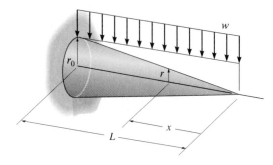

Prob. 11–32

11–33. Determine the variation in the depth of a cantilevered beam that supports a concentrated force **P** at its end so that it has a constant maximum bending stress σ_{allow} throughout its length. The beam has a constant width b_0.

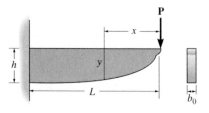

Prob. 11–33

11–34. The beam is made from a plate that has a constant thickness b. If it is simply supported and carries a uniform load w, determine the variation of its depth as a function of x so that it maintains a constant maximum bending stress σ_{allow} throughout its length.

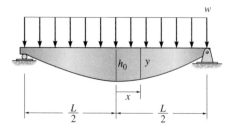

Prob. 11–34

11–35. The beam is made from a plate having a constant thickness t and a width that varies as shown. If it supports a concentrated force **P** at its center, determine the absolute maximum bending stress in the beam and specify its location $x, 0 < x < L/2$.

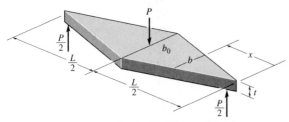

Prob. 11–35

*11–36. The motor at A is turning the shaft at a constant rate such that the torque **T** it develops creates horizontal reactions on the gears at C and D of 250 lb and 350 lb, respectively. The shaft is supported by a bearing in the motor at A and a bearing at B, which exert force components only in the x and z directions on the shaft. If the allowable shear stress for the shaft is $\tau_{\text{allow}} = 10$ ksi, determine to the nearest $\frac{1}{8}$ in. the smallest diameter of the shaft that will support the loading. Use the maximum-shear-stress theory of failure.

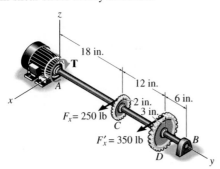

Prob. 11–36

11–37. The tapered simply supported beam supports the concentrated force **P** at its center. Determine the absolute maximum bending stress in the beam.

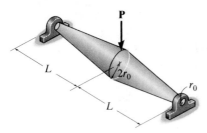

Prob. 11–37

11–38. Solve Prob. 11–36 using the maximum-distortion-energy theory of failure with $\sigma_{\text{allow}} = 20$ ksi.

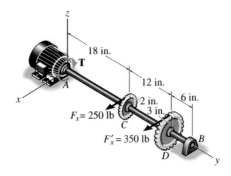

Prob. 11–38

11–39. The bearings at A and D exert only y and z components of force on the shaft. If $\tau_{\text{allow}} = 60$ MPa, determine to the nearest millimeter the smallest-diameter shaft that will support the loading. Use the maximum-shear-stress theory of failure.

*11–40. Solve Prob. 11–39 using the maximum-distortion-energy theory of failure. $\sigma_{\text{allow}} = 130$ MPa.

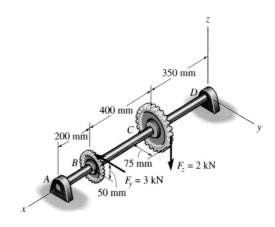

Probs. 11–39/11–40

11–41. The pulleys attached to the shaft are loaded as shown. If the bearings at A and B exert only horizontal and vertical forces on the shaft, determine the required diameter of the shaft to the nearest $\frac{1}{8}$ in. using the maximum-shear-stress theory of failure. $\tau_{\text{allow}} = 12$ ksi.

11–43. The shaft is supported by bearings at A and B that exert force components only in the x and z directions on the shaft. If the allowable normal stress for the shaft is $\sigma_{\text{allow}} = 15$ ksi, determine to the nearest $\frac{1}{8}$ in. the smallest diameter of the shaft that will support the gear loading. Use the maximum-distortion-energy theory of failure.

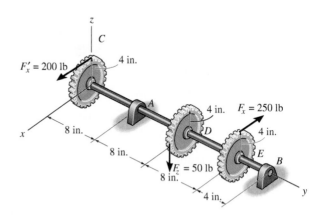

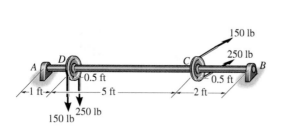

Prob. 11–41

Prob. 11–43

11–42. The pulleys attached to the shaft are loaded as shown. If the bearings at A and B exert only horizontal and vertical forces on the shaft, determine the required diameter of the shaft to the nearest $\frac{1}{8}$ in. using the maximum-distortion-energy theory of failure. $\sigma_{\text{allow}} = 20$ ksi.

*11–44. Determine to the nearest millimeter the diameter of the solid shaft if it is subjected to the gear loading. The bearings at A and B exert force components only in the y and z directions on the shaft. Base the design on the maximum-distortion-energy theory of failure with $\sigma_{\text{allow}} = 150$ MPa.

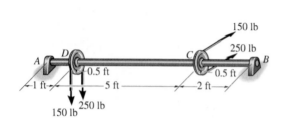

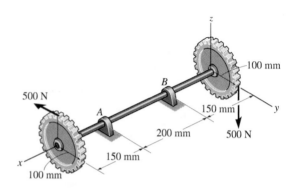

Prob. 11–42

Prob. 11–44

11–45. The shaft is supported on journal bearings that do not offer resistance to axial load. If the allowable normal stress for the shaft is $\sigma_{\text{allow}} = 80$ MPa, determine to the nearest millimeter the smallest diameter of the shaft that will support the loading. Use the maximum-distortion-energy theory of failure.

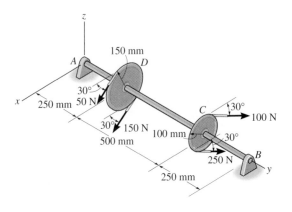

Prob. 11–45

11–46. The shaft is supported on journal bearings that do not offer resistance to axial load. If the allowable shear stress for the shaft is $\tau_{\text{allow}} = 35$ MPa, determine to the nearest millimeter the smallest diameter of the shaft that will support the loading. Use the maximum-shear-stress theory of failure.

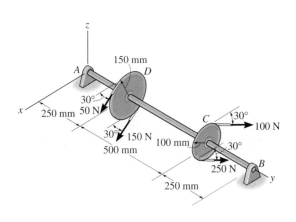

Prob. 11–46

11–47. The end gear connected to the shaft is subjected to the loading shown. If the bearings at A and B exert only y and z components of force on the shaft, determine the equilibrium torque T at gear C and then determine the smallest diameter of the shaft to the nearest millimeter that will support the loading. Use the maximum-shear-stress theory of failure with $\tau_{\text{allow}} = 60$ MPa.

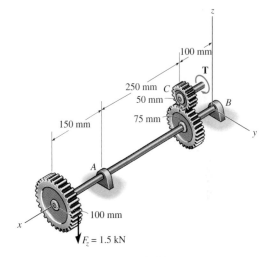

Prob. 11–47

*11–48. Solve Prob. 11–47 using the maximum-distortion-energy theory of failure with $\sigma_{\text{allow}} = 80$ MPa.

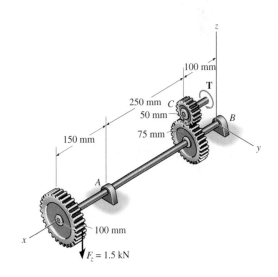

Prob. 11–48

REVIEW PROBLEMS

11–49. Draw the shear and moment diagrams for the shaft, and then determine its required diameter to the nearest millimeter if $\sigma_{allow} = 140$ MPa and $\tau_{allow} = 80$ MPa. The bearings at A and B exert only vertical reactions on the shaft.

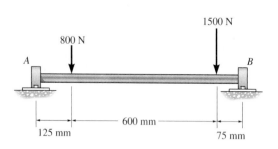

Prob. 11–49

11–51. The cantilevered beam has a circular cross section. If it supports a force **P** at its end, determine its radius y as a function of x so that it is subjected to a constant maximum bending stress σ_{allow} throughout its length.

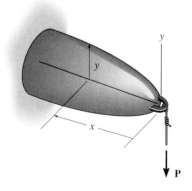

Prob. 11–51

11–50. The beam is constructed from three boards as shown. If each nail can support a shear force of 50 lb, determine the maximum spacing of the nails, s, s', and s'', to the nearest $\frac{1}{8}$ in., for regions AB, BC, and CD, respectively.

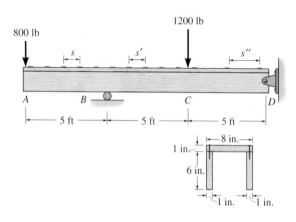

Prob. 11–50

*11–52. Select the lightest-weight steel wide-flange beam from Appendix B that will safely support the loading shown. The allowable bending stress is $\sigma_{allow} = 22$ ksi, and the allowable shear stress is $\tau_{allow} = 12$ ksi.

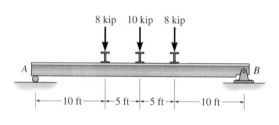

Prob. 11–52

11–53. Draw the shear and moment diagrams for the beam. Then select the lightest-weight steel wide-flange beam from Appendix B that will safely support the loading. Take σ_{allow} = 22 ksi, and τ_{allow} = 12 ksi.

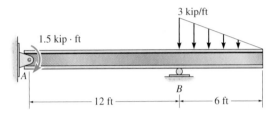

3 kip/ft

1.5 kip · ft

A

B

12 ft

6 ft

Prob. 11–53

11–55. The tapered beam supports a uniform distributed load w. If it is made from a plate and has a constant width b, determine the absolute maximum bending stress in the beam.

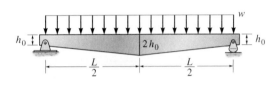

w

h_0

$2h_0$

h_0

$\dfrac{L}{2}$

$\dfrac{L}{2}$

Prob. 11–55

11–54. Select the lightest-weight steel wide-flange overhanging beam from Appendix B that will safely support the loading. Assume the support at A is a pin and the support at B is a roller. The allowable bending stress is σ_{allow} = 24 ksi and the allowable shear stress is τ_{allow} = 14 ksi.

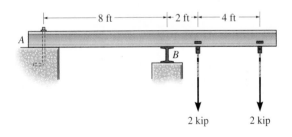

8 ft

2 ft

4 ft

A

B

2 kip

2 kip

Prob. 11–54

***11–56.** The 10-mm-wide bracket is used to support a force of 70 N at its end A. Determine the absolute maximum bending stress in the bracket, and specify its location x.

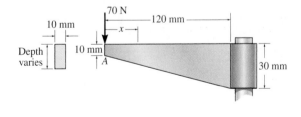

70 N

10 mm

120 mm

Depth varies

10 mm

A

x

30 mm

Prob. 11–56

The shelf loading causes noticeable deflection of the supporting beam, which can be calculated using the methods of this chapter.

12 DEFLECTION OF BEAMS AND SHAFTS

CHAPTER OBJECTIVES

Often limits must be placed on the amount of deflection a beam or shaft may undergo when it is subjected to a load, and so in this chapter we will discuss various methods for determining the deflection and slope at specific points on beams and shafts. The analytical methods include the integration method, the use of discontinuity functions, and the method of superposition. Also, a semigraphical technique, called the moment-area method, will be presented. At the end of the chapter, we will use these methods to solve for the support reactions on a beam or shaft that is statically indeterminate.

12.1 THE ELASTIC CURVE

Before the slope or the displacement at a point on a beam (or shaft) is determined, it is often helpful to sketch the deflected shape of the beam when it is loaded, in order to "visualize" any computed results and thereby partially check these results. The deflection diagram of the longitudinal axis that passes through the centroid of each cross-sectional area of the beam is called the *elastic curve*. For most beams the elastic curve can be sketched without much difficulty. When doing so, however, it is necessary to know how the slope or displacement is restricted at various types of supports. In general, supports that resist a *force*, such as a pin, restrict *displacement*, and those that resist a *moment*, such as a fixed wall, restrict *rotation* or *slope* as well as displacement. With this in mind, two typical examples of the elastic curves for loaded beams (or shafts), sketched to a greatly exaggerated scale, are shown in Fig. 12–1.

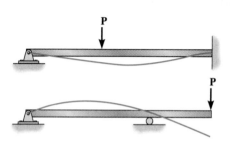

Fig. 12–1

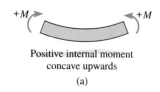

Positive internal moment
concave upwards

(a)

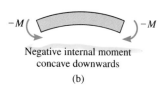

Negative internal moment
concave downwards

(b)

Fig. 12–2

If the elastic curve for a beam seems difficult to establish, it is suggested that the moment diagram for the beam be drawn first. Using the beam sign convention established in Sec. 6.1, a positive internal moment tends to bend the beam concave upward, Fig. 12–2a. Likewise, a negative moment tends to bend the beam concave downward, Fig. 12–2b. Therefore, if the moment diagram is *known*, it will be easy to construct the elastic curve. For example, consider the beam in Fig. 12–3a with its associated moment diagram shown in Fig. 12–3b. Due to the roller and pin supports, the displacement at *B* and *D* must be zero. Within the region of negative moment, *AC*, Fig. 12–3b, the elastic curve must be concave downward, and within the region of positive moment, *CD*, the elastic curve must be concave upward. Hence, there must be an *inflection point* at point *C*, where the curve changes from concave up to concave down, since this is a point of zero moment. Using these facts, the beam's elastic curve is sketched to a greatly exaggerated scale in Fig. 12–3c. It should also be noted that the displacements Δ_A and Δ_E are especially critical. At point *E* the *slope* of the elastic curve is *zero*, and there the beam's *deflection* may be a *maximum*. Whether Δ_E is actually greater than Δ_A depends on the relative magnitudes of $\mathbf{P}_1$ and $\mathbf{P}_2$ and the location of the roller at *B*.

Following these same principles, note how the elastic curve in Fig. 12–4 was constructed. Here the beam is cantilevered from a fixed support at *A* and therefore the elastic curve must have both zero displacement and zero slope at this point. Also, the largest displacement will occur either at *D*, where the slope is zero, or at *C*.

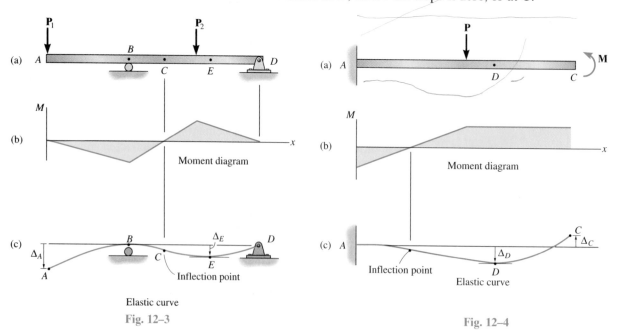

Fig. 12–3

Elastic curve

Fig. 12–4

Moment–Curvature Relationship.   We will now develop an important relationship between the internal moment in the beam and the radius of curvature ρ (rho) of the elastic curve at a point. The resulting equation will be used throughout the chapter as a basis for establishing each of the methods presented for finding the slope and displacement of the elastic curve for a beam (or shaft).

The following analysis, here and in the next section, will require the use of three coordinates. As shown in Fig. 12–5a, the x axis extends positive to the right, along the initially straight longitudinal axis of the beam. It is used to locate the differential element, having an undeformed width dx. The v axis extends *positive upward* from the x axis. It measures the *displacement* of the centroid on the cross-sectional area of the element. With these two coordinates, we will later define the equation of the elastic curve, v, as a function of x. Lastly, a "localized" y coordinate is used to specify the position of a fiber in the beam element. It is measured *positive upward* from the neutral axis, as shown in Fig. 12–5b. Recall that this same sign convention for x and y was used in the derivation of the flexure formula.

To derive the relationship between the internal moment and ρ, we will limit the analysis to the most common case of an initially straight beam that is elastically deformed by loads applied perpendicular to the beam's x axis and lying in the $x-v$ plane of symmetry for the beam's cross-sectional area. Due to the loading, the deformation of the beam is caused by both the internal shear force and bending moment. If the beam has a length that is much greater than its depth, the greatest deformation will be caused by bending, and therefore we will direct our attention to its effects. Deflections caused by shear will be discussed later in the chapter.

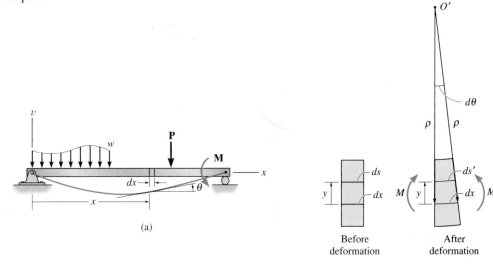

(a)

Before deformation

After deformation

(b)

Fig. 12–5

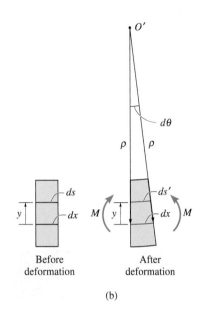

Before deformation

After deformation

(b)

Fig. 12–5

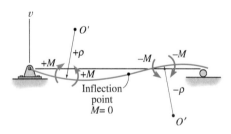

Fig. 12–6

When the internal moment M deforms the element of the beam, the angle between the cross sections becomes $d\theta$, Fig. 12–5b. The arc dx represents a portion of the elastic curve that intersects the neutral axis for each cross section. The *radius of curvature* for this arc is defined as the distance ρ, which is measured from the *center of curvature* O' to dx. Any arc on the element other than dx is subjected to a normal strain. For example, the strain in arc ds, located at a position y from the neutral axis, is $\epsilon = (ds' - ds)/ds$. However, $ds = dx = \rho\, d\theta$ and $ds' = (\rho - y)\, d\theta$, and so $\epsilon = [(\rho - y)d\theta - \rho d\theta]/\rho d\theta$ or

$$\frac{1}{\rho} = -\frac{\epsilon}{y} \tag{12–1}$$

If the material is homogeneous and behaves in a linear-elastic manner, then Hooke's law applies, $\epsilon = \sigma/E$. Also, since the flexure formula applies, $\sigma = -My/I$. Combining these equations and substituting into the above equation, we have

$$\frac{1}{\rho} = \frac{M}{EI} \tag{12–2}$$

where

ρ = the radius of curvature at a specific point on the elastic curve ($1/\rho$ is referred to as the *curvature*)

M = the internal moment in the beam at the point where ρ is to be determined

E = the material's modulus of elasticity

I = the beam's moment of inertia computed about the neutral axis

The product EI in this equation is referred to as the *flexural rigidity*, and it is always a positive quantity. The sign for ρ therefore depends on the direction of the moment. As shown in Fig. 12–6, when M is *positive*, ρ extends *above* the beam, i.e., in the positive v direction; when M is *negative*, ρ extends *below* the beam, or in the negative v direction.

Using the flexure formula, $\sigma = -My/I$, we can also express the curvature in terms of the stress in the beam, namely,

$$\frac{1}{\rho} = -\frac{\sigma}{Ey} \tag{12–3}$$

Both Eqs. 12–2 and 12–3 are valid for either small or large radii of curvature. However, the value of ρ is almost always calculated as a *very large quantity*. For example, consider an A-36 steel beam made from a W14 × 53 (Appendix B), where $E_{st} = 29(10^3)$ ksi and $\sigma_Y = 36$ ksi. When the material at the outer fibers, $y = \pm 7$ in., is about to *yield*, then, from Eq. 12–3, $\rho = \pm 5639$ in. Values of ρ calculated at other points along the beam's elastic curve may be even *larger*, since σ cannot exceed σ_Y at the outer fibers.

12.2   SLOPE AND DISPLACEMENT BY INTEGRATION

The elastic curve for a beam can be expressed mathematically as $v = f(x)$. To obtain this equation, we must first represent the curvature $(1/\rho)$ in terms of v and x. In most calculus books it is shown that this relationship is

$$\frac{1}{\rho} = \frac{d^2v/dx^2}{[1 + (dv/dx)^2]^{3/2}}$$

Substituting into Eq. 12–2, we get

$$\frac{d^2v/dx^2}{[1 + (dv/dx)^2]^{3/2}} = \frac{M}{EI} \qquad (12\text{–}4)$$

This equation represents a nonlinear second-order differential equation. Its solution, which is called the *elastica*, gives the exact shape of the elastic curve, assuming, of course, that beam deflections occur only due to bending. Through the use of higher mathematics, elastica solutions have been obtained only for simple cases of beam geometry and loading.

In order to facilitate the solution of a greater number of deflection problems, Eq. 12–4 can be modified. Most engineering design codes specify *limitations* on deflections for tolerance or esthetic purposes, and as a result the elastic deflections for the majority of beams and shafts form a shallow curve. Consequently, the slope of the elastic curve which is determined from dv/dx will be *very small*, and its square will be negligible compared with unity.* Therefore the curvature, as defined above, can be approximated by $1/\rho = d^2v/dx^2$. Using this simplification, Eq. 12–4 can now be written as

The moment of inertia of this bridge support varies along its length and this must be taken into account when computing its deflection.

$$\frac{d^2v}{dx^2} = \frac{M}{EI} \qquad (12\text{–}5)$$

It is also possible to write this equation in two alternative forms. If we differentiate each side with respect to x and substitute $V = dM/dx$ (Eq. 6–2), we get

$$\frac{d}{dx}\left(EI\frac{d^2v}{dx^2}\right) = V(x) \qquad (12\text{–}6)$$

Differentiating again, using $-w = dV/dx$ (Eq. 6–1), yields

$$\frac{d^2}{dx^2}\left(EI\frac{d^2v}{dx^2}\right) = -w(x) \qquad (12\text{–}7)$$

---

*See Example 12–1.

For most problems the flexural rigidity will be constant along the length of the beam. Assuming this to be the case, the above results may be reordered into the following set of equations:

$$EI\frac{d^4v}{dx^4} = -w(x) \tag{12-8}$$

$$EI\frac{d^3v}{dx^3} = V(x) \tag{12-9}$$

$$EI\frac{d^2v}{dx^2} = M(x) \tag{12-10}$$

Solution of any of these equations requires successive integrations to obtain the deflection v of the elastic curve. For each integration it is necessary to introduce a "constant of integration" and then solve for all the constants to obtain a unique solution for a particular problem. For example, if the distributed load is expressed as a function of x and Eq. 12–8 is used, then four constants of integration must be evaluated; however, if the internal moment M is determined and Eq. 12–10 is used, only two constants of integration must be found. The choice of which equation to start with depends on the problem. Generally, however, it is easier to determine the internal moment M as a function of x, integrate twice, and evaluate only two integration constants.

Recall from Sec. 6.1 that if the loading on a beam is discontinuous, that is, consists of a series of several distributed and concentrated loads, then several functions must be written for the internal moment, each valid within the region between the discontinuities. Also, for convenience in writing each moment expression, *the origin for each x coordinate can be selected arbitrarily*. For example, consider the beam shown in Fig. 12–7a. The internal moment in regions AB, BC, and CD can be written in terms of the x_1, x_2, and x_3 coordinates selected, as shown in either Fig. 12–7b or 12–7c, or in fact in any manner that will yield $M = f(x)$ in as simple a form as possible. Once these functions are integrated through the use of Eq. 12–10 and the constants of integration determined, the functions will give the slope and deflection (elastic curve) for each region of the beam for which they are valid.

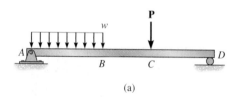

(a)

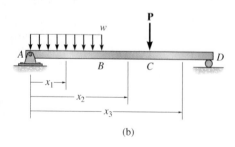

(b)

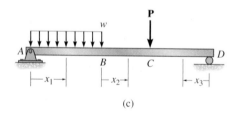

(c)

Fig. 12–7

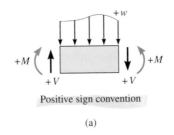

Positive sign convention

(a)

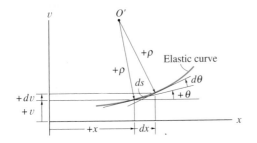

Positive sign convention

(b)

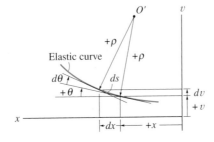

Positive sign convention

(c)

Fig. 12–8

Sign Convention and Coordinates. When applying Eqs. 12–8 through 12–10, it is important to use the proper signs for M, V, or w as established by the sign convention that was used in the derivation of these equations. For review, these terms are shown in their *positive directions* in Fig. 12–8a. Furthermore, recall that *positive deflection, v, is upward*, and as a result, the *positive slope angle θ* will be measured *counterclockwise* from the x axis when x is *positive to the right*. The reason for this is shown in Fig. 12–8b. Here positive increases dx and dv in x and v create an increased θ that is counterclockwise. On the other hand, if *positive x is directed to the left*, then θ will be *positive clockwise*, Fig. 12–8c.

It should be pointed out that by assuming dv/dx to be very small, the original horizontal length of the beam's axis and the arc of its elastic curve will be about the same. In other words, ds in Fig. 12–8b and 12–8c is approximately equal to dx, since $ds = \sqrt{(dx)^2 + (dv)^2} = \sqrt{1 + (dv/dx)^2}\ dx \approx dx$. As a result, points on the elastic curve are assumed to be *displaced vertically*, and not horizontally. Also, since the slope angle θ will be *very small*, its value in radians can be determined directly from $\theta \approx \tan\theta = dv/dx$.

The design of a roof system requires a careful consideration of deflection. For example, rain can accumulate on areas of the roof, which then causes ponding, leading to further deflection and possible failure of the roof.

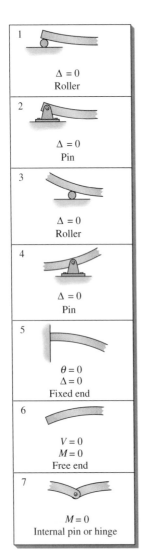

Table 12–1

Boundary and Continuity Conditions.   The constants of integration are determined by evaluating the functions for shear, moment, slope, or displacement at a particular point on the beam where the value of the function is known. These values are called **boundary conditions**. Several possible boundary conditions that are often used to solve beam (or shaft) deflection problems are listed in Table 12–1. For example, if the beam is supported by a roller or pin (1, 2, 3, 4), then it is required that the displacement be *zero* at these points. Furthermore, if these supports are located at the *ends of the beam* (1, 2), the internal moment in the beam must also be zero. At the fixed support (5), the slope and displacement are both zero, whereas the free-ended beam (6) has both zero moment and zero shear. Lastly, if two segments of a beam are connected by an "internal" pin or hinge (7), the moment must be zero at this connection.

If a single x coordinate cannot be used to express the equation for the beam's slope or the elastic curve, then **continuity conditions** must be used to evaluate some of the integration constants. For example, consider the beam in Fig. 12–9a. Here the x coordinates are both chosen with origins at A. Each is valid only within the regions $0 \le x_1 \le a$ and $a \le x_2 \le (a + b)$. Once the functions for the slope and deflection are obtained, they must give the *same values* for the slope and deflection at point B so the elastic curve is physically *continuous*. Expressed mathematically, this requires that $\theta_1(a) = \theta_2(a)$ and $v_1(a) = v_2(a)$. These equations can then be used to evaluate two constants of integration. On the other hand, if the elastic curve is expressed in terms of the coordinates $0 \le x_1 \le a$ and $0 \le x_2 \le b$, shown in Fig. 12–9b, then the continuity of slope and deflection at B requires $\theta_1(a) = -\theta_2(b)$ and $v_1(a) = v_2(b)$. In this particular case, a *negative* sign is necessary to match the slopes at B since x_1 extends positive to the right, whereas x_2 extends positive to the left. Consequently, θ_1 is positive counterclockwise, and θ_2 is positive clockwise. See Fig. 12–8b and 12–8c.

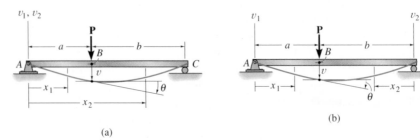

(a)   (b)

Fig. 12–9

Procedure for Analysis

The following procedure provides a method for determining the slope and deflection of a beam (or shaft) using the method of integration.

Elastic Curve.

• Draw an exaggerated view of the beam's elastic curve. Recall that zero slope and zero displacement occur at all fixed supports, and zero displacement occurs at all pin and roller supports.

• Establish the x and v coordinate axes. The x axis must be parallel to the undeflected beam and can have an origin at any point along the beam, with a positive direction either to the right or to the left.

• If several discontinuous loads are present, establish x coordinates that are valid for each region of the beam between the discontinuities. Choose these coordinates so that they will simplify subsequent algebraic work.

• In all cases, the associated positive v axis should be directed upward.

Load or Moment Function.

• For each region in which there is an x coordinate, express the loading w or the internal moment M as a function of x. In particular, *always* assume that M acts in the *positive direction* when applying the equation of moment equilibrium to determine $M = f(x)$.

Slope and Elastic Curve.

• Provided EI is constant, apply either the load equation $EI\, d^4v/dx^4 = -w(x)$, which requires four integrations to get $v = v(x)$, or the moment equation $EI\, d^2v/dx^2 = M(x)$, which requires only two integrations. For each integration it is important to include a constant of integration.

• The constants are evaluated using the boundary conditions for the supports (Table 12–1) and the continuity conditions that apply to slope and displacement at points where two functions meet. Once the constants are evaluated and substituted back into the slope and deflection equations, the slope and displacement at *specific points* on the elastic curve can then be determined.

• The numerical values obtained can be checked graphically by comparing them with the sketch of the elastic curve. Realize that *positive* values for *slope* are *counterclockwise* if the x axis extends *positive* to the *right*, and *clockwise* if the x axis extends *positive* to the *left*. In either of these cases, *positive displacement* is *upward*.

EXAMPLE 12-1

The cantilevered beam shown in Fig. 12–10a is subjected to a vertical load **P** at its end. Determine the equation of the elastic curve. *EI* is constant.

SOLUTION I

Elastic Curve. The load tends to deflect the beam as shown in Fig. 12–10a. By inspection, the internal moment can be represented throughout the beam using a single *x* coordinate.

Moment Function. From the free-body diagram, with **M** acting in the *positive direction*, Fig. 12–10b, we have

$$M = -Px$$

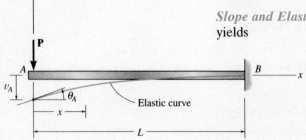

Slope and Elastic Curve. Applying Eq. 12–10 and integrating twice yields

$$EI \frac{d^2v}{dx^2} = -Px \tag{1}$$

$$EI \frac{dv}{dx} = -\frac{Px^2}{2} + C_1 \tag{2}$$

$$EI \, v = -\frac{Px^3}{6} + C_1 x + C_2 \tag{3}$$

(a)

Using the boundary conditions $dv/dx = 0$ at $x = L$ and $v = 0$ at $x = L$, Eqs. 2 and 3 become

$$0 = -\frac{PL^2}{2} + C_1$$

$$0 = -\frac{PL^3}{6} + C_1 L + C_2$$

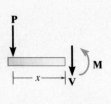

(b)

Thus, $C_1 = PL^2/2$ and $C_2 = -PL^3/3$. Substituting these results into Eqs. 2 and 3 with $\theta = dv/dx$, we get

$$\theta = \frac{P}{2EI}(L^2 - x^2)$$

$$v = \frac{P}{6EI}(-x^3 + 3L^2x - 2L^3) \qquad \textit{Ans.}$$

Fig. 12–10

Maximum slope and displacement occur at $A(x = 0)$, for which

$$\theta_A = \frac{PL^2}{2EI} \tag{4}$$

$$v_A = -\frac{PL^3}{3EI} \tag{5}$$

The *positive* result for θ_A indicates *counterclockwise* rotation and the *negative* result for v_A indicates that v_A is *downward*. This agrees with the results sketched in Fig. 12–10a.

In order to obtain some idea as to the actual *magnitude* of the slope and displacement at the end A, consider the beam in Fig. 12–10a to have a length of 15 ft, support a load of $P = 6$ kip, and be made of A-36 steel having $E_{st} = 29(10^3)$ ksi. Using the methods of Sec. 11.3, if this beam was designed without a factor of safety by assuming the allowable normal stress is equal to the yield stress $\sigma_{allow} = 36$ ksi, then a W12 × 26 would be found to be adequate ($I = 204$ in.^{4}). From Eqs. 4 and 5 we get

$$\theta_A = \frac{6 \text{ kip}(15 \text{ ft})^2(12 \text{ in./ft})^2}{2[29(10^3) \text{ kip/in.}^2](204 \text{ in.}^4)} = 0.0164 \text{ rad}$$

$$v_A = -\frac{6 \text{ kip}(15 \text{ ft})^3(12 \text{ in./ft})^3}{3[29(10^3) \text{ kip/in.}^2](204 \text{ in.}^4)} = -1.97 \text{ in.}$$

Since $\theta_A^2 = (dv/dx)^2 = 0.000270 \text{ rad}^2 \ll 1$, this justifies the use of Eq. 12–10, rather than applying the more exact Eq. 12–4, for computing the deflection of beams. Also, since this numerical application is for a *cantilevered beam*, we have obtained *larger values* for θ and v than would have been obtained if the beam was supported using pins, rollers, or other fixed supports.

SOLUTION II

This problem can also be solved using Eq. 12–8, $EI \, d^4v/dx^4 = -w(x)$. Here $w(x) = 0$ for $0 \le x \le L$, Fig. 12–10a, so that upon integrating once we get the form of Eq. 12–9, i.e.,

$$EI \frac{d^4v}{dx^4} = 0$$

$$EI \frac{d^3v}{dx^3} = C_1' = V$$

The shear constant C_1' can be evaluated at $x = 0$, since $V_A = -P$ (negative according to the beam sign convention, Fig. 12–8a.) Thus, $C_1' = -P$. Integrating again yields the form of Eq. 12–10, i.e.,

$$EI \frac{d^3v}{dx^3} = -P$$

$$EI \frac{d^2v}{dx^2} = -Px + C_2' = M$$

Here $M = 0$ at $x = 0$, so $C_2' = 0$, and as a result one obtains Eq. 1 and the solution proceeds as before.

EXAMPLE 12–2

The simply supported beam shown in Fig. 12–11a supports the triangular distributed loading. Determine its maximum deflection. EI is constant.

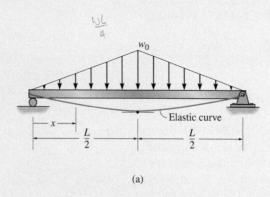

(a)

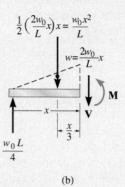

(b)

Fig. 12–11

SOLUTION I

Elastic Curve. Due to symmetry, only one x coordinate is needed for the solution, in this case $0 \leq x \leq L/2$. The beam deflects as shown in Fig. 12–11a. Notice that maximum deflection occurs at the center since the slope is zero at this point.

Moment Function. The distributed load acts downward, and therefore it is positive according to our sign convention. A free-body diagram of the segment on the left is shown in Fig. 12–11b. The equation for the distributed loading is

$$w = \frac{2w_0}{L}x \tag{1}$$

Hence,

$$\zeta + \Sigma M_{NA} = 0; \quad M + \frac{w_0 x^2}{L}\left(\frac{x}{3}\right) - \frac{w_0 L}{4}(x) = 0$$

$$M = -\frac{w_0 x^3}{3L} + \frac{w_0 L}{4}x$$

Slope and Elastic Curve.   Using Eq. 12–10 and integrating twice, we have

$$EI\frac{d^2v}{dx^2} = M = -\frac{w_0}{3L}x^3 + \frac{w_0L}{4}x \qquad (2)$$

$$EI\frac{dv}{dx} = -\frac{w_0}{12L}x^4 + \frac{w_0L}{8}x^2 + C_1$$

$$EIv = -\frac{w_0}{60L}x^5 + \frac{w_0L}{24}x^3 + C_1x + C_2$$

The constants of integration are obtained by applying the boundary condition $v = 0$ at $x = 0$ and the symmetry condition that $dv/dx = 0$ at $x = L/2$. This leads to

$$C_1 = -\frac{5w_0L^3}{192} \qquad C_2 = 0$$

Hence,

$$EI\frac{dv}{dx} = -\frac{w_0}{12L}x^4 + \frac{w_0L}{8}x^2 - \frac{5w_0L^3}{192}$$

$$EIv = -\frac{w_0}{60L}x^5 + \frac{w_0L}{24}x^3 - \frac{5w_0L^3}{192}x$$

Determining the maximum deflection at $x = L/2$, we have

$$v_{max} = -\frac{w_0L^4}{120EI} \qquad\qquad \textit{Ans.}$$

SOLUTION II

Starting with the distributed loading, Eq. 1, and applying Eq. 12–8, we have

$$EI\frac{d^4v}{dx^4} = -\frac{2w_0}{L}x$$

$$EI\frac{d^3v}{dx^3} = V = -\frac{w_0}{L}x^2 + C_1'$$

Since $V = +w_0L/4$ at $x = 0$, then $C_1' = w_0L/4$. Integrating again yields

$$EI\frac{d^3v}{dx^3} = V = -\frac{w_0}{L}x^2 + \frac{w_0L}{4}$$

$$EI\frac{d^2v}{dx^2} = M = -\frac{w_0}{3L}x^3 + \frac{w_0L}{4}x + C_2'$$

Here $M = 0$ at $x = 0$, so $C_2' = 0$. This yields Eq. 2. The solution now proceeds as before.

EXAMPLE 12-3

The simply supported beam shown in Fig. 12–12*a* is subjected to the concentrated force **P**. Determine the maximum deflection of the beam. *EI* is constant.

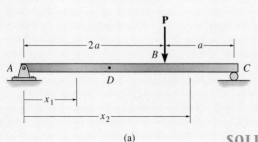

(a)

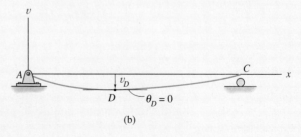

(b)

SOLUTION

Elastic Curve. The beam deflects as shown in Fig. 12–12*b*. Two coordinates must be used, since the moment becomes discontinuous at *P*. Here we will take x_1 and x_2, having the *same origin* at *A*, so that $0 \leq x_1 < 2a$ and $2a < x_2 \leq 3a$.

Moment Function. From the free-body diagrams shown in Fig. 12–12*c*,

$$M_1 = \frac{P}{3}x_1$$

$$M_2 = \frac{P}{3}x_2 - P(x_2 - 2a) = \frac{2P}{3}(3a - x_2)$$

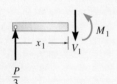

Slope and Elastic Curve. Applying Eq. 12–10 for M_1 and integrating twice yields

$$EI\frac{d^2v_1}{dx_1^2} = \frac{P}{3}x_1$$

$$EI\frac{dv_1}{dx^1} = \frac{P}{6}x_1^2 + C_1 \tag{1}$$

$$EIv_1 = \frac{P}{18}x_1^3 + C_1x_1 + C_2 \tag{2}$$

Likewise for M_2,

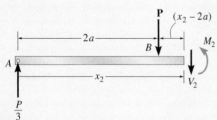

(c)

Fig. 12–12

$$EI\frac{d^2v_2}{dx_2^2} = \frac{2P}{3}(3a - x_2)$$

$$EI\frac{dv_2}{dx_2} = \frac{2P}{3}\left(3ax_2 - \frac{x_2^2}{2}\right) + C_3 \tag{3}$$

$$EIv_2 = \frac{2P}{3}\left(\frac{3}{2}ax_2^2 - \frac{x_2^3}{6}\right) + C_3x_2 + C_4 \tag{4}$$

The four constants are evaluated using *two* boundary conditions, namely, $x_1 = 0$, $v_1 = 0$ and $x_2 = 3a$, $v_2 = 0$. Also, *two* continuity conditions must be applied at B, that is, $dv_1/dx_1 = dv_2/dx_2$ at $x_1 = x_2 = 2a$ and $v_1 = v_2$ at $x_1 = x_2 = 2a$. Substitution as specified results in the following four equations:

$v_1 = 0$ at $x_1 = 0$; $0 = 0 + 0 + C_2$

$v_2 = 0$ at $x_2 = 3a$; $0 = \dfrac{2P}{3}\left(\dfrac{3}{2}a(3a)^2 - \dfrac{(3a)^3}{6}\right) + C_3(3a) + C_4$

$\dfrac{dv_1(2a)}{dx_1} = \dfrac{dv_2(2a)}{dx_2}$; $\dfrac{P}{6}(2a)^2 + C_1 = \dfrac{2P}{3}\left(3a(2a) - \dfrac{(2a)^2}{2}\right) + C_3$

$v_1(2a) = v_2(2a)$; $\dfrac{P}{18}(2a)^3 + C_1(2a) + C_2 = \dfrac{2P}{3}\left(\dfrac{3}{2}a(2a)^2 - \dfrac{(2a)^3}{6}\right) + C_3(2a) + C_4$

Solving these equations we get

$$C_1 = -\frac{4}{9}Pa^2 \qquad C_2 = 0$$
$$C_3 = -\frac{22}{9}Pa^2 \qquad C_4 = \frac{4}{3}Pa^3$$

Thus Eqs. 1–4 become

$$\frac{dv_1}{dx_1} = \frac{P}{6EI}x_1^2 - \frac{4}{9}\frac{Pa^2}{EI} \tag{5}$$

$$v_1 = \frac{P}{18EI}x_1^3 - \frac{4}{9}\frac{Pa^2}{EI}x_1 \tag{6}$$

$$\frac{dv_2}{dx_2} = \frac{2Pa}{EI}x_2 - \frac{P}{3EI}x_2^2 - \frac{22}{9}\frac{Pa^2}{EI} \tag{7}$$

$$v_2 = \frac{Pa}{EI}x_2^2 - \frac{P}{9EI}x_2^3 - \frac{22}{9}\frac{Pa^2}{EI}x_2 + \frac{4}{3}\frac{Pa^3}{EI} \tag{8}$$

By inspection of the elastic curve, Fig. 12–12b, the maximum deflection occurs at D, somewhere within region AB. Here the slope must be zero. From Eq. 5,

$$\frac{1}{6}x_1^2 - \frac{4}{9}a^2 = 0$$
$$x_1 = 1.633a$$

Substituting into Eq. 6,

$$v_{max} = -0.484\frac{Pa^3}{EI} \qquad\qquad Ans.$$

The negative sign indicates that the deflection is downward.

EXAMPLE 12-4

The beam in Fig. 12–13a is subjected to a load **P** at its end. Determine the displacement at C. EI is constant.

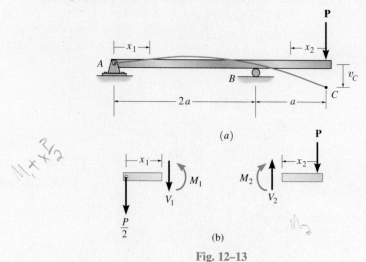

(a)

(b)

Fig. 12–13

SOLUTION

Elastic Curve. The beam deflects into the shape shown in Fig. 12–13a. Due to the loading, two x coordinates will be considered, namely, $0 \le x_1 < 2a$ and $0 \le x_2 < a$, where x_2 is directed to the left from C, since the internal moment is easy to formulate.

Moment Functions. Using the free-body diagrams shown in Fig. 12–13b, we have

$$M_1 = -\frac{P}{2}x_1 \qquad M_2 = -Px_2$$

Slope and Elastic Curve. Applying Eq. 12–10,

for $0 \le x_1 < 2a$, $EI \dfrac{d^2v_1}{dx_1{}^2} = -\dfrac{P}{2}x_1$

$$EI \frac{dv_1}{dx_1} = -\frac{P}{4}x_1{}^2 + C_1 \tag{1}$$

$$EIv_1 = -\frac{P}{12}x_1{}^3 + C_1x_1 + C_2 \tag{2}$$

For $0 \le x_2 < a$, $EI \dfrac{d^2v_2}{dx_2{}^2} = -Px_2$

$$EI \frac{dv_2}{dx_2} = -\frac{P}{2}x_2{}^2 + C_3 \tag{3}$$

$$EIv_2 = -\frac{P}{6}x_2{}^3 + C_3x_2 + C_4 \tag{4}$$

The *four* constants of integration are determined using *three* boundary conditions, namely, $v_1 = 0$ at $x_1 = 0$, $v_1 = 0$ at $x_1 = 2a$, and $v_2 = 0$ at $x_2 = a$ and *one* continuity equation. Here the continuity of slope at the roller requires $dv_1/dx_1 = -dv_2/dx_2$ at $x_1 = 2a$ and $x_2 = a$. Why is there a negative sign in this equation? (Note that continuity of displacement at B has been indirectly considered in the boundary conditions, since $v_1 = v_2 = 0$ at $x_1 = 2a$ and $x_2 = a$.) Applying these four conditions yields

$$v_1 = 0 \text{ at } x_1 = 0; \qquad 0 = 0 + 0 + C_2$$

$$v_1 = 0 \text{ at } x_1 = 2a; \qquad 0 = -\frac{P}{12}(2a)^3 + C_1(2a) + C_2$$

$$v_2 = 0 \text{ at } x_2 = a; \qquad 0 = -\frac{P}{6}a^3 + C_3 a + C_4$$

$$\frac{dv_1(2a)}{dx_1} = -\frac{dv_2(a)}{dx_2}; \qquad -\frac{P}{4}(2a)^2 + C_1 = -\left(-\frac{P}{2}(a)^2 + C_3\right)$$

Solving, we obtain

$$C_1 = \frac{Pa^2}{3} \qquad C_2 = 0 \qquad C_3 = \frac{7}{6}Pa^2 \qquad C_4 = -Pa^3$$

Substituting C_3 and C_4 into Eq. 4 gives

$$v_2 = -\frac{P}{6EI}x_2^3 + \frac{7Pa^2}{6EI}x_2 - \frac{Pa^3}{EI}$$

The displacement at C is determined by setting $x_2 = 0$. We get

$$v_C = -\frac{Pa^3}{EI} \qquad\qquad \textit{Ans.}$$

PROBLEMS

12–1. An A-36 steel strap having a thickness of 10 mm and a width of 20 mm is bent into a circular arc of radius $\rho = 10$ m. Determine the maximum bending stress in the strap.

12–2. A picture is taken of a man performing a pole vault, and the minimum radius of curvature of the pole is estimated by measurement to be 4.5 m. If the pole is 40 mm in diameter and it is made of a glass-reinforced plastic for which $E_g = 131$ GPa, determine the maximum bending stress in the pole.

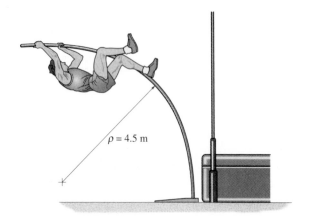

Prob. 12–2

12–3. Determine the equations of the elastic curve using the x_1 and x_2 coordinates. EI is constant.

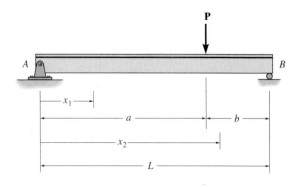

Prob. 12–3

*12–4. Determine the equation of the elastic curve for the beam using the x coordinate that is valid for $0 \le x < L/2$. Specify the slope at A and the beam's maximum deflection. EI is constant.

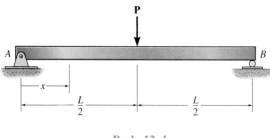

Prob. 12–4

12–5. Determine the equations of the elastic curve for the beam using the x_1 and x_2 coordinates. Specify the slope at A and the maximum deflection. EI is constant.

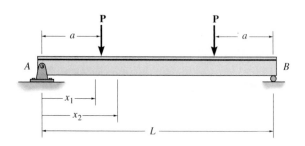

Prob. 12–5

12–6. Determine the equations of the elastic curve for the beam using the x_1 and x_2 coordinates. Specify the beam's maximum deflection. EI is constant.

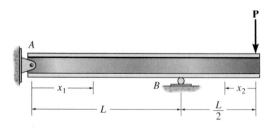

Prob. 12–6

12–7. The shaft is supported at A by a journal bearing that exerts only vertical reactions on the shaft, and at C by a thrust bearing that exerts horizontal and vertical reactions on the shaft. Determine the equations of the elastic curve using the coordinates x_1 and x_2. EI is constant.

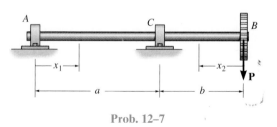

Prob. 12–7

12–10. Determine the maximum deflection of the beam and the slope at A. EI is constant.

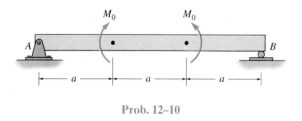

Prob. 12–10

*12–8. The fence board weaves between the three smooth fixed posts. If the posts remain along the same line, determine the maximum bending stress in the board. The board has a width of 6 in. and a thickness of 0.5 in. $E_w = 1.60(10^3)$ksi. Assume the displacement of each end of the board relative to its center is 3 in.

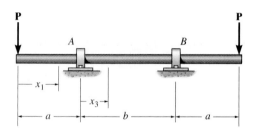

Prob. 12–8

12–11. The A-36 steel beam has a depth of 10 in. and is subjected to a constant moment M_0, which causes the stress at the outer fibers to become $\sigma_Y = 36$ ksi. Determine the radius of curvature of the beam and the maximum slope and deflection.

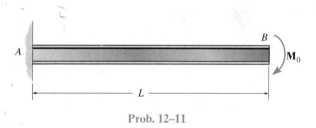

Prob. 12–11

12–9. Determine the equations of the elastic curve for the shaft using the x_1 and x_3 coordinates. Specify the slope at A and the deflection at the center of the shaft. EI is constant.

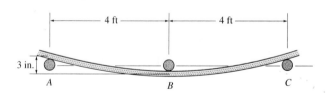

Prob. 12–9

*12–12. The shaft is supported at A by a journal bearing that exerts only vertical reactions on the shaft and at B by a thrust bearing that exerts horizontal and vertical reactions on the shaft. Draw the bending-moment diagram for the shaft and then, from this diagram, sketch the deflection or elastic curve for the shaft's centerline. Determine the equations of the elastic curve using the coordinates x_1 and x_2. EI is constant.

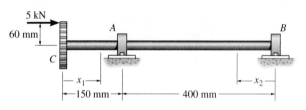

Prob. 12–12

12–13. Determine the elastic curve for the cantilevered beam, which is subjected to the couple moment M_0. Also compute the maximum slope and maximum deflection of the beam. EI is constant.

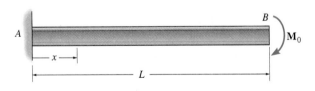

Prob. 12–13

12–14. Determine the maximum slope and maximum deflection of the simply-supported beam which is subjected to the couple moment M_0. EI is constant.

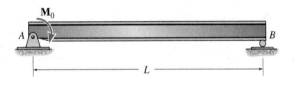

Prob. 12–14

12–15. Determine the elastic curve for the simply supported beam, which is subjected to the couple moments M_0. Also, compute the maximum slope and the maximum deflection of the beam. EI is constant.

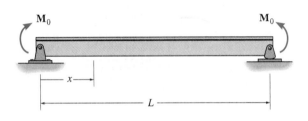

Prob. 12–15

*12–16. Determine the equation of the elastic curve using the coordinate x, and specify the slope at point A and the deflection at point C. EI is constant.

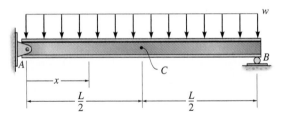

Prob. 12–16

12–17. The bar is supported by a roller constraint at B, which allows vertical displacement but resists axial load and moment. If the bar is subjected to the loading shown, determine the slope at A and the deflection at C. EI is constant.

12–18. Determine the deflection at B of the bar in Prob. 12–17.

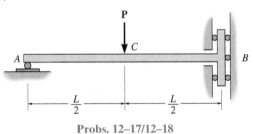

Probs. 12–17/12–18

12–19. Wooden posts used for a retaining wall have a diameter of 3 in. If the soil pressure along a post varies uniformly from zero at the top A to a maximum of 300 lb/ft at the bottom B, determine the slope and displacement at the top of the post. $E_w = 1.6(10^3)$ksi.

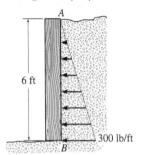

Prob. 12–19

*12–20.   The tapered beam has a rectangular cross section. Determine the deflection of its end in terms of the load **P**, length L, modulus of elasticity E, and the moment of inertia I_0 of its end.

12–22.   The beam is made of a material having a specific weight of γ. Determine the displacement and slope at its end A due to its weight. The modulus of elasticity for the material is E.

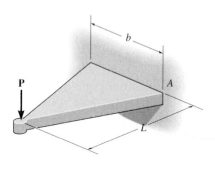

Prob. 12–22

Prob. 12–20

12–23.   The beam is made from a plate that has a constant thickness t and a width that varies linearly. The plate is cut into strips to form a series of leaves that are stacked to make a leaf spring consisting of n leaves. Determine the deflection at its end when loaded. Neglect friction between the leaves.

12–21.   The tapered beam has a rectangular cross section. Determine the deflection of its center in terms of the load P, length L, modulus of elasticity E, and the moment of inertia I_c of its center.

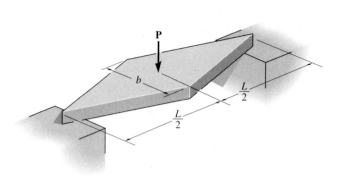

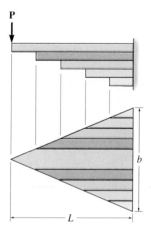

Prob. 12–21

Prob. 12–23

*12.3 DISCONTINUITY FUNCTIONS

The method of integration, used to find the equation of the elastic curve for a beam or shaft, is convenient if the load or internal moment can be expressed as a *continuous function* throughout the beam's *entire length*. If several different loadings act on the beam, however, the method becomes more tedious to apply, because separate loading or moment functions must be written for each region of the beam. Furthermore, integration of these functions requires the evaluation of integration constants using boundary conditions and/or continuity conditions. For example, the beam shown in Fig. 12–14 requires four moment functions to be written. They describe the moment in regions AB, BC, CD, and DE. When applying the moment-curvature relationship, $EI\ d^2v/dx^2 = M$, and integrating each moment equation twice, we must evaluate *eight* constants of integration. These involve *two* boundary conditions that require zero displacement at points A and E, and *six* continuity conditions for both slope and displacement at points B, C, and D.

In this section we will discuss a method for finding the equation of the elastic curve for a *multiply loaded beam* using a *single expression*, either formulated from the loading on the beam, $w = w(x)$, or the beam's internal moment, $M = M(x)$. If the expression for w is substituted into $EI\ d^4v/dx^4 = -w(x)$ and integrated four times, or if the expression for M is substituted into $EI\ d^2v/dx^2 = M(x)$, and integrated twice, the constants of integration will be determined only from the boundary conditions. Since the continuity equations will not be involved, the analysis will be greatly simplified.

Discontinuity Functions. In order to express the load on the beam or the internal moment within it using a single expression, we will use two types of mathematical operators known as *discontinuity functions*.

For safety purposes these cantilevered beams must be designed for both strength and a restricted amount of deflection.

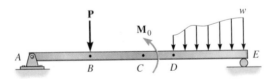

Fig. 12–14

Loading	Loading Function $w=w(x)$	Shear $V=-\int w(x)dx$	Moment $M=\int V dx$
1 M_0	$w = M_0\langle x-a\rangle^{-2}$	$V = -M_0\langle x-a\rangle^{-1}$	$M = -M_0\langle x-a\rangle^{0}$
2 P	$w = P\langle x-a\rangle^{-1}$	$V = -P\langle x-a\rangle^{0}$	$M = -P\langle x-a\rangle^{1}$
3 w	$w = w_0\langle x-a\rangle^{0}$	$V = -w_0\langle x-a\rangle^{1}$	$M = -\dfrac{w_0}{2}\langle x-a\rangle^{2}$
4 slope $= m$	$w = m\langle x-a\rangle^{1}$	$V = \dfrac{-m}{2}\langle x-a\rangle^{2}$	$M = \dfrac{-m}{6}\langle x-a\rangle^{3}$

Table 12–2

Macaulay Functions. For purposes of beam or shaft deflection, Macaulay functions, named after the mathematician W. H. Macaulay, can be used to describe ***distributed loadings***. They can be written in general form as

$$\langle x - a\rangle^{n} = \begin{cases} 0 & \text{for } x < a \\ (x - a)^{n} & \text{for } x \geq a \end{cases} \qquad (12\text{--}11)$$
$$n \geq 0$$

Here x represents the coordinate position of a point along the beam, and a is the location on the beam where a "discontinuity" occurs, namely the point where a distributed loading *begins*. Note that the Macaulay function $\langle x - a\rangle^{n}$ is written with angle brackets to distinguish it from the ordinary function $(x - a)^{n}$, written with parentheses. As stated by the equation, only when $x \geq a$ is $\langle x - a\rangle^{n} = (x - a)^{n}$, otherwise it is zero. Furthermore, these functions are valid only for exponential values $n \geq 0$. Integration of Macaulay functions follows the same rules as for ordinary functions, i.e.,

$$\int \langle x - a\rangle^{n} \, dx = \frac{\langle x - a\rangle^{n+1}}{n + 1} + C \qquad (12\text{--}12)$$

Note how the Macaulay functions describe both the *uniform load* w_0 ($n = 0$) and *triangular load* ($n = 1$), shown in Table 12–2, items 3 and 4. This type of description can, of course, be extended to distributed loadings having other forms. Also, it is possible to use superposition with the uniform and triangular loadings to create the Macaulay function for a trapezoidal loading. Using integration, the Macaulay functions for shear, $V = -\int w(x)\, dx$, and moment, $M = \int V\, dx$, are also shown in the table.

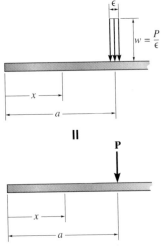

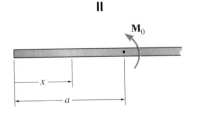

Fig. 12–15

Singularity Functions. These functions are used only to describe the point location of concentrated forces or couple moments acting on a beam or shaft. Specifically, a **concentrated force P** can be considered as a special case of a distributed loading, where the intensity of the loading is $w = P/\epsilon$ such that its width is ϵ, where $\epsilon \to 0$, Fig. 12–15. The area under this loading diagram is equivalent to P, *positive downward*, and so we will use the singularity function

$$w = P\langle x - a \rangle^{-1} = \begin{cases} 0 & \text{for } x \neq a \\ P & \text{for } x = a \end{cases} \qquad (12\text{–}13)$$

to describe the force P. Note that here $n = -1$ so that the units for w are force per length, as it should be. Furthermore, the function takes on the value of **P** only at the point $x = a$ where the load occurs, otherwise it is zero.

In a similar manner, a couple moment $\mathbf{M}_0$, considered *positive counterclockwise*, is a limitation as $\epsilon \to 0$ of two distributed loadings as shown in Fig. 12–16. Here the following function describes its value.

$$w = M_0 \langle x - a \rangle^{-2} = \begin{cases} 0 & \text{for } x \neq a \\ M_0 & \text{for } x = a \end{cases} \qquad (12\text{–}14)$$

The exponent $n = -2$, in order to ensure that the units of w, force per length, are maintained.

Integration of the above two singularity functions follow the rules of operational calculus and yields results that are *different* from those of Macaulay functions. Specifically,

$$\int \langle x - a \rangle^n = \langle x - a \rangle^{n+1}, \, n = -1, -2 \qquad (12\text{–}15)$$

Here, only the exponent n increases by one, and no constant of integration will be associated with this operation. Using this formula, notice how M_0 and P, described in Table 12–2, items 1 and 2, are integrated once, then twice, to obtain the internal shear and moment in the beam.

Application of Eqs. 12–11 through 12–15 provides a rather direct means for expressing the loading or the internal moment in a beam as a function of x. When doing so, close attention must be paid to the signs of the external loadings. As stated above, and as shown in Table 12–2, *concentrated forces and distributed loads are positive downward, and couple moments are positive counterclockwise.* If this sign convention is followed, then the internal shear and moment are in accordance with the beam sign convention established in Sec. 6.1.

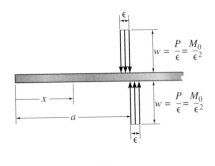

Fig. 12–16

As an example of how to apply discontinuity functions to describe the loading or internal moment in a beam, we will consider the beam loaded as shown in Fig. 12–17a. Here the reactive force $\mathbf{R}_1$ created by the pin, Fig. 12–17b, is negative since it acts upward, and $\mathbf{M}_0$ is negative since it acts clockwise. Using Table 12–2, the loading at any point x on the beam is therefore,

$$w = -R_1\langle x - 0\rangle^{-1} + P\langle x - a\rangle^{-1} - M_0\langle x - b\rangle^{-2} + w_0\langle x - c\rangle^0$$

The reactive force at the roller is not included here since x is never greater than L, and furthermore, this value is of no consequence in computing slope or deflection. Note that when $x = a$, $w = P$, all other terms being zero. Also, when $x > c$, $w = w_0$, etc.

Integrating this equation twice yields the expression that describes the internal moment in the beam. The constants of integration will be ignored here since the boundary conditions, or the end shear and moment, have been calculated ($V = R_1$ and $M = 0$) and these values are incorporated into the beam loading w. One can also obtain this result directly from Table 12–2. In either case,

$$M = R_1\langle x - 0\rangle - P\langle x - a\rangle + M_0\langle x - b\rangle^0 - \tfrac{1}{2}w_0\langle x - c\rangle^2 \quad (12\text{–}16)$$

The validity of this expression may be checked by using the method of sections, say, within the region $b < x < c$, Fig. 12–17b. Moment equilibrium requires that

$$M = R_1 x - P(x - a) + M_0 \quad\quad (12\text{–}17)$$

This result agrees with that obtained from the discontinuity functions, since by Eqs. 12–11, 12–13 and 12–14 only the last term in Eq. 12–16 is zero when $x < c$.

As a second example, consider the beam in Fig. 12–18a. The support reaction at A has been computed in Fig. 12–18b, and the trapezoidal loading has been separated into triangular and uniform loadings. From Table 12–2, the loading is therefore

$$w = -2.75\ \text{kN}\langle x - 0\rangle^{-1} - 1.5\ \text{kN} \cdot \text{m}\langle x - 3\ \text{m}\rangle^{-2} + 3\ \text{kN/m}\langle x - 3\ \text{m}\rangle^0 + 1\ \text{kN/m}^2\langle x - 3\ \text{m}\rangle^1$$

We can determine the moment expression directly from Table 12–2, rather than integrating this expression twice. In either case,

$$M = 2.75\ \text{kN}\langle x - 0\rangle^1 + 1.5\ \text{kN} \cdot \text{m}\langle x - 3\ \text{m}\rangle^0 - \frac{3\ \text{kN/m}}{2}\langle x - 3\ \text{m}\rangle^2 - \frac{1\ \text{kN/m}^2}{6}\langle x - 3\ \text{m}\rangle^3$$

$$= 2.75x + 1.5\langle x - 3\rangle^0 - 1.5\langle x - 3\rangle^2 - \frac{1}{6}\langle x - 3\rangle^3$$

The deflection of the beam can now be determined after this equation is integrated two successive times and the constants of integration are evaluated using the boundary conditions of zero displacement at A and B.

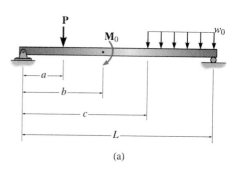

(a)

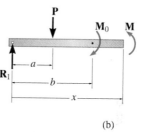

(b)

Fig. 12–17

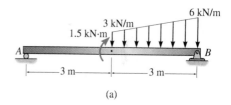

(a)

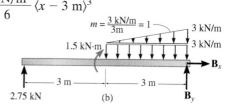

(b)

Fig. 12–18

PROCEDURE FOR ANALYSIS

The following procedure provides a method for using discontinuity functions to determine a beam's elastic curve. This method is particularly advantageous for solving problems involving beams or shafts subjected to *several loadings*, since the constants of integration can be evaluated by using *only* the boundary conditions, while the compatibility conditions are automatically satisfied.

Elastic Curve.

• Sketch the beam's elastic curve and identify the boundary conditions at the supports.

• Zero displacement occurs at all pin and roller supports, and zero slope and zero displacement occur at fixed supports.

• Establish the x axis so that it extends to the right and has its origin at the beam's left end.

Load or Moment Function.

• Calculate the support reactions and then use the discontinuity functions in Table 12–2 to express either the loading w or the internal moment M as a function of x. Make sure to follow the sign convention for each loading as it applies for this equation.

• Note that the distributed loadings must extend all the way to the beam's right end to be valid. If this does not occur, use the method of superposition, which is illustrated in Example 12–5.

Slope and Elastic Curve.

• Substitute w into $EI \, d^4v/dx^4 = -w(x)$ or M into the moment-curvature relation $EI \, d^2v/dx^2 = M$, and integrate to obtain the equations for the beam's slope and deflection.

• Evaluate the constants of integration using the boundary conditions, and substitute these constants into the slope and deflection equations to obtain the final results.

• When the slope and deflection equations are evaluated at any point on the beam, a *positive slope* is *counterclockwise*, and a *positive displacement* is *upward*.

EXAMPLE 12-5

Determine the equation of the elastic curve for the cantilevered beam shown in Fig. 12–19a. EI is constant.

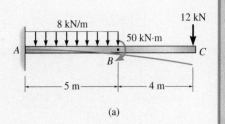

(a)

SOLUTION

Elastic Curve. The loads cause the beam to deflect as shown in Fig. 12–19a. The boundary conditions require zero slope and displacement at A.

Loading Function. The support reactions at A have been calculated by statics and are shown on the free-body diagram in Fig. 12–19b. Since the distributed loading in Fig. 12–19a does not extend to C as required, we can use the superposition of loadings shown in Fig. 12–19b to represent the same effect. By our sign convention, the 50-kN · m couple moment, the 52-kN force at A, and the portion of distributed loading from B to C on the bottom of the beam are all negative. The beam's loading is therefore

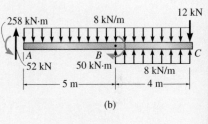

(b)

Fig. 12–19

$$w = -52 \text{ kN}\langle x - 0 \rangle^{-1} + 258 \text{ kN} \cdot \text{m}\langle x - 0 \rangle^{-2} + 8 \text{ kN/m}\langle x - 0 \rangle^{0}$$
$$- 50 \text{ kN} \cdot \text{m}\langle x - 5 \text{ m} \rangle^{-2} - 8 \text{ kN/m}\langle x - 5 \text{ m} \rangle^{0}$$

The 12-kN load is *not included* here, since x cannot be greater than 9 m. Because $dV/dx = -w(x)$, then by integrating, neglecting the constant of integration since the reactions are included in the load function, we have

$$V = 52\langle x - 0 \rangle^{0} - 258\langle x - 0 \rangle^{-1} - 8\langle x - 0 \rangle^{1} + 50\langle x - 5 \rangle^{-1} + 8\langle x - 5 \rangle^{1}$$

Furthermore, $dM/dx = V$, so that integrating again yields

$$M = -258\langle x - 0 \rangle^{0} + 52 \langle x - 0 \rangle^{1} - \frac{1}{2}(8)\langle x - 0 \rangle^{2} + 50\langle x - 5 \rangle^{0} + \frac{1}{2}(8) \langle x - 5 \rangle^{2}$$

$$= (-258 + 52x - 4x^2 + 4\langle x - 5 \rangle^2 + 50\langle x - 5 \rangle^0) \text{ kN} \cdot \text{m}$$

This same result can be obtained *directly* from Table 12–2.

Slope and Elastic Curve. Applying Eq. 12–10 and integrating twice, we have

$$EI \frac{d^2v}{dx^2} = -258 + 52x - 4x^2 + 50\langle x - 5 \rangle^0 + 4\langle x - 5 \rangle^2$$

$$EI \frac{dv}{dx} = -258x + 26x^2 - \frac{4}{3}x^3 + 50\langle x - 5 \rangle^1 + \frac{4}{3}\langle x - 5 \rangle^3 + C_1$$

$$EIv = -129x^2 + \frac{26}{3}x^3 - \frac{1}{3}x^4 + 25\langle x - 5 \rangle^2 + \frac{1}{3}\langle x - 5 \rangle^4 + C_1 x + C_2$$

Since $dv/dx = 0$ at $x = 0$, $C_1 = 0$; and $v = 0$ at $x = 0$, so $C_2 = 0$. Thus,

$$v = \frac{1}{EI}\left(-129x^2 + \frac{26}{3}x^3 - \frac{1}{3}x^4 + 25\langle x - 5 \rangle^2 + \frac{1}{3}\langle x - 5 \rangle^4\right) \text{ m} \qquad Ans.$$

EXAMPLE 12-6

Determine the maximum deflection of the beam shown in Fig. 12–20a. *EI* is constant.

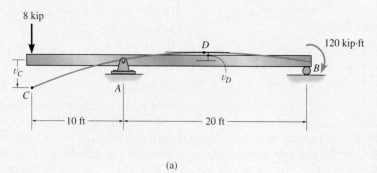

(a)

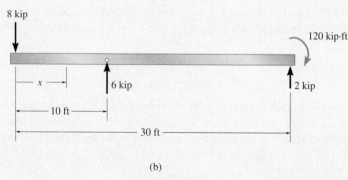

(b)

Fig. 12–20

SOLUTION

Elastic Curve.   The beam deflects as shown in Fig. 12–20a. The boundary conditions require zero displacement at *A* and *B*.

Loading Function.   The reactions have been calculated and are shown on the free-body diagram in Fig. 12–20b. The loading function for the beam can be written as

$$w = 8 \text{ kip}\langle x - 0\rangle^{-1} - 6 \text{ kip}\langle x - 10 \text{ ft}\rangle^{-1}$$

The couple moment and force at *B* are not included here, since they are located at the right end of the beam, and *x* cannot be greater than 30 ft. Applying $dV/dx = -w(x)$, we get

$$V = -8\langle x - 0\rangle^0 + 6\langle x - 10\rangle^0$$

In a similar manner, $dM/dx = V$ yields

$$M = -8\langle x - 0\rangle^1 + 6\langle x - 10\rangle^1$$
$$= (-8x + 6\langle x - 10\rangle^1) \text{ kip} \cdot \text{ft}$$

Notice how this equation can also be established *directly* using the results of Table 12–2 for moment.

Slope and Elastic Curve.   Integrating twice yields

$$EI \frac{d^2v}{dx^2} = -8x + 6\langle x - 10 \rangle^1$$

$$EI \frac{dv}{dx} = -4x^2 + 3\langle x - 10 \rangle^2 + C_1$$

$$EIv = -\frac{4}{3}x^3 + \langle x - 10 \rangle^3 + C_1 x + C_2 \qquad (1)$$

From Eq. 1, the boundary condition $v = 0$ at $x = 10$ ft and $v = 0$ at $x = 30$ ft gives

$$0 = -1333 + (10 - 10)^3 + C_1(10) + C_2$$
$$0 = -36\,000 + (30 - 10)^3 + C_1(30) + C_2$$

Solving these equations simultaneously for C_1 and C_2, we get $C_1 = 1333$ and $C_2 = -12\,000$. Thus,

$$EI \frac{dv}{dx} = -4x^2 + 3\langle x - 10 \rangle^2 + 1333 \qquad (2)$$

$$EIv = -\frac{4}{3}x^3 + \langle x - 10 \rangle^3 + 1333x - 12\,000 \qquad (3)$$

From Fig. 12–20a, maximum displacement may occur at either C or D, where the slope $dv/dx = 0$. To obtain the displacement of C, set $x = 0$ in Eq. 3. We get

$$v_C = -\frac{12\,000 \text{ kip} \cdot \text{ft}^3}{EI}$$

The *negative* sign indicates that the displacement is *downward* as shown in Fig. 12–20a. To locate point D, use Eq. 2 with $x > 10$ ft and $dv/dx = 0$. This gives

$$0 = -4x_D^2 + 3(x_D - 10)^2 + 1333$$
$$x_D^2 + 60x_D - 1633 = 0$$

Solving for the positive root,

$$x_D = 20.3 \text{ ft}$$

Hence, from Eq. 3,

$$EIv_D = -\frac{4}{3}(20.3)^3 + (20.3 - 10)^3 + 1333(20.3) - 12\,000$$

$$v_D = \frac{5000 \text{ kip} \cdot \text{ft}^3}{EI} \qquad\qquad\qquad Ans.$$

Comparing this value with v_C, we see that $v_{\max} = v_C$.

PROBLEMS

*12–24. The shaft supports the two pulley loads shown. Determine the slope of the shaft at the bearings A and B. The bearings exert only vertical reactions on the shaft. EI is constant.

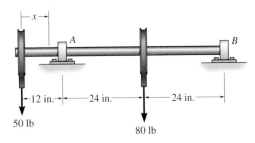

Prob. 12–24

12–25. The beam is subjected to the load shown. Determine the equation of the elastic curve. EI is constant.

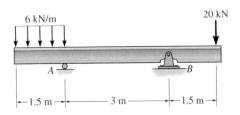

Prob. 12–25

12–26. The shaft is made of 304 stainless steel and has a diameter of 15 mm. Determine its maximum deflection. The bearings at A and B exert only vertical reactions on the shaft.

12–27. The shaft is made of 304 stainless steel and has a diameter of 15 mm. Determine the equation of the elastic curve and find the slopes at the bearings A and B. The bearings exert only vertical reactions on the shaft.

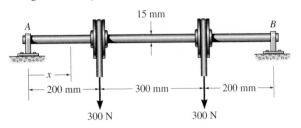

Probs. 12–26/12–27

*12–28. The beam is subjected to the load shown. Determine the equation of the elastic curve. EI is constant.

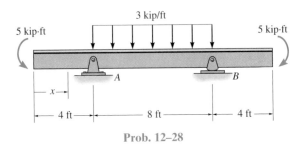

Prob. 12–28

12–29. The shaft supports the three pulley loads shown. Determine the equation of the elastic curve. The bearings at A and B exert only vertical reactions on the shaft. What is the maximum deflection? EI is constant.

12–30. The shaft supports the three pulley loads shown. Determine the deflection of the shaft at its center and its slopes at the bearings A and B. The bearings exert only vertical reactions on the shaft. EI is constant.

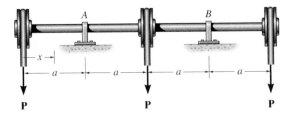

Probs. 12–29/12–30

12–31. The beam is subjected to the load shown. Determine the equation of the elastic curve. EI is constant.

*12–32. The beam is subjected to the load shown. Determine the displacement at $x = 7$ m and the slope at A. EI is constant.

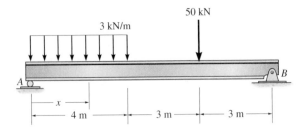

Probs. 12–31/12–32

12–33. The beam is subjected to the load shown. Determine the equations of the slope and elastic curve. EI is constant.

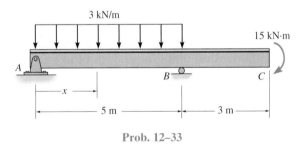

3 kN/m

15 kN·m

A B C

x

5 m   3 m

Prob. 12–33

12–34. The beam is subjected to the load shown. Determine the equation of the elastic curve. EI is constant.

12–35. The beam is subjected to the load shown. Determine the slopes at A and B and the displacement at C. EI is constant.

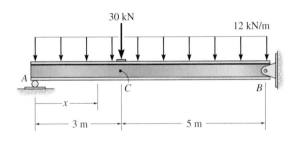

30 kN

12 kN/m

A C B

x

3 m   5 m

Probs. 12–34/12–35

*12–36. Determine the equation of the elastic curve. Specify the slope at A and the displacement at C. EI is constant.

12–37. Determine the equation of the elastic curve. Specify the slopes at A and B. EI is constant.

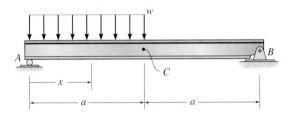

w

A C B

x

a a

Probs. 12–36/12–37

12–38. The beam is subjected to the load shown. Determine the equation of the elastic curve. EI is constant.

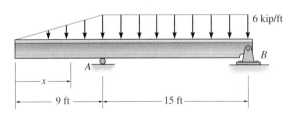

6 kip/ft

A B

x

9 ft   15 ft

Prob. 12–38

12–39. The wooden beam is subjected to the load shown. Determine the equation of the elastic curve. If $E_w = 12$ GPa, determine the deflection and the slope at end B.

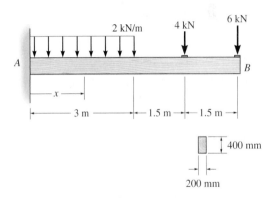

2 kN/m   4 kN   6 kN

A B

x

3 m   1.5 m   1.5 m

400 mm

200 mm

Prob. 12–39

*12–40. The beam is subjected to the load shown. Determine the equation of the elastic curve.

12–41. Determine the displacement at C and the slope at A of the beam.

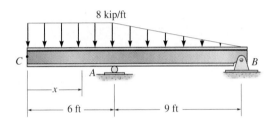

8 kip/ft

C A B

x

6 ft   9 ft

Probs. 12–40/12–41

*12.4 SLOPE AND DISPLACEMENT BY THE MOMENT-AREA METHOD

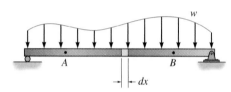

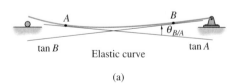

tan B

Elastic curve

(a)

The moment-area method provides a semigraphical technique for finding the slope and displacement at *specific points* on the elastic curve of a beam or shaft. Application of the method requires computing areas associated with the beam's moment diagram; so if this diagram consists of simple shapes, the method is very convenient to use. Normally this is the case when the beam is loaded with concentrated forces and couple moments.

To develop the moment-area method we will make the same assumptions as we used for the method of integration: The beam is initially straight, it is elastically deformed by the loads, such that the slope and deflection of the elastic curve are very small, and the deformations are caused by bending. The moment-area method is based on two theorems used to determine the slope and displacement at a point on the elastic curve.

Theorem 1. Consider the simply supported beam with its associated elastic curve, Fig. 12–21a. A differential segment dx of the beam is isolated in Fig. 12–21b. It is seen that the beam's internal moment M deforms the element such that the *tangents* to the elastic curve at each side of the element intersect at an angle $d\theta$. This angle can be determined from Eq. 12–10, written as

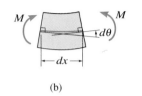

(b)

$$EI \frac{d^2v}{dx^2} = EI \frac{d}{dx}\left(\frac{dv}{dx}\right) = M$$

Since the *slope* is *small*, $\theta = dv/dx$, and therefore

$$d\theta = \frac{M}{EI} dx \tag{12-18}$$

If the moment diagram for the beam is constructed and divided by both the beam's moment of inertia I and modulus of elasticity E, Fig. 12–21c, then Eq. 12–18 indicates that $d\theta$ is equal to the *area* under the "M/EI diagram" for the beam segment dx. Integrating from a selected point A on the elastic curve to another point B, we have

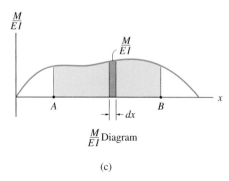

$\frac{M}{EI}$ Diagram

(c)

Fig. 12–21

$$\theta_{B/A} = \int_A^B \frac{M}{EI} dx \tag{12-19}$$

This equation forms the basis for the first moment-area theorem.

Theorem 1: *The angle between the tangents at any two points on the elastic curve equals the area under the M/EI diagram between these two points.*

The notation $\theta_{B/A}$ is referred to as the angle of the tangent at B measured *with respect to* the tangent at A. From the proof it should be evident that this angle is measured *counterclockwise*, from tangent A to tangent B, if the area under the M/EI diagram is *positive*. Conversely, if the area is *negative*, or lies below the x axis, the angle $\theta_{B/A}$ is measured clockwise from tangent A to tangent B. Furthermore, from the dimensions of Eq. 12–19, $\theta_{B/A}$ will be *measured* in *radians*.

Theorem 2.   The second moment-area theorem is based on the relative deviation of tangents to the elastic curve. Shown in Fig. 12–22a is a greatly exaggerated view of the vertical deviation dt of the tangents on each side of the differential element dx. This deviation is caused by the curvature of the element and has been measured along a vertical line passing through point A located on the elastic curve. Since the slope of the elastic curve and its deflection are assumed to be very small, it is satisfactory to approximate the length of each tangent line by x and the arc ds' by dt. Using the circular-arc formula $s = \theta r$, where r is the length x and s is dt, we can write $dt = x\, d\theta$. Substituting Eq. 12–18 into this equation and integrating from A to B, the vertical deviation of the tangent at A *with respect to* the tangent at B can then be determined; i.e.,

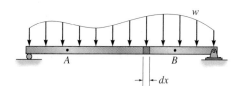

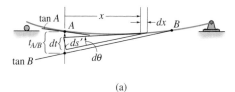

$$t_{A/B} = \int_A^B x\, \frac{M}{EI}\, dx \qquad (12\text{–}20)$$

Since the centroid of an area is found from $\bar{x} \int dA = \int x\, dA$, and $\int (M/EI)dx$ represents the area under the M/EI diagram, we can also write

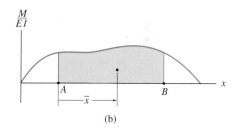

$$t_{A/B} = \bar{x} \int_A^B \frac{M}{EI}\, dx \qquad (12\text{–}21)$$

Here $\bar{x}$ is the distance from A to the *centroid* of the area under the M/EI diagram between A and B, Fig. 12–22b.

The second moment-area theorem can now be stated as follows:

Theorem 2: *The vertical deviation of the tangent at a point (A) on the elastic curve with respect to the tangent extended from another point (B) equals the moment of the area under the M/EI diagram between these two points (A and B). This moment is computed about point (A) where the vertical deviation ($t_{A/B}$) is to be determined.*

The distance $t_{A/B}$ used in the theorem can also be interpreted as the vertical displacement from the point located on the extended tangent drawn from B to the point A on the elastic curve. Note that $t_{A/B}$ is *not* equal to $t_{B/A}$, which is shown in Fig. 12–22c. Specifically, the moment of the area under the M/EI diagram between A and B is computed about point A to determine $t_{A/B}$, Fig. 12–22b, and it is computed about point B to determine $t_{B/A}$, Fig. 12–22c.

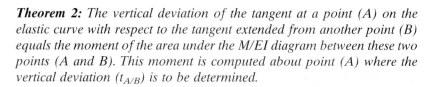

If the moment of a *positive M/EI* area from A to B is found for $t_{B/A}$, it indicates that point B is *above* the tangent extended from point A, Fig. 12–22a. Similarly, *negative M/EI* areas indicate that point B is *below* the tangent extended from point A. This same rule applies for $t_{A/B}$.

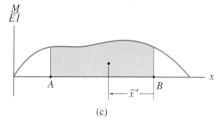

Fig. 12–22

PROCEDURE FOR ANALYSIS

The following procedure provides a method that may be used to apply the two moment-area theorems.

M/EI Diagram.

• Determine the support reactions and draw the beam's M/EI diagram. If the beam is loaded with concentrated forces, the M/EI diagram will consist of a series of straight line segments, and the areas and their moments required for the moment-area theorems will be relatively easy to compute. If the loading consists of a series of distributed loads, the M/EI diagram will consist of parabolic or perhaps higher-order curves, and it is suggested that the table on the inside front cover be used to locate the area and centroid under each curve.

Elastic Curve.

• Draw an exaggerated view of the beam's elastic curve. Recall that points of zero slope and zero displacement always occur at a fixed support, and zero displacement occurs at all pin and roller supports.

• If it becomes difficult to draw the general shape of the elastic curve, use the moment (or M/EI) diagram. Realize that when the beam is subjected to a *positive moment*, the beam bends *concave up*, whereas *negative moment* bends the beam *concave down*. Furthermore, an inflection point or change in curvature occurs where the moment in the beam (or M/EI) is zero.

• The unknown displacement and slope to be determined should be indicated on the curve.

• Since the moment-area theorems apply *only between two tangents*, attention should be given as to which tangents should be constructed so that the angles or deviations between them will lead to the solution of the problem. In this regard, *the tangents at the supports should be considered*, since the beam usually has zero displacement and/or zero slope at the supports.

Moment-Area Theorems.

• Apply Theorem 1 to determine the *angle* between any two tangents on the elastic curve and Theorem 2 to determine the *tangential deviation*.

• The algebraic sign of the answer can be checked from the angle or deviation indicated on the elastic curve.

• A *positive* $\theta_{B/A}$ represents a *counterclockwise* rotation of the tangent at B with respect to the tangent at A, and a *positive* $t_{B/A}$ indicates that point B on the elastic curve lies *above* the extended tangent from point A.

EXAMPLE 12–7

Determine the slope of the beam shown in Fig. 12–23a at points B and C. EI is constant.

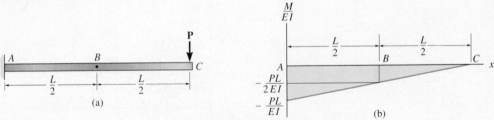

Fig. 12–23

SOLUTION

M/EI Diagram.   See Fig. 12–23b.

Elastic Curve.   The force **P** causes the beam to deflect as shown in Fig. 12–23c. (The elastic curve is concave downward, since M/EI is negative.) The tangents at B and C are indicated since we are required to find θ_B and θ_C. Also, the tangent at the support (A) is shown. This tangent has a *known* zero slope. By the construction, the angle between tan A and tan B, that is, $\theta_{B/A}$, is equivalent to θ_B, or

$$\theta_B = \theta_{B/A}$$

Also

$$\theta_C = \theta_{C/A}$$

Moment-Area Theorem.   Applying Theorem 1, $\theta_{B/A}$ is equal to the area under the M/EI diagram between points A and B; that is,

$$\theta_B = \theta_{B/A} = \left(-\frac{PL}{2EI}\right)\left(\frac{L}{2}\right) + \frac{1}{2}\left(-\frac{PL}{2EI}\right)\left(\frac{L}{2}\right)$$

$$= -\frac{3PL^2}{8EI} \qquad \text{Ans.}$$

The *negative sign* indicates that the angle measured from the tangent at A to the tangent at B is *clockwise*. This checks, since the beam slopes downward at B.

   In a similar manner, the area under the M/EI diagram between points A and C equals $\theta_{C/A}$. We have

$$\theta_C = \theta_{C/A} = \frac{1}{2}\left(-\frac{PL}{EI}\right)L$$

$$= -\frac{PL^2}{2EI} \qquad \text{Ans.}$$

EXAMPLE 12–8

Determine the displacement of points B and C of the beam shown in Fig. 12–24a. EI is constant.

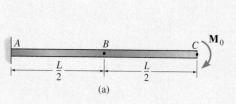

(a)

Fig. 12–24

(b)

SOLUTION

M/EI Diagram. See Fig. 12–24b.

Elastic Curve. The couple moment at C causes the beam to deflect as shown in Fig. 12–24c. The tangents at B and C are indicated since we are required to find Δ_B and Δ_C. Also, the tangent at the support (A) is shown since it is horizontal. The required displacements can now be related directly to the deviations between the tangents at B and A and C and A. Specifically, Δ_B is equal to the deviation of tan A from tan B; that is,

$$\Delta_B = t_{B/A}$$

$$\Delta_C = t_{C/A}$$

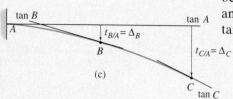

(c)

Moment-Area Theorem. Applying Theorem 2, $t_{B/A}$ is equal to the moment of the shaded area under the M/EI diagram between A and B computed about point B (the point on the elastic curve), since this is the point where the tangential deviation is to be determined. Hence, from Fig. 12–24b,

$$\Delta_B = t_{B/A} = \left(\frac{L}{4}\right)\left[\left(-\frac{M_0}{EI}\right)\left(\frac{L}{2}\right)\right] = -\frac{M_0 L^2}{8EI} \qquad \textit{Ans.}$$

Likewise, for $t_{C/A}$ we must determine the moment of the area under the *entire M/EI* diagram from A to C about point C (the point on the elastic curve). We have

$$\Delta_C = t_{C/A} = \left(\frac{L}{2}\right)\left[\left(-\frac{M_0}{EI}\right)(L)\right] = -\frac{M_0 L^2}{2EI} \qquad \textit{Ans.}$$

Since both answers are *negative*, they indicate that points B and C lie *below* the tangent at A. This checks with Fig. 12–24c.

EXAMPLE 12–9

Determine the slope at point C of the beam in Fig. 12–25a. EI is constant.

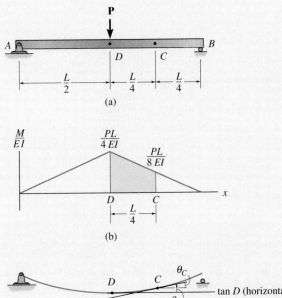

(a)

(b)

(c)

Fig. 12–25

SOLUTION

M/EI Diagram.   See Fig. 12–25b.

Elastic Curve.   Since the loading is applied symmetrically to the beam, the elastic curve is symmetric, and the tangent at D is horizontal, Fig. 12–25c. Also the tangent at C is drawn, since we must find the slope θ_C. By the construction, the angle $\theta_{C/D}$ between tan D and tan C is equal to θ_C; that is,

$$\theta_C = \theta_{C/D}$$

Moment-Area Theorem.   Using Theorem 1, $\theta_{C/D}$ is equal to the shaded area under the M/EI diagram between points D and C. We have

$$\theta_C = \theta_{C/D} = \left(\frac{PL}{8EI}\right)\left(\frac{L}{4}\right) + \frac{1}{2}\left(\frac{PL}{4EI} - \frac{PL}{8EI}\right)\left(\frac{L}{4}\right) = \frac{3PL^2}{64EI} \qquad Ans.$$

What does the positive result indicate?

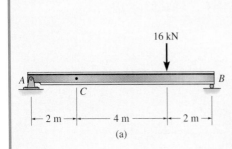

16 kN

2 m | 4 m | 2 m

(a)

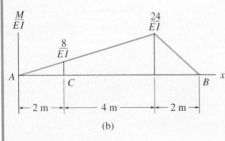

$\dfrac{M}{EI}$

$\dfrac{24}{EI}$

$\dfrac{8}{EI}$

A   C   B   x

2 m | 4 m | 2 m

(b)

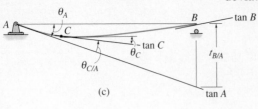

θ_A

θ_C

$\theta_{C/A}$

tan B

tan C

$t_{B/A}$

tan A

(c)

Fig. 12–26

EXAMPLE 12-10

Determine the slope at point C for the steel beam in Fig. 12–26a. Take $E_{st} = 200$ GPa, $I = 17(10^6)$ mm^4.

SOLUTION

M/EI Diagram.   See Fig. 12–26b.

Elastic Curve.   The elastic curve is shown in Fig. 12–26c. The tangent at C is shown since we are required to find θ_C. Tangents at the *supports*, A and B, are also constructed as shown. Angle $\theta_{C/A}$ is the angle between the tangents at A and C. The slope at A, θ_A, in Fig. 12–26c can be found using $|\theta_A| = |t_{B/A}|/L_{AB}$. This equation is valid since $t_{B/A}$ is actually very small, so that θ_A in radians can be approximated by the length of a circular arc defined by a radius of $L_{AB} = 8$ m and a sweep of θ_A. (Recall that $s = \theta r$.) From the geometry of Fig. 12–26c, we have

$$|\theta_C| = |\theta_A| - |\theta_{C/A}| = \left|\frac{t_{B/A}}{8}\right| - |\theta_{C/A}| \qquad (1)$$

Note that Example 12–9 could also be solved using this method.

Moment-Area Theorems.   Using Theorem 1, $\theta_{C/A}$ is equivalent to the area under the *M/EI* diagram between points A and C; that is,

$$\theta_{C/A} = \frac{1}{2}(2 \text{ m})\left(\frac{8 \text{ kN} \cdot \text{m}}{EI}\right) = \frac{8 \text{ kN} \cdot \text{m}^2}{EI}$$

Applying Theorem 2, $t_{B/A}$ is equivalent to the moment of the area under the *M/EI* diagram between B and A about point B (the point on the elastic curve), since this is the point where the tangential deviation is to be determined. We have

$$t_{B/A} = \left(2 \text{ m} + \frac{1}{3}(6 \text{ m})\right)\left[\frac{1}{2}(6 \text{ m})\left(\frac{24 \text{ kN} \cdot \text{m}}{EI}\right)\right]$$
$$+ \left(\frac{2}{3}(2 \text{ m})\right)\left[\frac{1}{2}(2 \text{ m})\left(\frac{24 \text{ kN} \cdot \text{m}}{EI}\right)\right]$$
$$= \frac{320 \text{ kN} \cdot \text{m}^3}{EI}$$

Substituting these results into Eq. 1, we get

$$\theta_C = \frac{320 \text{ kN} \cdot \text{m}^2}{(8\text{m})EI} - \frac{8 \text{ kN} \cdot \text{m}^2}{EI} = \frac{32 \text{ kN} \cdot \text{m}^2}{EI} \, \downarrow$$

We have calculated this result in units of kN and m, so converting EI into these units, we have

$$\theta_C = \frac{32 \text{ kN} \cdot \text{m}^2}{200(10^6) \text{ kN/m}^2 \, 17(10^{-6}) \text{ m}^4} = 0.00941 \text{ rad } \downarrow \qquad \textit{Ans.}$$

EXAMPLE 12–11

Determine the displacement at C for the beam shown in Fig. 12–27a. EI is constant.

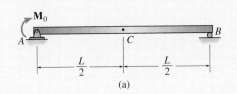

(a)

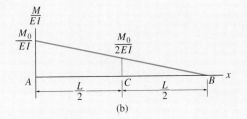

(b)

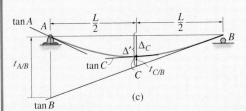

(c)

Fig. 12–27

SOLUTION

M/EI Diagram.   See Fig. 12–27b.

Elastic Curve.   The tangent at C is drawn on the elastic curve since we are required to find Δ_C, Fig. 12–27c. (Note that C is *not* the location of the maximum deflection of the beam, because the loading and hence the elastic curve are *not symmetric.*) Also indicated in Fig. 12–27c are the tangents at the supports A and B. It is seen that $\Delta_C = \Delta' - t_{C/B}$. If $t_{A/B}$ is determined, then Δ' can be found from proportional triangles, that is, $\Delta'/(L/2) = t_{A/B}/L$ or $\Delta' = t_{A/B}/2$. Hence,

$$\Delta_C = \frac{t_{A/B}}{2} - t_{C/B} \qquad (1)$$

Moment-Area Theorem.   Applying Theorem 2 to determine $t_{A/B}$ and $t_{C/B}$, we have

$$t_{A/B} = \left(\frac{1}{3}(L)\right)\left[\frac{1}{2}(L)\left(\frac{M_0}{EI}\right)\right] = \frac{M_0 L^2}{6EI}$$

$$t_{C/B} = \left(\frac{1}{3}\left(\frac{L}{2}\right)\right)\left[\frac{1}{2}\left(\frac{L}{2}\right)\left(\frac{M_0}{2EI}\right)\right] = \frac{M_0 L^2}{48EI}$$

Substituting these results into Eq. 1 gives

$$\Delta_C = \frac{1}{2}\left(\frac{M_0 L^2}{6EI}\right) - \left(\frac{M_0 L^2}{48EI}\right)$$

$$= \frac{M_0 L^2}{16EI} \downarrow \qquad\qquad\qquad Ans.$$

EXAMPLE 12–12

Determine the displacement at point C for the steel overhanging beam shown in Fig. 12–28a. Take $E_{st} = 29(10^3)$ ksi, $I = 125$ in^4.

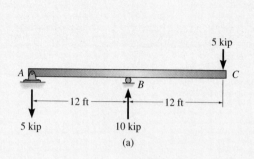

(a)

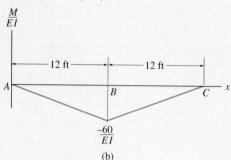

Fig. 12–28

(b)

SOLUTION

M/EI Diagram.   See Fig. 12–28b.

Elastic Curve.   The loading causes the beam to deflect as shown in Fig. 12–28c. We are required to find Δ_C. By constructing tangents at C and at the supports A and B, it is seen that $\Delta_C = |t_{C/A}| - \Delta'$. However, Δ' can be related to $t_{B/A}$ by proportional triangles; that is, $\Delta'/24 = |t_{B/A}|/12$ or $\Delta' = 2|t_{B/A}|$. Hence

$$\Delta_C = |t_{C/A}| - 2|t_{B/A}| \qquad (1)$$

Moment-Area Theorem.   Applying Theorem 2 to determine $t_{C/A}$ and $t_{B/A}$, we have

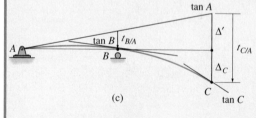

(c)

$$t_{C/A} = (12 \text{ ft})\left(\frac{1}{2}(24 \text{ ft})\left(-\frac{60 \text{ kip} \cdot \text{ft}}{EI}\right)\right)$$

$$= -\frac{8640 \text{ kip} \cdot \text{ft}^3}{EI}$$

$$t_{B/A} = \left(\frac{1}{3}(12 \text{ ft})\right)\left[\frac{1}{2}(12 \text{ ft})\left(-\frac{60 \text{ kip} \cdot \text{ft}}{EI}\right)\right] = -\frac{1440 \text{ kip} \cdot \text{ft}^3}{EI}$$

Why are these terms negative? Substituting the results into Eq. 1 yields

$$\Delta_C = \frac{8640 \text{ kip} \cdot \text{ft}^3}{EI} - 2\left(\frac{1440 \text{ kip} \cdot \text{ft}^3}{EI}\right) = \frac{5760 \text{ kip} \cdot \text{ft}^3}{EI} \downarrow$$

Realizing that the computations were made in units of kip and ft, we have

$$\Delta_C = \frac{5760 \text{ kip} \cdot \text{ft}^3 (1728 \text{ in.}^3/\text{ft}^3)}{[29(10^3) \text{ kip/in.}^2](125 \text{ in.}^4)} = 2.75 \text{ in.} \downarrow \qquad Ans.$$

PROBLEMS

12–42. Determine the slope and deflection at *C*. *EI* is constant.

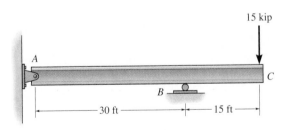

Prob. 12–42

12–43. If the bearings exert only vertical reactions on the shaft, determine the slope at the bearings and the maximum deflection of the shaft. *EI* is constant.

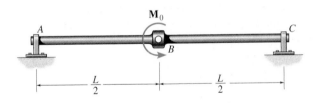

Prob. 12–43

*12–44. Determine the slope and deflection at *B* if the A-36 steel beam is (*a*) a solid rod having a diameter of 3 in., (*b*) a tube having an outer diameter of 3 in. and thickness of 0.25 in.

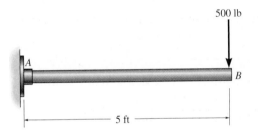

Prob. 12–44

12–45. Determine the slope at *B* and the deflection at *C*. *EI* is constant.

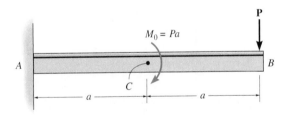

Prob. 12–45

12–46. Determine the slope at *C* and the deflection at *B*. *EI* is constant.

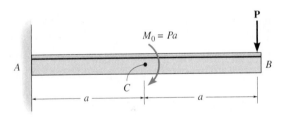

Prob. 12–46

12–47. The shaft is subjected to the loading shown. If the bearings at *A* and *B* only exert vertical reactions on the shaft, determine the slope at *A* and the displacement at *C*. *EI* is constant.

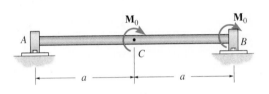

Prob. 12–47

*12–48. The beam is subjected to the loading shown. Determine the slope at *A* and the displacement at *C*. Assume the support at *A* is a pin and *B* is a roller. *EI* is constant.

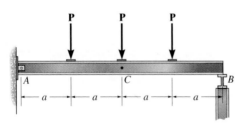

Prob. 12–48

12–49. The rod is constructed from two shafts for which the moment of inertia of *AB* is *I* and of *BC* is *2I*. Determine the maximum slope and deflection of the rod due to the loading. The modulus of elasticity is *E*.

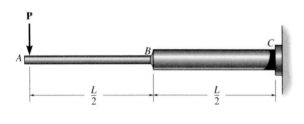

Prob. 12–49

12–50. Determine the value of *a* so that the slope at *A* is equal to zero. *EI* is constant.

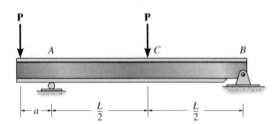

Prob. 12–50

12–51. Determine the deflection at *C* and the slope of the beam at *A*, *B*, and *C*. *EI* is constant.

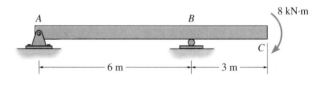

Prob. 12–51

*12–52. The beam is made of a ceramic material. In order to obtain its modulus of elasticity, it is subjected to the elastic loading shown. If the moment of inertia is *I* and the beam has a measured maximum deflection Δ, determine *E*. The supports at *A* and *D* exert only vertical reactions on the beam.

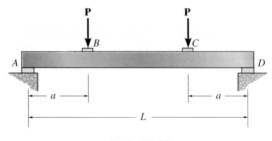

Prob. 12–52

12–53. The beam is subjected to the two loads. Determine the slope and displacement at points *A* and *B*. *EI* is constant.

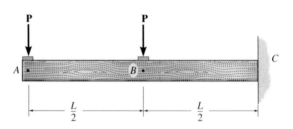

Prob. 12–53

12–54. At what distance a should the bearing supports at A and B be placed so that the deflection at the center of the shaft is equal to the deflection at its ends? The bearings exert only vertical reactions on the shaft. EI is constant.

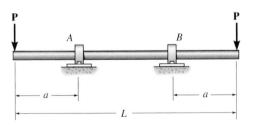

Prob. 12–54

12–55. Determine the maximum deflection of the 50-mm-diameter A-36 steel shaft. It is supported by bearings at its ends A and B which only exert vertical reactions on the shaft.

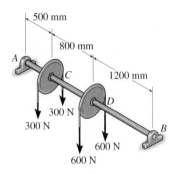

Prob. 12–55

*12–56. Determine the slope of the 50-mm-diameter A-36 steel shaft at the bearings at A and B. The bearings exert only vertical reactions on the shaft.

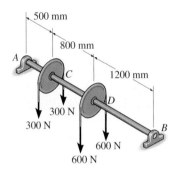

Prob. 12–56

12–57. Determine the maximum deflection of the shaft. EI is constant. The bearings exert only vertical reactions on the shaft.

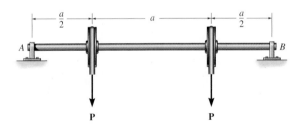

Prob. 12–57

12–58. Determine the displacement of the 20-mm-diameter A-36 steel shaft at the pulley D. The bearings at A and B exert only vertical reactions on the shaft.

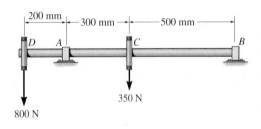

Prob. 12–58

12–59. Determine the slope at B and deflection at C. EI is constant.

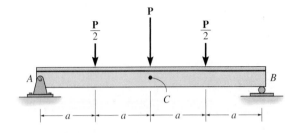

Prob. 12–59

*12–60. The simply supported shaft has a moment of inertia of $2I$ for region BC and a moment of inertia I for regions AB and CD. Determine the maximum deflection of the shaft due to the load **P**. The modulus of elasticity is E.

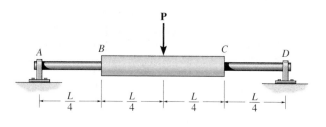

Prob. 12–60

12–61. The beam is subjected to the loading shown. Determine the slope at B and deflection at C. EI is constant.

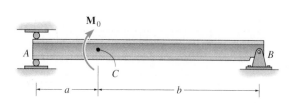

Prob. 12–61

12–62. Determine the slope at B and the displacement at C. The bearings at A and B exert only vertical reactions on the shaft. EI is constant.

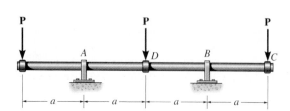

Prob. 12–62

12–63. A beam having a constant EI is supported as shown. Attached to the beam at A is a pointer, free of load. Both the beam and pointer are originally horizontal when no load is applied to the beam. Determine the distance between the end of the beam and the pointer after each has been displaced by the loading shown.

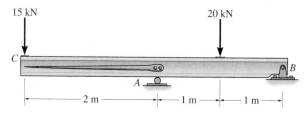

Prob. 12–63

*12–64. Determine the slope of the shaft at A and the displacement at D. The bearings at A and B exert only vertical reactions on the shaft. EI is constant.

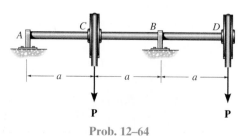

Prob. 12–64

12–65. The beam is subjected to the load **P** as shown. Determine the magnitude of force **F** that must be applied at the end of the overhang C so that the displacement at C is zero. EI is constant.

12–66. The beam is subjected to the load **P** as shown. If $\mathbf{F} = \mathbf{P}$, determine the displacement at D. EI is constant.

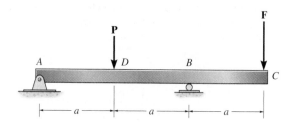

Probs. 12–65/12–66

12–67. Determine the displacement at C, D, and E. The bearings at A and B exert only vertical reactions on the shaft. EI is constant.

*12–68. Determine the slope of the shaft at the bearings at A and B, which exert only vertical reactions on the shaft. EI is constant.

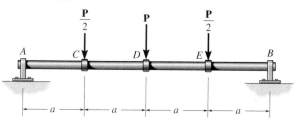

Probs. 12–67/12–68

12–69. The two A-36 steel bars have a thickness of 1 in. and a width of 4 in. They are designed to act as a spring for the machine which exerts a force of 4 kip on them at A and B. If the supports exert only vertical forces on the bars, determine the maximum deflection of the bottom bar.

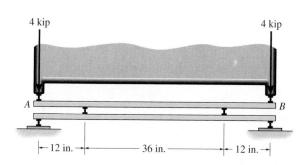

Prob. 12–69

12–70. The cantilevered beam is subjected to the loading shown. Determine the slope and displacement at C. Assume the support at A is fixed. EI is constant.

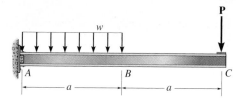

Prob. 12–70

12–71. The A-36 steel shaft is used to support a rotor that exerts a uniform load of 5 kN/m within the region CD of the shaft. Determine the slope of the shaft at the bearings A and B. The bearings exert only vertical reactions on the shaft.

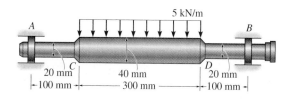

Prob. 12–71

*12–72. Determine the slope at B and the displacement at C. The member is an A-36 steel structural tee for which $I = 76.8$ in[4].

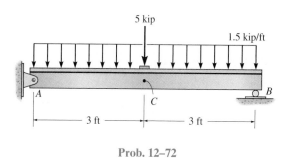

Prob. 12–72

12–73. Determine the slope at C and displacement at B. EI is constant.

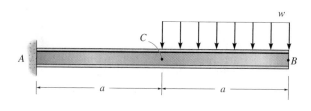

Prob. 12–73

12.5   METHOD OF SUPERPOSITION

The differential equation $EId^4v/dx^4 = -w(x)$ satisfies the two necessary requirements for applying the principle of superposition; i.e., the load $w(x)$ is linearly related to the deflection $v(x)$, and the load is assumed not to change significantly the original geometry of the beam or shaft. As a result, the deflections for a series of separate loadings acting on a beam may be superimposed. For example, if v_1 is the deflection for one load and v_2 is the deflection for another load, the total deflection for both loads acting together is the algebraic sum $v_1 + v_2$. Using tabulated results for various beam loadings, such as the ones listed in Appendix C, or those found in various engineering handbooks, it is therefore possible to find the slope and displacement at a point on a beam subjected to several different loadings by algebraically adding the effects of its various component parts.

The following examples illustrate how to use the method of superposition to solve deflection problems, where the deflection is caused not only by beam deformations, but also by rigid-body displacements, which can occur when the beam is supported by springs or portions of a segmented beam are supported by hinges.

The resultant deflection at any point on this beam can be determined from the superposition of the deflections caused by each of the separate loadings acting on the beam.

EXAMPLE 12–13

Determine the displacement at point C and the slope at the support A of the beam shown in Fig. 12–29a. EI is constant.

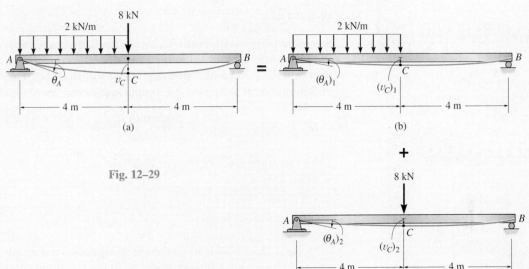

Fig. 12–29

SOLUTION

The loading can be separated into two component parts as shown in Figs. 12–29b and 12–29c. The displacement at C and slope at A are found using the table in Appendix C for each part.

For the distributed loading,

$$(\theta_A)_1 = \frac{3wL^3}{128EI} = \frac{3(2\ \text{kN/m})(8\ \text{m})^3}{128EI} = \frac{24\ \text{kN} \cdot \text{m}^2}{EI}\ \downarrow$$

$$(v_C)_1 = \frac{5wL^4}{768EI} = \frac{5(2\ \text{kN/m})(8\ \text{m})^4}{768EI} = \frac{53.33\ \text{kN} \cdot \text{m}^3}{EI}\ \downarrow$$

For the 8-kN concentrated force,

$$(\theta_A)_2 = \frac{PL^2}{16EI} = \frac{8\ \text{kN}(8\ \text{m})^2}{16EI} = \frac{32\ \text{kN} \cdot \text{m}^2}{EI}\ \downarrow$$

$$(v_C)_2 = \frac{PL^3}{48EI} = \frac{8\ \text{kN}(8\ \text{m})^3}{48EI} = \frac{85.33\ \text{kN} \cdot \text{m}^3}{EI}\ \downarrow$$

The total displacement at C and the slope at A are the algebraic sums of these components. Hence

$(^+\!\downarrow)$ $\theta_A = (\theta_A)_1 + (\theta_A)_2 = \dfrac{56\ \text{kN} \cdot \text{m}^2}{EI}\ \downarrow$ *Ans.*

$(^+\!\downarrow)$ $v_C = (v_C)_1 + (v_C)_2 = \dfrac{139\ \text{kN} \cdot \text{m}^3}{EI}\ \downarrow$ *Ans.*

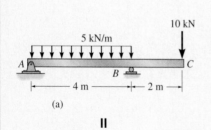

(a)

=

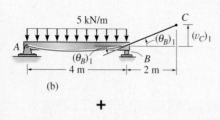

(b)

+

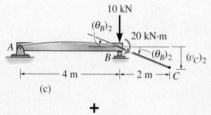

(c)

+

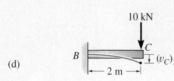

(d)

Fig. 12–30

EXAMPLE 12–14

Determine the displacement at the end C of the overhanging beam shown in Fig. 12–30a. EI is constant.

SOLUTION

Since the table in Appendix C *does not* include beams with overhangs, the beam will be separated into a simply supported and a cantilevered portion. First we will calculate the slope at B, as caused by the distributed load acting on the simply supported span, Fig. 12–30b.

$$(\theta_B)_1 = \frac{wL^3}{24EI} = \frac{5 \text{ kN/m}(4 \text{ m})^3}{24EI} = \frac{13.33 \text{ kN} \cdot \text{m}^2}{EI} \; \text{↖}$$

Since this angle is *small*, $(\theta_B)_1 \approx \tan(\theta_B)_1$, and the vertical displacement at point C is

$$(v_C)_1 = (2 \text{ m})\left(\frac{13.33 \text{ kN} \cdot \text{m}^2}{EI}\right) = \frac{26.67 \text{ kN} \cdot \text{m}^3}{EI} \; \text{↑}$$

Next, the 10-kN load on the overhang causes a statically equivalent force of 10 kN and couple moment of 20 kN · m at the support B of the simply supported span, Fig. 12–30c. The 10-kN force does not cause a displacement or slope at B; however, the 20-kN · m couple moment does cause a slope. The slope at B due to this moment is

$$(\theta_B)_2 = \frac{M_0 L}{3EI} = \frac{20 \text{ kN} \cdot \text{m}(4 \text{ m})}{3EI} = \frac{26.67 \text{ kN} \cdot \text{m}^2}{EI} \; \text{↘}$$

So that the extended point C is displaced

$$(v_C)_2 = (2 \text{ m})\left(\frac{26.7 \text{ kN} \cdot \text{m}^2}{EI}\right) = \frac{53.33 \text{ kN} \cdot \text{m}^3}{EI} \; \text{↓}$$

Finally, the cantilevered portion BC is displaced by the 10-kN force, Fig. 12–30d. We have

$$(v_C)_3 = \frac{PL^3}{3EI} = \frac{10 \text{ kN}(2 \text{ m})^3}{3EI} = \frac{26.67 \text{ kN} \cdot \text{m}^3}{EI} \; \text{↓}$$

Summing these results algebraically, we obtain the final displacement of point C,

$$(+\downarrow) \qquad v_C = -\frac{26.7}{EI} + \frac{53.3}{EI} + \frac{26.7}{EI} = \frac{53.3 \text{ kN} \cdot \text{m}^3}{EI} \; \text{↓} \qquad Ans.$$

EXAMPLE 12–15

Determine the displacement at the end C of the cantilever beam shown in Fig. 12–31. EI is constant.

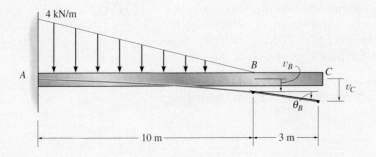

Fig. 12–31

SOLUTION

Using the table in Appendix C for the triangular loading, the slope and displacement at point B are

$$\theta_B = \frac{w_0 L^3}{24\,EI} = \frac{4\ \text{kN/m}\ (10\text{m})^3}{24\,EI} = \frac{166.67\ \text{kN} \cdot \text{m}^2}{EI}$$

$$v_B = \frac{w_0 L^4}{30\,EI} = \frac{4\ \text{kN/m}\ (10\text{m})^4}{30\,EI} = \frac{1333.33\ \text{kN} \cdot \text{m}^3}{EI}$$

The unloaded region BC of the beam remains straight, as shown in Fig. 12–31. Since θ_B is small, the displacement at C becomes

$$(+\downarrow) \qquad v_C = v_B + \theta_B\,(3\ \text{m})$$

$$= \frac{1333.33\ \text{kN} \cdot \text{m}^3}{EI} + \frac{166.67\ \text{kN} \cdot \text{m}^2}{EI}\,(3\ \text{m})$$

$$= \frac{1833\ \text{kN} \cdot \text{m}^3}{EI} \quad \downarrow \qquad\qquad\qquad \textit{Ans.}$$

EXAMPLE 12–16

The steel bar shown in Fig. 12–32a is supported by two springs at its ends A and B. Each spring has a stiffness of $k = 15$ kip/ft and is originally unstretched. If the bar is loaded with a force of 3 kip at point C, determine the vertical displacement of the force. Neglect the weight of the bar and take $E_{st} = 29(10^3)$ ksi, $I = 12$ in^4.

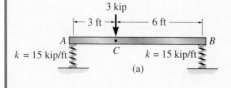

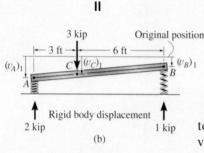

Fig. 12–32

SOLUTION

The end reactions at A and B are computed and shown in Fig. 12–32b. Each spring deflects by an amount

$$(v_A)_1 = \frac{2\ \text{kip}}{15\ \text{kip/ft}} = 0.1333\ \text{ft}$$

$$(v_B)_1 = \frac{1\ \text{kip}}{15\ \text{kip/ft}} = 0.0667\ \text{ft}$$

If the bar is considered to be *rigid*, these displacements cause it to move into the position shown in Fig. 12–32b. For this case, the vertical displacement at C is

$$(v_C)_1 = (v_B)_1 + \frac{6\ \text{ft}}{9\ \text{ft}}[(v_A)_1 - (v_B)_1]$$

$$= 0.0667\ \text{ft} + \frac{2}{3}[0.1333\ \text{ft} - 0.0667\ \text{ft}] = 0.1111\ \text{ft} \downarrow$$

We can find the displacement at C caused by the *deformation* of the bar, Fig. 12–32c, by using the table in Appendix C. We have

$$(v_C)_2 = \frac{Pab}{6EIL}(L^2 - b^2 - a^2)$$

$$= \frac{3\ \text{kip}(3\ \text{ft})(6\ \text{ft})[(9\ \text{ft})^2 - (6\ \text{ft})^2 - (3\ \text{ft})^2]}{6[29(10^3)]\ \text{kip/in.}^2(144\ \text{in.}^2/1\ \text{ft}^2)12\ \text{in.}^4(1\ \text{ft}^4/20\ 736\ \text{in.}^4)(9\ \text{ft})}$$

$$= 0.0149\ \text{ft} \downarrow$$

Adding the two displacement components, we get

$$(+\downarrow) \qquad v_C = 0.1111\ \text{ft} + 0.0149\ \text{ft} = 0.126\ \text{ft} = 1.51\ \text{in.} \downarrow \qquad \textit{Ans.}$$

PROBLEMS

12–74. The W8 × 48 cantilevered beam is made of A-36 steel and is subjected to the loading shown. Determine the deflection at its end A.

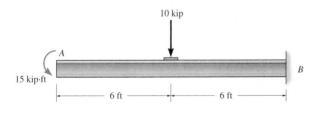

Prob. 12–74

12–75. The W12 × 45 simply supported beam is made of A-36 steel and is subjected to the loading shown. Determine the deflection at its center C.

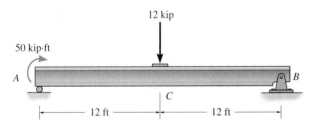

Prob. 12–75

*12–76. Determine the moment M_0 in terms of the load P and dimension a so that the deflection at the center of the beam is zero. EI is constant.

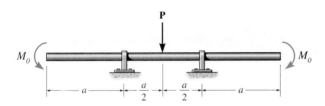

Prob. 12–76

12–77. The W24 × 104 beam is made of A-36 steel and is subjected to the loading shown. Determine the displacement at its end A.

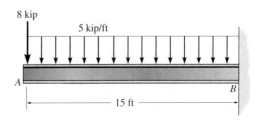

Prob. 12–77

12–78. The beam supports the loading shown. Code restrictions, due to a plaster ceiling, require the maximum deflection not to exceed 1/360 of the span length. Select the lightest-weight A-36 steel wide-flange beam from Appendix B that will satisfy this requirement and safely support the load. The allowable bending stress is $\sigma_{allow} = 24$ ksi and the allowable shear stress is $\tau_{allow} = 14$ ksi. Assume A is a roller and B is a pin.

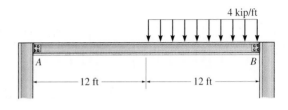

Prob. 12–78

12–79. The relay switch consists of a thin metal strip or armature AB that is made of red brass C83400 and is attracted to the solenoid S by a magnetic field. Determine the smallest force F required to attract the armature at C in order that contact is made at the free end B. Also, what should the distance a be for this to occur? The armature is fixed at A and has a moment of inertia of $I = 0.18(10^{-12})$ m^4.

12–81. Determine the vertical deflection at the end A of the bracket. Assume that the bracket is fixed supported at its base B and neglect axial deflection. EI is constant.

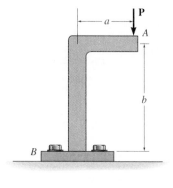

Prob. 12–81

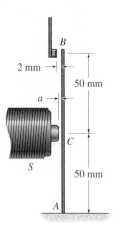

Prob. 12–79

*12–80. The simply supported beam carries a uniform load of 2 kip/ft. Code restrictions, due to a plaster ceiling, require the maximum deflection not to exceed 1/360 of the span length. Select the lightest-weight A-36 steel wide-flange beam from Appendix B that will satisfy this requirement and safely support the load. The allowable bending stress is $\sigma_{\text{allow}} = 24$ ksi and the allowable shear stress is $\tau_{\text{allow}} = 14$ ksi. Assume A is a pin and B a roller support.

12–82. The rod is pinned at its end A and attached to a torsional spring having a stiffness k, which measures the torque per radian of rotation of the spring. If a force $\mathbf{P}$ is always applied perpendicular to the end of the rod, determine the displacement of the force. EI is constant.

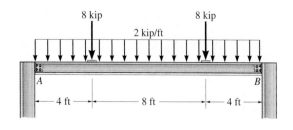

Prob. 12–80

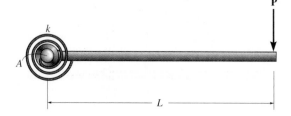

Prob. 12–82

12–83. The pipe assembly consists of three equal-sized pipes with flexibility stiffness EI and torsional stiffness GJ. Determine the vertical deflection at point A.

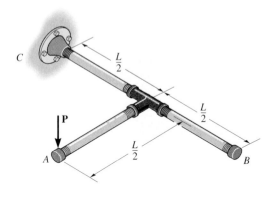

Prob. 12–83

12–85. Determine the vertical deflection and slope at the end A of the bracket. Assume that the bracket is fixed supported at its base, and neglect the axial deformation of segment AB. EI is constant.

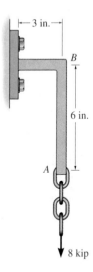

Prob. 12–85

*__12–84.__ Determine the slope at the end A of the bracket. Assume that the bracket is fixed supported at its base C, and neglect axial deformation of segment BC. EI is constant.

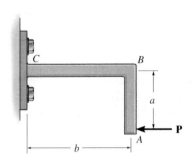

Prob. 12–84

12–86. The W24 × 104 A-36 steel beam is used to support the uniform distributed load and a concentrated force which is applied at its end. If the force acts at an angle with the vertical as shown, determine the horizontal and vertical displacement at point A.

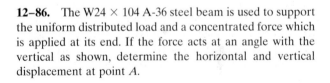

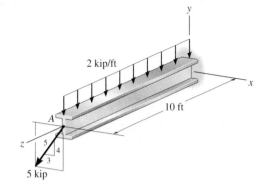

Prob. 12–86

12–87. The pipe assembly consists of two equal-sized pipes with flexibility stiffness EI and torsional stiffness GJ. Determine the vertical displacement at point A

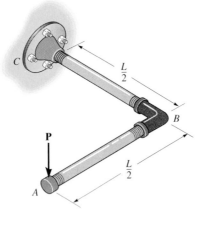

Prob. 12–87

*12–88. The framework consists of two A-36 steel cantilevered beams CD and BA and a simply supported beam CB. If each beam is made of steel and has a moment of inertia about its principal axis of $I_x = 118$ in^4, determine the deflection at the center G of beam CB.

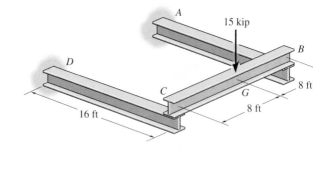

Prob. 12–88

12.6 STATICALLY INDETERMINATE BEAMS AND SHAFTS

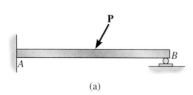

(a)

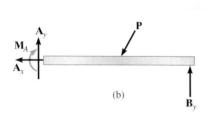

(b)

Fig. 12–33

The analysis of statically indeterminate axially loaded bars and torsionally loaded shafts has been discussed in Secs. 4.4 and 5.5, respectively. In this section we will illustrate a general method for determining the reactions on statically indeterminate beams and shafts. Specifically, a member of any type is classified as *statically indeterminate* if the number of unknown reactions *exceeds* the available number of equilibrium equations.

The additional support reactions on the beam or shaft that are *not needed* to keep it in stable equilibrium are called *redundants*. The number of these redundants is referred to as the *degree of indeterminacy*. For example, consider the beam shown in Fig. 12–33a. If the free-body diagram is drawn, Fig. 12–33b, there will be four unknown support reactions, and since three equilibrium equations are available for solution, the beam is classified as being indeterminate to the first degree. Either $\mathbf{A}_y$, $\mathbf{B}_y$, or $\mathbf{M}_A$ can be classified as the redundant, for if any one of these reactions is removed, the beam remains stable and in equilibrium ($\mathbf{A}_x$ cannot be classified as the redundant, for if it were removed, $\Sigma F_x = 0$ would not be satisfied.) In a similar manner, the *continuous beam* in Fig. 12–34a is indeterminate to the second degree, since there are five unknown reactions and only three available equilibrium equations, Fig. 12–34b. Here the two redundant support reactions can be chosen among $\mathbf{A}_y$, $\mathbf{B}_y$, $\mathbf{C}_y$, and $\mathbf{D}_y$.

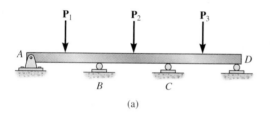

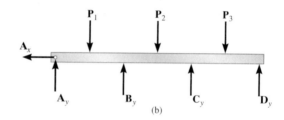

(a)

(b)

Fig. 12–34

An example of a statically indeterminate beam used to support a bridge deck.

To determine the reactions on a beam (or shaft) that is statically indeterminate, it is first necessary to specify the redundant reactions. We can determine these redundants from conditions of geometry known as *compatibility conditions*. Once determined, the redundants are then applied to the beam, and the remaining reactions are determined from the equations of equilibrium.

In the following sections we will illustrate this procedure for solution using the method of integration, Sec. 12.7; the moment-area method, Sec. 12.8; and the method of superposition, Sec. 12.9.

12.7   STATICALLY INDETERMINATE BEAMS AND SHAFTS—METHOD OF INTEGRATION

The method of integration, discussed in Sec. 12.2, requires two integrations of the differential equation $d^2v/dx^2 = M/EI$ once the internal moment M in the beam is expressed as a function of position x. If the beam is statically indeterminate, however, M can also be expressed in terms of the *unknown* redundants. After integrating this equation twice, there will be two constants of integration and the redundants to be determined. Although this is the case, these unknowns can always be found from the boundary and/or continuity conditions for the problem. For example, the beam in Fig. 12–35a has one redundant. It can either be A_y, M_A, or B_y, Fig. 12–35b. Once it is chosen the internal moment M can be written in terms of the redundant, and integrating the moment–displacement relationship, we can then determine the two constants of integration and the redundant from the *three* boundary conditions $v = 0$ at $x = 0$, $dv/dx = 0$ at $x = 0$, and $v = 0$ at $x = L$.

The following example problems illustrate specific applications of this method using the procedure for analysis outlined in Sec. 12.2.

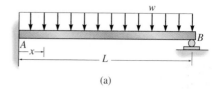

(a)

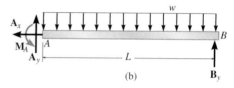

(b)

Fig. 12–35

EXAMPLE 12-17

The beam is subjected to the distributed loading shown in Fig. 12–36a. Determine the reactions at A. EI is constant.

SOLUTION

Elastic Cu ... Only one coordinate ... ed to the right, since ...

Moment F ... legree as indicated fr ... press the internal mc ... sing the segment sh ...

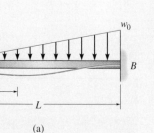

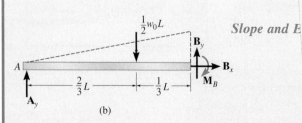

Slope and E ...

Fig. 12–36

$$EIv = \frac{1}{6}A_y x^3 - \frac{1}{120}w_0\frac{x^5}{L} + C_1 x + C_2$$

The three unknowns A_y, C_1, and C_2 are determined from the boundary conditions $x = 0$, $v = 0$; $x = L$, $dv/dx = 0$; and $x = L$, $v = 0$. Applying these conditions yields

$$x = 0, v = 0; \qquad 0 = 0 - 0 + 0 + C_2$$

$$x = L, \frac{dv}{dx} = 0; \qquad 0 = \frac{1}{2}A_y L^2 - \frac{1}{24}w_0 L^3 + C_1$$

$$x = L, v = 0; \qquad 0 = \frac{1}{6}A_y L^3 - \frac{1}{120}w_0 L^4 + C_1 L + C_2$$

Solving,

$$A_y = \frac{1}{10}w_0 L \qquad\qquad\qquad Ans.$$

$$C_1 = -\frac{1}{120}w_0 L^3 \quad C_2 = 0$$

Using the result for A_y, the reactions at B can be determined from the equations of equilibrium, Fig. 12–36b. Show that $B_x = 0$, $B_y = 2w_0 L/5$, and $M_B = w_0 L^2/15$.

EXAMPLE 12–18

The beam in Fig. 12–37*a* is fixed supported at both ends and is subjected to the uniform loading shown. Determine the reactions at the supports. Neglect the effect of axial load.

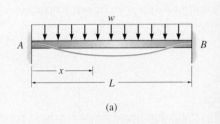

(a)

SOLUTION

Elastic Curve.   The beam deflects as shown in Fig. 12–37*a*. As in the previous problem, only one *x* coordinate is necessary for the solution since the loading is continuous across the span.

Moment Function.   From the free-body diagram, Fig. 12–37*b*, the respective shear and moment reactions at *A* and *B* must be equal, since there is symmetry of both loading and geometry. Because of this, the equation of equilibrium, $\Sigma F_y = 0$, requires

$$V_A = V_B = \frac{wL}{2} \qquad \text{Ans.}$$

The beam is indeterminate to the first degree, where M' is redundant. Using the beam segment shown in Fig. 12–37*c*, the internal moment **M** can be expressed in terms of M' as follows:

$$M = \frac{wL}{2}x - \frac{w}{2}x^2 - M'$$

Slope and Elastic Curve.   Applying Eq. 12–10, we have

$$EI\frac{d^2v}{dx} = \frac{wL}{2}x - \frac{w}{2}x^2 - M'$$

$$EI\frac{dv}{dx} = \frac{wL}{4}x^2 - \frac{w}{6}x^3 - M'x + C_1$$

$$EIv = \frac{wL}{12}x^3 - \frac{w}{24}x^4 - \frac{M'}{2}x^2 + C_1 x + C_2$$

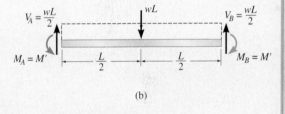

(b)

The three unknowns, M', C_1, and C_2, can be determined from the *three* boundary conditions $v = 0$ at $x = 0$, which yields $C_2 = 0$; $dv/dx = 0$ at $x = 0$, which yields $C_1 = 0$; and $v = 0$ at $x = L$, which yields

$$M' = \frac{wL^2}{12} \qquad \text{Ans.}$$

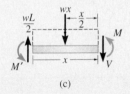

(c)

Using these results, notice that because of symmetry the remaining boundary condition $dv/dx = 0$ at $x = L$ is automatically satisfied.

Fig. 12–37

   It should be realized that this method of solution is generally suitable when only one *x* coordinate is needed to describe the elastic curve. If several *x* coordinates are needed, equations of continuity must be written, thus complicating the solution process.

PROBLEMS

12–89.   Determine the reactions at the supports *A* and *B*, then draw the shear and moment diagrams. Use discontinuity functions. *EI* is constant.

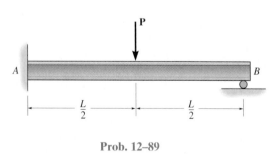

Prob. 12–89

12–90.   Determine the reactions at the supports, then draw the shear and moment diagram. *EI* is constant.

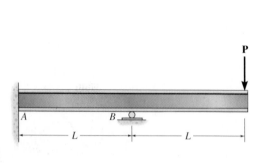

Prob. 12–90

12–91.   Determine the reactions at the supports *A* and *B*, then draw the shear and moment diagrams. *EI* is constant.

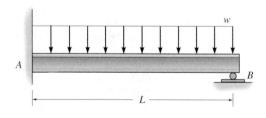

Prob. 12–91

*12–92.   The loading on a floor beam used in the airplane is shown. Use discontinuity functions and determine the reactions at the supports *A* and *B*, and then draw the moment diagram for the beam. The beam is made of aluminum and has a moment of inertia of $I = 320$ in^4.

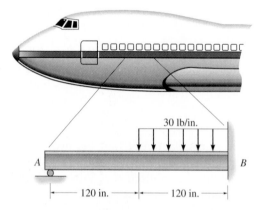

Prob. 12–92

12–93. Determine the reactions at the supports, then draw the shear and moment diagrams. *EI* is constant.

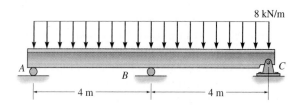

8 kN/m

A   *B*   *C*

4 m   4 m

Prob. 12–93

12–94. Determine the reactions at the supports, then draw the shear and moment diagrams. *EI* is constant.

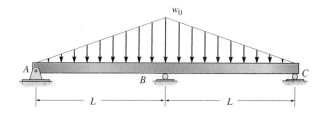

w_0

A   *B*   *C*

L   *L*

Prob. 12–94

12–95. Determine the reactions at the supports *A* and *B*. *EI* is constant.

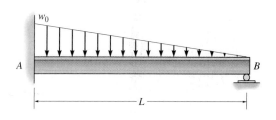

w_0

A   *B*

L

Prob. 12–95

*12–96. Determine the reactions at the supports, then draw the shear and moment diagrams. *EI* is constant.

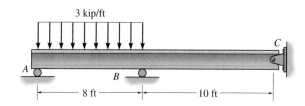

3 kip/ft

A   *B*   *C*

8 ft   10 ft

Prob. 12–96

12–97. The beam has a constant E_1I_1 and is supported by the fixed wall at *B* and the rod *AC*. If the rod has a cross-sectional area A_2 and the material has a modulus of elasticity E_2, determine the force in the rod.

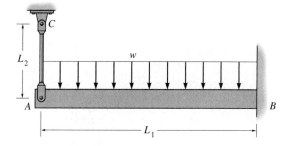

C

L_2

w

A   *B*

L_1

Prob. 12–97

*12.8 STATICALLY INDETERMINATE BEAMS AND SHAFTS—MOMENT-AREA METHOD

If the moment-area method is used to determine the unknown redundants of a statically indeterminate beam or shaft, then the M/EI diagram must be drawn such that the redundants are represented as *unknowns* on this diagram. Once the M/EI diagram is established, the two moment-area theorems can then be applied to obtain the proper relationships between the tangents on the elastic curve in order to meet the conditions of displacement and/or slope at the supports of the beam or shaft. In all cases the number of these compatibility conditions will be equivalent to the number of redundants, and so a solution for the redundants can be obtained.

Moment Diagrams Constructed by the Method of Superposition. Since application of the moment-area theorems requires calculation of both the area under the M/EI diagram and the centroidal location of this area, it is often convenient to use *separate M/EI* diagrams for *each* of the known loads and redundants rather than using the resultant diagram to compute these geometric quantities. This is especially true if the resultant moment diagram has a complicated shape. The method for drawing the moment diagram in parts is based on the principle of superposition.

Most loadings on beams or shafts will be a combination of the four loadings shown in Fig. 12–38. Construction of the associated moment diagrams, also shown in this figure, has been discussed in the examples of Chapter 6. Based on these results, we will now show how to use the method of superposition to represent the resultant moment diagram for the cantilevered beam shown in Fig. 12–39a by a series of separate moment diagrams. To do this, we will first replace the loads by a system of statically equivalent loads. For example, the three cantilevered beams shown in Fig. 12–39a are statically equivalent to the resultant beam, since

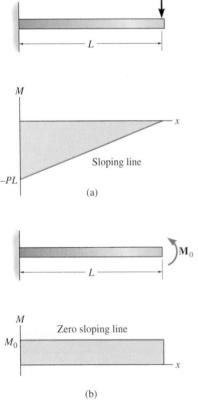
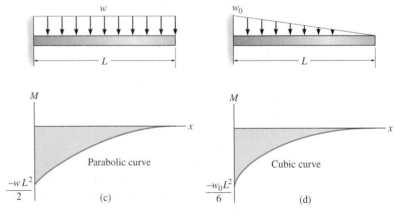

Fig. 12–38

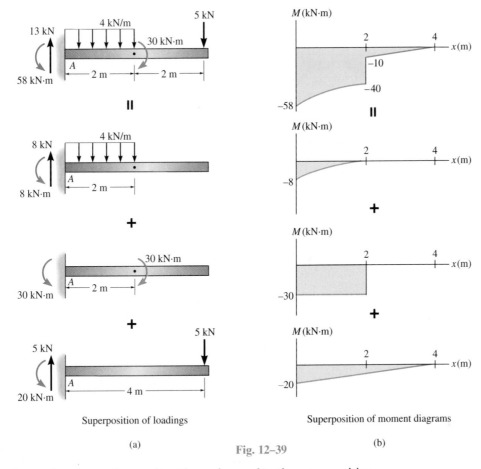

Superposition of loadings

(a)

Superposition of moment diagrams

(b)

Fig. 12–39

the load at each point on the resultant beam is equal to the superposition or addition of the loadings on the three separate beams. Indeed, the shear reaction at end A is 13 kN when the reactions on the separate beams are added together. In the same manner, the internal moment at any point on the resultant beam is equal to the sum of the internal moments at any point on the separate beams. Thus, if the moment diagrams for each separate beam are drawn, Fig. 12–39b, the superposition of these diagrams will yield the moment diagram for the resultant beam, shown at the top. For example, from each of the separate moment diagrams, the moment at end A is $M_A = -8 \text{ kN} \cdot \text{m} - 30 \text{ kN} \cdot \text{m} - 20 \text{ kN} \cdot \text{m} = -58 \text{ kN} \cdot \text{m}$, as verified by the top moment diagram. This example demonstrates that it is sometimes easier to construct a series of separate statically equivalent moment diagrams for the beam, *rather* than constructing its more complicated resultant moment diagram. Obviously, the area and location of the centroid for each part are easier to establish than those of the centroid for the resultant diagram.

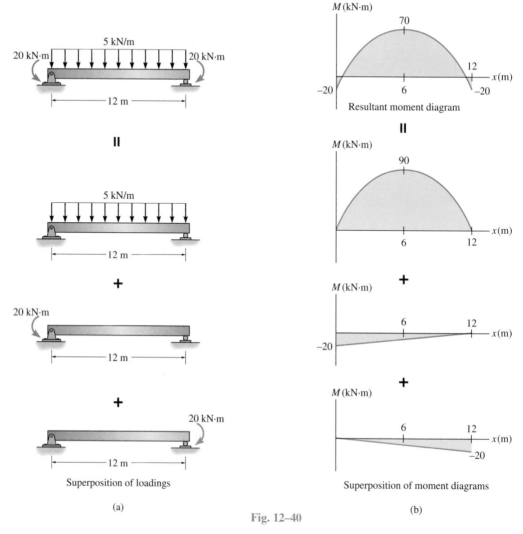

Fig. 12–40

In a similar manner, we can also represent the resultant moment diagram for a beam by using a superposition of moment diagrams for a series of simply supported beams. For example, the beam loading shown at the top of Fig. 12–40a is equivalent to the sum of the beam loadings shown below it. Consequently, the sum of the moment diagrams for each of these three loadings can be used rather than the resultant moment diagram shown at the top of Fig. 12–40b. For complete understanding, these results should be verified.

The examples that follow should also clarify some of these points and illustrate how to use the moment-area theorems to obtain the redundant reactions on statically indeterminate beams and shafts. The solutions follow the procedure for analysis outlined in Sec. 12.4.

EXAMPLE 12–19

The beam is subjected to the concentrated loading shown in Fig. 12–41*a*. Determine the reactions at the supports. *EI* is constant.

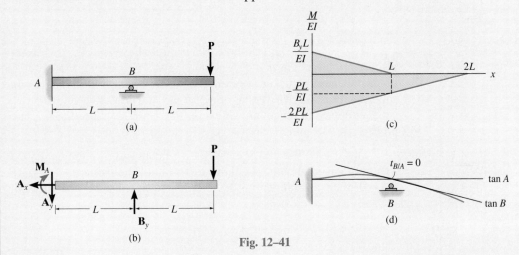

Fig. 12–41

SOLUTION

M/EI Diagram.   The free-body diagram is shown in Fig. 12–41*b*. Using the method of superposition, the separate *M/EI* diagrams for the redundant reaction $\mathbf{B}_y$ and the load $\mathbf{P}$ are shown in Fig. 12–41*c*.

Elastic Curve.   The elastic curve for the beam is shown in Fig. 12–41*d*. The tangents at the supports *A* and *B* have been constructed. Since $\Delta_B = 0$, then

$$t_{B/A} = 0$$

Moment-Area Theorem.   Applying Theorem 2, we have

$$t_{B/A} = \left(\frac{2}{3}L\right)\left[\frac{1}{2}\left(\frac{B_y L}{EI}\right)L\right] + \left(\frac{L}{2}\right)\left[\frac{-PL}{EI}(L)\right]$$

$$+ \left(\frac{2}{3}L\right)\left[\frac{1}{2}\left(\frac{-PL}{EI}\right)(L)\right] = 0$$

$$B_y = 2.5P \qquad\qquad\qquad \textit{Ans.}$$

Equations of Equilibrium.   Using this result, the reactions at *A* on the free-body diagram, Fig. 12–41*b*, are determined as follows:

$$\xrightarrow{+} \Sigma F_x = 0; \qquad\qquad A_x = 0 \qquad\qquad\qquad \textit{Ans.}$$

$$+\uparrow \Sigma F_y = 0; \qquad -A_y + 2.5P - P = 0$$

$$A_y = 1.5P \qquad\qquad\qquad \textit{Ans.}$$

$$\curvearrowright + \Sigma M_A = 0; \qquad -M_A + 2.5P(L) - P(2L) = 0$$

$$M_A = 0.5PL \qquad\qquad\qquad \textit{Ans.}$$

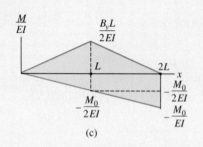

(a)

(b)

(c)

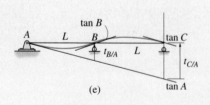

(d)

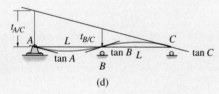

(e)

Fig. 12–42

EXAMPLE 12–20

The beam is subjected to the couple moment at its end C as shown in Fig. 12–42a. Determine the reaction at B. EI is constant.

SOLUTION

M/EI Diagram. The free-body diagram is shown in Fig. 12–42b. By inspection, the beam is indeterminate to the first degree. In order to obtain a direct solution, we will choose $\mathbf{B}_y$ as the redundant. Using superposition, the M/EI diagrams for $\mathbf{B}_y$ and $\mathbf{M}_0$, each applied to a simply supported beam, are shown in Fig. 12–42c. (Note that for such a beam A_x, A_y, and C_y do not contribute an M/EI diagram.)

Elastic Curve. The elastic curve for the beam is shown in Fig. 12–42d. The tangents at A, B, and C have been established. Since $\Delta_A = \Delta_B = \Delta_C = 0$, then the tangential deviations shown must be proportional; i.e.,

$$t_{B/C} = \frac{1}{2} t_{A/C}$$

From Fig. 12–42c, we have

$$t_{B/C} = \left(\frac{1}{3}L\right)\left[\frac{1}{2}\left(\frac{B_y L}{2EI}\right)(L)\right] + \left(\frac{2}{3}L\right)\left[\frac{1}{2}\left(\frac{-M_0}{2EI}\right)(L)\right]$$
$$+ \left(\frac{L}{2}\right)\left[\left(\frac{-M_0}{2EI}\right)(L)\right]$$

$$t_{A/C} = (L)\left[\frac{1}{2}\left(\frac{B_y L}{2EI}\right)(2L)\right] + \left(\frac{2}{3}(2L)\right)\left[\frac{1}{2}\left(\frac{-M_0}{EI}\right)(2L)\right]$$

Substituting into Eq. 1 and simplifying yields

$$B_y = \frac{3M_0}{2L} \qquad \qquad \textit{Ans.}$$

Equations of Equilibrium. The reactions at A and C can now be determined from the equations of equilibrium, Fig. 12–42b. Show that $A_x = 0$, $C_y = 5M_0/4L$, and $A_y = M_0/4L$.

Note from Fig. 12–42e that this problem can also be worked in terms of the tangential deviations,

$$t_{B/A} = \frac{1}{2} t_{C/A}$$

PROBLEMS

12–98. Determine the moment reactions at the supports A and B. EI is constant.

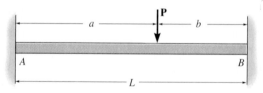

Prob. 12–98

12–99. Determine the value of a for which the maximum positive moment has the same magnitude as the maximum negative moment. EI is constant.

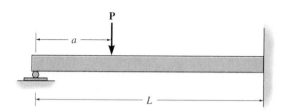

Prob. 12–99

*12–100. Determine the reactions at the supports. EI is constant.

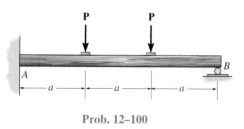

Prob. 12–100

12–101. Determine the reactions at the supports, then draw the shear and moment diagrams. EI is constant.

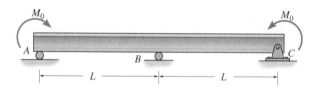

Prob. 12–101

12–102. The rod is fixed at A, and the connection at B consists of a roller constraint which allows vertical displacement but resists axial load and moment. Determine the moment reactions at these supports. EI is constant.

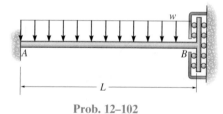

Prob. 12–102

12–103. Determine the moment reactions at the supports A and B, then draw the shear and moment diagrams. EI is constant.

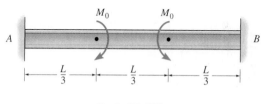

Prob. 12–103

*12–104. Determine the reactions at the supports, then draw the shear and moment diagrams. EI is constant. Support B is a thrust bearing.

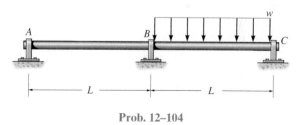

Prob. 12–104

12–105. Determine the moment reactions at the supports A and B. EI is constant.

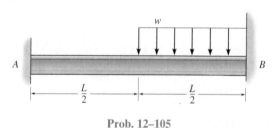

Prob. 12–105

12.9   STATICALLY INDETERMINATE BEAMS AND SHAFTS—METHOD OF SUPERPOSITION

The method of superposition has been used previously to solve for the redundant loadings on axially loaded bars and torsionally loaded shafts. In order to apply this method to the solution of statically indeterminate beams (or shafts), it is first necessary to identify the redundant support reactions as explained in Sec. 12.6. By *removing* them from the beam we obtain the so-called *primary beam*, which is statically determinate and stable, and is subjected *only* to the external load. If we add to this beam a succession of similarly supported beams, each loaded with a *separate* redundant, then by the principle of superposition, we obtain the actual loaded beam. Finally, in order to solve for the redundants, we must write the *conditions of compatibility* that exist at the supports where each of the redundants act. Since the redundant forces are determined directly in this manner, this method of analysis is sometimes called the **force method**. Once the redundants are obtained, the other reactions on the beam are then determined from the three equations of equilibrium.

To clarify these concepts, consider the beam shown in Fig. 12–43a. If we choose the reaction $\mathbf{B}_y$ at the roller as the redundant, then the primary beam is shown in Fig. 12–43b, and the beam with the redundant $\mathbf{B}_y$ acting on it is shown in Fig. 12–43c. The displacement at the roller is to be zero, and since the displacement of point B on the primary beam is v_B, and $\mathbf{B}_y$ causes point B to be displaced upward v_B', we can write the compatibility equation at B as

$$(+\uparrow) \qquad\qquad 0 = -v_B + v_B'$$

The displacements v_B and v_B' can be obtained using any one of the methods discussed in Secs. 12.2 through 12.5. Here we will obtain them directly from the table in Appendix C. We have

$$v_B = \frac{5PL^3}{48EI} \qquad \text{and} \qquad v_B' = \frac{B_y L^3}{3EI}$$

Substituting into the compatibility equation, we get

$$0 = -\frac{5PL^3}{48EI} + \frac{B_y L^3}{3EI}$$

$$B_y = \frac{5}{16}P$$

Now that $\mathbf{B}_y$ is known, the reactions at the wall are determined from the three equations of equilibrium applied to the entire beam, Fig. 12–43d. The results are

$$A_x = 0 \qquad A_y = \frac{11}{16}P$$

$$M_A = \frac{3}{16}PL$$

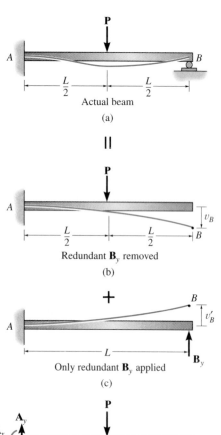

Actual beam

(a)

||

Redundant $\mathbf{B}_y$ removed

(b)

+

Only redundant $\mathbf{B}_y$ applied

(c)

(d)

Fig. 12–43

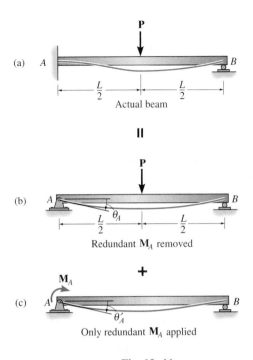

(a)

Actual beam

‖

(b)

Redundant $\mathbf{M}_A$ removed

+

(c)

Only redundant $\mathbf{M}_A$ applied

Fig. 12–44

As stated in Sec. 12.6, choice of the redundant is *arbitrary*, provided the primary beam remains stable. For example, the moment at A for the beam in Fig. 12–44a can also be chosen as the redundant. In this case the capacity of the beam to resist $\mathbf{M}_A$ is removed, and so the primary beam is then pin supported at A, Fig. 12–44b. Also, the redundant at A acts alone on this beam, Fig. 12–44c. Referring to the slope at A caused by the load $\mathbf{P}$ as θ_A, and the slope at A caused by the redundant $\mathbf{M}_A$ as θ_A', the compatibility equation for the slope at A requires

$(\circlearrowleft+)$ $0 = \theta_A + \theta_A'$

Again using the table in Appendix C, we have

$$\theta_A = \frac{PL^2}{16EI} \quad \text{and} \quad \theta_A' = \frac{M_A L}{3EI}$$

Thus

$$0 = \frac{PL^2}{16EI} + \frac{M_A L}{3EI}$$

$$M_A = -\frac{3}{16}PL$$

This is the same result computed previously. Here the negative sign for M_A simply means that $\mathbf{M}_A$ acts in the opposite sense of direction of that shown in Fig. 12–44c.

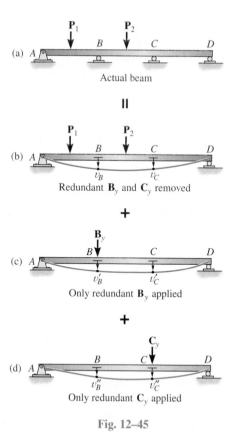

(a) Actual beam

‖

(b) Redundant $\mathbf{B}_y$ and $\mathbf{C}_y$ removed

+

(c) Only redundant $\mathbf{B}_y$ applied

+

(d) Only redundant $\mathbf{C}_y$ applied

Fig. 12–45

Another example that illustrates this method is given in Fig. 12–45a. In this case the beam is indeterminate to the second degree and therefore *two* compatibility equations will be necessary for the solution. We will choose the forces at the roller supports B and C as redundants. The primary (statically determinate) beam deforms as shown in Fig. 12–45b when the redundants are removed. Each redundant force deforms this beam as shown in Figs. 12–45c and 12–45d, respectively. By superposition, the compatibility equations for the displacements at B and C are

$(+\downarrow)$ $$0 = v_B + v_B' + v_B''$$

$(+\downarrow)$ $$0 = v_C + v_C' + v_C''$$

(12–22)

Here the displacement components v_B' and v_C' will be expressed in terms of the unknown $\mathbf{B}_y$, and the components v_B'' and v_C'' will be expressed in terms of the unknown $\mathbf{C}_y$. When these displacements have been determined and substituted into Eq. 12–22, these equations may then be solved simultaneously for the two unknowns $\mathbf{B}_y$ and $\mathbf{C}_y$.

PROCEDURE FOR ANALYSIS

The following procedure provides a means for applying the method of superposition (or the force method) to determine the reactions on statically indeterminate beams or shafts.

Elastic Curve.

• Specify the unknown redundant forces or moments that must be removed from the beam in order to make it statically determinate and stable.

• Using the principle of superposition, draw the statically indeterminate beam and show it equal to a sequence of corresponding *statically determinate beams*.

• The first of these beams, the primary beam, supports the same external loads as the statically indeterminate beam, and each of the other beams "added" to the primary beam shows the beam loaded with a separate redundant force or moment.

• Sketch the deflection curve for each beam and indicate symbolically the displacement or slope at the point of each redundant force or moment.

Compatibility Equations.

• Write a compatibility equation for the displacement or slope at each point where there is a redundant force or moment.

• Determine all the displacements or slopes using an appropriate method as explained in Secs. 12.2 through 12.5.

• Substitute the results into the compatibility equations and solve for the unknown redundants.

• If a numerical value for a redundant is *positive*, it has the *same sense of direction* as originally assumed. Similarly, a *negative* numerical value indicates the redundant acts *opposite* to its assumed *sense of direction*.

Equilibrium Equations.

• Once the redundant forces and/or moments have been determined, the remaining unknown reactions can be found from the equations of equilibrium applied to the loadings shown on the beam's free-body diagram.

The following examples illustrate application of this procedure. For brevity, all displacements and slopes have been found using the table in Appendix C.

EXAMPLE 12-21

Determine the reactions at the roller support B of the beam shown in Fig. 12–46a, then draw the shear and moment diagrams. EI is constant.

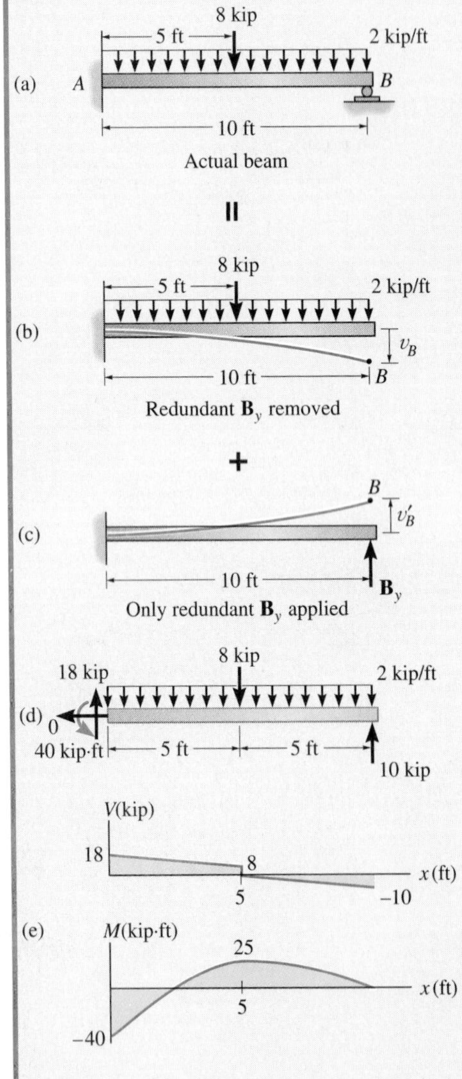

(a)

Actual beam

=

(b)

Redundant $\mathbf{B}_y$ removed

+

(c)

Only redundant $\mathbf{B}_y$ applied

(d)

(e)

Fig. 12–46

SOLUTION

Principle of Superposition. By inspection, the beam is statically indeterminate to the first degree. The roller support at B will be chosen as the redundant so that $\mathbf{B}_y$ will be determined *directly*. Figures 12–46b and 12–46c show application of the principle of superposition. Here we have assumed that $\mathbf{B}_y$ acts upward on the beam.

Compatibility Equation. Taking positive displacement as downward, the compatibility equation at B is

$$(+\downarrow) \qquad\qquad 0 = v_B - v_B' \qquad\qquad (1)$$

These displacements can be obtained directly from the table in Appendix C.

$$v_B = \frac{wL^4}{8EI} + \frac{5PL^3}{48EI}$$

$$= \frac{2\text{ kip/ft}(10\text{ ft})^4}{8EI} + \frac{5(8\text{ kip})(10\text{ ft})^3}{48EI} = \frac{3333\text{ kip}\cdot\text{ft}^3}{EI} \downarrow$$

$$v_B' = \frac{PL^3}{3EI} = \frac{B_y(10\text{ ft})^3}{3EI} = \frac{333.3\text{ ft}^3 B_y}{EI} \uparrow$$

Substituting into Eq. 1 and solving yields

$$0 = \frac{3333}{EI} - \frac{333.3 B_y}{EI}$$

$$B_y = 10\text{ kip} \qquad\qquad\qquad Ans.$$

Equilibrium Equations. Using this result and applying the three equations of equilibrium, we obtain the results shown on the beam's free-body diagram in Fig. 12–46d. The shear and moment diagrams are shown in Fig. 12–46e.

EXAMPLE 12-22

Determine the reactions on the beam shown in Fig. 12–47a. Due to the loading and poor construction, the roller support at B settles 12 mm. Take $E = 200$ GPa and $I = 80 \, (10^6) \, \text{mm}^4$.

SOLUTION

Principle of Superposition.   By inspection, the beam is indeterminate to the first degree. The roller support at B will be chosen as the redundant. The principle of superposition is shown in Figs. 12–47b and 12–47c. Here $\mathbf{B}_y$ is assumed to act upward on the beam.

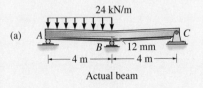

(a)

24 kN/m

A B 12 mm C

4 m   4 m

Actual beam

||

Compatibility Equation.   With reference to point B, using units of meters, we require

$$(+\downarrow) \qquad\qquad 0.012 \text{ m} = v_B - v'_B \qquad\qquad (1)$$

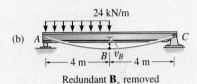

(b)

24 kN/m

A B v_B C

4 m   4 m

Redundant $\mathbf{B}_y$ removed

+

Using the table in Appendix C, the displacements are

$$v_B = \frac{5wL^4}{768EI} = \frac{5(24 \text{ kN/m})(8 \text{ m})^4}{768EI} = \frac{640 \text{ kN} \cdot \text{m}^3}{EI} \downarrow$$

$$v'_B = \frac{PL^3}{48EI} = \frac{B_y(8 \text{ m})^3}{48EI} = \frac{10.67 \text{ m}^3 B_y}{EI} \uparrow$$

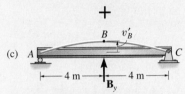

(c)

B v'_B

A C

4 m   4 m

$\mathbf{B}_y$

Only redundant $\mathbf{B}_y$ applied

Thus Eq. 1 becomes

$$0.012EI = 640 - 10.67B_y$$

Expressing E and I in units of kN/m^2 and m^4, respectively, we have

$$0.012(200)(10^6)[80(10^{-6})] = 640 - 10.67B_y$$
$$B_y = 42.0 \text{ kN} \uparrow \qquad\qquad \textit{Ans.}$$

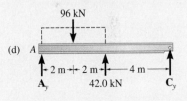

(d)

96 kN

A

2 m   2 m   4 m

A_y 42.0 kN C_y

Fig. 12–47

Equilibrium Equations. Applying this result to the beam, Fig. 12–47d, we can calculate the reactions at A and C using the equations of equilibrium. We obtain

$$\zeta+ \ \Sigma M_A = 0; \qquad -96 \text{ kN}(2 \text{ m}) + 42.0 \text{ kN}(4 \text{ m}) + C_y(8 \text{ m}) = 0$$
$$C_y = 3.00 \text{ kN} \uparrow \qquad\qquad \textit{Ans.}$$

$$+\uparrow \ \Sigma F_y = 0; \qquad A_y - 96 \text{ kN} + 42.0 \text{ kN} + 3.00 \text{ kN} = 0$$
$$A_y = 51 \text{ kN} \uparrow \qquad\qquad \textit{Ans.}$$

EXAMPLE 12–23

The beam in Fig. 12–48a is fixed supported to the wall at A and pin connected to a $\frac{1}{2}$-in.-diameter rod BC. If $E = 29(10^3)$ ksi for both members, determine the force developed in the rod due to the loading. The moment of inertia of the beam about its neutral axis is $I = 475$ in^4.

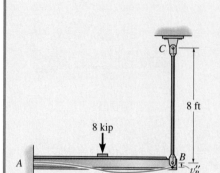

Actual beam and rod

(a)

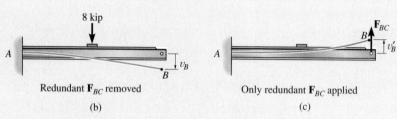

Redundant $\mathbf{F}_{BC}$ removed

(b)

Only redundant $\mathbf{F}_{BC}$ applied

(c)

Fig. 12–48

SOLUTION I

Principle of Superposition. By inspection, this problem is indeterminate to the first degree. Here B will undergo an unknown displacement v_B'', since the rod will stretch. The rod will be treated as the redundant and hence the force of the rod is removed from the beam at B, Fig. 12–48b, and then reapplied, Fig. 12–48c.

Compatibility Equation.   At point B we require

$$(+\downarrow) \qquad\qquad v_B'' = v_B - v_B' \qquad\qquad (1)$$

The displacements v_B and v_B' are determined from the table in Appendix C. v_B'' is calculated from Eq. 4–2. Working in kilopounds and inches, we have

$$v_B'' = \frac{PL}{AE} = \frac{F_{BC}(8\text{ ft})(12\text{ in./ft})}{(\pi/4)(\frac{1}{2}\text{ in.})^2[29(10^3)\text{ kip/in}^2]} = 0.01686F_{BC}\downarrow$$

$$v_B = \frac{5PL^3}{48EI} = \frac{5(8\text{ kip})(10\text{ ft})^3(12\text{ in./ft})^3}{48[29(10^3)\text{ kip/in.}^2](475\text{ in.}^4)} = 0.1045\text{ in.}\downarrow$$

$$v_B' = \frac{PL^3}{3EI} = \frac{F_{BC}(10\text{ ft})^3(12\text{ in./ft})^3}{3[29(10^3)\text{ kip/in.}^2](475\text{ in.}^4)} = 0.04181F_{BC}\uparrow$$

Thus, Eq. 1 becomes

$$(+\downarrow) \qquad\qquad 0.01686F_{BC} = 0.1045 - 0.04181F_{BC}$$

$$F_{BC} = 1.78\text{ kip} \qquad\qquad\textit{Ans.}$$

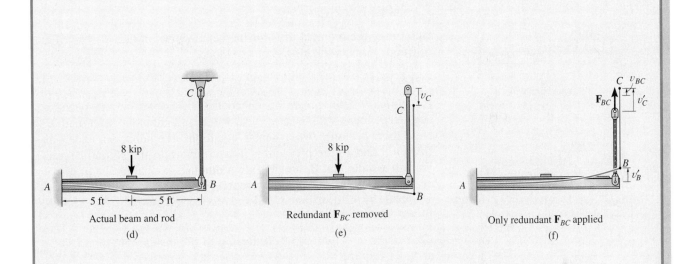

Actual beam and rod
(d)

Redundant $\mathbf{F}_{BC}$ removed
(e)

Only redundant $\mathbf{F}_{BC}$ applied
(f)

SOLUTION II

Principle of Superposition.   We can also solve this problem by removing the pin support at C and keeping the rod attached to the beam. In this case the 8-kip load will cause points B and C to be displaced downward the *same amount* v_C, Fig. 12–48e, since no force exists in rod BC. When the redundant force $\mathbf{F}_{BC}$ is applied at point C, it causes the end C of the rod to be displaced upward v_C' and the end B of the beam to be displaced upward v_B', Fig. 12–48f. The difference in these two displacements, v_{BC}, represents the stretch of the rod due to $\mathbf{F}_{BC}$, so that $v_C' = v_{BC} + v_B'$. Hence, from Figs. 12–48d, 12–48e, and 12–48f, the compatibility of displacement at point C is

$$(+\downarrow) \qquad\qquad 0 = v_C - (v_{BC} + v_B') \qquad\qquad (2)$$

From Solution I, we have

$$v_C = v_B = 0.1045 \text{ in.} \downarrow$$
$$v_{BC} = v_B'' = 0.01686 F_{BC} \uparrow$$
$$v_B' = 0.04181 F_{BC} \uparrow$$

Therefore, Eq. 2 becomes

$$(+\downarrow) \qquad 0 = 0.1045 - (0.01686 F_{BC} + 0.04181 F_{BC})$$
$$F_{BC} = 1.78 \text{ kip} \qquad\qquad\qquad ***Ans.***$$

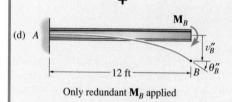

3 kip/ft

(a) A B

6 ft   6 ft

Actual beam

$=$

3 kip/ft

(b) A

6 ft   6 ft v_B

B θ_B

Redundants $\mathbf{M}_B$ and $\mathbf{B}_y$ removed

$+$

$\mathbf{B}_y$

(c) A v_B'

12 ft B θ_B'

Only redundant $\mathbf{B}_y$ applied

$+$

$\mathbf{M}_B$

(d) A v_B''

12 ft B θ_B''

Only redundant $\mathbf{M}_B$ applied

Fig. 12–49

EXAMPLE 12–24

Determine the moment at B for the beam shown in Fig. 12–49a. EI is constant. Neglect the effects of axial load.

SOLUTION

Principle of Superposition. Since the axial load on the beam is neglected, there will be a vertical force and moment at A and B. Here there are only two available equations of equilibrium ($\Sigma M = 0$, $\Sigma F_y = 0$), and so the problem is indeterminate to the second degree. We will assume that $\mathbf{B}_y$ and $\mathbf{M}_B$ are redundant, so that by the principle of superposition, the beam is represented as a cantilever, loaded *separately* by the distributed load and reactions $\mathbf{B}_y$ and $\mathbf{M}_B$, Figs. 12–49b, 12–49c, and 12–49d.

Compatibility Equations. Referring to the displacement and slope at B, we require

$(\stackrel{+}{\curvearrowright})$ $0 = \theta_B + \theta_B' + \theta_B''$ (1)

$(+\downarrow)$ $0 = v_B + v_B' + v_B''$ (2)

Using the table in Appendix C to compute the slopes and displacements, we have

$$\theta_B = \frac{wL^3}{48EI} = \frac{3 \text{ kip/ft}(12 \text{ ft})^3}{48EI} = \frac{108}{EI} \downarrow$$

$$v_B = \frac{7wL^4}{384EI} = \frac{7(3 \text{ kip/ft})(12 \text{ ft})^4}{384EI} = \frac{1134}{EI} \downarrow$$

$$\theta_B' = \frac{PL^2}{2EI} = \frac{B_y(12 \text{ ft})^2}{2EI} = \frac{72B_y}{EI} \downarrow$$

$$v_B' = \frac{PL^3}{3EI} = \frac{B_y(12 \text{ ft})^3}{3EI} = \frac{576B_y}{EI} \downarrow$$

$$\theta_B'' = \frac{ML}{EI} = \frac{M_B(12 \text{ ft})}{EI} = \frac{12M_B}{EI} \downarrow$$

$$v_B'' = \frac{ML^2}{2EI} = \frac{M_B(12 \text{ ft})^2}{2EI} = \frac{72M_B}{EI} \downarrow$$

Substituting these values into Eqs. 1 and 2 and canceling out the common factor EI, we get

$(\stackrel{+}{\curvearrowright})$ $0 = 108 + 72B_y + 12M_B$

$(+\downarrow)$ $0 = 1134 + 576B_y + 72M_B$

Solving these equations simultaneously gives

$$B_y = -3.375 \text{ kip}$$

$$M_B = 11.25 \text{ kip} \cdot \text{ft} \qquad\qquad \textit{Ans.}$$

PROBLEMS

12–106. The assembly consists of a steel and an aluminum bar, each of which is 1 in. thick, fixed at its ends A and B, and pin connected to the *rigid* short link CD. If a horizontal force of 80 lb is applied to the link as shown, determine the moments created at A and B. $E_{st} = 29(10^3)$ ksi, $E_{al} = 10(10^3)$ ksi.

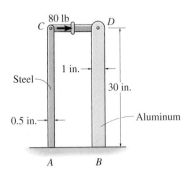

Prob. 12–106

12–107. The A-36 steel beam and rod are used to support the load of 8 kip. If it is required that the allowable normal stress for the steel is $\sigma_{allow} = 18$ ksi, and the maximum deflection not exceed 0.05 in., determine the smallest diameter rod that should be used. The beam is rectangular, having a height of 5 in. and a thickness of 3 in.

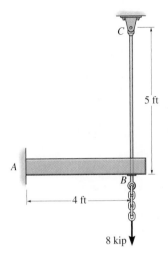

Prob. 12–107

*12–108. Determine the reactions at the supports A, B, and C, then draw the shear and moment diagrams. EI is constant.

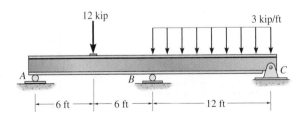

Prob. 12–108

12–109. The beam is used to support the 20-kip load. Determine the reactions at the supports. Assume A is fixed and B is a roller.

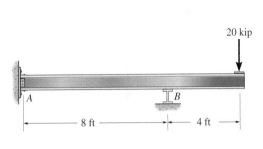

Prob. 12–109

12–110. The beam has a constant E_1I_1 and is supported by the fixed wall at B and the rod AC. If the rod has a cross-sectional area A_2 and the material has a modulus of elasticity E_2, determine the force in the rod.

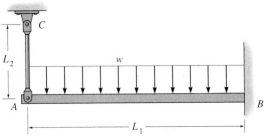

Prob. 12–110

12–111. Determine the reactions at support C. EI is constant for both beams.

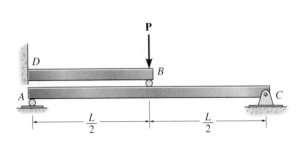

Prob. 12–111

*12–112. Determine the reactions at the supports A and B. EI is constant.

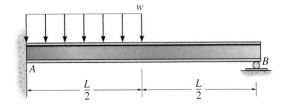

Prob. 12–112

12–113. The assembly consists of three simply supported beams for which the bottom of the top beam rests on the top of the bottom two. If a uniform load of 3 kN/m is applied to the top beam, determine the vertical reactions at each of the supports. EI is constant.

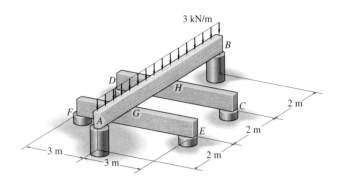

Prob. 12–113

12–114. Determine the reactions at the supports, then draw the shear and moment diagrams. *EI* is constant.

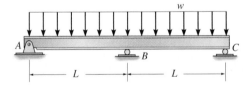

Prob. 12–114

12–115. The beam is supported by a pin at *A*, a spring having a stiffness *k* at *B*, and a roller at *C*. Determine the force the spring exerts on the beam. *EI* is constant.

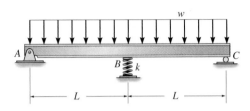

Prob. 12–115

12–116. Each of the two members is made from 6061-T6 aluminum and has a square cross section 1 in. × 1 in. They are pin connected at their ends and a jack is placed between them and opened until the force it exerts on each member is 50 lb. Determine the greatest force *P* that can be applied to the center of the top member without causing either of the two members to yield. For the analysis neglect the axial force in each member. Assume the jack is rigid.

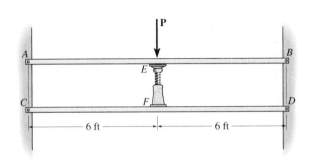

Prob. 12–116

12–117. The 1-in.-diameter A-36 steel shaft is supported by unyielding bearings at *A* and *C*. The bearing at *B* rests on a simply supported steel wide-flange beam having a moment of inertia of *I* = 500 in⁴. If the belt loads on the pulley are 400 lb each, determine the vertical reactions at *A*, *B*, and *C*.

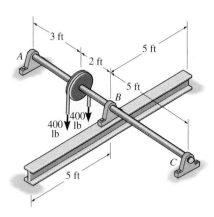

Prob. 12–117

12–118. Determine the force in the spring. *EI* is constant.

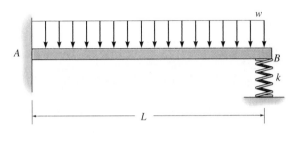

Prob. 12–118

12–119. The beam is made from a soft elastic material having a constant EI. If it is originally a distance Δ from the surface of its end support, determine the distance a at which it rests on this support when it is subjected to the uniform load w_0, which is great enough to cause this to happen.

*12–120. The rim on the flywheel has a thickness t, width b, and specific weight γ. If the flywheel is rotating at a constant rate of ω, determine the maximum moment developed in the rim. Assume that the spokes do not deform. *Hint:* Due to symmetry of the loading, the slope of the rim at each spoke is zero. Consider the radius to be sufficiently large so that the segment AB can be considered as a straight beam fixed at both ends and loaded with a uniform centrifugal force per unit length. Show that this force is $w = bt\gamma\omega^2 r/g$.

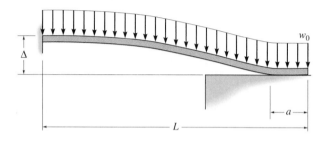

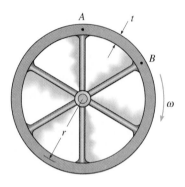

Prob. 12–119

Prob. 12–120

REVIEW PROBLEMS

12–121. The shaft supports the two pulley loads shown. Using discontinuity functions, determine the equation of the elastic curve. The bearings at A and B exert only vertical reactions on the shaft. EI is constant.

12–122. Use discontinuity functions to determine the slope at B and the displacement at C for the W10 × 45. $E_{st} = 29(10^3)$ ksi.

12–123. Solve Prob. 12–122 using the moment-area theorems.

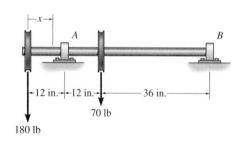

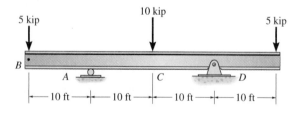

Prob. 12–121

Probs. 12–122/12–123

*12–124. The wooden beam is subjected to the loading shown. Assume the support at A is a pin and B is a roller. Determine the slope at A and the displacement at C. Use the moment-area theorems. EI is constant.

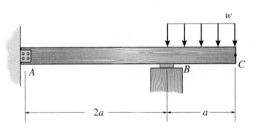

Prob. 12–124

12–125. The bearing supports A, B, and C exert only vertical reactions on the shaft. Determine these reactions, then draw the shear and moment diagrams. EI is constant. Use the moment-area theorems.

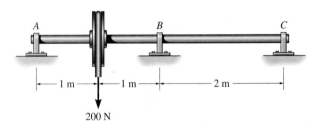

Prob. 12–125

12–126. If the bearings at A and B exert only vertical reactions on the shaft, determine the slope at B and the displacement at C. Use the moment-area theorems. EI is constant.

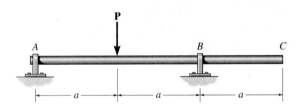

Prob. 12–126

12–127. Determine the value of a so that the slope at A is equal to zero. EI is constant. Use the moment-area theorems.

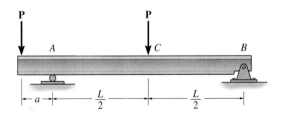

Prob. 12–127

*12–128. Determine the value of a so that the deflection at C is equal to zero. EI is constant. Use the moment-area theorems.

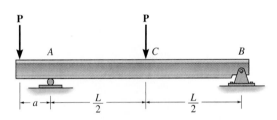

Prob. 12–128

12–129. Determine the moment reactions at the supports A and B. Use the method of integration. EI is constant.

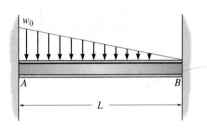

Prob. 12–129

The columns for this building are used to support the floor loading. Engineers design these members to resist the possibility of buckling. (Chris Baker/Tony Stone Images.)

13 BUCKLING OF COLUMNS

CHAPTER OBJECTIVES

In this chapter we will discuss the behavior of columns and indicate some of the methods used for their design. The chapter begins with a general discussion of buckling, followed by a determination of the axial load needed to buckle a so-called ideal column. Afterwards a more realistic analysis is considered, which accounts for any bending of the column. Also, inelastic buckling of a column is presented as a special topic. At the end of the chapter we will discuss some of the methods used to design both concentric and eccentric loaded columns made from common engineering materials.

13.1 CRITICAL LOAD

Whenever a member is designed, it is necessary that it satisfy specific strength, deflection, and stability requirements. In the preceding chapters we have discussed some of the methods used to determine a member's strength and deflection, while assuming that the member was always in stable equilibrium. Some members, however, may be subjected to compressive loadings, and if these members are long and slender the loading may be large enough to cause the member to deflect laterally or sideway. To be specific, long slender members subjected to an axial compressive force are called *columns*, and the lateral deflection that occurs is called *buckling*. Quite often the buckling of a column can lead to a sudden and dramatic failure of a structure or mechanism, and as a result, special attention must be given to the design of columns so that they can safely support their intended loadings without buckling.

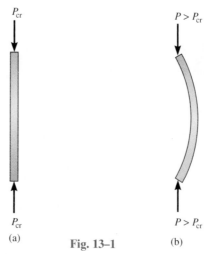

Fig. 13–1

(a)   (b)

The maximum axial load that a column can support when it is on the *verge* of buckling is called the ***critical load***, P_{cr}, Fig. 13–1a. Any additional loading will cause the column to buckle and therefore deflect laterally as shown in Fig. 13–1b. In order to better understand the nature of this instability, consider a two-bar mechanism consisting of weightless bars that are rigid and pin connected at their ends, Fig. 13–2a. When the bars are in the vertical position, the spring, having a stiffness k, is unstretched, and a *small* vertical force **P** is applied at the top of one of the bars. We can upset this equilibrium position by displacing the pin at A by a small amount Δ, Fig. 13–2b. As shown on the free-body diagram of the pin when the bars are displaced, Fig. 13–2c, the spring will produce a restoring force $F = k\Delta$, while the applied load **P** develops two horizontal components, $P_x = P \tan \theta$, which tend to push the pin (and the bars) further out of equilibrium. Since θ is small, $\Delta = \theta(L/2)$ and $\tan \theta \approx \theta$. Thus the *restoring* spring force becomes $F = k\theta L/2$, and the *disturbing* force is $2P_x = 2P\theta$.

If the restoring force is greater than the disturbing force, that is, $k\theta L/2 > 2P\theta$, then, noticing that θ cancels out, we can solve for P, which gives

$$P < \frac{kL}{4} \qquad \text{stable equilibrium}$$

This is a condition for *stable equilibrium* since the force developed by the spring would be adequate to restore the bars back to their vertical position. On the other hand, if $kL\theta/2 < 2P\theta$, or

$$P > \frac{kL}{4} \qquad \text{unstable equilibrium}$$

then the mechanism would be in *unstable equilibrium*. In other words, if this load P is applied, and a slight displacement occurs at A, the mechanism will tend to move out of equilibrium and not be restored to its original position.

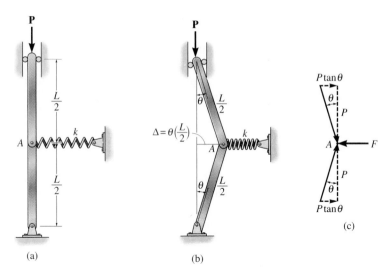

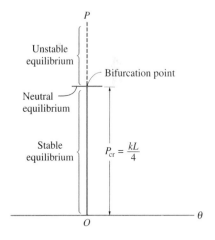

Fig. 13–2

The intermediate value of P, defined by requiring $kL\theta/2 = 2P\theta$, is the *critical load*. Here

$$P_{cr} = \frac{kL}{4} \qquad \text{neutral equilibrium}$$

This loading represents a case of the mechanism being in *neutral equilibrium*. Since P_{cr} is *independent* of the (small) displacement θ of the bars, any slight disturbance given to the mechanism will not cause it to move further out of equilibrium, nor will it be restored to its original position. Instead, the bars will *remain* in the deflected position.

These three different states of equilibrium are represented graphically in Fig. 13–3. The transition point where the load is equal to the critical value $P = P_{cr}$ is called the *bifurcation point*. At this point the mechanism will be in equilibrium for any *small value* of θ, measured either to the right or to the left of the vertical. Physically, P_{cr} represents the load for which the mechanism is on the verge of buckling. It is quite valid to determine this value by assuming *small displacements* as done here; however, it should be understood that P_{cr} may *not* be the largest value of P that the mechanism can support. Indeed, if a larger load is placed on the bars, then the mechanism may have to undergo a further deflection before the spring is compressed or elongated enough to hold the mechanism in equilibrium.

Like the two-bar mechanism just discussed, the critical buckling loads on columns supported in various ways can be obtained, and the method used to do this will be explained in the next section. Although in engineering design the critical load may be considered to be the largest load the column can support, realize that, like the two-bar mechanism in the deflected or buckled position, a column may actually support an

Fig. 13–3

even greater load than P_{cr}. Unfortunately, however, this loading may require the column to undergo a *large* deflection, which is generally not tolerated in engineering structures or machines. For example, it may take only a few newtons of force to buckle a meterstick, but the additional load it may support can be applied only after the stick undergoes a relatively large lateral deflection.

13.2   IDEAL COLUMN WITH PIN SUPPORTS

In this section we will determine the critical buckling load for a column that is pin supported as shown in Fig. 13–4a. The column to be considered is an ***ideal column***, meaning one that is perfectly straight before loading, is made of homogeneous material, and upon which the load is applied through the centroid of the cross section. It is further assumed that the material behaves in a linear-elastic manner and that the column buckles or bends in a single plane. In reality, the conditions of column straightness and load application are never accomplished; however, the analysis to be performed on an "ideal column" is similar to that used to analyze initially crooked columns or those having an eccentric load application. These more realistic cases will be discussed later in this chapter.

   Since an ideal column is straight, theoretically the axial load P could be increased until failure occurs by either fracture or yielding of the material. However, when the critical load P_{cr} is reached, the column is on the verge of becoming unstable, so that a small lateral force F, Fig. 13–4b, will cause the column to remain in the deflected position when F is removed, Fig. 13–4c. Any slight reduction in the axial load P from P_{cr} will allow the column to straighten out, and any slight increase in P, beyond P_{cr}, will cause further increases in lateral deflection.

Some pin-connected members used in moving machinery, such as this short link, are subjected to compressive loads and thus act as columns.

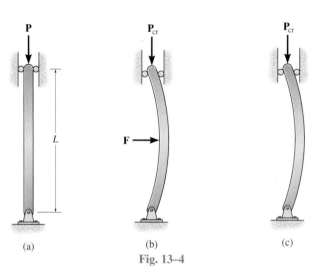

(a)   (b)   (c)

Fig. 13–4

Whether or not a column will remain stable or become unstable when subjected to an axial load will depend on its ability to restore itself, which is based on its resistance to bending. Hence, in order to determine the critical load and the buckled shape of the column, we will apply Eq. 12–10, which relates the internal moment in the column to its deflected shape, i.e.,

$$EI \frac{d^2v}{dx^2} = M \qquad (13\text{–}1)$$

Recall that this equation assumes that the slope of the elastic curve is small* and that deflections occur only by bending. When the column is in its deflected position, Fig. 13–5a, the internal bending moment can be determined by using the method of sections. The free-body diagram of a segment in the deflected position is shown in Fig. 13–5b. Here both the deflection v and the internal moment M are shown in the *positive direction* according to the sign convention used to establish Eq. 13–1. Summing moments, the internal moment is $M = -Pv$. Thus Eq. 13–1 becomes

$$EI \frac{d^2v}{dx^2} = -Pv$$

$$\frac{d^2v}{dx^2} + \left(\frac{P}{EI} \right) v = 0 \qquad (13\text{–}2)$$

This is a homogeneous, second-order, linear differential equation with constant coefficients. It can be shown by using the methods of differential equations, or by direct substitution into Eq. 13–2, that the general solution is

$$v = C_1 \sin\left(\sqrt{\frac{P}{EI}} x \right) + C_2 \cos\left(\sqrt{\frac{P}{EI}} x \right) \qquad (13\text{–}3)$$

The two constants of integration are determined from the boundary conditions at the ends of the column. Since $v = 0$ at $x = 0$, then $C_2 = 0$. And since $v = 0$ at $x = L$, then

$$C_1 \sin\left(\sqrt{\frac{P}{EI}} L \right) = 0$$

This equation is satisfied if $C_1 = 0$; however, then $v = 0$, which is a *trivial solution* that requires the column to always remain straight, even though the load causes the column to become unstable. The other possibility is for

$$\sin\left(\sqrt{\frac{P}{EI}} L \right) = 0$$

which is satisfied if

$$\sqrt{\frac{P}{EI}} L = n\pi$$

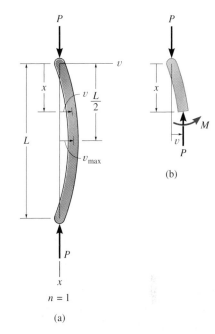

Fig. 13–5 a,b

*If large deflections are to be considered, the more accurate differential equation, Eq. 12–4, $EI(d^2v/dx^2)/[1 + (dv/dx)^2]^{3/2} = M$ must be used.

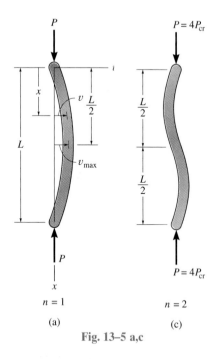

$n = 1$ $n = 2$

(a)   (c)

Fig. 13–5 a,c

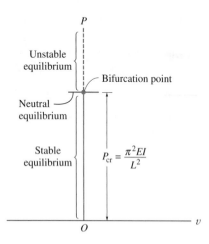

Fig. 13–6

or

$$P = \frac{n^2 \pi^2 EI}{L^2} \quad n = 1, 2, 3, \ldots \quad (13\text{–}4)$$

The *smallest value* of P is obtained when $n = 1$, so the *critical load* for the column is therefore

$$P_{cr} = \frac{\pi^2 EI}{L^2}$$

This load is sometimes referred to as the *Euler load*, after the Swiss mathematician Leonhard Euler, who originally solved this problem in 1757. The corresponding buckled shape is defined by the equation

$$v = C_1 \sin \frac{\pi x}{L}$$

Here the constant C_1 represents the maximum deflection, v_{max}, which occurs at the midpoint of the column, Fig. 13–5a. Specific values for C_1 cannot be obtained, since the exact deflected form for the column is unknown once it has buckled. It has been assumed, however, that this deflection is small.

Realize that n in Eq. 13–4 represents the number of waves in the deflected shape of the column. For example, if $n = 2$, then from Eqs. 13–3 and 13–4, *two waves* will appear in the buckled shape, Fig. 13–5c, and the column will support a critical load that is $4P_{cr}$ just prior to buckling. Since this value is four times the critical load and the deflected shape is unstable, this form of buckling, practically speaking, will not exist.

Like the two-bar mechanism discussed in Sec. 13–1, we can represent the load-deflection characteristics of the ideal column by the graph shown in Fig. 13–6. The bifurcation point represents the state of *neutral equilibrium*, at which point the *critical load* acts on the column. Here the column is on the verge of impending buckling.

It should be noted that the critical load is independent of the strength of the material; rather it depends only on the column's dimensions (I and L) and the material's stiffness or modulus of elasticity E. For this reason, as far as elastic buckling is concerned, columns made, for example, of high-strength steel offer no advantage over those made of lower-strength steel, since the modulus of elasticity for both is approximately the same. Also note that the load-carrying capacity of a column will increase as the moment of inertia of the cross section increases. Thus, efficient columns are designed so that most of the column's cross-sectional area is located as far away as possible from the principal centroidal axes for the section. This is why hollow sections such as tubes are more economical than solid sections. Furthermore, wide-flange sections, and columns that are "built up" from channels, angles, plates, etc., are better than sections that are solid and rectangular.

It is also important to realize that a column will buckle about the principal axis of the cross section having the **least moment of inertia** (the weakest axis). For example, a column having a rectangular cross section, like a meter stick, as shown in Fig. 13–7, will buckle about the $a-a$ axis, not the $b-b$ axis. As a result, engineers usually try to achieve a balance, keeping the moments of inertia the same in all directions. Geometrically, then, circular tubes would make excellent columns. Also, square tubes or those shapes having $I_x \approx I_y$ are often selected for columns.

Summarizing the above discussion, the buckling equation for a pin-supported long slender column can be rewritten, and the terms defined as follows:

$$P_{cr} = \frac{\pi^2 EI}{L^2} \qquad (13\text{–}5)$$

where

P_{cr} = critical or maximum axial load on the column just before it begins to buckle. This load must *not* cause the stress in the column to exceed the proportional limit

E = modulus of elasticity for the material

I = *least* moment of inertia for the column's cross-sectional area

L = unsupported length of the column, whose ends are pinned

Fig. 13–7

For purposes of design, Eq. 13–5 can also be written in a more useful form by expressing $I = Ar^2$, where A is the cross-sectional area and r is the **radius of gyration** of the cross-sectional area. Thus,

$$P_{cr} = \frac{\pi^2 E(Ar^2)}{L^2}$$

$$\left(\frac{P}{A}\right)_{cr} = \frac{\pi^2 E}{(L/r)^2}$$

or

$$\sigma_{cr} = \frac{\pi^2 E}{(L/r)^2} \qquad (13\text{–}6)$$

Typical interior steel pipe columns used to support the roof of a single story building.

Here

σ_{cr} = critical stress, which is an *average stress* in the column just before the column buckles. This stress is an *elastic stress* and therefore $\sigma_{cr} \leq \sigma_Y$

E = modulus of elasticity for the material

L = unsupported length of the column, whose ends are pinned

r = *smallest* radius of gyration of the column, determined from $r = \sqrt{I/A}$, where I is the *least* moment of inertia of the column's cross-sectional area A

The geometric ratio L/r in Eq. 13–6 is known as the **slenderness ratio**. It is a measure of the column's flexibility, and as will be discussed later, it serves to classify columns as long, intermediate, or short.

It is possible to graph Eq. 13–6 using axes that represent the critical stress versus the slenderness ratio. Examples of this graph for columns made of a typical structural steel and aluminum alloy are shown in Fig. 13–8. Note that the curves are hyperbolic and are valid only for critical stresses below the material's yield point (proportional limit), since the material must behave elastically. For the steel the yield stress is $(\sigma_Y)_{st} = 36$ ksi $[E_{st} = 29(10^3)$ ksi$]$, and for the aluminum it is $(\sigma_Y)_{al} = 27$ ksi $[E_{al} = 10(10^3)$ ksi$]$. Substituting $\sigma_{cr} = \sigma_Y$ into Eq. 13–6, the *smallest* acceptable slenderness ratios for the steel and aluminum columns are therefore $(L/r)_{st} = 89$ and $(L/r)_{al} = 60.5$, respectively. Thus, for a steel column, if $(L/r)_{st} \geq 89$, Euler's formula can be used to determine the buckling load since the stress in the column remains elastic. On the other hand, if $(L/r)_{st} < 89$, the column's stress will exceed the yield point before buckling can occur, and therefore the Euler formula is not valid in this case.

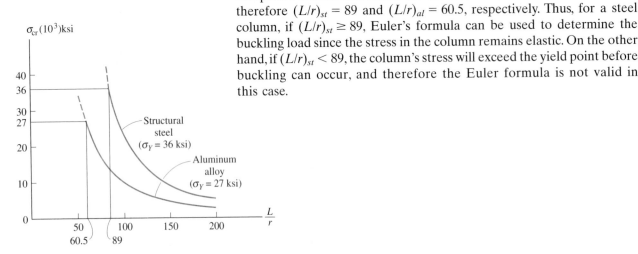

Fig. 13–8

IMPORTANT POINTS

• *Columns* are long slender members that are subjected to axial loads.

• The *critical load* is the maximum axial load that a column can support when it is on the verge of buckling. This loading represents a case of *neutral equilibrium*.

• An *ideal column* is initially perfectly straight, made of homogeneous material, and the load is applied through the centroid of the cross section.

• A pin-connected column will buckle about the principal axis of the cross section having the *least* moment of inertia.

• The *slenderness ratio* is L/r, where r is the smallest radius of gyration of the cross section. Buckling will occur about the axis where this ratio gives the greatest value.

EXAMPLE 13-1

A 24-ft-long A-36 steel tube having the cross section shown in Fig. 13–9 is to be used as a pin-ended column. Determine the maximum allowable axial load the column can support so that it does not buckle.

SOLUTION

Using Eq. 13–5 to obtain the critical load with $E_{st} = 29(10^3)$ ksi,

$$P_{cr} = \frac{\pi^2 EI}{L^2}$$

$$= \frac{\pi^2 [29(10^3) \text{ kip/in}^2](\frac{1}{4}\pi(3)^4 - \frac{1}{4}\pi(2.75)^4) \text{ in}^4}{[24 \text{ ft}(12 \text{ in./ft})]^2}$$

$$= 64.5 \text{ kip} \qquad\qquad\qquad \textit{Ans.}$$

This force creates an average compressive stress in the column of

$$\sigma_{cr} = \frac{P_{cr}}{A} = \frac{64.5 \text{ kip}}{[\pi(3)^2 - \pi(2.75)^2] \text{ in}^2} = 14.3 \text{ ksi}$$

Since $\sigma_{cr} < \sigma_Y = 36$ ksi, application of Euler's equation is appropriate.

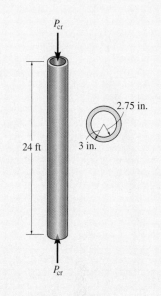

Fig. 13–9

EXAMPLE 13-2

The A-36 steel $W\ 8 \times 31$ member shown in Fig. 13–10 is to be used as a pin-connected column. Determine the largest axial load it can support before it either begins to buckle or the steel yields.

SOLUTION

From the table in Appendix B, the column's cross-sectional area and moments of inertia are $A = 9.13$ in^2, $I_x = 110$ in^4, and $I_y = 37.1$ in^4. By inspection, buckling will occur about the $y-y$ axis. Why? Applying Eq. 13–5, we have

$$P_{cr} = \frac{\pi^2 EI}{L^2} = \frac{\pi^2 [29(10^3) \text{ kip/in}^2](37.1 \text{ in}^4)}{[12 \text{ ft}(12 \text{ in./ft})]^2} = 512 \text{ kip}$$

When fully loaded, the average compressive stress in the column is

$$\sigma_{cr} = \frac{P_{cr}}{A} = \frac{512 \text{ kip}}{9.13 \text{ in}^2} = 56.1 \text{ ksi}$$

Since this stress exceeds the yield stress (36 ksi), the load P is determined from simple compression:

$$36 \text{ ksi} = \frac{P}{9.13 \text{ in}^2}; \qquad\qquad P = 329 \text{ kip} \qquad\qquad \textit{Ans.}$$

In actual practice, a factor of safety would be placed on this loading.

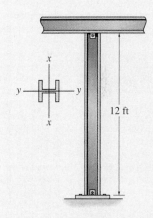

Fig. 13–10

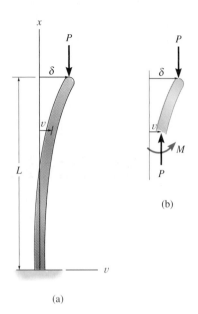

Fig. 13-11

The tubular columns used to support this water tank have been braced at three locations along their length to prevent them from buckling.

13.3 COLUMNS HAVING VARIOUS TYPES OF SUPPORTS

In Sec. 13.2 we derived the Euler load for a column that is pin connected or free to rotate at its ends. Oftentimes, however, columns may be supported in some other way. For example, consider the case of a column fixed at its base and free at the top, Fig. 13–11a. Determination of the buckling load on this column follows the same procedure as that used for the pinned column. From the free-body diagram in Fig. 13–11b, the internal moment at the arbitrary section is $M = P(\delta - v)$. Consequently, the differential equation for the deflection curve is

$$EI \frac{d^2v}{dx^2} = P(\delta - v)$$

$$\frac{d^2v}{dx^2} + \frac{P}{EI}v = \frac{P}{EI}\delta \qquad (13-7)$$

Unlike Eq. 13–2, this equation is nonhomogeneous because of the nonzero term on the right side. The solution consists of both a complementary and particular solution, namely,

$$v = C_1 \sin\left(\sqrt{\frac{P}{EI}}x\right) + C_2 \cos\left(\sqrt{\frac{P}{EI}}x\right) + \delta$$

The constants are determined from the boundary conditions. At $x = 0, v = 0$, so that $C_2 = -\delta$. Also,

$$\frac{dv}{dx} = C_1\sqrt{\frac{P}{EI}} \cos\left(\sqrt{\frac{P}{EI}}x\right) - C_2\sqrt{\frac{P}{EI}} \sin\left(\sqrt{\frac{P}{EI}}x\right)$$

At $x = 0, dv/dx = 0$, so that $C_1 = 0$. The deflection curve is therefore

$$v = \delta\left[1 - \cos\left(\sqrt{\frac{P}{EI}}x\right)\right] \qquad (13-8)$$

Since the deflection at the top of the column is δ, that is, at $x = L$, $v = \delta$, we require

$$\delta \cos\left(\sqrt{\frac{P}{EI}}L\right) = 0$$

The trivial solution $\delta = 0$ indicates that no buckling occurs, regardless of the load P. Instead,

$$\cos\left(\sqrt{\frac{P}{EI}}L\right) = 0 \qquad \text{or} \qquad \sqrt{\frac{P}{EI}}L = \frac{n\pi}{2}$$

The smallest critical load occurs when $n = 1$, so that

$$P_{cr} = \frac{\pi^2 EI}{4L^2} \qquad (13-9)$$

By comparison with Eq. 13–5 it is seen that a column fixed-supported at its base will carry only one-fourth the critical load that can be applied to a pin-supported column.

Other types of supported columns are analyzed in much the same way and will not be covered in detail here.* Instead, we will tabulate the results for the most common types of column support and show how to apply these results by writing Euler's formula in a general form.

Effective Length.   As stated previously, the Euler formula, Eq. 13–5, was developed for the case of a column having ends that are pinned or free to rotate. In other words, L in the equation represents the unsupported distance between the points of zero moment. If the column is supported in other ways, then Euler's formula can be used to determine the critical load provided "L" represents the distance between the zero-moment points. This distance is called the column's **effective length**, L_e. Obviously, for a pin-ended column $L_e = L$, Fig. 13–12a. For the fixed and free-ended column analyzed above, the deflection curve was found to be one-half that of a column that is pin-connected and has a length of $2L$, Fig. 13–12b. Thus the effective length between the points of zero moment is $L_e = 2L$. Examples for two other columns with different end supports are also shown in Fig. 13–12. The column fixed at its ends, Fig. 13–12c, has inflection points or points of zero moment $L/4$ from each support. The effective length is therefore represented by the middle half of its length, that is, $L_e = 0.5L$. Lastly, the pin- and fixed-ended column, Fig. 13–12d, has an inflection point at approximately $0.7L$ from its pinned end, so that $L_e = 0.7L$.

Rather than specifying the column's effective length, many design codes provide column formulas that employ a dimensionless coefficient K called the **effective-length factor**. K is defined from

$$L_e = KL \tag{13–10}$$

Specific values of K are also given in Fig. 13–12. Based on this generality, we can therefore write Euler's formula as

$$P_{cr} = \frac{\pi^2 EI}{(KL)^2} \tag{13–11}$$

or

$$\sigma_{cr} = \frac{\pi^2 E}{(KL/r)^2} \tag{13–12}$$

Here (KL/r) is the column's **effective-slenderness ratio**. For example, note that for the column fixed at its base and free at its end, we have $K = 2$, and therefore Eq. 13–11 gives the same result as Eq. 13–9.

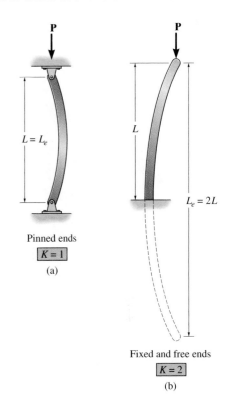

P

$L = L_e$

Pinned ends

$\boxed{K = 1}$

(a)

P

L

$L_e = 2L$

Fixed and free ends

$\boxed{K = 2}$

(b)

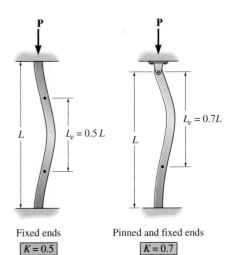

P

L

$L_e = 0.5L$

Fixed ends

$\boxed{K = 0.5}$

(c)

P

L

$L_e = 0.7L$

Pinned and fixed ends

$\boxed{K = 0.7}$

(d)

Fig. 13–12

*See Problems 13–42, 13–43, 13–44, and 13–45.

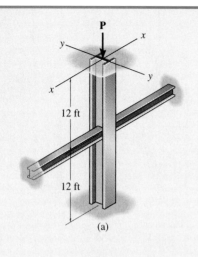

12 ft

12 ft

(a)

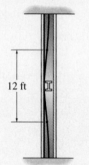

12 ft

x–x axis buckling

(b)

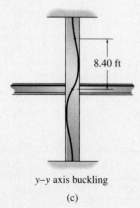

8.40 ft

y–y axis buckling

(c)

Fig. 13–13

EXAMPLE 13–3

A $W\,6 \times 15$ steel column is 24 ft long and is fixed at its ends as shown in Fig. 13–13a. Its load-carrying capacity is increased by bracing it about the $y-y$ (weak) axis using struts that are assumed to be pin-connected to its midheight. Determine the load it can support so that the column does not buckle nor the material exceed the yield stress. Take $E_{st} = 29(10^3)$ ksi and $\sigma_Y = 60$ ksi.

SOLUTION

The buckling behavior of the column will be *different* about the x and y axes due to the bracing. The buckled shape for each of these cases is shown in Figs. 13–13b and 13–13c. From Fig. 13–13b, the effective length for buckling about the $x-x$ axis is $(KL)_x = 0.5(24\ \text{ft}) = 12\ \text{ft} = 144$ in., and from Fig. 13–13c, for buckling about the $y-y$ axis, $(KL)_y = 0.7(24\ \text{ft}/2) = 8.40\ \text{ft} = 100.8$ in. The moments of inertia for a $W\,6 \times 15$ are determined from the table in Appendix B. We have $I_x = 29.1$ in^4, $I_y = 9.32$ in^4.

Applying Eq. 13–11, we have

$$(P_{cr})_x = \frac{\pi^2 E I_x}{(KL)_x^2} = \frac{\pi^2 [29(10^3)\ \text{ksi}]29.1\ \text{in}^4}{(144\ \text{in.})^2} = 401.7\ \text{kip} \qquad (1)$$

$$(P_{cr})_y = \frac{\pi^2 E I_y}{(KL)_y^2} = \frac{\pi^2 [29(10^3)\ \text{ksi}]9.32\ \text{in}^4}{(100.8\ \text{in.})^2} = 262.5\ \text{kip} \qquad (2)$$

By comparison, buckling will occur about the $y-y$ axis.

The area of the cross section is 4.43 in^2, so the average compressive stress in the column will be

$$\sigma_{cr} = \frac{P_{cr}}{A} = \frac{262.5\ \text{kip}}{4.43\ \text{in}^2} = 59.3\ \text{ksi}$$

Since this stress is less than the yield stress, buckling will occur before the material yields. Thus,

$$P_{cr} = 263\ \text{kip} \qquad \qquad \textit{Ans.}$$

Note: From Eq. 13–11 it can be seen that buckling will always occur about the column axis having the *largest* slenderness ratio, since a large slenderness ratio will give a small critical load. Thus, using the data for the radius of gyration from the table in Appendix B, we have

$$\left(\frac{KL}{r}\right)_x = \frac{144\ \text{in.}}{2.56\ \text{in.}} = 56.2$$

$$\left(\frac{KL}{r}\right)_y = \frac{100.8\ \text{in.}}{1.46\ \text{in.}} = 69.0$$

Hence, $y-y$ axis buckling will occur, which is the same conclusion reached by comparing Eqs. 1 and 2.

EXAMPLE 13–4

The aluminum column is fixed at its bottom and is braced at its top by cables so as to prevent movement at the top along the x axis, Fig. 13–14a. If it is assumed to be fixed at its base, determine the largest allowable load P that can be applied. Use a factor of safety for buckling of F.S. = 3.0. Take $E_{al} = 70$ GPa, $\sigma_Y = 215$ MPa, $A = 7.5(10^{-3})$ m^2, $I_x = 61.3(10^{-6})$ m^4, $I_y = 23.2(10^{-6})$ m^4.

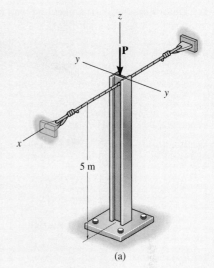

(a)

SOLUTION

Buckling about the x and y axes is shown in Fig. 13–14b and 13–14c, respectively. Using Fig. 13–12, for x–x axis buckling, $K = 2$, so $(KL)_x = 2(5$ m$) = 10$ m. Also, for y–y axis buckling, $K = 0.7$, so $(KL)_y = 0.7(5$ m$) = 3.5$ m.

Applying Eq. 13–11, the critical loads for each case are

$$(P_{cr})_x = \frac{\pi^2 EI_x}{(KL)_x^2} = \frac{\pi^2[70(10^9) \text{ N/m}^2](61.3(10^{-6}) \text{ m}^4)}{(10 \text{ m})^2}$$

$$= 424 \text{ kN}$$

$$(P_{cr})_y = \frac{\pi^2 EI_y}{(KL)_y^2} = \frac{\pi^2[70(10^9) \text{ N/m}^2](23.2(10^{-6}) \text{ m}^4)}{(3.5 \text{ m})^2}$$

$$= 1.31 \text{ MN}$$

By comparison, as P is increased the column will buckle about the x–x axis. The allowable load is therefore

$$P_{allow} = \frac{P_{cr}}{\text{F.S.}} = \frac{424 \text{ kN}}{3.0} = 141 \text{ kN} \qquad Ans.$$

Since

$$\sigma_{cr} = \frac{P_{cr}}{A} = \frac{424 \text{ kN}}{7.5(10^{-3}) \text{ m}^2} = 56.5 \text{ MPa} < 215 \text{ MPa}$$

Euler's equation can be applied.

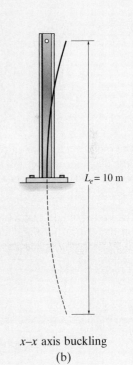

$L_e = 10$ m

x–x axis buckling

(b)

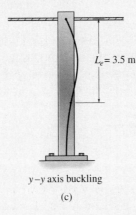

$L_e = 3.5$ m

y–y axis buckling

(c)

Fig. 13–14

PROBLEMS

13–1. Determine the critical buckling load for the column. The material can be assumed rigid.

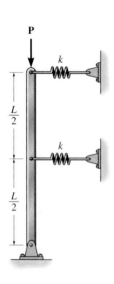

Prob. 13–1

13–2. The rod is made from an A-36 steel rod. Determine the smallest diameter of the rod, to the nearest $\frac{1}{16}$ in., that will support the load of $P = 5$ kip without buckling. The ends are roller supported.

13–3. The rod is made from a 1-in.-diameter steel rod. Determine the critical buckling load if the ends are roller supported. $E_{st} = 29(10^3)$ ksi, $\sigma_Y = 50$ ksi.

Probs. 13–2/13–3

*13–4. The column consists of a rigid member that is pinned at its bottom and attached to a spring at its top. If the spring is unstretched when the column is in the vertical position, determine the critical load that can be placed on the column.

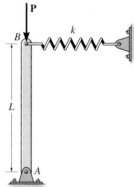

Prob. 13–4

13–5. The aircraft link is made from an A-36 steel rod. Determine the smallest diameter of the rod, to the nearest $\frac{1}{16}$ in., that will support the load of 4 kip without buckling. The ends are pin connected.

Prob. 13–5

13–6. An A-36 steel column has a length of 4 m and is pinned at both ends. If the cross sectional area has the dimensions shown, determine the critical load.

13–7. Solve Prob. 13–6 if the column is fixed at its bottom and pinned at its top.

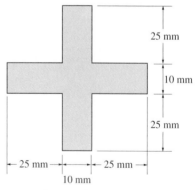

Probs. 13–6/13–7

*13–8. The W8 × 67 is used as a structural A-36 steel column that can be assumed fixed at its base and pinned at its top. Determine the largest axial force P that can be applied without causing it to buckle.

13–9. Solve Prob. 13–8 if the column is assumed fixed at its bottom and free at its top.

25 ft

Probs. 13–8/13–9

13–10. A steel column has a length of 9 m and is fixed at both ends. If the cross-sectional area has the dimensions shown, determine the critical load. $E_{st} = 200$ GPa, $\sigma_Y = 250$ MPa.

13–11. Solve Prob. 13–10 if the column is pinned at its top and bottom.

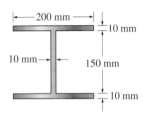

200 mm
10 mm
10 mm
150 mm
10 mm

Probs. 13–10/13–11

*13–12. A rod made from polyurethane has a stress–strain diagram in compression as shown. If the rod is pinned at its ends and is 37 in. long, determine its smallest diameter so it does not fail from elastic buckling.

13–13. A rod made from polyurethane has a stress–strain diagram in compression as shown. If the rod is pinned at its top and fixed at its base, and is 37 in. long, determine its smallest diameter so it does not fail from elastic buckling.

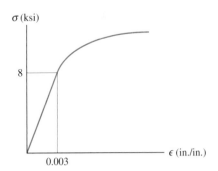

σ (ksi)

8

0.003

ϵ (in./in.)

Probs. 13–12/13–13

13–14. The 10-ft wooden rectangular column has the dimensions shown. Determine the critical load if the ends are assumed to be pin connected. $E_w = 1.6(10^3)$ ksi, $\sigma_Y = 5$ ksi.

13–15. The 10-ft column has the dimensions shown. Determine the critical load if the bottom is fixed and the top is pinned. $E_w = 1.6(10^3)$ ksi, $\sigma_Y = 5$ ksi.

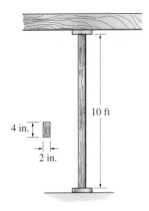

10 ft

4 in.

2 in.

Probs. 13–14/13–15

*13–16. An L-2 tool steel link in a forging machine is pin connected to the forks at its ends as shown. Determine the maximum load P it can carry without buckling. Use a factor of safety with respect to buckling of F.S. = 1.75. Note from the figure on the left that the ends are pinned for buckling, whereas from the figure on the right the ends are fixed.

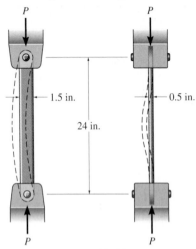

1.5 in.   0.5 in.
24 in.

Prob. 13–16

13–17. The W 12 × 87 structural A-36 steel column has a length of 12 ft. If its bottom end is fixed supported while its top is free, and it is subjected to an axial load of $P = 380$ kip, determine the factor of safety with respect to buckling.

13–18. The W 12 × 87 structural A-36 steel column has a length of 12 ft. If its bottom end is fixed supported while its top is free, determine the largest axial load it can support. Use a factor of safety with respect to buckling of 1.75.

P
12 ft

Probs. 13–17/13–18

13–19. The handle is used to operate a simple press used to crush cans. Determine the maximum force P that can be applied to the handle so that the rod BC does not buckle. The rod is made of steel and has a diameter of 0.5 in. It is pin connected at its ends. $E_{st} = 29(10^3)$ ksi, $\sigma_Y = 36$ ksi.

*13–20. The handle is used to operate a simple press used to crush cans. Determine the smallest diameter steel rod BC, to the nearest $\frac{1}{8}$ in., that can be used if the maximum force P applied to the handle is $P = 60$ lb. The rod is pin connected at its ends. $E_{st} = 29(10^3)$ ksi $\sigma_Y = 36$ ksi.

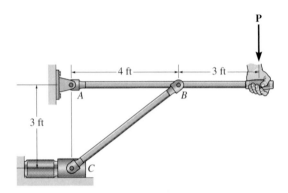

P
4 ft   3 ft
A   B
3 ft
C

Probs. 13–19/13–20

13–21. The 12-ft A-36 steel pipe column has an outer diameter of 3 in. and a thickness of 0.25 in. Determine the critical load if the ends are assumed to be pin connected.

13–22. The 12-ft A-36 steel column has an outer diameter of 3 in. and a thickness of 0.25 in. Determine the critical load if the bottom is fixed and the top is pinned.

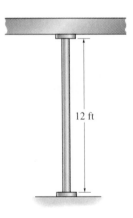

12 ft

Probs. 13–21/13–22

13–23. The A-36 steel pipe has an outer diameter of 2 in. and a thickness of 0.5 in. If it is held in place by a guywire, determine the largest horizontal force P that can be applied without causing the pipe to buckle. Assume that the ends of the pipe are pin connected.

*13–24. The A-36 steel pipe has an outer diameter of 2 in. If it is held in place by a guywire, determine the pipe's required inner diameter to the nearest $\frac{1}{8}$ in., so that it can support a maximum horizontal load of $P = 4$ kip without causing the pipe to buckle. Assume the ends of the pipe are pin connected.

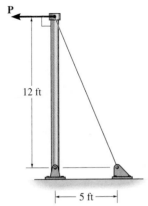

Probs. 13–23/13–24

13–25. The truss is made from A-36 steel bars, each of which has a circular cross section with a diameter of 1.5 in. Determine the maximum force P that can be applied without causing any of the members to buckle. The members are pin connected at their ends.

13–26. The truss is made from A-36 steel bars, each of which has a circular cross section. If the applied load $P = 10$ kip, determine the diameter of member AB to the nearest $\frac{1}{8}$ in. that will prevent this member from buckling. The members are pin supported at their ends.

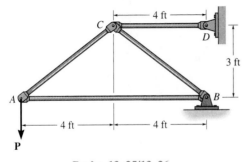

Probs. 13–25/13–26

13–27. The members of the truss are assumed to be pin connected. If member GF is an A-36 steel rod having a diameter of 2 in., determine the greatest magnitude of load **P** that can be supported by the truss without causing this member to buckle.

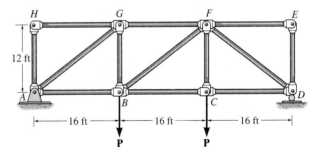

Prob. 13–27

*13–28. The members of the truss are assumed to be pin connected. If member AG is an A-36 steel rod having a diameter of 2 in., determine the greatest magnitude of load **P** that can be supported by the truss without causing this member to buckle.

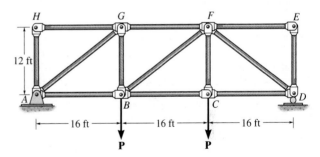

Prob. 13–28

13–29. Determine the maximum force P that can be applied to the handle so that the A-36 steel control rod BC does not buckle. The rod has a diameter of 25 mm.

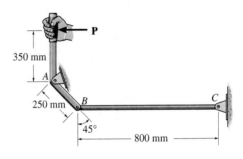

Prob. 13–29

13–30. The linkage is made using two A-36 steel rods, each having a circular cross section. If each rod has a diameter of $\frac{3}{4}$ in., determine the largest load it can support without causing any rod to buckle. Assume that the rods are pin connected at their ends.

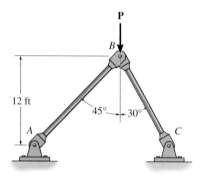

Prob. 13–30

13–31. The A-36 steel bar AB has a square cross section. If it is pin-connected at its ends, determine the maximum allowable load P that can be applied to the frame. Use a factor of safety with respect to buckling of 2.

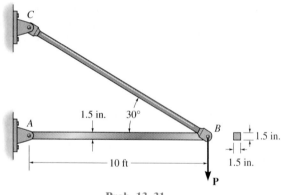

Prob. 13–31

*13–32. The 50-mm diameter C86100 bronze rod is fixed supported at A and has a gap of 2 mm from the wall at B. Determine the increase in temperature ΔT that will cause the rod to buckle. Assume that the contact at B acts as a pin.

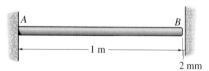

Prob. 13–32

13–33. Determine if the frame can support a load of $w = 6$ kN/m if the factor of safety with respect to buckling of member AB is 3. Assume that AB is made of steel and is pinned at its ends for $x-x$ axis buckling and fixed at its ends for $y-y$ axis buckling. $E_{st} = 200$ GPa, $\sigma_Y = 360$ MPa.

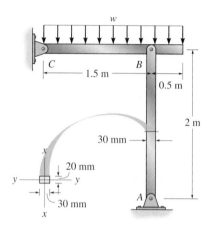

Prob. 13–33

13–34. The A-36 steel bar AB of the frame is pin connected at its ends. If $P = 30$ kip, determine the factor of safety with respect to buckling about the $y-y$ axis due to the applied loading.

13–35. The A-36 steel bar AB of the frame is pin connected at its ends. Determine the largest load P that can be applied to the frame without causing it to buckle about the $y-y$ axis.

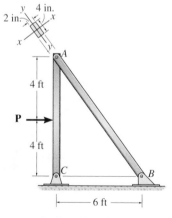

Probs. 13–34/13–35

*13–36. The deck is supported by the two 40-mm-square columns. Column *AB* is pinned at *A* and fixed at *B*, whereas *CD* is pinned at *C* and *D*. If the deck is prevented from sidesway, determine the greatest weight of the load that can be applied without causing the deck to collapse. The center of gravity of the load is located at $d = 2$ m. Both columns are made from Douglas fir.

13–37. The deck is supported by the two 40-mm-square columns. Column *AB* is pinned at *A* and fixed at *B*, whereas *CD* is pinned at *C* and *D*. If the deck is prevented from sidesway, determine the position *d* of the center of gravity of the load and the load's greatest magnitude without causing the deck to collapse. Both columns are made from Douglas fir.

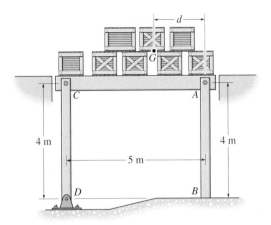

Probs. 13–36/13–37

13–38. The steel bar *AC* of the frame is pin connected at its ends. Determine the factor of safety with respect to buckling about the *y–y* axis due to the applied loading of $P = 15$ kN. $E_{st} = 200$ GPa, $\sigma_Y = 360$ MPa.

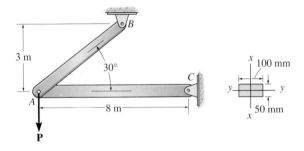

Prob. 13–38

13–39. The beam is supported by the three pin-connected suspender bars, each having a diameter of 0.5 in. and made from A-36 steel. Determine the greatest uniform load *w* that can be applied to the beam without causing *AB* and *CB* to buckle.

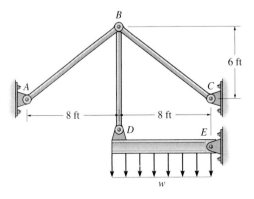

Prob. 13–39

*13–40. Determine the maximum distributed loading that can be applied to the wide-flange beam so that the brace *CD* does not buckle. The brace is an A-36 steel rod having a diameter of 50 mm.

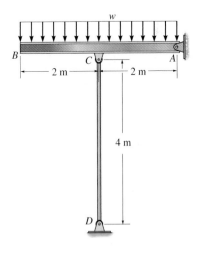

Prob. 13–40

13–41. The column is supported at B by a support that does not permit rotation but allows vertical deflection. Determine the critical load P_{cr}. EI is constant.

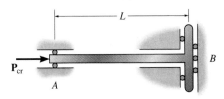

Prob. 13–41

13–42. Consider an ideal column as in Fig. 13–12c, having both ends fixed. Show that the critical load on the column is given by $P_{cr} = 4\pi^2EI/L^2$. *Hint:* Due to the vertical deflection of the top of the column, a constant moment $\mathbf{M}'$ will be developed at the supports. Show that $d^2v/dx^2 + (P/EI)v = M'/EI$. The solution is of the form $v = C_1 \sin(\sqrt{P/EI}x) + C_2 \cos(\sqrt{P/EI}x) + M'/P$.

13–43. Consider an ideal column as in Fig. 13–12d, having one end fixed and the other pinned. Show that the critical load on the column is given by $P_{cr} = 20.19\, EI/L^2$. *Hint:* Due to the vertical deflection at the top of the column, a constant moment $\mathbf{M}'$ will be developed at the fixed support and horizontal reactive forces $\mathbf{R}'$ will be developed at both supports. Show that $d^2v/dx^2 + (P/EI)v = (R'/EI)(L - x)$. The solution is of the form $v = C_1 \sin(\sqrt{P/EI}x) + C_2 \cos(\sqrt{P/EI}x) + (R'/P)(L - x)$. After application of the boundary conditions show that $\tan(\sqrt{P/EI}\,L) = \sqrt{P/EI}\,L$. Solve by trial and error for the smallest root.

*__13–44.__ The ideal column is subjected to the force $\mathbf{F}$ at its midpoint and the axial load $\mathbf{P}$. Determine the maximum displacement and the maximum moment in the column at midspan. EI is constant. *Hint:* Establish the differential equation for deflection Eq. 13–1. The general solution is $v = A \sin kx + B \cos kx - c^2x/k^2$, where $c^2 = F/2EI$, $k^2 = P/EI$.

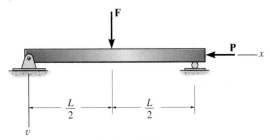

Prob. 13–44

13–45. The ideal column has a weight w (force/length) and rests in the horizontal position when it is subjected to the axial load $\mathbf{P}$. Determine the maximum displacement and the maximum moment in the column at midspan. EI is constant. *Hint:* Establish the differential equation for deflection Eq. 13–1, with the origin at the midspan. The general solution is $v = A \sin kx + B \cos kx + (w/2P)x^2 - (wL/2P)x - (wEI/P^2)$, where $k^2 = P/EI$.

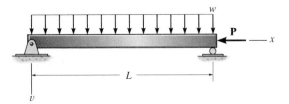

Prob. 13–45

*13.4 THE SECANT FORMULA

The Euler formula was derived with the assumptions that the load P is always applied through the centroid of the column's cross-sectional area and that the column is perfectly straight. This is actually quite unrealistic, since manufactured columns are *never* perfectly straight, nor is the application of the load known with great accuracy. In reality, then, columns never *suddenly buckle*; instead they begin to bend, although ever so slightly, immediately upon application of the load. As a result, the actual criterion for load application will be limited either to a specified deflection of the column or by not allowing the maximum stress in the column to exceed an allowable stress.

To study this effect, we will apply the load P to the column at a short *eccentric distance e* from the centroid of the cross section, Fig. 13–15a. This loading on the column is statically equivalent to the axial load P and bending moment $M' = Pe$ shown in Fig. 13–15b. As shown, in both cases, the ends A and B are supported so that they are free to rotate (pin-supported). As before, we will only consider small slopes and deflections and linear-elastic material behavior. Furthermore, the $x-v$ plane is a plane of symmetry for the cross-sectional area.

From the free-body diagram of the arbitrary section, Fig. 13–15c, the internal moment in the column is

$$M = -P(e + v) \qquad (13\text{–}13)$$

The differential equation for the deflection curve is therefore

$$EI \frac{d^2v}{dx^2} = -P(e + v)$$

These timber columns can be considered pinned at their bottom and fixed connected to the beams at their tops. Deflection of the beams will cause the columns to be eccentrically loaded.

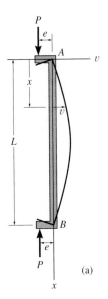

(a)

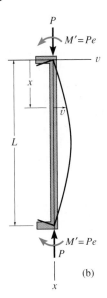

(b)

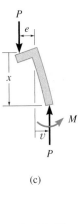

(c)

Fig. 13–15

or

$$\frac{d^2v}{dx^2} + \frac{P}{EI}v = -\frac{P}{EI}e$$

This equation is similar to Eq. 13–7 and has a general solution consisting of the complementary and particular solutions, namely,

$$v = C_1 \sin \sqrt{\frac{P}{EI}}x + C_2 \cos \sqrt{\frac{P}{EI}}x - e \qquad (13\text{--}14)$$

To evaluate the constants we must apply the boundary conditions. At $x = 0, v = 0$, so $C_2 = e$. And at $x = L, v = 0$, which gives

$$C_1 = \frac{e[1 - \cos(\sqrt{P/EI}\ L)]}{\sin(\sqrt{P/EI}\ L)}$$

Since $1 - \cos(\sqrt{P/EI}\ L) = 2 \sin^2(\sqrt{P/EI}\ L/2)$ and $\sin(\sqrt{P/EI}\ L) = 2 \sin(\sqrt{P/EI}\ L/2)\cos(\sqrt{P/EI}\ L/2)$, we have

$$C_1 = e \tan\left(\sqrt{\frac{P}{EI}}\ \frac{L}{2}\right)$$

Hence, the deflection curve, Eq. 13–14, can be written as

$$v = e\left[\tan\left(\sqrt{\frac{P}{EI}}\ \frac{L}{2}\right) \sin\left(\sqrt{\frac{P}{EI}}x\right) + \cos\left(\sqrt{\frac{P}{EI}}x\right) - 1\right] \qquad (13\text{--}15)$$

Maximum Deflection.   Due to symmetry of loading, both the maximum deflection and maximum stress occur at the column's midpoint. Therefore, when $x = L/2, v = v_{max}$, so

$$v_{max} = e\left[\sec\left(\sqrt{\frac{P}{EI}}\ \frac{L}{2}\right) - 1\right] \qquad (13\text{--}16)$$

Notice that if e approaches zero, then v_{max} approaches zero. However, if the terms in the brackets approach infinity as e approaches zero, then v_{max} will have a nonzero value. Mathematically, this would represent the behavior of an axially loaded column at failure when subjected to the critical load P_{cr}. Therefore, to find P_{cr} we require

$$\sec\left(\sqrt{\frac{P_{cr}}{EI}}\ \frac{L}{2}\right) = \infty$$

$$\sqrt{\frac{P_{cr}}{EI}}\ \frac{L}{2} = \frac{\pi}{2}$$

$$P_{cr} = \frac{\pi^2 EI}{L^2} \qquad (13\text{--}17)$$

which is the same result found from the Euler formula, Eq. 13–5.

If Eq. 13–16 is plotted as load P versus deflection v_{max} for various values of eccentricity e, the family of colored curves shown in Fig. 13–16 results. Here the critical load becomes an asymptote to the curves, and of course represents the unrealistic case of an ideal column ($e = 0$). As

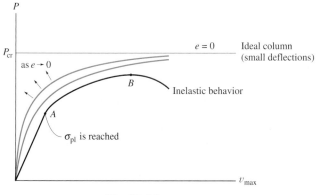

Fig. 13–16

stated earlier, *e* is *never zero* due to imperfections in initial column straightness and load application; however, as $e \to 0$, the curves tend to approach the ideal case. Furthermore, these curves are appropriate only for *small deflections*, since the curvature was approximated by d^2v/dx^2 when Eq. 13–16 was developed. Had a more exact analysis been performed, all these curves would tend to turn upward, intersecting and then rising above the line $P = P_{cr}$. This, of course, indicates that a larger load *P* is needed to create larger column deflections. We have not considered this analysis here, however, since most often engineering design restricts the deflection of columns to small values.

It should also be noted that the colored curves in Fig. 13–16 apply only for linear-elastic material behavior. Such is the case if the column is long and slender. However, if a short or intermediate-length stocky column is considered, then the applied load, as it is increased, may eventually cause the material to yield, and the column will begin to behave in an *inelastic manner*. This occurs at point *A* for the black curve in Fig. 13–16. As the load is further increased, the curve never reaches the critical load, and instead the load reaches a maximum value at *B*. Afterwards, a sudden decrease in load-carrying capacity occurs as the column continues to deflect by larger amounts.

Lastly, the colored curves in Fig. 13–16 also illustrate that a *nonlinear* relationship occurs between the load *P* and the deflection *v*. As a result, the principle of superposition *cannot be used* to determine the total deflection of a column caused by applying *successive loads* to the column. Instead, the loads must first be added, and then the corresponding deflection due to their resultant can be determined. Physically, the reason that successive loads and deflections cannot be superimposed is that the column's internal moment *depends* on both the load *P* and the *deflection v*, that is, $M = -P(e + v)$, Eq. 13–13. In other words, any *deflection* caused by a component load increases the moment. This behavior is unlike beam bending, where the actual deflection caused by the load *does not* increase the internal moment.

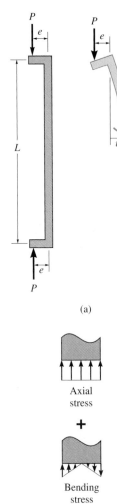

(a)

Axial
stress

+

Bending
stress

=

σ_{max}

Resultant
stress

(b)

Fig. 13–17

The Secant Formula.   The maximum stress in the column can be determined by realizing that it is caused by both the axial load and the moment, Fig. 13–17a. Maximum moment occurs at the column's midpoint, and using Eqs. 13–13 and 13–16, it has a magnitude of

$$M = |P(e + v_{max})| \qquad M = Pe \sec\left(\sqrt{\frac{P}{EI}}\,\frac{L}{2}\right) \qquad (13\text{–}18)$$

As shown in Fig. 13–17b, the maximum stress in the column is compressive, and it has a value of

$$\sigma_{max} = \frac{P}{A} + \frac{Mc}{I}; \qquad \sigma_{max} = \frac{P}{A} + \frac{Pec}{I}\sec\left(\sqrt{\frac{P}{EI}}\,\frac{L}{2}\right)$$

Since the radius of gyration is defined as $r^2 = I/A$, the above equation can be written in a form called the *secant formula*:

$$\sigma_{max} = \frac{P}{A}\left[1 + \frac{ec}{r^2}\sec\left(\frac{L}{2r}\sqrt{\frac{P}{EA}}\right)\right] \qquad (13\text{–}19)$$

Here

σ_{max} = maximum *elastic stress* in the column, which occurs at the inner concave side at the column's midpoint. This stress is compressive

P = vertical load applied to the column. $P < P_{cr}$ unless $e = 0$, then $P = P_{cr}$ (Eq. 13–5)

e = eccentricity of the load P, measured from the neutral axis of the column's cross-sectional area to the line of action of P

c = distance from the neutral axis to the outer fiber of the column where the maximum compressive stress σ_{max} occurs

A = cross-sectional area of the column

L = unsupported length of the column *in the plane of bending.* For supports other than pins, the effective length L_e should be used. See Fig. 13–12

E = modulus of elasticity for the material

r = radius of gyration, $r = \sqrt{I/A}$, where I is computed about the neutral or bending axis.

Like Eq. 13–16, Eq. 13–19 indicates that there is a nonlinear relationship between the load and the stress. Hence, the principle of superposition does not apply, and therefore the loads have to be added *before* the stress is determined. Furthermore, due to this nonlinear relationship, any factor of safety used for design purposes applies to the load and not to the stress.

For a given value of σ_{max}, graphs of Eq. 13–19 can be plotted as KL/r versus P/A for various values of the *eccentricity ratio* ec/r^2. A specific set of graphs for a structural-grade A-36 steel having a yield point of $\sigma_{max} = \sigma_Y = 36$ ksi and a modulus of elasticity of $E_{st} = 29(10^3)$ ksi is shown in Fig. 13–18. Here the abscissa and ordinate represent the slenderness ratio KL/r and the average stress P/A, respectively. Note that when $e \to 0$, or when $ec/r^2 \to 0$, Eq. 13–19 gives $\sigma_{max} = P/A$, where P is the critical load on the column, defined by Euler's formula. This results in Eq. 13–6, which has been plotted in Fig. 13–8 and repeated in Fig. 13–18. Since both Eqs. 13–6 and 13–19 are valid only for elastic loadings, the stresses shown in Fig. 13–18 cannot exceed $\sigma_Y = 36$ ksi, represented here by the horizontal line.

The curves in Fig. 13–18 indicate that differences in the eccentricity ratio have a marked effect on the load-carrying capacity of columns that have small slenderness ratios. On the other hand, columns that have large slenderness ratios tend to fail at or near the Euler critical load regardless of the eccentricity ratio. When using Eq. 13–19 for design purposes, it is therefore important to have a somewhat accurate value for the eccentricity ratio for shorter-length columns.

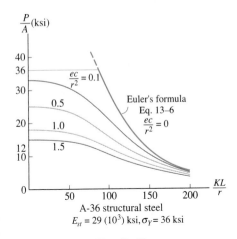

Fig. 13–18

Design.   Once the eccentricity ratio has been determined, the column data can be substituted into Eq. 13–19. If a value of $\sigma_{max} = \sigma_Y$ is chosen, then the corresponding load P_Y can be determined from a trial-and-error procedure, since the equation is transcendental and cannot be solved explicitly for P_Y. As a design aid, computer software, or graphs such as those in Fig. 13–18, can also be used to determine P_Y directly.

Realize that P_Y is the load that will cause the column to develop a maximum compressive stress of σ_Y at its inner concave fibers. Due to the eccentric application of P_Y, this load will *always be smaller* than the critical load P_{cr}, which is determined from the Euler formula that assumes unrealistically that the column is axially loaded. Once P_Y is obtained, an appropriate factor of safety can then be applied in order to specify the column's safe load.

IMPORTANT POINTS

- Due to imperfections in manufacturing or application of the load, a column will never suddenly buckle, instead it begins to bend.

- The load applied to a column is related to its deflection in a non-linear manner, and so the principle of superposition does not apply.

- As the slenderness ratio increases, eccentrically loaded columns tend to fail at or near the Euler buckling load.

EXAMPLE 13–5

The steel column shown in Fig. 13–19 is assumed to be pin-connected at its top and base. Determine the allowable eccentric load P that can be applied. Also, what is the maximum deflection of the column due to this loading? Due to bracing, assume buckling does not occur about the y axis. Take $E_{st} = 29(10^3)$ ksi, $\sigma_Y = 36$ ksi.

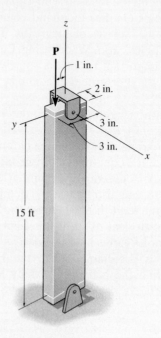

Fig. 13–19

SOLUTION

Computing the necessary geometrical properties, we have

$$I_x = \frac{1}{12}\,(2\text{ in.})(6\text{ in.})^3 = 36\text{ in.}^4$$

$$A = (2\text{ in.})(6\text{ in.}) = 12\text{ in.}^2$$

$$r_x = \sqrt{\frac{36\text{ in.}^4}{12\text{ in.}^2}} = 1.732\text{ in.}$$

$$e = 1\text{ in.}$$

$$KL = 1(15\text{ ft})(12\text{ in./ft}) = 180\text{ in.}$$

$$\frac{KL}{r_x} = \frac{180\text{ in.}}{1.732\text{ in.}} = 104$$

Since the curves in Fig. 13–18 have been established for $E_{st} =$ 29(10³) ksi, and $\sigma_Y = 36$ ksi, we can use them to determine the value of P/A and thus avoid a trial-and-error solution of the secant formula. Here $KL/r_x = 104$. Using the curve defined by the eccentricity ratio $ec/r^2 = 1$ in.(3 in.)/(1.732 in.)² = 1, we get

$$\frac{P}{A} \approx 12 \text{ ksi}$$

$$P = (12 \text{ ksi})(12 \text{ in.}^2) = 144 \text{ kip} \qquad \textit{Ans.}$$

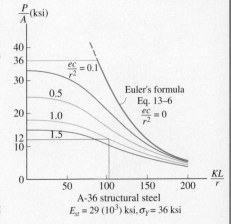

A-36 structural steel
$E_{st} = 29 (10^3)$ ksi, $\sigma_Y = 36$ ksi

Fig. 13–18

We can check this value by showing that it satisfies the secant formula, Eq. 13–19:

$$\sigma_{max} = \frac{P}{A}\left[1 + \frac{ec}{r^2} \sec\left(\frac{L}{2r}\sqrt{\frac{P}{EA}}\right)\right]$$

$$36 \overset{?}{=} \frac{144 \text{ kip}}{12 \text{ in.}^2}\left[1 + (1) \sec\left(\frac{180 \text{ in.}}{2(1.732 \text{ in.})}\sqrt{\frac{144 \text{ kip}}{29(10^3) \text{ ksi}(12 \text{ in}^2)}}\right)\right]$$

$$36 \overset{?}{=} 12[1 + \sec(1.0570 \text{ rad})]$$

$$36 \overset{?}{=} 12(1 + \sec 60.56°)$$

$$36 \approx 36.4$$

The maximum deflection occurs at the column's center, where $\sigma_{max} = 36$ ksi. Applying Eq. 13–16, we have

$$v_{max} = e\left[\sec\left(\sqrt{\frac{P}{EI}}\frac{L}{2}\right) - 1\right]$$

$$= 1 \text{ in.}\left[\sec\left(\sqrt{\frac{144 \text{ kip}}{29(10^3) \text{ ksi}(36 \text{ in}^4)}}\frac{180 \text{ in.}}{2}\right) - 1\right]$$

$$= 1 \text{ in.}[\sec 1.057 \text{ rad} - 1]$$

$$= 1 \text{ in.}[\sec 60.56° - 1]$$
$$= 1.03 \text{ in.} \qquad \textit{Ans.}$$

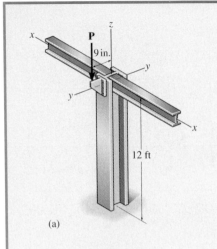

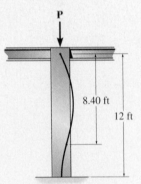

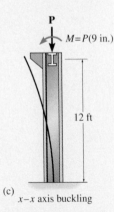

(a)

(b)   *y–y axis buckling*

(c)   *x–x axis buckling*

Fig. 13–20

EXAMPLE 13-6

The W8 × 40 A-36 steel column shown in Fig. 13–20a is fixed at its base and braced at the top so that it is fixed from displacement, yet free to rotate about the $y–y$ axis. Also, it can sway to the side in the $y–z$ plane. Determine the maximum eccentric load the column can support before it either begins to buckle or the steel yields.

SOLUTION

From the support conditions it is seen that about the $y–y$ axis the column behaves as if it were pinned at its top and fixed at the bottom and subjected to an axial load P, Fig. 13–20b. About the $x–x$ axis the column is free at the top and fixed at the bottom, and it is subjected to both an axial load P and moment $M = P(9 \text{ in.})$, Fig. 13–20c.

y–y Axis Buckling.   From Fig. 13–12d the effective length factor is $K_y = 0.7$, so $(KL)_y = 0.7(12)$ ft $= 8.40$ ft $= 100.8$ in. Using the table in Appendix B to determine I_y for the W8 × 40 section and applying Eq. 13–11, we have

$$(P_{cr})_y = \frac{\pi^2 EI_y}{(KL)_y^2} = \frac{\pi^2[29(10^3) \text{ ksi}](49.1 \text{ in.}^4)}{(100.8 \text{ in.})^2} = 1383 \text{ kip}$$

x–x Axis Yielding.   From Fig. 13–12b, $K_x = 2$, so $(KL)_x = 2(12)$ ft $= 24$ ft $= 288$ in. Again using the table in Appendix B to determine $A = 11.7 \text{ in}^2$, $c = 8.25 \text{ in.}/2 = 4.125$ in., and $r_x = 3.53$ in., and applying the secant formula, we have

$$\sigma_Y = \frac{P_x}{A}\left[1 + \frac{ec}{r_x^2} \sec\left(\frac{(KL)_x}{2r_x}\sqrt{\frac{P_x}{EA}}\right)\right]$$

or

$$421.2 = P_x[1 + 2.979 \sec(0.0700\sqrt{P_x})]$$

Solving for P_x by trial and error, noting that the argument for sec is in radians, we get

$$P_x = 88.4 \text{ kip} \qquad\qquad \textit{Ans.}$$

Since this value is less than $(P_{cr})_y = 1383$ kip, failure will occur about the $x–x$ axis. Also, $\sigma = 88.4 \text{ kip}/11.7 \text{ in}^2 = 7.56 \text{ ksi} < \sigma_Y = 36 \text{ ksi}$.

*13.5   INELASTIC BUCKLING

In engineering practice, columns are generally classified according to the type of stresses developed within the column at the time of failure. *Long slender columns* will become unstable when the compressive stress remains elastic. The failure that occurs is referred to as *elastic instability*. *Intermediate columns* fail due to *inelastic instability*, meaning that the compressive stress at failure is greater than the material's proportional limit. And *short columns*, sometimes called *posts*, do not become unstable; rather the material simply yields or fractures.

Application of the Euler equation requires that the stress in the column remain *below* the material's yield point (actually the proportional limit) when the column buckles, and so this equation applies only to long columns. In practice, however, most columns are selected to have intermediate lengths. The behavior of these columns can be studied by modifying the Euler equation so that it applies for inelastic buckling. To show how this can be done, consider the material to have a stress–strain diagram as shown in Fig. 13–21a. Here the proportional limit is σ_{pl}, and the modulus of elasticity, or slope of the line AB, is E. A plot of Euler's hyperbola, Fig. 13–8, is shown in Fig. 13–21b. This equation is valid for a column having a slenderness ratio as small as $(KL/r)_{pl}$, since at this point the axial stress in the column becomes $\sigma_{cr} = \sigma_{pl}$.

If the column has a slenderness ratio that is *less than* $(KL/r)_{pl}$, then the critical stress in the column must be greater than σ_Y. For example, suppose a column has a slenderness ratio of $(KL/r)_1 < (KL/r)_{pl}$, with corresponding critical stress $\sigma_D > \sigma_{pl}$ needed to cause instability. When the column is *about to buckle*, the change in strain that occurs in the column is within a *small range* $\Delta\epsilon$, so that the modulus of elasticity or stiffness for the material can be taken as the ***tangent modulus*** E_t, defined as the slope of the σ–ϵ diagram at point D, Fig. 13–21a. In other words, at the time of failure, the column behaves as if it were made from a material that has a *lower stiffness* than when it behaves elastically, $E_t < E$.

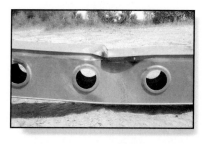

This crane boom failed by buckling caused by an overload. Note the region of localized collapse.

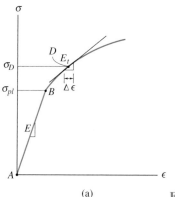

(a)   **Fig. 13–21**

In general, therefore, as the slenderness ratio decreases, the *critical stress* for a column continues to rise; and from the σ–ϵ diagram, the *tangent modulus* for the material *decreases*. Using this idea, we can modify Euler's equation to include these cases of inelastic buckling by substituting the material's tangent modulus E_t for E, so that

$$\sigma_{cr} = \frac{\pi^2 E_t}{(KL/r)^2} \tag{13–20}$$

This is the so-called *tangent modulus* or *Engesser equation*, proposed by F. Engesser in 1889. A plot of this equation for intermediate and short-length columns of a material defined by the σ–ϵ diagram in Fig. 13–21*a* is shown in Fig. 13–21*b*.

No *actual column* can be considered to be either perfectly straight or loaded along its centroidal axis, as assumed here, and therefore it is indeed very difficult to develop an expression that will provide a full analysis of this phenomenon. It should also be pointed out that other methods of describing the inelastic buckling of columns have been considered. One of these methods was developed by the aeronautical engineer F. R. Shanley and is called the *Shanley theory* of inelastic buckling. Although it provides a better description of the phenomenon than the tangent modulus theory, as explained here, experimental testing of a large number of columns, each of which approximates the ideal column, has shown that Eq. 13–20 is *reasonably accurate* in predicting the column's critical stress. Furthermore, the tangent modulus approach to modeling inelastic column behavior is relatively easy to apply.

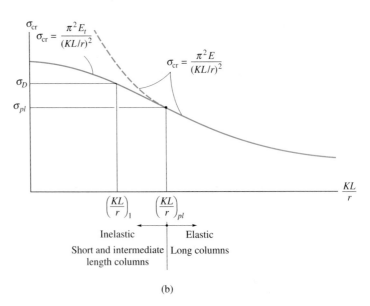

(b)

Fig. 13–21

EXAMPLE 13–7

A solid rod has a diameter of 30 mm and is 600 mm long. It is made of a material that can be modeled by the stress–strain diagram shown in Fig. 13–22. If it is used as a pin-supported column, determine the critical load.

SOLUTION

The radius of gyration is

$$r = \sqrt{\frac{I}{A}} = \sqrt{\frac{(\pi/4)(15)^4}{\pi(15)^2}} = 7.5 \text{ mm}$$

and therefore the slenderness ratio is

$$\frac{KL}{r} = \frac{1(600 \text{ mm})}{7.5 \text{ mm}} = 80$$

Applying Eq. 13–20 yields

$$\sigma_{\text{cr}} = \frac{\pi^2 E_t}{(KL/r)^2} = \frac{\pi^2 E_t}{(80)^2} = 1.542(10^{-3})E_t \qquad (1)$$

First we will assume that the critical stress is elastic. From Fig. 13–22,

$$E = \frac{150 \text{ MPa}}{0.001} = 150 \text{ GPa}$$

Thus, Eq. 1 becomes

$$\sigma_{\text{cr}} = 1.542(10^{-3})[150(10^3)] \text{ MPa} = 231.3 \text{ MPa}$$

Since $\sigma_{\text{cr}} > \sigma_{pl} = 150$ MPa, inelastic buckling occurs.
From the second line segment of the σ–ϵ diagram, Fig. 13–22, we have

$$E_t = \frac{\Delta\sigma}{\Delta\epsilon} = \frac{270 \text{ MPa} - 150 \text{ MPa}}{0.002 - 0.001} = 120 \text{ GPa}$$

Applying Eq. 1 yields

$$\sigma_{\text{cr}} = 1.542(10^{-3})[120(10^3)] \text{ MPa} = 185.1 \text{ MPa}$$

Since this value falls within the limits of 150 MPa and 270 MPa, it is indeed the critical stress.
The critical load on the rod is therefore

$$P_{\text{cr}} = \sigma_{\text{cr}}A = 185.1 \text{ MPa}[\pi(0.015 \text{ m})^2] = 131 \text{ kN} \qquad \textit{Ans.}$$

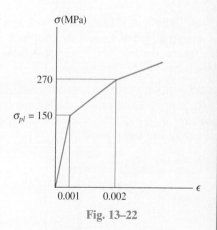

Fig. 13–22

PROBLEMS

13–46.   The W10 × 12 structural A-36 steel column is used to support a load of 4 kip. If the column is fixed at the base and free at the top, determine the deflection at the top of the column due to the loading.

13–47.   The W10 × 12 structural A-36 steel column is used to support a load of 4 kip. If the column is fixed at its base and free at its top, determine the maximum stress in the column due to this loading.

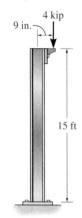

Probs. 13–46/13–47

*13–48.   The W14 × 26 structural A-36 steel member is used as a 20-ft-long column that is assumed to be fixed at its top and fixed at its bottom. If the 15-kip load is applied at an eccentric distance of 10 in., determine the maximum stress in the column.

13–49.   The W14 × 26 structural A-36 steel member is used as a column that is assumed to be fixed at its top and pinned at its bottom. If the 15-kip load is applied at an eccentric distance of 10 in., determine the maximum stress in the column.

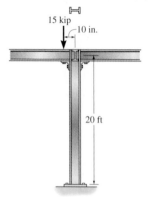

Probs. 13–48/13–49

13–50.   A W14 × 30 structural A-36 steel member is to be used as a pin-connected 20-ft column. Determine the maximum eccentric load P that can be applied so the column does not buckle or yield. Compare this value with an axial critical load P' applied through the centroid of the column.

13–51.   Solve Prob. 13–50 if the column is fixed connected at its ends.

*13–52.   Solve Prob. 13–50 if the column is fixed at its bottom and free at its top.

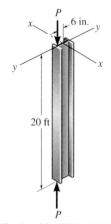

Probs. 13–50/13–51/13–52

13–53.   The tube is made of copper and has an outer diameter of 35 mm and a wall thickness of 7 mm. Using a factor of safety with respect to buckling and yielding of F.S. = 2.5, determine the allowable eccentric load P. The tube is pin supported at its ends. E_{cu} = 120 GPa, σ_Y = 750 MPa.

13–54.   The tube is made of copper and has an outer diameter of 35 mm and a wall thickness of 7 mm. Using a factor of safety with respect to buckling and yielding of F.S. = 2.5, determine the allowable eccentric load P that it can support without failure. The tube is fixed supported at its ends. E_{cu} = 120 GPa, σ_Y = 750 MPa.

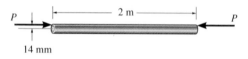

Probs. 13–53/13–54

13–55. The brass rod is fixed at one end and free at the other end. If the eccentric load $P = 200$ kN is applied, determine the greatest allowable length L of the rod so that it does not buckle or yield. $E_{br} = 101$ GPa, $\sigma_Y = 69$ MPa.

*13–56. The brass rod is fixed at one end and free at the other end. If the length of the rod is $L = 2$ m, determine the greatest allowable load P that can be applied so that the rod does not buckle or yield. Also, determine the largest sidesway deflection of the rod due to the loading. $E_{br} = 101$ GPa, $\sigma_Y = 69$ MPa.

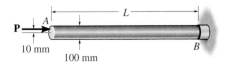

Probs. 13–55/13–56

13–57. The W10 × 30 structural A-36 steel column is pinned at its top and bottom. If it is subjected to the eccentric load of 85 kip, determine the factor of safety with respect to yielding.

13–58. The W10 × 30 structural A-36 steel column is fixed at its bottom and free at its top. If it is subjected to the eccentric load of 85 kip, determine if the column fails by yielding. The column is braced so that it does not buckle about the $y-y$ axis.

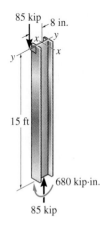

Probs. 13–57/13–58

13–59. Determine the load P required to cause the steel W12 × 50 structural A-36 steel column to fail either by buckling or by yielding. The column is fixed at its bottom and the cables at its top act as a pin to hold it.

*13–60. Solve Prob. 13–59 if the column is an A-36 steel W12 × 16 section.

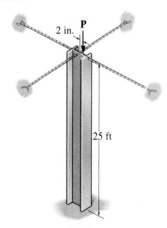

Probs. 13–59/13–60

13–61. A W12 × 26 structural A-36 steel column is fixed connected at its ends and has a length $L = 23$ ft. Determine the maximum eccentric load P that can be applied so the column does not buckle or yield.

13–62. A W14 × 30 structural A-36 steel column is fixed connected at its ends and has a length $L = 20$ ft. Determine the maximum eccentric load P that can be applied so the column does not buckle or yield.

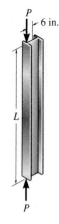

Probs. 13–61/13–62

13–63. The aluminum column has the cross section shown. If it is fixed at the bottom and free at the top, determine the maximum force P that can be applied at A without causing it to buckle or yield. Use a factor of safety of 3 with respect to buckling and yielding. $E_{al} = 70$ GPa, $\sigma_Y = 95$ MPa.

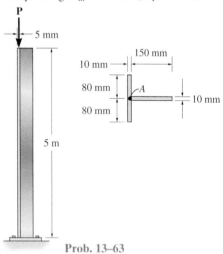

Prob. 13–63

*****13–64.** The steel column supports the two eccentric loadings. If it is assumed to be pinned at its top, fixed at the bottom, and fully braced against buckling about the y–y axis, determine the maximum deflection of the column and the maximum stress in the column. $E_{st} = 200$ GPa, $\sigma_Y = 360$ MPa.

13–65. The steel column supports the two eccentric loadings. If it is assumed to be fixed at its top and bottom, and braced against buckling about the y–y axis, determine the maximum deflection of the column and the maximum stress in the column. $E_{st} = 200$ GPa, $\sigma_Y = 360$ MPa.

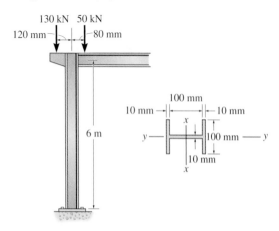

Probs. 13–64/13/65

13–66. A column of intermediate length buckles when the compressive stress is 40 ksi. If the slenderness ratio is 60, determine the tangent modulus.

13–67. The stress–strain diagram for a material can be approximated by the two line segments shown. If a bar having a diameter of 60 mm and a length of 2 m is made from this material, determine the critical load if both ends are pinned. Assume that the load acts through the axis of the bar. Use Engesser's equation.

*****13–68.** The stress–strain diagram for a material can be approximated by the two line segments shown. If a bar having a diameter of 60 mm and a length of 2 m is made from this material, determine the critical load provided the ends are fixed. Assume that the load acts through the axis of the bar. Use Engesser's equation.

13–69. The stress–strain diagram for a material can be approximated by the two line segments shown. If a bar having a diameter of 60 mm and length of 2 m is made from this material, determine the critical load provided one end is free and the other is fixed. Assume that the load acts through the axis of the bar. Use Engesser's equation.

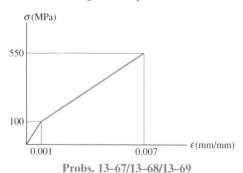

Probs. 13–67/13–68/13–69

13–70. Construct the buckling curve, P/A versus L/r, for a column that has a bilinear stress–strain curve in compression as shown.

Prob. 13–70

*13.6 DESIGN OF COLUMNS FOR CONCENTRIC LOADING

The theory presented thus far applies to columns that are perfectly straight, made of homogeneous material, and originally stress-free. Practically speaking, though, as stated previously, columns are not perfectly straight, and most have residual stresses in them, primarily due to nonuniform cooling during manufacture. Also, the supports for columns are less than exact, and the points of application and directions of loads are not known with absolute certainty. In order to compensate for these effects, which actually vary from one column to the next, many design codes specify the use of column formulas that are empirical. By performing experimental tests on a large number of *axially loaded columns*, the results may be plotted and a design formula developed by curve-fitting the mean of the data.

An example of such tests for wide-flange steel columns is shown in Fig. 13–23. Notice the similarity between these results and those of the family of curves determined from the secant formula, Fig. 13–18. The reason for this similarity has to do with the influence of an "accidental" eccentricity ratio on the column's strength. As stated in Sec. 13.4, this ratio has more of an effect on the strength of short and intermediate-length columns than on those that are long. Tests have indicated that ec/r^2 can range from 0.1 to 0.6 for most axially loaded columns.

In order to account for the behavior of different-length columns, design codes usually specify several formulas that will best fit the data within the short, intermediate, and long column range. Hence, each formula will apply only for a specific *range* of slenderness ratios, and so it is important that the engineer carefully observe the KL/r limits for which a particular formula is valid. Examples of design formulas for steel, aluminum, and wood columns that are currently in use will now be discussed. The purpose is to give some idea as to how columns are designed in practice. These formulas should not, however, be used for the design of actual columns, unless the code from which they are referenced is consulted.

These long unbraced timber columns are used to support the roof of this building.

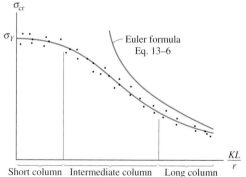

Fig. 13–23

Steel Columns.   Columns made of structural steel are currently designed on the basis of formulas proposed by the Structural Stability Research Council (SSRC). Factors of safety have been applied to these formulas and adopted as specifications for building construction by the American Institute of Steel Construction (AISC). Basically these specifications provide two formulas for column design, each of which gives the maximum allowable stress in the column for a specific range of slenderness ratios. For long columns the Euler formula is proposed, i.e., $\sigma_{max} = \pi^2 E/(KL/r)^2$.

Application of this formula requires that a factor of safety F.S. = $\frac{23}{12} \approx 1.92$ be applied. Thus, for design,

$$\sigma_{allow} = \frac{12\pi^2 E}{23(KL/r)^2} \qquad \left(\frac{KL}{r}\right)_c \leq \frac{KL}{r} \leq 200 \qquad (13\text{–}21)$$

As stated, this equation is applicable for a slenderness ratio bounded by 200 and $(KL/r)_c$. A specific value of $(KL/r)_c$ is obtained by requiring the Euler formula to be used only for elastic material behavior. Through experiments it has been determined that compressive residual stresses can exist in rolled-formed steel sections that may be as much as one-half the yield stress. Consequently, if the stress in the Euler formula is greater than $\frac{1}{2}\sigma_Y$, the equation will not apply. Therefore the value of $(KL/r)_c$ can be determined as follows:

$$\frac{1}{2}\sigma_Y = \frac{\pi^2 E}{(KL/r)_c^2}$$

$$\left(\frac{KL}{r}\right)_c = \sqrt{\frac{2\pi^2 E}{\sigma_Y}} \qquad (13\text{–}22)$$

Columns having slenderness ratios less than $(KL/r)_c$ are designed on the basis of an empirical formula that is parabolic and has the form

$$\sigma_{max} = \left[1 - \frac{(KL/r)^2}{2(KL/r)_c^2}\right]\sigma_Y$$

Since there is more uncertainty in the use of this formula for longer columns, it is divided by a factor of safety defined as follows:

$$\text{F.S.} = \frac{5}{3} + \frac{3}{8}\frac{(KL/r)}{(KL/r)_c} - \frac{(KL/r)^3}{8(KL/r)_c^3}$$

Here it is seen that F.S. = $\frac{5}{3} \approx 1.67$ at $KL/r = 0$ and increases to F.S. = $\frac{23}{12} \approx 1.92$ at $(KL/r)_c$. Hence, for design purposes,

$$\sigma_{allow} = \frac{\left[1 - \frac{(KL/r)^2}{2(KL/r)_c^2}\right]\sigma_Y}{\{(5/3) + [(3/8)(KL/r)/(KL/r)_c] - [(KL/r)^3/8(KL/r)_c^3]\}} \qquad (13\text{–}23)$$

Equations 13–21 and 13–23 are plotted in Fig. 13–24. When applying any of these equations, either FPS or SI units can be used for the calculations.

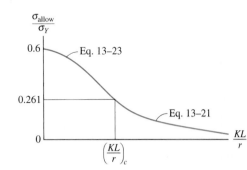

Fig. 13–24

Aluminum Columns. Column design for structural aluminum is specified by the Aluminum Association using three equations, each applicable for a specific range of slenderness ratios. Since several types of aluminum alloy exist, there is a unique set of formulas for each type. For a common alloy (2014-T6) used in building construction, the formulas are

$$\sigma_{allow} = 28 \text{ ksi} \qquad 0 \le \frac{KL}{r} \le 12 \qquad (13\text{–}24)$$

$$\sigma_{allow} = \left[30.7 - 0.23\left(\frac{KL}{r}\right)\right]\text{ksi} \qquad 12 < \frac{KL}{r} < 55 \qquad (13\text{–}25)$$

$$\sigma_{allow} = \frac{54\,000 \text{ ksi}}{(KL/r)^2} \qquad 55 \le \frac{KL}{r} \qquad (13\text{–}26)$$

These equations are plotted in Fig. 13–25. As shown, the first two represent straight lines and are used to model the effects of columns in the short and intermediate range. The third formula has the same form as the Euler formula and is used for long columns.

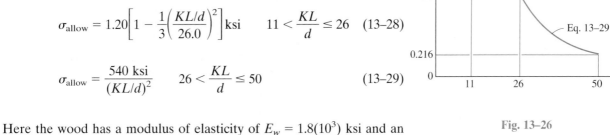

Timber Columns. Columns used in timber construction are designed on the basis of formulas published by the National Forest Products Association (NFPA) or the American Institute of Timber Construction (AITC). For example, the NFPA formulas for the allowable stress in short, intermediate, and long columns having a rectangular cross section of dimensions b and d, where d is the *smallest* dimension of the cross section, are

$$\sigma_{allow} = 1.20 \text{ ksi} \qquad 0 \le \frac{KL}{d} \le 11 \qquad (13\text{–}27)$$

$$\sigma_{allow} = 1.20\left[1 - \frac{1}{3}\left(\frac{KL/d}{26.0}\right)^2\right]\text{ksi} \qquad 11 < \frac{KL}{d} \le 26 \qquad (13\text{–}28)$$

$$\sigma_{allow} = \frac{540 \text{ ksi}}{(KL/d)^2} \qquad 26 < \frac{KL}{d} \le 50 \qquad (13\text{–}29)$$

Here the wood has a modulus of elasticity of $E_w = 1.8(10^3)$ ksi and an allowable compressive stress of 1.2 ksi parallel to the grain. In particular, Eq. 13–29 is simply Euler's equation having a factor of safety of 3. These three equations are plotted in Fig. 13–26.

PROCEDURE FOR ANALYSIS

Column Analysis.

• When using any formula to *analyze* a column, that is, to find its allowable load, it is first necessary to calculate the slenderness ratio in order to determine which column formula applies.

• Once the average allowable stress has been computed, the allowable load in the column is determined from $P = \sigma_{\text{allow}}A$.

Column Design.

• If a formula is used to *design* a column, that is, to determine the column's cross-sectional area for a given loading and effective length, then a trial-and-check procedure generally must be followed if the column has a composite shape, such as a wide-flange section.

• One possible way to apply a trial-and-check procedure would be to *assume* the column's cross-sectional area, A', and calculate the corresponding stress $\sigma' = P/A'$. Also, with A' use an appropriate design formula to determine the allowable stress σ_{allow}. From this, calculate the *required* column area $A_{\text{req'd}} = P/\sigma_{\text{allow}}$.

• If $A' > A_{\text{req'd}}$, the design is safe. When making the comparison, it is practical to require A' to be close to but greater than $A_{\text{req'd}}$, usually within 2–3%. A redesign is necessary if $A' < A_{\text{req'd}}$.

• Whenever a trial-and-check procedure is repeated, the choice of an area is determined by the previously calculated required area. In engineering practice this method for design is usually shortened through the use of computer software or published tables and graphs.

EXAMPLE 13-8

An A-36 steel W10 × 100 member is used as a pin-supported column, Fig. 13–27. Using the AISC column design formulas, determine the largest load that it can safely support.

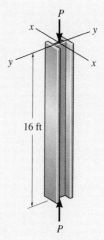

Fig. 13–27

SOLUTION

The following data for a W10 × 100 is taken from the table in Appendix B.

$$A = 29.4 \text{ in}^2 \quad r_x = 4.60 \text{ in.} \quad r_y = 2.65 \text{ in.}$$

Since $K = 1$ for both x and y axis buckling, the slenderness ratio is largest if r_y is used. Thus,

$$\frac{KL}{r} = \frac{1(16 \text{ ft})(12 \text{ in./ft})}{(2.65 \text{ in.})} = 72.45$$

From Eq. 13–22, we have

$$\left(\frac{KL}{r}\right)_c = \sqrt{\frac{2\pi^2 E}{\sigma_Y}}$$

$$= \sqrt{\frac{2\pi^2[29(10^3) \text{ ksi}]}{36 \text{ ksi}}}$$

$$= 126.1$$

Here $0 < KL/r < (KL/r)_c$, so Eq. 13–23 applies.

$$\sigma_{\text{allow}} = \frac{\left[1 - \dfrac{(KL/r)^2}{2(KL/r)_c^2}\right]\sigma_Y}{\{(5/3) + [(3/8)(KL/r)/(KL/r)_c] - [(KL/r)^3/8(KL/r_c)^3]\}}$$

$$= \frac{[1 - (72.45)^2/2(126.1)^2]36 \text{ ksi}}{\{(5/3) + [(3/8)(72.45/126.1)] - [(72.45)^3/8(126.1)^3]\}}$$

$$= 16.17 \text{ ksi}$$

The allowable load P on the column is therefore

$$\sigma_{\text{allow}} = \frac{P}{A}; \qquad 16.17 \text{ kip/in.}^2 = \frac{P}{29.4 \text{ in.}^2}$$

$$P = 476 \text{ kip} \qquad\qquad \textit{Ans.}$$

EXAMPLE 13-9

The steel rod in Fig. 13–28 is to be used to support an axial load of 18 kip. If $E_{st} = 29(10^3)$ ksi and $\sigma_Y = 50$ ksi, determine the smallest diameter of the rod as allowed by the AISC specification. The rod is fixed at both ends.

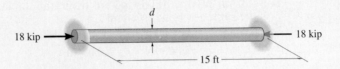

Fig. 13–28

SOLUTION

For a circular cross section the radius of gyration becomes

$$r = \sqrt{\frac{I}{A}} = \sqrt{\frac{(1/4)\pi(d/2)^4}{(1/4)\pi d^2}} = \frac{d}{4}$$

Applying Eq. 13–22, we have

$$\left(\frac{KL}{r}\right)_c = \sqrt{\frac{2\pi^2 E}{\sigma_Y}} = \sqrt{\frac{2\pi^2[29(10^3)\text{ ksi}]}{50\text{ ksi}}} = 107.0$$

Since the rod's radius of gyration is unknown, KL/r is unknown, and therefore a choice must be made as to whether Eq. 13–21 or Eq. 13–23 applies. We will consider Eq. 13–21. For a fixed-end column $K = 0.5$, so

$$\sigma_{\text{allow}} = \frac{12\pi^2 E}{23(KL/r)^2}$$

$$\frac{18\text{ kip}}{(1/4)\pi d^2} = \frac{12\pi^2[29(10^3)\text{ kip/in}^2]}{23[0.5(15\text{ ft})(12\text{ in./ft})/(d/4)]^2}$$

$$\frac{22.92}{d^2} = 1.152d^2$$

$$d = 2.11\text{ in.}$$

Use

$$d = 2.25\text{ in.} = 2\tfrac{1}{4}\text{ in.} \qquad \textit{Ans.}$$

For this design, we must check the slenderness-ratio limits; i.e.,

$$\frac{KL}{r} = \frac{0.5(15)(12)}{(2.25/4)} = 160$$

Since $107.0 < 160 < 200$, use of Eq. 13–21 is appropriate.

EXAMPLE 13-10

A bar having a length of 30 in. is used to support an axial compressive load of 12 kip, Fig. 13–29. It is pin supported at its ends and made from a 2014-T6 aluminum alloy. Determine the dimensions of its cross-sectional area if its width is to be twice its thickness.

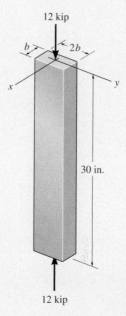

12 kip

30 in.

12 kip

Fig. 13–29

SOLUTION

Since $KL = 30$ in. is the same for both $x-x$ and $y-y$ axis buckling, the largest slenderness ratio is determined using the smallest radius of gyration, i.e., using $I_{min} = I_y$:

$$\frac{KL}{r_y} = \frac{KL}{\sqrt{I_y/A}} = \frac{1(30)}{\sqrt{(1/12)2b(b)3/[2b(b)]}} = \frac{103.9}{b} \qquad (1)$$

Here we must apply Eq. 13–24, 13–25, or 13–26. Since we do not as yet know the slenderness ratio, we will begin by using Eq. 13–24.

$$\frac{P}{A} = 28 \text{ ksi}$$

$$\frac{12 \text{ kip}}{2b(b)} = 28 \text{ kip/in}^2$$

$$b = 0.463 \text{ in.}$$

Checking the slenderness ratio, we have

$$\frac{KL}{r} = \frac{103.9}{0.463} = 224.5 > 12$$

Try Eq. 13–26, which is valid for $KL/r \geq 55$:

$$\frac{P}{A} = \frac{54\,000 \text{ ksi}}{(KL/r)^2}$$

$$\frac{12}{2b(b)} = \frac{54\,000}{(103.9/b)^2}$$

$$b = 1.05 \text{ in.} \qquad \qquad \textit{Ans.}$$

From Eq. 1

$$\frac{KL}{r} = \frac{103.9}{1.05} = 99.3 > 55 \quad \text{OK}$$

Note: It would be satisfactory to choose the cross section with dimensions 1 in. by 2 in.

EXAMPLE 13-11

A board having cross-sectional dimensions of 5.5 in. by 1.5 in. is used to support an axial load of 5 kip, Fig. 13–30. If the board is assumed to be pin supported at its top and bottom, determine its *greatest* allowable length L as specified by the NFPA.

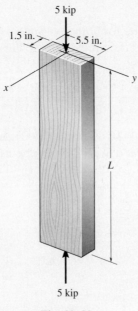

Fig. 13–30

SOLUTION

By inspection, the board will buckle about the y axis. In the NFPA equations, $d = 1.5$ in. Assuming that Eq. 13–29 applies, we have

$$\frac{P}{A} = \frac{540 \text{ ksi}}{(KL/d)^2}$$

$$\frac{5 \text{ kip}}{(5.5 \text{ in.})(1.5 \text{ in.})} = \frac{540 \text{ ksi}}{(1 \, L/1.5 \text{ in.})^2}$$

$$L = 44.8 \text{ in.} \qquad \textit{Ans.}$$

Here

$$\frac{KL}{d} = \frac{1(44.8 \text{ in.})}{1.5 \text{ in.}} = 29.8$$

Since $26 < KL/d \le 50$, the solution is valid.

PROBLEMS

13–71. Determine the largest length of a W10 × 12 structural A-36 steel section if it is fixed supported and is subjected to an axial load of 28 kip. Use the AISC equations.

*13–72. Determine the largest length of a W8 × 31 structural A-36 steel section if it is pin supported and is subjected to an axial load of 130 kip. Use the AISC equations.

13–73. Using the AISC equations, check if a W6 × 9 structural A-36 steel column that is 10 ft long can support an axial load of 40 kip. The ends are fixed.

13–74. Solve Prob. 13–73 if the ends are pin supported.

13–75. Using the AISC equations, select from Appendix B the lightest-weight structural A-36 steel column that is 30 ft long and supports an axial load of 200 kip. The ends are fixed.

*13–76. Using the AISC equations, select from Appendix B the lightest-weight structural A-36 steel column that is 24 ft long and supports an axial load of 100 kip. The ends are fixed.

13–77. Determine the largest length of a W6 × 16 structural A-36 steel column if it is pin supported and subjected to an axial load of 70 kip. Use the AISC equations.

13–78. Determine the largest length of a W8 × 48 structural A-36 steel column if it is pin supported and subjected to an axial load of 55 kip. Use the AISC equations.

13–79. Using the AISC equations, check if a column having the cross section shown can support an axial force of 1500 kN. The column has a length of 4 m, is made from A-36 steel, and its ends are pinned.

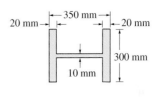

Prob. 13–79

*13–80. Determine the largest length of a W8 × 31 structural A-36 steel column if it is to support an axial load of 35 kip. The ends are pinned.

13–81. Using the AISC equations, select from Appendix B the lightest-weight structural A-36 steel column that is 20 ft long and supports an axial load of 40 kip. The ends are pinned.

13–82. Determine the largest length of a W10 × 19 structural A-36 steel column if it is to support an axial load of 50 kip. The ends are fixed supported.

■13–83. Determine the largest length of a W10 × 45 structural steel column if it is pin supported and subjected to an axial load of 290 kip. $E_{st} = 29(10^3)$ ksi, $\sigma_Y = 50$ ksi. Use the AISC equations.

*13–84. The beam and column arrangement is used in a railroad yard for loading and unloading cars. If the maximum anticipated hoist load is 12 kip, determine if the W8 × 31 structural A-36 steel column is adequate for supporting the load. The hoist travels along the bottom flange of the beam, 1 ft ≤ x ≤ 25 ft, and has negligible size. Assume the beam is pinned to the column at B and roller supported at A. The column is also pinned at C and is braced so it will not buckle out of the plane of the loading.

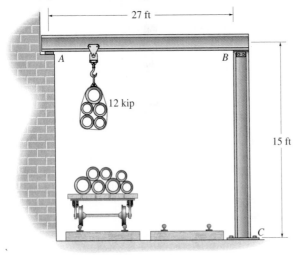

Prob. 13–84

13–85. The 1-in.-diameter rod is used to support an axial load of 5 kip. Determine its greatest allowable length L if it is made of 2014-T6 aluminum. Assume that the ends are pin connected.

13–86. The 1-in.-diameter rod is used to support an axial load of 5 kip. Determine its greatest allowable length L if it is made of 2014-T6 aluminum. Assume that the ends are fixed connected.

Probs 13–85/13–86

13–87. The 2-in.-diameter rod is used to support an axial load of 8 kip. Determine its greatest allowable length L if it is made of 2014-T6 aluminum. Assume that the ends are pin connected.

*13–88. The 2-in.-diameter rod is used to support an axial load of 8 kip. Determine its greatest allowable length L if it is made of 2014-T6 aluminum. Assume that the ends are fixed connected.

Prob. 13–87/13–88

13–89. A 5-ft-long rod is used in a machine to transmit an axial compressive load of 3 kip. Determine its diameter if it is pin connected at its ends and is made of a 2014-T6 aluminum alloy.

13–90. Solve Prob. 13–89 if the rod is fixed connected at its ends.

13–91. The tube is 0.25 in. thick, is made of a 2014-T6 aluminum alloy, and is fixed at its bottom and pinned at its top. Determine the largest axial load that it can support.

*13–92. The tube is 0.25 in. thick, is made of a 2014-T6 aluminum alloy, and is fixed connected at its ends. Determine the largest axial load that it can support.

13–93. The tube is 0.25 in. thick, is made of a 2014-T6 aluminum alloy, 2014-T6, and is pin connected at its ends. Determine the largest axial load it can support.

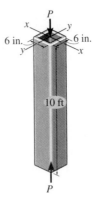

Probs 13–91/13–92/13–93

13–94. A 6-ft-long rod is used in a machine to transmit an axial compressive load of 3 kip. Determine its diameter if it is fixed at one end and pinned at the other. The material is 2014-T6 aluminum alloy.

13–95. The timber column has a square cross section and is assumed to be pin connected at its top and bottom. If it supports an axial load of 20 kip, determine its side dimensions a to the nearest $\frac{1}{2}$ in. Use the NFPA formulas.

***13–96.** Solve Prob. 13–95 if the column is assumed to be fixed connected at its top and bottom.

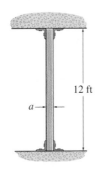

Probs. 13–95/13–96

13–97. The timber column has a length of 20 ft and is pin connected at its ends. Use the NFPA formulas to determine the largest axial force P that it can support.

13–98. The timber column has a length of 20 ft and is fixed connected at its ends. Use the NFPA formulas to determine the largest axial force P that it can support.

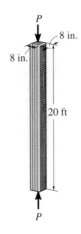

Prob. 13–97/13–98

13–99. The column is made of wood. It is fixed at its bottom and free at its top. Use the NFPA formulas to determine its greatest allowable length if it supports an axial load of $P = 6$ kip.

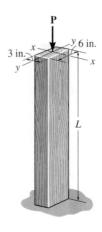

Prob. 13–99

***13–100.** The column is made of wood. It is fixed at its bottom and free at its top. Use the NFPA formulas to determine the largest allowable axial load P that it can support if it has a length $L = 6$ ft.

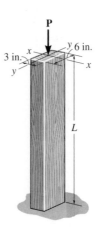

Prob. 13–100

*13.7   DESIGN OF COLUMNS FOR ECCENTRIC LOADING

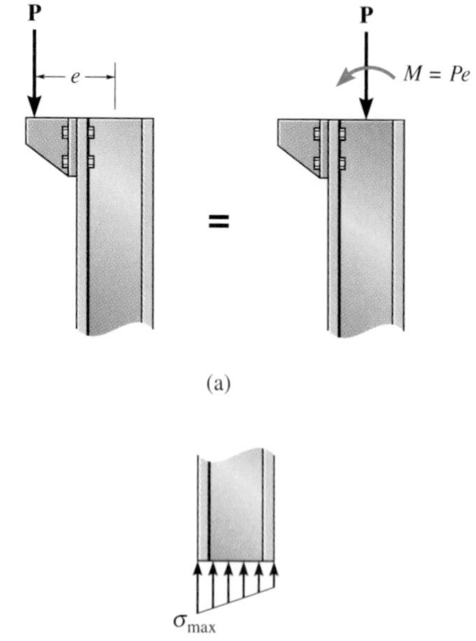

(a)

σ_{max}

(b)

Fig. 13–31

Occasionally a column may be required to support a load acting either at its edge or on an angle bracket attached to its side, such as shown in Fig. 13–31a. The bending moment $M = Pe$, which is caused by the eccentric loading, must be accounted for when the column is designed. There are several acceptable ways in which this is done in engineering practice. We will discuss two of the most common methods.

Use of Available Column Formulas. The stress distribution acting over the cross-sectional area of the column shown in Fig. 13–31a is determined from both the axial force P and the bending moment $M = Pe$. In particular, the maximum compressive stress is

$$\sigma_{max} = \frac{P}{A} + \frac{Mc}{I} \qquad (13\text{–}30)$$

A typical stress profile is shown in Fig. 13–31b. If we conservatively *assume* that the entire cross section is subjected to the uniform stress σ_{max} as determined from Eq. 13–30, then we can compare σ_{max} with σ_{allow}, which is determined using the formulas given in Sec. 13–6. Calculation of σ_{allow} is usually done using the *largest* slenderness ratio for the column, regardless of the axis about which the column experiences bending. This requirement is normally specified in design codes and will in most cases lead to a conservative design. If

$$\sigma_{max} \leq \sigma_{allow}$$

then the column can carry the specified loading. If this inequality does not hold, then the column's area A must be increased, and a new σ_{max} and σ_{allow} must be calculated. This method of design is rather simple to apply and works well for columns that are short or of intermediate length.

Interaction Formula. When *designing* an eccentrically loaded column it is desirable to see how the bending and axial loads *interact*, so that a balance between these two effects can be achieved. To do this, we will consider the separate contributions made to the total column area by the axial force and moment. If the allowable stress for the axial load is $(\sigma_a)_{allow}$, then the required area for the column needed to support the load P is

$$A_a = \frac{P}{(\sigma_a)_{allow}}$$

Similarly, if the allowable bending stress is $(\sigma_b)_{allow}$, then since $I = Ar^2$, the required area of the column needed to support the eccentric moment is determined from the flexure formula, that is,

$$A_b = \frac{Mc}{(\sigma_b)_{\text{allow}}r^2}$$

The total area A for the column needed to resist *both* the axial load and moment requires that

$$A_a + A_b = \frac{P}{(\sigma_a)_{\text{allow}}} + \frac{Mc}{(\sigma_b)_{\text{allow}}r^2} \leq A$$

or

$$\frac{P/A}{(\sigma_a)_{\text{allow}}} + \frac{Mc/Ar^2}{(\sigma_b)_{\text{allow}}} \leq 1$$

$$\frac{\sigma_a}{(\sigma_a)_{\text{allow}}} + \frac{\sigma_b}{(\sigma_b)_{\text{allow}}} \leq 1 \qquad (13\text{--}31)$$

Here

> σ_a = axial stress caused by the force P and determined from $\sigma_a = P/A$, where A is the cross-sectional area of the column
>
> σ_b = bending stress caused by an eccentric load or applied moment M; σ_b is found from $\sigma_b = Mc/I$, where I is the moment of inertia of the cross-sectional area computed about the bending or neutral axis
>
> $(\sigma_a)_{\text{allow}}$ = allowable axial stress as defined by formulas given in Sec. 13.6 or by other design code specifications. For this purpose, always use the *largest* slenderness ratio for the column, regardless of the axis about which the column experiences bending
>
> $(\sigma_b)_{\text{allow}}$ = allowable bending stress as defined by code specifications.

In particular, if the column is subjected only to an axial load, then the bending-stress ratio in Eq. 13–31 would be equal to zero and the design will be based only on the allowable axial stress. Likewise, when no axial load is present, the axial-stress ratio is zero and the stress requirement will be based on the allowable bending stress. Hence, each stress ratio indicates the contribution of axial load or bending moment. Since Eq. 13–31 shows how these loadings interact, this equation is sometimes referred to as the ***interaction formula***. This design approach requires a trial-and-check procedure, where it is required that the designer *pick* an available column and then check to see if the inequality is satisfied. If it is not, a larger section is then picked and the process repeated. An economical choice is made when the left side is close to but less than 1.

The interaction method is often specified in codes for the design of members made of steel, aluminum, or timber. In particular, the American Institute of Steel Construction specifies the use of this equation only when the axial-stress ratio $\sigma_a/(\sigma_a)_{\text{allow}} \leq 0.15$. For other values of this ratio, a modified form of Fig. 13–31 is used.

The following examples illustrate the above methods for design and analysis of eccentrically loaded columns.

Typical example of a column used to support an eccentric roof loading.

EXAMPLE 13–12

The column in Fig. 13–32 is made of aluminum alloy 2014-T6 and is used to support an eccentric load **P**. Determine the magnitude of **P** that can be supported if the column is fixed at its base and free at its top. Use Eq. 13–30.

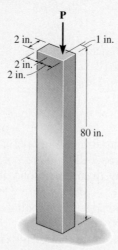

P

2 in.

1 in.

2 in.

2 in.

80 in.

Fig. 13–32

SOLUTION

From Fig. 13–12b, $K = 2$. The largest slenderness ratio for the column is therefore

$$\frac{KL}{r} = \frac{2(80 \text{ in.})}{\sqrt{[(1/12)(4 \text{ in.})(2 \text{ in.})^3]/[(2 \text{ in.})4 \text{ in.}]}} = 277.1$$

By inspection, Eq. 13–26 must be used $(277.1 > 55)$. Thus,

$$\sigma_{\text{allow}} = \frac{54\,000 \text{ ksi}}{(KL/r)^2} = \frac{54\,000 \text{ ksi}}{(277.1)^2} = 0.703 \text{ ksi}$$

The actual maximum compressive stress in the column is determined from the combination of axial load and bending. We have

$$\sigma_{\text{max}} = \frac{P}{A} + \frac{(Pe)c}{I}$$

$$= \frac{P}{2 \text{ in.}(4 \text{ in.})} + \frac{P(1 \text{ in.})(2 \text{ in.})}{(1/12)(2 \text{ in.})(4 \text{ in.})^3}$$

$$= 0.3125P$$

Assuming that this stress is *uniform* over the cross section, instead of just at the outer boundary, we require

$$\sigma_{\text{allow}} = \sigma_{\text{max}}; \qquad 0.703 = 0.3125P$$

$$P = 2.25 \text{ kip} \qquad \qquad \textit{Ans.}$$

EXAMPLE 13-13

The A-36 steel W6 × 20 column in Fig. 13–33 is pin connected at its ends and is subjected to the eccentric load **P**. Determine the maximum allowable value of P using the interaction method if the allowable bending stress is $(\sigma_b)_{\text{allow}} = 22$ ksi.

SOLUTION

Here $K = 1$. The necessary geometric properties for the W6 × 20 are taken from the table in Appendix B.

$$A = 5.87 \text{ in}^2 \qquad I_x = 41.4 \text{ in}^4 \qquad r_y = 1.50 \text{ in.} \qquad d = 6.20 \text{ in.}$$

We will consider r_y because this will lead to the *largest* value of the slenderness ratio. Also, I_x is needed since bending occurs about the x axis ($c = 6.20$ in./2 = 3.10 in.). To determine the allowable compressive stress, we have

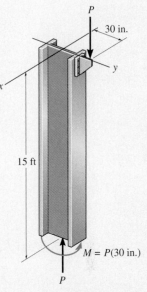

$$\frac{KL}{r} = \frac{1(15 \text{ ft})(12 \text{ in./ft})}{1.50 \text{ in.}} = 120$$

Since

$$\left(\frac{KL}{r}\right)_c = \sqrt{\frac{2\pi^2 E}{\sigma_Y}} = \sqrt{\frac{2\pi^2[29(10^3) \text{ ksi}]}{36 \text{ ksi}}} = 126.1$$

Fig. 13–33

then $KL/r < (KL/r)_c$ and so Eq. 13–23 must be used.

$$\sigma_{\text{allow}} = \frac{[1 - (KL/r)^2/2(KL/r)_c^2]\sigma_Y}{\{(5/3) + [(3/8)(KL/r)/(KL/r)_c] - [(KL/r)^3/8(KL/r)_c^3]\}}$$

$$= \frac{[1 - (120)^2/2(126.1)^2]36 \text{ ksi}}{\{(5/3) + [(3/8)(120)/(126.1)] - [(120)^3/8(126.1)^3]\}}$$

$$= 10.28 \text{ ksi}$$

Applying the interaction Eq. 13–31 yields

$$\frac{\sigma_a}{(\sigma_a)_{\text{allow}}} + \frac{\sigma_b}{(\sigma_b)_{\text{allow}}} \leq 1$$

$$\frac{P/5.87 \text{ in.}^2}{10.28 \text{ ksi}} + \frac{P(30 \text{ in.})(3.10 \text{ in.})/(41.4 \text{ in}^4)}{22 \text{ ksi}} = 1$$

$$P = 8.43 \text{ kip} \qquad\qquad Ans.$$

Checking the application of the interaction method for the steel section, we require

$$\frac{\sigma_a}{(\sigma_a)_{\text{allow}}} = \frac{(8.43 \text{ kip})/(5.87 \text{ in.})}{10.28 \text{ kip/in}^2} = 0.140 < 0.15 \qquad \text{OK}$$

EXAMPLE 13–14

The timber column in Fig. 13–34 is made from two boards nailed together so that the cross section has the dimensions shown. If the column is fixed at its base and free at its top, use Eq. 13–30 to determine the eccentric load **P** that can be supported.

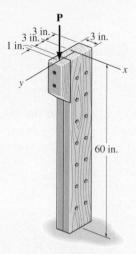

Fig. 13–34

SOLUTION

From Fig. 13–12b, $K = 2$. Here we must calculate KL/d to determine which equation from Eqs. 13–27 through 13–29 should be used. Since σ_{allow} is determined using the largest slenderness ratio, we choose $d = 3$ in. This is done to make this ratio as large as possible, and thereby yields the lowest possible allowable axial stress. This is done even though bending due to P is about the x axis. We have

$$\frac{KL}{d} = \frac{2(60 \text{ in.})}{3 \text{ in.}} = 40$$

The allowable axial stress is determined using Eq. 13–29 since $26 < KL/d < 50$. Thus,

$$\sigma_{\text{allow}} = \frac{540 \text{ ksi}}{(KL/d)^2} = \frac{540 \text{ ksi}}{(40)^2} = 0.3375 \text{ ksi}$$

Applying Eq. 13–30 with $\sigma_{\text{allow}} = \sigma_{\text{max}}$, we have

$$\sigma_{\text{allow}} = \frac{P}{A} + \frac{Mc}{I}$$

$$0.3375 \text{ ksi} = \frac{P}{3 \text{ in.}(6 \text{ in.})} + \frac{P(4 \text{ in.})(3 \text{ in.})}{(1/12)(3 \text{ in.})(6 \text{ in.})^3}$$

$$P = 1.22 \text{ kip} \qquad \textit{Ans.}$$

PROBLEMS

13–101. The W8 × 15 structural A-36 steel column is fixed at its top and bottom. If it supports end moments of $M = 5$ kip · ft, determine the axial force P that can be applied. Bending is about the $x-x$ axis. Use the AISC equations of Sec. 13.6 and Eq. 13–30.

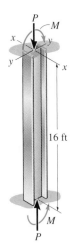

16 ft

Prob. 13–101

13–102. The W8 × 15 structural A-36 steel column is fixed at its top and bottom. If it supports end moments of $M = 23$ kip · ft, determine the axial force P that can be applied. Bending is about the $x-x$ axis. Use the interaction formula with $(\sigma_b)_{\text{allow}} = 24$ ksi.

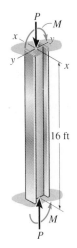

16 ft

Prob. 13–102

13–103. The W12 × 22 structural A-36 steel column is fixed at its bottom and free at its top. Determine the greatest eccentric load P that can be applied using Eq. 13–30 and the AISC equations of Sec. 13.6.

*13–104. The W10 × 15 structural A-36 steel column is fixed at its bottom and free at its top. Determine the greatest eccentric load P that can be applied using Eq. 13–30 and the AISC equations of Sec. 13.6.

13–105. The W10 × 15 structural A-36 steel column is fixed at its bottom and free at its top. If it is subjected to a load of $P = 2$ kip, determine if it is safe based on the AISC equations of Sec. 13.6 and Eq. 13–30.

13–106. The W12 × 22 structural A-36 steel column is fixed at its bottom and free at its top. If it is subjected to a load of $P = 4$ kip, determine if it is safe based on the AISC equations of Sec. 13.6 and Eq. 13–30.

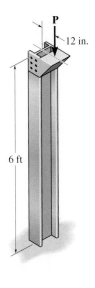

P

12 in.

6 ft

Probs. 13–103/13–104/13–105/13–106

13–107. The W8 × 15 structural A-36 steel column is assumed to be pinned at its top and bottom. Determine the largest eccentric load P that can be applied using Eq. 13–30 and the AISC equations of Sec. 13.6.

***13–108.** Solve Prob. 13–107 if the column is fixed at its top and bottom.

13–109. Solve Prob. 13–107 if the column is fixed at its bottom and pinned at its top.

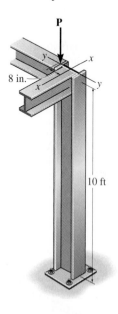

Probs. 13–107/13–108/13–109

13–110. A 20-ft-long column is made of aluminum alloy 2014-T6. If it is pinned at its top and bottom, and a compressive load **P** is applied at point A, determine the maximum allowable magnitude of **P** using the equations of Sec. 13.6 and Eq. 13–30.

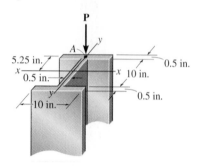

Prob. 13–110

13–111. A 20-ft-long column is made of aluminum alloy 2014-T6. If it is pinned at its top and bottom, and a compressive load **P** is applied at point A, determine the maximum allowable magnitude of **P** using the equations of Sec. 13.6 and the interaction formula with $(\sigma_b)_{allow} = 20$ ksi.

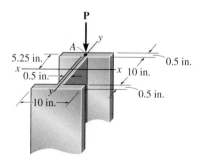

Prob. 13–111

***13–112.** The W14 × 43 structural A-36 steel column is fixed at its bottom and free at its top. Determine the greatest eccentric load P that can be applied using Eq. 13–30 and the AISC equations of Sec. 13.6.

13–113. The W10 × 45 structural A-36 steel column is fixed at its bottom and free at its top. If it is subjected to a load of $P = 2$ kip, determine if it is safe based on the AISC equations of Sec. 13.6 and Eq. 13–30.

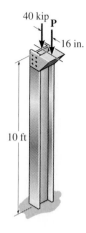

Probs. 13–112/13–113

13–114. Check if the wood column is adequate for supporting the eccentric load of $P = 800$ lb applied at its top. It is fixed at its base and free at its top. Use the NFPA equations of Sec. 13.6 and Eq. 13–30.

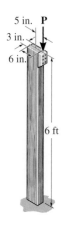

5 in.   P
3 in.
6 in.

6 ft

Prob. 13–114

*13–116.** Check if the wood column is adequate for supporting the eccentric load of $P = 600$ lb applied at its top. It is fixed at its base and free at its top. Use the NFPA equations of Sec. 13.6 and Eq. 13–30.

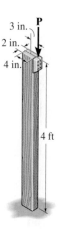

3 in.   P
2 in.
4 in.

4 ft

Prob. 13–116

13–115. Determine the maximum allowable eccentric load P that can be applied to the wood column. The column is fixed at its base and free at its top. Use the NFPA equations of Sec. 13.6 and Eq. 13–30.

5 in.   P
3 in.
6 in.

6 ft

Prob. 13–115

13–117. Determine the maximum allowable eccentric load P that can be applied to the wood column. The column is fixed at its base and free at its top. Use the NFPA equations of Sec. 13.6 and Eq. 13–30.

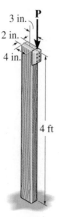

3 in.   P
2 in.
4 in.

4 ft

Prob. 13–117

REVIEW PROBLEMS

13–118.   The member has a symmetric cross section. If it is pin connected at its ends, determine the largest force it can support. It is made of 2014-T6 aluminum alloy.

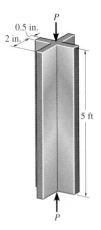

Prob. 13–118

13–119.   The wood column is 4 m long and is required to support the axial load of 25 kN. If the cross section is square, determine the dimension a of each of its sides using a factor of safety against buckling of F.S. = 2.5. The column is assumed to be pinned at its top and bottom. Use the Euler equation. $E_w = 11$ GPa, $\sigma_Y = 10$ MPa.

*****13–120.**   The wood column is 4 m long and is required to support the axial load of 25 kN. If the cross section is square, determine the dimension a of each of its sides using a factor of safety against buckling of F.S. = 1.5. The column is assumed to be fixed at its top and bottom. Use the Euler equation, $E_w = 11$ GPa, $\sigma_Y = 10$ MPa.

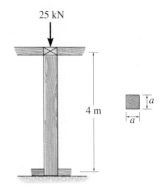

Probs. 13–119/13–120

13–121.   The steel column is assumed to be pin connected at its top and bottom and fully braced against buckling about the y–y axis. If it is subjected to an axial load of 200 kN, determine the maximum moment M that can be applied to its ends without causing it to yield. $E_{st} = 200$ GPa, $\sigma_Y = 250$ MPa.

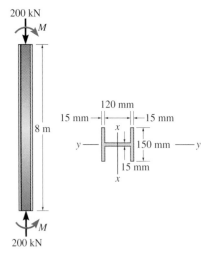

Prob. 13–121

13–122.   The steel column is assumed to be fixed connected at its top and bottom and braced against buckling about the y–y axis. If it is subjected to an axial load of 200 kN, determine the maximum moment M that can be applied to its ends without causing it to yield. $E_{st} = 200$ GPa, $\sigma_Y = 250$ MPa.

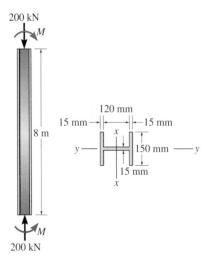

Prob. 13–122

13–123. The steel bar AB has a rectangular cross section. If it is pin connected at its ends, determine the maximum allowable intensity w of the distributed load that can be applied to BC without causing bar AB to buckle. Use a factor of safety with respect to buckling of F.S. = 1.5. E_{st} = 200 GPa, σ_Y = 360 MPa.

*13–124. The steel bar AB has a rectangular cross section. If it is assumed to be pin connected at its ends, determine if member AB will buckle if the distributed load w = 2 kN/m. Use a factor of safety with respect to buckling of F.S. = 1.5. E_{st} = 200 GPa. σ_Y = 360 MPa.

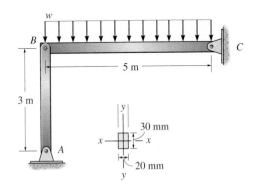

Probs. 13–123/13–124

13–125. Use the AISC equations and check if the W6 × 15 A-36 steel column can support the axial load of 16 kip. The column is fixed at its base and free at its top.

13–126. Use the AISC equations and check if the W12 × 45 A-36 steel column can support the axial load of 16 kip. The column is fixed at its base and free at its top.

Probs. 13–125/13–126

13–127. The steel pipe is fixed supported at its ends. If it is 4 m long and has an outer diameter of 50 mm, determine its required thickness so that it can support an axial load of P = 100 kN without buckling. E_{st} = 200 GPa, σ_Y = 250 MPa.

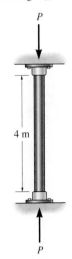

Prob. 13–127

*13–128. The distributed loading is supported by two pin-connected columns, each having a solid circular cross section. If AB is made of aluminum and CD of steel, determine the required diameter of each column so that both will be on the verge of buckling at the same time. E_{st} = 200 GPa, E_{al} = 70 GPa, $(\sigma_Y)_{st}$ = 250 MPa, $(\sigma_Y)_{al}$ = 100 MPa.

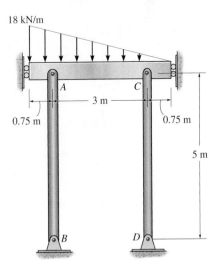

Prob. 13–128

As piles are driven in place, their ends are subjected to impact loading. The nature of impact and the energy derived from it must be understood in order to determine the stress developed within the pile. (Courtesy of Manitowoc Engineering Company.)

14 ENERGY METHODS

CHAPTER OBJECTIVES

In this chapter we will show how to apply energy methods to solve problems involving deflection. The chapter begins with a discussion of work and strain energy, followed by a development of the principle of conservation of energy. Using this principle, the stress and deflection of a member are determined when the member is subjected to impact. The method of virtual work and Castigliano's theorem are then developed, and these methods are used to determine the displacement and slope at points on structural members and mechanical elements.

14.1 EXTERNAL WORK AND STRAIN ENERGY

Before developing any of the energy methods that will be used throughout this chapter, we will first define the work caused by an external force and couple moment and show how to express the work in terms of a body's strain energy. The formulations to be presented here and in the next section will provide the basis for applying the work and energy methods that follow throughout the chapter.

Work of a Force.   In mechanics, a force does *work* when it undergoes a displacement dx that is in the *same direction* as the force. The work done is a scalar, defined as $dU_e = F\,dx$. If the total displacement is x, the work becomes

$$U_e = \int_0^x F\,dx \qquad (14\text{–}1)$$

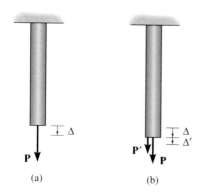

(a)   (b)

To show how to apply this equation, we will calculate the work done by an axial force applied to the end of the bar shown in Fig. 14–1a. As the magnitude of **F** is *gradually* increased from zero to some limiting value $F = P$, the final displacement of the end of the bar becomes Δ. If the material behaves in a linear-elastic manner, then the force will be directly proportional to the displacement; that is, $F = (P/\Delta)x$. Substituting into Eq. 14–1 and integrating from 0 to Δ, we get

$$U_e = \frac{1}{2}P\Delta \qquad (14\text{–}2)$$

Therefore, as the force is gradually applied to the bar, its magnitude builds from zero to some value P, and consequently, the work done is equal to the *average force magnitude*, $P/2$, times the total displacement Δ. We can represent this graphically as the light color-shaded area of the triangle in Fig. 14–1c.

Suppose, however, that **P** is already applied to the bar and that *another force* **P′** is now applied, so that the end of the bar is displaced *further* by an amount Δ', Fig. 14–1b. The work done by **P** (not **P′**) when the bar undergoes this further displacement Δ' is then

$$U_e' = P\Delta' \qquad (14\text{–}3)$$

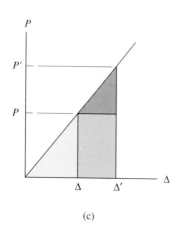

(c)

Fig. 14–1

Here the work represents the dark color-shaded *rectangular area* in Fig. 14–1c. In this case **P** does not change its magnitude, since the bar's displacement Δ' is caused only by **P′**. Therefore, work here is simply the force magnitude P times the displacement Δ'.

In summary, then, when a force **P** is applied to the bar, followed by application of the force **P′**, the total work done by both forces is represented by the area of the entire triangle in Fig. 14–1c. The light-colored triangular area represents the work of **P** that is caused by its displacement Δ. The light-gray shaded triangular area represents the work of **P′**, since this force is displaced Δ'; and lastly, the dark color-shaded rectangular area represents the additional work done by **P** when **P** is displaced Δ', as caused by **P′**.

Work of a Couple Moment.   A couple moment **M** does work when it undergoes a rotational displacement $d\theta$ along its line of action. The work done is defined as $dU_e = M\, d\theta$, Fig. 14–2. If the total angle of rotational displacement is θ rad, the work becomes

$$U_e = \int_0^\theta M\, d\theta \qquad (14\text{–}4)$$

As in the case of force, if the couple moment is applied to a *body* having linear-elastic material behavior, such that its magnitude is increased gradually from zero at $\theta = 0$ to M at θ, then the work is

$$U_e = \frac{1}{2}M\theta \qquad (14\text{–}5)$$

However, if the couple moment is already applied to the body and other loadings further rotate the body by an amount θ', then the work is

$$U'_e = M\theta'$$

Strain Energy.   When loads are applied to a body, they will deform the material. Provided no energy is lost in the form of heat, the external work done by the loads will be converted into internal work called ***strain energy***. This energy, which is *always positive*, is stored in the body and is caused by the action of either normal or shear stress.

Normal Stress.   If the volume element shown in Fig. 14–3 is subjected to the normal stress σ_z, then the force created on the top and bottom faces is $dF_z = \sigma_z\, dA = \sigma_z\, dx\, dy$. If this force is applied gradually to the element, like the force **P** discussed previously, its magnitude is increased from zero to dF_z, while the element undergoes a displacement $d\Delta_z = \epsilon_z\, dz$. The work done by dF_z is therefore $dU_i = \frac{1}{2}dF_z\, d\Delta_z = \frac{1}{2}[\sigma_z\, dx\, dy]\epsilon_z\, dz$. Since the volume of the element is $dV = dx\, dy\, dz$, we have

$$dU_i = \frac{1}{2}\sigma_z\epsilon_z\, dV \qquad (14\text{–}6)$$

Notice that U_i is *always positive*, even if σ_z is compressive, since σ_z and ϵ_z will always be in the same direction.

In general then, if the body is subjected only to a uniaxial *normal stress* σ, acting in a specified direction, the strain energy in the body is then

$$U_i = \int_V \frac{\sigma\epsilon}{2}\, dV \qquad (14\text{–}7)$$

Also, if the material behaves in a linear-elastic manner, Hooke's law applies, $\sigma = E\epsilon$, and therefore we can express the strain energy in terms of the normal stress as

$$\boxed{U_i = \int_V \frac{\sigma^2}{2E}\, dV} \qquad (14\text{–}8)$$

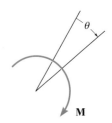

M

Fig. 14–2

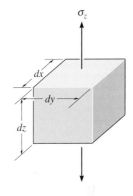

Fig. 14–3

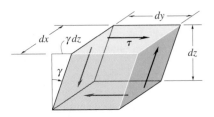

Fig. 14–4

Shear Stress.   A strain-energy expression similar to that for normal stress can also be established for the material when it is subjected to shear stress. Consider the volume element shown in Fig. 14–4. Here the shear stress causes the element to deform such that only the shear force $dF = \tau(dx\ dy)$, acting on the top face of the element, is displaced $\gamma\ dz$ relative to the bottom face. The *vertical faces* only rotate, and therefore the shear forces on these faces do no work. Hence, the strain energy stored in the element is

$$dU_i = \frac{1}{2}[\tau(dx\ dy)]\gamma\ dz$$

or

$$dU_i = \frac{1}{2}\ \tau\gamma\ dV \tag{14–9}$$

where $dV = dx\ dy\ dz$ is the volume of the element.

Integrating over the body's entire volume to obtain the strain energy stored in the body, we have

$$U_i = \int_V \frac{\tau\gamma}{2}\ dV \tag{14–10}$$

Like the case of normal stress, shear strain energy is always positive since τ and γ are always in the same direction. If the material is linear-elastic, then, applying Hooke's law, $\gamma = \tau/G$, we can express the strain energy in terms of the shear stress as

$$U_i = \int_V \frac{\tau^2}{2G}\ dV \tag{14–11}$$

In the next section we will use Eqs. 14–8 and 14–11 to obtain formal expressions for the strain energy stored in members subjected to several types of loads. Once this is done we will then be able to develop the energy methods necessary to determine the displacement and slope at points on a body.

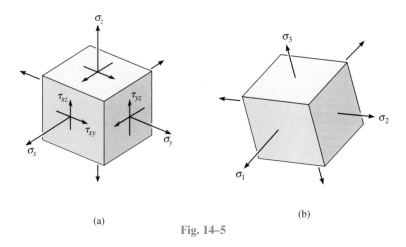

(a)

(b)

Fig. 14–5

Multiaxial Stress. The previous development may be expanded to determine the strain energy in a body when it is subjected to a general state of stress, Fig. 14–5a. The strain energies associated with each of the normal and shear stress components can be obtained from Eqs. 14–6 and 14–9. Since energy is a scalar, the total strain energy in the body is therefore

$$U_i = \int_V \left[\frac{1}{2} \sigma_x \epsilon_x + \frac{1}{2} \sigma_y \epsilon_y + \frac{1}{2} \sigma_z \epsilon_z \right.$$

$$\left. + \frac{1}{2} \tau_{xy} \gamma_{xy} + \frac{1}{2} \tau_{yz} \gamma_{yz} + \frac{1}{2} \tau_{xz} \gamma_{xz} \right] dV \qquad (14\text{–}12)$$

The strains can be eliminated by using the generalized form of Hooke's law given by Eqs. 10–18 and 10–19. After substituting and combining terms, we have

$$U_i = \int_V \left[\frac{1}{2E} (\sigma_x^2 + \sigma_y^2 + \sigma_z^2) - \frac{\nu}{E} (\sigma_x \sigma_y + \sigma_y \sigma_z + \sigma_x \sigma_z) \right.$$

$$\left. + \frac{1}{2G} (\tau_{xy}^2 + \tau_{yz}^2 + \tau_{xz}^2) \right] dV \qquad (14\text{–}13)$$

If only the principal stresses $\sigma_1, \sigma_2, \sigma_3$ act on the element, Fig. 14–5b, this equation reduces to a simpler form, namely,

$$U_i = \int_V \left[\frac{1}{2E} (\sigma_1^2 + \sigma_2^2 + \sigma_3^2) - \frac{\nu}{E} (\sigma_1 \sigma_2 + \sigma_2 \sigma_3 + \sigma_1 \sigma_3) \right] dV \quad (14\text{–}14)$$

Recall that we used this equation in Sec. 10.7 as a basis for developing the maximum-distortion-energy theory.

14.2   ELASTIC STRAIN ENERGY FOR VARIOUS TYPES OF LOADING

Using the equations for elastic strain energy developed in the previous section, we will now formulate the strain energy stored in a member when it is subjected to an axial load, bending moment, transverse shear, and torsional moment. Examples will be given to show how to calculate the strain energy in members subjected to each of these loadings.

Axial Load.   Consider a bar of variable yet slightly tapered cross section, which is subjected to an axial load coincident with the bar's centroidal axis, Fig. 14–6. The *internal axial force* at a section located a distance x from one end is N. If the cross-sectional area at this section is A, then the normal stress on the section is $\sigma = N/A$. Applying Eq. 14–8, we have

$$U_i = \int_V \frac{\sigma_x^{\,2}}{2E}\, dV = \int_V \frac{N^2}{2EA^2}\, dV$$

Fig. 14–6

If we choose an element or differential slice having a volume $dV = A\ dx$, the general formula for the strain energy in the bar is therefore

$$\boxed{U_i = \int_0^L \frac{N^2}{2AE}\, dx} \qquad (14\text{–}15)$$

For the more common case of a prismatic bar of constant cross-sectional area A, length L, and constant axial load N, Fig. 14–7, Eq. 14–15, when integrated, gives

Fig. 14–7

$$\boxed{U_i = \frac{N^2L}{2AE}} \qquad (14\text{–}16)$$

From this equation it can be seen that the bar's elastic strain energy will *increase* if the length of the bar is increased, or if the modulus of elasticity or cross-sectional area is decreased. For example, an aluminum rod $[E_{al} = 10(10^3)$ ksi$]$ will store approximately three times as much energy as a steel rod $[E_{st} = 29(10^3)$ ksi$]$ having the same size and subjected to the same load. On the other hand, doubling the cross-sectional area of a given rod will decrease its ability to store energy by one-half. The following example illustrates this point numerically.

EXAMPLE 14–1

One of the two high-strength steel bolts A and B shown in Fig. 14–8 is to be chosen to support a sudden tensile loading. For the choice it is necessary to determine the greatest amount of elastic strain energy that each bolt can absorb. Bolt A has a diameter of 0.875 in. for 2 in. of its length and a root (or smallest) diameter of 0.731 in. within the 0.25-in. threaded region. Bolt B has "upset" threads, such that the diameter throughout its 2.25-in. length can be taken as 0.731 in. In both cases, neglect the extra material that makes up the threads. Take $E_{st} = 29(10^3)$ ksi, $\sigma_Y = 44$ ksi.

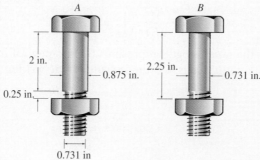

Fig. 14–8

SOLUTION

Bolt A.   If the bolt is subjected to its maximum tension, the maximum stress of $\sigma_Y = 44$ ksi will occur within the 0.25-in. region. This tension force is

$$P_{max} = \sigma_Y A = 44 \text{ ksi}\left[\pi\left(\frac{0.731 \text{ in.}}{2}\right)^2\right] = 18.47 \text{ kip}$$

Applying Eq. 14–16 to each region of the bolt, we have

$$U_i = \sum \frac{N^2 L}{2AE}$$

$$= \frac{(18.47 \text{ kip})^2(2 \text{ in.})}{2[\pi(0.875 \text{ in.}/2)^2][29(10^3) \text{ ksi}]} + \frac{(18.47 \text{ kip})^2(0.25 \text{ in.})}{2[\pi(0.731 \text{ in.}/2)^2]29(10^3) \text{ ksi}}$$

$$= 0.0231 \text{ in.} \cdot \text{kip} \qquad\qquad \textit{Ans.}$$

Bolt B.   Here the bolt is assumed to have a uniform diameter of 0.731 in. throughout its 2.25-in. length. Also, from the calculation above, it can support a maximum tension force of $P_{max} = 18.47$ kip. Thus,

$$U_i = \frac{N^2 L}{2AE} = \frac{(18.47 \text{ kip})^2(2.25 \text{ in.})}{2[\pi(0.731 \text{ in.}/2)^2][29(10^3) \text{ ksi}]} = 0.0315 \text{ in.} \cdot \text{kip} \quad \textit{Ans.}$$

By comparison, bolt B can absorb 36% more elastic energy than bolt A, even though it has a smaller cross section along its shank.

Bending Moment.   Since a bending moment applied to a straight prismatic member develops *normal stress* in the member, we can use Eq. 14–8 to determine the strain energy stored in the member due to bending. For example, consider the axisymmetric beam shown in Fig. 14–9. Here the internal moment is M, and the normal stress acting on the arbitrary element a distance y from the neutral axis is $\sigma = My/I$. If the volume of the element is $dV = dA\ dx$, where dA is the area of its exposed face and dx is its length, the elastic strain energy in the beam is

$$U_i = \int_V \frac{\sigma^2}{2E}\, dV = \int_V \frac{1}{2E}\left(\frac{My}{I}\right)^2 dA\ dx$$

or

$$U_i = \int_0^L \frac{M^2}{2EI^2}\left(\int_A y^2\, dA\right) dx$$

Realizing that the area integral represents the moment of inertia of the beam about the neutral axis, the final result can be written as

$$U_i = \int_0^L \frac{M^2\, dx}{2EI} \qquad (14\text{–}17)$$

To evaluate the strain energy, therefore, we must first express the internal moment as a function of its position x along the beam, and then perform the integration over the beam's entire length.* The following examples illustrate this procedure.

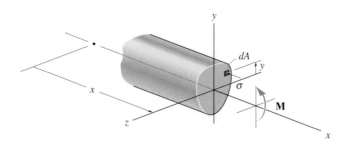

Fig. 14–9

---

*Recall that the flexure formula, as used here, can also be used with justifiable accuracy to determine the stress in slightly tapered beams. (See Sec. 6.4.) So in the general sense, I in Eq. 14–17 may also have to be expressed as a function of x.

EXAMPLE 14-2

Determine the elastic strain energy due to bending of the cantilevered beam if the beam is subjected to the uniform distributed load w, Fig. 14–10a. EI is constant.

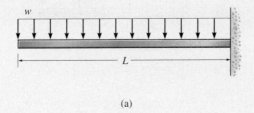

(a)

Fig. 14–10

SOLUTION

The internal moment in the beam is determined by establishing the x coordinate with origin at the left side. The left segment of the beam is shown in Fig. 14–10b. We have

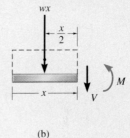

$$\zeta + \Sigma M_{NA} = 0; \qquad M + wx\left(\frac{x}{2}\right) = 0$$

$$M = -w\left(\frac{x^2}{2}\right)$$

(b)

Applying Eq. 14–17 yields

$$U_i = \int_0^L \frac{M^2\, dx}{2EI} = \int_0^L \frac{[-w(x^2/2)]^2\, dx}{2EI} = \frac{w^2}{8EI}\int_0^L x^4\, dx$$

or

$$U_i = \frac{w^2 L^5}{40EI} \qquad\qquad Ans.$$

We can also obtain the strain energy using an x coordinate having its origin at the right side of the beam and extending positive to the left, Fig. 14–10c. In this case,

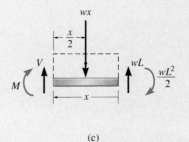

$$\zeta + \Sigma M_{NA} = 0; \quad -M - wx\left(\frac{x}{2}\right) + wL(x) - \frac{wL^2}{2} = 0$$

$$M = -\frac{wL^2}{2} + wLx - w\left(\frac{x^2}{2}\right)$$

(c)

Applying Eq. 14–17, we obtain the same result as before.

EXAMPLE 14-3

Determine the bending strain energy in region AB of the beam shown in Fig. 14–11a. EI is constant.

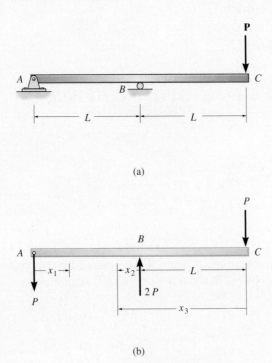

(a)

(b)

Fig. 14–11

SOLUTION

A free-body diagram of the beam is shown in Fig. 14–11b. To obtain the answer we can express the internal moment in terms of any one of the indicated three "x" coordinates and then apply Eq. 14–17. Each of these solutions will now be considered.

$0 \leq x_1 \leq L$.   From the free-body diagram of the section in Fig. 14–11c, we have

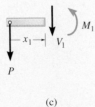

(c)

$$\zeta + \Sigma M_{NA} = 0; \qquad M_1 + Px_1 = 0$$
$$M_1 = -Px_1$$

$$U_i = \int \frac{M^2\, dx}{2EI} = \int_0^L \frac{(-Px_1)^2\, dx_1}{2EI}$$

$$= \frac{P^2 L^3}{6EI} \qquad\qquad Ans.$$

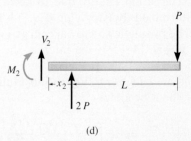

(d)

$0 \le x_2 \le L.$ Using the free-body diagram of the section in Fig. 14–11d gives

$\zeta^+ \Sigma M_{NA} = 0;$ $-M_2 + 2P(x_2) - P(x_2 + L) = 0$

$$M_2 = P(x_2 - L)$$

$$U_i = \int \frac{M^2 \, dx}{2EI} = \int_0^L \frac{[P(x_2 - L)]^2 \, dx_2}{2EI}$$

$$= \frac{P^2 L^3}{6EI} \qquad\qquad\qquad Ans.$$

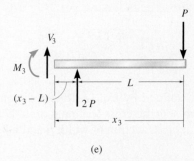

(e)

$L \le x_3 \le 2L.$ From the free-body diagram in Fig. 14–11e, we have

$\zeta^+ \Sigma M_{NA} = 0;$ $-M_3 + 2P(x_3 - L) - P(x_3) = 0$

$$M_3 = P(x_3 - 2L)$$

$$U_i = \int \frac{M^2 \, dx}{2EI} = \int_L^{2L} \frac{[P(x_3 - 2L)]^2 \, dx_3}{2EI}$$

$$= \frac{P^2 L^3}{6EI} \qquad\qquad\qquad Ans.$$

This and the previous example indicate that the strain energy for the beam can be computed using *any* suitable x coordinate. It is only necessary to integrate over the range of the coordinate where the internal energy is to be determined. Here the choice of x_1 provides the simplest solution.

Transverse Shear.   The strain energy due to shear stress in a beam element can be determined by applying Eq. 14–11. Here we will consider the beam to be prismatic and to have an axis of symmetry about the y axis as shown in Fig. 14–12. If the internal shear at the section x is V, then the shear stress acting on the volume element of material, having a length dx and area dA, is $\tau = VQ/It$. Substituting into Eq. 14–11, the strain energy for shear becomes

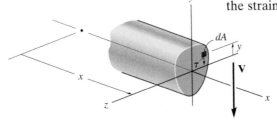

$$U_i = \int_V \frac{\tau^2}{2G}\, dV = \int_V \frac{1}{2G}\left(\frac{VQ}{It}\right)^2 dA\, dx$$

$$U_i = \int_0^L \frac{V^2}{2GI^2}\left(\int_A \frac{Q^2}{t^2}\, dA\right) dx$$

Fig. 14–12

The integral in parentheses is evaluated over the beam's cross-sectional area. To simplify this expression we will define the *form factor* for shear as

$$f_s = \frac{A}{I^2}\int_A \frac{Q^2}{t^2}\, dA \qquad (14\text{--}18)$$

Substituting into the above equation, we get

$$U_i = \int_0^L \frac{f_s V^2\, dx}{2GA} \qquad (14\text{--}19)$$

The form factor defined by Eq. 14–18 is a dimensionless number that is unique for each specific cross-sectional area. For example, if the beam has a rectangular cross section of width b and height h, Fig. 14–13, then

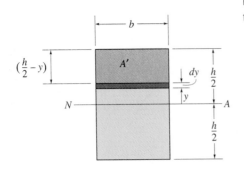

$$t = b$$

$$A = bh$$

$$I = \frac{1}{12}bh^3$$

$$Q = \bar{y}'A' = \left(y + \frac{(h/2) - y}{2}\right) b\left(\frac{h}{2} - y\right) = \frac{b}{2}\left(\frac{h^2}{4} - y^2\right)$$

Substituting these terms into Eq. 14–18, we get

$$f_s = \frac{bh}{(\frac{1}{12}bh^3)^2}\int_{-h/2}^{h/2} \frac{b^2}{4b^2}\left(\frac{h^2}{4} - y^2\right)^2 b\, dy = \frac{6}{5} \qquad (14\text{--}20)$$

Fig. 14–13

The form factor for other sections can be determined in a similar manner. Once obtained, this number is substituted into Eq. 14–19 and the strain energy for transverse shear can then be evaluated.

EXAMPLE 14-4

Determine the strain energy in the cantilevered beam due to shear if the beam has a square cross section and is subjected to a uniform distributed load w, Fig. 14–14a. EI and G are constant.

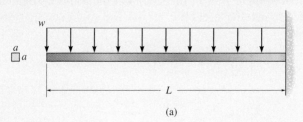

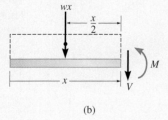

(a)   (b)

Fig. 14–14

SOLUTION

From the free-body diagram of an arbitrary section, Fig. 14–14b, we have

$$+\uparrow \Sigma F_y = 0; \qquad\qquad -V - wx = 0$$
$$V = -wx$$

Since the cross section is square, the form factor $f_s = \frac{6}{5}$ (Eq. 14–20) and therefore Eq. 14–19 becomes

$$(U_i)_s = \int_0^L \frac{\frac{6}{5}(-wx)^2\, dx}{2GA} = \frac{3w^2}{5GA} \int_0^L x^2\, dx$$

or

$$(U_i)_s = \frac{w^2 L^3}{5GA} \qquad\qquad Ans.$$

Using the results of Example 14–2, with $A = a^2$, $I = \frac{1}{12}a^4$, the ratio of shear to bending strain energy is

$$\frac{(U_i)_s}{(U_i)_b} = \frac{w^2 L^3/5Ga^2}{w^2 L^5/40E(\frac{1}{12}a^4)} = \frac{2}{3}\left(\frac{a}{L}\right)^2 \frac{E}{G}$$

Since $G = E/2(1 + v)$ and $v \leq \frac{1}{2}$ (Sec. 10.6), then as an *upper bound*, $E = 3G$, so that

$$\frac{(U_i)_s}{(U_i)_b} = 2\left(\frac{a}{L}\right)^2$$

It can be seen that this ratio will increase as L decreases. However, even for very short beams, where, say, $L = 5a$, the contribution due to shear strain energy is only 8% of the bending strain energy. For this reason, the shear strain energy stored in beams is usually neglected in engineering analysis.

Torsional Moment. To determine the internal strain energy in a circular shaft or tube due to an applied torsional moment, we must apply Eq. 14–11. Consider the slightly tapered shaft in Fig. 14–15. A section of the shaft taken a distance x from one end is subjected to an internal torque T. The shear stress distribution that causes this torque varies linearly from the center of the shaft. On the arbitrary element of length dx and area dA, the stress is $\tau = T\rho/J$. The strain energy stored in the shaft is thus

$$U_i = \int_V \frac{\tau^2}{2G}\, dV = \int_V \frac{1}{2G}\left(\frac{T\rho}{J}\right)^2 dA\, dx$$

$$= \int_0^L \frac{T^2}{2GJ^2}\left(\int_A \rho^2\, dA\right) dx$$

Fig. 14–15

Since the area integral represents the polar moment of inertia J for the shaft at the section, the final result can be written as

$$\boxed{U_i = \int_0^L \frac{T^2}{2GJ}\, dx} \qquad (14\text{–}21)$$

The most common case occurs when the shaft (or tube) has a constant cross-sectional area and the applied torque is constant, Fig. 14–16. Integration of Eq. 14–21 then gives

$$\boxed{U_i = \frac{T^2 L}{2GJ}} \qquad (14\text{–}22)$$

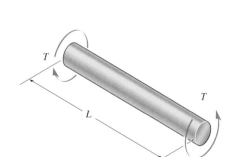

Fig. 14–16

From this equation we may conclude that, like an axially loaded member, the energy-absorbing capacity of a torsionally loaded shaft is *decreased* by increasing the diameter of the shaft, since this increases J.

If the cross section of the shaft is some shape other than circular or tubular, Eq. 14–22 must be modified. For example, if it is rectangular, having dimensions $h > b$, then using a mathematical analysis based on the theory of elasticity, it can be shown that the strain energy in the shaft is determined from

$$U_i = \frac{T^2 L}{2Cb^3 hG} \qquad (14\text{–}23)$$

where

$$C = \frac{hb^3}{16}\left[\frac{16}{3} - 3.336\frac{b}{h}\left(1 - \frac{b^4}{12h^4}\right)\right] \qquad (14\text{–}24)$$

The following example illustrates how to determine the strain energy in a shaft due to a torsional loading.

EXAMPLE 14-5

The tubular shaft in Fig. 14–17a is fixed at the wall and subjected to two torques as shown. Determine the strain energy stored in the shaft due to this loading. $G = 75$ GPa.

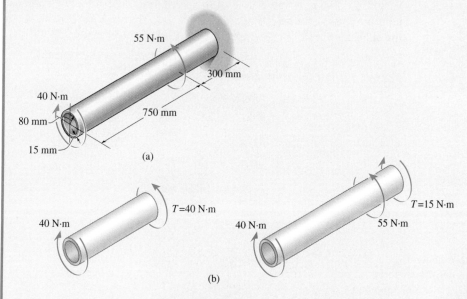

Fig. 14–17

SOLUTION

Using the method of sections, the internal torque is first determined within the two regions of the shaft where it is constant, Fig. 14–17b. Although these torques (40 N · m and 15 N · m) are in opposite directions, this will be of no consequence in determining the strain energy, since the torque is squared in Eq. 14–22. In other words, the strain energy is always positive. The polar moment of inertia for the shaft is

$$J = \frac{\pi}{2}[(0.08 \text{ m})^4 - (0.065 \text{ m})^4] = 36.30(10^{-6}) \text{ m}^4$$

Applying Eq. 14–22, we have

$$U_i = \sum \frac{T^2 L}{2GJ}$$

$$= \frac{(40 \text{ N} \cdot \text{m})^2(0.750 \text{ m})}{2[75(10^9) \text{ N/m}^2]36.30(10^{-6}) \text{ m}^4} + \frac{(15 \text{ N} \cdot \text{m})^2(0.300 \text{ m})}{2[75(10^9) \text{ N/m}^2]36.30(10^{-6}) \text{ m}^4}$$

$$= 233 \text{ } \mu\text{J} \qquad\qquad\qquad\qquad\qquad Ans.$$

IMPORTANT POINTS

• A *force* does work when it moves through a *displacement*. If the force is increased gradually in magnitude from zero to F, the work is $U = (F/2)\Delta$, whereas if the force is constant when the displacement occurs then $U = F\Delta$.

• A *couple moment* does work when it moves through a *rotation*.

• *Strain energy* is caused by the internal work of the normal and shear stresses. It is always a *positive* quantity.

• The strain energy can be related to the resultant internal loadings N, V, M, and T.

• As the beam becomes longer, the strain energy due to bending becomes much larger than the strain energy due to shear. For this reason, the *shear strain energy* in beams can generally be *neglected*.

PROBLEMS

14–1. A material is subjected to a general state of plane stress. Express the strain energy density in terms of the elastic constants E, G, and v and the stress components σ_x, σ_y, and τ_{xy}.

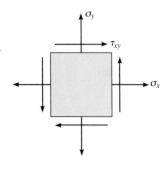

Prob. 14–1

14–2. The strain energy density must be the same whether the state of stress is represented by σ_x, σ_y, and τ_{xy}, or by the principal stresses σ_1 and σ_2. Equate the strain energy expressions for each of these two cases and show that $G = E/[2(1 + v)]$.

14–3. If σ_x and σ_y are applied to an element, for what ratio of σ_x to σ_y is the strain energy a minimum?

***14–4.** Determine the torsional strain energy in the A-36 steel shaft. The shaft has a radius of 30 mm.

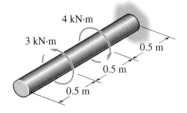

Prob. 14–4

14–5. Determine the bending strain energy in the A-36 structural steel W10 × 12 beam. Obtain the answer using the coordinates (a) x_1 and x_4, and (b) x_2 and x_3.

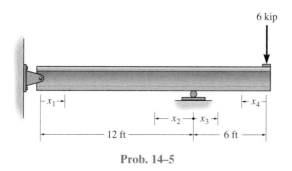

6 kip

12 ft   6 ft

Prob. 14–5

14–6. Determine the bending strain energy in the beam due to the loading shown. EI is constant.

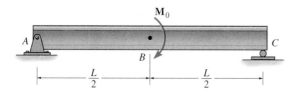

$\mathbf{M}_0$

A   B   C

$\dfrac{L}{2}$ $\dfrac{L}{2}$

Prob. 14–6

14–7. Determine the bending strain energy in the beam. EI is constant.

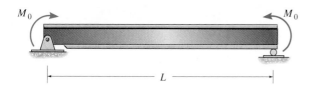

M_0 M_0

L

Prob. 14–7

*14–8. Determine the bending strain energy in the cantilevered beam due to a uniform load w. Solve the problem two ways. (a) Apply Eq. 14–17. (b) The load $w\,dx$ acting on a segment dx of the beam is displaced a distance y, where $y = w(-x^4 + 4L^3x - 3L^4)/(24EI)$, the equation of the elastic curve. Hence the internal strain energy in the differential segment dx of the beam is equal to the external work, i.e., $dU_i = \frac{1}{2}(w\,dx)(-y)$. Integrate this equation to obtain the total strain energy in the beam. EI is constant.

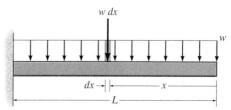

$w\,dx$ w

dx x

L

Prob. 14–8

14–9. Determine the bending strain energy in the simply supported beam due to a uniform load w. Solve the problem two ways. (a) Apply Eq. 14–17. (b) The load $w\,dx$ acting on the segment dx of the beam is displaced a distance y, where $y = w(-x^4 + 2Lx^3 - L^3x)/(24EI)$, the equation of the elastic curve. Hence the internal strain energy in the differential segment dx of the beam is equal to the external work, i.e., $dU_i = \frac{1}{2}(w\,dx)(-y)$. Integrate this equation to obtain the total strain energy in the beam. EI is constant.

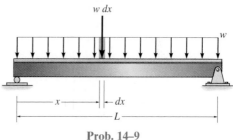

$w\,dx$ w

x dx

L

Prob. 14–9

14–10. Determine the bending strain energy in the beam. EI is constant.

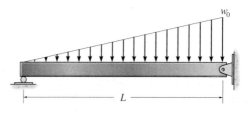

w_0

L

Prob. 14–10

14–11.   Determine the total axial and bending strain energy in the A-36 steel beam. $A = 2300$ mm^2, $I = 9.5(10^6)$ mm^4.

14–13.   The bolt has a diameter of 10 mm, and the link AB has a rectangular cross section that is 12 mm wide by 7 mm thick. Determine the strain energy in the link due to bending and in the bolt due to axial force. The bolt is tightened so that it has a tension of 500 N. Both members are made of A-36 steel. Neglect the hole in the link.

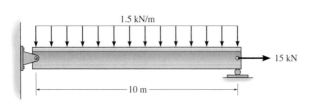

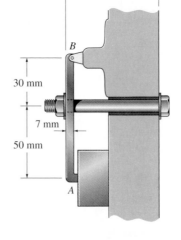

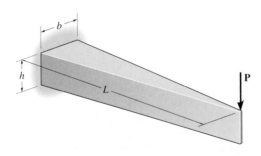

Prob. 14–11

Prob. 14–13

*14–12.   The beam shown is tapered along its width. If a force **P** is applied to its end, determine the strain energy in the beam and compare this result with that of a beam that has a constant rectangular cross section of width b and height h.

14–14.   The steel beam is supported on two springs, each having a stiffness of $k = 8$ MN/m. Determine the strain energy in each of the springs and the bending strain energy in the beam. $E_{st} = 200$ GPa, $I = 5(10^6)$ mm^4.

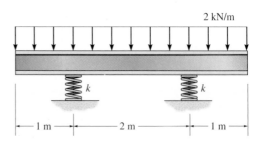

Prob. 14–12

Prob. 14–14

14–15.   Determine the total axial and bending strain energy in the A-36 structural steel W8 × 58 beam.

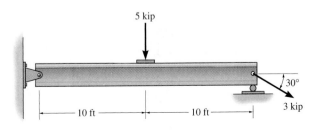

5 kip

30°

3 kip

10 ft

10 ft

Prob. 14–15

14–17.   Determine the bending strain energy in the beam due to the distributed load. EI is constant.

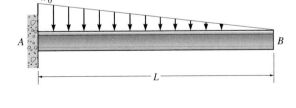

w_0

A

B

L

Prob. 14–17

*14–16.**   Determine the bending strain energy in the beam and the axial strain energy in each of the two posts. All members are made of aluminum and have a square cross section 50 mm by 50 mm. Assume the posts only support an axial load. $E_{al} = 70$ GPa.

14–18.   The A-36 steel bar consists of two segments, one of circular cross section of radius r, and one of square cross section. If it is subjected to the axial loading of P, determine the dimensions a of the square segment so that the strain energy within the square segment is the same as in the circular segment.

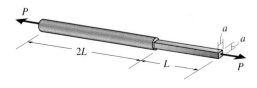

P

$2L$

L

a

a

P

Prob. 14–18

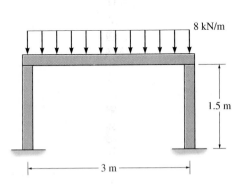

8 kN/m

1.5 m

3 m

Prob. 14–16

14–19.   Consider the thin-walled tube of Fig. 5–30. Use the formula for shear stress, $\tau_{avg} = T/2tA_m$, Eq. 5–18, and the general equation of shear strain energy, Eq. 14–11, to show that the twist of the tube is given by Eq. 5–20. *Hint:* Equate the work done by the torque T to the strain energy in the tube, determined from integrating the strain energy for a differential element, Fig. 14–4, over the volume of material.

14.3   Conservation of Energy

All energy methods used in mechanics are based on a balance of energy, often referred to as the conservation of energy. In this chapter, only mechanical energy will be considered in the energy balance; that is, the energy developed by heat, chemical reactions, and electromagnetic effects will be neglected. As a result, if a loading is applied *slowly* to a body, so that kinetic energy can also be neglected, then physically the external loads tend to deform the body so that the loads do *external work* U_e as they are displaced. This external work caused by the loads is transformed into *internal work* or strain energy U_i, which is stored in the body. Furthermore, when the loads are removed, the strain energy restores the body back to its original undeformed position, provided the material's elastic limit is not exceeded. The conservation of energy for the body can therefore be stated mathematically as

$$\boxed{U_e = U_i} \qquad (14\text{--}25)$$

We will now show three examples of how this equation can be applied to determine the displacement of a point on a deformable member or structure. As the first example, consider the truss in Fig. 14–18 subjected to the known load **P**. Provided **P** is applied gradually, the external work done by **P** is determined from Eq. 14–2, that is, $U_e = \frac{1}{2}P\Delta$, where Δ is the vertical displacement of the truss at the joint where **P** is applied. Assuming that **P** develops an axial force **N** in a particular member, the strain energy stored in this member is determined from Eq. 14–16, that is, $U_i = N^2L/2AE$. Summing the strain energies for all the members of the truss, we can write Eq. 14–25 as

$$\frac{1}{2}\,P\Delta = \sum \frac{N^2L}{2AE} \qquad (14\text{--}26)$$

Once the internal forces (N) in all the members of the truss are determined and the terms on the right calculated, it is then possible to determine the unknown displacement Δ.

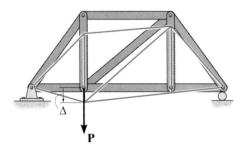

Fig. 14–18

As a second example, consider finding the vertical displacement Δ under the known load **P** acting on the beam in Fig. 14–19. Again, the external work is $U_e = \frac{1}{2}P\Delta$. However, in this case the strain energy would be the result of internal shear and moment loadings caused by **P**. In particular, the contribution of strain energy due to shear is generally *neglected* in most beam deflection problems unless the beam is short and supports a very large load. (See Example 14–4.) Consequently, the beam's strain energy will be determined only by the internal bending moment M, and therefore, using Eq. 14–17, Eq. 14–25 can be written symbolically as

$$\frac{1}{2}P\Delta = \int_0^L \frac{M^2}{2EI}\,dx \tag{14–27}$$

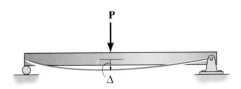

Fig. 14–19

Once M is expressed as a function of position and the integral is evaluated, Δ can then be determined.

As the last example, we will consider a beam loaded by a couple moment **M**$_0$ as shown in Fig. 14–20. This moment causes the rotational displacement θ at the point of application of the couple moment. Since the couple moment only does work when it *rotates*, using Eq. 14–5, the external work is $U_e = \frac{1}{2}M_0\theta$. Therefore Eq. 14–25 becomes

$$\frac{1}{2}M_0\theta = \int_0^L \frac{M^2}{2EI}\,dx \tag{14–28}$$

Fig. 14–20

Here the strain energy is determined as a result of the internal bending moment M caused by application of the couple moment **M**$_0$. Once M has been expressed as a function of x and the strain energy evaluated, then θ can be calculated.

In each of the above examples, it should be noted that application of Eq. 14–25 is *quite limited*, because only a *single* external force or couple moment must act on the member or structure. In other words, the displacement can *only* be calculated at the point and in the direction of the external force or couple moment. If more than one external force or couple moment were applied, then the external work of each loading would involve its associated unknown displacement. As a result, *all* these unknown displacements could not be determined, since only the single Eq. 14–25 is available for the solution. Although application of the conservation of energy as described here has these restrictions, it does serve as an introduction to more general energy methods, which we will consider throughout the rest of this chapter. Specifically, it will be shown in later sections of this chapter that by modifying the method for applying the conservation-of-energy principle, we will be able to perform a completely general deflection analysis of a member or structure.

EXAMPLE 14-6

The three-bar truss in Fig. 14–21a is subjected to a horizontal force of 5 kip. If the cross-sectional area of each member is 0.20 in^2, determine the horizontal displacement at point B. $E = 29(10^3)$ ksi.

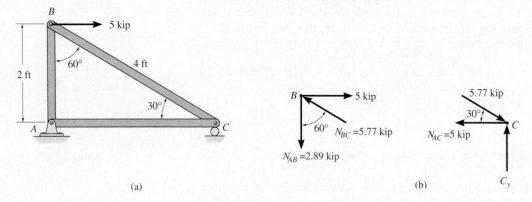

(a)

(b)

Fig. 14–21

SOLUTION

We can apply the conservation of energy to solve this problem because only a *single* external force acts on the truss and the required displacement is in the *same direction* as the force. Furthermore, the reactive forces on the truss do no work since they are not displaced.

Using the method of joints, the force in each member is determined as shown on the free-body diagrams of the pins at B and C, Fig. 14–21b.

Applying Eq. 14–26, we have

$$\frac{1}{2} P\Delta = \sum \frac{N^2 L}{2AE}$$

$$\frac{1}{2} (5 \text{ kip})(\Delta_B)_h = \frac{(2.89 \text{ kip})^2 (2 \text{ ft})}{2AE} + \frac{(-5.77 \text{ kip})^2 (4 \text{ ft})}{2AE}$$

$$+ \frac{(-5 \text{ kip})^2 (3.46 \text{ ft})}{2AE}$$

$$(\Delta_B)_h = \frac{47.32 \text{ kip} \cdot \text{ft}}{AE}$$

Notice that since N is squared, it does not matter if a particular member is in tension or compression. Substituting in the numerical data for A and E and solving, we get

$$(\Delta_B)_h = \frac{47.32 \text{ kip} \cdot \text{ft}(12 \text{ in./ft})}{(0.2 \text{ in}^2)[29(10^3) \text{ kip/in}^2]}$$

$$= 0.0979 \text{ in.} \quad \rightarrow \qquad \textit{Ans.}$$

EXAMPLE 14–7

The cantilevered beam in Fig. 14–22a has a rectangular cross section and is subjected to a load **P** at its end. Determine the displacement of the load. EI is constant.

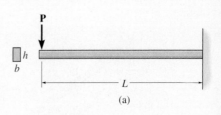

(a)

SOLUTION

The internal shear and moment in the beam as a function of x are determined using the method of sections, Fig. 14–22b.

When applying Eq. 14–25 we will consider the strain energy due to shear and bending. Using Eqs. 14–19 and 14–17, we have

$$\frac{1}{2} P\Delta = \int_0^L \frac{f_s V^2 \, dx}{2GA} + \int_0^L \frac{M^2 \, dx}{2EI}$$

$$= \int_0^L \frac{(\tfrac{6}{5})(-P)^2 \, dx}{2GA} + \int_0^L \frac{(-Px)^2 \, dx}{2EI} = \frac{3P^2 L}{5GA} + \frac{P^2 L^3}{6EI} \qquad (1)$$

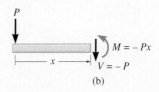

(b)

Fig. 14–22

The first term on the right side of this equation represents the strain energy due to shear, while the second is the strain energy due to bending. As stated in Example 14–4, for most beams the shear strain energy is much smaller than the bending strain energy. To show when this is the case for the beam in Figure 14–22, we require

$$\frac{3}{5} \frac{P^2 L}{GA} \overset{?}{<} \frac{P^2 L^3}{6EI}$$

$$\frac{3}{5} \frac{P^2 L}{G(bh)} \overset{?}{<} \frac{P^2 L^3}{6E[\frac{1}{12}(bh^3)]}$$

$$\frac{3}{5G} \overset{?}{<} \frac{2L^2}{Eh^2}$$

Since $E \leq 3G$ (see Example 14–4), then

$$0.9 \overset{?}{<} \left(\frac{L}{h}\right)^2$$

Hence if h is small and L relatively long (compared with h), the beam becomes slender and the shear strain energy can be neglected. In other words, the *shear strain energy* becomes important *only* for *short, deep beams.* For example, beams for which $L = 5h$ have more than 28 times more bending strain energy than shear strain energy, so neglecting the shear strain energy represents an error of about 3.6%. With this in mind, Eq. 1 can be simplified to

$$\frac{1}{2} P\Delta = \frac{P^2 L^3}{6EI}$$

So that

$$\Delta = \frac{PL^3}{3EI} \qquad \qquad \textit{Ans.}$$

PROBLEMS

*14–20. Determine the vertical displacement of joint C. AE is constant.

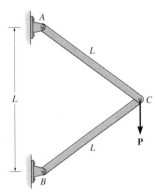

Prob. 14–20

14–21. Determine the horizontal displacement of joint D. AE is constant.

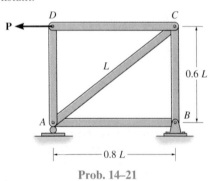

Prob. 14–21

14–22. Determine the vertical displacement of joint D. AE is constant.

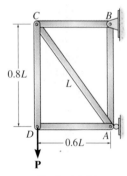

Prob. 14–22

14–23. The cantilevered beam has a rectangular cross-sectional area A, a moment of inertia I, and a modulus of elasticity E. If a load P acts at point B as shown, determine the displacement at B in the direction of P, accounting for bending, axial force, and shear.

Prob. 14–23

*14–24. Determine the displacement of point B on the A-36 steel beam. $I = 80(10^6)$ mm^4.

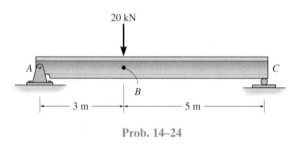

20 kN

Prob. 14–24

14–25. Determine the displacement of point B on the 2014-T6 aluminum beam.

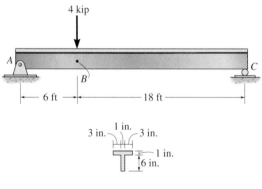

Prob. 14–25

14–26.   Determine the slope at point C of the A-36 steel beam. $I = 9.50(10^6)$ mm^4.

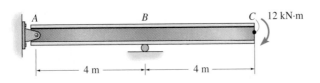

A   B   C   12 kN·m

4 m   4 m

Prob. 14–26

14–27.   Determine the deflection of the beam at its center caused by shear. The shear modulus is G.

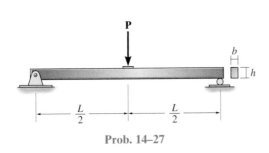

P

$\frac{L}{2}$ $\frac{L}{2}$

b   h

Prob. 14–27

*14–28.   The A-36 steel bars are pin connected at B. If each has a square cross section, determine the vertical displacement at B.

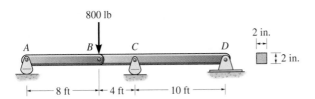

800 lb

A   B   C   D

2 in.

2 in.

8 ft   4 ft   10 ft

Prob. 14–28

14–29.   Determine the slope at point A of the beam. EI is constant.

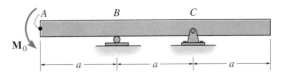

A   B   C

$\mathbf{M}_0$

a   a   a

Prob. 14–29

14–30.   The rod has a circular cross section with a moment of inertia I. If a vertical force $\mathbf{P}$ is applied at A, determine the vertical displacement at this point. Only consider the strain energy due to bending. The modulus of elasticity is E.

r

A

P

Prob. 14–30

14–31.   The coiled spring has n coils and is made of a material having a shear modulus G. Determine the stretch of the spring when it is subjected to the load $\mathbf{P}$. Assume that the coils are close to each other so that $\theta \approx 0°$ and the deflection is caused entirely by the torsional stress in the coil.

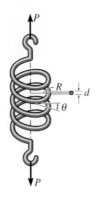

P

R   d

θ

P

Prob. 14–31

14.4 IMPACT LOADING

Throughout this text we have considered all loadings to be applied to a body in a gradual manner, such that when they reach a maximum value they remain constant or static. Some loadings, however, are dynamic; that is, they vary with time. A typical example would be caused by the collision of objects. This is called an impact loading. Specifically, **impact** occurs when one object strikes another, such that large forces are developed between the objects during a very short period of time.

If we assume no energy is lost during impact, we can study the mechanics of impact using the conservation of energy. To show how this is done, we will first analyze the motion of a simple block-and-spring system as shown in Fig. 14–23. When the block is released from rest, it falls a distance h, striking the spring and compressing it a distance Δ_{max} before momentarily coming to rest. If we neglect the mass of the spring and assume that the spring responds *elastically*, then the conservation of energy requires that the energy of the falling block be transformed into stored (strain) energy in the spring; or in other words, the work done by the block's weight, falling $h + \Delta_{max}$, is equal to the work needed to displace the end of the spring by an amount Δ_{max}. Since the force in a spring is related to Δ_{max} by the equation $F = k\Delta_{max}$, where k is the spring stiffness, then applying the conservation of energy and Eq. 14–2, we have

$$U_e = U_i$$

$$W(h + \Delta_{max}) = \frac{1}{2}(k\Delta_{max})\,\Delta_{max}$$

$$W(h + \Delta_{max}) = \frac{1}{2}k\Delta_{max}^2 \tag{14–29}$$

$$\Delta_{max}^2 - \frac{2W}{k}\Delta_{max} - 2\left(\frac{W}{k}\right)h = 0$$

This quadratic equation may be solved for Δ_{max}. The maximum root is

$$\Delta_{max} = \frac{W}{k} + \sqrt{\left(\frac{W}{k}\right)^2 + 2\left(\frac{W}{k}\right)h}$$

If the weight W is applied statically (or gradually) to the spring, the end displacement of the spring is $\Delta_{st} = W/k$. Using this simplification, the above equation becomes

$$\Delta_{max} = \Delta_{st} + \sqrt{(\Delta_{st})^2 + 2\Delta_{st}h}$$

or

$$\Delta_{max} = \Delta_{st}\left[1 + \sqrt{1 + 2\left(\frac{h}{\Delta_{st}}\right)}\right] \tag{14–30}$$

Once Δ_{max} is computed, the maximum force applied to the spring can be determined from

Fig. 14–23

The impact loading of slow moving vehicles is intended to be absorbed by this crash barrier.

$$F_{max} = k\Delta_{max} \qquad (14\text{--}31)$$

It should be realized, however, that this force and associated displacement occur only at an *instant*. Provided the block does not rebound off the spring, it will continue to vibrate until the motion dampens out and the block assumes the static position, Δ_{st}. Note also that if the block is held just above the spring, $h = 0$, and *dropped*, then, from Eq. 14–30, the maximum displacement of the block is

$$\Delta_{max} = 2\Delta_{st}$$

In other words, when the block is dropped from the top of the spring (dynamically applied load), the displacement is *twice* what it would be if it were set on the spring (statically applied load).

Using a similar analysis, it is also possible to determine the maximum displacement of the end of the spring if the block is sliding on a smooth horizontal surface with a known velocity **v** just before it collides with the spring, Fig. 14–24. Here the block's kinetic energy,* $\frac{1}{2}(W/g)v^2$, is transformed into stored energy in the spring. Hence,

$$U_e = U_i$$

$$\frac{1}{2}\left(\frac{W}{g}\right)v^2 = \frac{1}{2}k\,\Delta_{max}^2$$

$$\Delta_{max} = \sqrt{\frac{Wv^2}{gk}} \qquad (14\text{--}32)$$

Since the static displacement at the top of the spring caused by the weight W resting on it is $\Delta_{st} = W/k$, then

$$\Delta_{max} = \sqrt{\frac{\Delta_{st}v^2}{g}} \qquad (14\text{--}33)$$

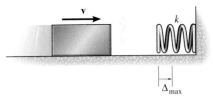

Fig. 14–24

The results of this simplified analysis can be used to determine both the approximate deflection and the stress developed in a deformable member when it is subjected to impact. To do this we must make the necessary assumptions regarding the collision, so that the behavior of the colliding bodies is similar to the response of the block-and-spring models discussed above. Hence we will consider the moving body to be *rigid* like the block and the stationary body to be deformable like the spring. It is assumed that the material behaves in a linear-elastic manner. Furthermore, during collision no energy is lost due to heat, sound, or localized plastic deformations. When collision occurs, the bodies remain in contact until the elastic body reaches its maximum deformation, and during the motion the inertia or mass of the elastic body is neglected. Realize that each of these assumptions will lead to a *conservative* estimate of both the stress and deflection of the elastic body. In other words, their values will be larger than those that actually occur.

---

*Recall from physics that kinetic energy is "energy of motion." For the translation of a body it is determined from $\frac{1}{2}mv^2$, where m is the body's mass, $m = W/g$.

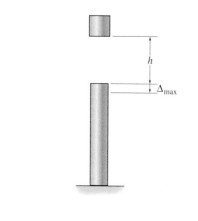

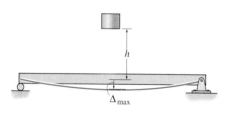

Fig. 14–25

The members of this crash guard must be designed to resist a prescribed impact loading in order to arrest the motion of a rail car.

A few examples of when the above theory can be applied are shown in Fig. 14–25. Here a known weight (block) is dropped onto a post or beam, causing it to deform a maximum amount Δ_{max}. The energy of the falling block is transformed momentarily into axial strain energy in the post and bending strain energy in the beam.* Although vibrations are established in each member after impact, they will tend to dissipate as time passes. In order to determine the deformation Δ_{max}, we could use the same approach as the block–spring system, and that is to write the conservation-of-energy equation for the block and post or block and beam, and then solve for Δ_{max}. However, we can also solve these problems in a more direct manner by modeling the post and beam by an *equivalent spring*. For example, if a force **P** displaces the top of the post $\Delta = PL/AE$, then a spring having a stiffness $k = AE/L$ would be displaced the same amount by **P**, that is, $\Delta = P/k$. In a similar manner, from Appendix C, a force **P** applied to the center of a simply supported beam displaces the center $\Delta = PL^3/48EI$, and therefore an equivalent spring would have a stiffness of $k = 48EI/L^3$. It is not necessary, however, to actually find the equivalent spring stiffness to apply Eq. 14–30 or 14–32. All that is needed to determine the dynamic displacement, Δ_{max}, is to calculate the *static displacement*, Δ_{st}, due to the weight W of the block resting on the member.

Once Δ_{max} is determined, the maximum dynamic force can then be calculated from $P_{max} = k\Delta_{max}$. If we consider P_{max} to be an *equivalent static load* then the maximum stress in the member can be determined using statics and the theory of mechanics of materials. Recall that this stress acts only for an *instant*. In reality, vibrational waves pass through the material, and the stress in the post or the beam, for example, does not remain constant.

The ratio of the equivalent static load P_{max} to the load W is called the *impact factor*, n. Since $P_{max} = k\Delta_{max}$ and $W = k\Delta_{st}$, then from Eq. 14–30, we can express it as

$$n = 1 + \sqrt{1 + 2\left(\frac{h}{\Delta_{st}}\right)} \qquad (14\text{–}34)$$

This factor represents the magnification of a statically applied load so that it can be treated dynamically. Using Eq. 14–34, n can be computed for any member that has a linear relationship between load and deflection. For a complicated system of connected members, however, impact factors are determined by experience or by experimental testing. Once n is determined, the dynamic stress and deflections are easily found from the static stress σ_{st} and static deflection Δ_{st} caused by the load W, that is, $\sigma_{max} = n\sigma_{st}$ and $\Delta_{max} = n\Delta_{st}$.

---

*Strain energy due to shear is neglected for reasons discussed in Example 14–4.

IMPORTANT POINTS

- *Impact* occurs when a large force is developed between two objects which strike one another during a short period of time.

- We can analyze the effects of impact by assuming the moving body is rigid, the material of the stationary body is linear elastic, no energy is lost in the collision, the bodies remain in contact during collision, and the inertia of the elastic body is neglected.

- The dynamic load on a body can be treated as a statically applied load by multiplying the static load by a *magnification factor*.

EXAMPLE 14-8

The aluminum pipe shown in Fig. 14–26 is used to support a load of 150 kip. Determine the maximum displacement at the top of the pipe if the load is (a) applied gradually, and (b) applied suddenly by releasing it from the top of the pipe at $h = 0$. Take $E_{al} = 10(10^3)$ ksi and assume that the aluminum behaves elastically.

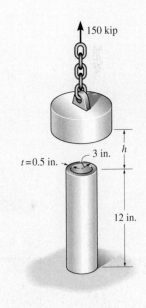

Fig. 14–26

SOLUTION

Part (a).   When the load is applied gradually, the work done by the weight is transformed into elastic strain energy in the pipe. Applying the conservation of energy, we have

$$U_e = U_i$$

$$\frac{1}{2} W\Delta_{st} = \frac{W^2 L}{2AE}$$

$$\Delta_{st} = \frac{WL}{AE} = \frac{150 \text{ kip}(12 \text{ in.})}{\pi[(3 \text{ in.})^2 - (2.5 \text{ in.})^2]10(10^3) \text{ kip/in}^2}$$

$$= 0.02083 \text{ in.} = 0.0208 \text{ in.} \qquad \textit{Ans}$$

Part (b).   Here Eq. 14–30 can be applied, with $h = 0$. Hence,

$$\Delta_{max} = \Delta_{st}\left[1 + \sqrt{1 + 2\left(\frac{h}{\Delta_{st}}\right)}\right]$$

$$= 2\Delta_{st} = 2(0.02083 \text{ in.})$$

$$= 0.0417 \text{ in.} \qquad \textit{Ans.}$$

Hence, the displacement of the weight is twice as great as when the load is applied statically. In other words, the impact factor is $n = 2$, Eq. 14–34.

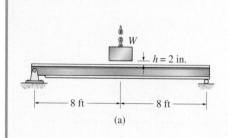

(a)

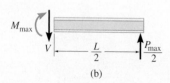

(b)

Fig. 14–27

EXAMPLE 14–9

The A-36 steel beam shown in Fig. 14–27a is a W10 × 39. Determine the maximum bending stress in the beam and the beam's maximum deflection if the weight $W = 1.50$ kip is dropped from a height $h = 2$ in. onto the beam. $E_{st} = 29(10^3)$ ksi.

SOLUTION I

We will apply Eq. 14–30. First, however, we must calculate Δ_{st}. Using the table in Appendix C, and the data in Appendix B for the properties of a W10 × 39, we have

$$\Delta_{st} = \frac{WL^3}{48EI} = \frac{1.50 \text{ kip}(16 \text{ ft})^3(12 \text{ in./ft})^3}{48[29(10^3) \text{ ksi}](209 \text{ in}^4)} = 0.0365 \text{ in.}$$

$$\Delta_{max} = \Delta_{st}\left[1 + \sqrt{1 + 2\left(\frac{h}{\Delta_{st}}\right)}\right]$$

$$= 0.0365 \text{ in.}\left[1 + \sqrt{1 + 2\left(\frac{2 \text{ in.}}{0.0365 \text{ in.}}\right)}\right] = 0.420 \text{ in.} \qquad Ans.$$

This deflection is caused by an equivalent static load P_{max}, determined from $P_{max} = (48EI/L^3)\,\Delta_{max}$.

The internal moment caused by this load is maximum at the center of the beam, such that by the method of sections, Fig. 14–27b, $M_{max} = P_{max}L/4$. Applying the flexure formula to determine the bending stress, we have

$$\sigma_{max} = \frac{M_{max}c}{I} = \frac{P_{max}Lc}{4I} = \frac{12E\Delta_{max}c}{L^2}$$

$$= \frac{12[29(10^3) \text{ kip/in}^2](0.420 \text{ in.})(9.92 \text{ in.}/2)}{(16 \text{ ft})^2(12 \text{ in./ft})^2} = 19.7 \text{ ksi} \qquad Ans.$$

SOLUTION II

It is also possible to obtain the dynamic or maximum deflection Δ_{max} from first principles. The external work of the falling weight W is $U_e = W(h + \Delta_{max})$. Since the beam deflects Δ_{max}, and $P_{max} = 48EI\Delta_{max}/L^3$, then

$$U_e = U_i$$

$$W(h + \Delta_{max}) = \frac{1}{2}\left(\frac{48EI\Delta_{max}}{L^3}\right)\Delta_{max}$$

$$(1.50 \text{ kip})(2 \text{ in.} + \Delta_{max}) = \frac{1}{2}\left[\frac{48[29(10^3) \text{ kip/in}^2]209 \text{ in}^4}{(16 \text{ ft})^3(12 \text{ in./ft})^3}\right]\Delta_{max}^2$$

$$20.55\Delta_{max}^2 - 1.50\Delta_{max} - 3.00 = 0$$

Solving and choosing the positive root yields

$$\Delta_{max} = 0.420 \text{ in.} \qquad Ans.$$

EXAMPLE 14–10

A railroad car that is assumed to be rigid and has a mass of 80 Mg is moving forward at a speed of $v = 0.2$ m/s when it strikes a steel 200-mm by 200-mm post at A, Fig. 14–28a. If the post is fixed to the ground at C, determine the maximum horizontal displacement of its top B due to the impact. Take $E_{st} = 200$ GPa.

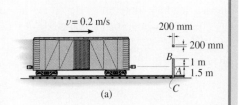

$v = 0.2$ m/s

200 mm

200 mm

B

1 m

A 1.5 m

C

(a)

SOLUTION

Here the kinetic energy of the railroad car is transformed into internal bending strain energy only for region AC of the post. Assuming that point A is displaced $(\Delta_A)_{max}$, then the force P_{max} that causes this displacement can be determined from the table in Appendix C. We have

$$P_{max} = \frac{3EI(\Delta_A)_{max}}{L_{AC}^3} \qquad (1)$$

$$U_e = U_i; \quad \frac{1}{2}mv^2 = \frac{1}{2}P_{max}(\Delta_A)_{max}$$

$$\frac{1}{2}mv^2 = \frac{1}{2}\frac{3EI}{L_{AC}^3}(\Delta_A)^2_{max}; \quad (\Delta_A)_{max} = \sqrt{\frac{mv^2 L_{AC}^3}{3EI}}$$

Substituting in the numerical data yields

$$(\Delta_A)_{max} = \sqrt{\frac{80(10^3)\ kg(0.2\ m/s)^2(1.5\ m)^3}{3[200(10^9)\ N/m^2][\frac{1}{12}(0.2\ m)^4]}} = 0.0116\ m = 11.6\ mm$$

Using Eq. 1, the force P_{max} is therefore

$$P_{max} = \frac{3[200(10^9)\ N/m^2][\frac{1}{12}(0.2\ m)^4](0.0116\ m)}{(1.5\ m)^3} = 275.4\ kN$$

With reference to Fig. 14–28b, segment AB of the post remains straight. To determine the maximum displacement at B, we must first determine the slope at A. Using the appropriate formula from the table in Appendix C to determine θ_A, we have

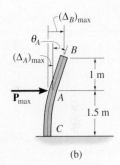

$(\Delta_B)_{max}$

θ_A

B

$(\Delta_A)_{max}$

P_{max} A

1 m

1.5 m

C

(b)

Fig. 14–28

$$\theta_A = \frac{P_{max}L_{AC}^2}{2EI} = \frac{275.4(10^3)\ N\ (1.5\ m)^2}{2[200(10^9)\ N/m^2][\frac{1}{12}\ (0.2\ m)^4]} = 0.01162\ rad$$

The maximum displacement at B is thus

$$(\Delta_B)_{max} = (\Delta_A)_{max} + \theta_A L_{AB}$$
$$= 11.62\ mm + (0.01162\ rad)\ 1(10^3)\ mm = 23.2\ mm \qquad Ans.$$

PROBLEMS

*14–32. A bar is 8 ft long and has a diameter of 0.5 in. If it is to be used to absorb energy in tension from an impact loading, determine the total amount of elastic energy that it can absorb if it is made of red brass C83400.

14–33. Solve Prob. 14–32 if the bar is made of aluminum 2014-T6.

14–34. Determine the diameter of a red brass C83400 bar that is 8 ft long if it is to be used to absorb 800 ft · lb of energy in tension from an impact loading. No yielding occurs.

14–35. The composite aluminum bar is made from two segments having diameters of 5 mm and 10 mm. Determine the maximum axial stress developed in the bar if the 5-kg collar is dropped from a height of $h = 100$ mm. $E_{al} = 70$ GPa, $\sigma_Y = 410$ MPa.

*14–36. The composite aluminum bar is made from two segments having diameters of 5 mm and 10 mm. Determine the maximum height h from which the 5-kg collar should be dropped so that it produces a maximum axial stress in the bar of $\sigma_{max} = 300$ MPa. $E_{al} = 70$ GPa, $\sigma_Y = 410$ MPa.

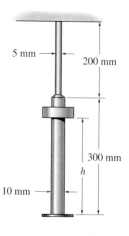

Probs. 14–35/14–36

14–37. Determine the speed v of the 50-Mg mass when it is just over the top of the steel post, if after impact, the maximum stress developed in the post is 550 MPa. The post has a length of $L = 1$ m and a cross-sectional area of 0.01 m². $E_{st} = 200$ GPa, $\sigma_Y = 600$ MPa.

Prob. 14–37

14–38. The A-36 steel bolt is required to absorb the energy of a 2-kg mass that falls from rest $h = 30$ mm. If the bolt has a diameter of 4 mm, determine its required length L so the stress in the bolt does not exceed 150 MPa.

14–39. The A-36 steel bolt is required to absorb the energy of a 2-kg mass that falls from rest $h = 30$ mm. If the bolt has a diameter of 4 mm and a length of $L = 200$ mm, determine if the stress in the bolt will exceed 175 MPa.

*14–40. The A-36 steel bolt is required to absorb the energy of a 2-kg mass that falls from rest along the 4-mm-diameter bolt shank that is $L = 150$ mm long. Determine the maximum height h of release so the stress in the bolt does not exceed 150 MPa.

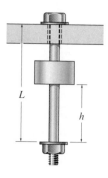

Probs. 14–38/14–39/14–40

14–41. The 50-lb weight is falling at 3 ft/s at the instant it is 2 ft above the spring and post assembly. Determine the maximum stress in the post if the spring has a stiffness of $k = 200$ kip/in. The post has a diameter of 3 in. and a modulus of elasticity of $E = 6.80(10^3)$ ksi. Assume the material will not yield.

Prob. 14–41

14–42. The composite aluminum 2014-T6 bar is made from two segments having diameters of 7.5 mm and 15 mm. Determine the maximum axial stress developed in the bar if the 10-kg collar is dropped from a height of $h = 100$ mm.

14–43. The composite aluminum 2014-T6 bar is made from two segments having diameters of 7.5 mm and 15 mm. Determine the maximum height h from which the 10-kg collar should be dropped so that it produces a maximum axial stress in the bar of $\sigma_{max} = 300$ MPa.

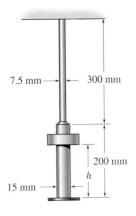

Probs. 14–42/14–43

*14–44. A cylinder having the dimensions shown is made of magnesium Am 1004-T61. If it is struck by a rigid block having a weight of 800 lb and traveling at 2 ft/s, determine the maximum stress in the cylinder. Neglect the mass of the cylinder.

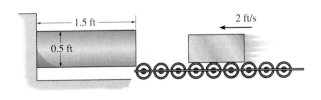

Prob. 14–44

14–45. The sack of cement has a weight of 90 lb. If it is dropped from rest at a height of $h = 4$ ft onto the center of the W10 × 39 structural steel A-36 beam, determine the maximum bending stress developed in the beam due to the impact. Also, what is the impact factor?

14–46. The sack of cement has a weight of 90 lb. Determine the maximum height h from which it can be dropped from rest onto the center of the W10 × 39 structural steel A-36 beam so that the maximum bending stress due to impact does not exceed 30 ksi.

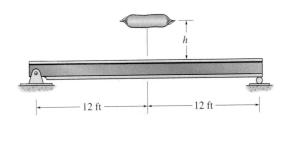

Probs. 14–45/14–46

14–47. The 200-lb block has a downward velocity of 4 ft/s when it is 3 ft from the top of the wooden beam. Determine the maximum stress in the beam due to the impact and compute the maximum deflection of its end C. $E_w = 1.9(10^3)$ ksi, $\sigma_Y = 6$ ksi.

*14–48. The 100-lb block has a downward velocity of 4 ft/s when it is 3 ft from the top of the wooden beam. Determine the maximum stress in the beam due to the impact and compute the maximum deflection of point B. $E_w = 1.9(10^3)$ ksi, $\sigma_Y = 8$ ksi.

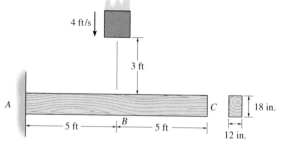

Probs. 14–47/14–48

14–49. The diver weighs 150 lb. While holding himself rigid he strikes the end of a wooden diving board with a downward velocity of 4 ft/s when $h = 0$. Determine the maximum bending stress developed in the board. The board has a thickness of 1.5 in. and width of 1.5 ft. $E_w = 1.8(10^3)$ ksi. Assume the material does not yield.

14–50. The diver weighs 150 lb. While holding himself rigid he strikes the end of the wooden diving board. Determine the maximum height h from which he can jump onto the board so that the maximum bending stress in the wood does not exceed 6 ksi. The board has a thickness of 1.5 in. and width of 1.5 ft. $E_w = 1.8(10^3)$ ksi.

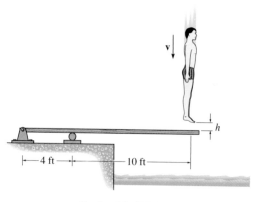

Probs. 14–49/14–50

14–51. The 75-lb block has a downward velocity of 2 ft/s when it is 3 ft from the top of the wooden beam. Determine the maximum bending stress in the beam due to the impact, and compute the maximum deflection of its end D. $E_w = 1.9(10^3)$ ksi. Assume the material will not yield.

*14–52. The 75-lb block has a downward velocity of 2 ft/s when it is 3 ft from the top of the wood beam. Determine the maximum bending stress in the beam due to the impact, and compute the maximum deflection of point B. $E_w = 1.9(10^3)$ ksi.

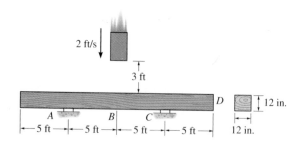

Probs. 14–51/14–52

14–53. Determine the maximum height h from which an 80-lb weight can be dropped onto the end of the A-36 steel W6 × 12 beam without exceeding the maximum elastic stress.

14–54. The 80-lb weight is dropped from rest at a height of $h = 4$ ft onto the end of the A-36 steel W6 × 12 beam. Determine the maximum bending stress developed in the beam.

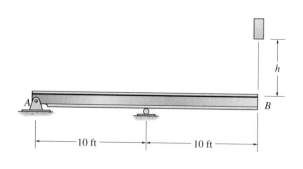

Probs. 14–53/14–54

14–55. The tugboat has a weight of 120 000 lb and is traveling forward at 2 ft/s when it strikes the 12-in.-diameter fender post AB used to protect a bridge pier. If the post is made from treated white spruce and is assumed fixed at the river bed, determine the maximum horizontal distance the top of the post will move due to the impact. Assume the tugboat is rigid and neglect the effect of the water.

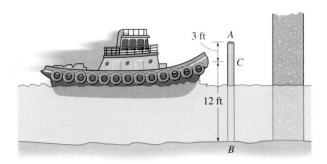

Prob. 14–55

14–57. The 150-lb weight has a velocity of 4 ft/s at a height of 4 ft from the top of the A-36 steel beam. Determine the maximum deflection and maximum bending stress in the beam if the supporting springs at A and B each have a stiffness of $k = 500$ lb/in.

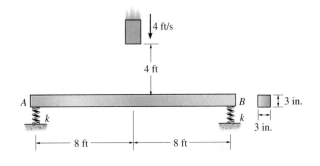

Prob. 14–57

*__14–56.__ The car bumper is made of polycarbonate-polybutylene terephthalate. If $E = 2.0$ GPa, determine the maximum deflection and maximum stress in the bumper if it strikes the rigid post when the car is coasting at $v = 0.75$ m/s. The car has a mass of 1.80 Mg, and the bumper can be considered simply supported on two spring supports connected to the rigid frame of the car. For the bumper take $I = 300(10^6)$ mm^4, $c = 75$ mm, $\sigma_Y = 30$ MPa, and $k = 1.5$ MN/m.

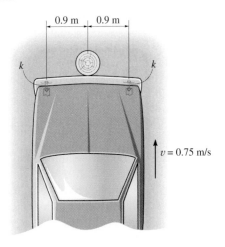

Prob. 14–56

14–58. The 150-lb weight has a velocity of 4 ft/s at a height of 4 ft from the top of the A-36 steel beam. Determine the load factor n if the supporting springs at A and B each have a stiffness of $k = 300$ lb/in.

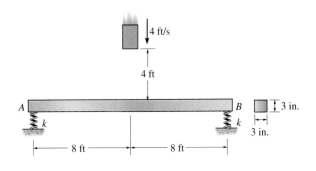

Prob. 14–58

*14.5   Principle of Virtual Work

The principle of virtual work was developed by John Bernoulli in 1717, and like other energy methods of analysis, it is based on the conservation of energy. Although the principle of virtual work has many applications in mechanics, in this text we will use it to obtain the displacement and slope at various points on a deformable body. Before doing this, however, we will need to make some preliminary remarks that apply to the development of this method.

Whenever a body is fixed from moving, it is necessary that the loadings satisfy the equilibrium conditions and the displacements satisfy the compatibility conditions. Specifically, *equilibrium conditions* require the external loads to be uniquely related to the internal loads, and the *compatibility conditions* require the external displacements to be uniquely related to the internal deformations. For example, if we consider a deformable body of any shape or size and apply a series of external loads **P** to it, these loadings will cause internal loadings u within the body. Here the external and internal loads are related by the equations of equilibrium. Furthermore, since the body is deformable, the external loads will be displaced Δ, and the internal loadings will undergo displacements δ. In general, the material does *not* have to behave elastically, and so the displacements may *not* be related to the loads. However, if the external displacements are known, the corresponding internal displacements are uniquely defined since the body is continuous. For this case, the conservation of energy states that

$$U_e = U_i; \qquad\qquad \Sigma\, P\, \Delta = \Sigma\, u\, \delta \qquad\qquad (14\text{--}35)$$

Based on this concept, we will now develop the principle of virtual work so that it can be used to determine the displacement and slope at *any point* on a body. To do this, we will consider the body to be of arbitrary shape as shown in Fig. 14–29b, and to be subjected to the "real loads" $\mathbf{P}_1$, $\mathbf{P}_2$, and $\mathbf{P}_3$. It is to be understood that these loads cause no movement of the supports; however, in general they can strain the material *beyond* the elastic limit. Suppose that it is necessary to determine the displacement Δ of point A on the body caused by these loads. To do this we will consider applying the conservation-of-energy principle, Eq. 14–35. In this case, however, there is no force acting at A, and so the unknown displacement Δ will *not* be included as an external "work term" in the equation.

In order to get around this limitation, we will place an *imaginary* or "virtual" force $\mathbf{P}'$ on the body at point A, such that $\mathbf{P}'$ acts in the *same direction* as Δ. Furthermore, this load is applied to the body *before* the real loads are applied, Fig. 14–29a. For convenience, which will be made clear later, we will choose $\mathbf{P}'$ to have a "unit" magnitude; that is, $P' = 1$. It is to be emphasized that the term "***virtual***" is used to describe the load because it is *imaginary* and does not actually exist as part of the

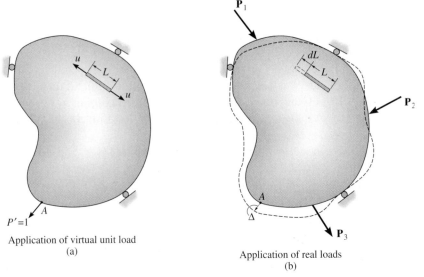

Application of virtual unit load
(a)

Application of real loads
(b)

Fig. 14–29

real loading. This external virtual load, however, does create an internal virtual load **u** in a representative element or fiber of the body, as shown in Fig. 14–29*a*. As expected, *P′* and *u* can be related by the equations of equilibrium. Also, because of *P′* and *u*, the body and the element will each undergo a virtual displacement, although we will *not* be concerned with their magnitudes. Once the virtual load is applied and *then* the body is subjected to the *real loads* P_1, P_2, and P_3, point *A* will be displaced a real amount Δ, which causes the element to be displaced *dL*, Fig. 14–29*b*. As a result, the external virtual force **P′** and internal virtual load **u** "ride along" by Δ and *dL*, respectively; consequently these loads perform *external virtual work* of $1 \cdot \Delta$ on the body and *internal virtual work* of $u \cdot dL$ on the element. Considering *only* the conservation of *virtual* energy, the external virtual work is then equal to the internal virtual work done on all the elements of the body. Therefore, we can write the virtual-work equation as

$$\underbrace{1 \cdot \Delta = \Sigma u \cdot dL}_{\text{real displacements}}^{\text{virtual loadings}} \qquad (14\text{–}36)$$

Here

$P' = 1 =$ external virtual unit load acting in the direction of Δ

$u =$ internal virtual load acting on the element

$\Delta =$ external displacement caused by the real loads

$dL =$ internal displacement of the element in the direction of **u**, caused by the real loads

By choosing $P' = 1$, it can be seen that the solution for Δ follows directly, since $\Delta = \Sigma u\, dL$.

In a similar manner, if the rotational displacement or slope of the tangent at a point on the body is to be determined, a virtual *couple moment* **M′**, having a "unit" magnitude, is applied at the point. As a consequence, this couple moment causes a virtual load u_θ in one of the elements of the body. Assuming that the real loads deform the element an amount dL, the rotation θ can be found from the virtual-work equation

virtual loadings

$$1 \cdot \theta = \Sigma u_\theta\, dL \tag{14–37}$$

real displacements

Here

$M' = 1 =$ external virtual unit couple moment acting in the direction of θ

$u_\theta =$ internal virtual load acting on an element

$\theta =$ external rotational displacement in radians caused by the real loads

$dL =$ internal displacement of the element in the direction of u_θ, caused by the real loads

This method for applying the principle of virtual work is often referred to as the *method of virtual forces*, since a *virtual force* is applied, resulting in a calculation of an external *real displacement*. The equation of virtual work in this case represents a statement of *compatibility requirements* for the body. Although it is not important here, realize that we can also apply the principle of virtual work as a *method of virtual displacements*. In this case, *virtual displacements* are imposed on the body when the body is subjected to *real loadings*. This method can be used to determine the external reactive force on the body or an unknown internal loading in the body. When it is used in this manner, the equation of virtual work is a statement of the *equilibrium requirements* for the body.*

Internal Virtual Work.   The terms on the right-hand side of Eqs. 14–36 and 14–37 represent the internal virtual work developed in the body. The real internal displacements dL in these terms can be produced in several different ways. For example, these displacements may result from geometric fabrication errors, from temperature, or more commonly from stress. In particular, no restriction has been placed on the magnitude of the external loading, so the stress may be large enough to cause yielding or even strain hardening of the material.

---

*See *Engineering Mechanics: Statics*, 9th edition, R.C. Hibbeler, Prentice Hall, Inc., 2001.

Deformation caused by	Strain energy	Internal virtual work
Axial load N	$\displaystyle\int_0^L \frac{N^2}{2EA}\,dx$	$\displaystyle\int_0^L \frac{nN}{EA}\,dx$
Shear V	$\displaystyle\int_0^L \frac{f_s V^2}{2GA}\,dx$	$\displaystyle\int_0^L \frac{f_s vV}{GA}\,dx$
Bending moment M	$\displaystyle\int_0^L \frac{M^2}{2EI}\,dx$	$\displaystyle\int_0^L \frac{mM}{EI}\,dx$
Torsional moment T	$\displaystyle\int_0^L \frac{T^2}{2GJ}\,dx$	$\displaystyle\int_0^L \frac{tT}{GJ}\,dx$

Table 14–1

If we assume that the material behavior is linear-elastic and the stress does not exceed the proportional limit, we can formulate the expressions for internal virtual work caused by stress using the equations of elastic strain energy developed in Sec. 14.2. They are listed in the center column of Table 14–1. Recall that each of these expressions assumes that the stress resultant $\mathbf{N}$, $\mathbf{V}$, $\mathbf{M}$, or $\mathbf{T}$ was applied gradually from zero to its full value. As a result, the work done by the stress resultant is shown in these expressions as *one-half* the product of the stress resultant and its displacement. In the case of the virtual-force method, however, the "full" virtual loading is applied *before* the real loads cause displacements, and therefore the work of the internal virtual loading is simply the product of the internal virtual load and its real displacement. Referring to these internal virtual loadings (u) by the corresponding lowercase symbols n, v, m, and t, the virtual work due to axial load, shear, bending moment, and torsional moment is listed in the right-hand column of Table 14–1. Using these results, the virtual-work equation for a body subjected to a general loading can therefore be written as

$$1 \cdot \Delta = \int \frac{nN}{AE}\,dx + \int \frac{mM}{EI}\,dx + \int \frac{f_s vV}{GA}\,dx + \int \frac{tT}{GJ}\,dx \quad (14\text{–}38)$$

In the following sections we will apply the above equation to problems involving the deflections of trusses, beams, and mechanical elements. We will also include a discussion of how to handle the effects of fabrication errors and differential temperature. For application it is important that a consistent set of units be used for all the terms. For example, if the real loads are expressed in kilonewtons and the body's dimensions are in meters, a 1-kN virtual force or 1-kN · m virtual couple should be applied to the body. By doing so a calculated displacement Δ will be in meters, and a calculated slope will be in radians.

*14.6 METHOD OF VIRTUAL FORCES APPLIED TO TRUSSES

In this section we will apply the method of virtual forces to determine the displacement of a truss joint. To illustrate the principles, the vertical displacement of joint A of the truss shown in Fig. 14–30b will be determined. This displacement is caused by the "real loads" $\mathbf{P}_1$ and $\mathbf{P}_2$, and since these loads cause only axial force in the members, it is only necessary to consider the internal virtual work due to axial load, Table 14–1. To obtain this virtual work, we will assume that each member has a constant cross-sectional area A, and the virtual load n and real load N are constant throughout the member's length. As a result, the internal virtual work for a member is

$$\int_0^L \frac{nN}{AE}\, dx = \frac{nNL}{AE}$$

And the virtual-work equation for the entire truss is therefore

$$1 \cdot \Delta = \sum \frac{nNL}{AE} \qquad (14\text{–}39)$$

Here

1 = external virtual unit load acting on the truss joint in the stated direction of Δ

Δ = joint displacement caused by the real loads on the truss

n = internal virtual force in a truss member caused by the external virtual unit load

N = internal force in a truss member caused by the real loads

L = length of a member

A = cross-sectional area of a member

E = modulus of elasticity of a member

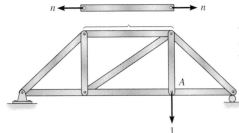

Application of virtual unit load

(a)

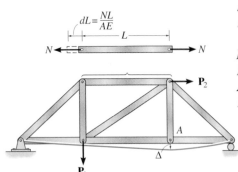

Application of real loads

(b)

Fig. 14–30

The formulation of this equation follows naturally from the development in Sec. 14.5. Here the external virtual unit load creates internal virtual "n" forces in each of the truss members, Fig. 14–30a. When the real loads are applied to the truss, they cause the truss joint to be displaced Δ in the same direction as the virtual unit load, Fig. 14–30b, and each member undergoes a displacement NL/AE, in the same direction as its respective n force. Consequently, the external virtual work $1 \cdot \Delta$ equals the internal virtual work or the internal (virtual) strain energy stored in *all* the truss members, i.e., Eq. 14–39.

Temperature Change. Truss members can change their length due to a change in temperature. If α is the coefficient of thermal expansion for a member and ΔT is the change in temperature, the change in length of a member is $\Delta L = \alpha\, \Delta T L$ (Eq. 4–4). Hence, we can determine the displacement of a selected truss joint due to this temperature change from Eq. 14–36, written as

$$1 \cdot \Delta = \Sigma n\alpha\, \Delta T L \qquad\qquad (14\text{--}40)$$

Here

1 = external virtual unit load acting on the truss joint in the stated direction of Δ

n = internal virtual force in a truss member caused by the external virtual unit load

Δ = external joint displacement caused by the temperature change

α = coefficient of thermal expansion of member

ΔT = change in temperature of member

L = length of member

Fabrication Errors. Occasionally errors in fabricating the lengths of the members of a truss may occur. If this happens, the displacement in a particular direction of a truss joint from its expected position can be determined from direct application of Eq. 14–36 written as

$$1 \cdot \Delta = \Sigma n\, \Delta L \qquad\qquad (14\text{--}41)$$

Here

1 = external virtual unit load acting on the truss joint in the stated direction of Δ

n = internal virtual force in a truss member caused by the external virtual unit load

Δ = external joint displacement caused by the fabrication errors

ΔL = difference in length of the member from its intended length caused by a fabrication error

A combination of the right-hand sides of Eqs. 14–39 through 14–41 will be necessary if external loads act on the truss and some of the members undergo a temperature change or have been fabricated with the wrong dimensions.

PROCEDURE FOR ANALYSIS

The following procedure provides a method that may be used to determine the displacement of any joint on a truss using the method of virtual force.

Virtual Forces n.

• Place the virtual unit load on the truss at the joint where the desired displacement is to be determined. The load should be directed along the line of action of the displacement.

• With the unit load so placed and all the real loads *removed* from the truss, calculate the internal *n* force in each truss member. Assume that tensile forces are positive and compressive forces are negative.

Real Forces N.

• Determine the *N* forces in each member. These forces are caused only by the real loads acting on the truss. Again, assume that tensile forces are positive and compressive forces are negative.

Virtual-Work Equation.

• Apply the equation of virtual work to determine the desired displacement. It is important to retain the algebraic sign for each of the corresponding *n* and *N* forces when substituting these terms into the equation.

• If the resultant sum $\Sigma nNL/AE$ is positive, the displacement Δ is in the same direction as the virtual unit load. If a negative value results, Δ is opposite to the virtual unit load.

• When applying $1 \cdot \Delta = \Sigma n\alpha \, \Delta TL$, realize that if any of the members undergo an *increase* in temperature, ΔT will be *positive*; whereas a *decrease* in temperature will result in a *negative* value for ΔT.

• For $1 \cdot \Delta = \Sigma n \, \Delta L$, when a fabrication error *increases* the length of a member, ΔL is *positive*, whereas a *decrease* in length is *negative*.

• When applying this method, attention should be paid to the units of each numerical quantity. Notice, however, that the virtual unit load can be assigned any arbitrary unit: pounds, kips, newtons, etc., since the *n* forces will have these *same* units, and as a result, the units for both the virtual unit load and the *n* forces will cancel from both sides of the equation.

EXAMPLE 14–11

Determine the vertical displacement of joint C of the steel truss shown in Fig. 14–31a. The cross-sectional area of each member is $A = 400$ mm^2 and $E_{st} = 200$ GPa.

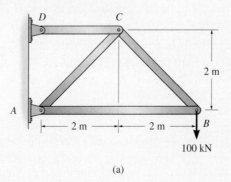

(a)

SOLUTION

Virtual Forces n. Since the vertical displacement at joint C is to be determined, *only* a vertical 1-kN virtual load is placed at joint C; and the force in each member is calculated using the method of joints. The results of this analysis are shown in Fig. 14–31b. Using our sign convention, positive numbers indicate tensile forces and negative numbers indicate compressive forces.

Real Forces N. The applied load of 100 kN causes forces in the members that can be calculated using the method of joints. The results of this analysis are shown in Fig. 14–31c.

Virtual-Work Equation. Arranging the data in tabular form, we have

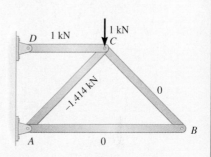

Virtual forces

(b)

Member	n	N	L	nNL
AB	0	−100	4	0
BC	0	141.4	2.828	0
AC	−1.414	−141.4	2.828	565.7
CD	1	200	2	400
				Σ 965.7 kN$^2 \cdot$ m

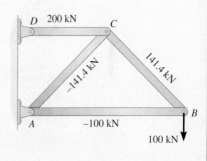

Real forces

(c)

Fig. 14–31

Thus

$$1 \text{ kN} \cdot \Delta_{C_v} = \sum \frac{nNL}{AE} = \frac{965.7 \text{ kN}^2 \cdot \text{m}}{AE}$$

Substituting the numerical values for A and E, we have

$$1 \text{ kN} \cdot \Delta_{C_v} = \frac{965.7 \text{ kN}^2 \cdot \text{m}}{[400(10^{-6}) \text{ m}^2]200(10^6) \text{ kN/m}^2}$$

$$\Delta_{C_v} = 0.01207 \text{ m} = 12.1 \text{ mm} \qquad \textbf{\textit{Ans.}}$$

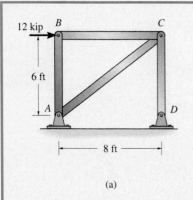

(a)

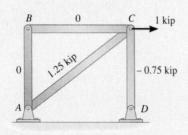

Virtual forces

(b)

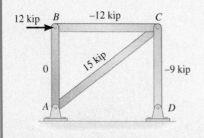

Real forces

(c)

Fig. 14-32

EXAMPLE 14-12

The cross-sectional area of each member of the steel truss shown in Fig. 14-32a is $A = 0.5$ in², and the modulus of elasticity for the steel members is $E_{st} = 29(10^3)$ ksi. (a) Determine the horizontal displacement of joint C if a force of 12 kip is applied to the truss at B. (b) If no external loads act on the truss, what is the horizontal displacement of joint C if member AC is fabricated 0.25 in. too short?

SOLUTION

Part (a).

Virtual Forces n. Since the *horizontal displacement* of joint C is to be determined, a horizontal virtual force of 1 kip is applied at C. The n force in each member is determined by the method of joints and shown on the truss in Fig. 14-32b. As usual, a positive number represents a tensile force and a negative number represents a compressive force.

Real Forces N. The force in each member as caused by the externally applied 12-kip force is shown in Fig. 14-32c.

Virtual-Work Equation. Since AE is constant, ΣnNL is computed as follows:

Member	n	N	L	nNL
AB	0	0	6	0
AC	1.25	15	10	187.5
CB	0	−12	8	0
CD	−0.75	−9	6	40.5
				$\Sigma\, 228.0$ kip² · ft

$$1 \text{ kip} \cdot \Delta_{C_h} = \sum \frac{nNL}{AE} = \frac{228.0 \text{ kip}^2 \cdot \text{ft}}{AE}$$

$$1 \text{ kip} \cdot \Delta_{C_h} = \frac{228.0 \text{ kip}^2 \cdot \text{ft}(12 \text{ in./ft})}{(0.5 \text{ in}^2)29(10^3) \text{ kip/in}^2}$$

$$\Delta_{C_h} = 0.189 \text{ in.} \qquad\qquad Ans.$$

Part (b). Here we must apply Eq. 14-41. Since the horizontal displacement of C is to be determined, we can use the results of Fig. 14-32b. Realizing that member AC is *shortened* by $\Delta L = -0.25$ in., we have

$$1 \cdot \Delta = \Sigma n\Delta L; \quad 1 \text{ kip} \cdot \Delta_{C_h} = (1.25 \text{ kip})(-0.25 \text{ in.})$$

$$\Delta_{C_h} = -0.312 \text{ in.} = 0.312 \text{ in.} \leftarrow \qquad Ans.$$

The negative sign indicates that joint C is displaced to the left, opposite to the 1-kip horizontal load.

EXAMPLE 14-13

Determine the horizontal displacement of joint B of the truss shown in Fig. 14–33a. Due to radiant heating, member AB is subjected to an *increase* in temperature of $\Delta T = +60°C$. The members are made of steel, for which $\alpha_{st} = 12(10^{-6})/°C$ and $E_{st} = 200$ GPa. The cross-sectional area of each member is 250 mm^2.

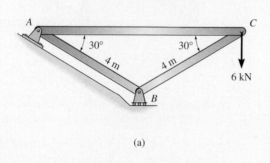

(a)

SOLUTION

Virtual Forces n. A horizontal 1-kN virtual load is applied to the truss at joint B, and the forces in each member are computed, Fig. 14–33b.

Real Forces N. Since the n forces in members AC and BC are *zero*, the N forces in these members do *not* have to be determined. Why? For completeness, though, the entire "real" force analysis is shown in Fig. 14–33c.

Virtual-Work Equation. Both loads and temperature affect the deformation; therefore, Eqs. 14–39 and 14–40 are combined, which gives

$$1 \text{ kN} \cdot \Delta_{B_h} = \sum \frac{nNL}{AE} + \Sigma n\alpha \, \Delta TL$$

$$= 0 + 0 + \frac{(-1.155 \text{ kN})(-12 \text{ kN})(4 \text{ m})}{[250(10^{-6}) \text{ m}^2][200(10^6) \text{ kN/m}^2]}$$

$$+ 0 + 0 + (-1.155 \text{ kN})[12(10^{-6})/°C](60°C)(4 \text{ m})$$

$$\Delta_{B_h} = -0.00222 \text{ m}$$

$$= 2.22 \text{ mm} \rightarrow \qquad \qquad \textit{Ans.}$$

The negative sign indicates that roller B moves to the right, opposite to the direction of the virtual load, Fig. 14–33b.

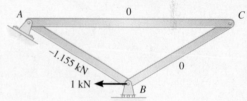

Virtual forces

(b)

Real forces

(c)

Fig. 14–33

PROBLEMS

14–59.   Determine the vertical displacement of joint B of the truss. Each A-36 steel member has a cross-sectional area of 300 mm².

*14–60.   Determine the horizontal displacement of joint B of the truss. Each A-36 steel member has a cross-sectional area of 300 mm².

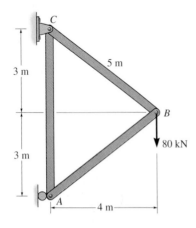

Probs. 14–59/14–60

14–61.   Determine the horizontal displacement of point B. Each A-36 steel member has a cross-sectional area of 2 in².

14–62.   Determine the vertical displacement of point B. Each A-36 steel member has a cross-sectional area of 2 in².

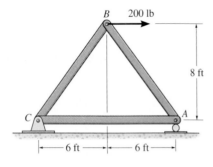

Probs. 14–61/14–62

14–63.   Determine the vertical displacement of joint B. For each A-36 steel member $A = 1.5$ in².

*14–64.   Determine the vertical displacement of joint E. For each A-36 steel member $A = 1.5$ in².

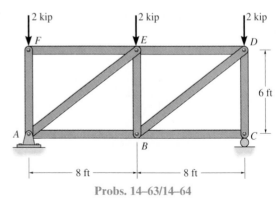

Probs. 14–63/14–64

14–65.   Determine the horizontal displacement of point C. Each A-36 steel member has a cross-sectional area of 400 mm².

14–66.   Determine the vertical displacement of point D. Each A-36 steel member has a cross-sectional area of 400 mm².

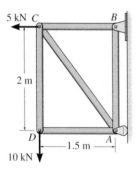

Probs. 14–65/14–66

14–67.   Determine the vertical displacement of point A. Each A-36 steel member has a cross-sectional area of 3 in².

*14–68.   Determine the vertical displacement of point B. Each A-36 steel member has a cross-sectional area of 3 in².

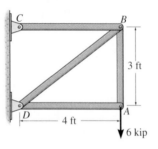

Probs. 14–67/14–68

14–69. Determine the vertical displacement of point *A*. Each A-36 steel member has a cross-sectional area of 400 mm².

14–70. Determine the vertical displacement of point *B*. Each A-36 steel member has a cross-sectional area of 400 mm².

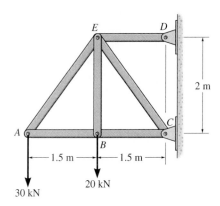

Probs. 14–69/14–70

14–71. Determine the horizontal displacement of point *D*. Each A-36 steel member has a cross-sectional area of 300 mm².

***14–72.** Determine the horizontal displacement of point *E*. Each A-36 steel member has a cross-sectional area of 300 mm².

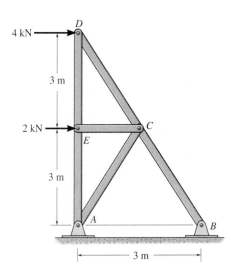

Probs. 14–71/14–72

14–73. Determine the vertical displacement of point *B*. Each A-36 steel member has a cross-sectional area of 4.5 in², $E_{st} = 29(10^3)$ ksi.

14–74. Determine the vertical displacement of point *E*. Each A-36 steel member has a cross-sectional area of 4.5 in².

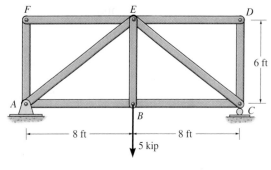

Probs. 14–73/14–74

14–75. Determine the vertical displacement of joint *C*. Each A-36 steel member has a cross-sectional area of 4.5 in².

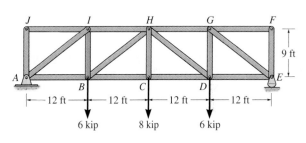

Prob. 14–75

***14–76.** Determine the vertical displacement of joint *H*. Each A-36 steel member has a cross-sectional area of 4.5 in².

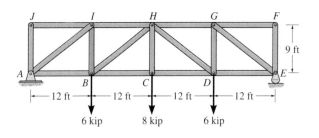

Prob. 14–76

*14.7   METHOD OF VIRTUAL FORCES APPLIED TO BEAMS

In this section we will apply the method of virtual forces to determine the displacement and slope at a point on a beam. To illustrate the principles, the displacement Δ of point A on the beam shown in Fig. 14–34b will be determined. This displacement is caused by the "real distributed load" w, and since this load causes both a shear and moment within the beam, we must actually consider the internal virtual work due to both of these loadings. In Example 14–7, however, it was shown that beam deflections due to shear are negligible compared with those caused by bending, particularly if the beam is long and slender. Since this type of beam is most often used in practice, we will consider only the virtual strain energy due to bending, Table 14–1. Applying Eq. 14–36, the virtual-work equation for the beam is therefore

$$1 \cdot \Delta = \int_0^L \frac{mM}{EI} \, dx \qquad\qquad (14\text{–}42)$$

Here

1 = external virtual unit load acting on the beam in the direction of Δ
Δ = displacement caused by the real loads acting on the beam
m = internal virtual moment in the beam, expressed as a function of x and caused by the external virtual unit load
M = internal moment in the beam, expressed as a function of x and caused by the real loads
E = modulus of elasticity of the material
I = moment of inertia of the cross-sectional area, computed about the neutral axis

In a similar manner, if the slope θ of the tangent at a point on the beam's elastic curve is to be determined, a virtual unit couple moment must be applied at the point, and the corresponding internal virtual moment m_θ has to be determined. If we apply Eq. 14–37 for this case and neglect the effect of shear deformations, we have

$$1 \cdot \theta = \int_0^L \frac{m_\theta M}{EI} \, dx \qquad\qquad (14\text{–}43)$$

Notice that the formulation of the above equations follows naturally from the development in Sec. 14.5. For example, the external virtual unit load creates an internal virtual moment m in the beam at position x, Fig. 14–34a. When the real load w is applied, it causes the element dx at x to deform or rotate by an angle $d\theta$, Fig. 14–34b. Provided the material responds elastically, then $d\theta$ equals $(M/EI) \, dx$. Consequently, the external

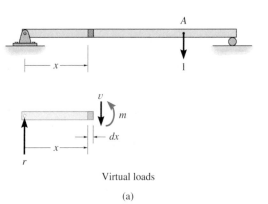

Virtual loads

(a)

Fig. 14–34

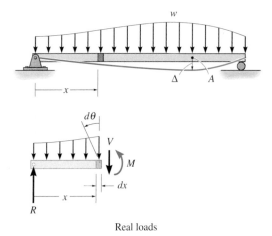

Real loads

(b)

virtual work $1 \cdot \Delta$ equals the internal virtual work for the entire beam, $\int m(M/EI) \, dx$, Eq. 14–42.

Unlike beams, as discussed here, some members may also be subjected to significant virtual strain energy caused by axial load, shear, and torsional moment. When this is the case, we must include in the above equations the energy terms for these loadings as formulated in Eq. 14–38.

When applying Eqs. 14–42 and 14–43, it is important to realize that the integrals on the right side represent the amount of virtual bending strain energy that is *stored* in the beam. If concentrated forces or couple moments act on the beam or the distributed load is discontinuous, a single integration *cannot* be performed across the beam's entire length. Instead, separate x coordinates must be chosen within regions that have no discontinuity of loading. Also, it is not necessary that each x have the same origin; however, the x selected for determining the real moment M in a particular region must be the *same* x as that selected for determining the virtual moment m or m_θ within the same region. For example, consider the beam shown in Fig. 14–35. In order to determine the displacement at D, we can use x_1 to determine the strain energy in region AB, x_2 for region BC, x_3 for region DE, and x_4 for region DC. In any case, each x coordinate should be selected so that both M and m (or m_θ) can easily be formulated.

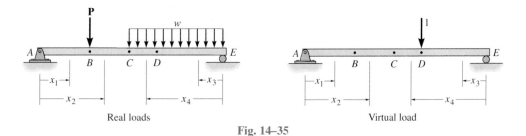

Real loads   Virtual load

Fig. 14–35

Procedure for Analysis

The following procedure provides a method that may be used to determine the displacement and slope at a point on the elastic curve of a beam using the method of virtual work.

Virtual Moments m or m_θ.

• Place a *virtual unit load* on the beam at the point and directed along the line of action of the desired displacement.

• If the slope is to be determined, place a virtual *unit couple moment* at the point.

• Establish appropriate x coordinates that are valid within regions of the beam where there is no discontinuity of real or virtual load.

• With the virtual load in place, and all the real loads *removed* from the beam, calculate the internal moment m or m_θ as a function of each x coordinate.

• Assume that m or m_θ acts in the positive direction according to the established beam sign convention, Fig. 6–3.

Real Moments.

• Using the *same x* coordinates as those established for m or m_θ, determine the internal moments M caused by the real loads.

• Since positive m or m_θ was assumed to act in the conventional "positive direction," it is important that positive M acts in this *same direction*. This is necessary since positive or negative internal virtual work depends on the directional sense of both the virtual load, defined by $\pm m$ or $\pm m_\theta$, and displacement, caused by $\pm M$.

Virtual-Work Equation.

• Apply the equation of virtual work to determine the desired displacement Δ or slope θ. It is important to retain the algebraic sign of each integral calculated within its specified region.

• If the algebraic sum of all the integrals for the entire beam is positive, Δ or θ is in the same direction as the virtual unit load or virtual unit couple moment, respectively. If a negative value results, Δ or θ is opposite to the virtual unit load or couple moment.

EXAMPLE 14-14

Determine the displacement of point B on the beam shown in Fig. 14–36a. EI is constant.

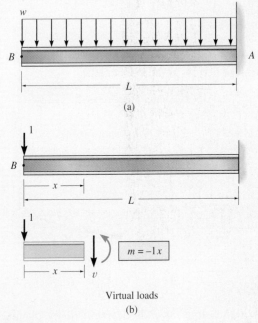

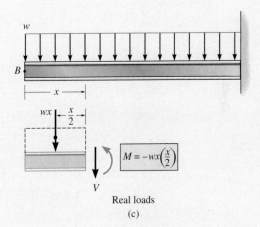

Virtual loads
(b)

Real loads
(c)

SOLUTION

Fig. 14–36

Virtual Moment m.   The vertical displacement of point B is obtained by placing a virtual unit load at B, Fig. 14–36b. By inspection, there are no discontinuities of loading on the beam for *both* the real and virtual loads. Thus, a *single x* coordinate can be used to determine the virtual strain energy. This coordinate will be selected with its origin at B, since then the reactions at A do not have to be determined in order to find the internal moments m and M. Using the method of sections, the internal moment m is computed as shown in Fig. 14–36b.

Real Moment M.   Using the *same x* coordinate, the internal moment M is computed as shown in Fig. 14–36c.

Virtual-Work Equation.   The vertical displacement at B is thus

$$1 \cdot \Delta_B = \int \frac{mM}{EI}\, dx = \int_0^L \frac{(-1x)(-wx^2/2)\, dx}{EI}$$

$$\Delta_B = \frac{wL^4}{8EI} \qquad\qquad\qquad Ans.$$

EXAMPLE 14–15

Determine the slope at point B of the beam shown in Fig. 14–37a. EI is constant.

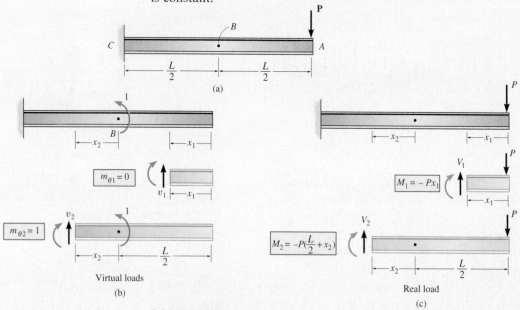

(a)

Virtual loads
(b)

Real load
(c)

Fig. 14–37

SOLUTION

Virtual Moments m_θ. The slope at B is determined by placing a virtual unit couple moment at B, Fig. 14–37b. Two x coordinates must be selected in order to determine the total virtual strain energy in the beam. Coordinate x_1 accounts for the strain energy within segment AB, and coordinate x_2 accounts for the strain energy in segment BC. The internal moments m_θ within each of these segments are computed using the method of sections as shown in Fig. 14–37b.

Real Moments M. Using the *same coordinates* x_1 and x_2 (Why?), the internal moments M are computed as shown in Fig. 14–37c.

Virtual-Work Equation. The slope at B is thus

$$1 \cdot \theta_B = \int \frac{m_\theta M}{EI}\, dx$$

$$= \int_0^{L/2} \frac{0(-Px_1)\, dx_1}{EI} + \int_0^{L/2} \frac{1\{-P[(L/2) + x_2]\}\, dx_2}{EI}$$

$$\theta_B = -\frac{3PL^2}{8EI} \qquad\qquad Ans.$$

The *negative sign* indicates that θ_B is *opposite* to the direction of the virtual couple moment shown in Fig. 14–37b.

EXAMPLE 14–16

Determine the displacement of point A of the steel beam shown in Fig. 14–38a. $I = 450$ in^4, $E_{st} = 29(10^3)$ ksi.

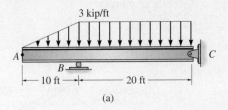

(a)

SOLUTION

Virtual Moments m.   The beam is subjected to the virtual unit load at A and the reactions are computed, Fig. 14–38b. By inspection, two coordinates x_1 and x_2 must be chosen to cover all regions of the beam. For purposes of integration it is simplest to use origins at A and C. Using the method of sections, the internal moments m are shown in Fig. 14–38b.

Real Moments M.   The reactions on the real beam are found first, Fig. 14–38a. Then, using the *same x* coordinates as those used for m, the internal moments M are determined.

Virtual-Work Equation.   Applying the equation of virtual work to the beam, we have

$$1 \text{ kip} \cdot \Delta_A = \int \frac{mM}{EI} \, dx = \int_0^{10} \frac{(-1x_1)(-0.05x_1{}^3) \, dx_1}{EI}$$

$$+ \int_0^{20} \frac{(-0.5x_2)(27.5x_2 - 1.5x_2{}^2) \, dx_2}{EI}$$

or

$$1 \text{ kip} \cdot \Delta_A = \frac{0.010(10)^5}{EI} - \frac{4.583(20)^3}{EI} + \frac{0.1875(20)^4}{EI}$$

$$\Delta_A = \frac{-5666.7 \text{ kip} \cdot \text{ft}^3}{EI}$$

Substituting in the data for E and I, we get

$$\Delta_A = \frac{-5666.7 \text{ kip} \cdot \text{ft}^3(12 \text{ in./ft})^3}{[29(10^3) \text{ kip/in}^2] \, 450 \text{ in}^4}$$

$$= -0.750 \text{ in.} \qquad \textit{Ans.}$$

The negative sign indicates that point A is displaced upward.

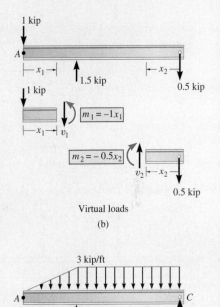

Virtual loads

(b)

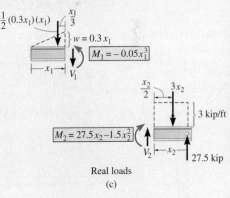

Real loads

(c)

Fig. 14–38

PROBLEMS

14–77. Determine the displacement at point C. EI is constant.

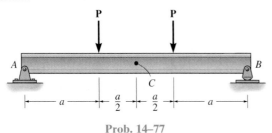

Prob. 14–77

14–78. Determine the displacement of point C. EI is constant.

14–79. Determine the slope at point C. EI is constant.

*14–80.** Determine the slope at point A. EI is constant.

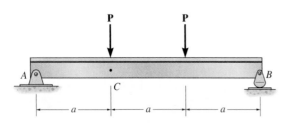

Probs. 14–78/14–79/14–80

14–81. Determine the slope of the 60-mm-diameter A-36 steel shaft at the bearing support A.

14–82. Determine the displacement at C of the 60-mm-diameter A-36 steel shaft.

14–83. Determine the tilt of the pulley at C if the A-36 steel shaft has a 60-mm diameter.

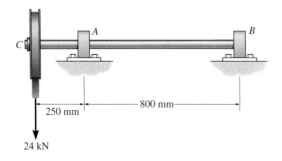

Probs. 14–81/14–82/14–83

*14–84.** Determine the displacement of collar B. The A-36 steel shaft has a diameter of 60 mm.

14–85. Determine the slope at C. The A-36 steel shaft has a diameter of 60 mm.

14–86. Determine the slope at A. The A-36 steel shaft has a diameter of 60 mm.

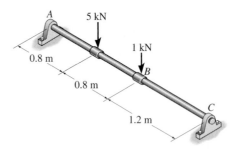

Probs. 14–84/14–85/14–86

14–87. The A-36 steel beam has a moment of inertia of $I = 125(10^6)$ mm^4. Determine the displacement at D.

*14–88.** The A-36 steel beam has a moment of inertia of $I = 125(10^6)$ mm^4. Determine the slope at E.

14–89. The A-36 structural steel beam has a moment of inertia of $I = 125(10^6)$ mm^4. Determine the slope of the beam at B.

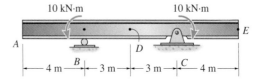

Probs. 14–87/14–88/14–89

14–90. Determine the displacement at point B. The moment of inertia of the center portion DG of the shaft is 2I, whereas the end segments AD and GC have a moment of inertia I. The modulus of elasticity for the material is E.

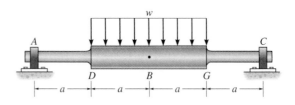

Prob. 14–90

14–91. Determine the displacement of point C of the W14 × 26 beam made from A-36 steel.

*14–92. Determine the slope at A of the W14 × 26 beam made from A–36 steel.

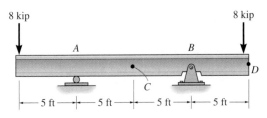

Probs. 14–91/14–92

14–93. Determine the displacement at C of the A-36 steel beam. $I = 70(10^6)$ mm^4.

14–94. Determine the slope at A of the A-36 steel beam. $I = 70(10^6)$ mm^4.

14–95. Determine the slope at B of the A-36 steel beam. $I = 70(10^6)$ mm^4.

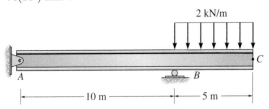

Probs. 14–93/14–94/14–95

*14–96. The beam is made of Douglas fir. Determine the displacement at A.

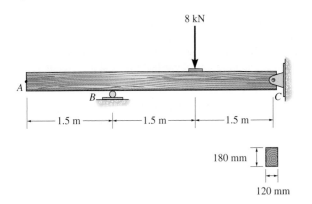

Prob. 14–96

14–97. The beam is made of Douglas fir. Determine the slope at C.

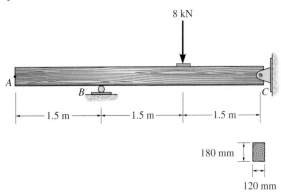

Prob. 14–97

14–98. The beam is made of oak, for which $E_o = 11$ GPa. Determine the slope and displacement at A.

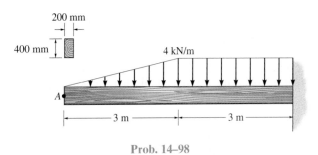

Prob. 14–98

14–99. Determine the displacement at point C. EI is constant.

*14–100. Determine the slope at B. EI is constant.

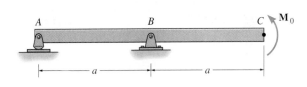

Probs. 14–99/14–100

14–101. Beam AB has a square cross section of 100 mm by 100 mm. Bar CD has a diameter of 10 mm. If both members are made of A-36 steel, determine the vertical displacement of point B due to the loading of 10 kN.

14–102. Beam AB has a square cross section of 100 mm by 100 mm. Bar CD has a diameter of 10 mm. If both members are made of A-36 steel, determine the slope at A due to the loading of 10 kN.

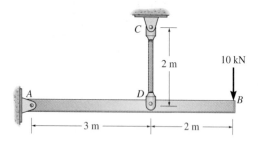

Probs. 14–101/14–102

14–103. Bar ABC has a rectangular cross section of 300 mm by 100 mm. Attached rod DB has a diameter of 20 mm. If both members are made of A-36 steel, determine the vertical displacement of point C due to the loading. Consider only the effect of bending in ABC and axial force in DB.

*14–104. Bar ABC has a rectangular cross section of 300 mm by 100 mm. Attached rod DB has a diameter of 20 mm. If both members are made of A-36 steel, determine the slope at A due to the loading. Consider only the effect of bending in ABC and axial force in DB.

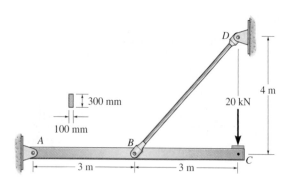

Probs. 14–103/14–104

14–105. The L-shaped frame is made from two segments, each of length L and flexural stiffness EI. If it is subjected to the uniform distributed load, determine the horizontal displacement of the end C.

14–106. The L-shaped frame is made from two segments, each of length L and flexural stiffness EI. If it is subjected to the uniform distributed load, determine the vertical displacement of point B.

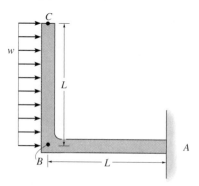

Probs. 14–105/14–106

14–107. Determine the horizontal and vertical displacements of point C. There is a fixed support at A. EI is constant.

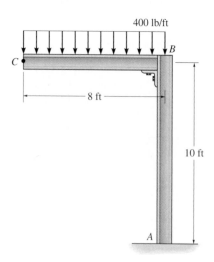

Prob. 14–107

*14–108. Determine the horizontal displacement of point *A* on the angle bracket due to the concentrated force **P**. The bracket is fixed connected to its support. *EI* is constant. Consider only the effect of bending.

14–110. The frame is made from two segments, each of length *L* and flexural stiffness *EI*. If it is subjected to the uniform distributed load, determine the horizontal displacement of point *B*. Consider only the effect of bending.

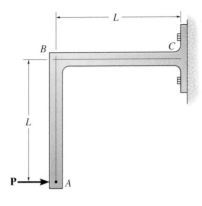

Prob. 14–108

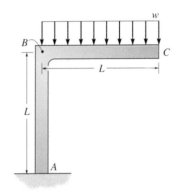

Prob. 14–110

14–109. The frame is made from two segments, each of length *L* and flexural stiffness *EI*. If it is subjected to the uniform distributed load, determine the vertical displacement of point *C*. Consider only the effect of bending.

14–111. The semi-circular rod has a moment of inertia *I* and modulus of elasticity *E*. Determine the horizontal deflection at the roller due to bending.

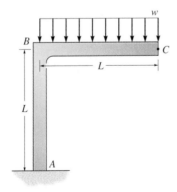

Prob. 14–109

Prob. 14–111

*14.8   CASTIGLIANO'S THEOREM

In 1879 Alberto Castigliano, an Italian railroad engineer, published a book in which he outlined a method for determining the displacement and slope at a point in a body. This method, which is referred to as *Castigliano's second theorem*, applies only to bodies that have constant temperature and material with linear-elastic behavior. If the displacement at a point is to be determined, the theorem states that the displacement is equal to the first partial derivative of the strain energy in the body with respect to a force acting at the point and in the direction of displacement. In a similar manner, the slope of the tangent at a point in a body is equal to the first partial derivative of the strain energy in the body with respect to a couple moment acting at the point and in the direction of the slope angle.

To derive Castigliano's second theorem, consider a body of any arbitrary shape, which is subjected to a series of n forces $\mathbf{P}_1, \mathbf{P}_2, \ldots, \mathbf{P}_n$, Fig. 14–39. Since the external work done by these forces is equal to the internal strain energy stored in the body, we can apply the conservation of energy, i.e.,

$$U_i = U_e$$

However, the external work is a function of the external loads, $U_e = \Sigma \int P \, dx$, Eq. 14–1, so the internal work is also a function of the external loads. Thus,

$$U_i = U_e = f(P_1, P_2, \ldots, P_n) \tag{14–44}$$

Now, if any one of the external forces, say P_j, is increased by a differential amount dP_j, the internal work will also be increased, such that the strain energy becomes

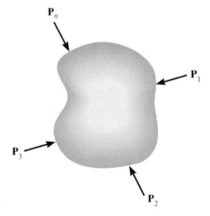

Fig. 14–39

$$U_i + dU_i = U_i + \frac{\partial U_i}{\partial P_j} dP_j \qquad (14\text{--}45)$$

This value, however, should not depend on the sequence in which the n forces are applied to the body. For example, we could apply $d\mathbf{P}_j$ to the body *first*, then apply the loads $\mathbf{P}_1, \mathbf{P}_2, \ldots, \mathbf{P}_n$. In this case, $d\mathbf{P}_j$ would cause the body to be displaced a differential amount $d\Delta_i$ in the direction of $d\mathbf{P}_j$. By Eq. 14–2 ($U_e = \frac{1}{2}P_j\Delta_j$), the increment of strain energy would be $\frac{1}{2} dP_j \, d\Delta_j$. This quantity, however, is a second-order differential and may be neglected. Further application of the loads $\mathbf{P}_1, \mathbf{P}_2, \ldots, \mathbf{P}_n$ causes $d\mathbf{P}_j$ to move through the displacement Δ_j so that now the strain energy becomes

$$U_i + dU_j = U_i + dP_j \, \Delta_j \qquad (14\text{--}46)$$

Here, as above, U_i is the internal strain energy in the body, caused by the loads $\mathbf{P}_1, \mathbf{P}_2, \ldots, \mathbf{P}_n$, and $dU_j = dP_j \, \Delta_j$ is the *additional* strain energy caused by $d\mathbf{P}_j$.

In summary, Eq. 14–45 represents the strain energy in the body determined by first applying the loads $\mathbf{P}_1, \mathbf{P}_2, \ldots, \mathbf{P}_n$, then $d\mathbf{P}_j$; Eq. 14–46 represents the strain energy determined by first applying $d\mathbf{P}_j$ and then the loads $\mathbf{P}_1, \mathbf{P}_2, \ldots, \mathbf{P}_n$. Since these two equations must be equal, we require

$$\Delta_i = \frac{\partial U_i}{\partial P_j} \qquad (14\text{--}47)$$

which proves the theorem; i.e., the displacement Δ_j in the direction of $\mathbf{P}_j$ is equal to the first partial derivative of the strain energy with respect to $\mathbf{P}_j$.

It should be noted that Eq. 14–47 is a statement regarding the body's *compatibility requirements*, since it is a condition related to displacement. Also, the above derivation requires that *only conservative forces* be considered for the analysis. These forces can be applied in any order, and furthermore, they do work that is independent of the path and therefore create no energy loss. As long as the material has linear-elastic behavior, the applied forces will be conservative and the theorem is valid. It should also be mentioned that Castigliano's first theorem is similar to his second theorem; however, it relates the load P_j to the partial derivative of the strain energy with respect to the corresponding displacement, that is, $P_j = \partial U_i/\partial \Delta_j$. The proof is similar to that given above. This theorem is another way of expressing the *equilibrium requirements* for the body; however, it has limited application and therefore it will not be discussed here.

*14.9   Castigliano's Theorem Applied to Trusses

Since a truss member is subjected to an axial load, the strain energy is given by Eq. 14–16, $U_i = N^2 L / 2AE$. Substituting this equation into Eq. 14–47 and omitting the subscript i, we have

$$\Delta = \frac{\partial}{\partial P} \sum \frac{N^2 L}{2AE}$$

It is generally easier to perform the differentiation prior to summation. Also, L, A, and E are constant for a given member, and therefore we may write

$$\boxed{\Delta = \sum N \left(\frac{\partial N}{\partial P} \right) \frac{L}{AE}} \tag{14–48}$$

Here

Δ = joint displacement of the truss

P = external force of *variable magnitude* applied to the truss joint in the direction of Δ

N = internal axial force in a member caused by *both* the force **P** and the loads on the truss

L = length of a member

A = cross-sectional area of a member

E = modulus of elasticity of the material

In order to determine the partial derivative $\partial N/\partial P$, it will be necessary to treat P as a *variable*, not a specific numerical quantity. In other words, each internal axial force N must be expressed as a function of P.

By comparison, Eq. 14–48 is similar to that used for the method of virtual work, Eq. 14–39 ($1 \cdot \Delta = \Sigma nNL/AE$), except that n is replaced by $\partial N/\partial P$. These terms, n and $\partial N/\partial P$, will, however, be the *same*, since they represent the rate of change of the internal axial force with respect to the load P or, in other words, the axial force per unit load.

Procedure for Analysis

The following procedure provides a method that may be used to determine the displacement of any joint on a truss using Castigliano's second theorem.

External Force P.

• Place a force **P** on the truss at the joint where the desired displacement is to be determined. This force is assumed to have a *variable magnitude* and should be directed along the line of action of the displacement.

Internal Forces N.

• Determine the force N in each member caused by both the real (numerical) loads and the variable force P. Assume that tensile forces are positive and compressive forces are negative.

• Find the respective partial derivative $\partial N/\partial P$ for each member.

• After N and $\partial N/\partial P$ have been determined, assign P its numerical value if it has actually replaced a real force on the truss. Otherwise, set P equal to zero.

Castigliano's Second Theorem.

• Apply Castigliano's theorem to determine the desired displacement Δ. It is important to retain the algebraic signs for corresponding values of N and $\partial N/\partial P$ when substituting these terms into the equation.

• If the resultant sum $\Sigma N(\partial N/\partial P)L/AE$ is positive, Δ is in the same direction as **P**. If a negative value results, Δ is opposite to **P**.

EXAMPLE 14–17

Determine the horizontal displacement of joint C of the steel truss shown in Fig. 14–40a. The cross-sectional area of each member is indicated in the figure. Take $E_{st} = 29(10^3)$ ksi.

SOLUTION

*External Force **P***.   Since the horizontal displacement of C is to be determined, a horizontal *variable* force **P** is applied to joint C, Fig. 14–40b. Later this force will be set equal to the fixed value of 8 kip.

Internal Forces N.   Using the method of joints, the force N in each member is found. The results are shown in Fig. 14–40b. Arranging the data in tabular form, we have

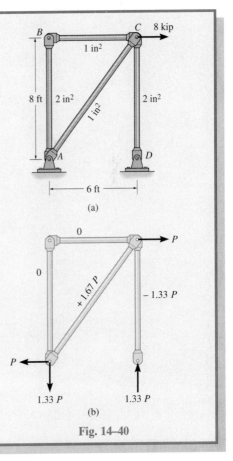

(a)

(b)

Fig. 14–40

Member	N	$\dfrac{\partial N}{\partial P}$	$N(P = 8\text{ kip})$	L	$N\left(\dfrac{\partial N}{\partial P}\right)L$
AB	0	0	0	8	0
BC	0	0	0	6	0
AC	$1.67P$	1.67	13.33	10	222.2
CD	$-1.33P$	-1.33	-10.67	8	113.8

Castigliano's Second Theorem.   Applying Eq. 14–48 gives

$$\Delta_{Ch} = \Sigma N\left(\frac{\partial N}{\partial P}\right)\frac{L}{AE}$$

$$= 0 + 0 + \frac{(222.2\text{ kip}\cdot\text{ft})(12\text{ in./ft})}{(1\text{ in}^2)\,29(10^3)\text{ kip/in}^2} + \frac{(113.8\text{ kip}\cdot\text{ft})(12\text{ in./ft})}{(2\text{ in}^2)\,29(10^3)\text{ kip/in}^2}$$

$$= 0.115\text{ in.}\qquad\qquad Ans.$$

EXAMPLE 14-18

Determine the vertical displacement of joint C of the steel truss shown in Fig. 14–41a. The cross-sectional area of each member is $A = 400$ mm^2, and $E_{st} = 200$ GPa.

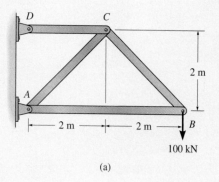

(a)

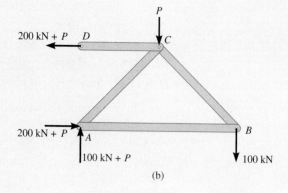

(b)

Fig. 14–41

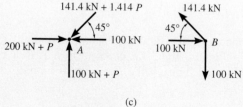
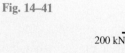

(c)

SOLUTION

External Force P.   A vertical force **P** is applied to the truss at joint C, since this is where the vertical displacement is to be determined, Fig. 14–41b.

Internal Forces N.   The reactions at the truss supports A and D are calculated and the results are shown in Fig. 14–41b. Using the method of joints, the **N** forces in each member are determined, Fig. 14–41c.* For convenience, these results along with the partial derivatives $\partial N/\partial P$ are listed in tabular form. Note that since **P** does not actually exist as a real load on the truss, we require $P = 0$.

Member	N	$\dfrac{\partial N}{\partial P}$	$N(P = 0)$	L	$N\left(\dfrac{\partial N}{\partial P}\right)L$
AB	-100	0	-100	4	0
BC	141.4	0	141.4	2.828	0
AC	$-141.4 - 1.414\,P$	-1.414	-141.4	2.828	565.7
CD	$200 + P$	1	200	2	400
					Σ 965.7 kN $\cdot$ m

Castigliano's Second Theorem.   Applying Eq. 14–48, we have

$$\Delta_{C_v} = \Sigma N\left(\frac{\partial N}{\partial P}\right)\frac{L}{AE} = \frac{965.7 \text{ kN} \cdot \text{m}}{AE}$$

Substituting the numerical values for A and E, we get

$$\Delta_{C_v} = \frac{965.7 \text{ kN} \cdot \text{m}}{[400(10^{-6}) \text{ m}^2] \, 200(10^6) \text{ kN/m}^2}$$

$$= 0.01207 \text{ m} = 12.1 \text{ mm} \qquad \textit{Ans.}$$

This solution should be compared with that of Example 14–11, using the virtual-work method.

---

*It may be more convenient to analyze the truss with just the 100-kN load on it, then analyze the truss with the **P** load on it. The results can then be added together to give the **N** forces.

PROBLEMS

*14–112. Solve Prob. 14–60 using Castigliano's theorem.

14–113. Solve Prob. 14–59 using Castigliano's theorem.

14–114. Solve Prob. 14–61 using Castigliano's theorem.

14–115. Solve Prob. 14–62 using Castigliano's theorem.

*14–116. Solve Prob. 14–64 using Castigliano's theorem.

14–117. Solve Prob. 14–63 using Castigliano's theorem.

14–118. Solve Prob. 14–65 using Castigliano's theorem.

14–119. Solve Prob. 14–66 using Castigliano's theorem.

*14–120. Solve Prob. 14–68 using Castigliano's theorem.

14–121. Solve Prob. 14–67 using Castigliano's theorem.

14–122. Solve Prob. 14–69 using Castigliano's theorem.

14–123. Solve Prob. 14–70 using Castigliano's theorem.

*14–124. Solve Prob. 14–72 using Castigliano's theorem.

14–125. Solve Prob. 14–71 using Castigliano's theorem.

14–126. Solve Prob. 14–73 using Castigliano's theorem.

14–127. Solve Prob. 14–74 using Castigliano's theorem.

*14–128. Solve Prob. 14–76 using Castigliano's theorem.

14–129. Solve Prob. 14–75 using Castigliano's theorem.

*14.10   CASTIGLIANO'S THEOREM APPLIED TO BEAMS

The internal strain energy for a beam is caused by both bending and shear. However, as pointed out in Example 14–7, if the beam is long and slender, the strain energy due to shear can be neglected compared with that of bending. Assuming this to be the case, the internal strain energy for a beam is given by $U_i = \int M^2\, dx/2EI$, Eq. 14–17. Substituting into $\Delta_i = \partial U_i/\partial P_i$, Eq. 14–47, and omitting the subscript i, we have

$$\Delta = \frac{\partial}{\partial P} \int_0^L \frac{M^2\, dx}{2EI}$$

Rather than squaring the expression for internal moment M, integrating, and then taking the partial derivative, it is generally easier to differentiate prior to integration. Provided E and I are constant, we have

$$\Delta = \int_0^L M\left(\frac{\partial M}{\partial P}\right)\frac{dx}{EI} \tag{14–49}$$

where

Δ = displacement of the point caused by the real loads acting on the beam

P = external force of variable magnitude applied to the beam in the direction of Δ

M = internal moment in the beam, expressed as a function of x and caused by both the force P and the loads on the beam

E = modulus of elasticity of the material

I = moment of inertia of cross-sectional area computed about the neutral axis

If the slope of the tangent θ at a point on the elastic curve is to be determined, the partial derivative of the internal moment M with respect to an *external couple moment* M' acting at the point must be found. For this case,

$$\theta = \int_0^L M\left(\frac{\partial M}{\partial M'}\right)\frac{dx}{EI} \tag{14–50}$$

The above equations are similar to those used for the method of virtual work, Eqs. 14–42 and 14–43, except m and m_θ replace $\partial M/\partial P$ and $\partial M/\partial M'$, respectively.

It should be mentioned that if the loading on a member causes significant strain energy within the member due to axial load, shear, bending moment, and torsional moment, then the effects of all these loadings should be included when applying Castigliano's theorem. To do this we must use the strain-energy functions developed in Sec. 14.2, along with their associated partial derivatives. The result is

$$\Delta = \Sigma N\left(\frac{\partial N}{\partial P}\right)\frac{L}{AE} + \int_0^L f_s V\left(\frac{\partial V}{\partial P}\right)\frac{dx}{GA} + \int_0^L M\left(\frac{\partial M}{\partial P}\right)\frac{dx}{EI} + \int_0^L T\left(\frac{\partial T}{\partial P}\right)\frac{dx}{GJ} \quad (14\text{-}15)$$

The method of applying this general formulation is similar to that used to apply Eqs. 14–49 and 14–50.

PROCEDURE FOR ANALYSIS

The following procedure provides a method that may be used to apply Castigliano's second theorem.

External Force P or Couple Moment M′.

• Place a force **P** on the beam at the point and directed along the line of action of the desired displacement.

• If the slope of the tangent is to be determined, place a couple moment **M′** at the point.

• Assume that both **P** and **M′** have a variable magnitude.

Internal Moments M.

• Establish appropriate x coordinates that are valid within regions of the beam where there is no discontinuity of force, distributed load, or couple moment.

• Calculate the internal moments M as a function of P or $M′$ and the partial derivatives $\partial M/\partial P$ or $\partial M/\partial M′$ for each coordinate x.

• After M and $\partial M/\partial P$ or $\partial M/\partial M′$ have been determined, assign P or $M′$ its numerical value if it has actually replaced a real force or couple moment. Otherwise, set P or $M′$ equal to zero.

Castigliano's Second Theorem.

• Apply Eq. 14–49 or 14–50 to determine the desired displacement Δ or θ. It is important to retain the algebraic signs for corresponding values of M and $\partial M/\partial P$ or $\partial M/\partial M′$.

• If the resultant sum of all the definite integrals is positive, Δ or θ is in the same direction as **P** or **M′**. If a negative value results, Δ or θ is opposite to **P** or **M′**.

EXAMPLE 14–19

Determine the displacement of point B on the beam shown in Fig. 14–42a. EI is constant.

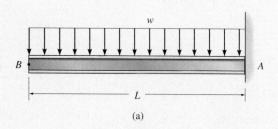

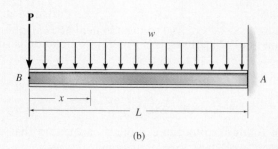

(a)

(b)

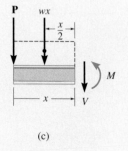

(c)

Fig. 14–42

SOLUTION

External Force **P**.   A vertical force **P** is placed on the beam at B as shown in Fig. 14–42b.

Internal Moments M.   A single x coordinate is needed for the solution, since there are no discontinuities of loading between A and B. Using the method of sections, Fig. 14–42c, the internal moment and partial derivative are determined as follows:

$$\zeta^+ \Sigma M_{NA} = 0; \qquad M + wx\left(\frac{x}{2}\right) + P(x) = 0$$

$$M = -\frac{wx^2}{2} - Px$$

$$\frac{\partial M}{\partial P} = -x$$

Setting $P = 0$ gives

$$M = \frac{-wx^2}{2} \qquad \text{and} \qquad \frac{\partial M}{\partial P} = -x$$

Castigliano's Second Theorem.   Applying Eq. 14–49, we have

$$\Delta_B = \int_0^L M\left(\frac{\partial M}{\partial P}\right)\frac{dx}{EI} = \int_0^L \frac{(-wx^2/2)(-x)\,dx}{EI}$$

$$= \frac{wL^4}{8EI} \qquad\qquad\qquad\qquad\qquad\qquad \textit{Ans.}$$

The similarity between this solution and that of the virtual-work method, Example 14–14, should be noted.

EXAMPLE 14–20

Determine the slope at point B of the beam shown in Fig. 14–43a. EI is constant.

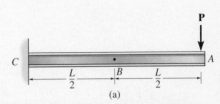

(a)

SOLUTION

External Couple Moment **M'**. Since the slope at point B is to be determined, an external couple moment **M'** is placed on the beam at this point, Fig. 14–43b.

Internal Moments M. Two coordinates, x_1 and x_2, must be used to determine the internal moments within the beam since there is a discontinuity, **M'**, at B. As shown in Fig. 14–43b, x_1 ranges from A to B and x_2 ranges from B to C. Using the method of sections, Fig. 14–43c, the internal moments and the partial derivatives are determined as follows:

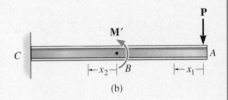

(b)

For x_1,

$$\zeta + \Sigma M_{NA} = 0; \qquad -M_1 - Px_1 = 0$$

$$M_1 = -Px_1$$

$$\frac{\partial M_1}{\partial M'} = 0$$

For x_2,

$$\zeta + \Sigma M_{NA} = 0; \qquad -M_2 + M' - P\left(\frac{L}{2} + x_2\right) = 0$$

$$M_2 = M' - P\left(\frac{L}{2} + x_2\right)$$

$$\frac{\partial M_2}{\partial M'} = 1$$

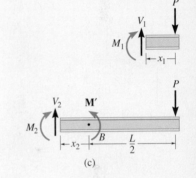

(c)

Fig. 14–43

Castigliano's Second Theorem. Setting $M' = 0$ and applying Eq. 14–50, we have

$$\theta_B = \int_0^L M\left(\frac{\partial M}{\partial M'}\right)\frac{dx}{EI}$$

$$= \int_0^{L/2} \frac{(-Px_1)(0)\ dx_1}{EI} + \int_0^{L/2} \frac{-P[(L/2) + x_2](1)\ dx_2}{EI}$$

$$= -\frac{3PL^2}{8EI} \qquad\qquad Ans.$$

The negative sign indicates that θ_B is opposite to the direction of the couple moment **M'**. Note the similarity between this solution and that of Example 14–15.

EXAMPLE 14–21

Determine the vertical displacement of point C of the steel beam shown in Fig. 14–44a. Take $E_{st} = 200$ GPa, $I = 125(10^{-6})$ m^4.

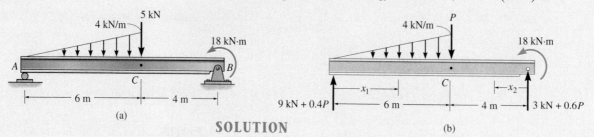

(a)

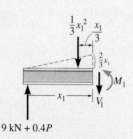

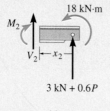

(c)

Fig. 14–44

SOLUTION

(b)

External Force **P**.   A vertical force **P** is applied at point C, Fig. 14–44b. Later this force will be set equal to the fixed value of 5 kN.

Internal Moments M.   In this case two x coordinates are needed for the integration since the load is discontinuous at C. Using the method of sections, Fig. 14–44c, the internal moments and partial derivatives are determined as follows:

For x_1,

$$+\Sigma M_{NA} = 0; \quad M_1 + \frac{1}{3}x_1^2\left(\frac{x_1}{3}\right) - (9 + 0.4P)x_1 = 0$$

$$M_1 = (9 + 0.4P)x_1 - \frac{1}{9}x_1^3$$

$$\frac{\partial M_1}{\partial P} = 0.4x_1$$

For x_2,

$$\downarrow^+ \Sigma M_{NA} = 0; \quad -M_2 + 18 + (3 + 0.6P)x_2 = 0$$
$$M_2 = 18 + (3 + 0.6P)x_2$$

$$\frac{\partial M_2}{\partial P} = 0.6x_2$$

Castigliano's Second Theorem.   Setting $P = 5$ kN and applying Eq. 14–49, we have

$$\Delta_{C_v} = \int_0^L M\left(\frac{\partial M}{\partial P}\right)\frac{dx}{EI}$$

$$= \int_0^6 \frac{(11x_1 - \frac{1}{9}x_1^3)(0.4x_1)\,dx_1}{EI} + \int_0^4 \frac{(18 + 6x_2)(0.6x_2)\,dx_2}{EI}$$

$$= \frac{410.9 \text{ kN} \cdot \text{m}^3}{[200(10^6) \text{ kN/m}^2]\,125(10^{-6}) \text{ m}^4}$$

$$= 0.0164 \text{ m} = 16.4 \text{ mm} \qquad\qquad\qquad Ans.$$

PROBLEMS

14–130. Solve Prob. 14–77 using Castigliano's theorem.

14–131. Solve Prob. 14–78 using Castigliano's theorem.

*14–132. Solve Prob. 14–84 using Castigliano's theorem.

14–133. Solve Prob. 14–83 using Castigliano's theorem.

14–134. Solve Prob. 14–88 using Castigliano's theorem.

14–135. Solve Prob. 14–94 using Castigliano's theorem.

*14–136. Solve Prob. 14–92 using Castigliano's theorem.

14–137. Solve Prob. 14–102 using Castigliano's theorem.

14–138. Solve Prob. 14–98 using Castigliano's theorem.

14–139. Solve Prob. 14–99 using Castigliano's theorem.

*14–140. Solve Prob. 14–96 using Castigliano's theorem.

14–141. Solve Prob. 14–103 using Castigliano's theorem.

14–142. Solve Prob. 14–106 using Castigliano's theorem.

14–143. Solve Prob. 14–107 using Castigliano's theorem.

*14–144. Solve Prob. 14–108 using Castigliano's theorem.

14–145. Solve Prob. 14–109 using Castigliano's theorem.

14–146. Solve Prob. 14–111 using Castigliano's theorem.

REVIEW PROBLEMS

14–147. The cantilevered beam is subjected to a couple moment M_0 applied at its end. Determine the displacement of the beam at B. EI is constant. Use the method of virtual work.

*14–148. Solve Prob. 14–147 using Castigliano's theorem.

14–149. Use the method of virtual work to determine the displacement at point A. EI is constant.

14–150. Solve Prob. 14–149 using Castigliano's theorem.

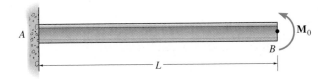

Probs. 14–147/14–148

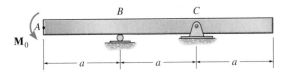

Probs. 14–149/14–150

14–151. Determine the slope of point C of the A-36 steel beam. $I = 9.50(10^6)$ mm^4. Use the method of virtual work.

*14–152. Solve Prob. 14–151 using Castigliano's theorem.

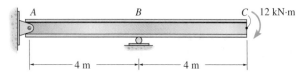

Probs. 14–151/14–152

14–153. Determine the vertical displacement of joint E. For each member $A = 400$ mm^2, $E = 200$ GPa. Use the method of virtual work.

14–154. Solve Prob. 14–153 using Castigliano's theorem.

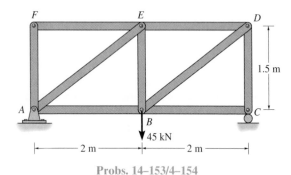

Probs. 14–153/4–154

14–155. Determine the strain energy in the *horizontal* curved bar due to torsion. There is a *vertical* force **P** acting at its end. JG is constant.

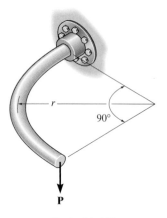

Prob. 14–155

*14–156. Determine the total strain energy in the A-36 steel assembly. Consider the axial strain energy in the two 0.5-in.-diameter rods and the bending strain energy in the beam for which $I = 43.4$ in^4.

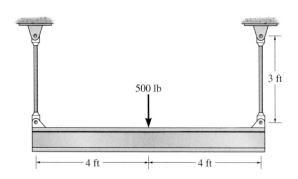

Prob. 14–156

14–157. The W10 × 12 beam is made of A-36 steel and is cantilevered from the wall at B. The spring mounted on the beam has a stiffness of $k = 1000$ lb/in. If a weight of 8 lb is dropped onto the spring from a height of $h = 3$ ft, determine the maximum bending stress developed in the beam.

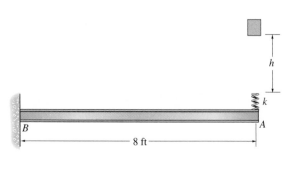

Prob. 14–157

A GEOMETRIC PROPERTIES OF AN AREA

A.1 CENTROID OF AN AREA

The *centroid* of an area refers to the point that defines the geometric center for the area. If the area has an arbitrary shape, as shown in Fig. A–1a, the x and y coordinates defining the location of the centroid C are determined using the formulas

$$\bar{x} = \frac{\int_A x \, dA}{\int_A dA} \qquad \bar{y} = \frac{\int_A y \, dA}{\int_A dA} \qquad (A–1)$$

The numerators in these equations are formulations of the "first moment" of the area element dA about the y and the x axis, respectively, Fig. A–1b; the denominators represent the total area A of the shape.

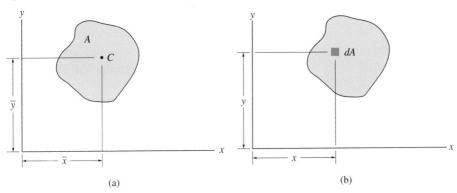

(a) (b)

Fig. A–1

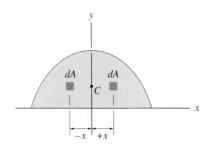

Fig. A–2

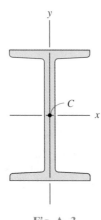

Fig. A–3

It should be noted that the location of the centroid for some areas may be partially or completely specified by using symmetry conditions. In cases where the area has an axis of symmetry, the centroid for the area will lie along that axis. For example, the centroid C for the area shown in Fig. A–2 must lie along the y axis, since for every elemental area dA a distance $+x$ to the right of the y axis, there is an identical element a distance $-x$ to the left. The total moment for all the elements about the axis of symmetry will therefore cancel; that is, $\int x \, dA = 0$ (Eq. A–1), so that $\bar{x} = 0$. In cases where a shape has two axes of symmetry, it follows that the centroid lies at the intersection of these axes, Fig. A–3. Based on the principle of symmetry, or using Eq. A–1, the locations of the centroid for common area shapes are listed on the inside front cover of this book.

Composite Areas.   Often an area can be sectioned or divided into several parts having simpler shapes. Provided the area and location of the centroid of each of these "composite shapes" are known, one can eliminate the need for integration to determine the centroid for the entire area. In this case, equations analogous to Eq. A–1 must be used, except that finite summation signs replace the integrals; i.e.,

$$\bar{x} = \frac{\Sigma \tilde{x} A}{\Sigma A} \qquad \bar{y} = \frac{\Sigma \tilde{y} A}{\Sigma A} \tag{A–2}$$

Here $\tilde{x}$ and $\tilde{y}$ represent the *algebraic distances* or x, y coordinates for the centroid of each composite part, and ΣA represents the sum of the areas of the composite parts or simply the *total area*. In particular, if a hole, or a geometric region having no material, is located within a composite part, the hole is considered as an additional composite part having a *negative* area. Also, as discussed above, if the total area is symmetrical about an axis, the centroid of the area lies on the axis.

The following example illustrates application of Eq. A–2.

EXAMPLE A-1

Locate the centroid C of the cross-sectional area for the T-beam shown in Fig. A–4a.

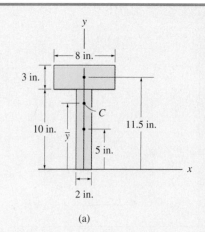

(a)

SOLUTION I

The y axis is placed along the axis of symmetry so that $\bar{x} = 0$, Fig. A–4a. To obtain $\bar{y}$ we will establish the x axis (reference axis) through the base of the area. The area is segmented into two rectangles as shown, and the centroidal location $\tilde{y}$ for each is established. Applying Eq. A–2, we have

$$\bar{y} = \frac{\Sigma \tilde{y}A}{\Sigma A} = \frac{[5 \text{ in.}](10 \text{ in.})(2 \text{ in.}) + [11.5 \text{ in.}](3 \text{ in.})(8 \text{ in.})}{(10 \text{ in.})(2 \text{ in.}) + (3 \text{ in.})(8 \text{ in.})}$$

$$= 8.55 \text{ in.} \qquad\qquad\qquad Ans.$$

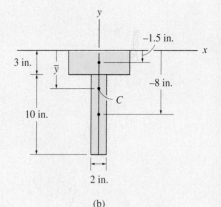

(b)

SOLUTION II

Using the same two segments, the x axis can be located at the top of the area as shown in Fig. A–4b. Here

$$\bar{y} = \frac{\Sigma \tilde{y}A}{\Sigma A} = \frac{[-1.5 \text{ in.}](3 \text{ in.})(8 \text{ in.}) + [-8 \text{ in.}](10 \text{ in.})(2 \text{ in.})}{(3 \text{ in.})(8 \text{ in.}) + (10 \text{ in.})(2 \text{ in.})}$$

$$= -4.45 \text{ in.} \qquad\qquad\qquad Ans.$$

The negative sign indicates that C is located *below* the origin, which is to be expected. Also note that from the two answers 8.55 in. + 4.45 in. = 13.0 in., which is the depth of the beam.

SOLUTION III

It is also possible to consider the cross-sectional area to be one large rectangle *less* two small rectangles, Fig. A–4c. Here we have

$$\bar{y} = \frac{\Sigma \tilde{y}A}{\Sigma A} = \frac{[6.5 \text{ in.}](13 \text{ in.})(8 \text{ in.}) - 2[5 \text{ in.}](10 \text{ in.})(3 \text{ in.})}{(13 \text{ in.})(8 \text{ in.}) - 2(10 \text{ in.})(3 \text{ in.})}$$

$$= 8.55 \text{ in.} \qquad\qquad\qquad Ans.$$

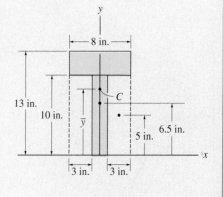

(c)

Fig. A–4

A.2 MOMENT OF INERTIA FOR AN AREA

When calculating the centroid for an area, we considered the first moment of the area about an axis; i.e., for the computation it was necessary to evaluate an integral of the form $\int x \, dA$. There are some topics in mechanics of materials that require evaluation of an integral of the second moment of an area, that is, $\int x^2 dA$. This integral is referred to as the **moment of inertia** for an area. To show formally how it is defined, consider the area A, shown in Fig. A–5, which lies in the $x-y$ plane. By definition, the moments of inertia of the differential element dA about the x and y axes are $dI_x = y^2 dA$ and $dI_y = x^2 dA$, respectively. For the entire area, the moment of inertia is determined by integration, i.e.,

$$I_x = \int_A y^2 dA$$

$$I_y = \int_A x^2 dA$$

(A–3)

We can also formulate the second moment of the differential element about the pole O or z axis, Fig. A–5. This is referred to as the *polar moment of inertia*, $dJ_O = r^2 dA$. Here r is the perpendicular distance from the pole (z axis) to the element dA. For the entire area, the polar moment of inertia is

$$J_O = \int_A r^2 dA = I_x + I_y$$

(A–4)

The relationship between J_O and I_x, I_y is possible since $r^2 = x^2 + y^2$, Fig. A–5.

From the above formulations it is seen that I_x, I_y, and J_O will *always* be *positive*, since they involve the product of distance squared and area. Furthermore, the units for moment of inertia involve length raised to the fourth power, e.g., m^4, mm^4, or ft^4, in^4.

Using the above equations, the moments of inertia for some common area shapes have been calculated about their *centroidal axes* and are listed on the inside front cover.

Parallel-Axis Theorem for an Area.   If the moment of inertia for an area is known about a centroidal axis, we can determine the moment of inertia of the area about a corresponding parallel axis using the *parallel-axis theorem*. To derive this theorem, consider finding the moment of inertia of the shaded area shown in Fig. A–6 about the x axis. In this case, a differential element dA is located at an arbitrary distance y' from the centroidal x' axis, whereas the *fixed distance* between the

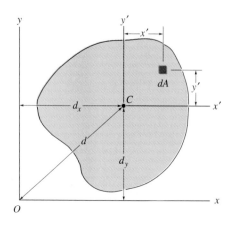

Fig. A–5

Fig. A–6

parallel x and x' axes is defined as d_y. Since the moment of inertia of dA about the x axis is $dI_x = (y' + d_y)^2 dA$, then for the entire area,

$$I_x = \int_A (y' + d_y)^2 dA = \int_A y'^2 dA + 2d_y \int_A y'dA + d_y^2 \int_A dA$$

The first term on the right represents the moment of inertia of the area about the x' axis, $\bar{I}_{x'}$. The second term is zero since the x' axis passes through the area's centroid C, that is, $\int y'dA = \bar{y}A = 0$ since $\bar{y} = 0$. The final result is therefore

$$\boxed{I_x = \bar{I}_{x'} + Ad_y^2} \tag{A–5}$$

A similar expression can be written for I_y, that is,

$$\boxed{I_y = \bar{I}_{y'} + Ad_x^2} \tag{A–6}$$

And finally, for the polar moment of inertia about an axis perpendicular to the x–y plane and passing through the pole O (z axis), Fig. A–6, we have

$$\boxed{J_O = \bar{J}_C + Ad^2} \tag{A–7}$$

The form of each of the above equations states that the moment of inertia of an area about an axis is equal to the area's moment of inertia about a parallel axis passing through the "centroid" plus the product of the area and the square of the perpendicular distance between the axes.

Composite Areas.   Many cross-sectional areas consist of a series of connected simpler shapes, such as rectangles, triangles, and semicircles. Provided the moment of inertia of each of these shapes is known, or can be found about a common axis, then the moment of inertia of the "composite area" can be determined as the *algebraic sum* of the moments of inertia of its composite parts.

In order to properly determine the moment of inertia of such an area about a specified axis, it is first necessary to divide the area into its composite parts and indicate the perpendicular distance from the axis to the parallel centroidal axis for each part. Using the table on the inside front cover of the book, the moment of inertia of each part is computed about the centroidal axis. If this axis does not coincide with the specified axis, the parallel-axis theorem, $I = \bar{I} + Ad^2$, should be used to determine the moment of inertia of the part about the specified axis. The moment of inertia of the entire area about this axis is then determined by summing the results of its composite parts. In particular, if a composite part has a "hole," the moment of inertia for the composite is found by "subtracting" the moment of inertia for the hole from the moment of inertia of the entire area including the hole.

The following examples illustrate application of this method.

EXAMPLE A–2

Determine the moment of inertia of the cross-sectional area of the T-beam shown in Fig. A–7a about the centroidal x' axis.

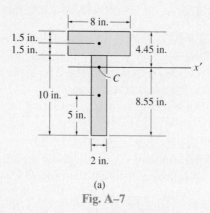

(a)

Fig. A–7

SOLUTION I

The area is segmented into two rectangles as shown in Fig. A–7a, and the distance from the x' axis and each centroidal axis is determined. Using the table on the inside front cover, the moment of inertia of a rectangle about its centroidal axis is $I = \frac{1}{12}bh^3$. Applying the parallel-axis theorem, Eq. A–5, to each rectangle and adding the results, we have

$$I = \Sigma \bar{I}_{x'} + A d_y^2$$

$$= \left[\frac{1}{12}(2 \text{ in.})(10 \text{ in.})^3 + (2 \text{ in.})(10 \text{ in.})(8.55 \text{ in.} - 5 \text{ in.})^2 \right]$$

$$+ \left[\frac{1}{12}(8 \text{ in.})(3 \text{ in.})^3 + (8 \text{ in.})(3 \text{ in.})(4.45 \text{ in.} - 1.5 \text{ in.})^2 \right]$$

$$I = 646 \text{ in}^4 \qquad\qquad Ans.$$

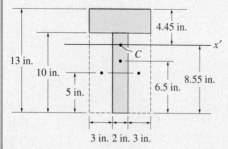

(b)

SOLUTION II

The area can be considered as one large rectangle less two small rectangles, shown dashed in Fig. A–7b. We have

$$I = \Sigma \bar{I}_{x'} + A d_y^2$$

$$= \left[\frac{1}{12}(8 \text{ in.})(13 \text{ in.})^3 + (8 \text{ in.})(13 \text{ in.})(8.55 \text{ in.} - 6.5 \text{ in.})^2 \right]$$

$$- 2\left[\frac{1}{12}(3 \text{ in.})(10 \text{ in.})^3 + (3 \text{ in.})(10 \text{ in.})(8.55 \text{ in.} - 5 \text{ in.})^2 \right]$$

$$I = 646 \text{ in}^4 \qquad\qquad Ans.$$

EXAMPLE A-3

Determine the moments of inertia of the beam's cross-sectional area shown in Fig. A–8a about the x and y centroidal axes.

SOLUTION

The cross section can be considered as three composite rectangular areas A, B, and D shown in Fig. A–8b. For the calculation, the centroid of each of these rectangles is located in the figure. From the table on the inside front cover, the moment of inertia of a rectangle about its centroidal axis is $I = \frac{1}{12}bh^3$. Hence, using the parallel-axis theorem for rectangles A and D, the computations are as follows:

Rectangle A:

$$I_x = \bar{I}_{x'} + A d_y^2 = \frac{1}{12}(100 \text{ mm})(300 \text{ mm})^3 + (100 \text{ mm})(300 \text{ mm})(200 \text{ mm})^2$$

$$= 1.425(10^9) \text{ mm}^4$$

$$I_y = \bar{I}_{y'} + A d_x^2 = \frac{1}{12}(300 \text{ mm})(100 \text{ mm})^3 + (100 \text{ mm})(300 \text{ mm})(250 \text{ mm})^2$$

$$= 1.90(10^9) \text{ mm}^4$$

Rectangle B:

$$I_x = \frac{1}{12}(600 \text{ mm})(100 \text{ mm})^3 = 0.05(10^9) \text{ mm}^4$$

$$I_y = \frac{1}{12}(100 \text{ mm})(600 \text{ mm})^3 = 1.80(10^9) \text{ mm}^4$$

Rectangle D:

$$I_x = \bar{I}_{x'} + A d_y^2 = \frac{1}{12}(100 \text{ mm})(300 \text{ mm})^3 + (100 \text{ mm})(300 \text{ mm})(200 \text{ mm})^2$$

$$= 1.425(10^9) \text{ mm}^4$$

$$I_y = \bar{I}_{y'} + A d_x^2 = \frac{1}{12}(300 \text{ mm})(100 \text{ mm})^3 + (100 \text{ mm})(300 \text{ mm})(250 \text{ mm})^2$$

$$= 1.90(10^9) \text{ mm}^4$$

The moments of inertia for the entire cross section are thus

$$I_x = 1.425(10^9) + 0.05(10^9) + 1.425(10^9)$$
$$= 2.90(10^9) \text{ mm}^4 \qquad \qquad \textit{Ans.}$$
$$I_y = 1.90(10^9) + 1.80(10^9) + 1.90(10^9)$$
$$= 5.60(10^9) \text{ mm}^4 \qquad \qquad \textit{Ans.}$$

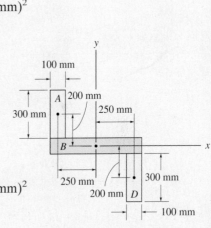

(a)

(b)

Fig. A–8

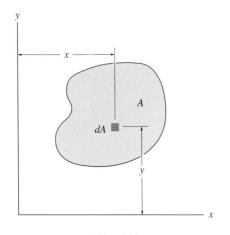

Fig. A–9

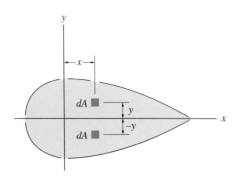

Fig. A–10

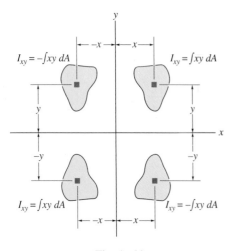

Fig. A–11

A.3 PRODUCT OF INERTIA FOR AN AREA

In general, the moment of inertia for an area is different for every axis about which it is computed. In some applications of mechanical or structural design it is necessary to know the orientation of those axes that give, respectively, the maximum and minimum moments of inertia for the area. The method for determining this is discussed in Sec. A.4. To use this method, however, one must first compute the product of inertia for the area as well as its moments of inertia for given x, y axes.

The ***product of inertia*** for the differential element dA in Fig. A–9, which is located at point (x, y), is defined as $dI_{xy} = xy\, dA$. Thus, for the entire area A, the product of inertia is

$$I_{xy} = \int_A xy\, dA \qquad\qquad (A-8)$$

Like the moment of inertia, the product of inertia has units of length raised to the fourth power, e.g., m^4, mm^4 or ft^4, in^4. However, since x or y may be a negative quantity, while the element of area is always positive, the product of inertia may be positive, negative, or zero, depending on the location and orientation of the coordinate axes. For example, the product of inertia I_{xy} for an area will be *zero* if either the x or y axis is an axis of *symmetry* for the area. To show this, consider the shaded area in Fig. A–10, where for every element dA located at point (x, y) there is a corresponding element dA located at $(x, -y)$. Since the products of inertia for these elements are, respectively, $xy\, dA$ and $-xy\, dA$, their algebraic sum or the integration of all the elements of area chosen in this way will cancel each other. Consequently, the product of inertia for the total area becomes zero. It also follows from the definition of I_{xy} that the "sign" of this quantity depends on the quadrant where the area is located. As shown in Fig. A–11, the sign of I_{xy} will change as the area is rotated from one quadrant to the next.

Parallel-Axis Theorem.   Consider the shaded area shown in Fig. A–12, where x' and y' represent a set of centroidal axes, and x and y represent a corresponding set of parallel axes. Since the product of inertia of dA with respect to the x and y axes is $dI_{xy} = (x' + d_x)(y' + d_y)\, dA$, then for the entire area,

$$I_{xy} = \int_A (x' + d_x)(y' + d_y)\, dA$$
$$= \int_A x'y'\, dA + d_x \int_A y'\, dA + d_y \int_A x'\, dA + d_x d_y \int_A dA$$

The first term on the right represents the product of inertia of the area with respect to the centroidal axis, $\bar{I}_{x'y'}$. The second and third terms are zero since the moments of the area are taken about the centroidal axis. Realizing that the fourth integral represents the total area A, we therefore have as the final result

$$\boxed{I_{xy} = \bar{I}_{x'y'} + Ad_xd_y} \qquad (A\text{-}9)$$

The similarity between this equation and the parallel-axis theorem for moments of inertia should be noted. In particular, it is important that the *algebraic signs* for d_x and d_y be maintained when applying Eq. A–9. As illustrated in the following example, the parallel-axis theorem finds important application in determining the product of inertia of a *composite area* with respect to a set of x, y axes.

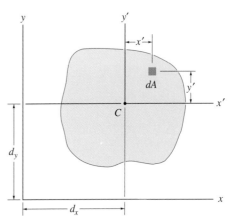

Fig. A–12

EXAMPLE A-4

Determine the product of inertia of the beam's cross-sectional area, shown in Fig. A–13a, about the x and y centroidal axes.

SOLUTION

As in Example A–3, the cross section can be considered as three composite rectangular areas A, B, and D, Fig. A–13b. The coordinates for the centroid of each of these rectangles are shown in the figure. Due to symmetry, the product of inertia of *each rectangle is zero* about a set of x', y' axes that pass through the rectangle's centroid. Hence, application of the parallel-axis theorem to each of the rectangles yields

Rectangle A:
$$I_{xy} = \bar{I}_{x'y'} + Ad_xd_y$$
$$= 0 + (300 \text{ mm})(100 \text{ mm})(-250 \text{ mm})(200 \text{ mm})$$
$$= -1.50(10^9) \text{ mm}^4$$

Rectangle B:
$$I_{xy} = \bar{I}_{x'y'} + Ad_xd_y$$
$$= 0 + 0$$
$$= 0$$

Rectangle D:
$$I_{xy} = \bar{I}_{x'y'} + Ad_xd_y$$
$$= 0 + (300 \text{ mm})(100 \text{ mm})(250 \text{ mm})(-200 \text{ mm})$$
$$= -1.50(10^9) \text{ mm}^4$$

The product of inertia for the entire cross section is thus
$$I_{xy} = [-1.50(10^9)] + 0 + [-1.50(10^9)]$$
$$= -3.00(10^9) \text{ mm}^4 \qquad \qquad Ans.$$

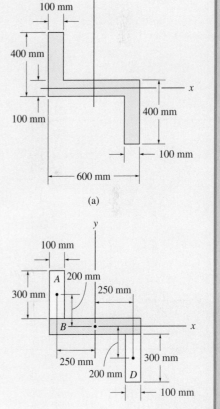

(a)

(b)

Fig. A–13

A.4   MOMENTS OF INERTIA FOR AN AREA ABOUT INCLINED AXES

In mechanical or structural design, it is sometimes necessary to calculate the moments and product of inertia $I_{x'}, I_{y'}$ and $I_{x'y'}$ for an area with respect to a set of inclined x' and y' axes when the values for θ, I_x, I_y, and I_{xy} are *known*. As shown in Fig. A–14, the coordinates to the area element dA from the two coordinate systems are related by the *transformation equations*

$$x' = x \cos \theta + y \sin \theta$$
$$y' = y \cos \theta - x \sin \theta$$

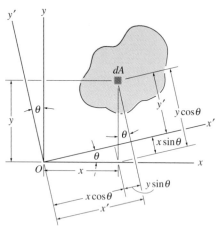

Fig. A–14

Using these equations, the moments and product of inertia of dA about the x' and y' axes become

$$dI_{x'} = y'^2 \, dA = (y \cos \theta - x \sin \theta)^2 \, dA$$
$$dI_{y'} = x'^2 \, dA = (x \cos \theta + y \sin \theta)^2 \, dA$$
$$dI_{x'y'} = x'y' \, dA = (x \cos \theta + y \sin \theta)(y \cos \theta - x \sin \theta) \, dA$$

Expanding each expression and integrating, realizing that $I_x = \int y^2 \, dA$, $I_y = \int x^2 \, dA$, and $I_{xy} = \int xy \, dA$, we obtain

$$I_{x'} = I_x \cos^2 \theta + I_y \sin^2 \theta - 2I_{xy} \sin \theta \cos \theta$$
$$I_{y'} = I_x \sin^2 \theta + I_y \cos^2 \theta + 2I_{xy} \sin \theta \cos \theta$$
$$I_{x'y'} = I_x \sin \theta \cos \theta - I_y \sin \theta \cos \theta + I_{xy}(\cos^2 \theta - \sin^2 \theta)$$

These equations may be simplified by using the trigonometric identities $\sin 2\theta = 2 \sin \theta \cos \theta$ and $\cos 2\theta = \cos^2 \theta - \sin^2 \theta$, in which case

$$I_{x'} = \frac{I_x + I_y}{2} + \frac{I_x - I_y}{2} \cos 2\theta - I_{xy} \sin 2\theta$$

$$I_{y'} = \frac{I_x + I_y}{2} - \frac{I_x - I_y}{2} \cos 2\theta + I_{xy} \sin 2\theta \qquad \text{(A–10)}$$

$$I_{x'y'} = \frac{I_x - I_y}{2} \sin 2\theta + I_{xy} \cos 2\theta$$

Note that if the first and second equations are added together, it is seen that the polar moment of inertia about the z axis passing through point O is *independent* of the orientation of the x' and y' axes; i.e.,

$$J_O = I_{x'} + I_{y'} = I_x + I_y$$

Principal Moments of Inertia.   From Eq. A–10, it may be seen that $I_{x'}$, $I_{y'}$, and $I_{x'y'}$ depend on the angle of inclination, θ, of the x', y' axes. We will now determine the orientation of these axes about which the moments of inertia for the area, $I_{x'}$ and $I_{y'}$, are maximum and minimum. This particular set of axes is called the **principal axes** of inertia for the area, and the corresponding moments of inertia with respect to these axes are called the *principal moments of inertia*. In general, there is a set of principal axes for every chosen origin O; however, the area's centroid is normally the most important location for O.

The angle $\theta = \theta_p$, which defines the orientation of the principal axes for the area, may be found by differentiating the first of Eq. A–10 with respect to θ and setting the result equal to zero. Thus,

$$\frac{dI_{x'}}{d\theta} = -2\left(\frac{I_x - I_y}{2}\right)\sin 2\theta - 2\,I_{xy}\cos 2\theta = 0$$

Therefore, at $\theta = \theta_p$,

$$\tan 2\theta_p = \frac{-I_{xy}}{(I_x - I_y)/2} \tag{A–11}$$

This equation has two roots, θ_{p_1} and θ_{p_2}, which are $90°$ apart and so specify the inclination of each principal axis.

The sine and cosine of $2\theta_{p_1}$ and $2\theta_{p_2}$ can be obtained from the triangles shown in Fig. A–15, which are based on Eq. A–11. If these trigonometric relations are substituted into the first or second of Eq. A–10 and simplified, the result is

$$I_{\substack{max \\ min}} = \frac{I_x + I_y}{2} \pm \sqrt{\left(\frac{I_x - I_y}{2}\right)^2 + I_{xy}^2} \tag{A–12}$$

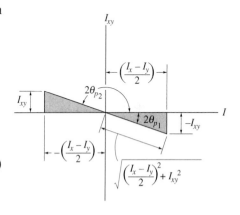

Fig. A–15

Depending on the sign chosen, this result gives the maximum or minimum moment of inertia for the area. Furthermore, if the above trigonometric relations for θ_{p_1} and θ_{p_2} are substituted into the third of Eq. A–10, it will be seen that $I_{x'y'} = 0$; that is, the *product of inertia with respect to the principal axes is zero*. Since it was indicated in Sec. A.3 that the product of inertia is zero with respect to any symmetrical axis, it therefore follows that *any symmetrical axis represents a principal axis of inertia for the area*.

Also, notice that the equations derived in this section are similar to those for stress and strain transformation developed in Chapters 9 and 10, respectively. The following example illustrates their application.

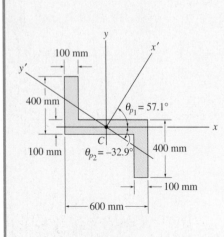

Fig. A–16

EXAMPLE A–5

Determine the principal moments of inertia for the beam's cross-sectional area shown in Fig. A–16 with respect to an axis passing through the centroid C.

SOLUTION

The moments and product of inertia of the cross section with respect to the x, y axes have been computed in Examples A–3 and A–4. The results are

$$I_x = 2.90(10^9) \text{ mm}^4 \quad I_y = 5.60(10^9) \text{ mm}^4 \quad I_{xy} = -3.00(10^9) \text{ mm}^4$$

Using Eq. A–11, the angles of inclination of the principal axes x' and y' are

$$\tan 2\theta_p = \frac{-I_{xy}}{(I_x - I_y)/2} = \frac{3.00(10^9)}{[2.90(10^9) - 5.60(10^9)]/2} = -2.22$$

$$2\theta_{p_1} = 114.2° \quad \text{and} \quad 2\theta_{p_2} = -65.8°$$

Thus, as shown in Fig. A–16,

$$\theta_{p_1} = 57.1° \quad \text{and} \quad \theta_{p_2} = -32.9°$$

The principal moments of inertia with respect to the x' and y' axes are determined by using Eq. A–12. Hence,

$$I_{\substack{max \\ min}} = \frac{I_x + I_y}{2} \pm \sqrt{\left(\frac{I_x - I_y}{2}\right)^2 + I_{xy}^2}$$

$$= \frac{2.90(10^9) + 5.60(10^9)}{2} \pm \sqrt{\left[\frac{2.90(10^9) - 5.60(10^9)}{2}\right]^2 + [-3.00(10^9)]^2}$$

$$= 4.25(10^9) \pm 3.29(10^9)$$

or

$$I_{max} = 7.54(10^9) \text{ mm}^4 \quad I_{min} = 0.960(10^9) \text{ mm}^4 \qquad \textit{Ans.}$$

Specifically, the maximum moment of inertia, $I_{max} = 7.54(10^9)$ mm^4, occurs with respect to the x' axis (major axis), since *by inspection* most of the cross-sectional area is farthest away from this axis. To prove this, substitute the data with $\theta = 57.1°$ into the first of Eq. A–10.

A.5 MOHR'S CIRCLE FOR MOMENTS OF INERTIA

Equations A–10 through A–12 have a graphical solution that is convenient to use and generally easy to remember. Squaring the first and third of Eq. A–10 and adding, it is found that

$$\left(I_{x'} - \frac{I_x + I_y}{2}\right)^2 + I_{x'y'}^2 = \left(\frac{I_x - I_y}{2}\right)^2 + I_{xy}^2 \qquad (A–13)$$

In any given problem, $I_{x'}$ and $I_{x'y'}$ are *variables*, and I_x, I_y, and I_{xy} are *known constants*. Thus, Eq. A–13 may be written in compact form as

$$(I_{x'} - a)^2 + I_{x'y'}^2 = R^2$$

When this equation is plotted, the resulting graph represents a *circle* of radius

$$R = \sqrt{\left(\frac{I_x - I_y}{2}\right)^2 + I_{xy}^2}$$

having its center located at point $(a, 0)$, where $a = (I_x + I_y)/2$. The circle so constructed is called *Mohr's circle*. Its application is similar to that used for stress and strain transformation developed in Chapters 9 and 10, respectively.

PROCEDURE FOR ANALYSIS

The main purpose for using Mohr's circle here is to have a convenient means of transforming I_x, I_y, and I_{xy} into the principal moments of inertia. The following procedure provides a method for doing this.

Compute I_x, I_y, I_{xy}. Establish the x, y axes for the area, with the origin located at the point P of interest, usually the centroid, and determine I_x, I_y, and I_{xy}, Fig. A–17a.

Construct the Circle. Establish a rectangular coordinate system such that the abscissa represents the moment of inertia I, and the ordinate represents the product of inertia I_{xy}, Fig. A–17b. Determine the center of the circle, C, which is located a distance $(I_x + I_y)/2$ from the origin, and plot the "reference point" A having coordinates (I_x, I_{xy}). By definition, I_x is always positive, whereas I_{xy} will be either positive or negative. Connect the reference point A with the center of the circle, and determine the distance CA by trigonometry. This distance represents the radius of the circle, Fig. A–17b. Finally, draw the circle.

Principal Moments of Inertia. The points where the circle intersects the abscissa give the values of the principal moments of inertia I_{min} and I_{max}. Notice that the *product of inertia will be zero at these points*, Fig. A–17b.

To find the direction of the major principal axis, determine by trigonometry the angle $2\theta_{p_1}$, *measured from the radius CA to the positive I axis*, Fig. A–17b. This angle represents twice the angle from the x axis to the axis of maximum moment of inertia I_{max}, Fig. A–17a. Both the angle on the circle, $2\theta_{p_1}$, and the angle on the area, θ_{p_1}, *must be measured in the same sense*, as shown in Fig. A–17. The minor axis is for minimum moment of inertia I_{min}, which is perpendicular to the major axis defining I_{max}.

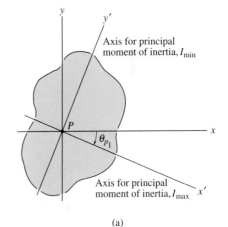

(a)

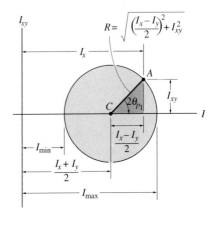

(b)

Fig. A–17

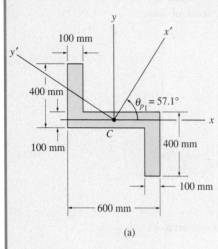

(a)

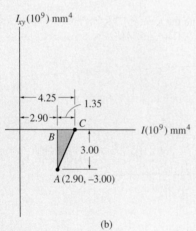

(b)

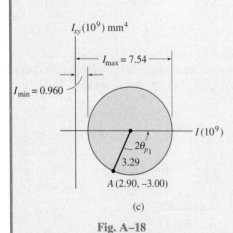

(c)

Fig. A–18

EXAMPLE A-6

Use Mohr's circle to determine the principal moments of inertia for the beam's cross-sectional area, shown in Fig. A–18a, with respect to axes passing through the centroid C.

SOLUTION

Compute I_x, I_y, I_{xy}. The moments of inertia and the product of inertia have been determined in Examples A–3 and A–4 with respect to the x, y axes shown in Fig. A–18a. The results are $I_x = 2.90(10^9)$ mm^4, $I_y = 5.60(10^9)$ mm^4, and $I_{xy} = -3.00(10^9)$ mm^4.

Construct the Circle. The I and I_{xy} axes are shown in Fig. A–18b. The center of the circle, C, lies at a distance $(I_x + I_y)/2 = (2.90 + 5.60)/2 = 4.25$ from the origin. When the reference point $A(2.90, -3.00)$ is connected to point C, the radius CA is determined from the shaded triangle CBA using the Pythagorean theorem:

$$CA = \sqrt{(1.35)^2 + (-3.00)^2} = 3.29$$

The circle is constructed in Fig. A–18c.

Principal Moments of Inertia. The circle intersects the I axis at points $(7.54, 0)$ and $(0.960, 0)$. Hence

$$I_{max} = 7.54(10^9) \text{ mm}^4 \qquad \text{Ans.}$$
$$I_{min} = 0.960(10^9) \text{ mm}^4 \qquad \text{Ans.}$$

As shown in Fig. A–18c, the angle $2\theta_{p_1}$ is determined from the circle by measuring *counterclockwise* from CA to the direction of the *positive I* axis. Hence,

$$2\theta_{p_1} = 180° - \tan^{-1}\left(\frac{|BA|}{|BC|}\right) = 180° - \tan^{-1}\left(\frac{3.00}{1.35}\right) = 114.2°$$

The major principal axis (for $I_{max} = 7.54(10^9)$ mm^4) is therefore oriented at an angle $\theta_{p_1} = 57.1°$, measured *counterclockwise*, from *positive x* axis. The minor axis is perpendicular to this axis. The results are shown in Fig. A–18a.

PROBLEMS

A–1. Determine the distance $\bar{y}$ to the centroid C of the beam's cross-sectional area. The beam is symmetric with respect to the y axis.

A–2. Determine I_x and I_y for the beam's cross-sectional area.

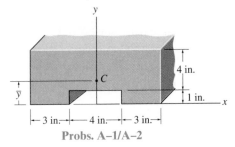

Probs. A–1/A–2

A–3. Determine the distance $\bar{y}$ to the centroid C of the beam's cross-sectional area, and then find $\bar{I}_{x'}$.

*A–4. Determine $\bar{I}_y$ for the beam's cross-sectional area.

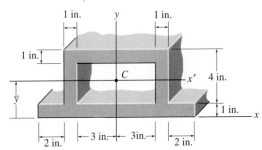

Probs. A–3/A–4

A–5. Determine $\bar{y}$, which locates the centroid, and then find the moments of inertia $\bar{I}_{x'}$ and $\bar{I}_y$ for the T-beam.

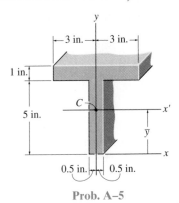

Prob. A–5

A–6. Determine $\bar{y}$, which locates the centroid C, and then compute the moments of inertia $\bar{I}_{x'}$ and $\bar{I}_{y'}$ for the cross-sectional area.

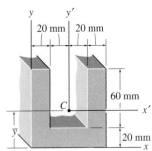

Prob. A–6

A–7. Determine the location $(\bar{x}, \bar{y})$ of the centroid C of the angle's cross-sectional area, and then find the moments of inertia $\bar{I}_{x'}$ and $\bar{I}_{y'}$.

*A–8. Determine the location $(\bar{x}, \bar{y})$ of the centroid C of the angle's cross-sectional area, and then find the product of inertia $\bar{I}_{x'y'}$ with respect to the x' and y' axes.

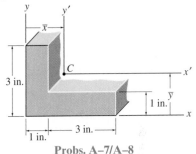

Probs. A–7/A–8

A–9. Determine the location $(\bar{x}, \bar{y})$ of the centroid C of the angle's cross-sectional area, and then compute the moments of inertia $\bar{I}_{x'}$ and $\bar{I}_{y'}$.

A–10. Determine the location $(\bar{x}, \bar{y})$ of the centroid C of the angle's cross-sectional area, and then compute the product of inertia $\bar{I}_{x'y'}$ with respect to the x' and y' axes.

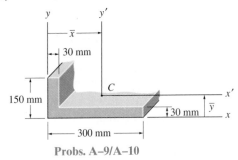

Probs. A–9/A–10

A–11. Determine the location $(\bar{x}, \bar{y})$ of the channel's cross-sectional area, and then determine the moments of inertia $\bar{I}_{x'}$ and $\bar{I}_{y'}$.

*A–12. Determine the location $(\bar{x}, \bar{y})$ of the channel's cross-sectional area, and then determine the product of inertia $\bar{I}_{x'y'}$ with respect to the x' and y' axes.

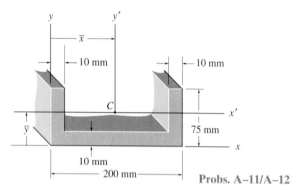

Probs. A–11/A–12

A–13. Determine the moments of inertia $\bar{I}_x$ and $\bar{I}_y$ of the Z-section. The origin of coordinates is at the centroid C.

A–14. Determine the product of inertia $\bar{I}_{xy}$ of the cross-sectional area of the Z-section. The origin of coordinates is at the centroid C.

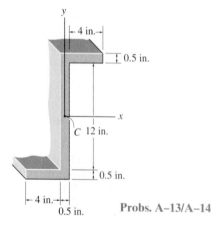

Probs. A–13/A–14

A–15. Determine the location $(\bar{x}, \bar{y})$ of the centroid C of the cross-sectional area, and then determine the moments of inertia $\bar{I}_{x'}$ and $\bar{I}_{y'}$ with respect to the x' and y' axes that have their origin located at the centroid C.

*A–16. Determine the location $(\bar{x}, \bar{y})$ of the centroid C of the cross-sectional area, then determine the product of inertia $\bar{I}_{x'y'}$ with respect to the x' and y' axes.

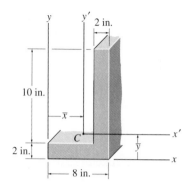

Probs. A–15/A–16

A–17. Determine the product of inertia I_{xy} of the cross-sectional area with respect to the x and y axes.

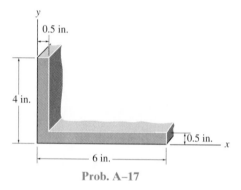

Prob. A–17

A–18. Compute the moments of inertia $I_{x'}$ and $I_{y'}$ and the product of inertia $I_{x'y'}$ of the cross-sectional area. Use the equations of Sec. A.4.

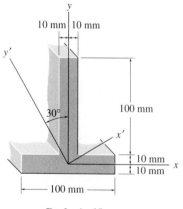

Prob. A–18

A–19. Determine the moments of inertia $I_{x'}$ and $I_{y'}$ and the product of inertia $I_{x'y'}$ for the rectangular area. The x' and y' axes pass through the centroid C. Use the equations of Sec. A.4.

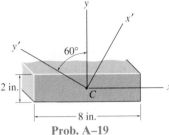

Prob. A–19

*A–20. Determine the location $(\bar{x}, \bar{y})$ of the centroid C of the angle's cross-sectional area, then find the product of inertia $I_{x'y'}$. Use the equations of Sec. A.4.

A–21. Determine the location $(\bar{x}, \bar{y})$ of the centroid C of the angle's cross-sectional area, then find the principal moments of inertia and the orientation of the principal axes of inertia with respect to the x and y axes.

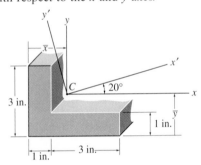

Probs. A–20/A–21

A–22. Determine the moments of inertia $I_{x'}$ and $I_{y'}$ and the product of inertia $I_{x'y'}$ for the semicircular area.

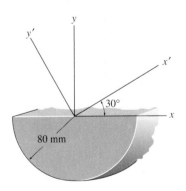

Prob. A–22

A–23. Determine the principal moments of inertia and the orientation of the principal axes of inertia for the angle's cross-sectional area with respect to a set of principal axes that have their origin located at the centroid C. Use the equations developed in Sec. A.4.

*A–24. Solve Prob. A–23 using Mohr's circle.

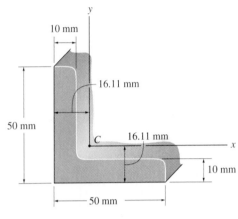

Probs. A–23/A–24

A–25. Determine the principal moments of inertia and the orientation of the principal axes of inertia of the cross-sectional area that have their origin located at the centroid C. Use the equations developed in Sec. A.4.

A–26. Solve Prob. A–25 using Mohr's circle.

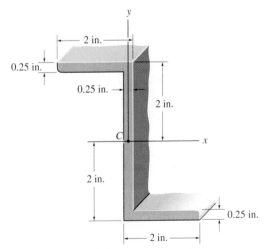

Probs. A–25/A–26

B Geometric Properties of Structural Shapes

Wide-Flange Sections or W Shapes   FPS Units

Designation	Area A	Depth d	Web thickness t_w	Flange width b_f	Flange thickness t_f	$x-x$ axis I	$x-x$ axis S	$x-x$ axis r	$y-y$ axis I	$y-y$ axis S	$y-y$ axis r
in. $\times$ lb/ft	in^2	in.	in.	in.	in.	in^4	in^3	in.	in^4	in^3	in.
W24 $\times$ 104	30.6	24.06	0.500	12.750	0.750	3100	258	10.1	259	40.7	2.91
W24 $\times$ 94	27.7	24.31	0.515	9.065	0.875	2700	222	9.87	109	24.0	1.98
W24 $\times$ 84	24.7	24.10	0.470	9.020	0.770	2370	196	9.79	94.4	20.9	1.95
W24 $\times$ 76	22.4	23.92	0.440	8.990	0.680	2100	176	9.69	82.5	18.4	1.92
W24 $\times$ 68	20.1	23.73	0.415	8.965	0.585	1830	154	9.55	70.4	15.7	1.87
W24 $\times$ 62	18.2	23.74	0.430	7.040	0.590	1550	131	9.23	34.5	9.80	1.38
W24 $\times$ 55	16.2	23.57	0.395	7.005	0.505	1350	114	9.11	29.1	8.30	1.34
W18 $\times$ 65	19.1	18.35	0.450	7.590	0.750	1070	117	7.49	54.8	14.4	1.69
W18 $\times$ 60	17.6	18.24	0.415	7.555	0.695	984	108	7.47	50.1	13.3	1.69
W18 $\times$ 55	16.2	18.11	0.390	7.530	0.630	890	98.3	7.41	44.9	11.9	1.67
W18 $\times$ 50	14.7	17.99	0.355	7.495	0.570	800	88.9	7.38	40.1	10.7	1.65
W18 $\times$ 46	13.5	18.06	0.360	6.060	0.605	712	78.8	7.25	22.5	7.43	1.29
W18 $\times$ 40	11.8	17.90	0.315	6.015	0.525	612	68.4	7.21	19.1	6.35	1.27
W18 $\times$ 35	10.3	17.70	0.300	6.000	0.425	510	57.6	7.04	15.3	5.12	1.22
W16 $\times$ 57	16.8	16.43	0.430	7.120	0.715	758	92.2	6.72	43.1	12.1	1.60
W16 $\times$ 50	14.7	16.26	0.380	7.070	0.630	659	81.0	6.68	37.2	10.5	1.59
W16 $\times$ 45	13.3	16.13	0.345	7.035	0.565	586	72.7	6.65	32.8	9.34	1.57
W16 $\times$ 36	10.6	15.86	0.295	6.985	0.430	448	56.5	6.51	24.5	7.00	1.52
W16 $\times$ 31	9.12	15.88	0.275	5.525	0.440	375	47.2	6.41	12.4	4.49	1.17
W16 $\times$ 26	7.68	15.69	0.250	5.500	0.345	301	38.4	6.26	9.59	3.49	1.12
W14 $\times$ 53	15.6	13.92	0.370	8.060	0.660	541	77.8	5.89	57.7	14.3	1.92
W14 $\times$ 43	12.6	13.66	0.305	7.995	0.530	428	62.7	5.82	45.2	11.3	1.89
W14 $\times$ 38	11.2	14.10	0.310	6.770	0.515	385	54.6	5.87	26.7	7.88	1.55
W14 $\times$ 34	10.0	13.98	0.285	6.745	0.455	340	48.6	5.83	23.3	6.91	1.53
W14 $\times$ 30	8.85	13.84	0.270	6.730	0.385	291	42.0	5.73	19.6	5.82	1.49
W14 $\times$ 26	7.69	13.91	0.255	5.025	0.420	245	35.3	5.65	8.91	3.54	1.08
W14 $\times$ 22	6.49	13.74	0.230	5.000	0.335	199	29.0	5.54	7.00	2.80	1.04

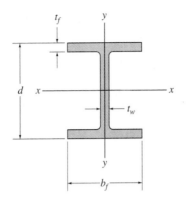

Wide-Flange Sections or W Shapes   FPS Units

Designation	Area	Depth	Web thickness	Flange width	Flange thickness	$x-x$ axis			$y-y$ axis		
	A	d	t_w	b_f	t_f	I	S	r	I	S	r
in. $\times$ lb/ft	in^2	in.	in.	in.	in.	in^4	in^3	in.	in^4	in^3	in.
W12 $\times$ 87	25.6	12.53	0.515	12.125	0.810	740	118	5.38	241	39.7	3.07
W12 $\times$ 50	14.7	12.19	0.370	8.080	0.640	394	64.7	5.18	56.3	13.9	1.96
W12 $\times$ 45	13.2	12.06	0.335	8.045	0.575	350	58.1	5.15	50.0	12.4	1.94
W12 $\times$ 26	7.65	12.22	0.230	6.490	0.380	204	33.4	5.17	17.3	5.34	1.51
W12 $\times$ 22	6.48	12.31	0.260	4.030	0.425	156	25.4	4.91	4.66	2.31	0.847
W12 $\times$ 16	4.71	11.99	0.220	3.990	0.265	103	17.1	4.67	2.82	1.41	0.773
W12 $\times$ 14	4.16	11.91	0.200	3.970	0.225	88.6	14.9	4.62	2.36	1.19	0.753
W10 $\times$ 100	29.4	11.10	0.680	10.340	1.120	623	112	4.60	207	40.0	2.65
W10 $\times$ 54	15.8	10.09	0.370	10.030	0.615	303	60.0	4.37	103	20.6	2.56
W10 $\times$ 45	13.3	10.10	0.350	8.020	0.620	248	49.1	4.32	53.4	13.3	2.01
W10 $\times$ 39	11.5	9.92	0.315	7.985	0.530	209	42.1	4.27	45.0	11.3	1.98
W10 $\times$ 30	8.84	10.47	0.300	5.810	0.510	170	32.4	4.38	16.7	5.75	1.37
W10 $\times$ 19	5.62	10.24	0.250	4.020	0.395	96.3	18.8	4.14	4.29	2.14	0.874
W10 $\times$ 15	4.41	9.99	0.230	4.000	0.270	68.9	13.8	3.95	2.89	1.45	0.810
W10 $\times$ 12	3.54	9.87	0.190	3.960	0.210	53.8	10.9	3.90	2.18	1.10	0.785
W8 $\times$ 67	19.7	9.00	0.570	8.280	0.935	272	60.4	3.72	88.6	21.4	2.12
W8 $\times$ 58	17.1	8.75	0.510	8.220	0.810	228	52.0	3.65	75.1	18.3	2.10
W8 $\times$ 48	14.1	8.50	0.400	8.110	0.685	184	43.3	3.61	60.9	15.0	2.08
W8 $\times$ 40	11.7	8.25	0.360	8.070	0.560	146	35.5	3.53	49.1	12.2	2.04
W8 $\times$ 31	9.13	8.00	0.285	7.995	0.435	110	27.5	3.47	37.1	9.27	2.02
W8 $\times$ 24	7.08	7.93	0.245	6.495	0.400	82.8	20.9	3.42	18.3	5.63	1.61
W8 $\times$ 15	4.44	8.11	0.245	4.015	0.315	48.0	11.8	3.29	3.41	1.70	0.876
W6 $\times$ 25	7.34	6.38	0.320	6.080	0.455	53.4	16.7	2.70	17.1	5.61	1.52
W6 $\times$ 20	5.87	6.20	0.260	6.020	0.365	41.4	13.4	2.66	13.3	4.41	1.50
W6 $\times$ 16	4.74	6.28	0.260	4.030	0.405	32.1	10.2	2.60	4.43	2.20	0.966
W6 $\times$ 15	4.43	5.99	0.230	5.990	0.260	29.1	9.72	2.56	9.32	3.11	1.46
W6 $\times$ 12	3.55	6.03	0.230	4.000	0.280	22.1	7.31	2.49	2.99	1.50	0.918
W6 $\times$ 9	2.68	5.90	0.170	3.940	0.215	1.64	5.56	2.47	2.19	1.11	0.905

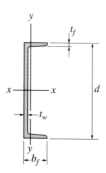

American Standard Channels or C Shapes   FPS Units

Designation	Area A	Depth d	Web thickness t_w		Flange width b_f		Flange thickness t_f		$x-x$ axis I	S	r	$y-y$ axis I	S	r
in. × lb/ft	in²	in.	in.		in.		in.		in⁴	in³	in.	in⁴	in³	in.
C15 × 50	14.7	15.00	0.716	$^{11}/_{16}$	3.716	$3\frac{3}{4}$	0.650	$\frac{5}{8}$	404	53.8	5.24	11.0	3.78	0.867
C15 × 40	11.8	15.00	0.520	$\frac{1}{2}$	3.520	$3\frac{1}{2}$	0.650	$\frac{5}{8}$	349	46.5	5.44	9.23	3.37	0.886
C15 × 33.9	9.96	15.00	0.400	$\frac{3}{8}$	3.400	$3\frac{3}{8}$	0.650	$\frac{5}{8}$	315	42.0	5.62	8.13	3.11	0.904
C12 × 30	8.82	12.00	0.510	$\frac{1}{2}$	3.170	$3\frac{1}{8}$	0.501	$\frac{1}{2}$	162	27.0	4.29	5.14	2.06	0.763
C12 × 25	7.35	12.00	0.387	$\frac{3}{8}$	3.047	3	0.501	$\frac{1}{2}$	144	24.1	4.43	4.47	1.88	0.780
C12 × 20.7	6.09	12.00	0.282	$^{5}/_{16}$	2.942	3	0.501	$\frac{1}{2}$	129	21.5	4.61	3.88	1.73	0.799
C10 × 30	8.82	10.00	0.673	$^{11}/_{16}$	3.033	3	0.436	$^{7}/_{16}$	103	20.7	3.42	3.94	1.65	0.669
C10 × 25	7.35	10.00	0.526	$\frac{1}{2}$	2.886	$2\frac{7}{8}$	0.436	$^{7}/_{16}$	91.2	18.2	3.52	3.36	1.48	0.676
C10 × 20	5.88	10.00	0.379	$\frac{3}{8}$	2.739	$2\frac{3}{4}$	0.436	$^{7}/_{16}$	78.9	15.8	3.66	2.81	1.32	0.692
C10 × 15.3	4.49	10.00	0.240	$\frac{1}{4}$	2.600	$2\frac{5}{8}$	0.436	$^{7}/_{16}$	67.4	13.5	3.87	2.28	1.16	0.713
C9 × 20	5.88	9.00	0.448	$^{7}/_{16}$	2.648	$2\frac{5}{8}$	0.413	$^{7}/_{16}$	60.9	13.5	3.22	2.42	1.17	0.642
C9 × 15	4.41	9.00	0.285	$^{5}/_{16}$	2.485	$2\frac{1}{2}$	0.413	$^{7}/_{16}$	51.0	11.3	3.40	1.93	1.01	0.661
C9 × 13.4	3.94	9.00	0.233	$\frac{1}{4}$	2.433	$2\frac{3}{8}$	0.413	$^{7}/_{16}$	47.9	10.6	3.48	1.76	0.962	0.669
C8 × 18.75	5.51	8.00	0.487	$\frac{1}{2}$	2.527	$2\frac{1}{2}$	0.390	$\frac{3}{8}$	44.0	11.0	2.82	1.98	1.01	0.599
C8 × 13.75	4.04	8.00	0.303	$^{5}/_{16}$	2.343	$2\frac{3}{8}$	0.390	$\frac{3}{8}$	36.1	9.03	2.99	1.53	0.854	0.615
C8 × 11.5	3.38	8.00	0.220	$\frac{1}{4}$	2.260	$2\frac{1}{4}$	0.390	$\frac{3}{8}$	32.6	8.14	3.11	1.32	0.781	0.625
C7 × 14.75	4.33	7.00	0.419	$^{7}/_{16}$	2.299	$2\frac{1}{4}$	0.366	$\frac{3}{8}$	27.2	7.78	2.51	1.38	0.779	0.564
C7 × 12.25	3.60	7.00	0.314	$^{5}/_{16}$	2.194	$2\frac{1}{4}$	0.366	$\frac{3}{8}$	24.2	6.93	2.60	1.17	0.703	0.571
C7 × 9.8	2.87	7.00	0.210	$^{3}/_{16}$	2.090	$2\frac{1}{8}$	0.366	$\frac{3}{8}$	21.3	6.08	2.72	0.968	0.625	0.581
C6 × 13	3.83	6.00	0.437	$^{7}/_{16}$	2.157	$2\frac{1}{8}$	0.343	$^{5}/_{16}$	17.4	5.80	2.13	1.05	0.642	0.525
C6 × 10.5	3.09	6.00	0.314	$^{5}/_{16}$	2.034	2	0.343	$^{5}/_{16}$	15.2	5.06	2.22	0.866	0.564	0.529
C6 × 8.2	2.40	6.00	0.200	$^{3}/_{16}$	1.920	$1\frac{7}{8}$	0.343	$^{5}/_{16}$	13.1	4.38	2.34	0.693	0.492	0.537
C5 × 9	2.64	5.00	0.325	$^{5}/_{16}$	1.885	$1\frac{7}{8}$	0.320	$^{5}/_{16}$	8.90	3.56	1.83	0.632	0.450	0.489
C5 × 6.7	1.97	5.00	0.190	$^{3}/_{16}$	1.750	$1\frac{3}{4}$	0.320	$^{5}/_{16}$	7.49	3.00	1.95	0.479	0.378	0.493
C4 × 7.25	2.13	4.00	0.321	$^{5}/_{16}$	1.721	$1\frac{3}{4}$	0.296	$^{5}/_{16}$	4.59	2.29	1.47	0.433	0.343	0.450
C4 × 5.4	1.59	4.00	0.184	$^{3}/_{16}$	1.584	$1\frac{5}{8}$	0.296	$^{5}/_{16}$	3.85	1.93	1.56	0.319	0.283	0.449
C3 × 6	1.76	3.00	0.356	$\frac{3}{8}$	1.596	$1\frac{5}{8}$	0.273	$\frac{1}{4}$	2.07	1.38	1.08	0.305	0.268	0.416
C3 × 5	1.47	3.00	0.258	$\frac{1}{4}$	1.498	$1\frac{1}{2}$	0.273	$\frac{1}{4}$	1.85	1.24	1.12	0.247	0.233	0.410
C3 × 4.1	1.21	3.00	0.170	$^{3}/_{16}$	1.410	$1\frac{3}{8}$	0.273	$\frac{1}{4}$	1.66	1.10	1.17	0.197	0.202	0.404

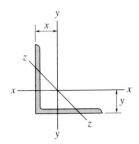

Angles Having Equal Legs   FPS Units

Size and thickness	Weight per foot	Area A	x–x axis					y–y axis				z–z axis
			I	S	r	y	I	S	r	x	r
in.	lb	in^2	in^4	in^3	in.	in.	in^4	in^3	in.	in.	in.
L 8 × 8 × 1	51.0	15.0	89.0	15.8	2.44	2.37	89.0	15.8	2.44	2.37	1.56
L 8 × 8 × ¾	38.9	11.4	69.7	12.2	2.47	2.28	69.7	12.2	2.47	2.28	1.58
L 8 × 8 × ½	26.4	7.75	48.6	8.36	2.50	2.19	48.6	8.36	2.50	2.19	1.59
L 6 × 6 × 1	37.4	11.0	35.5	8.57	1.80	1.86	35.5	8.57	1.80	1.86	1.17
L 6 × 6 × ¾	28.7	8.44	28.2	6.66	1.83	1.78	28.2	6.66	1.83	1.78	1.17
L 6 × 6 × ½	19.6	5.75	19.9	4.61	1.86	1.68	19.9	4.61	1.86	1.68	1.18
L 6 × 6 × ⅜	14.9	4.36	15.4	3.53	1.88	1.64	15.4	3.53	1.88	1.64	1.19
L 5 × 5 × ¾	23.6	6.94	15.7	4.53	1.51	1.52	15.7	4.53	1.51	1.52	0.975
L 5 × 5 × ½	16.2	4.75	11.3	3.16	1.54	1.43	11.3	3.16	1.54	1.43	0.983
L 5 × 5 × ⅜	12.3	3.61	8.74	2.42	1.56	1.39	8.74	2.42	1.56	1.39	0.990
L 4 × 4 × ¾	18.5	5.44	7.67	2.81	1.19	1.27	7.67	2.81	1.19	1.27	0.778
L 4 × 4 × ½	12.8	3.75	5.56	1.97	1.22	1.18	5.56	1.97	1.22	1.18	0.782
L 4 × 4 × ⅜	9.8	2.86	4.36	1.52	1.23	1.14	4.36	1.52	1.23	1.14	0.788
L 4 × 4 × ¼	6.6	1.94	3.04	1.05	1.25	1.09	3.04	1.05	1.25	1.09	0.795
L 3½ × 3½ × ½	11.1	3.25	3.64	1.49	1.06	1.06	3.64	1.49	1.06	1.06	0.683
L 3½ × 3½ × ⅜	8.5	2.48	2.87	1.15	1.07	1.01	2.87	1.15	1.07	1.01	0.687
L 3½ × 3½ × ¼	5.8	1.69	2.01	0.794	1.09	0.968	2.01	0.794	1.09	0.968	0.694
L 3 × 3 × ½	9.4	2.75	2.22	1.07	0.898	0.932	2.22	1.07	0.898	0.932	0.584
L 3 × 3 × ⅜	7.2	2.11	1.76	0.833	0.913	0.888	1.76	0.833	0.913	0.888	0.587
L 3 × 3 × ¼	4.9	1.44	1.24	0.577	0.930	0.842	1.24	0.577	0.930	0.842	0.592
L 2½ × 2½ × ½	7.7	2.25	1.23	0.724	0.739	0.806	1.23	0.724	0.739	0.806	0.487
L 2½ × 2½ × ⅜	5.9	1.73	0.984	0.566	0.753	0.762	0.984	0.566	0.753	0.762	0.487
L 2½ × 2½ × ¼	4.1	1.19	0.703	0.394	0.769	0.717	0.703	0.394	0.769	0.717	0.491
L 2 × 2 × ⅜	4.7	1.36	0.479	0.351	0.594	0.636	0.479	0.351	0.594	0.636	0.389
L 2 × 2 × ¼	3.19	0.938	0.348	0.247	0.609	0.592	0.348	0.247	0.609	0.592	0.391
L 2 × 2 × ⅛	1.65	0.484	0.190	0.131	0.626	0.546	0.190	0.131	0.626	0.546	0.398

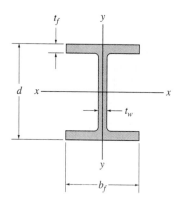

Wide-Flange Sections or W Shapes   SI Units

Designation	Area A	Depth d	Web thickness t_w	Flange width b_f	Flange thickness t_f	$x-x$ axis I	$x-x$ axis S	$x-x$ axis r	$y-y$ axis I	$y-y$ axis S	$y-y$ axis r
mm × kg/m	mm^2	mm	mm	mm	mm	10^6 mm^4	10^3 mm^3	mm	10^6 mm^4	10^3 mm^3	mm
W610 × 155	19 800	611	12.70	324.0	19.0	1 290	4 220	255	108	667	73.9
W610 × 140	17 900	617	13.10	230.0	22.2	1 120	3 630	250	45.1	392	50.2
W610 × 125	15 900	612	11.90	229.0	19.6	985	3 220	249	39.3	343	49.7
W610 × 113	14 400	608	11.20	228.0	17.3	875	2 880	247	34.3	301	48.8
W610 × 101	12 900	603	10.50	228.0	14.9	764	2 530	243	29.5	259	47.8
W610 × 92	11 800	603	10.90	179.0	15.0	646	2 140	234	14.4	161	34.9
W610 × 82	10 500	599	10.00	178.0	12.8	560	1 870	231	12.1	136	33.9
W460 × 97	12 300	466	11.40	193.0	19.0	445	1 910	190	22.8	236	43.1
W460 × 89	11 400	463	10.50	192.0	17.7	410	1 770	190	20.9	218	42.8
W460 × 82	10 400	460	9.91	191.0	16.0	370	1 610	189	18.6	195	42.3
W460 × 74	9 460	457	9.02	190.0	14.5	333	1 460	188	16.6	175	41.9
W460 × 68	8 730	459	9.14	154.0	15.4	297	1 290	184	9.41	122	32.8
W460 × 60	7 590	455	8.00	153.0	13.3	255	1 120	183	7.96	104	32.4
W460 × 52	6 640	450	7.62	152.0	10.8	212	942	179	6.34	83.4	30.9
W410 × 85	10 800	417	10.90	181.0	18.2	315	1 510	171	18.0	199	40.8
W410 × 74	9 510	413	9.65	180.0	16.0	275	1 330	170	15.6	173	40.5
W410 × 67	8 560	410	8.76	179.0	14.4	245	1 200	169	13.8	154	40.2
W410 × 53	6 820	403	7.49	177.0	10.9	186	923	165	10.1	114	38.5
W410 × 46	5 890	403	6.99	140.0	11.2	156	774	163	5.14	73.4	29.5
W410 × 39	4 960	399	6.35	140.0	8.8	126	632	159	4.02	57.4	28.5
W360 × 79	10 100	354	9.40	205.0	16.8	227	1 280	150	24.2	236	48.9
W360 × 64	8 150	347	7.75	203.0	13.5	179	1 030	148	18.8	185	48.0
W360 × 57	7 200	358	7.87	172.0	13.1	160	894	149	11.1	129	39.3
W360 × 51	6 450	355	7.24	171.0	11.6	141	794	148	9.68	113	38.7
W360 × 45	5 710	352	6.86	171.0	9.8	121	688	146	8.16	95.4	37.8
W360 × 39	4 960	363	6.48	128.0	10.7	102	578	143	3.75	58.6	27.5
W360 × 33	4 190	349	5.84	127.0	8.5	82.9	475	141	2.91	45.8	26.4

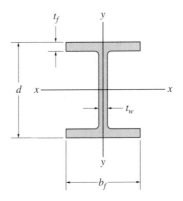

Wide-Flange Sections or W Shapes   SI Units

Designation	Area A	Depth d	Web thickness t_w	Flange width b_f	Flange thickness t_f	$x-x$ axis I	$x-x$ axis S	$x-x$ axis r	$y-y$ axis I	$y-y$ axis S	$y-y$ axis r
mm × kg/m	mm^2	mm	mm	mm	mm	10^6 mm^4	10^3 mm^3	mm	10^6 mm^4	10^3 mm^3	mm
W310 × 129	16 500	318	13.10	308.0	20.6	308	1 940	137	100	649	77.8
W310 × 74	9 480	310	9.40	205.0	16.3	165	1 060	132	23.4	228	49.7
W310 × 67	8 530	306	8.51	204.0	14.6	145	948	130	20.7	203	49.3
W310 × 39	4 930	310	5.84	165.0	9.7	84.8	547	131	7.23	87.6	38.3
W310 × 33	4 180	313	6.60	102.0	10.8	65.0	415	125	1.92	37.6	21.4
W310 × 24	3 040	305	5.59	101.0	6.7	42.8	281	119	1.16	23.0	19.5
W310 × 21	2 680	303	5.08	101.0	5.7	37.0	244	117	0.986	19.5	19.2
W250 × 149	19 000	282	17.30	263.0	28.4	259	1 840	117	86.2	656	67.4
W250 × 80	10 200	256	9.40	255.0	15.6	126	984	111	43.1	338	65.0
W250 × 67	8 560	257	8.89	204.0	15.7	104	809	110	22.2	218	50.9
W250 × 58	7 400	252	8.00	203.0	13.5	87.3	693	109	18.8	185	50.4
W250 × 45	5 700	266	7.62	148.0	13.0	71.1	535	112	7.03	95	35.1
W250 × 28	3 620	260	6.35	102.0	10.0	39.9	307	105	1.78	34.9	22.2
W250 × 22	2 850	254	5.84	102.0	6.9	28.8	227	101	1.22	23.9	20.7
W250 × 18	2 280	251	4.83	101.0	5.3	22.5	179	99.3	0.919	18.2	20.1
W200 × 100	12 700	229	14.50	210.0	23.7	113	987	94.3	36.6	349	53.7
W200 × 86	11 000	222	13.00	209.0	20.6	94.7	853	92.8	31.4	300	53.4
W200 × 71	9 100	216	10.20	206.0	17.4	76.6	709	91.7	25.4	247	52.8
W200 × 59	7 580	210	9.14	205.0	14.2	61.2	583	89.9	20.4	199	51.9
W200 × 46	5 890	203	7.24	203.0	11.0	45.5	448	87.9	15.3	151	51.0
W200 × 36	4 570	201	6.22	165.0	10.2	34.4	342	86.8	7.64	92.6	40.9
W200 × 22	2 860	206	6.22	102.0	8.0	20.0	194	83.6	1.42	27.8	22.3
W150 × 37	4 730	162	8.13	154.0	11.6	22.2	274	68.5	7.07	91.8	38.7
W150 × 30	3 790	157	6.60	153.0	9.3	17.1	218	67.2	5.54	72.4	38.2
W150 × 22	2 860	152	5.84	152.0	6.6	12.1	159	65.0	3.87	50.9	36.8
W150 × 24	3 060	160	6.60	102.0	10.3	13.4	168	66.2	1.83	35.9	24.5
W150 × 18	2 290	153	5.84	102.0	7.1	9.19	120	63.3	1.26	24.7	23.5
W150 × 14	1 730	150	4.32	100.0	5.5	6.84	91.2	62.9	0.912	18.2	23.0

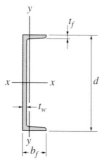

American Standard Channels or C Shapes   SI Units

Designation	Area A	Depth d	Web thickness t_w	Flange width b_f	Flange thickness t_f	x–x axis I	x–x axis Z	x–x axis S	x–x axis r	y–y axis I	y–y axis Z	y–y axis S	y–y axis r
mm × kg/m	mm²	mm	mm	mm	mm	10^6 mm⁴	10^3 mm³	10^3 mm³	mm	10^6 mm⁴	10^3 mm³	10^3 mm³	mm
C380 × 74	9 480	381.0	18.20	94.4	16.50	168	1 120	882	133	4.58	134	61.8	22.0
C380 × 60	7 610	381.0	13.20	89.4	16.50	145	937	761	138	3.84	113	55.1	22.5
C380 × 50	6 430	381.0	10.20	86.4	16.50	131	826	688	143	3.38	102	50.9	22.9
C310 × 45	5 690	305.0	13.00	80.5	12.70	67.4	551	442	109	2.14	71.0	33.8	19.4
C310 × 37	4 740	305.0	9.83	77.4	12.70	59.9	479	393	112	1.86	62.9	30.9	19.8
C310 × 31	3 930	305.0	7.16	74.7	12.70	53.7	416	352	117	1.61	57.2	28.3	20.2
C250 × 45	5 690	254.0	17.10	77.0	11.10	42.9	436	338	86.8	1.61	61.9	27.1	17.0
C250 × 37	4 740	254.0	13.40	73.3	11.10	38.0	377	299	89.5	1.40	52.3	24.3	17.2
C250 × 30	3 790	254.0	9.63	69.6	11.10	32.8	316	258	93.0	1.17	44.4	21.6	17.6
C250 × 23	2 900	254.0	6.10	66.0	11.10	28.1	259	221	98.4	0.949	38.5	19.0	18.1
C230 × 30	3 790	229.0	11.40	67.3	10.50	25.3	275	221	81.7	1.01	40.5	19.2	16.3
C230 × 22	2 850	229.0	7.24	63.1	10.50	21.2	221	185	86.2	0.803	33.6	16.7	16.8
C230 × 20	2 540	229.0	5.92	61.8	10.50	19.9	205	174	88.5	0.733	32.0	15.8	17.0
C200 × 28	3 550	203.0	12.40	64.2	9.90	18.3	226	180	71.8	0.824	35.6	16.5	15.2
C200 × 20	2 610	203.0	7.70	59.5	9.90	15.0	179	148	75.8	0.637	28.3	14.0	15.6
C200 × 17	2 180	203.0	5.59	57.4	9.90	13.6	156	134	79.0	0.549	25.9	12.8	15.9
C180 × 22	2 790	178.0	10.60	58.4	9.30	11.3	159	127	63.6	0.574	26.9	12.8	14.3
C180 × 18	2 320	178.0	7.98	55.7	9.30	10.1	138	113	66.0	0.487	23.4	11.5	14.5
C180 × 15	1 850	178.0	5.33	53.1	9.30	8.87	117	99.7	69.2	0.403	20.6	10.2	14.8
C150 × 19	2 470	152.0	11.10	54.8	8.70	7.24	119	95.3	54.1	0.437	22.3	10.5	13.3
C150 × 16	1 990	152.0	7.98	51.7	8.70	6.33	101	83.3	56.4	0.360	18.8	9.22	13.5
C150 × 12	1 550	152.0	5.08	48.8	8.70	5.45	84.1	71.7	59.3	0.288	16.3	8.04	13.6
C130 × 13	1 700	127.0	8.25	47.9	8.10	3.70	71.4	58.3	46.7	0.263	15.0	7.35	12.4
C130 × 10	1 270	127.0	4.83	44.5	8.10	3.12	57.5	49.1	49.6	0.199	12.5	6.18	12.5
C100 × 11	1 370	102.0	8.15	43.7	7.50	1.91	46.0	37.5	37.3	0.180	11.4	5.62	11.5
C100 × 8	1 030	102.0	4.67	40.2	7.50	1.60	37.0	31.4	39.4	0.133	9.32	4.65	11.4
C75 × 9	1 140	76.2	9.04	40.5	6.90	0.862	28.2	22.6	27.5	0.127	8.91	4.39	10.6
C75 × 7	948	76.2	6.55	38.0	6.90	0.770	24.6	20.2	28.5	0.103	7.64	3.83	10.4
C75 × 6	781	76.2	4.32	35.8	6.90	0.691	21.3	18.1	29.8	0.082	6.57	3.32	10.2

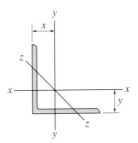

Angles Having Equal Legs   SI Units

Size and thickness	Mass per Meter	Area	x–x axis				y–y axis				z–z axis
			I	S	r	y	I	S	r	x	r
mm	kg	mm²	10^6 mm⁴	10^6 mm⁴	mm	mm	10^6 mm⁴	10^6 mm⁴	mm	mm	mm
⌐ 203 × 203 × 25.4	75.9	9 680	36.9	258	61.7	60.1	36.9	258	61.7	60.1	39.6
⌐ 203 × 203 × 19.0	57.9	7 380	28.9	199	62.6	57.8	28.9	199	62.6	57.8	40.1
⌐ 203 × 203 × 12.7	39.3	5 000	20.2	137	63.6	55.5	20.2	137	63.6	55.5	40.4
⌐ 152 × 152 × 25.4	55.7	7 100	14.6	139	45.3	47.2	14.6	139	45.3	47.2	29.7
⌐ 152 × 152 × 19.0	42.7	5 440	11.6	108	46.2	45.0	11.6	108	46.2	45.0	29.7
⌐ 152 × 152 × 12.7	29.2	3 710	8.22	75.1	47.1	42.7	8.22	75.1	47.1	42.7	30.0
⌐ 152 × 152 × 9.5	22.2	2 810	6.35	57.4	47.5	41.5	6.35	57.4	47.5	41.5	30.2
⌐ 127 × 127 × 19.0	35.1	4 480	6.54	73.9	38.2	38.7	6.54	73.9	38.2	38.7	24.8
⌐ 127 × 127 × 12.7	24.1	3 060	4.68	51.7	39.1	36.4	4.68	51.7	39.1	36.4	25.0
⌐ 127 × 127 × 9.5	18.3	2 330	3.64	39.7	39.5	35.3	3.64	39.7	39.5	35.3	25.1
⌐ 102 × 102 × 19.0	27.5	3 510	3.23	46.4	30.3	32.4	3.23	46.4	30.3	32.4	19.8
⌐ 102 × 102 × 12.7	19.0	2 420	2.34	32.6	31.1	30.2	2.34	32.6	31.1	30.2	19.9
⌐ 102 × 102 × 9.5	14.6	1 840	1.84	25.3	31.6	29.0	1.84	25.3	31.6	29.0	20.0
⌐ 102 × 102 × 6.4	9.8	1 250	1.28	17.3	32.0	27.9	1.28	17.3	32.0	27.9	20.2
⌐ 89 × 89 × 12.7	16.5	2 100	1.52	24.5	26.9	26.9	1.52	24.5	26.9	26.9	17.3
⌐ 89 × 89 × 9.5	12.6	1 600	1.20	19.0	27.4	25.8	1.20	19.0	27.4	25.8	17.4
⌐ 89 × 89 × 6.4	8.6	1 090	0.840	13.0	27.8	24.6	0.840	13.0	27.8	24.6	17.6
⌐ 76 × 76 × 12.7	14.0	1 770	0.915	17.5	22.7	23.6	0.915	17.5	22.7	23.6	14.8
⌐ 76 × 76 × 9.5	10.7	1 360	0.726	13.6	23.1	22.5	0.726	13.6	23.1	22.5	14.9
⌐ 76 × 76 × 6.4	7.3	927	0.514	9.39	23.5	21.3	0.514	9.39	23.5	21.3	15.0
⌐ 64 × 64 × 12.7	11.5	1 450	0.524	12.1	19.0	20.6	0.524	12.1	19.0	20.6	12.4
⌐ 64 × 64 × 9.5	8.8	1 120	0.420	9.46	19.4	19.5	0.420	9.46	19.4	19.5	12.4
⌐ 64 × 64 × 6.4	6.1	766	0.300	6.59	19.8	18.2	0.300	6.59	19.8	18.3	12.5
⌐ 51 × 51 × 9.5	7.0	877	0.202	5.82	15.2	16.2	0.202	5.82	15.2	16.2	9.88
⌐ 51 × 51 × 6.4	4.7	605	0.146	4.09	15.6	15.1	0.146	4.07	15.5	15.1	9.93
⌐ 51 × 51 × 3.2	2.5	312	0.080	2.16	16.0	13.9	0.080	2.16	16.0	13.9	10.1

C Slopes and Deflections of Beams

Simply Supported Beam Slopes and Deflections

Beam	Slope	Deflection	Elastic Curve
	$\theta_{max} = \dfrac{-PL^2}{16EI}$	$v_{max} = \dfrac{-PL^3}{48EI}$	$v = \dfrac{-Px}{48EI}(3L^2 - 4x^2)$ $0 \le x \le L/2$
	$\theta_1 = \dfrac{-Pab(L+b)}{6EIL}$ $\theta_2 = \dfrac{Pab(L+a)}{6EIL}$	$v\Big|_{x=a} = \dfrac{-Pba}{6EIL}(L^2 - b^2 - a^2)$	$v = \dfrac{-Pbx}{6EIL}(L^2 - b^2 - x^2)$ $0 \le x \le a$
	$\theta_1 = \dfrac{-M_0 L}{3EI}$ $\theta_2 = \dfrac{M_0 L}{6EI}$	$v_{max} = \dfrac{-M_0 L^2}{\sqrt{243}EI}$	$v = \dfrac{-M_0 x}{6EIL}(x^2 - 3Lx + 2L^2)$
	$\theta_{max} = \dfrac{-wL^3}{24EI}$	$v_{max} = \dfrac{-5wL^4}{384EI}$	$v = \dfrac{-wx}{24EI}(x^3 - 2Lx^2 + L^3)$
	$\theta_1 = \dfrac{-3wL^3}{128EI}$ $\theta_2 = \dfrac{7wL^3}{384EI}$	$v\Big|_{x=L/2} = \dfrac{-5wL^4}{768EI}$ $v_{max} = -0.006563\dfrac{wL^4}{EI}$ at $x = 0.4598L$	$v = \dfrac{-wx}{384EI}(16x^3 - 24Lx^2 + 9L^3)$ $0 \le x \le L/2$ $v = \dfrac{-wL}{384EI}(8x^3 - 24Lx^2$ $\qquad\qquad +17L^2 x - L^3)$ $L/2 \le x < L$
	$\theta_1 = \dfrac{-7w_0 L^3}{360EI}$ $\theta_2 = \dfrac{w_0 L^3}{45EI}$	$v_{max} = -0.00652\dfrac{w_0 L^4}{EI}$ at $x = 0.5193$	$v = \dfrac{-w_0 x}{360EIL}(3x^4 - 10L^2 x^2 + 7L^4)$

Cantilevered Beam Slopes and Deflections

Beam	Slope	Deflection	Elastic Curve
	$\theta_{max} = \dfrac{-PL^2}{2EI}$	$v_{max} = \dfrac{-PL^3}{3EI}$	$v = \dfrac{-Px^2}{6EI}(3L - x)$
	$\theta_{max} = \dfrac{-PL^2}{8EI}$	$v_{max} = \dfrac{-5PL^3}{48EI}$	$v = \dfrac{-Px^2}{6EI}(\tfrac{3}{2}L - x) \qquad 0 \le x \le L/2$ $v = \dfrac{-PL^2}{24EI}(3x - \tfrac{1}{2}L) \qquad L/2 \le x \le L$
	$\theta_{max} = \dfrac{-wL^3}{6EI}$	$v_{max} = \dfrac{-wL^4}{8EI}$	$v = \dfrac{-wx^2}{24EI}(x^2 - 4Lx + 6L^2)$
	$\theta_{max} = \dfrac{M_0 L}{EI}$	$v_{max} = \dfrac{M_0 L^2}{2EI}$	$v = \dfrac{M_0 x^2}{2EI}$
	$\theta_{max} = \dfrac{-wL^3}{48EI}$	$v_{max} = \dfrac{-7wL^4}{384EI}$	$v = \dfrac{-wx^2}{24EI}(x^2 - 2Lx + \tfrac{3}{2}L^2)$ $\qquad\qquad\qquad\qquad 0 \le x \le L/2$ $v = \dfrac{-wL^3}{192EI}(4x - L/2)$ $\qquad\qquad\qquad\qquad L/2 \le x \le L$
	$\theta_{max} = \dfrac{-w_0 L^3}{24EI}$	$v_{max} = \dfrac{-w_0 L^4}{30EI}$	$v = \dfrac{-w_0 x^2}{120EIL}(10L^3 - 10L^2 x + 5Lx^2 - x^3)$

D REVIEW FOR THE FUNDAMENTALS OF ENGINEERING EXAMINATION

The Fundamentals of Engineering (FE) exam is given semiannually by the National Council of Engineering Examiners (NCEE), and is one of the requirements for obtaining a Professional Engineering License. A portion of this exam contains problems in mechanics of materials, and this appendix provides a review of the subject matter most often asked on this exam.

Before solving any of the problems, you should review the sections indicated in each chapter in order to become familiar with the boldfaced definitions and the procedures used to solve various types of problems. Also, review the example problems in these sections. The following problems are arranged in the same sequence as the topics in each chapter. Partial solutions to *all the problems* are given at the back of this appendix.

Chapter 1—Review All Sections

D–1 Determine the resultant internal moment in the member of the frame at point F.

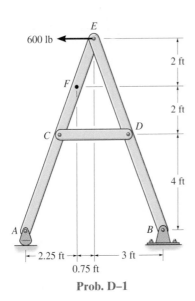

Prob. D–1

D–2 The beam is supported by a pin at A and a link BC. Determine the resultant internal shear in the beam at point D.

D–3 The beam is supported by a pin at A and a link BC. Determine the average shear stress in the pin at B if it has a diameter of 20 mm and is in double shear.

D–4 The beam is supported by a pin at A and a link BC. Determine the average shear stress in the pin at A if it has a diameter of 20 mm and is in single shear.

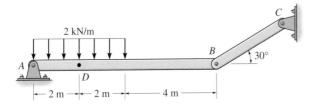

Probs. D–2/D–3/D–4

D–5 How many independent stress components are there in three dimensions?

D–6 The bars of the truss each have a cross-sectional area of 2 in². Determine the average normal stress in member CB.

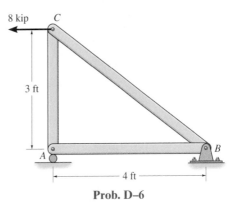

Prob. D–6

D–7 The frame supports the loading shown. The pin at A has a diameter of 0.25 in. If it is subjected to double shear, determine the average shear stress in the pin.

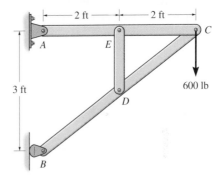

Prob. D–7

D–8 The uniform beam is supported by two rods AB and CD that have cross-sectional areas of 10 mm² and 15 mm², respectively. Determine the intensity w of the distributed load so that the average normal stress in each rod does not exceed 300 kPa.

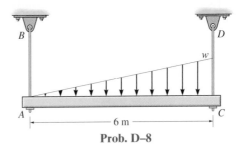

Prob. D–8

D–9 The bolt is used to support the load of 3 kip. Determine its diameter d to the nearest $\frac{1}{8}$ in. The allowable normal stress for the bolt is $\sigma_{\text{allow}} = 24$ ksi.

Prob. D–9

D–10 The two rods support the vertical force of $P = 30$ kN. Determine the diameter of rod AB if the allowable tensile stress for the material is $\sigma_{\text{allow}} = 150$ MPa.

D–11 The rods AB and AC have diameters of 15 mm and 12 mm, respectively. Determine the largest vertical force **P** that can be applied. The allowable tensile stress for the rods is $\sigma_{\text{allow}} = 150$ MPa.

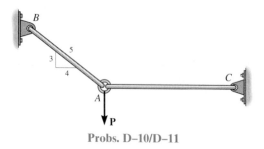

Probs. D–10/D–11

D–12 The allowable bearing stress for the material under the supports A and B is $\sigma_{\text{allow}} = 500$ psi. Determine the maximum uniform distributed load w that can be applied to the beam. The bearing plates at A and B have square cross sections of 3 in. $\times$ 3 in. and 2 in. $\times$ 2 in., respectively.

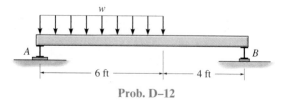

Prob. D–12

Chapter 2—Review All Sections

D–13 A rubber band has an unstretched length of 9 in. If it is stretched around a pole having a diameter of 3 in., determine the average normal strain in the band.

D–14 The rigid rod is supported by a pin at A and wires BC and DE. If the maximum allowable normal strain in each wire is $\epsilon_{\text{allow}} = 0.003$, determine the maximum vertical displacement of the load **P**.

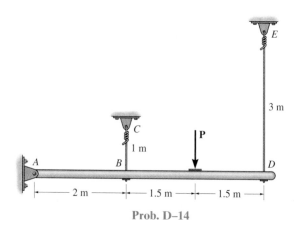

Prob. D–14

D–15 The load **P** causes a normal strain of 0.0045 in./in. in cable AB. Determine the angle of rotation of the rigid beam due to the loading if the beam is originally horizontal before it is loaded.

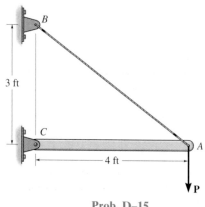

Prob. D–15

D–16 The square piece of material is deformed into the dashed position. Determine the shear strain at corner C.

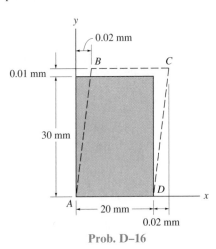

Prob. D–16

Chapter 3—Review Sections 3.1–3.7
D–17 Define homogeneous material.

D–18 Indicate the points on the stress–strain diagram which represent the proportional limit and the ultimate stress.

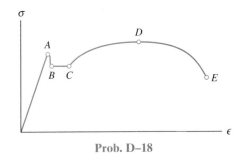

Prob. D–18

D–19 Define the modulus of elasticity E.

D–20 At room temperature, mild steel is a ductile material. True or false.

D–21 Engineering stress and strain are calculated using the *actual* cross-sectional area and length of the specimen. True or false.

D–22 If a rod is subjected to an axial load, there is only strain in the material in the direction of the load. True or false.

D–23 A 100-mm long rod has a diameter of 15 mm. If an axial tensile load of 100 kN is applied, determine its change in length. $E = 200$ GPa.

D–24 A bar has a length of 8 in. and cross-sectional area of 12 in^2. Determine the modulus of elasticity of the material if it is subjected to an axial tensile load of 10 kip and stretches 0.003 in. The material has linear-elastic behavior.

D–25 A 10-mm-diameter brass rod has a modulus of elasticity of $E = 100$ GPa. If it is 4 m long and subjected to an axial tensile load of 6 kN, determine its elongation.

D–26 A 100-mm long rod has a diameter of 15 mm. If an axial tensile load of 10 kN is applied to it, determine its change in diameter. $E = 70$ GPa, $\nu = 0.35$.

Chapter 4—Review Sections 4.1–4.6
D–27 What is Saint-Venant's principle?

D–28 What are the two conditions for which the principle of superposition is valid?

D–29 Determine the displacement of end A with respect to end C of the shaft. The cross-sectional area is 0.5 in^2 and $E = 29(10^3)$ ksi.

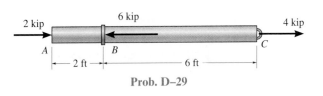

Prob. D–29

D–30 Determine the displacement of end A with respect to C of the shaft. The diameters of each segment are indicated in the figure. $E = 200$ GPa.

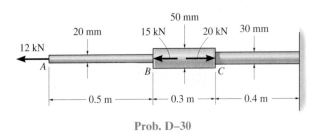

Prob. D–30

D–31 Determine the angle of tilt of the rigid beam when it is subjected to the load of 5 kip. Before the load is applied the beam is horizontal. Each rod has a diameter of 0.5 in., and $E = 29(10^3)$ ksi.

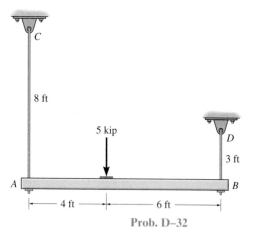

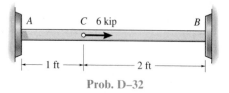

Prob. D–32

D–32 The uniform bar is subjected to the load of 6 kip. Determine the horizontal reactions at the supports A and B.

Prob. D–32

D–33 The cylinder is made from steel and has an aluminum core. If its ends are subjected to the axial force of 300 kN, determine the average normal stress in the steel. The cylinder has an outer diameter of 100 mm and an inner diameter of 80 mm. $E_{st} = 200$ GPa, $E_{al} = 73.1$ GPa.

$P = 300$ kN

Prob. D–33

D–34 The column is constructed from concrete and six steel reinforcing rods. If it is subjected to an axial force of 20 kip, determine the force supported by the concrete. Each rod has a diameter of 0.75 in. $E_{conc} = 4.20(10^3)$ ksi, $E_{st} = 29(10^3)$ ksi.

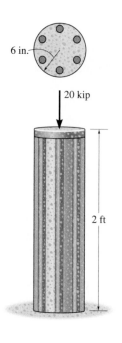

Prob. D–34

D–35 Two bars, each made of a different material, are connected and placed between two walls when the temperature is $T_1 = 15°C$. Determine the force exerted on the (rigid) supports when the temperature becomes $T_2 = 25°C$. The material properties and cross-sectional area of each bar are given in the figure.

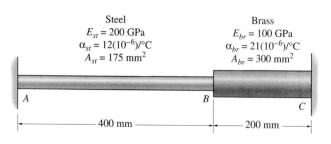

Prob. D–35

D–36 The aluminum rod has a diameter of 0.5 in. and is attached to the rigid supports at A and B when $T_1 = 80°F$. If the temperature becomes $T_2 = 100°F$, and an axial force of $P = 1200$ lb is applied to the rigid collar as shown, determine the reactions at A and B. $\alpha_{al} = 12.8(10^{-6})/°F$, $E_{al} = 10.6(10^6)$ psi.

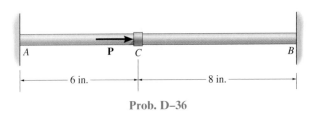

Prob. D–36

D–37 The aluminum rod has a diameter of 0.5 in. and is attached to the rigid supports at A and B when $T_1 = 80°F$. Determine the force **P** that must be applied to the rigid collar so that, when $T_2 = 50°F$, the reaction at B is zero. $\alpha_{al} = 12.8(10^{-6})/°F$, $E_{al} = 10.6(10^3)$ ksi.

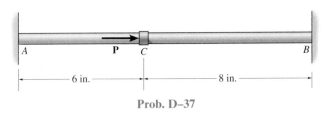

Prob. D–37

Chapter 5—Review Sections 5.1–5.5
D–38 Can the torsion formula, $\tau = Tc/J$, be used if the cross section is noncircular?

D–39 The solid 0.75-in.-diameter shaft is used to transmit the torques shown. Determine the absolute maximum shear stress developed in the shaft.

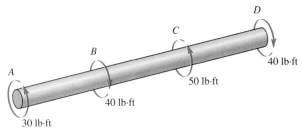

Prob. D–39

D–40 The solid 1.5-in.-diameter shaft is used to transmit the torques shown. Determine the shear stress developed in the shaft at point B.

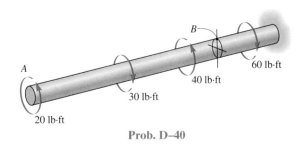

Prob. D–40

D–41 The solid shaft is used to transmit the torques shown. Determine the absolute maximum shear stress developed in the shaft.

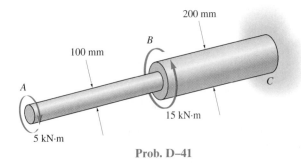

Prob. D–41

D–42 The shaft is subjected to the torques shown. Determine the angle of twist of end A with respect to end B. The shaft has a diameter of 1.5 in. $G = 11(10^3)$ ksi.

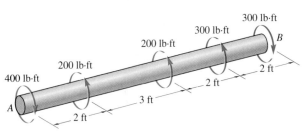

Prob. D–42

D–43 Determine the angle of twist of the 1-in.-diameter shaft at end A when it is subjected to the torsional loading shown. $G = 11(10^3)$ ksi.

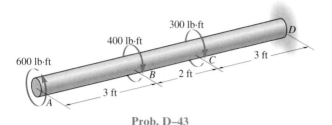

Prob. D–43

D–44 The shaft consists of a solid section AB with a diameter of 30 mm, and a tube BD with an inner diameter of 25 mm and outer diameter of 50 mm. Determine the angle of twist at its end A when it is subjected to the torsional loading shown. $G = 75$ GPa.

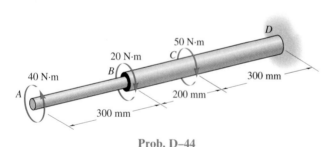

Prob. D–44

D–45 A motor delivers 200 hp to a steel shaft, which is tubular and has an outer diameter of 1.75 in. If it is rotating at 150 rad/s, determine its largest inner diameter to the nearest $\frac{1}{8}$ in. if the allowable shear stress for the material is $\tau_{allow} = 20$ ksi.

D–46 A motor delivers 300 hp to a steel shaft, which is tubular and has an outer diameter of 2.5 in. and an inner diameter of 2 in. Determine the smallest angular velocity at which it can rotate if the allowable shear stress for the material is $\tau_{allow} = 20$ ksi.

D–47 The shaft is made from a steel tube having a brass core. If it is fixed to the rigid support, determine the angle of twist that occurs at its end. $G_{st} = 75$ GPa and $G_{br} = 37$ GPa.

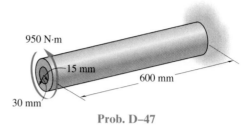

Prob. D–47

D–48 Determine the absolute maximum shear stress in the shaft. JG is constant.

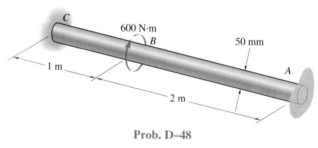

Prob. D–48

Chapter 6—Review Sections 6.1–6.5

D–49 Determine the internal moment in the beam as a function of x, where $2 \text{ m} \le x < 3 \text{ m}$.

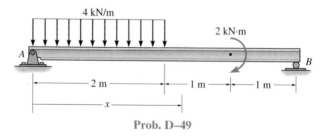

Prob. D–49

D–50 Determine the internal moment in the beam as a function of x, where $0 \le x \le 3$ m.

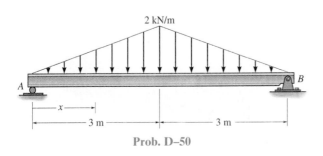

Prob. D–50

D–51 Determine the maximum moment in the beam.

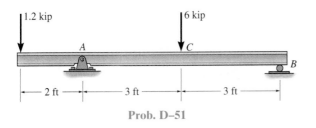

Prob. D–51

D–52 Determine the absolute maximum bending moment in the beam.

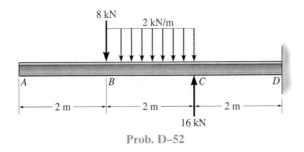

Prob. D–52

D–53 Determine the maximum moment in the beam.

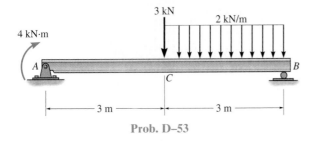

Prob. D–53

D–54 Determine the maximum moment in the beam.

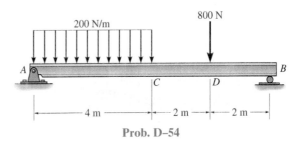

Prob. D–54

D–55 Determine the absolute maximum bending stress in the beam.

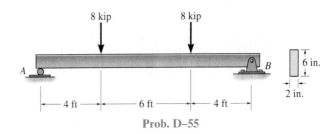

Prob. D–55

D–56 Determine the maximum bending stress in the 50-mm-diameter rod at C. There is a journal bearing at A.

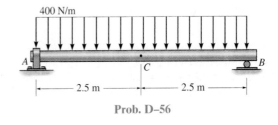

Prob. D–56

D–57 What is the strain in a beam at the neutral axis?

D–58 Determine the moment M that should be applied to the beam in order to create a compressive stress at point D of 10 ksi.

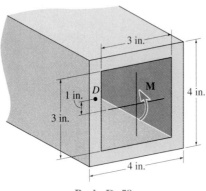

Prob. D–58

D–59 Determine the maximum bending stress in the beam.

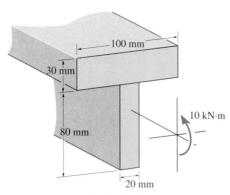

Prob. D–59

D–60 Determine the maximum load P that can be applied to the beam that is made from a material having an allowable bending stress of $\sigma_{allow} = 12$ MPa.

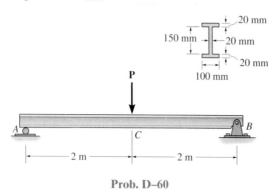

Prob. D–60

D–61 Determine the maximum stress in the beam's cross section.

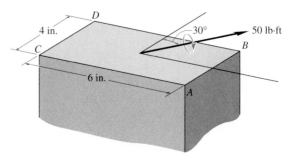

Prob. D–61

Chapter 7—Review Sections 7.1–7.4
D–62 Determine the maximum shear stress in the beam.

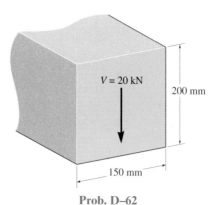

Prob. D–62

D–63 The beam has a rectangular cross section and is subjected to a shear of $V = 2$ kip. Determine the maximum shear stress in the beam.

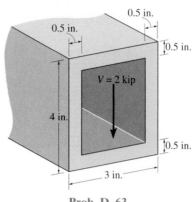

Prob. D–63

D–64 Determine the absolute maximum shear stress in the shaft having a diameter of 60 mm. The supports at A and B are journal bearings.

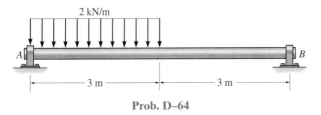

2 kN/m

A B

3 m   3 m

Prob. D–64

D–65 Determine the shear stress in the beam at point A, which is located at the top of the web.

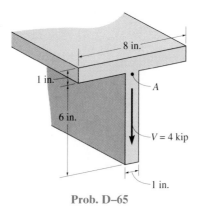

8 in.

1 in.

A

6 in.

$V = 4$ kip

1 in.

Prob. D–65

D–66 The beam is made from two boards fastened together at the top and bottom with nails spaced every 2 in. If an internal shear force of $V = 150$ lb is applied to the boards, determine the shear force resisted by each nail.

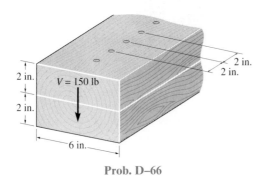

2 in.

2 in.

2 in.

2 in.

$V = 150$ lb

6 in.

Prob. D–66

D–67 The beam is made from four boards fastened together at the top and bottom with two rows of nails spaced every 4 in. If an internal shear force of $V = 400$ lb is applied to the boards, determine the shear force resisted by each nail.

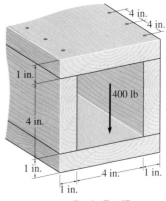

4 in.

4 in.

1 in.

400 lb

4 in.

1 in.

4 in.

1 in.

1 in.

Prob. D–67

Chapter 8—Review All Sections

D–68 A cylindrical tank is subjected to an internal pressure of 80 psi. If the internal diameter of the tank is 30 in., and the wall thickness is 0.3 in., determine the maximum normal stress in the material.

D–69 A pressurized spherical tank is to be made of 0.25-in.-thick steel. If it is subjected to an internal pressure of $p = 150$ psi, determine its inner diameter if the maximum normal stress is not to exceed 10 ksi.

D–70 Determine the magnitude of the load P that will cause a maximum normal stress of $\sigma_{max} = 30$ ksi in the link along section a–a.

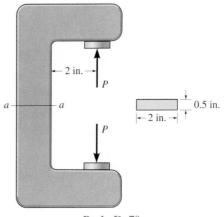

2 in.

P

a a

0.5 in.

2 in.

P

Prob. D–70

D–71 Determine the maximum normal stress in the horizontal portion of the bracket. The bracket has a thickness of 1 in. and a width of 0.75 in.

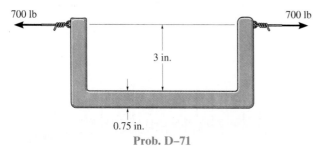

Prob. D–71

D–72 Determine the maximum load P that can be applied to the rod so that the normal stress in the rod does not exceed $\sigma_{max} = 30$ MPa.

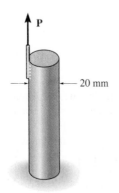

Prob. D–72

D–73 The beam has a rectangular cross section and is subjected to the loading shown. Determine the components of stress σ_x, σ_y, and τ_{xy} at point B.

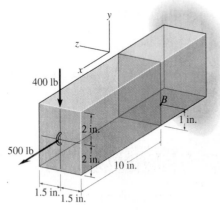

Prob. D–73

D–74 The solid cylinder is subjected to the loading shown. Determine the components of stress at point B.

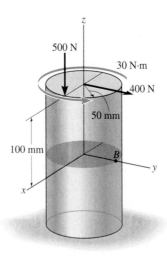

Prob. D–74

Chapter 9—Review Sections 9.1–9.3

D–75 When the state of stress at a point is represented by the principal stress, no shear stress will act on the element. True or false.

D–76 The state of stress at a point is shown on the element. Determine the maximum principal stress.

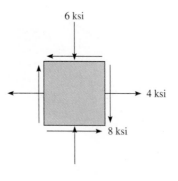

Prob. D–76

D–77 The state of stress at a point is shown on the element. Determine the maximum in-plane shear stress.

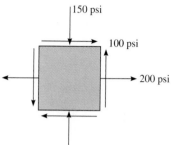

Prob. D–77

D–78 The state of stress at a point is shown on the element. Determine the maximum in-plane shear stress.

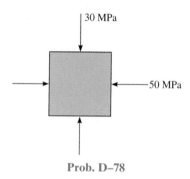

Prob. D–78

D–79 The beam is subjected to the load at its end. Determine the maximum principal stress at point B.

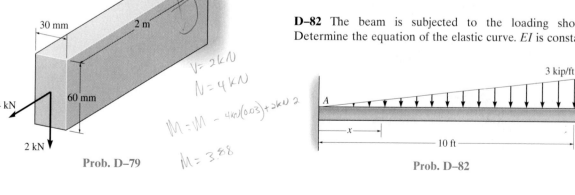

Prob. D–79

D–80 The beam is subjected to the loading shown. Determine the principal stress at point C.

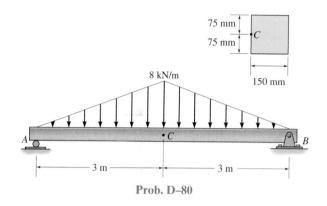

Prob. D–80

Chapter 12—Review Sections 12.1–12.2, 12.5

D–81 The beam is subjected to the loading shown. Determine the equation of the elastic curve. EI is constant.

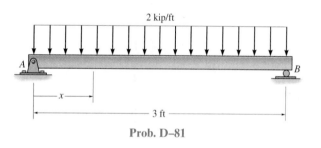

Prob. D–81

D–82 The beam is subjected to the loading shown. Determine the equation of the elastic curve. EI is constant.

Prob. D–82

D–83 Determine the displacement at point C of the beam shown. Use the method of superposition. EI is constant.

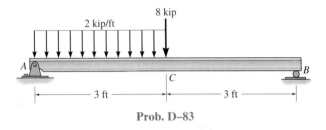

2 kip/ft

8 kip

A

C

B

3 ft

3 ft

Prob. D–83

D–84 Determine the slope at point A of the beam shown. Use the method of superposition. EI is constant.

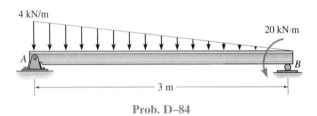

4 kN/m

20 kN·m

A

B

3 m

Prob. D–84

Chapter 13—Review Sections 13.1–13.3
D–85 The critical load is the maximum axial load that a column can support when it is on the verge of buckling. This loading represents a case of neutral equilibrium. True or false.

D–86 A 50-in.-long rod is made from a 1-in.-diameter steel rod. Determine the critical buckling load if the ends are fixed supported. $E = 29(10^3)$ ksi, $\sigma_Y = 36$ ksi.

D–87 A 12-ft wooden rectangular column has the dimensions shown. Determine the critical load if the ends are assumed to be pin-connected. $E = 1.6(10^3)$ ksi. Yielding does not occur.

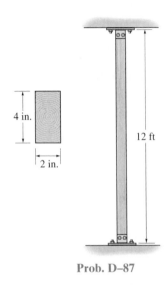

4 in.

2 in.

12 ft

Prob. D–87

D–88 A steel pipe is fixed-supported at its ends. If it is 5 m long and has an outer diameter of 50 mm and a thickness of 10 mm, determine the maximum axial load P that it can carry without buckling. $E_{st} = 200$ GPa, $\sigma_Y = 250$ MPa.

D–89 A steel pipe is pin-supported at its ends. If it is 6 ft long and has an outer diameter of 2 in., determine its smallest thickness so that it can support an axial load of $P = 40$ kip without buckling. $E_{st} = 29(10^3)$ ksi, $\sigma_Y = 36$ ksi.

D–90 Determine the smallest diameter of a solid 40-in.-long steel rod, to the nearest $\frac{1}{16}$ in., that will support an axial load of $P = 3$ kip without buckling. The ends are pin connected. $E_{st} = 29(10^3)$ ksi, $\sigma_Y = 36$ ksi.

Partial Solutions and Answers

D-1 Entire frame:
$\Sigma M_B = 0; A_y = 800$ lb
CD is a two-force member
Member AE:
$\Sigma M_E = 0; F_{CD} = 600$ lb
Segment ACF:
$\Sigma M_F = 0; M_F = 600$ lb $\cdot$ ft   *Ans.*

D-2 BC is a two-force member.
Beam AB:
$\Sigma M_B = 0; A_y = 6$ kN
Segment AD:
$\Sigma F_y = 0; V = 2$ kN   *Ans.*

D-3 BC is a two-force member.
Beam AB:
$\Sigma M_A = 0; T_{BC} = 4$ kN
Pin B:
$\tau_B = \dfrac{T_{BC}/2}{A} = \dfrac{4/2}{\frac{\pi}{4}(0.02)^2} = 6.37$ MPa   *Ans.*

D-4 BC is a two-force member
Beam AB:
$\Sigma M_A = 0; T_{BC} = 4$ kN
$\Sigma F_x = 0; A_x = 3.464$ kN
$\Sigma F_y = 0; A_y = 6$ kN
$F_A = \sqrt{(3.464)^2 + (6)^2} = 6.928$ kN
$\tau_A = \dfrac{F_A}{A} = \dfrac{6.928}{\frac{\pi}{4}(0.02)^2} = 22.1$ MPa   *Ans.*

D-5 6: $\sigma_x, \sigma_y, \sigma_z, \tau_{xy}, \tau_{yz}, \tau_{zx}$ *Ans.*

D-6 Joint C:
$\xrightarrow{+}\Sigma F_x = 0; T_{CB} = 10$ kip
$\sigma = \dfrac{T_{CB}}{A} = \dfrac{10}{2} = 5$ ksi   *Ans.*

D-7 Entire frame
$\Sigma F_y = 0; A_y = 600$ lb
$\Sigma M_B = 0; A_x = 800$ lb
$F_A = \sqrt{(600)^2 + (800)^2} = 1000$ lb
$\tau_A = \dfrac{F_A/2}{A} = \dfrac{1000/2}{\frac{\pi}{4}(0.25)^2} = 10.2$ ksi   *Ans.*

D-8 Beam:
$\Sigma M_A = 0; T_{CD} = 2\,w$
$\Sigma F_y = 0; T_{AB} = w$
Rod AB:
$\sigma = \dfrac{P}{A}; 300(10^3) = \dfrac{w}{10}; w = 3$ MN/m
Rod CD:
$\sigma = \dfrac{P}{A}; 300(10^3) = \dfrac{2w}{15}; w = 2.25$ MN/m   *Ans.*

D-9 $\sigma = \dfrac{P}{A}; 24 = \dfrac{3}{\frac{\pi}{4}d^2}; d = 0.3989$ in.
use $d = 0.5$ in.   *Ans.*

D-10 Joint A:
$\Sigma F_y = 0; F_{AB} = 50$ kN
$\sigma = \dfrac{P}{A}; 150(10^6) = \dfrac{50(10^3)}{\frac{\pi}{4}d^2}; d = 20.6$ mm   *Ans.*

D-11 Joint A:
$\Sigma F_y = 0; F_{AB} = 1.667P$
$\Sigma F_x = 0; F_{AC} = 1.333\,P$
Rod AB:
$\sigma = \dfrac{P}{A}; 150(10^6) = \dfrac{1.667\,P}{\frac{\pi}{4}(0.015)^2}; P = 15.9$ kN
Rod AC:
$\sigma = \dfrac{P}{A}; 150(10^6) = \dfrac{1.333\,P}{\frac{\pi}{4}(0.012)^2}; P = 12.7$ kN   *Ans.*

D-12 Beam:
$\Sigma M_A = 0; B_y = 1.8\,w$
$\Sigma F_y = 0; A_y = 4.2\,w$
At A:
$\sigma = \dfrac{P}{A}; 500 = \dfrac{4.2\,w}{(3)(3)}; w = 1.07$ kip/ft   *Ans.*
At B:
$\sigma = \dfrac{P}{A}; 500 = \dfrac{1.8\,w}{(2)(2)}; w = 1.11$ kip/ft

D-13 $\epsilon = \dfrac{l - l_0}{l_0} = \dfrac{\pi(3) - 9}{9} = 0.0472$ in./in.   *Ans.*

D-14 $(\delta_{DE})_{max} = \epsilon_{max}\, l_{DE} = 0.003(3) = 0.009$ m
By proportion from A,
$\delta_{BC} = 0.009\left(\frac{2}{5}\right) = 0.0036$ m
$(\delta_{BC})_{max} = \epsilon_{max}\, l_{BC} = 0.003(1) = 0.003$ m < 0.0036 m
Use $\delta_{BC} = 0.003$ m. By proportion from A,
$\delta_P = 0.003\left(\dfrac{3.5}{2}\right) = 0.00525$ m $= 5.25$ mm   *Ans.*

D–15 $l_{AB} = \sqrt{(4)^2 + (3)^2} = 5$ ft

$l'_{AB} = 5 + 5(0.0045) = 5.0225$ ft

The angle BCA was originally $\theta = 90°$. Using the cosine law, the new angle BCA (θ') is

$5.0225 = \sqrt{(3)^2 + (4)^2 - 2(3)(4)\cos\theta}$

$\theta = 90.538°$

Thus

$\Delta\theta' = 90.538° - 90° = 0.538°$ *Ans.*

D–16 $\angle BCD = \angle BAD = \tan^{-1}\dfrac{30.01}{0.02} = 89.962°$

$\gamma_{xy} = (90° - 89.962°)\dfrac{\pi}{180°} = 0.666(10^{-3})$ rad *Ans.*

D–17 Material has uniform properties throughout. *Ans.*

D–18 Proportional limit is A. *Ans.*
Ultimate stress is D. *Ans.*

D–19 The initial slope of the $\sigma - \epsilon$ diagram. *Ans.*

D–20 True. *Ans.*

D–21 False. Use the *original* cross-sectional area and length. *Ans.*

D–22 False. There is also strain in the perpendicular directions due to the Poisson effect. *Ans.*

D–23 $\epsilon = \dfrac{\sigma}{E} = \epsilon\dfrac{P}{AE}$

$\delta = \epsilon L = \dfrac{PL}{AE} = \dfrac{100(10^3)(0.100)}{\frac{\pi}{4}(0.015)^2\,200(10^9)} = 0.283$ mm

Ans.

D–24 $\epsilon = \dfrac{\sigma}{E} = \dfrac{P}{AE}$

$\delta = \epsilon L = \dfrac{PL}{AE}; 0.003 = \dfrac{(10\,000)(8)}{12\,E}$

$E = 2.22(10^6)$ psi *Ans.*

D–25 $\epsilon = \dfrac{\sigma}{E} = \dfrac{P}{AE}$

$\delta = \epsilon L = \dfrac{PL}{AE} = \dfrac{6(10^3)\,4}{\frac{\pi}{4}(0.01)^2\,100(10^9)} = 3.06$ mm *Ans.*

D–26 $\sigma = \dfrac{P}{A} = \dfrac{10(10^3)}{\frac{\pi}{4}(0.015)^2} = 56.59$ MPa

$\epsilon_{long} = \dfrac{\sigma}{E} = \dfrac{56.59(10^6)}{70(10^9)} = 0.808(10^{-3})$

$\epsilon_{lat} = -\nu\epsilon_{long} = -0.35(0.808(10^{-3})) = -0.283(10^{-3})$

$\delta d = \epsilon_{lat}(15\text{ mm}) = -4.24$ mm *Ans.*

D–27 Stress distributions tend to smooth out on sections further removed from the load. *Ans.*

D–28 1) Linear-elastic material.
2) No large deformations. *Ans.*

D–29 $\delta_{A/C} = \sum\dfrac{PL}{AE} = \dfrac{-2(2)(12)}{0.5(29(10^3))} + \dfrac{4(6)(12)}{0.5(29(10^3))}$

$= 0.0166$ in. *Ans.*

D–30 $\delta_{A/C} = \sum\dfrac{PL}{AE} = \dfrac{12(10^3)(0.5)}{\frac{\pi}{4}(0.02)^2\,200(10^9)}$

$+ \dfrac{27(10^3)(0.3)}{\frac{\pi}{4}(0.05)^2\,200(10^9)} = 0.116$ mm *Ans.*

D–31 Beam AB:

$\Sigma M_A = 0; F_{BD} = 2$ kip

$\Sigma F_y = 0; F_{AC} = 3$ kip

$\delta_A = \dfrac{PL}{AE} = \dfrac{3(8)(12)}{\frac{\pi}{4}(0.5)^2\,29(10^3)} = 0.0506$ in. $\downarrow$

$\delta_B = \dfrac{PL}{AE} = \dfrac{2(3)(12)}{\frac{\pi}{4}(0.5)^2\,29(10^3)} = 0.01264$ in. $\downarrow$

$\theta = \dfrac{\Delta\delta}{l_{AB}} = \dfrac{\delta_A - \delta_B}{l_{AB}} = \dfrac{0.0506 - 0.01264}{10(12)}$

$= 0.000315$ rad $= 0.0181°$ *Ans.*

D–32 Equilibrium:

$F_A + F_B = 6$

Compatibility:

$\delta_{C/A} = \delta_{C/B}; \dfrac{F_A\,(1)}{AE} = \dfrac{F_B\,(2)}{AE}$

$F_A = 4$ kip, $F_B = 2$ kip *Ans.*

D–33 Equilibrium:

$P_{st} + P_{al} = 300\,(10^3)$

Compatibility:

$\delta_{st} = \delta_{al}; \dfrac{P_{st}\,L}{[\frac{\pi}{4}(0.1)^2 - \frac{\pi}{4}(0.08)^2]\,200(10^9)}$

$= \dfrac{P_{al}\,L}{[\frac{\pi}{4}(0.08)^2]\,73.1\,(10^9)}$

$P_{st} = 181.8$ kN
$P_{al} = 118$ kN

$\sigma_{st} = \dfrac{P_{st}}{A} = \dfrac{181.8}{[\frac{\pi}{4}(0.1)^2 - \frac{\pi}{4}(0.08)^2]} = 64.3$ MPa *Ans.*

D–34 Equilibrium:

$$P_{conc} + P_{st} = 20$$

Compatibility:

$$\delta_{conc} = \delta_{st}; \quad \frac{P_{conc}\,(2)}{[\pi\,(6)^2 - 6(\frac{\pi}{4})(0.75)^2]\,4.20(10^3)}$$

$$= \frac{P_{st}\,(2)}{6(\frac{\pi}{4})\,(0.75)^2\,29\,(10^3)}$$

$$P_{conc} = 17.2 \text{ kip} \hfill Ans.$$
$$P_{st} = 2.84 \text{ kip}$$

D–35 $\delta_{temp} = \Sigma \alpha \Delta TL$

$$\delta_{load} = \Sigma \frac{PL}{AE}$$

Compatibility: $\delta_{temp} + \delta_{load} = 0$

$$12(10^{-6})(25 - 15)\,(0.4) + 21(10^{-6})(25 - 15)(0.2) +$$

$$\frac{-F(0.4)}{175(10^{-6})(200(10^9))} - \frac{F(0.2)}{300(10^{-6})(100(10^9))} = 0$$

$$F = 4.97 \text{ kN} \hfill Ans.$$

D–36 Equilibrium:

$$F_A + F_B = 1200$$

Compatibility

Remove support at B. Require

$$\delta_B = (\delta_{B/A})_{temp} + (\delta_{B/A})_{load} = 0$$

$$\alpha \Delta TL + \Sigma \frac{PL}{AE} = 0$$

$$12.8(10^{-6})(100 - 80)(14) +$$

$$\frac{1200(6)}{\frac{\pi}{4}\,(0.5)^2\,10.6\,(10^6)} - \frac{F_B(14)}{\frac{\pi}{4}\,(0.5)^2\,10.6\,(10^6)} = 0$$

$$F_B = 1.05 \text{ kip} \hfill Ans.$$

$$F_A = 153 \text{ lb} \hfill Ans.$$

D–37 Equilibrium:

$$F_A + F_B = P$$

Since $F_B = 0$, $F_A = P$

Compatibility

Remove support at B. Require

$$\delta_B = (\delta_{B/A})_{temp} + (\delta_{B/A})_{load} = 0$$

$$\alpha \Delta TL + \frac{PL}{AE} = 0$$

$$12.8\,(10^{-6})\,(50 - 80)\,(14) + \frac{P(6)}{\frac{\pi}{4}\,(0.5)^2\,10.6\,(10^6)} = 0$$

$$P = 1.86 \text{ kip} \hfill Ans.$$

D–38 No, it is only valid for circular cross sections. Non-circular cross sections will warp. *Ans.*

D–39 $T_{max} = T_{CD} = 40 \text{ lb} \cdot \text{ft}$

$$\tau_{max} = \frac{Tc}{J} = \frac{40(12)(0.375)}{\frac{\pi}{2}\,(0.375)^4} = 5.79 \text{ ksi} \hfill Ans.$$

D–40 Equilibrium of segment AB:

$$T_B = 30 \text{ lb} \cdot \text{ft}$$

$$T_B = \frac{Tc}{J} = \frac{30(12)(0.75)}{\frac{\pi}{2}\,(0.75)^4} = 543 \text{ psi} \hfill Ans.$$

D–41 Segment AB:

$$\tau_{max} = \frac{Tc}{J} = \frac{5(10^3)(0.05)}{\frac{\pi}{2}\,(0.05)^4} = 25.5 \text{ MPa} \hfill Ans.$$

Segment BC:

$$\tau_{max} = \frac{Tc}{J} = \frac{10(10^3)(0.1)}{\frac{\pi}{2}\,(0.1)^4} = 6.37 \text{ MPa}$$

D–42 $\phi_{A/B} = \Sigma \dfrac{TL}{JG} = \dfrac{-400(12)(2)(12)}{\frac{\pi}{2}\,(0.75)^4\,11(10^6)}$

$$- \frac{200(12)(3)(12)}{\frac{\pi}{2}\,(0.75)^4\,11(10^6)} + 0 + \frac{300(12)(2)(12)}{\frac{\pi}{2}\,(0.75)^4\,11(10^6)}$$

$$= -0.0211 \text{ rad} = 0.0211 \text{ rad clockwise when viewed}$$
from A. *Ans.*

D–43 $\phi_A = \Sigma \dfrac{TL}{JG} = \dfrac{600(12)(3)(12)}{\frac{\pi}{2}\,(0.5)^4 11(10^6)}$

$$+ \frac{200(12)(2)(12)}{\frac{\pi}{2}\,(0.5)^4\,11(10^6)} - \frac{100(12)(3)(12)}{\frac{\pi}{2}\,(0.5)^4\,11\,(10^6)}$$

$$= 0.253 \text{ rad counterclockwise when viewed from } A.$$
Ans.

D–44 $\phi_A = \Sigma \dfrac{TL}{JG} = \dfrac{40(0.3)}{\frac{\pi}{2}\,(0.015)^4\,75(10^9)}$

$$+ \frac{20(0.2)}{\frac{\pi}{2}\,[(0.025)^4 - (0.0125)^4]\,75(10^9)}$$

$$- \frac{30(0.3)}{\frac{\pi}{2}\,[(0.025)^4 - (0.0125)^4]\,75(10^9)}$$

$$= 1.90\,(10^{-3}) \text{ rad counterclockwise when viewed}$$
from A towards D. *Ans.*

D–45 $P = 200 \text{ hp} \left(\dfrac{550 \text{ ft} \cdot \text{lb/s}}{1 \text{ hp}} \right) = 110\,000 \text{ ft} \cdot \text{lb/s}$

$T = \dfrac{P}{\omega} = \dfrac{110\,000}{150} = 733.33 \text{ lb} \cdot \text{ft} = 8800 \text{ lb} \cdot \text{in.}$

$\tau_{\text{allow}} = \dfrac{Tc}{J}; \ 20(10^3) = \dfrac{8800\,(0.875)}{\frac{\pi}{2}\,[(0.875)^4 - r_i^4]}$

$r_i = 0.764 \text{ in.}$
$d_i = 1.53 \text{ in. use } d_i = 1.625 \text{ in.} = 1\frac{5}{8} \text{ in.}$ *Ans.*

D–46 $P = 300 \text{ hp} \left(\dfrac{550 \text{ ft} \cdot \text{lb/s}}{1 \text{ hp}} \right) = 165\,000 \text{ ft} \cdot \text{lb/s}$

$T = \dfrac{P}{\omega} = \dfrac{165\,000}{\omega}$

$\tau_{\text{max}} = \dfrac{Tc}{J}; \ 20(10^3) = \dfrac{\frac{165\,000}{\omega}\,(12)(1.25)}{\frac{\pi}{2}\,[(1.25)^4 - (1)^4]}$

$\omega = 54.7 \text{ rad/s}$ *Ans.*

D–47 Equilibrium:
$T_{st} + T_{br} = 950$
Compatibility: $\phi_{st} = \phi_{br};$

$\dfrac{T_{st}\,(0.6)}{\frac{\pi}{2}\,[(0.03)^4 - (0.015)^4]\,75\,(10^9)} = \dfrac{T_{br}\,(0.6)}{\frac{\pi}{2}\,(0.015)^4\,37(10^9)}$

$T_{br} = 30.25 \text{ N} \cdot \text{m}$
$T_{st} = 919.8 \text{ N} \cdot \text{m}$

$\phi = \phi_{br} = \dfrac{30.25(0.6)}{\frac{\pi}{2}\,(0.015)^4\,37(10^9)} = 0.00617 \text{ rad}$ *Ans.*

D–48 Equilibrium:
$T_A + T_C = 600$
Compatibility:

$\delta_{B/C} = \delta_{B/A}; \ \dfrac{T_C\,(1)}{JG} = \dfrac{T_A\,(2)}{JG}$

$T_A = 200 \text{ N} \cdot \text{m}$
$T_C = 400 \text{ N} \cdot \text{m}$

$\tau_{\text{max}} = \dfrac{Tc}{J} = \dfrac{400(0.025)}{\frac{\pi}{2}\,(0.025)^4} = 16.3 \text{ MPa}$ *Ans.*

D–49 $A_y = 5.5 \text{ kN}$
Use section of length x.
$\lightning + \Sigma M = 0; \ -5.5\,x + 4\,(2)(x - 1) + M = 0$
$M = 8 - 2.5\,x$ *Ans.*

D–50 $A_y = 3 \text{ kN}$
Use section of length x.
Intensity of $w = \dfrac{2}{3}x$ at x.
$\lightning + \Sigma M = 0; \ -3x + \left(\dfrac{1}{3}\,x \right)\left[\dfrac{1}{2}(x)\left(\dfrac{2}{3}x \right) \right] + M = 0$
$M = 3x - \dfrac{x^3}{9}$ *Ans.*

D–51 $B_y = 2.6 \text{ kip}$
$A_y = 4.6 \text{ kip}$
Draw M-diagram
$M_{\text{max}} = 7.80 \text{ kip} \cdot \text{ft (at } C)$ *Ans.*

D–52 Draw M-diagram
$M_{\text{max}} = 20 \text{ kN} \cdot \text{m (at } C)$ *Ans.*

D–53 $A_y = 2.33 \text{ kN}$
$B_y = 6.617 \text{ kN}$
Draw M-diagram
$M_{\text{max}} = 11 \text{ kN} \cdot \text{m (at } C)$ *Ans.*

D–54 $A_y = B_y = 800 \text{ N}$
Draw M-diagram
$M_{\text{max}} = 1600 \text{ N} \cdot \text{m (within } CD)$ *Ans.*

D–55 $A_y = B_y = 8 \text{ kip}$
$M_{\text{max}} = 8\,(4) = 32 \text{ kip} \cdot \text{ft}$
$\sigma = \dfrac{Mc}{I} = \dfrac{32(12)(3)}{\frac{1}{12}\,(2)(6)^3} = 32 \text{ ksi}$ *Ans.*

D–56 $A_y = B_y = 1000 \text{ N}$
$M_{\text{max}} = 1250 \text{ N} \cdot \text{m}$
$\sigma_{\text{max}} = \dfrac{Mc}{I} = \dfrac{1250(0.025)}{\frac{\pi}{4}\,(0.025)^4} = 102 \text{ MPa}$ *Ans.*

D–57 $\epsilon = 0$ *Ans.*

D–58 $\sigma = \dfrac{My}{I}; \ 10(10^3) = \dfrac{M(1)}{[\frac{1}{12}\,(4)(4)^3 - \frac{1}{12}\,(3)(3)^3]}$

$M = 145.8 \text{ kip} \cdot \text{in.} = 12.2 \text{ kip} \cdot \text{ft}$ *Ans.*

D–59 From bottom of cross section

$$\bar{y} = \frac{\Sigma \bar{y}A}{\Sigma A} = \frac{40(80)(20) + 95(30)(100)}{80(20) + 30(100)} = 75.870 \text{ mm}$$

$$I = \frac{1}{12} (20)(80)^3 + 20(80)(75.870 - 40)^2$$

$$+ \frac{1}{12} (100)(30)^3 + 100(30)(95 - 75.870)^2 = 4.235(10^{-6}) \text{ m}$$

$$\sigma_{max} = \frac{Mc}{I} = \frac{10(10^3)(0.075870)}{4.235 (10^{-6})} = 179 \text{ MPa} \quad \textit{Ans.}$$

D–60 $A_y = P/2$

$M_{max} = P/2 \ (2) = P \text{ (at } C)$

$$I = \frac{1}{12} (0.02)(0.150)^3 + 2\left[\frac{1}{12} (0.1)(0.02)^3 \right.$$

$$+ (0.1)(0.02)(0.085)^2 \Big]$$

$$= 34.66 (10^{-6}) \text{ m}^4$$

$$\sigma = \frac{Mc}{I}; \ 12 \ (10^6) = \frac{P \ (0.095)}{34.66 \ (10^{-6})}$$

$$P = 4.38 \text{ kN} \quad \textit{Ans.}$$

D–61 Maximum stress occurs at D or A.

$$(\sigma_{max})_D = \frac{(50 \cos 30°) \ 12 \ (3)}{\frac{1}{12} \ (4) \ (6)^3}$$

$$+ \frac{(50 \sin 30°) \ 12 \ (2)}{\frac{1}{12} \ (6)(4)^3} = 40.4 \text{ psi} \quad \textit{Ans.}$$

D–62 Q is upper or lower half of cross section.

$$\tau_{max} = \frac{VQ}{It} = \frac{20(10^3) \ [(0.05)(0.1)(0.15)]}{[\frac{1}{12} (0.150)(0.2)^3](0.15)} = 1 \text{ MPa}$$

$$\textit{Ans.}$$

D–63 $I = \frac{1}{12} (3)(4)^3 - \frac{1}{12} (2)(3)^3 = 11.5 \text{ in}^4$

Q is upper or lower half of cross section.

$Q = (1) \ (2) \ (3) - (0.75) \ (1.5) \ (2) = 3.75 \text{ in}^3$

$$\tau_{max} = \frac{VQ}{It} = \frac{2(3.75)}{11.5 \ (1)} = 0.652 \text{ ksi} \quad \textit{Ans.}$$

D–64 $A_y = 4.5 \text{ kN}, B_y = 1.5 \text{ kN}$

$V_{max} = 4.5 \text{ kN (at } A)$

Q is upper half of cross section.

$$\tau_{max} = \frac{VQ}{It} = \frac{4.5(10^3) \ [(\frac{4(0.03)}{3\pi}) \frac{1}{2} \pi (0.03)^2]}{[\frac{1}{4} \pi (0.03)^4] \ (0.06)}$$

$$= 2.12 \text{ MPa} \quad \textit{Ans.}$$

D–65 From the bottom:

$$\bar{y} = \frac{\Sigma \bar{y}A}{\Sigma A} = \frac{3(6)(1) + 6.5(1)(8)}{6(1) + 1(8)} = 5 \text{ in.}$$

$$I = \frac{1}{12} (1)(6)^3 + 6(1)(5 - 3)^2 + \frac{1}{12} (8)(1)^3 +$$

$$8(1) \ (6.5 - 5)^2 = 60.67 \text{ in}^4$$

$$\tau = \frac{VQ}{It} = \frac{4(10^3)[8 \ (1)(6.5 - 5)]}{60.67 \ (1)} = 791 \text{ psi} \quad \textit{Ans.}$$

D–66 $I = \frac{1}{12} (6)(4)^3 = 32 \text{ in}^4$

$$q = \frac{VQ}{I} = \frac{150[(1)(6)(2)]}{32} = 56.25 \text{ lb/in.}$$

$$F = qs = (56.25 \text{ lb/in.}) \ (2 \text{ in.}) = 112.5 \text{ lb} \quad \textit{Ans.}$$

D–67 $I = \frac{1}{12} (6)(6)^3 - \frac{1}{12} (4)(4)^3 = 86.67 \text{ in}^4$

$$q = \frac{VQ}{I} = \frac{400(2.5)(6)(1)}{86.67} = 69.23 \text{ lb/in.}$$

For one nail

$q = 69.23/2 = 34.62 \text{ lb/in.}$

$F = qs = 34.62 \text{ lb/in.} \ (4 \text{ in.}) = 138 \text{ lb} \quad \textit{Ans.}$

D–68 $\sigma = \frac{pr}{t} = \frac{80(15)}{0.3} = 4000 \text{ psi} = 4 \text{ ksi} \quad \textit{Ans.}$

D–69 $\sigma = \frac{pr}{2t}; \ 10 \ (10^3) = \frac{150 \ r}{2 \ (0.25)}$

$r = 33.3 \text{ in.}$

$d = 66.7 \text{ in.} \quad \textit{Ans.}$

D–70 At the section through centroidal axis

$N = P$

$V = 0$

$M = (2 + 1)P = 3P$

$$\sigma = \frac{P}{A} + \frac{Mc}{I}$$

$$30 = \frac{P}{2(0.5)} + \frac{(3P)(1)}{\frac{1}{12} (0.5)(2)^3}$$

$P = 3 \text{ kip} \quad \textit{Ans.}$

D–71 At a section through the center of bracket on centroidal axis.

$N = 700 \text{ lb}$

$V = 0$

$M = 700(3 + 0.375) = 2362.5 \text{ lb} \cdot \text{in.}$

$$\sigma = \frac{P}{A} + \frac{Mc}{I} = \frac{700}{0.75(1)} + \frac{2362.5(0.375)}{[\frac{1}{12} (1)(0.75)^3]}$$

$$= 26.1 \text{ ksi} \quad \textit{Ans.}$$

D–72 At a cross section
$$N = P, M = P\,(0.01)$$
$$\sigma_{max} = \frac{P}{A} + \frac{Mc}{I}$$
$$30(10^6) = \frac{P}{\pi\,(0.01)^2} + \frac{P\,(0.01)(0.01)}{\frac{1}{4}\,\pi\,(0.01)^4}$$

$$P = 1.88\ kN \hspace{2cm} Ans.$$

D–73 At section through B:
$$N = 500\ lb, V = 400\ lb$$
$$M = 400\,(10) = 4000\ lb \cdot in.$$
Axial load:
$$\sigma_x = \frac{P}{A} = \frac{500}{4\,(3)} = 41.667\ psi\ (T)$$
Shear load:
$$\tau_{xy} = \frac{VQ}{It} = \frac{400[(1.5)(3)(1)]}{[\frac{1}{12}(3)(4)^3]\,3} = 37.5\ psi$$

Bending moment:
$$\sigma_x = \frac{My}{I} = \frac{4000(1)}{\frac{1}{12}\,(3)(4)^3} = 250\ psi\ (C)$$

Thus
$$\sigma_x = 41.667 - 250 = 208\ psi\ (C) \hspace{1cm} Ans.$$
$$\sigma_y = 0 \hspace{1cm} Ans.$$
$$\tau_{xy} = 37.5\ psi \hspace{1cm} Ans.$$

D–74 At section B:
$$N_z = 500\ N, V_y = 400\ N, M_x = 400\,(0.1) = 40\ N \cdot m$$
$$M_y = 500\,(0.05) = 25\ N \cdot m, T_z = 30\ N \cdot m$$
Axial load:
$$\sigma_z = \frac{P}{A} = \frac{500}{\pi\,(0.05)^2} = 63.66\ kPa\ (C)$$
Shear load:
$$\tau_{zy} = 0\ \text{since at}\ B, Q = 0.$$
Moment about x axis:
$$\sigma_z = \frac{Mc}{I} = \frac{40(0.05)}{\frac{\pi}{4}\,(0.05)^4} = 407.4\ kPa\ (C)$$

Moment about y axis:
$$\sigma_z = 0\ \text{since}\ B\ \text{is on neutral axis.}$$
Torque:
$$\tau_{zx} = \frac{Tc}{J} = \frac{30(0.05)}{\frac{\pi}{2}\,(0.05)^4} = 153\ kPa$$

Thus
$$\sigma_x = 0 \hspace{1cm} Ans.$$
$$\sigma_y = 0 \hspace{1cm} Ans.$$
$$\sigma_z = -63.66 - 407.4 = -471\ kPa \hspace{1cm} Ans.$$
$$\tau_{xy} = 0 \hspace{1cm} Ans.$$
$$\tau_{zy} = 0 \hspace{1cm} Ans.$$
$$\tau_{zx} = -153\ kPa \hspace{1cm} Ans.$$

D–75 True. $\hspace{4cm} Ans.$

D–76 $\sigma_x = 4\ ksi, \sigma_y = -6\ ksi, \tau_{xy} = -8\ ksi$
Apply Eq. 9-5, $\sigma_1 = 8.43\ ksi, \sigma_2 = -10.4\ ksi \hspace{0.5cm} Ans.$

D–77 $\sigma_x = 200\ psi, \sigma_y = -150\ psi, \tau_{xy} = 100\ psi$
Apply Eq. 9-7, $\tau_{max\atop \text{in-plane}} = 202\ psi \hspace{1cm} Ans.$

D–78 $\sigma_x = -50\ MPa, \sigma_y = -30\ MPa, \tau_{xy} = 0$
Use Eq. 9-7, $\tau_{max\atop \text{in-plane}} = 10\ MPa$

D–79 At the cross section through B:
$$N = 4\ kN, V = 2\ kN, M = 2\,(2) = 4\ kN \cdot m$$
$$\sigma_B = \frac{P}{A} + \frac{Mc}{I} = \frac{4(10^3)}{0.03(0.06)} + \frac{4(10^3)(0.03)}{\frac{1}{12}(0.03)(0.06)^3}$$
$$= 224\ MPa\ (T)$$
Note $\tau_B = 0$ since $Q = 0$.
Thus
$$\sigma_1 = 224\ MPa \hspace{1cm} Ans.$$
$$\sigma_2 = 0 \hspace{1cm} Ans.$$

D–80 $A_y = B_y = 12\ kN$
Segment AC:
$$V_C = 0, M_C = 24\ kN \cdot m$$
$$\tau_C = 0\ (\text{since}\ V_C = 0)$$
$$\sigma_C = 0\ (\text{since}\ C\ \text{is on neutral axis})$$
$$\sigma_1 = \sigma_2 = 0 \hspace{1cm} Ans.$$

D–81 $A_y = 3\ kip$
Use section having a length x.
$$\zeta + \Sigma M = 0;\ -3x + 2x\left(\frac{x}{2}\right) + M = 0$$
$$M = 3x - x^2$$
$$EI\,\frac{d^2v}{dx^2} = 3x - x^2$$
Integrate twice, use
$v = 0$ at $x = 0, v = 0$ at $x = 3$ m
$$v = \frac{1}{EI}\left(-\tfrac{1}{12}\,x^4 + 0.5\,x^3 - 2.25x\right) \hspace{0.5cm} Ans.$$

D–82 $A_y = 15\ kip$
$M_A = 100\ kip \cdot ft$
Use section having a length x.
Intensity of $w = \left(\frac{3}{10}\right)x$ at x.
$$\zeta + \Sigma M = 0;$$
$$-15x + 100 + \left(\frac{1}{3}x\right)\left[\frac{1}{2}\left(\frac{3}{10}x\right)(x)\right] + M = 0$$
$$M = 15x - 0.05\,x^3 - 100$$
$$EI\,\frac{d^2v}{dx^2} = 15x - 0.05\,x^3 - 100$$
Integrate twice, use
$v = 0$ at $x = 0, dv/dx = 0$ at $x = 0$
$$v = \frac{1}{EI}\,(2.5x^3 - 0.0025\,x^5 - 50\,x^2) \hspace{0.5cm} Ans.$$

D–83 From Appendix C, consider distributed and concentrated loads separately.

$$\Delta_C = \frac{5wL^4}{768\ EI} + \frac{PL^3}{48\ EI}$$

$$= \frac{5(2)(6)^4}{768\ EI} + \frac{8\ (6)^3}{48\ EI} = \frac{52.875\ \text{kip} \cdot \text{ft}^3}{EI} \downarrow \qquad Ans.$$

D–84 From Appendix C, consider distributed load and couple moment separately.

$$\theta_A = \frac{w_0 L^3}{45\ EI} + \frac{ML}{6\ EI}$$

$$= \frac{4\ (3)^3}{45\ EI} + \frac{20\ (3)}{6\ EI} = \frac{12.4\ \text{kN} \cdot \text{m}^2}{EI} \qquad \downarrow \qquad Ans.$$

D–85 True. $\qquad\qquad Ans.$

D–86 $P = \dfrac{\pi^2 EI}{(KL)^2} = \dfrac{\pi^2\ (29\ (10^3))\ (\frac{\pi}{4}\ (0.5)^4)}{[0.5\ (50)]^2} = 22.5\ \text{kip}$
$$Ans.$$

$$\sigma = \frac{P}{A} = \frac{22.5}{\pi(0.5)^2} = 28.6\ \text{ksi} < \sigma_Y \ \text{OK}$$

D–87 $P = \dfrac{\pi^2 EI}{(KL)^2} = \dfrac{\pi^2\ (1.6)(10^3)\ [\frac{1}{12}\ (4)(2)^3]}{[1\ (12)(12)]^2} = 2.03\ \text{kip}$
$$Ans.$$

D–88 $A = \pi((0.025)^2 - (0.015)^2) = 1.257\ (10^{-3})\ \text{m}^2$

$I = \frac{1}{4}\ \pi\ ((0.025)^4 - (0.015)^4) = 267.04\ (10^{-9})\ \text{m}^4$

$$P = \frac{\pi^2\ EI}{(KL)^2} = \frac{\pi^2\ (200\ (10^9))(267.04)(10^{-9})}{[0.5(5)]^2}$$
$$= 84.3\ \text{kN} \qquad\qquad Ans.$$

$$\sigma = \frac{P}{A} = \frac{84.3\ (10^3)}{1.257\ (10^{-3})} = 67.1\ \text{MPa} < 250\ \text{MPa}\ \ \text{OK}$$

D–89 $P = \dfrac{\pi^2\ EI}{(KL)^2};\quad 40 = \dfrac{\pi^2\ 29(10^3)[\frac{\pi}{4}\ (1^4 - r_2^4)]}{[1\ (6)(12)]^2}$

$r_2 = 0.528\ \text{in.}$

$$\sigma = \frac{P}{A} = \frac{40}{\pi\ [(1)^2 - (0.528)^2]} = 17.6\ \text{ksi} < 36\ \text{ksi OK}$$

Thus $t = 1 - 0.528 = 0.472\ \text{in.}$ $\qquad Ans.$

D–90 $P = \dfrac{\pi^2 EI}{(KL)^2};\quad 3 = \dfrac{\pi^2\ 29(10^3)\ (\frac{\pi}{4}\ r^4)}{[1\ (40)]^2}$

$r = 0.382\ \text{in.}$

$$\sigma = \frac{P}{A} = \frac{3}{\pi\ (0.382)^2} = 6.53\ \text{ksi} < 36\ \text{ksi OK}$$

$d = 2r = 0.765$

Use $d = \dfrac{13}{16}$ in. (0.8125 in.) $\qquad\qquad Ans.$

ANSWERS

Chapter 1

1–1. $T_B = 150 \text{ lb} \cdot \text{ft}, T_C = 500 \text{ lb} \cdot \text{ft}$

1–2. $N_A = 77.3 \text{ N}, V_A = 20.7 \text{ N},$
$M_A = -0.555 \text{ N} \cdot \text{m},$
$M_A = -0.555 \text{ N} \cdot \text{m}$

1–3. $N_C = 0, V_C = 3.50 \text{ kip},$
$M_C = -47.5 \text{ kip} \cdot \text{ft},$
$N_D = 0, V_D = 0.240 \text{ kip}, M_D = -0.360 \text{ kip} \cdot \text{ft}$

1–5. $N_C = 0, M_C = 11.2 \text{ kip} \cdot \text{ft}$

1–6. $N_E = 0, V_E = 0.450 \text{ kip}, M_E = -0.675 \text{ kip} \cdot \text{ft},$
$N_D = 0, V_D = 0.930 \text{ kip}, M_D = 11.0 \text{ kip} \cdot \text{ft}$

1–7. $N_D = 0.703 \text{ kN}, V_D = 0.3125 \text{ kN},$
$M_D = 0.3125 \text{ kN} \cdot \text{m}$

1–9. $N_A = 0, V_A = 450 \text{ lb},$
$M_A = -1125 \text{ lb} \cdot \text{ft} = -1.125 \text{ kip} \cdot \text{ft},$
$N_B = 0, V_B = 850 \text{ lb},$
$M_B = -6325 \text{ lb} \cdot \text{ft} = -6.325 \text{ kip} \cdot \text{ft},$
$V_C = 0, N_C = -1200 \text{ lb} = -1.20 \text{ kip},$
$M_C = -8125 \text{ lb} \cdot \text{ft} = -8.125 \text{ kip} \cdot \text{ft}$

1–10. $N_C = -45.0 \text{ kip}, V_C = 0, M_C = 9.00 \text{ kip} \cdot \text{ft}$

1–11. $N_B = 0, V_B = 288 \text{ lb},$
$M_B = -1152 \text{ lb} \cdot \text{ft} = -1.15 \text{ kip} \cdot \text{ft}$

1–13. $V_D = 17.3 \text{ N}, N_D = 10 \text{ N},$
$M_D = 1.60 \text{ N} \cdot \text{m},$

1–14. $N_B = -0.4 \text{ kip}, V_B = 0.960 \text{ kip},$
$M_B = -3.12 \text{ kip} \cdot \text{ft}$

1–15. $N_C = -0.4 \text{ kip}, V_C = 1.08 \text{ kip},$
$M_C = -6.18 \text{ kip} \cdot \text{ft},$
$N_D = 0, V_D = 1.45 \text{ kip}, M_D = -15.7 \text{ kip}$

1–17. $P = 0.5333 \text{ kN} = 0.533 \text{ kN}, N_C = -2.00 \text{ kN},$
$V_C = -0.533 \text{ kN}, M_C = 0.400 \text{ kN} \cdot \text{m}$

1–18. $N_D = 18 \text{ kN}, V_D = 90 \text{ kN},$
$M_D = 21.6 \text{ kN} \cdot \text{m},$
$N_E = 0, V_E = 90 \text{ kN}, M_E = 18 \text{ kN} \cdot \text{m}$

1–19. $N_C = -18.2 \text{ N}, V_C = 10.5 \text{ N}, M_C = -9.46 \text{ N} \cdot \text{m}$

1–21. $N_E = 0, V_E = 120 \text{ N}, M_E = 48.0 \text{ N} \cdot \text{m},$
$V = 0, N = 1.39 \text{ kN}, M = 0$

1–22. $N_D = 0, V_D = 0.750 \text{ kip}, M_D = 13.5 \text{ kip} \cdot \text{ft},$
$N_E = 0, V_E = -9.00 \text{ kip}, M_E = -24.0 \text{ kip} \cdot \text{ft}$

1–23. $N_{a-a} = -2338 \text{ lb} = -2.34 \text{ kip},$
$V_{a-a} = -900 \text{ lb} = -0.900 \text{ kip},$
$M_{a-a} = 3600 \text{ lb} \cdot \text{ft} = 3.60 \text{ kip} \cdot \text{ft}$

1–25. $V_{b-b} = 2475 \text{ lb} = 2.475 \text{ kip},$
$N_{b-b} = 389.7 \text{ lb} = 0.390 \text{ kip},$
$M_{b-b} = 3600 \text{ lb} \cdot \text{ft} = 3.60 \text{ kip} \cdot \text{ft}$

1–26. $(V_C)_x = -250 \text{ N}, (N_C)_y = 0, (V_C)_z = -240 \text{ N},$
$(M_C)_x = -108 \text{ N} \cdot \text{m},$
$(T_C)_y = 0, (M_C)_z = -138 \text{ N} \cdot \text{m}$

1–27. $(N_D)_x = 0, (V_D)_y = 154 \text{ N},$
$(V_D)_z = -171 \text{ N}, (T_D)_x = 0,$
$(M_D)_y = -94.3 \text{ N} \cdot \text{m}, (M_D)_z = -149 \text{ N} \cdot \text{m}$

1–29. $(V_A)_x = 0$, $(V_A)_y = 7.20$ kip, $(N_A)_z = 3.30$ kip,
$(M_A)_x = -99.3$ kip $\cdot$ ft, $(M_A)_y = -9.75$ kip $\cdot$ ft,
$(T_A)_z = 46.8$ kip $\cdot$ ft

1–30. $V_B = 0.785\ w\ r$,
$N_B = 0$, $T_B = 0.0783\ w\ r^2$, $M_B = -0.293\ w\ r^2$

1–31. $N_A = P\cos\theta$, $V_A = P\sin\theta$, $M_A = Pr(1 - \cos\theta)$

1–33. $\sigma = 1.82$ MPa

1–34. $P_{\text{allow}} = 6240$ lb $= 6.24$ kip

1–35. $\sigma = 15.4$ psi

1–37. $\sigma_{\max} = 357$ psi

1–38. $\theta = 47.4°$, $F_{AB} = 34.66$ lb, $F_{AC} = 44.37$ lb,
$\sigma_{AC} = 353$ psi, $\sigma_{AB} = 177$ psi

1–39. $\sigma_B = 151$ kPa, $\sigma_C = 32.5$ kPa, $\sigma_D = 25.5$ kPa

1–41. $F = 36$ kN, $d = 110$ mm

1–42. $\tau_{\text{avg}} = 119$ kPa

1–43. $(\tau_B)_{\text{avg}} = 6053$ psi $= 6.05$ ksi

1–45. $(\tau_D)_{\text{avg}} = 6621$ psi $= 6.62$ ksi,
$(\tau_E)_{\text{avg}} = 6217$ psi $= 6.22$ ksi

1–46. $(\tau_D)_{\text{avg}} = 13.2$ ksi, $(\tau_E)_{\text{avg}} = 12.4$ ksi

1–47. $\bar{x} = 4$ in., $\bar{y} = 4$ in., $\sigma = 9.26$ psi

1–49. $x = 0.4$ m, $\sigma = 306$ MPa

1–50. $\tau_A = 0$, $(\tau_B)_{\text{avg}} = 11.0$ ksi

1–51. $\sigma_D = 13.3$ MPa (C), $\sigma_E = 70.7$ MPa (T)

1–53. $\sigma_{a-a} = 90.0$ kPa, $\tau_{a-a} = 52.0$ kPa

1–54. 3.71 ksi

1–55. 1.59 ksi

1–57. $\tau_{\text{avg}} = 16$ psi

1–58. $\sigma = \dfrac{P}{A}\sin^2\theta$, $\tau_{\text{avg}} = \dfrac{P}{2A}\sin 2\theta$

1–59. 22.9 ksi

1–61. $P = 6.82$ kip

1–62. $\sigma = \dfrac{w_0}{2aA}(2a^2 - x^2)$

1–63. $\sigma = \dfrac{w_0}{2aA}(2a - x)^2$

1–65. $\sigma = (47.5 - 20.0x)$ MPa

1–66. $\sigma = 102$ psi

1–67. $\sigma_{AB} = 417$ psi   (C), $\sigma_{BC} = 469$ psi (T),
$\sigma_{AC} = 833$ psi   (T)

1–69. $\sigma_{AB} = 16.7$ MPa, $\sigma_{BC} = 8.64$ MPa

1–70. $\tau_B = 324$ MPa, $F_A = 165$ kN, $\tau_A = 324$ MPa

1–71. 3.70 kN

1–73. $\sigma = 72.2$ kPa

1–74. $\sigma = 49.5$ kPa

1–75. $\sigma = (238 - 22.6z)$ kPa

1–77. $h = 2\dfrac{3}{4}$ in.

1–78. $P = 90$ kN, $A = 6.19\,(10^{-3})$m^2, $P_{\max} = 155$ kN

1–79. $d = \dfrac{5}{8}$ in., $h = \dfrac{3}{8}$ in.

1–81. $a = 0.253$ in.

1–82. $P = 19.8$ kip

1–83. $P = 55.0$ kN *(Controls !)*

1–85. $d_A = 0.441$ in., $d_B = 0.794$ in.

1–86. $A_{BC} = 0.577$ in^2, $d_A = 0.743$ in., $d_B = 0.525$ in.

1–87. $P = 0.491$ kip *(Controls !)*

1–89. $d_A = 0.155$ in., $d_B = 0.162$ in.

1–90. $h = 1.74$ in.

1–91. $d_1 = 44.6$ mm,
$d_3 = 26.4$ mm,
$t = 15.8$ mm

1–93. $w = 0.452$ kip/ft (Controls), $w = 0.576$ kip/ft

1–94. $t = 0.1667$ m $= 167$ mm
$b = 33.3$ mm

1–95. $d_B = 0.00611$ m, $= 6.11$ mm
$d_w = 0.0154$ m, $= 15.4$ mm

1–97. $3\dfrac{1}{2}$ in. $\times$ $3\dfrac{1}{2}$ in. plate, $4\dfrac{1}{2}$ in. $\times$ $4\dfrac{1}{2}$ in. plate

1–98. $t = 0.001778$ m $= 1.78$ mm,
$d_t = 0.01185$ m $= 11.9$ mm,
$d_r = 0.004120$ m $= 4.12$ mm

1–99. $w_1 = 5.33$ kip/in., $w_2 = 8.00$ kip/in., $d = 0.714$ in.

1–101. $F_{HI} = 20.0$ kN, $F_{BF} = F_{AG} = F = 15.0$ kN,
$d_{EF} = d_{CG} = 11.3$ mm

1–102. $N_D = -2.16$ kip, $V_D = 0$, $M_D = 2.16$ kip $\cdot$ ft,
$V_E = 0.540$ kip, $N_E = 4.32$ kip, $M_E = 2.16$ kip $\cdot$ ft

1–103. F.S. = 2.71, F.S. = 1.53

1–105. $d = 0.620$ in., $P_{max} = 7.25$ kip

1–106. $t = 0.800$ in., $P_{max} = 24.0$ kip, $w = 2.50$ in.

1–107. $\sigma = 267$ kPa

1–109. $\sigma = 215$ kPa, $\sigma = 192$ kPa, $\sigma = 170$ kPa

1–110. $\tau_{avg} = 61.3$ MPa

1–111. $\tau_{avg} = 79.6$ MPa

1–113. $\sigma = \gamma L^2/8s$

1–114. $d = 5\frac{1}{2}$ in.

1–115. $\tau_{avg} = 267$ psi

Chapter 2

2–1. $\varepsilon_{avg} = 0.250$ in./in.

2–2. $\varepsilon = 0.885$ in./in.

2–3. $(\varepsilon_{AD})_{avg} = 0.250(10^{-3})$ in./in.,
$(\varepsilon_{BE})_{avg} = 2.87(10^{-3})$ in./in.,
$(\varepsilon_{CF})_{avg} = 5.49(10^{-3})$ in./in.

2–5. $\Delta p = 11.2$ mm

2–6. $(\varepsilon_{avg})_{Approx} = 1.50\,\theta$, $(\varepsilon_{avg})_{approx} = 52.4(10^{-3})$ m/m

2–7. $\varepsilon_{AB} = 6.50\,(10^{-3})$ mm/mm

2–9. $\varepsilon_{AB} = 0.152$ in./in., $\varepsilon_{AC} = 0.0274$ in./in.

2–10. $\varepsilon_{AB} = 16.8(10^{-3})$ m/m

2–11. $\Delta L = \dfrac{kL^2}{2}$

2–13. $\gamma_{xy} = -0.0200$ rad

2–14. $\gamma_{x'y'} = -7.27\,(10^{-3})$ rad

2–15. $\varepsilon_x = -1.25\,(10^{-3})$ in./in., $\varepsilon_y = 2.50\,(10^{-3})$ in./in.,
$\varepsilon_{x'} = \varepsilon_{y'} = 0.627\,(10^{-3})$ in./in.

2–17. $(\gamma_B)_{xy} = 0.0116$ rad $= 11.6(10^{-3})$ rad,
$(\gamma_A)_{xy} = -0.0116$ rad $= -11.6(10^{-3})$ rad

2–18. $(\gamma_C)_{xy} = -11.6(10^{-3})$ rad, $(\gamma_D)_{xy} = 11.6(10^{-3})$ rad

2–19. $\varepsilon_{AC} = 1.60(10^{-3})$ mm/mm,
$\varepsilon_{DB} = 12.8(10^{-3})$ mm/mm

2–21. $\varepsilon_{AC} = 0.01665$ mm/mm $= 16.7(10^{-3})$ mm/mm,
$\varepsilon_{BD} = 0.01134$ mm/mm $= 11.3(10^{-3})$ mm/mm

2–22. $(\gamma_B)_{xy} = 5.24(10^{-3})$ rad

2–23. $\varepsilon_{AB} = 1.61\,(10^{-3})$ mm/mm
$\varepsilon_{CD} = 126\,(10^{-3})$ mm/mm

2–25. $\varepsilon_{AB} = 0.0381$ mm/mm $= 38.1(10^{-3})$ mm

2–26. $(\gamma_C)_{xy} = -0.137$ rad, $(\gamma_D)_{xy} = 0.137$ rad

2–27. $\Delta y = 2.03$ mm

2–29. $\Delta = \displaystyle\int \varepsilon\, ds = 1.69$ in.

2–30. $\varepsilon_{AB} = \dfrac{v_B \sin\theta}{L} - \dfrac{u_A \cos\theta}{L}$

Chapter 3

3–1. $E_{approx} = 3.275\,(10^3)$ ksi

3–2. $E = 55.3(10^3)$ ksi, $u_r = 9.96\,\dfrac{\text{in} \cdot \text{lb}}{\text{in}^3}$

3–3. $(u_t)_{approx} = 85.0\,\dfrac{\text{in} \cdot \text{lb}}{\text{in}^3}$

3–5. $(E)_{approx} = 229$ GPa, $(\sigma_u)_{approx} = 528$ MPa,
$(\sigma_f)_{approx} = 479$ MPa

3–6. $(u_t)_{approx} = 117$ MJ/m^3

3–7. The amount of elastic recovery = 0.00350 in.,
Permanent elongation = 0.1565 in.

3–9. $(u_t)_{approx} = 18.0\,\dfrac{\text{in} \cdot \text{kip}}{\text{in}^3}$, $u_r = 20$ in. $\cdot$ lb/in^3

3–10. $E_{approx} = 173$ GPa,
$\sigma_{pl} = 260$ MPa, $\sigma_u = 400$ MPa, $u_r = 195$ kJ/m^3,
Elastic recovery = 0.00208 mm/mm,
Permanent set = 0.0729 mm/mm

3–11. $L = 50.0123$ in.

3–13. $\Delta P = 50.05$ kip

3–14. copolymer

3–15. $E = 28.6(10^3)$ ksi

3–17. $A = 0.209$ in^2, $P = 1.62$ kip

3–18. $d_{AB} = 3.54$ mm, $d_{AC} = 3.23$ mm, $L_{AB} = 750.49$ mm

3–19. $\varepsilon_{BE} = 0.417(10^{-3})$ in./in., $w = 14.5$ lb,
$\varepsilon_{CD} = 0.250\,(10^{-3})$ in./in.

3–21. $T = 4.50$ kN

3–22. $P = 15.0$ kip

3–23. $A_{BC} = 0.8$ in^2, $A_{BA} = 0.2$ in^2, $P = 3.02$ kN

3–25. $\delta = 8.33$ mm

3–26. $d = 1.5034$ in.

3–27. $\delta L = -0.0173$ mm, $d = 20.00162$ mm

3–29. $\varepsilon_y = -0.0150$ in./in., $\varepsilon_x = 0.00540$ in./in.,
$\gamma_{xy} = -0.00524$ rad

3–30. $L = 50.0377$ mm, $d = 12.99608$ mm

3–31. $E = 32.5(10^3)$ ksi, $P = 2.45$ kip

3–33. $\delta = \dfrac{P}{2\pi h G} \ln \dfrac{r_o}{r_i}$

3–34. $\varepsilon = 0.000999$ in./in., If the nut is unscrewed, the load is zero. Therefore, the strain $\varepsilon = 0$

3–35. $\delta_h = 3.02$ mm

3–37. $P = 205$ lb

3–38. $\varepsilon_b = 0.00227$ mm/mm, $\varepsilon_s = 0.000884$ mm/mm

3–39. $E_{\text{approx}} = 32.0(10^3)$ ksi,
$(\sigma_y)_{\text{approx}} = 55.4$ ksi, $\sigma_u = 110$ ksi, $\sigma_f = 93.1$ ksi

3–41. $P = 11.3$ kN (*Controls !*), $P = 238.8$ kN

Chapter 4

4–1. $\delta_A = -3.64\,(10^{-3})$ mm

4–2. $\delta_A = -0.0603$ in.

4–3. $P_1 = 70.7$ kip, $P_2 = 141$ kip

4–5. $\delta_{A/C} = 0.0298$ in.

4–6. $\delta_B = -0.00632$ in., $\delta_A = 0.0670$ in.

4–7. $P_1 = 16.4$ kip, $P_2 = 35.3$ kip

4–9. $\delta_C = 0.00843$ in., $\delta_E = 0.00169$ in., $\delta_B = 0.0333$ in.

4–10. $\delta_{C_y} = 0.150$ mm

4–11. $P = 13.3$ kN

4–13. $\alpha = 0.00341°$, $\beta = 0.0295°$

4–14. $\delta_B = 0.0311$ in.

4–15. $P = 15.4$ kip

4–17. $W = 9.69$ kN

4–18. $\delta_F = 0.0113$ in.

4–19. $0.00878°$

4–21. $x = 4.24$ ft, $w = 1.02$ kip/ft

4–22. $P = 6.80$ kip

4–23. $P = 11.8$ kip

4–25. $P = 300$ kip, $\delta = 12.9$ ft

4–26. $F = 12.0$ kN, $\delta_{A/B} = -0.864$ mm

4–27. $F = 17.0$ kN, $\delta_{A/B} = -1.03$ mm

4–29. $\delta = \delta_W + \delta_P$

$$= \frac{PL}{\pi E r_2 r_1} + \frac{\gamma L^2(r_2 + r_1)}{6E(r_2 - r_1)} - \frac{\gamma L^2 r_1^2}{3E r_2(r_2 - r_1)}$$

4–30. 2.37 mm

4–31. $\Delta = 0.0128$ in.

4–33. $\delta = \dfrac{0.511P}{\pi r_0 E}$

4–34. $\delta = \dfrac{\gamma L^2}{6E}$

4–35. $x = \dfrac{h}{1 + \dfrac{E_A}{E_B}} = \dfrac{E_B}{E_A + E_B} h$

4–37. $\sigma_{\text{st}} = 1.66$ ksi, $\sigma_{\text{con}} = 0.240$ ksi, $\delta = 0.0055$ in.

4–38. $A_{\text{st}} = 18.2$ in^2, $\delta = 0.00545$ in.

4–39. $\sigma_{al} = 27.5$ MPa, $\sigma_{st} = 79.9$ MPa

4–41. $d = 2.39$ in.

4–42. $\sigma_{\text{st}} = 65.9$ MPa, $\sigma_{\text{con}} = 8.24$ MPa

4–43. $d = 36.3$ mm

4–45. $F_B = \dfrac{P}{3}$, $F_D = \dfrac{2}{3}P$

4–46. $P_s = 35.61$ kN, $P_b = 14.4$ kN

4–47. $T_{AC} = 0.170$ kip, $T_{AB} = 1.33$ kip

4–49. $\sigma_D = 13.4$ MPa, $\sigma_{BC} = 9.55$ MPa

4–50. $\theta = 63.7\,(10^{-6})$ rad

4–51. $F_{AC} = 727$ N,
$F_{AB} = F_{AD} = 465$ N

4–53. $\sigma_A = 189$ MPa, $\sigma_B = 21.4$ MPa

4–54. $P = 1.16$ kN

4–55. $a = 0.120$ mm

4–57. $\sigma_{AB} = \dfrac{7P}{12A}$, $\sigma_{CD} = \dfrac{P}{3A}$, $\sigma_{EF} = \dfrac{P}{12A}$

4–58. $\delta_B = 0.00257$ in.

4–59. $F_{CD} = 0.211$ kip, $F_{EF} = 1.26$ kip

4–61. $\delta_A = 2.27$ mm

4–62. $F_1 = \left(\dfrac{A_1 E_1}{2A_1 E_1 + A_2 E_2}\right)P$, $F_2 = \left(\dfrac{A_2 E_2}{2A_1 E_1 + A_2 E_2}\right)P$

4–63. $A_1' = \left(\dfrac{E_1}{E_2}\right)A_1$

4–65. $d = 4.90$ m

4–66. $s = 0.133(10^{-3})$ in.

4–67. $F_B = 16.9$ kN,

$\qquad$ $F_A = 16.9$ kN

4–69. $F_A = 25.6$ kN,

$\qquad$ $F_{sp} = 48\ 836.5$ N

4–70. $\delta_A = 4.42$ mm

4–71. $F_A = 4.09$ kip, $F_B = 2.91$ kip

4–73. $F_A = 14.0$ kN

4–74. $F = 4.20$ kN

4–75. $F = 116$ kip

4–77. $\delta = 0.348$ in., $F = 19.5$ kip

4–78. $F = 2.54$ kip, $F = 2.54$ kip

4–79. $F = 1.68$ kip

4–81. $F = 7.66$ kip

4–82. $\varepsilon_{max} = 8.80(10^{-6})$ in./in., $\varepsilon_{min} = -4.40(10^{-6})$ in./in.

4–83. $\sigma = 5.74$ ksi

4–85. $T_2 = 116°F$, $\sigma_{al} = 11.6$ ksi

4–86. $\sigma_{al} = 25.4$ ksi, $L'_{al} = 8.00462$ in.

4–87. $\sigma = 19.1$ ksi

4–89. $\sigma_A = 2.85$ ksi, $\sigma_C = 4.30$ ksi

4–90. $F_{AC} = 10.0$ lb, $F_{AD} = 136$ lb

4–91. $\sigma_s = 40.1$ MPa, $\sigma_b = 29.5$ MPa

4–93. $T_2 = 244°$

4–94. $F = \dfrac{\alpha AE}{2}(T_B - T_A)$

4–95. $P = 5.05$ kN   (Controls)

4–97. $\sigma_{max} = 34.8$ ksi   (Controls !)

4–98. $\sigma_{max} = 88.3$ MPa   (Controls !)

4–99. $P = 10.9$ kN   (Controls!)

4–101. $P = 77142.86$ N $= 77.1$ kN, $\delta = 0.429$ mm

4–102. $P = 15$ kip, $K = 1.60$

4–103. $\sigma_{st} = 36.0$ ksi, $\sigma_{al} = 19.3$ ksi

4–105. $\delta_{Tot} = 0.432$ in.

4–106. **a)** $\delta_D = 0.375$ in., **b)** $\delta_D = 6.40$ in.

4–107. **a)** $P = 10.9$ kN, **b)** $P = 33.0$ kN

4–109. $\delta = \dfrac{\gamma^2 L^3}{3c^2}$

4–110. $w = 21.9$ kN/m, $\delta_G = 4.24$ mm

4–111. **a)** $w = 18.7$ kN/m, **b)** $w = 21.9$ kN/m

4–113. $T_2 = 244°C$

4–114. $F_1 = 602$ kN

4–115. $\theta = \dfrac{\delta_{EF} - \delta_{AB}}{2d} = \dfrac{3E_2 L(T_2 - T_1)(\alpha_2 - \alpha_1)}{d(5E_2 + E_1)}$

4–117. $\delta_{B/C} = -0.0278$ in.

4–118. $\delta_D = 1.17$ mm

4–119. $\delta_{B/A} = 0.0918$ mm

Chapter 5

5–1. **a)** $T = 7.95$ kip · in., **b)** $T' = 6.38$ kip · in.

5–2. **a)** $r' = 0.707\ r$, **b)** $r' = 0.707\ r$

5–3. $\tau_{max} = 6.62$ ksi

5–5. $\tau_C = 37.7$ MPa, $\tau_D = 75.5$ MPa

5–6. $\tau_{AB} = 7.82$ ksi, $\tau_{BC} = 2.36$ ksi

5–7. $T_1 = 215$ N · m

$\qquad$ $(\tau_{max})_{CD} = 4.00$ MPa, $(\tau_{max})_{DE} = 2.58$ MPa

5–9. $(\tau_{BC})_{max} = 5.07$ ksi, $(\tau_{DE})_{max} = 3.62$ ksi

5–10. $(\tau_{EF})_{max} = 0$, $(\tau_{CD})_{max} = 2.17$ ksi

5–11. $d = 2\frac{1}{2}$ in.

5–13. $\tau_A = 1.72$ ksi, $\tau_B = 3.02$ ksi

5–14. $\tau_{abs\ max} = 3.59$ ksi

5–15. $\tau_A = 9.43$ MPa, $\tau_B = 14.1$ MPa

5–17. $d = 0.03441$ m $= 34.4$ mm

5–18. $\tau_A = 6.88$ MPa, $\tau_B = 10.3$ MPa

5–19. $\tau_{abs\ max} = 49.7$ MPa

5–21. $\tau_{abs\ max} = 90.7$ MPa, in the region 0.3 m $> x > 0.4$ m

5–22. $d = 33.8$ mm

5–23. $d = 1\dfrac{1}{4}$ in.

5–25. $\tau_{max} = 856$ psi

5–26. $d = \dfrac{1}{2}$ in.

5–27. $n = \dfrac{2\ r^3}{Rd^2}$

5–29. $d_i = 1\dfrac{5}{8}$ in.

5–30. $\tau_{max} = 3.44$ MPa

5–31. $d_2 = 2\dfrac{3}{8}$ in.

5–33. $\tau_{max} = 6018$ psi $= 6.02$ ksi

5–34. $c = (2.98\ x)$ mm

5–35. $T_0 = 670$ N $\cdot$ m, $\tau_{abs\ max} = 6.66$ MPa

5–37. $\tau_{max} = \dfrac{T\ c}{J} = \dfrac{T\ r}{\pi\ r^4} = \dfrac{2T}{\pi\ r^3}$

$= \dfrac{2\ T}{\pi\left[\dfrac{r_A\ (L - x) + r_B x}{L}\right]^3}$

$= \dfrac{2TL^3}{\pi[r_A(L - x) + r_B x]^3}$

5–38. $t = 0.174$ in.

5–39. $t = 0.104$ in.

5–41. $(\tau_{avg})_{max} = \dfrac{T}{2\pi r_i{}^2 h}$

5–42. $\tau_{max} = 44.3$ MPa, $\phi = 11.9°$

5–43. % increase in shear stress $= 6.67\%$,
% increase in $\phi = 6.67\%$

5–45. $\phi_{B/A} = 0.578°$

5–46. $\phi_{A/D} = 0.638°$

5–47. $\phi_{B/C} = 0.129°$

5–49. $\phi_{C/B} = 2.72°$

5–50. $\tau_{max} = 64.0$ MPa

5–51. $d = 1\dfrac{1}{4}$ in.

5–53. $t = 0.00753$ m $= 7.53$ mm

5–54. $\omega = 131$ rad/s

5–55. $\tau_{abs\ max} = 65.2$ MPa, $\phi_{C/F} = 11.4°$

5–57. $\phi_C = 0.113°$

5–58. $\phi_B = 1.53°$

5–59. $\phi_A = 1.78°$

5–61. $\phi_A = 2.70°$

5–62. $\phi_{A/C} = 5.45°$

5–63. $(\tau_{BC})_{max} = 10.2$ ksi, $(\tau_{BA})_{max} = 1.86$ ksi, $\phi_C = 2.66°$

5–65. $\phi = \dfrac{T}{2a\pi G}\ (1 - e^{-4aL})$

5–66. $\phi_A = 1.59°$

5–67. $\phi = \dfrac{7\ t_0\ L^2}{6\ \pi\ c^4 G}$

5–69. $\tau_o = 3.18$ MPa, $\tau_i = 1.59$ MPa, $\phi_{A/B} = 0.730°$

5–70. $\phi_A = |\ 0.432°\ |$

5–71. $\phi = \dfrac{7TL}{12\pi r^4 G}$

5–74. $\theta = \dfrac{T}{4\pi\ hG}\left[\dfrac{1}{r_i^2} - \dfrac{1}{r_o^2}\right]$

5–75. $(\tau_{AC})_{max} = 14.3$ MPa, $(\tau_{CB})_{max} = 9.55$ MPa

5–77. $F = 23.4$ lb

5–78. $\tau_{max} = 389$ psi

5–79. $T_D = 385$ N $\cdot$ m, $T_A = 115$ N $\cdot$ m

5–81. $\tau_{abs\ max} = 3.67$ ksi

5–82. $T_A = 19.4$ lb $\cdot$ ft, $T_B = 481$ lb $\cdot$ ft

5–83. $T = 1.29$ kip $\cdot$ ft

5–85. $T_B = 222$ N $\cdot$ m, $T_A = 55.6$ N $\cdot$ m

5–86. $\phi_E = 1.66°$

5–87. $\phi_C = 0.002019$ rad $= 0.116°$,
$(\tau_{st})_{max\ BC} = 394.63$ psi $= 395$ psi (Max),
$(\gamma_{st})_{max} = 34.3(10^{-6})$ rad, $(\tau_{br})_{max} = 96.1$ psi (Max),
$(\gamma_{br})_{max} = 17.2(10^{-6})$ rad

5–89. $T_B = \dfrac{37}{189}T, T_A = \dfrac{152}{189}T$

5–90. $T = 7.74$ N $\cdot$ m, $\tau_{max} = 37.2$ MPa $< \tau_Y$

5–91. $\tau_{max} = 24.6$ MPa, $\phi_{A/C} = -0.04894$ rad $= |\ 2.80°\ |$

5–93. $(\tau_{BC})_{max} = 0.955$ MPa, $(\tau_{AC})_{max} = 1.59$ MPa,
$\phi_{B/C} = -0.001123$ rad $= |\ 0.0643°\ |$

5–94. $(\tau_{max})_A = 308$ MPa

5–95. $a = 20.2$ mm,
$\phi_{A/B} = -0.1472$ rad $= |\ 8.44°\ |$

5–97. $T_B = 32$ lb $\cdot$ ft, $T_A = 48$ lb $\cdot$ ft, $\phi_C = 0.0925°$

5–98. $T = 0.0820$ N $\cdot$ m, $\phi = 25.5$ rad

5–99. $a = 0.0289$ m $= 28.9$ mm

5–101. $\tau_{avg} = 3.35$ ksi

5–102. $T = 3360$ N $\cdot$ m $= 3.36$ kN $\cdot$ m, $\phi = 11.6°$

5–103. $\tau_{avg} = 1.19$ MPa

5–105. $b = 0.803$ in.

5–106. $(\tau_{avg})_A = 9.62$ MPa

5–107. The factor of increase $= 2.85$

5–109. $(\tau_{avg})_A = 357$ kPa

5–110. $a = 12.7$ mm

5–111. $(\tau_{max})_f = 50.6$ MPa(Max)

5–113. $\tau_{max} = 3.98$ MPa

5–114. $r = 0.075$ in.

5–115. $T = 8.16$ N $\cdot$ m

5–117. $r = 0.15$ in.

5–118. $\rho_y = 1.16$ in.

5–119. $T = 32.46$ kip $\cdot$ in. $= 2.71$ kip $\cdot$ ft,
$T_P = 33.51$ kip $\cdot$ in. $= 2.79$ kip $\cdot$ ft

5–121. $T_P = 0.565$ N $\cdot$ m

5–122. $T_y = 1256.64$ N $\cdot$ m $= 1.26$ kN $\cdot$ m, $\phi = 3.58°$
$\phi = 4.86°$

5–123. $\rho_y = 0.01298$ m $= 13.0$ mm

5–125. $T = 3.27$ kN $\cdot$ m, $\phi = 68.8°$

5–126. $T_p = 6.98$ kN $\cdot$ m, $\phi_r = 9.11°$

5–127. $T = 20.8$ kN $\cdot$ m,
$\phi = 34.4°$, $\phi_r = 12.2°$

5–129. $T_r = 9.15$ MPa at $\rho = c$

5–130. $T = 2.43$ kN $\cdot$ m, $\phi_r = 7.61°$

5–131. $T = 3.27$ kN $\cdot$ m, $\phi = 34.4°$

5–133. $t = 1.60$ mm

5–134. $\omega = 99.6$ rad/s

5–135. $F = 26.2$ N, $\phi = 1.86°$

5–137. $\phi_D = 0.391°$

5–138. $\tau_{max} = 82.0$ MPa

5–139. $P = 1.10$ kW, $\tau_{max} = 825$ kPa

5–141. $T = 71.5$ N $\cdot$ m, $\tau_{max} = 23.3$ MPa

5–142. $(\tau_{avg})_{max} = 2.03$ MPa, $\phi = 0.258°$

Chapter 6

6–1. $x = 0.25^-, V = -24, M = -6.0$

6–2. $x = 18^-, V = -400, M = -7200$

6–3. $x = 1.75^-, V = -124, M = -194$

6–5. $x = 1^-, V = -60, M = -60$

6–6. $V = 15.6$ N, $M = (15.6x + 100)$ N $\cdot$ m

6–7. $x = 0.2^-, V = -4.33, M = -0.866$

6–9. $x = 0.4, V = -1, M = -0.4$

6–10. $P = 809$ lb, $C_y = 1501$ lb, $C_x = 404$ lb,
$x = 8^-; V = -701, M = -5604$

6–11. $x = 3^-, V = -2000, M = -6000$

6–13. $x = 6^-, V = -800, M = -4800$

6–14. $x = 8, V = -6400, M = 2560$

6–15. $V = 0, M = m(L - x)$

6–17. $x = 2^-, V = 1000, M = 2400$

6–18. $0 \le x < 6, V = 30 - 2x, M = -x^2 + 30x - 216;$
$6 < x \le 10, V = 8, M = 8x - 120$

6–19. $x = 1, V = 0, M = 2.50$

6–21. $0 < x < 8, V = \{7.60 - 0.800x\}$ kip,
$M = \{-0.400x^2 + 7.60x - 44.8\}$ kip $\cdot$ ft, $8 < x \le 16,$
$V = 1.20$ kip, $M = \{1.20x - 19.2\}$ kip $\cdot$ ft

6–22. $0 < x < 4, V = -250$ lb, $M = \{-250x\}$ lb $\cdot$ ft,
$4 < x < 10, V = \{1050 - 150x\}$ lb,
$M = \{-75x^2 + 1050x - 4000\}$ lb $\cdot$ ft; $10 < x < 14,$
$V = 250$ lb,
$M = \{250x - 3500\}$ lb $\cdot$ ft

6–23. $T = 120$ kip $\cdot$ in.

6–25. $x = 2.54, V = 0, M = 346$

6–26. $A_y = 9.375$ kip, $M_A = 18.6$ kip $\cdot$ ft, $A_x = 0$

6–27. $x = L/2, V = 0, M = 23w_0L^2/216$

6–29. $x = 6^-, V = -2, M = -12$

6–30. $x = L/2, V = 0, M = w_0L^2/12$

6–31. $V = \dfrac{w_0}{4}(3L - 4x),$
$M = \dfrac{w_0}{24}(-12x^2 + 18Lx - 7L^2),$
$V = \dfrac{w_0}{L}(L - x)^2, M = -\dfrac{w_0}{3L}(L - x)^3$

6–33. $w = 40.0$ lb/ft

6–34. $w_0 = 1.2$ kN/m

6–35. $x = 3^-, V = 20, M = -60$

6–37. $a = 0.207L$

6–38. $\%\left(\dfrac{M'}{M}\right) = 84.6\%$

6–39. $M = 36.5$ kN · m, $\sigma_{max} = 40.0$ MPa

6–41. $(\sigma_{max})_t = 2.40$ ksi, $(\sigma_{max})_c = 4.80$ ksi

6–42. $\sigma_{max} = 74.7$ MPa, % of effectiveness $= 53.0\%$

6–43. $M = 129.2$ kip · in. $= 10.8$ kip · ft

6–45. $F = 3296.5$ lb $= 3.30$ kip

6–46. $F = 5494.19$ lb $= 5.49$ kip

6–47. $\sigma_B = 3.61$ MPa, $\sigma_C = 1.55$ MPa

6–49. $\sigma_{max} = 86.5$ psi

6–50. $F = 846$ lb

6–51. $\sigma_A = 6.81$ MPa, $\sigma_B = 1.01$ MPa,
$\sigma_C = 4.14$ MPa

6–53. $F_{R_A} = 0, F_{R_B} = 1.50$ kN

6–54. $\sigma_{max} = 15.4$ ksi

6–55. $\sigma_D = 5.00$ ksi (Max), $F_A = 17.7$ kip, $F_B = 13.7$ kip

6–57. $\sigma_A = 214$ psi, $\sigma_B = 33.0$ psi, $\sigma_C = 115$ psi

6–58. **(a)** $\sigma_{max} = 95.0$ MPa, By comparison, section
(b) will have the least amount of bending stress.,
% of effectiveness $= 38.4\%$

6–59. $\sigma_{max} = 244$ MPa

6–61. $\sigma_{max} = 7.59$ ksi

6–62. $\sigma_{max} = 15.1$ ksi

6–63. $\sigma_{max} = 12.2$ ksi

6–65. $a = 66.9$ mm

6–66. $a = 3.19$ in.

6–67. $\sigma_{max} = 6.26$ MPa

6–69. $b = 1.23$ in., $h = 2b = 2.47$ in.

6–70. $d_i = 2.56$ in.

6–71. $\sigma_{max} = 181$ MPa

6–73. $P = 1.67$ kN

6–74. $\sigma_{max} = 9.00$ MPa

6–75. $\sigma = 13.6$ ksi

6–77. $a = 160$ mm

6–78. $P = 7.29$ kip (*Controls*), $P = 9042$ lb $= 9.04$ kip

6–79. $P = 5.97$ kip (*Controls*)

6–81. $d = 199$ mm

6–82. $\sigma_{max} = 2.70$ ksi

6–83. $a = 0, \sigma_{max} = \dfrac{3}{2}\dfrac{PL}{bd^2}$

6–85. $d_o = 188$ mm, $d_i = 0.8d_o = 151$ mm

6–86. $\sigma_{max} = 45.1$ ksi

6–87. $\varepsilon_{max} = 0.711(10^{-3})$ mm/mm

6–89. $d = 86.3$ mm

6–90. $\sigma_{max} = 6.94$ MPa

6–91. $\sigma_{max} = 25.8$ ksi

6–93. $\bar{y} = 0.04625$ m,
$I = 9.90885(10^{-6})$ m^4, $w = 2.61$ kN/m

6–94. $b = 4.02$ in.

6–95. $\sigma_{max} = 2.18$ ksi

6–97. $\sigma_A = -119$ kPa, $\sigma_B = 446$ kPa,
$\sigma_D = -446$ kPa, $\sigma_E = 119$ kPa

6–98. $\bar{y} = 4.83$ in., $M = 75.7$ kip.ft (*Controls!*)

6–99. $\bar{y} = 4.83$ in., $\sigma_A = 2.35$ ksi (C), $\sigma_B = 2.62$ ksi

6–101. $\bar{z} = 36.6$ mm, $\sigma_A = 4.38$ MPa, $\sigma_B = -1.13$ MPa,
$\sigma_D = -1.977$ MPa, $\sigma_E = 5.23$ MPa

6–102. $P = 14.2$ kN

6–103. $\sigma_A = 7.60$ MPa (T)   (Max),
$\sigma_B = 7.60$ MPa(C)   (Max)

6–105. $\sigma_{max} = 66.8$ ksi

6–106. $\sigma_{max} = 161$ MPa

6–107. $d = 28.9$ mm

6–109. $d = 62.9$ mm

6–110. $a = 0, b = -\left(\dfrac{M_z I_y + M_y I_{yz}}{I_y I_z - I_{yz}^2}\right), c = \dfrac{M_y I_z + M_z I_{yz}}{I_y I_z - I_{yz}^2},$
$\sigma_x = -\left(\dfrac{M_z I_y + M_y I_{yz}}{I_y I_z - I_{yz}^2}\right)y + \left(\dfrac{M_y I_z + M_z I_{yz}}{I_y I_z - I_{yz}^2}\right)z$

6–111. $\sigma_A = 293$ kPa (C)

6–113. $\sigma_A = 326$ kPa (T)

6–114. $h = 41.3$ mm,
$M = 6.60$ kN · m (*Controls!*), $M = 29.2$ kN.

6–115. $M = 6.41$ kN · m (*Controls!*), $M = 28.4$ kN · m

6–117. $M = 128$ kN · m (*Controls*)

6–118. $M = 12.4$ kN · m (*Controls!*)

6–119. $M = 19.9$ kN · m (*Controls!*)

6–121. $M = 16.4$ kip · ft (*Controls!*)

6–122. $(\sigma_{max})_{st} = 15.0$ ksi,

   $(\sigma_{max})_w = 0.578$ ksi, $(\sigma_B)_{st} = 12.0$ ksi

6–123. $M = 11.7$ kip $\cdot$ ft (*Controls!*)

6–125. $M = 97.5$ kip $\cdot$ ft (*Controls*)

6–126. $(\sigma_{max})_{st} = 9.42$ MPa, $(\sigma_{max})_{br} = 6.63$ MPa,

   $\sigma_{st} = 1.86$ MPa, $\sigma_{br} = 0.937$ MPa

6–127. $\bar{y} = 0.08355$ m, $I_{NA} = 57.62060(10^{-6})$ m^4,

   $M = 58.8$ kN $\cdot$ m (*Controls!*)

6–129. $M = 1.14$ kip $\cdot$ ft (*Controls*)

6–130. $\sigma_A = 3.82$ MPa (T), $\sigma_B = 9.73$ MPa (C)

6–131. $\sigma_C = 2.66$ MPa (T)

6–133. $P = 3.09$ N (*Controls!*)

6–134. $\sigma_A = 144$ psi (T), $\sigma_B = 106$ psi (C),

   $\sigma_C = 6.11$ psi (C)

6–135. $\sigma_B = 26.2$ MPa (C)   (Max)

6–137. $\sigma = \dfrac{M\bar{r}}{AI}\left(-\dfrac{yA}{r}\right) = -\dfrac{My}{I}$

6–138. $\sigma_A = 0.892$ MP a (T), $\sigma_B = 0.447$ MPa (C),

   No! This is a consequence of Saint - Venant's principle.

6–139. $M = 14.0$ kN $\cdot$ m (*Controls!*)

6–141. $\sigma_{max} = 54.4$ MPa

6–142. $\sigma_{max} = 384$ MPa

6–143. $M = 9.11$ N $\cdot$ m

6–145. $\sigma_{max} = 12.5$ ksi

6–146. $\sigma_{max} = 80.6$ MPa

6–147. $M = 347$ lb $\cdot$ ft

6–149. $\sigma_{max} = 13.8$ ksi

6–150. $r = 8.0$ mm

6–151. $M_Y = 14.3$ kN $\cdot$ m, $M_P = 25.6$ kN $\cdot$ m

6–153. $k = 1.27$

6–154. $\sigma'_{top} = 67.1$ MPa

6–155. $K = 1.57$

6–157. $M_Y = 330$ kip $\cdot$ ft, $M_P = 585$ kip $\cdot$ ft

6–158. $Z = 195$ in^3, $k = 1.78$

6–159. $M_P = 172$ kip $\cdot$ ft

6–161. $M_Y = 312.5$ kN $\cdot$ m, $M_P = 437.5$ kN $\cdot$ m

6–162. $Z = 2.75a^3$, $k = 1.71$

6–163. $M_Y = 38.7$ kip $\cdot$ ft, $M_P = 66.0$ kip $\cdot$ ft

6–165. $Z = 0.0976bh^2$, $k = 2.34$

6–166. $M_Y = 3.07$ kN $\cdot$ m, $M_P = 7.19$ kN $\cdot$ m

6–167. **a)** $P = 25.0$ kN, **b)** $P = 37.5$ kN

6–169. $P = 100$ lb

6–170. **a)** $M = 35.0$ kip $\cdot$ ft, **b)** $M = 59.8$ kip $\cdot$ ft

6–171. $M = \dfrac{nbh^2}{2(2n + 1)}\sigma_{max}$

6–173. $\sigma'_{top} = 25.1$ ksi, $y = 2.356$ in.

6–174. $x = 4^-$, $V = 11.7$ $M = 46.7$

6–175. $Z = \dfrac{2}{3}a^3$, $k = 2.00$

6–177. $(\sigma_{max}) = 3.43$ MPa (T), $(\sigma_{max}) = 1.62$ MPa (C)

6–178. $\sigma_{max} = 6.44$ ksi

6–179. $x = 0.2$, $V = 217$, $M = 43.3$

6–181. $\sigma = \dfrac{6M}{a^3}(\cos\theta + \sin\theta)$, $\theta = 45°$, $\alpha = -45°$

6–182. **a)** $\sigma_A = 2.08$ MPa (C),

   b) $\sigma = 2.08$ MPa (C)

6–183. **a)** $\sigma_{max} = 0.410$ MPa

   b) $\sigma_{max} = 0.410$ MPa

Chapter 7

7–1. $\tau_A = 74.7$ MPa

7–2. $\tau_B = 98.7$ MPa

7–3. $\tau_{max} = 0.499$ ksi,

   $(\tau_{AB})_F = 0.166$ ksi, $(\tau_{AB})_W = 0.498$ ksi

7–5. $Q_{max} = 88.23$ in^4,

   $\tau_{max} = 276$ psi, shear stress jump $= 156$ psi

7–6. $V_f = 3.05$ kip

7–7. $\tau_B = 4.41$ MPa

7–9. $V = 9.96$ kip

7–10. $V = 32.1$ kip

7–11. $\tau_{max} = 4.48$ ksi

7–13. $\tau_{max} = 190$ kN

7–14. $\tau_{max} = 3.16$ MPa,

   $(\tau_A)_w = 1.87$ MPa, $(\tau_A)_f = 1.24$ MPa

7–15. $\tau_{max} = \dfrac{4V}{3A}$

7–17. $P = 80.1$ kip

7–18. $\tau_{max} = 99.8$ psi

7–19. $\tau_{max} = 14.7$ MPa

7–21. $w = 5.69$ kip/ft

7–22. $w = 13.3$ kip/ft, $\tau_{max} = 625$ psi

7–23. $\tau_A = 35.7$ psi

7–25. $F_t = 512$ lb

7–26. $\tau_{max} = 280$ psi

7–33. $F = 675$ lb

7–34. $F = 5.31$ kN

7–35. $\tau_B = 646$ psi, $\tau_A = 592$ psi

7–37. $s = 8.66$ in., $s' = 1.21$ in.

7–38. $V = 34.5$ kip

7–39. $P = 238$ N

7–41. $w = 354$ lb/ft

7–42. $q = 141$ lb/in.

7–43. $\tau_{nail} = 159$ MPa

7–45. $\tau = 14.4$ MPa

7–46. $V = 499$ kN

7–47. $s = 13.8$ in.

7–51. $q_A = 196$ lb/in., $q_B = 452$ lb/in., $q_{max} = 641$ lb/in.

7–53. $q_A = 13.0$ kN/m, $q_B = 9.44$ kN/m

7–54. $q_C = 38.6$ kN/m

7–55. $q_B = 12.6$ kN/m, $q_{max} = 22.5$ kN/m

7–57. $q_A = 0$, $q_B = 417$ lb/in.

7–58. $q_{max} = 106$ kN/m

7–59. $q_A = 79.5$ kN/m

7–61. $e = 1.07$ in.

7–62. $e = \left(\dfrac{h_2^3}{h_1^3 + h_2^3}\right)b$

7–63. $e = \dfrac{3b^2}{2(d + 3b)}$

7–65. $e = 0$

7–66. $e = \dfrac{3(b_2^2 - b_1^2)}{h + 6(b_1 + b_2)}$

7–67. $e = \dfrac{b(6h_1\,h^2 + 3bh^2 - 8h_1^3)}{(h + 2h_1)^3 + 6b\,h^2}$

7–69. $e = \dfrac{15}{38}d$

7–70. $e = \dfrac{7}{10}a$

7–71. $e = \left(\dfrac{4 - \pi}{\pi}\right)r$

7–73. $e = 2\,r$

7–74. $e = \left(\dfrac{12}{3\pi + 4}\right)r$

7–75. $\tau_{max} = 928$ psi

7–77. $V = 4.10$ kip

7–78. $V = 749$ lb

7–79. $q_A = 0$, $q_B = 1.21$ kN/m, $q_C = 3.78$ kN/m

7–81. $\tau_A = 1.99$ MPa, $\tau_B = 1.65$ MPa

7–82. $\tau_A = 160$ psi, $\tau_C = 0$

7–83. $\tau_B = 213$ psi

Chapter 8

8–1. $t = 18.75$ mm

8–2. $\sigma_1 = 600$ psi, $\sigma_2 = 0$

8–3. $\sigma_1 = 600$ psi, $\sigma_2 = 300$ psi

8–5. $T = 18.2$ kip $\cdot$ ft, $P = 18.1$ kip, $F = 9.05$ kip

8–6. $\sigma_c = 2.69$ ksi

8–7. $p = 25$ psi, $\delta = 0.00140$ in.

8–9. $s = 33.3$ in.

8–10. $T_1 = 128°F$, $\sigma_1 = 12.1$ ksi, $p = 252$ psi

8–11. $p = \dfrac{E(r_2 - r_3)}{\dfrac{r_2^2}{r_2 - r_1} + \dfrac{r_3^2}{r_4 - r_3}}$

8–13.   **a)** $\sigma_1 = 127$ MPa, **b)** $\sigma_1' = 63.3$ MPa,
   c) $(\tau_{avg})_b = 258$ MPa

8–14. $\theta = 54.7°$ 322

8–15. $\sigma_{max} = 44.0$ ksi (T)

8–17.   2.13 ksi (T), 1.07 ksi (C)

8–18. $\sigma_A = 0.533$ ksi (T), $\sigma_B = 1.07$ ksi (C),
   $\tau_A = 0.600$ ksi, $\tau_B = 0$

8–19. $w = 79.7$ mm

8–21. $\sigma_{max} = 26.7$ MPa (C), $\sigma_{min} = 13.3$ MPa (T)

8–22. $\sigma_{max} = 53.3$ MPa (C), $\sigma_{min} = 40.0$ MPa (T)

8–23. $d = 13.1$ mm

8–25. $\sigma_A = 0.318$ MPa, $\tau_A = 0.735$ MPa

8–26. $\sigma_B = -21.7$ MPa, $\tau_B = 0$

8–27. $(\sigma_C)_{max} = 11.0$ MPa, $(\sigma_C)_{min} = 8.33$ MPa

8–29. $T = 2.16$ kip

8–30. $T = 2.16$ kip

8–31. $\sigma_B = 235.1$ psf (C), Since σ_B is in compression, the chimney is *safe*.

8–33. $\sigma_A = 25.0$ psi (C), $\sigma_B = 75.0$ psi (C)

8–34. $\sigma_A = 25.0$ psi (C), $\sigma_B = 75.0$ psi (C),
$\sigma_C = 25.0$ psi (C), $\sigma_D = 25.0$ psi (T)

8–35. $\theta = 0.286°$

8–37. $\sigma_D = -88.0$ MPa, $\tau_D = 0$

8–38. $\sigma_E = 57.8$ MPa, $\tau_E = 864$ kPa

8–39. $e_y = \dfrac{r}{4}$

8–41. $\sigma_B = 0.522$ MPa (C), $\tau_B = 0$

8–42. Point A: $\sigma_A = 107$ MPa, $\tau_A = 15.3$ MPa,
Point B : $\sigma_B = 0$, $\tau_B = 14.8$ MPa

8–43. Point C: $\sigma_C = 107$ MPa (C), $\tau_C = 15.3$ MPa,
Point D : $\sigma_D = 0$, $\tau_D = 15.8$ MPa

8–45. $\sigma_B = 81.3$ MPa (C), $(\tau_{xz})_B = 2.36$ MPa, $(\tau_{xy})_B = 0$

8–46. $\sigma_C = 103$ MPa (C), $(\tau_{xy})_C = 3.54$ MPa, $(\tau_{xz})_C = 0$

8–47. Point D: $(\sigma_D)_y = -178$ psi, $(\tau_D)_{yx} = (\tau_D)_{yz} = 0$,
Point E: $(\sigma_E)_y = 9.78$ psi, $(\tau_E)_{yx} = (\tau_E)_{yz} = 0$

8–49. $\sigma_A = 4.37$ MPa (C), $\sigma_B = 0.318$ MPa (C), $\tau_A = 0$,
$\tau_B = 0.477$ MPa

8–50. $\sigma_A = 10.1$ ksi (T), $\sigma_B = 10.3$ ksi (C), $\tau_A = \tau_B = 0$

8–51. $\sigma_E = 1.01$ MPa (C), $\sigma_F = 27.7$ MPa (C),
$\tau_E = 1.96$ MPa, $\tau_F = 0$

8–53. $\sigma_D = 21.4$ ksi (C), $\sigma_E = 10.8$ ksi (C), $\tau_D = \tau_E = 0$

8–54. $\sigma_A = 6.61$ ksi (T), $(\tau_{xz})_A = 1.39$ ksi, $(\tau_{xy})_A = 0$

8–55. $\sigma_B = 5.76$ ksi (C), $(\tau_{xy})_B = 1.36$ ksi, $(\tau_{xz})_B = 0$

8–57. $\sigma_A = 360$ kPa (T), $(\tau_{xz})_A = 39.1$ kPa, $(\tau_{xy})_A = 0$

8–58. $\sigma_C = 360$ kPa (C), $(\tau_{xz})_C = -9.05$ kPa, $(\tau_{xy})_C = 0$

8–59. $6e_y + 18e_z < 5a$

8–61. $\sigma_A = 9.88$ kPa (T), $\sigma_B = 49.4$ kPa (C),
$\sigma_C = 128$ kPa (C), $\sigma_D = 69.1$ kPa (C)

8–62. $(\sigma_t)_{max} = 49.0$ ksi (T), $(\sigma_c)_{max} = 40.8$ ksi (C)

8–63. $(\sigma_t)_{max} = 28.8$ ksi (T), $(\sigma_c)_{max} = 24.0$ ksi (C)

8–65. $(\sigma_t)_{max} = 15.5$ ksi (T), $(\sigma_c)_{max} = 9.31$ ksi (C)

8–66. $\sigma_{max} = \dfrac{1.33P}{a^2}$ (C), $\sigma_{min} = \dfrac{P}{3a^2}$ (T)

8–67. $\sigma_{max} = \dfrac{0.368P}{r^2}$ (C), $\sigma_{min} = \dfrac{0.0796P}{r^2}$ (T)

8–69. $p = 133.3$ kPa, $P = 848$ N

8–70. $p = 3.60$ MPa, $n = 113$ bolts

8–71. $\sigma_1 = 50.0$ MPa, $\sigma_2 = 25.0$ MPa, $F_b = 133$ kN

8–73. $\sigma_{max} = 26.7$ ksi

8–74. $\sigma_C = 11.6$ ksi, $\tau_C = 0$, $\sigma_D = -23.2$ ksi, $\tau_D = 0$

8–75. $\sigma_F = 6.40$ MPa (C), $\tau_F = 0$

8–77. $\sigma_A = 5.89$ ksi (C), $\sigma_B = 1.68$ ksi (T),
$\tau_A = 0$, $\tau_B = 0.397^-$ksi

Chapter 9

9–2. $\sigma_{x'} = -4.05$ ksi, $\tau_{x'y'} = -0.404$ ksi

9–3. $\sigma_{x'} = -4.05$ ksi, $\tau_{x'y'} = -0.404$ ksi

9–5. $\sigma_x = -2.71$ ksi, $\tau_{x'y'} = 4.17$ ksi

9–6. $\sigma_{x'} = -388$ psi, $\tau_{x'y'} = 455$ psi

9–7. $\sigma_{x'} = -388$ psi, $\tau_{x'y'} = 455$ psi

9–9. $\sigma_{x'} = 329$ psi, $\sigma_{y'} = -28.9$ psi, $\tau_{x'y'} = -69.9$ psi

9–10. **a)** $\sigma_1 = 10.1$ ksi, $\sigma_2 = -4.07$ ksi,
$\theta_{p_1} = -40.9°$, $\theta_{p_2} = 49.1°$

b) $\tau_{\substack{max \\ \text{in-plane}}} = 7.07$ ksi,

$\theta_s = 4.07°$ and $-85.9°$, $\sigma_{avg} = 3.00$ ksi

9–11. $\sigma_{x'} = -19.9$ ksi, $\tau_{x'y'} = 7.70$ ksi, $\sigma_{y'} = 9.89$ ksi

9–13. **a)** $\sigma_1 = 64.1$ ksi, $\sigma_2 = -14.1$ ksi
$\theta_{p1} = -64.9°$, $\theta_{p2} = 25.1°$

b) $\tau_{\substack{max \\ \text{in-plane}}} = 39.1$ ksi

$\sigma_{avg} = 25.0$ ksi, $\theta_s = -19.9°$ and $70.1°$

9–14. **a)** $\sigma_1 = 474$ psi, $\sigma_2 = -1034$ psi,
$\theta_{p1} = -55.9°$, $\theta_{p2} = 34.1°$,

b) $\tau_{\substack{max \\ \text{in-plane}}} = 754$ psi $\theta_s = -10.9°$ and $79.1°$,

$\sigma_{avg} = -280$ psi

9–15. $\sigma_1 = 40.0$ psi, $\sigma_2 = -40.0$ psi

9–18. $\sigma_x = 33.0$ MPa, $\sigma_y = 137$ MPa, $\tau_{xy} = -30$ MPa

9–19. $\tau_a = -1.96$ ksi, $\sigma_1 = 80.1$ ksi, $\sigma_2 = 19.9$ ksi

9–21. $\sigma_{x'} = 0.507$ MPa,

$\tau_{x'y'} = 0.958$ MPa

9–22. $\sigma_1 = 2.29$ MPa, $\sigma_2 = -7.20$ kPa

9–23. $\sigma_1 = 0$, $\sigma_2 = -192$ MPa, $\sigma_1 = 24.0$ MPa,

$\sigma_2 = -24.0$ MPa, $\theta_{p_1} = -45.0°$, $\theta_{p_2} = 45.0°$

9–25. $\tau_{\max}_{\text{in-plane}} = 5$ kPa, $\sigma_{\text{avg}} = 0$

9–26. $\sigma_1 = 152$ MPa, $\sigma_2 = 0$, $\sigma_1 = 0.229$ MPa,

$\sigma_2 = -196$ MPa, $\theta_{p_1} = 88.0°$, $\theta_{p_2} = 1.96°$

9–27. $\sigma_1 = 0$, $\sigma_2 = -87.1$ MPa $\sigma_1 = 93.9$ MPa,

$\sigma_2 = 0$, $\tau_{\max}_{\text{in-plane}} = 43.6$ MPa, $\tau_{\max}_{\text{in-plane}} = 47.0$ MPa

9–29. $\sigma_1 = 0$, $\sigma_2 = -118$ psi,

$\sigma_1 = 95.6$ psi, $\sigma_2 = 0$, $\tau_{\max}_{\text{in-plane}} = 58.9$ psi and 47.8 psi

9–30. $\sigma_1 = 150$ MPa, $\sigma_2 = -1.52$ MPa, $\sigma_1 = 1.60$ MPa,

$\sigma_2 = -143$ MPa

9–31. $\sigma_1 = \dfrac{2}{\pi d^2}\left(-F + \sqrt{F^2 + \dfrac{64T_0^2}{d^2}}\right)$,

$\sigma_2 = -\dfrac{2}{\pi d^2}\left(F + \sqrt{F^2 + \dfrac{64T_0^2}{d^2}}\right)$,

$\tau_{\max}_{\text{in-plane}} = \dfrac{2}{\pi d^2}\sqrt{F^2 + \dfrac{64T_0^2}{d^2}}$

9–33. $M = 8.73$ kip · in.

9–34. $\sigma_1 = 4.33$ MPa, $\sigma_2 = -13.0$ MPa

9–35. $\sigma_1 = 0$, $\sigma_2 = -77.4$ MPa, $\tau_{\max}_{\text{in-plane}} = 38.7$ MPa

9–37. $\sigma_1 = 54.6$ MPa, $\sigma_2 = -59.8$ MPa, $\tau_{\max}_{\text{in-plane}} = 57.2$ MPa

9–38. $\sigma_1 = 1.06$ ksi, $\sigma_2 = -0.850$ ksi, $\tau_{\max}_{\text{in-plane}} = 0.957$ ksi

9–39. $\sigma_1 = 0$, $\sigma_2 = -0.814$ ksi,

$\tau_{\max}_{\text{in-plane}} = 0.407$ ksi

9–41. $\sigma_1 = 0$, $\sigma_2 = 29.5$ MPa, $\sigma_1 = 0.541$ MPa,

$\sigma_2 = -1.04$ MPa, $\theta_{p_1} = -54.2°$, $\theta_{p_2} = 35.8°$

9–42. $\sigma_1 = 5.50$ MPa, $\sigma_2 = -0.611$ MPa

9–43. $\sigma_1 = 1.29$ MPa, $\sigma_2 = -1.29$ MPa

9–45. $\sigma_1 = 382$ psi, $\sigma_2 = -471$ psi, $\tau_{\max}_{\text{in-plane}} = 427$ psi

9–46. For point A: $\sigma_1 = 61.7$ psi, $\sigma_2 = 0$.

For point B: $\sigma_1 = 0$, $\sigma_2 = -46.3$ psi

9–49. $\sigma_{x'} = -4.05$ ksi, $\tau_{x'y'} = -0.404$ ksi

9–50. $\sigma_{x'} = -2.71$ ksi, $\tau_{x'y'} = 4.17$ ksi

9–51. $\sigma_{x'} = -388$ psi, $\tau_{x'y'} = 455$ psi

9–53. $\sigma_{\text{avg}} = 3.00$ ksi,

a) $\sigma_1 = 10.1$ ksi, $\sigma_2 = -4.07$ ksi, $\theta_{p_1} = 40.9°$

b) $\tau_{\max}_{\text{in-plane}} = -7.07$ ksi, $\theta_s = 4.07°$

9–54. $\sigma_{x'} = -19.9$ ksi, $\tau_{x'y'} = 7.70$ ksi,

$\sigma_{y'} = 9.89$ ksi

9–55. $\sigma_{\text{avg}} = 25.0$ ksi,

a) $\sigma_1 = 64.1$ ksi, $\sigma_2 = -14.1$ ksi, $\theta_{p_1} = 64.9°$

b) $\tau_{\max}_{\text{in-plane}} = -39.1$ ksi, $\theta_s = 19.9°$

9–57. $\sigma_{\text{avg}} = -280$ psi,

a) $\sigma_1 = 474$ psi, $\sigma_2 = -1034$ psi, $\theta_{p_1} = 55.9°$

b) $\tau_{\max}_{\text{in-plane}} = -754$ psi, $\theta_s = 10.9°$

9–58. $\sigma_{\text{avg}} = 3.50$ MPa, $\tau_{\max}_{\text{in-plane}} = 0.500$ MPa

9–59. $\sigma_{x'} = -56.3$ ksi, $\sigma_{y'} = 56.3$ ksi, $\tau_{x'y'} = 32.5$ ksi

9–61. $\sigma_{x'} = 736$ MPa, $\sigma_{y'} = -156$ MPa, $\tau_{x'y'} = 188$ MPa

9–62. $\sigma_{x'} = -421$ MPa, $\tau_{x'y'} = -354$ MPa,

$\sigma_{y'} = 421$ MPa

9–63. a) $\sigma_1 = 12.3$ ksi, $\sigma_2 = -17.3$ ksi, $\theta_p = 16.3°$

b) $\tau_{\max}_{\text{in-plane}} = 14.8$ ksi, $\sigma_{\text{avg}} = -2.5$ ksi, $\theta_s = 28.7°$

9–65. $\sigma_{\text{avg}} = 7.50$ ksi,

a) $\sigma_1 = 16.5$ ksi, $\sigma_2 = -1.51$ ksi, $\theta_{p_1} = 16.8°$

b) $\tau_{\max}_{\text{in-plane}} = -9.01$ ksi, $\theta_s = 28.2°$

9–66. a) $\sigma_{\text{avg}} = 100$ psi, $R = 700$ psi

b) $\sigma_{\text{avg}} = -1.00$ ksi, $R = 1.00$ ksi

c) $\sigma_{\text{avg}} = 0$, $R = 20.0$ MPa

9–67. $\sigma_x = 33.0$ MPa, $\sigma_y = 137$ MPa,

$\tau_{xy} = -30.0$ MPa

9–70. $\sigma_1 = 1.50$ ksi, $\sigma_2 = -0.0235$ ksi

9–71. $\sigma_1 = 0.0723$ ksi, $\sigma_2 = -0.683$ ksi

9–73. $\sigma_1 = 0$, $\sigma_2 = -87.1$ MPa, $\tau_{\max_{\text{in-plane}}} = 43.6$ MPa,

$\theta_s = 45.0°$, $\sigma_1 = 93.9$ MPa,

$\sigma_2 = 0$ MPa, $\tau_{\max_{\text{in-plane}}} = 47.0$ MPa,

$\theta_s = 45.0°$

9–74. $\sigma_1 = 4.15$ ksi, $\sigma_2 = 0$, $\tau_{\max_{\text{in-plane}}} = 2.075$ ksi,

$\theta_s = 45.0°$, $\sigma_1 = 0$,

$\sigma_2 = -3.95$ ksi, $\tau_{\max_{\text{in-plane}}} = 1.975$ ksi, $\theta_s = 45.0°$

9–75. $\sigma_{\text{avg}} = 4.80$ ksi

9–77. $\sigma_1 = 4.00$ ksi, $\sigma_2 = -0.0317$ ksi

9–78. $\sigma_1 = 0.711$ ksi, $\sigma_2 = -0.445$ ksi

9–79. $\sigma_1 = 4.71$ ksi, $\sigma_2 = -0.0262$ ksi

9–81. $\sigma_{x'} = -12.5$ kPa, $\tau_{x'y'} = 21.7$ kPa

9–82. $\sigma_1 = 141$ MPa, $\sigma_2 = 9.05$ MPa, $\tau_{\max_{\text{in-plane}}} = 66.0$ MPa

9–83. $\sigma_{x'} = 500$ MPa, $\tau_{x'y'} = 167$ MPa

9–85. **a)** $\sigma_{\max} = 6$ ksi, $\sigma_{\text{int}} = \sigma_{\min} = 0$

b) $\sigma_{\max} = 50$ MPa, $\sigma_{\text{int}} = 0$, $\sigma_{\min} = -40$ MPa

c) $\sigma_{\max} = 600$ psi, $\sigma_{\text{int}} = 200$ psi, $\sigma_{\min} = 100$ psi

d) $\sigma_{\max} = 0$, $\sigma_{\text{int}} = -7$ ksi, $\sigma_{\min} = -9$ ksi

e) $\sigma_{\max} = \sigma_{\text{int}} = \sigma_{\min} = -30$ MPa

9–86. *For x–y Plane:* $\tau_{\max_{\text{in-plane}}} = 4$ ksi, $\sigma_{\text{avg}} = 11$ ksi

For x–z Plane: $\tau_{\max_{\text{in-plane}}} = 2$ ksi, $\sigma_{\text{avg}} = 13$ ksi

For y–z Plane: $\tau_{\max_{\text{in-plane}}} = 2$ ksi, $\sigma_{\text{avg}} = 9$ ksi

9–87. $\sigma_{\max} = 7.00$ ksi, $\sigma_{\text{int}} = 5.00$ ksi,

$\sigma_{\min} = -5.00$ ksi, $\tau_{\text{abs}_{\max}} = 6.00$ ksi

9–89. For x–y plane: $\tau_{\max} = 6.0$ ksi, $\sigma_{\text{avg}} = -2.0$ ksi

For x–z plane: $\tau_{\max} = 3.0$ ksi, $\sigma_{\text{avg}} = 7.0$ ksi

For y–z plane: $\tau_{\max} = 9.0$ ksi, $\sigma_{\text{avg}} = 1.0$ ksi

9–90. $\sigma_1 = \sigma_2 = \sigma_3 = -p$

9–91. $\sigma_{\max} = 0.615$ kPa, $\sigma_{\text{int}} = 0$, $\sigma_{\min} = -4.62$ MPa,

$\tau_{\text{abs}_{\max}} = 2.31$ MPa

9–93. $\sigma_{\max} = 48.8$ ksi, $\sigma_{\text{int}} = 0$,

$\sigma_{\min} = -25.4$ ksi, $\tau_{\text{abs}_{\max}} = 37.1$ ksi, $\theta_s = 9.22°$

9–94. $\sigma_{\max} = 34.7$ ksi, $\sigma_{\text{int}} = 0$

$\sigma_{\min} = -34.7$ ksi, $\tau_{\text{abs}_{\max}} = 34.7$ ksi

9–95. $\sigma_1 = 233$ psi, $\sigma_2 = -774$ psi, $\tau_{\max_{\text{in-plane}}} = 503$ psi

9–97. $\sigma_{\text{avg}} = 62.5$ ksi,

a) $\sigma_1 = 153$ ksi, $\sigma_2 = -27.7$ ksi, $\theta_{p_1} = 23.1°$

b) $\tau_{\max_{\text{in-plane}}} = 90.2$ ksi, $\theta_s = 21.9°$

9–98. $\sigma_{\max} = 153$ ksi, $\sigma_{\text{int}} = 0$,

$\sigma_{\min} = -27.7$ ksi, $\tau_{\text{abs}_{\max}} = 90.2$ ksi

9–99. $M = 81.8$ lb $\cdot$ ft, $T = 231$ lb $\cdot$ ft

9–101. $\sigma_{x'} = -0.611$ ksi, $\tau_{x'y'} = 7.88$ ksi, $\sigma_{y'} = -3.39$ ksi

9–102. $\sigma_{\text{avg}} = 0$, $\tau_{\max_{\text{in-plane}}} = 5.00$ kPa

9–103. $\sigma_{x}' = -22.9$ kPa, $\tau_{x'y'} = -13.2$ kPa

Chapter 10

10–2. $\epsilon_{x'} = -309(10^{-6})$, $\epsilon_{y'} = -541(10^{-6})$,

$\gamma_{x'y'} = -423(10^{-6})$

10–3. $\epsilon_{x'} = -116(10^{-6})$, $\epsilon_{y'} = 466(10^{-6})$, $\gamma_{x'y'} = 393(10^{-6})$

10–5. $\epsilon_{x'} = 649(10^{-6})$, $\gamma_{x'y'} = -85.1(10^{-6})$, $\epsilon_{y'} = 201(10^{-6})$

10–6. **a)** $\epsilon_1 = 368(10^{-6})$, $\epsilon_2 = 182(10^{-6})$,

$\theta_{p_1} = -52.8°$, $\theta_{p_2} = 37.2°$

b) $\gamma_{\max_{\text{in-plane}}} = 187(10^{-6})$, $\theta_s = -7.76°$, and $82.2°$,

$\epsilon_{\text{avg}} = 275(10^{-6})$

10–7. **a)** $\epsilon_1 = 1039(10^{-6})$, $\epsilon_2 = 291(10^{-6})$,

$\theta_{p1} = 30.2°$, $\theta_{p2} = 120°$

b) $\gamma_{\max_{\text{in-plan}}} = 748(10^{-6})$, $\epsilon_{\text{avg}} = 665(10^{-6})$,

$\theta_s = -14.8°$ and $75.2°$, $\gamma_{x'y'} = 748(10^{-6})$

10–9. **a)** $\epsilon_1 = 385(10^{-6})$, $\epsilon_2 = 195(10^{-6})$

$\theta_{p_1} = 54.2°$, $\theta_{p2} = -35.8°$

b) $\gamma_{\max_{\text{in-plane}}} = 190(10^{-6})$,

$\theta_s = 9.22°, -80.8°$, $\epsilon_{\text{avg}} = 290(10^{-6})$

10–10. **a)** $\epsilon_1 = 138(10^{-6})$, $\epsilon_2 = -198(10^{-6})$,

$\theta_{p_1} = 13.3°$, $\theta_{p_2} = -76.7°$

b) $\gamma_{\max_{\text{in-plane}}} = 335(10^{-6})$, $\epsilon_{\text{avg}} = -30.0(10^{-6})$,

$\theta_s = 39.8°$ and $130°$, $\gamma_{x'y'} = 417(10^{-6})$

10–13. $\epsilon_{x'} = -116(10^{-6})$, $\gamma_{x'y'} = -393(10^{-6})$,

$\epsilon_{y'} = 466(10^{-6})$

10–14. $\epsilon_{x'} = 466(10^{-6})$, $\epsilon_{y'} = -116(10^{-6})$

$\gamma_{x'y'} = -393(10^{-6})$

10–15. $\epsilon_{x'} = 649(10^{-6})$, $\gamma_{x'y'} = -85.1(10^{-6})$

$\epsilon_{y'} = 201(10^{-6})$

10–17. $\epsilon_{avg} = 275(10^{-6})$, $\epsilon_1 = 368(10^{-6})$, $\epsilon_2 = 182(10^{-6})$,

$\theta_{p_1} = 52.8°$, $\gamma_{\substack{max \\ in\text{-}plane}} = -187(10^{-6})$,

$\theta_s = 7.76°$

10–18. $\epsilon_{avg} = 665(10^{-6})$, $\epsilon_1 = 1039(10^{-6})$, $\epsilon_2 = 291(10^{-6})$,

$\theta_{p_1} = 30.2°$, $\gamma_{\substack{max \\ in\text{-}plane}} = 748(10^{-6})$,

$\theta_s = 14.8°$

10–19. $\epsilon_{avg} = 290(10^{-6})$, $\epsilon_1 = 385(10^{-6})$,

$\epsilon_2 = 195(10^{-6})$, $\theta_{p_1} = 54.2°$

$\gamma_{\substack{max \\ in\text{-}plan}} = 190(10^{-6})$, $\theta_s = 9.22°$

10–21. $\gamma_{\substack{max \\ in\text{-}plan}} = -1015(10^{-6})$, $\epsilon_{max} = 418(10^{-6})$,

$\epsilon_{int} = 0$, $\epsilon_{min} = -598(10^{-6})$, $\gamma_{\substack{abs \\ max}} = 1015(10^{-6})$

10–22. $\cdot\gamma_{\substack{max \\ in\text{-}plane}} = 168(10^{-6})$, $\epsilon_{max} = 289(10^{-6})$,

$\epsilon_{int} = 121(10^{-6})$, $\epsilon_{min} = 0$, $\gamma_{\substack{abs \\ max}} = 289(10^{-6})$

10–23. $\gamma_{\substack{max \\ in\text{-}plane}} = 740(10^{-6})$, $\epsilon_{max} = 0$, $\epsilon_{int} = -65.1(10^{-6})$,

$\epsilon_{min} = -805(10^{-6})$, $\gamma_{\substack{abs \\ max}} = 805(10^{-6})$

10–25. $\gamma_{\substack{max \\ in\text{-}plane}} = 465(10^{-6})$, $\epsilon_{max} = 870(10^{-6})$,

$\epsilon_{int} = 405(10^{-6})$, $\epsilon_{min} = 0$, $\gamma_{\substack{abs \\ max}} = 870(10^{-6})$

10–26. $\gamma_{\substack{max \\ in\text{-}plane}} = -320(10^{-6})$, $\epsilon_{max} = 0$, $\epsilon_{int} = -140(10^{-6})$,

$\epsilon_{min} = -460(10^{-6})$, $\gamma_{\substack{abs \\ max}} = 460(10^{-6})$

10–27. **a)** $\epsilon_1 = 773(10^{-6})$, $\epsilon_2 = 76.8(10^{-6})$,

b) $\gamma_{\substack{max \\ in\text{-}plane}} = 696(10^{-6})$ **c)** $\gamma_{\substack{abs \\ max}} = 773(10^{-6})$

10–29. **a)** $\epsilon_1 = 192(10^{-6})$, $\epsilon_2 = -152(10^{-6})$,

b,c) $\gamma_{\substack{abs \\ max}} = 344(10^{-6})$

10–30. **a)** $\epsilon_1 = 1434(10^{-6})$, $\epsilon_2 = -304(10^{-6})$,

b) $\gamma_{\substack{max \\ in\text{-}plane}} = 1738(10^{-6})$, $\epsilon_{avg} = 565(10^{-6})$

10–31. $\epsilon_1 = 1046(10^{-6})$, $\epsilon_2 = -306(10^{-6})$, $\theta_p = 15.4°$

10–35. $k = 7.78(10^3)$ ksi

10–37. $\nu_{pvc} = 0.614$

10–38. $\epsilon_{max} = 30.5(10^{-6})$, $\epsilon_{int} = \epsilon_{min} = -10.7(10^{-6})$

10–39. $p = 0.967$ ksi, $\gamma_{\substack{max \\ in\text{-}plane}} = 1.30(10^{-3})$

10–41. $\epsilon_{max} = 0.755(10^{-3})$,

$\epsilon_{int} = 0.621(10^{-3})$, $\epsilon_{min} = -1.31(10^{-3})$

10–42. $E_p = 769$ ksi, $e = 0.254(10^{-3})$

10–43. $E = 17.4$ GPa, $\delta d = 12.65(10^{-6})$ mm

10–45. $\nu = 0.309$, $E = 28.1(10^3)$ ksi

10–46. $p = 3.43$ MPa, $\tau_{\substack{max \\ in\text{-}plane}} = 0$, $\tau_{\substack{abs \\ max}} = 85.7$ MPa

10–47. $\sigma_1 = 8.37$ ksi, $\sigma_2 = 6.26$ ksi

10–49. $\epsilon_{max} = 289(10^{-6})$, $\epsilon_{int} = 0$, $\epsilon_{min} = -289(10^{-6})$

10–50. $\Delta d = 0.800$ mm, $\sigma_{AB} = 315$ MPa

10–51. $p = 0.432$ ksi,

$\gamma_{\substack{max \\ in\text{-}plane}} = 0.294(10^{-3})$, $\gamma_{\substack{abs \\ max}} = 0.587(10^{-3})$

10–53. $\varepsilon_x = \varepsilon_y = 0$, $T = 41.3$ N $\cdot$ m

10–54. $\delta V = \dfrac{1 - 2\nu}{2E}(2PL + w_0 L^2)$

10–57. $k = 1.35$

10–59. $\varepsilon_1 = \dfrac{pr}{2Et}(2 - \nu)$, $\delta d = 2.72$ mm, $\varepsilon_2 = \dfrac{pr}{2Et}(1 - 2\nu)$,

$\delta L = 1.60$ mm

10–61. $\sigma_x = \sigma_y = -70.0$ ksi, $\sigma_z = -55.2$ ksi

10–62. $\varepsilon_x = \varepsilon_y = 0$, $\varepsilon_z = 5.44(10^{-3})$

10–63. $\Delta T = 67.7°$F

10–65. $\sigma_x^2 + \sigma_y^2 - \sigma_x\sigma_y + 3\tau_{xy}^2 = \sigma_Y^2$

10–66. Yes.

10–67. Yes.

10–69. $\sigma_1 = 11.6$ ksi

10–70. $\sigma_2 = 38.9$ ksi

10–71. $\sigma_2 = 23.9$ MPa

10–73. F.S. $= 1.59$

10–74. F.S. $= 1.80$

10–75. $d = 2.03$ in.

10–77. $a = 1.78$ in.

10–78. F.S. $= 5.38$

10–79. $\sigma_Y = 19.7$ ksi

10–81. **a)** $\tau = 32.5$ ksi, **b)** $\tau = 37.5$ ksi

10–82. $M_e = \sqrt{M^2 + T^2}$

10–83. $\sigma_Y = 94.3$ ksi

10–85. **a)** $\sigma_1 = 924$ MPa (*Controls!*), **b)** $\sigma_1 = 1067$ MPa

10–86. $d = 0.942$ in.

10–87. $d = 0.988$ in.

10–89. $d = 0.833$ in.

10–90. $d = 0.794$ in.

10–91. **a)** $p = \dfrac{t}{r}\sigma_Y$ (*Controls!*), **b)** $p = \dfrac{2t}{\sqrt{3}r}\sigma_Y$

10–93. $\sigma_Y = 32.0$ ksi

10–94. $d = 1.68$ in.

10–95. The material does not fail according to the maximum normal stress theory.

10–97. No.

10–98. **a)** $\varepsilon_1 = 482(10^{-6})$, $\varepsilon_2 = 168(10^{-6})$,

   b) $\gamma_{\substack{max \\ \text{lb·plane}}} = 313(10^{-6})$ **c)** $\gamma_{\substack{abs \\ max}} = 482(10^{-6})$

10–101. **a)** $\varepsilon_1 = 441(10^{-6})$, $\varepsilon_2 = -641(10^{-6})$,

   $\theta_{p_1} = -24.8°$, $\theta_{p_2} = 65.2°$

   b) $\gamma_{\substack{max \\ \text{in-plane}}} = 1.08(10^{-3})$, $\theta_s = 20.2$ and $-69.8°$,

   $\varepsilon_{avg} = -100(10^{-6})$

10–102. $\Delta b = 1.2(10^{-3})$in., $\Delta b = -1.2(10^{-3})$in.

10–103. $k_r = 3.33$ ksi, $k_g = 5.13(10^3)$ksi

10–105. $\varepsilon_{avg} = 83.3(10^{-6})$,

   a) $\varepsilon_1 = 880(10^{-6})$, $\varepsilon_2 = -713(10^{-6})$, $\theta_{p_1} = 54.8°$

   b) $\gamma_{\substack{max \\ \text{in-plane}}} = -1593(10^{-6})$, $\theta_s = 9.78°$

Chapter 11

11–1. $h = 6$ in.

11–2. $h = 4\frac{3}{4}$ in.

11–3. $b = 3.40$ in.

11–5. W12 × 16

11–6. W12 × 26

11–7. Yes.

11–9. $h = 5\frac{1}{4}$ in.

11–10. $h = 4\frac{1}{2}$ in.

11–11. 0.394 in.

11–13. No, the beam fails due to bending stress criteria.

11–14. W24 × 62

11–15. The wide flange section W12 × 14 fails due to the bending stress and will not safely support the loading.

11–17. 11.4 mm

11–18. 13.0 mm

11–19. $P = 12.5$ kip, $\tau_{\text{req'd}} = 466$ psi

11–21. $b = 4.00$ in.

11–22. $b = 5.00$ in.

11–23. $d = 1\frac{1}{4}$ in.

11–25. 4.25 in.

11–26. W12 × 22

11–27. 0.595 in.

11–29. $\sigma_{max} = \dfrac{3PL}{8bh_0^2}$

11–30. $w = \dfrac{w_0}{L}x$

11–31. $\sigma_{max} = \dfrac{16PL}{27\pi r_0^3}$

11–33. $d = h\sqrt{\dfrac{x}{L}}$

11–34. The beam has a semi-elliptical shape.

11–35. $\sigma = \dfrac{3PL}{2b_0 t^2}$, The bending stress is independent of x. Therefore, the stress is constant throughout the span.

11–37. $\sigma_{max} = \dfrac{8PL}{27\pi r_0^3}$

11–38. $d = 1\frac{1}{4}$ in.

11–39. $d = 36$ mm

11–41. $d = 1\frac{1}{2}$ in.

11–42. $d = 1\frac{5}{8}$ in.

11–43. $d = 1\frac{1}{8}$ in.

11–45. $d = 20$ mm

11–46. $d = 21$ mm

11–47. $T = 100$ N · m, $d = 29$ mm

11–49. $d = 21$ mm

11–50. $s = \dfrac{5}{8}$ in., $s' = \dfrac{1}{2}$ in., $s'' = 2\frac{3}{4}$ in.

11–51. $y = \left[\dfrac{4P}{\pi\sigma_{allow}}x\right]^{\frac{1}{3}}$

11–53. W10 × 12

11–54. W10 × 12

11–55. $\sigma_{max} = wL^2/4bh_0^2$

Chapter 12

12–1. $\sigma = 100$ MPa

12–2. $\sigma = 582$ MPa

12–3. $v_1 = \dfrac{Pb}{6EIL}(x_1^3 - (L^2 - b^2)x_1),$

$v_2 = \dfrac{Pa}{6EIL}[3x_2^2 L - x_2^3 - (2L^2 + a^2)x_2 + a^2 L]$

12–5. $\theta_A = \dfrac{Pa(a - L)}{2EI}, v_1 = \dfrac{Px_1}{6EI}[x_1^2 + 3a(a - L)],$

$v_2 = \dfrac{Pa}{6EI}[3x(x - L) + a^2],$

$v_{max} = \dfrac{Pa}{24EI}(4a^2 - 3L^2)$

12–6. $v_1 = \dfrac{Px_1}{12EI}(-x_1^2 + L^2),$

$v_2 = \dfrac{P}{24EI}(-4x_2^3 + 7L^2 x_2 - 3L^3), v_{max} = \dfrac{PL^3}{8EI}$

12–7. $v_1 = \dfrac{-Pb}{6aEI}[x_1^3 - a^2 x_1],$

$v_2 = \dfrac{P}{6EI}[-x_2^3 + (-2ab + 3b^2)x_2 - 2(b^3 + ab^2)]$

12–9. $\theta_A = \dfrac{Pab}{2EI},$

$v_1 = \dfrac{P}{6EI}[-x_1^3 + 3a(a + b)x_1 - a^2(2a + 3b)],$

$v_3 = \dfrac{Pax_3}{2EI}(-x_3 + b), v_C = \dfrac{Pab^2}{8EI}$

12–10. $\theta_A = \dfrac{M_0 a}{2EI}, v_{max} = -\dfrac{5M_0 a^2}{8EI}$

12–11. $\rho = 336$ ft, $\theta_{max} = \dfrac{M_0 L}{EI}, v_{max} = -\dfrac{M_0 L^2}{2EI}$

12–13. $\theta_{max} = \dfrac{-M_0 L}{EI}, v = -\dfrac{M_0 x^2}{2EI}, v_{max} = -\dfrac{M_0 L^2}{2EI}$

12–14. $\theta_{max} = \dfrac{M_0 L}{3EI}, v_{max} = -\dfrac{\sqrt{3}M_0 L^2}{27EI}$

12–15. $|\theta_{max}| = \dfrac{M_0 L}{2EI}, v = \dfrac{M_0 x}{2EI}(x - L), v_{max} = -\dfrac{M_0 L^2}{8EI}$

12–17. $\theta_A = -\dfrac{3}{8}\dfrac{PL^2}{EI}, v_C = \dfrac{-PL^3}{6EI}$

12–18. $v_B = \dfrac{-11PL^3}{48EI}$

12–19. $\theta_A = 0.0611$ rad, $v_A = -3.52$ in.

12–21. $v_C = -\dfrac{PL^3}{32EI_C}$

12–22. $\theta_A = \dfrac{\gamma L^3}{3r^2 E}, v_A|_{x=0} = -\dfrac{\gamma L^4}{6r^2 E}$

12–23. $v_{max} = \dfrac{6PL^3}{Ebt^3}$

12–25. $v = \dfrac{1}{EI}[-0.25x^4 +$

$0.208 <x - 1.5>^3 + 0.25 <x - 1.5>^4 +$

$4.625 <x - 4.5>^3 + 25.1x - 36.4]$ kN · m^3

12–26. $v_{max} = 6.83$ mm

12–27. $\theta_B = 0.0313$ rad, $v = \dfrac{1}{EI}\{50x^3 - 50 <x - 0.2>^3$

$- 50 <x - 0.5>^3 - 15.0x\}$ N · m^3

12–29. $v = \dfrac{P}{12EI}\{-2x^3 + 3 <x - a>^3$

$- 2 <x - 2a>^3 +$

$3 <x - 3a>^3 + 15a^2 x - 13a^3\},$

$v_{max} = -\dfrac{13Pa^3}{12EI}$

12–30. $\theta_B = \dfrac{3Pa^2}{4EI}, v|_{x=2a} = \dfrac{Pa^3}{3EI}$

12–31. $v = \dfrac{1}{EI}\{4.10x^3 - 0.125x^4 +$

$0.125 <x - 4>^4 - 8.33 <x - 7>^3 - 279x\}$ kN · m^3

12–33. $\dfrac{dv}{dx} = \dfrac{1}{EI}[2.25x^2 - 0.5x^3 + 5.25 <x - 5>^2 +$

$0.5 <x - 5>^3 - 3.125]$ kN · m^2,

$v = \dfrac{1}{EI}[0.75x^3 - 0.125x^4 + 1.75 <x - 5>^3$

$+ 0.125 <x - 5>^4 - 3.125x]$ kN · m^3

12–34. $v = \dfrac{1}{EI}\{11.1x^3 - 0.5x^4 - 5 <x - 3>^3$

$- 378x\}$ kN · m^3

12–35. $\theta_A = -\dfrac{378 \text{ kN · m}^2}{EI}, \theta_B = \dfrac{359 \text{ kN · m}^2}{EI},$

$v_C = -\dfrac{874 \text{ kN · m}^3}{EI}$

12–37. $\theta_A = -\dfrac{3wa^3}{16EI}$, $\theta_B = \dfrac{7wa^3}{48EI}$,

$v = \dfrac{w}{48EI}\{6ax^3 - 2x^4 + 2 <x - a>^4 - 9a^3x\}$

12–38. $v = \dfrac{1}{EI}[-0.00556x^5$

$+ 12.9 <x - 9>^3 + 0.00556 <x - 9>^5$

$- 256x + 2637] \text{ kip} \cdot \text{ft}^3$

12–39. $v = \dfrac{1}{EI}\Big[-31.5x^2 + \dfrac{8}{3}x^3 - \dfrac{x^4}{12} +$

$\dfrac{1}{12} <x - 3>^4 - \dfrac{2}{3} <x - 4.5>^3\Big] \text{kN} \cdot \text{m}^3$,

$\theta_B = -0.705°$, $v_B = -51.7$ mm

12–41. $\theta_A = \dfrac{302 \text{ kip} \cdot \text{ft}^2}{EI}$, $v_C = -\dfrac{3110 \text{ kip} \cdot \text{ft}^3}{EI}$

12–42. $\theta_C = \dfrac{3937.5}{EI}$, $\Delta_C = \dfrac{50\,625}{EI}$

12–43. $\theta_A = \dfrac{M_0L}{24EI}$, $\theta_C = \theta_A = \dfrac{M_0L}{24EI}$, $\Delta_{max} = \dfrac{\sqrt{3}M_0L^2}{216EI}$

12–45. $\theta_B = \dfrac{3Pa^2}{EI}$, $\Delta_C = \dfrac{4Pa^3}{3EI}$

12–46. $\theta_C = \dfrac{5Pa^2}{2EI}$, $\Delta_B = \dfrac{25Pa^3}{6EI}$

12–47. $\theta_A = \dfrac{5M_0a}{12EI}$, $\Delta_C = \dfrac{M_0a^2}{4EI}$

12–49. $\theta_{max} = \dfrac{5PL^2}{16EI}$, $\Delta_{max} = \dfrac{3PL^3}{16EI}$

12–50. $a = \dfrac{3}{16}L$

12–51. $\Delta_C = \dfrac{84}{EI}$, $\theta_A = \dfrac{8}{EI}$, $\theta_B = \dfrac{16}{EI}$, $\theta_C = \dfrac{40}{EI}$

12–53. $\theta_A = \dfrac{5PL^2}{8EI}$, $\theta_B = \dfrac{PL^2}{2EI}$, $\Delta_A = \dfrac{7PL^3}{16EI}$, $\Delta_B = \dfrac{7PL^3}{48EI}$

12–54. $a = 0.152L$

12–55. $\Delta_{max} = 8.16$ mm

12–57. $\Delta_{max} = \dfrac{11Pa^3}{48EI}$

12–58. $\Delta_D = 4.98$ mm

12–59. $\theta_B = \dfrac{7Pa^2}{4EI}$, $\Delta_C = \dfrac{9Pa^3}{4EI}$

12–61. $\theta_B = \dfrac{M_0(b^3 + 3ab^2 - 2a^3)}{6EI(a + b)^2}$, $\Delta_C = \dfrac{M_0a\,b(b - a)}{3EI(a + b)}$

12–62. $\theta_B = \dfrac{3Pa^2}{4EI}$, $\Delta_C = \dfrac{13Pa^3}{12EI}$

12–63. $t_{C/A} = \dfrac{40}{EI}$

12–65. $F = \dfrac{P}{4}$

12–66. $\Delta_D = \dfrac{Pa^3}{12EI}$

12–67. $\Delta_D = \dfrac{9Pa^3}{4EI}$, $\Delta_E = \dfrac{19Pa^3}{12EI}$

12–69. $\Delta_{max} = 2.12$ in.

12–70. $\theta_C = \dfrac{a^2}{6EI}(12P + wa)$, $\Delta_C = \dfrac{a^3}{24EI}(64P + 7wa)$

12–71. $\theta_B = \theta_A = 0.175°$

12–73. $\theta_C = \dfrac{wa^3}{EI}$, $\Delta_B = \dfrac{41wa^4}{24EI}$

12–74. $\Delta_A = 0.933$ in.

12–75. $\Delta_C = 0.895$ in.

12–77. $\Delta_A = 0.781$ in.

12–78. W16 × 50

12–79. $F = 0.349$ N, $a = 0.800$ mm

12–81. $\Delta_A = \dfrac{Pa^2(3b + a)}{3EI}$

12–82. $\Delta = PL^2\Big(\dfrac{1}{k} + \dfrac{L}{3EI}\Big)$

12–83. $\Delta_A = PL^3\Big(\dfrac{1}{12EI} + \dfrac{1}{8JG}\Big)$

12–85. $\Delta_A = \dfrac{72}{EI}$, $\theta_A = \dfrac{36}{EI}$

12–86. $(\Delta_A)_v = 0.0737$ in., $(\Delta_A)_h = 0.230$ in.

12–87. $(\Delta_A)_v = \dfrac{PL^3}{24}\Big(\dfrac{2}{EI} + \dfrac{3}{JG}\Big)$

12–89. $A_y = \dfrac{11}{16}P$, $M_A = \dfrac{3PL}{16}$

12–90. $A_x = 0$, $M_A = \dfrac{PL}{2}$, $A_y = \dfrac{3P}{2}$, $B_y = \dfrac{5P}{2}$

12–91. $B_y = \dfrac{3wL}{8}$, $A_y = \dfrac{5wL}{8}$, $M_A = \dfrac{wL^2}{8}$

12–93. $C_x = 0$, $A_y = 12.0$ kN, $B_y = 40.0$ kN, $C_y = 12.0$ kN

12–94. $A_x = 0$, $C_y = \dfrac{w_0L}{10}$, $B_y = \dfrac{4w_0L}{5}$, $A_y = \dfrac{w_0L}{10}$

12–95. $A_x = 0$, $B_y = \dfrac{w_0L}{10}$, $A_y = \dfrac{2w_0L}{5}$, $M_A = \dfrac{w_0L^2}{15}$

12–97. $T_{AC} = \dfrac{3A_2E_2wL_1^4}{8(A_2E_2L_1^3 + 3E_1I_1L_2)}$

12–98. $M_A = \dfrac{Pab^2}{L^2}$, $M_B = \dfrac{Pa^2b}{L^2}$

12–99. $a = 0.414L$

12–101. $A_y = \dfrac{3M_0}{2L}$, $C_y = \dfrac{3M_0}{2L}$, $B_y = \dfrac{3M_0}{L}$, $C_x = 0$

12–102. $M_B = \dfrac{wL^2}{6}$, $M_A = \dfrac{wL^2}{3}$.

12–103. $M_A = \dfrac{M_0}{3}$, $M_B = \dfrac{M_0}{3}$

12–105. $M_A = \dfrac{5wL^2}{192}$, $M_B = \dfrac{11wL^2}{192}$

12–106. $M_A = 0.639$ kip $\cdot$ in., $M_B = 1.76$ kip $\cdot$ in.

12–107. $d = 0.708$ in.

12–109. $A_x = 0$, $B_y = 35.0$ kip, $A_y = 15.0$ kip, $M_A = 40.0$ kip $\cdot$ ft

12–110. $T_{AC} = \dfrac{3wA_2E_2L_1^4}{8[3E_1I_1L_2 + A_2E_2L_1^3]}$

12–111. $C_x = 0$, $B_y = \dfrac{2P}{3}$, $C_y = \dfrac{P}{3}$

12–113. $A_y = 2.15$ kN

12–114. $A_x = 0$, $B_y = \dfrac{5wL}{4}$, $C_y = \dfrac{3wL}{8}$

12–115. $F_{sp} = \dfrac{5wkL^4}{4(6EI + kL^3)}$

12–117. $B_y = 634$ lb, $A_y = 243$ lb, $C_y = 76.8$ lb

12–118. $F_{sp} = \dfrac{3kwL^4}{24EI + 8kL^3}$

12–119. $R = \left(\dfrac{8\Delta EI}{9w_0}\right)^{\frac{1}{4}}$, $a = L - \left(\dfrac{72\Delta EI}{w_0}\right)^{\frac{1}{4}}$

12–121. $v = \dfrac{1}{EI}[-30x^3 + 46.25 <x - 12>^3$
$\qquad\qquad - 11.7 <x - 24>^3 + 38\,700x - 412\,560]$

12–122. $\theta_B = 0.0100$ rad, $v_C = 0.200$ in.

12–123. $\theta_B = 0.0100$ rad, $\Delta_C = 0.200$ in.

12–125. $B_y = 138$ N, $A_y = 81.3$ N, $C_y = 18.8$ N

12–126. $\theta_B = \dfrac{Pa^2}{4EI}$, $\Delta_C = \dfrac{Pa^3}{4EI}$

12–127. $a = \dfrac{3L}{16}$

12–129. $M_B = \dfrac{w_0L^2}{30}$, $B_y = \dfrac{3w_0L}{20}$, $M_A = \dfrac{w_0L^2}{20}$, $A_y = \dfrac{7w_0L}{20}$

Chapter 13

13–1. $P_{cr} = \dfrac{5\,k\,L}{4}$

13–2. $d = \dfrac{5}{8}$ in.

13–3. $P_{cr} = 35.1$ kip

13–5. $d = \dfrac{9}{16}$ in.

13–6. $P_{cr} = 22.7$ kN

13–7. $P_{cr} = 46.4$ kN

13–9. $P_{cr} = 70.4$ kip

13–10. $P_{cr} = 1.30$ MN

13–11. $P_{cr} = 325$ kN

13–13. $d = 1.81$ in.

13–14. $P_{cr} = 2.92$ kip

13–15. $P_{cr} = 5.97$ kip

13–17. F.S. $= 2.19$

13–18. $P = 475$ kip

13–19. $P = 83.6$ lb

13–21. $P_{cr} = 28.4$ kip

13–22. $P_{cr} = 58.0$ kip

13–23. $P = 4.23$ kip

13–25. $P = 5.79$ kip   (Controls)

13–26. $d = 1\dfrac{3}{4}$ in.

13–27. $P = 4.57$ kip

13–29. $P = 29.9$ kN

13–30. $P = 207$ lb   (Controls)

13–31. $P = 2.42$ kip

13–33. No, AB will fail.

13–34. F.S. $= 2.12$

13–35. $P = 63.6$ kip

13–37. $W = 5.24$ kN, $d = 1.64$ m

13–38. F.S. $= 1.24$

13–39. $w = 18.3$ lb/ft

13–41. $P_{cr} = \dfrac{\pi^2\, E\, I}{4\, L^2}$

13–45. $v_{max} = \dfrac{wEI}{P^2}\left[\sec\left(\sqrt{\dfrac{P}{EI}}\,\dfrac{L}{2}\right) - \dfrac{PL^2}{8EI} - 1\right]$

$M_{max} = \dfrac{wEI}{P}\left[\sec\left(\sqrt{\dfrac{P}{EI}}\,\dfrac{L}{2}\right) - 1\right]$

13–46. $v_{max} = 0.387$ in.

13–47. $\sigma_{max} = 4.57$ ksi

13–49. $\sigma_{max} = 6.24$ ksi

13–50. $P' = 97.4$ kip

13–51. $P = 139$ kip, the column would carry $2.29\left(\dfrac{P'}{P}\right)$ times more load if the load P acts concentrically.

13–53. $P = 6.75$ kN

13–54. $P = 20.1$ kN

13–55. $L = 1.71$ m (*Controls!*)

13–57. F.S. $= 1.12$

13–58. Yes.

13–59. $P_{max} = 343$ kip (*Controls!*)

13–61. $P = 113$ kip (*Controls*)

13–62. $P' = 390$ kip, $P = 139$ kip (*Controls*)

13–63. $P_{allow} = 7.89$ kN (*Controls*)

13–65. $\sigma_{max} = 178$ MPa, $v_{max} = 10.8$ mm

13–66. $E_t = 14.6(10^3)$ ksi

13–67. $P_{cr} = 157$ kN

13–69. $P_{cr} = 39.2$ kN

13–71. $L = 18.0$ ft

13–73. Adequate.

13–74. Not Adequate.

13–75. W8 × 48

13–77. $L = 6.86$ ft

13–78. $L = 33.9$ ft

13–79. Adequate.

13–81. W8 × 24

13–82. $L = 18.9$ ft

13–83. $L = 10.9$ ft

13–85. $L = 1.92$ ft

13–86. $L = 3.84$ ft

13–87. $L = 6.07$ ft

13–89. $d = 1.42$ in.

13–90. $d = 1.00$ in.

13–91. $P_{allow} = 129$ kip

13–93. $P_{allow} = 109$ kip

13–94. $d = 1.30$ in.

13–95. $a = 5\dfrac{1}{2}$ in.

13–97. $P = 38.4$ kip

13–98. $P = 68.3$ kip

13–99. $L = 5.03$ ft

13–101. $P = 29.6$ kip

13–102. $P = 1.48$ kip

13–103. $P = 0.967$ kip

13–105. Not adequate.

13–106. Not adequate.

13–107. $P = 8.83$ kip

13–109. $P = 15.0$ kip

13–110. $P = 33.1$ kip

13–111. $P = 57.7$ kip

13–113. Yes.

13–114. No.

13–115. $P = 703$ lb

13–117. $P = 341$ lb

13–118. $P_{allow} = 57.6$ kip

13–119. $a = 103$ mm

13–121. $M = 44.2$ kN · m

13–122. $M = 60.5$ kN · m

13–123. $w = 1.17$ kN/m

13–125. Adequate.

13–126. Adequate.

13–127. $t = 5.92$ mm

Chapter 14

14–1. $\dfrac{U_i}{V} = \dfrac{1}{2E}(\sigma_x^2 + \sigma_y^2 - 2\nu\sigma_x\sigma_y) + \dfrac{\tau_{xy}^2}{2G}$

14–3. $\dfrac{\sigma_x}{\sigma_y} = \nu$

14–5. **a)** $U_i = 4.31$ kip $\cdot$ in., **b)** $U_i = 4.31$ kip $\cdot$ in.

14–6. $U_i = \dfrac{M_0^2 L}{24EI}$

14–7. $U_i = \dfrac{M_0 L}{2EI}$

14–9. **a)** $U_i = \dfrac{w^2 L^5}{240EI}$, **b)** $U_i = \dfrac{w^2 L^5}{240EI}$

14–10. $U_i = \dfrac{w_0^2 L^5}{945EI}$

14–11. $U_i = 496$ J

14–13. $(U_i)_a = 0.477\,(10^{-3})$ J, $(U_i)_b = 0.0171$ J

14–14. $(U_i)_{sp} = 1.00$ J, $(U_i)_b = 0.400$ J

14–15. $U_i = 45.5$ ft $\cdot$ lb

14–17. $U_i = \dfrac{w_0^2 L^5}{504EI}$

14–18. $a = \sqrt{\dfrac{\pi}{2}} r$

14–21. $\Delta_D = \dfrac{3.50PL}{AE}$

14–22. $(\Delta_D)_v = \dfrac{3.50PL}{AE}$

14–23. $\Delta_B = \dfrac{PL}{15}\left(\dfrac{15\cos^2\theta}{AE} + \dfrac{5L^2\sin^2\theta}{EI} + \dfrac{18\sin^2\theta}{GA}\right)$

14–25. $\Delta_B = 1.82$ in.

14–26. $\theta_C = 1.93°$

14–27. $\Delta = \dfrac{3PL}{10bhG}$

14–29. $\theta_A = \dfrac{4M_0 a}{3EI}$

14–30. $\Delta_A = \dfrac{3\pi Pr^3}{2EI}$

14–31. $\Delta = \dfrac{64nPR^3}{d^4 G}$

14–33. $U_i = 267$ ft $\cdot$ lb

14–34. $d = 5.35$ in.

14–35. $\sigma_{max} = 359$ MPa

14–37. $v = 0.499$ m/s

14–38. $L = 850$ mm

14–39. Yes, σ_{max} exceeded 175 MPa.

14–41. $\sigma_{max} = 3.06$ ksi

14–42. $\sigma_{max} = 414$ MPa

14–43. $h = 95.6$ mm

14–45. $n = 115$, $\sigma_{max} = 17.7$ ksi

14–46. $h = 11.6$ ft

14–47. $\sigma_{max} = 4.55$ ksi $< \sigma_Y$, $\Delta_C = 0.799$ in.

14–49. $\sigma_{max} = 5.88$ ksi

14–50. $h = 3.73$ in.

14–51. $\sigma_{max} = 2.34$ ksi, $(\Delta_D)_{max} = 0.370$ in.

14–53. $h = 4.20$ ft

14–54. $\sigma_{max} = 35.2$ ksi

14–55. $(\Delta_A)_{max} = 15.4$ in.

14–57. $\Delta_{max} = 5.45$ in., $\sigma_{max} = 33.1$ ksi

14–58. $n = 17.8$

14–59. $\Delta_{B_v} = 11.3$ mm

14–61. $(\Delta_B)_h = 0.699(10^{-3})$ in.

14–62. $(\Delta_B)_v = 0.0931(10^{-3})$ in.

14–63. $\Delta_{B_v} = 0.0132$ in.

14–65. $(\Delta_C)_h = 0.234$ mm

14–66. $(\Delta_D)_v = 1.16$ mm

14–67. $(\Delta_A)_v = 0.0199$ in.

14–69. $\Delta_{C_v} = 6.23$ mm

14–70. $\Delta_{H_v} = 3.79$ mm

14–71. $(\Delta_D)_h = 4.12$ mm

14–73. $(\Delta_B)_v = 0.0124$ in.

14–74. $(\Delta_E)_v = 0.00966$ in.

14–75. $(\Delta_C)_v = 0.163$ in.

14–77. $\Delta_C = \dfrac{23Pa^3}{24EI}$

14–78. $\Delta_C = \dfrac{5Pa^3}{6EI}$

14–79. $\theta_C = \dfrac{Pa^2}{2EI}$

14–81. $\theta_A = 0.0126$ rad

14–82. $\Delta_C = 4.13$ mm

14–83. $\theta_C = 0.0185$ rad

14–85. $\theta_C = 0.0174$ rad

14–86. $\theta_B = 0.0216$ rad

14–87. $\Delta_D = 1.80$ mm

14–89. $\theta_B = 1.20(10^{-3})$ rad

14–90. $\Delta_B = \dfrac{65wa^4}{48EI}$

14–91. $\Delta_C = 0.122$ in.

14–93. $\Delta_C = 40.9$ mm

14–94. $\theta_A = 0.00298$ rad

14–95. $\theta_B = 0.00595$ rad

14–97. $\theta_C = 5.89(10^{-3})$ rad

14–98. $\Delta_A = 27.4$ m, $\theta_A = 5.75(10^{-3})$ rad

14–99. $\Delta_C = \dfrac{5M_0a^2}{6EI}$

14–101. $\Delta_B = 43.5$ mm

14–102. $\theta_A = 0.00529$ rad

14–103. $\Delta_C = 17.9$ mm

14–105. $\Delta_{C_h} = \dfrac{5wL^4}{8EI}$

14–106. $\Delta_{B_v} = \dfrac{wL^4}{4EI}$

14–107. $\Delta_{C_h} = \dfrac{640\,000 \text{ lb} \cdot \text{ft}^3}{EI}$, $\Delta_{C_v} = \dfrac{1\,228\,800 \text{ lb} \cdot \text{ft}^3}{EI}$

14–109. $(\Delta_C)_v = \dfrac{5wL^4}{8EI}$

14–110. $(\Delta_B)_h = \dfrac{wL^4}{4EI}$

14–111. $\Delta = \dfrac{\pi Pr^3}{2EI}$

14–113. $(\Delta_B)_v = 11.3$ mm

14–114. $(\Delta_B)_h = 0.699(10^{-3})$ in.

14–115. $0.0931(10^{-3})$ in.

14–117. $(\Delta_B)_v = 0.0132$ in.

14–118. $(\Delta_C)_h = 0.234$ mm

14–119. $(\Delta_D)_v = 1.16$ mm

14–121. $(\Delta_A)_v = 0.0199$ in.

14–122. $(\Delta_A)_v = 6.23$ mm

14–123. $(\Delta_B)_v = 3.79$ mm

14–125. $(\Delta_D)_h = 4.12$ mm

14–126. $(\Delta_B)_v = 0.0124$ in.

14–127. $(\Delta_E)_v = 0.00966$ in.

14–129. $(\Delta_C)_v = 0.163$ in.

14–130. $\Delta_C = \dfrac{23PL^3}{24EI}$

14–131. $\Delta_C = \dfrac{5Pa^3}{6EI}$

14–133. $\theta_C = 0.0185$ rad

14–134. $\theta_E = 0.00120$ rad

14–135. $\theta_A = 0.00298$ rad

14–137. $\theta_A = 0.00529$ rad

14–138. $\Delta_A = 27.4$ m, $\theta_A = 5.75(10^{-3})$ rad

14–139. $\Delta_C = \dfrac{5M_0a^2}{6EI}$

14–141. $\Delta_C = 17.9$ mm

14–142. $\Delta_B = \dfrac{wL^4}{4EI}$

14–143. $(\Delta_C)_h = \dfrac{640(10^3) \text{ lb} \cdot \text{ft}^3}{EI}$, $(\Delta_C)_v = \dfrac{1\,228\,800 \text{ lb} \cdot \text{ft}^3}{EI}$

14–145. $(\Delta_C)_v = \dfrac{5wL^4}{8EI}$

14–146. $\Delta_A = \dfrac{\pi Pr^3}{2EI}$

14–147. $\Delta_B = \dfrac{M_0L^2}{2EI}$

14–149. $\Delta_A = \dfrac{5M_0a^2}{6EI}$

14–150. $\Delta_A = \dfrac{5M_0a^2}{6EI}$

14–151. $\theta_C = 0.0337$ rad

14–153. $\Delta_{E_v} = 2.95$ mm

14–154. $\Delta_{E_v} = 2.95$ mm

14–155. $U_i = \dfrac{P^2r^3}{8GJ}(3\pi - 8)$

14–157. $\sigma_{max} = 6.20$ ksi

Appendix A

A–1. 2.67 in.

A–2. $I_x = 415\ \text{in}^4$, $I_y = 411\ \text{in}^4$

A–3. 2.19 in., 83.7 in^4

A–5. $\bar{y} = 4.14$ in., $\bar{I}_{x'} = 35.5\ \text{in}^4$, $\bar{I}_y = 18.4\ \text{in}^4$

A–6. 34.0 mm, $\bar{I}_{x'} = 2.31(10^6)\ \text{mm}^4$, $\bar{I}_y = 3.09(10^6)\ \text{mm}^4$

A–7. $\bar{y} = 1.00$ in., $\bar{x} = 1.50$ in., $\bar{I}_{x'} = 4.00\ \text{in}^4$,
$\bar{I}_{y'} = 8.50\ \text{in}^4$

A–9. $\bar{y} = 36.4$ mm, $\bar{x} = 111$ mm, $\bar{I}_{x'} = 19.5(10^6)\ \text{mm}^4$,
$\bar{I}_{y'} = 115(10^6)\ \text{mm}^4$

A–10. $\bar{y} = 36.4$ mm, $\bar{x} = 111$ mm, $\bar{I}_{x'y'} = -26.1(10^6)\ \text{mm}^4$

A–11. $\bar{y} = 19.8$ mm, $\bar{x} = 100$ mm, $\bar{I}_{x'} = 1.58(10^6)\ \text{mm}^4$,
$\bar{I}_{y'} = 18.4(10^6)\ \text{mm}^4$

A–13. $\bar{I}_x = 248\ \text{in}^4$, $\bar{I}_y = 25.7\ \text{in}^4$

A–14. $\bar{I}_{xy} = 56.25\ \text{in}^4$

A–15. $\bar{y} = 4.33$ in., $\bar{x} = 5.67$ in., $\bar{I}_{x'} = 492\ \text{in}^4$,
$\bar{I}_{y'} = 172\ \text{in}^4$

A–17. $\bar{I}_{xy} = 3.23\ \text{in}^4$

A–18. $I_{x'} = 7.13(10^6)\ \text{mm}^4$, $I_{y'} = 3.53(10^6)\ \text{mm}^4$,
$I_{x'y'} = 3.12(10^6)\ \text{mm}^4$

A–19. $I_{x'} = 65.3\ \text{in}^4$, $I_{y'} = 25.3\ \text{in}^4$, $I_{x'y'} = -34.6\ \text{in}^4$

A–21. $\bar{y} = 1.00$ in., $\bar{x} = 1.50$ in., $I_{\max} = 10.0\ \text{in}^4$,
$I_{\min} = 2.50\ \text{in}^4$, $\theta_{p1} = 63.4°$, $\theta_{p2} = -26.6°$

A–22. $I_{x'} = 16.1(10^6)\ \text{mm}^4$, $I_{y'} = 16.1(10^6)\ \text{mm}^4$, $I_{x'y'} = 0$

A–23. $I_{\max} = 308(10^3)\ \text{mm}^4$, $I_{\min} = 85.3(10^3)\ \text{mm}^4$,
$\theta_{p_1} = 45.0°$, $\theta_{p_2} = -45.0°$

A–25. $I_{\max} = 5.09\ \text{in}^4$, $I_{\min} = 0.428\ \text{in}^4$, $\theta_{p_1} = 22.4°$,
$\theta_{p_2} = -67.6°$

A–26. $I_{\max} = 5.09\ \text{in}^4$, $I_{\min} = 0.428\ \text{in}^4$, $\theta_{p_1} = 22.4°$

INDEX

Average Mechanical Properties of Typical Engineering Materials[a]
(U.S. Customary Units)

Materials	Specific Weight γ (lb/in³)	Modulus of Elasticity E (10³) ksi	Modulus of Rigidity G (10³) ksi	Yield Strength (ksi) σ_Y [b] Tens.	Comp.	Shear	Ultimate Strength (ksi) σ_u [b] Tens.	Comp.	Shear	% Elongation in 2 in. specimen	Poisson's Ratio ν	Coef. of Therm. Expansion α (10⁻⁶)/°F
Metallic												
Aluminum Wrought Alloys [2014-T6	0.101	10.6	3.9	60	60	25	68	68	42	10	0.35	12.8
6061-T6]	0.098	10.0	3.7	37	37	19	42	42	27	12	0.35	13.1
Cast Iron Alloys [Gray ASTM 20	0.260	10.0	3.9	–	–	–	26	97	–	0.6	0.28	6.70
Malleable ASTM A-197]	0.263	25.0	9.8	–	–	–	40	83	–	5	0.28	6.60
Copper Alloys [Red Brass C83400	0.316	14.6	5.4	11.4	11.4	–	35	35	–	35	0.35	9.80
Bronze C86100]	0.319	15.0	5.6	50	50	–	95	95	–	20	0.34	9.60
Magnesium Alloy [Am 1004-T61]	0.066	6.48	2.5	22	22	–	40	40	22	1	0.30	14.3
Steel Alloys [Structural A36	0.284	29.0	11.0	36	36	–	58	58	–	30	0.32	6.60
Stainless 304	0.284	28.0	11.0	30	30	–	75	75	–	40	0.27	9.60
Tool L2]	0.295	29.0	11.0	102	102	–	116	116	–	22	0.32	6.50
Titanium Alloy [Ti-6A1-4V]	0.160	17.4	6.4	134	134	–	145	145	–	16	0.36	5.20
Nonmetallic												
Concrete [Low Strength	0.086	3.20	–	–	–	1.8	–	–	–	–	0.15	6.0
High Strength]	0.086	4.20	–	–	–	5.5	–	–	–	–	0.15	6.0
Plastic Reinforced [Kevlar 49	0.0524	19.0	–	–	–	–	104	70	10.2	2.8	0.34	–
30% Glass]	0.0524	10.5	–	–	–	–	13	19	–	–	0.34	–
Wood Select Structural Grade [Douglas Fir	0.017	1.90	–	–	–	–	0.30[c]	3.78[d]	0.90[d]	–	0.29[e]	–
White Spruce]	0.130	1.40	–	–	–	–	0.36[c]	5.18[d]	0.97[d]	–	0.31[e]	–

[a] Specific values may vary for a particular material due to alloy or mineral composition, mechanical working of the specimen, or heat treatment. For a more exact value reference books for the material should be consulted.

[b] The yield and ultimate strengths for ductile materials can be assumed equal for both tension and compression.

[c] Measured perpendicular to the grain.

[d] Measured parallel to the grain.

[e] Deformation measured perpendicular to the grain when the load is applied along the grain.